Communications
in Computer and Informa

3

Tai-hoon Kim   Adrian Stoica
Ruay-Shiung Chang (Eds.)

Security-Enriched Urban Computing and Smart Grid

First International Conference, SUComS 2010
Daejeon, Korea, September 15-17, 2010
Proceedings

Volume Editors

Tai-hoon Kim
Hannam University
Daejeon, South Korea
E-mail: taihoonn@hannam.ac.kr

Adrian Stoica
Jet Propulsion Laboratory
Pasadena, CA, USA
E-mail: adrian.stoica@jpl.nasa.gov

Ruay-Shiung Chang
National Dong Hwa University
Shoufeng, Hualien, Taiwan, R.O.C.
E-mail: rschang@mail.ndhu.edu.tw

Library of Congress Control Number: 2010936201

CR Subject Classification (1998): C.2.4, J.3, C.2, H.4, I.4, I.5

ISSN 1865-0929
ISBN-10 3-642-16443-9 Springer Berlin Heidelberg New York
ISBN-13 978-3-642-16443-9 Springer Berlin Heidelberg New York

springer.com

Printed in Germany

Typesetting: Camera-ready by author, data conversion by Scientific Publishing Services, Chennai, India
Printed on acid-free paper 06/3180

Foreword

Security-enriched urban computing and smart grids are areas that attracted many academic and industry professionals to research and develop. The goal of this conference was to bring together researchers from academia and industry as well as practitioners to share ideas, problems and solutions relating to the multifaceted aspects of urban computing and the smart grid.

This conference includes the following special sessions: Signal Processing, Image Processing, Pattern Recognition and Communications (SIPC 2010), Networking, Fault-tolerance and Security For Distributed Computing Systems (NFSDCS 2010), Security Technology Application (STA 2010), Electric Transportation (ElecTrans 2010), Techniques of Bi-directional Power Computing in High Voltage Power Supply (TBPC 2010), Low Power IT and Applications (LPITA 2010), Computational Intelligence and Soft Computing (CISC 2010), Distributed Computing and Sensor Networks (DCSN 2010), Advanced Fusion IT (AFIT 2010), Social Media and Social Networking (SMSN 2010), Software Engineering and Medical Information Engineering (SEMIE 2010), Human-Centered Advanced Research/Education (HuCARE 2010), Database Integrity and Security (DIS 2010), Ubiquitous IT Application (UITA 2010) and Smart Grid Applications (SGA 2010).

We would like to express our gratitude to all of the authors of the submitted papers and to all attendees, for their contributions and participation. We believe in the need for continuing this undertaking in the future.

We acknowledge the great effort of all the Chairs and the members of advisory boards and Program Committees of the above-listed events, who selected 15% of over 570 submissions, following a rigorous peer-review process. Special thanks go to SERSC (Science & Engineering Research Support soCiety) for supporting these co-located conferences.

We are grateful in particular to the following speakers who kindly accepted our invitation and, in this way, helped to meet the objectives of the conference: Ruay-Shiung Chang (National Dong Hwa University), Mohamed Hamdi (University of November 7th at Carthage), Andres Iglesias Prieto (University of Cantabria), Osvaldo Gervasi (University of Perugia), Zita Vale (GGECAD / ISEP-IPP), Hussein Mustapha Khodr (GGECAD / ISEP-IPP), Goreti Marreiros (GGECAD / ISEP-IPP), Marek Ogiela (Computer Science and Electronics Insitute of Automatics), Hoon Ko (GGECAD / ISEP-IPP), Martin Drahansky (BUT, Faculty of Information Technology), Brian King (Indiana University—Purdue University) and Wael Adi (Institute of Computer and Network Engineering).

We would also like to thank Rosslin John Robles and Maricel O. Balitanas—graduate students of Hannam University—who helped with a great passion in editing the material.

July 2010

Adrian Stoica
Ruay-Shiung Chang
Tai-hoon Kim

Preface

This volume comprises the proceedings of the the First International Conference on Security-Enriched Urban Computing and Smart Grid (SUComS), which was held on September 15–17, 2010, at the Yousung Hotel, Daejeon, Korea.

SUComS 2010 was focused on various aspects of advances in urban computing and smart grids, particularly in their nexus with computational sciences, mathematics and information technology. It provided a chance for academic and industry professionals to discuss recent progress in the related areas. We expect that the conference and its publications will be a trigger for further related research and technology improvements in this important subject. We would like to acknowledge the great effort of all the Chairs and members of the Program Committee. Out of around 570 submissions to SUComS 2010, we accepted 81 papers to be included in the proceedings and presented during the conference. This constitutes an acceptance ratio firmly below 15%. All 85 accepted papers can be found in this volume.

We would like to express our gratitude to each of the authors of the submitted papers and to all the attendees, for their contributions and participation. We firmly believe in the need for continuing this undertaking in the future.

Once more, we would like to thank all the organizations and individuals who supported this event as a whole and ultimately, helped make SUComS 2010 a success.

July 2010

Tai-hoon Kim
Adrian Stoica
Ruay-Shiung Chang

Organization

Organizing Committee

General Chair	Adrian Stoica (NASA Jet Propulsion Laboratory, USA)
Program Co-chairs	Ruay-Shiung Chang (National Dong Hwa University, Taiwan) Tai-hoon Kim (Hannam University, Korea)
International Advisory Board	Tughrul Arslan (The University of Edinburgh, UK) Jianhua Ma (Hosei University, Japan) Sankar K. Pal (Indian Statistical Institute, India) Xiaofeng Song (Nanjing University of Aeronautics and Astronautics, China) Frode Eika Sandnes (Oslo University College, Norway)
Publicity Co-chair	Yang Xiao (The University of Alabama, USA) J.H. Abawajy (Deakin University, Australia) Robert C. Hsu (Chung Hua University, Taiwan)

Program Committee

Abdelwahab Hamou-Lhadj
Aboul Ella Hassanien
Agustinus Borgy Waluyo
Ai-Chun Pang
Albert Levi
Ami Marowka
Biplab K. Sarker
Brian King
Chantana Chantrapornchai
Chao-Tung Yang
Chu-Hsing Lin
Dan Liu
Eric Renault
Fangguo Zhang
Farrukh A. Khan
Fionn Murtagh
George A. Gravvanis
Georgios Kambourakis
Gerard Damm
Hong Sun
Hui Chen
Hyeong-Ok Lee
Hyun Sung Kim
Jalal Al-Muhtadi
Jiann-Liang
José Manuel Molina López
Juha Jaakko Röning
Kaiqi Xiong
Kendra M.L. Cooper
Khalil Drira
Larbi Esmahi
Luigi Buglione
Matthias Reuter
MalRey Lee
Ming Li
Mohammad Riaz Moghal
Paolo D'Arco
Petr Hanacek
Qi Shi
Ramin Yahyapour
Reinhard Schwarz
Robert C. Hsu
Rodrigo Mello
Schahram Dustdar
Serge Chaumette
Stuart J. Barnes
Sun-Yuan Hsieh
Swee-Huay Heng
Tae (Tom) Oh
Tatsuya Akutsu
Toshihiro Yamauchi
Wanquan Liu
Wei Zhong
Witold Pedrycz
Yali (Tracy) Liu
Yang Li
Yannis Stamatiou
Yao-Chung Chang
Yeong-Deok Kim

Table of Contents

Genetic-Annealing Algorithm in Grid Environment for Scheduling Problems*

Marco Antonio Cruz-Chávez[1], Abelardo Rodríguez-León[2], Erika Yesenia Ávila-Melgar[1], Fredy Juárez-Pérez[1], Martín H. Cruz-Rosales[3], and Rafael Rivera-López[2]

[1] CIICAp
[2] Technological Institute of Veracruz
Miguel Ángel de Quevedo 2779, Formando Hogar, 91860 Veracruz, Veracruz, México
{arleon,rrivera}@itver.edu.mx
[3] FC, Autonomous University of Morelos State
Av. Universidad 1001. Col. Chamilpa, C.P. 62209. Cuernavaca, Morelos, México
{mcruz,juarezfredy,erikay,mcr}@uaem.mx

Abstract. This paper presents a parallel hybrid evolutionary algorithm executed in a grid environment. The algorithm executes local searches using simulated annealing within a Genetic Algorithm to solve the job shop scheduling problem. Experimental results of the algorithm obtained in the "Tarantula MiniGrid" are shown. Tarantula was implemented by linking two clusters from different geographic locations in Mexico (Morelos-Veracruz). The technique used to link the two clusters and configure the Tarantula MiniGrid is described. The effects of latency in communication between the two clusters are discussed. It is shown that the evolutionary algorithm presented is more efficient working in Grid environments because it can carry out major exploration and exploitation of the solution space.

1 Introduction

The Job Shop Scheduling Problem (JSSP) deals with the resource assignment in manufacturing industries. According to the complexity theory, JSSP is a complex problem, and is classified as NP-Complete [2]. Throughout time, several methods have been proposed to solve complex problems. Finding the global optimum solution using an exact method would take years for an NP-Complete problem. Hence, finding solutions to complex problems with deterministic algorithms is not applicable. In order to deal with NP-Complete problems, the meta-heuristics [3], [4] have been successfully applied. The evolutionary algorithms are meta-heuristics that work efficiently by exploring the solution space to find the best solution. The drawback to these algorithms is that they generally do not completely exploit the solution space. One way to compensate for this problem is to codify a hybrid meta-heuristic that has the capability to completely exploit and explore the solution space, and thereby escape from local optimum solutions.

* This work was supported by project CUDI-CONACYT 2009, 2010.

T.-h. Kim, A. Stoica, and R.-S. Chang (Eds.): SUComS 2010, CCIS 78, pp. 1–9, 2010.

Although meta-heuristics are efficient techniques to deal with complex problems, generally the computational resources of a computer are not sufficient to obtain results in reasonable computational times with problems in the NP-complete class. It is necessary to make use of computational tools such as clusters or grids, which are a set of interconnected computers that enable distributed processing. Distributed processing allows the use of more computational resources, so that the execution of a polynomial time algorithm can be more efficient.

For the execution of the polynomial algorithm proposed in this paper, a proprietary high performance MiniGrid was created[1]. The MiniGrid was formed by linking two institutions located in different geographic places: the Autonomous University of Morelos State, located in Cuernavaca, Morelos, and the Technological Institute of Veracruz located in Veracruz, Veracruz. Each institution had its own cluster and they were linked to form the Tarantula MiniGrid. The experimental platform created allowed for the execution of parallel programs by means of a passing interface for messages, better known as an *MPI library* [4].

The principal contribution of this work is the presentation of a Parallel Hybrid Evolutionary Algorithm in a Grid Environment with low latency as used to solve the Job Shop Scheduling Problem.

The remainder of this paper is divided into the following sections: Section 2 explains the formation of the job shop scheduling problem. Section 3 presents the description of the hybrid evolutionary algorithm and the process of deployment in a parallel platform with MPI. Section 4 shows the test scenarios used to execute the parallel algorithm. Section 5 describes the experimental results of efficiency and latency of the parallel hybrid evolutionary algorithm execution in a grid environment. Finally, Section 6 presents the conclusions of this work.

2 Job Shop Scheduling Problem

The JSSP is an attempt to establish a scheduling of machines in a manufacturing shop to realize a set of jobs [1]. Each job requires a set of operations to be carried out. The solution to the problem is an optimum sequence of jobs to be executed by each machine, respecting precedence constraints while executing the operations in each job.

$$\text{Min}\left[\max_{j \in O}\left(s_j + p_j\right)\right] \tag{1}$$

$$\forall\ j \in \boldsymbol{O} \qquad s_j \geq 0 \tag{2}$$

$$\forall\ i, j \in \boldsymbol{O},\ (i \prec j) \in J_k \qquad s_i + p_i \leq s_j \tag{3}$$

$$\forall\ i, j \in \boldsymbol{O},\ \left(i, j \in M_k\right) \qquad s_i + p_i \leq s_j \vee s_j + p_j \leq s_i \tag{4}$$

Equation (1) is the objective function related to the maximum completion time needed to finish the last operation realized in the manufacturing workshop. Constraint (2) states that the start time of each operation j should be greater or equal to zero.

Constraint (3) represents the existing precedence between two operations of the same job. Constraint (4) defines the resource capacity between pairs of operations executed by the same machine.

3 Hybrid Evolutionary Algorithm Deployed in Grid Environment

Hybrid evolutionary algorithms are gaining popularity due to their capabilities in handling several real world problems involving complexity, noisy environments, imprecision, uncertainty, and vagueness [5]. Figure 1 shows the solution methodology used to solve the job shop scheduling problem with the proposed hybrid evolutionary algorithm. It presents a genetic algorithm containing a randomly generated feasible population, a selection operator "*the best*" (elitism) [13], a crossover operator Subsequence Exchange (*SXX*) [12] and the iterative mutation operator *SA* [6]. According to their fitness, the operator "the best" selects an average of the best individual values from a population and recombines them, thus producing offspring that will compose the next generation.

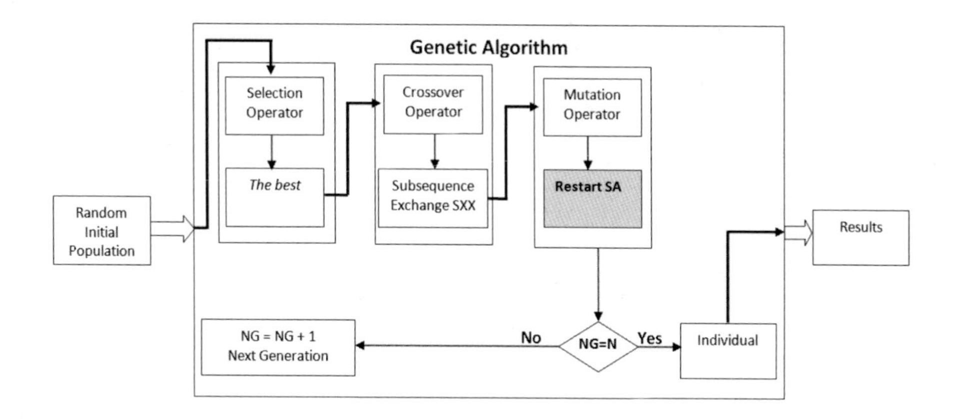

Fig. 1. Solution Methodology for the JSSP

Application of a crossover operator allows for exploration of the solution space. An iterative operator is used to mutate [6]. It optimizes the fitness function, applying a mutation in each individual. It helps to find the best individual by using an iterated local search with the Simulated Annealing process (SA). The previously described procedure is repeated for each generation of the population.

The hybrid evolutionary algorithm is implemented by using distributed processing and parallel techniques. The MiniGrid consists of two clusters. Each cluster contains some nodes, which have one or more processors. The processors are involved in the implementation of parallel algorithms using the Interface Passing Messages library (MPI). The parallel hybrid evolutionary algorithm implemented uses the master-slave paradigm, which maintains the sequence of the original algorithm [7].

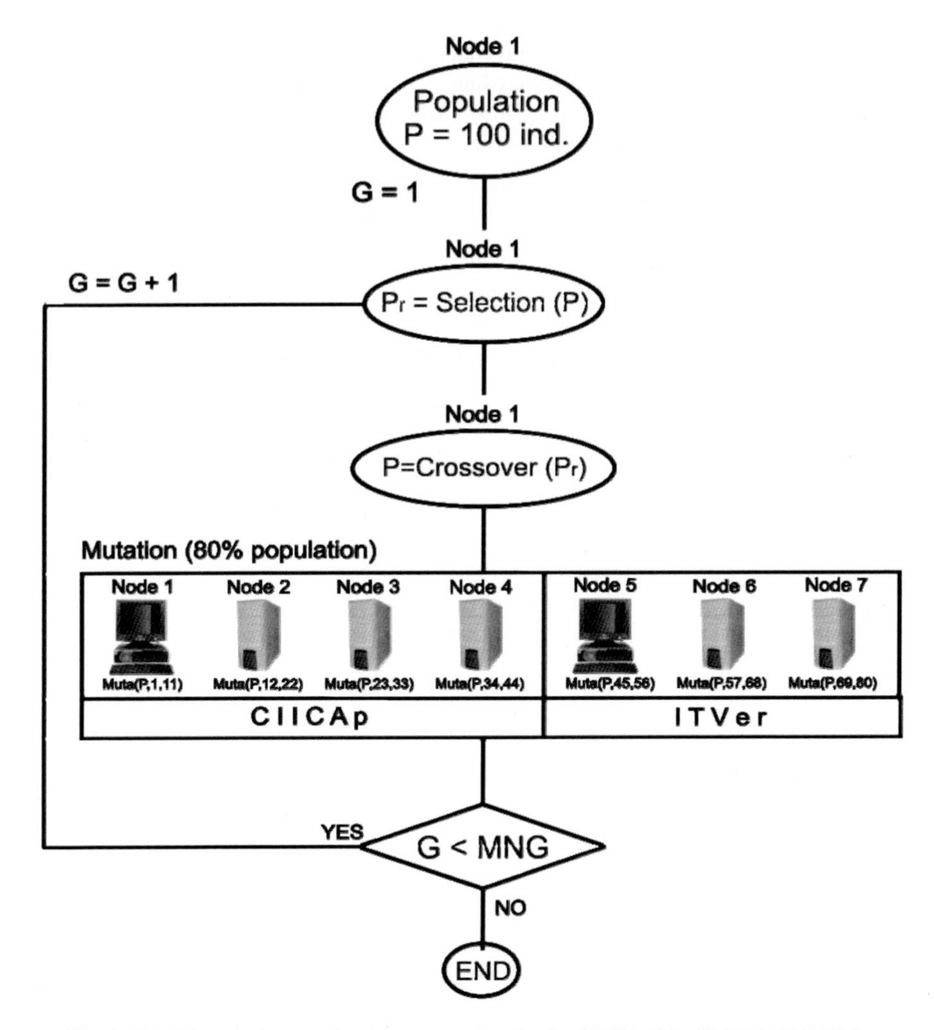

Fig. 2. Hybrid evolutionary algorithm execution for the JSSP in MiniGrid UAEM-ITVer

The main idea of the proposed parallelization of the evolutionary algorithm is to first allocate tasks to processors with lower communication requirements. The tasks are divided proportionally according to the available processors. Figure 2 shows the use of nodes in the MiniGrid for the algorithm. First, an initial feasible population is created in a master node. The master node carries out selection and crossover operations for genetic algorithms; it then sends a sub-population to the slaves' nodes on the MiniGrid. Based on experimental tests, the mutation operator has been identified as the most time consuming procedure. To decrease the execution time, the mutation function is divided proportionally according to the available nodes in the MiniGrid. That means that a process number is generated based on the number of processors available in the MiniGrid (1:1). Each processor executes a set of iterations using the Restart SA technique (iterative mutation). A Restart SA is executed for each individual of a sub-population that is to be mutated. The slave nodes return the newly mutated individuals to the master. Next, the master node collects all the individuals mutated by the slaves' nodes and uses them to create a new population. This process continues until an optimal solution is found or a given number of generations are created. For example, for a population of 100 with an 80% mutation rate, and 20 processors, each processor would mutate 4 individuals, as shown in Figure 2.

The MPI library was used for sending messages for the parallel hybrid evolutionary algorithm. The MPI library enables the transfer of complex data types. However, the MPI library has no mechanism to transfer dynamic data types. It only recognizes the primitive data types used in C language. For every primitive data type in C language as int, char, long, double, and so on, there is an equivalent MPI data type: MPI_INT, MPI_CHAR, MPI_LONG, MPI_DOUBLE. When using dynamic data types, there is not an equivalent MPI data type, so it is necessary to develop a conversion procedure, because the hybrid algorithm uses dynamic data types.

The conversion procedure necessary to transfer data is known in some platforms as serialization and de-serialization. It consists of serializing the dynamic data structure, sending it, receiving it, and reconstituting it. The conversion process is complex and tedious. Currently, an automatic tool called *Automap* [10] is proposed to do the conversion automatically. Once the data is converted to MPI data types, it is easy to send dynamic data structures through the grid.

4 Test Scenario

The MiniGrid is formed by joining two high-performance cluster computers (HPC), each computer belonging to a different domain. The clusters are integrated as a single parallel machine. They need for integration of the HPC is due to their distant geographic locations. The integration allows for local management of both clusters despite the fact that they are independent entities.

Table 1. Infrastructure of Hardware and Software for the Grid

<table>
<tr><th colspan="2">GRID TARANTULA ---- SOFTWARE ---</th></tr>
<tr><td colspan="2">Red Hat Enterprise Linux 4, Compiler gcc versión 3.4.3, OpenMPI 1.2.8, MPICH2-1.0.8, Ganglia 3.0.6
NIS ypserv-2.13-5, NFS nfs-utils-1.0.6-46, OpenVPN, Torque + Maui.</td></tr>
<tr><th>CLUSTER CIICAP</th><th>CLUSTER ITVER</th></tr>
<tr><td>Switch Cisco C2960 24/10/100</td><td>Switch 3com 8/10/100</td></tr>
<tr><td>Master node Pentium 4, 2793 MHz, 512 MB RAM, 80 GB HD, 2 network cards 10/100 Mb/s</td><td>Master node Pentium 4 Dual Core, 3200 Mhz, 1 GB RAM, 80 GC HD, 1 network card 10/100 Mb/s</td></tr>
<tr><td>9 slave nodes Intel® Celeron® Dual Core, 2000MHz, 2 GB RAM, 160GB HD, 1 network card 10/100 Mb/s</td><td>2 slave nodes Pentium 4 Dual Core, 3201 Mhz, 1 GB RAM, 80 GB HD,1 network card 10/100 Mb/s</td></tr>
</table>

The basic components in the construction of a high performance cluster with support for MPI programming were based on the installation and configuration of the following software and services on Red Hat Enterprise Linux 4. 1) Configuration of the private network. 2) Set up of a network file system using Network File System (NFS). 3) Set up of the Network Information Service (NIS). 4) Set up of public keys to enable the execution of remote jobs. 5) Set up of the MPI for use of OpenMPI or MPICH, which is the open source implementation of MPI-1 standard and MPI-2. 6) Configuration of Ganglia [11] for real time monitoring of the cluster. 7) Configuration of Torque settings to manage the cluster resources. 8) Set up of MAUI for planning jobs in the cluster.

The integration of the components shown in Table 1 is essential for the creation of high-performance clusters. The Mini-Grid consists of two or more high-performance clusters, which are geographically remote. An important element of the MiniGrid is a link that joins the two clusters by a *Virtual Private Network* (VPN, network-to-network). For the MiniGrid implemented in this study, an Open VPN was used instead of routers (via hardware). The joined clusters were in different subnets.

5 Experimental Results

The latency computation is based on the transfer rates between the two clusters: cluster.ciicap.edu.mx and nopal.itver.edu.mx. The process used to compute latency in the two clusters is described in detail in [9].

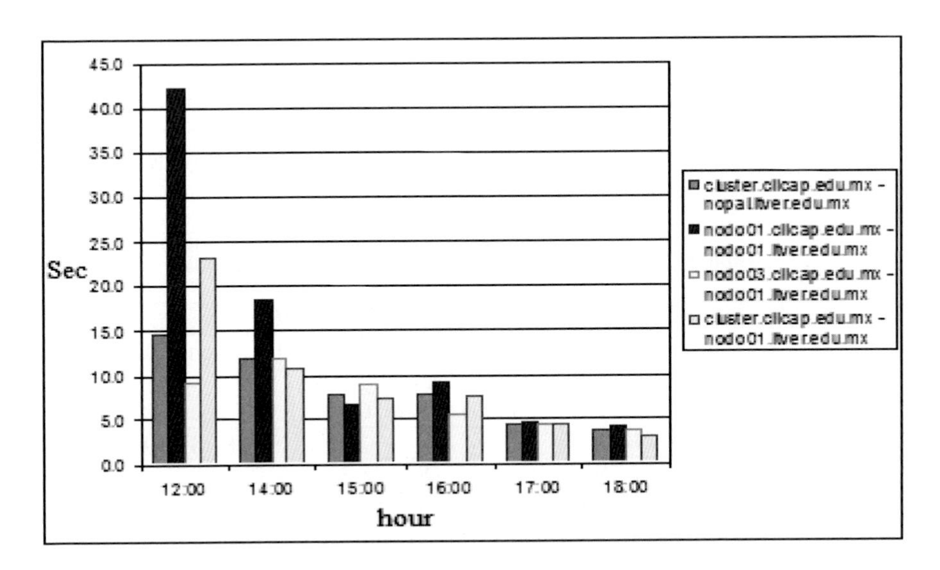

Fig. 3. Tarantula MiniGrid Latency

Figure 3 shows the latency obtained for the Tarantula MiniGrid throughout the day, from 12 AM to 6 PM. The results shown are the average values over a period of five days. It can be observed that the latency decreases as the day progresses. At 12 hours, higher latencies are reported, measuring between 45 and 40 seconds. At 18 hours, a latency of less than 0.4 seconds is observed. The latency tends to decrease as the day progresses. This indicates that in the afternoon the data transfer is more efficient.

In order to measure the efficiency of the hybrid Evolutionary Algorithm, it was executed for two benchmarks taken from the OR Library [8]. The algorithm was executed from the master node of the CIICAp cluster. The first executed benchmark was LA40 and it consisted of 15 jobs and 15 machines. The stop criteria were the number of generations created. The input data is presented in Table 2.

The obtained results of the algorithm execution in the MiniGrid are presented in Table 3. The average time of execution in the nodes (cores) of each cluster and the best makespan was presented after it finished 35 generations. Logical reasoning indicates that if the number of processors is increased then the population size could be increased, while requiring a similar amount of time. Increasing the number of processors would let the algorithm explore a larger portion of the solution space and find better solutions in less time. This shows the importance of having a large GRID that includes computational resources for high performance applications, such as the algorithm presented here.

Table 2. Input Data for the Hybrid Genetic Algorithm for the LA40 and YN1 Instance

Variable	**LA40**	**YN1**
Initial Temperature for Simulated Annealing	25	20
Final Temperature for Simulated Annealing	1.0	1.0
Coefficient Temperature for Simulated Annealing	0.99	0.955
Markov Chain Length	210	5000
Given Upper Bound for the JSSP	1222	885
Restarting number for each individual in Simulated Annealing	20	20
Good Population Selection (%)	40	40
Bad Population Selection (%)	5	5
Population to mutate (%)	80	80
Number of Generations for Genetic Algorithm	35	6

Table 2 indicates that the number of restarts for SA for each individual was 20. According to the restart number, in the worst case, the average execution time of the mutation in each processor for each generation was 345 seconds vs. 25 seconds for the worst latency registered in the MiniGrid. Based on this observation, it can be determined that latency is not significantly important because the time the processors were in use was greater than the communication time between clusters (93.2% vs. 6.8%).

The second executed benchmark was an instance of YN1 [8]. It had 20 jobs, 20 machines and 400 operations. Table 2 presents the input data and Table 5 shows the obtained results. The average latency detected was approximately 25 seconds. It indicates that, in the worst case, the total latency in execution time of the algorithm was

Table 3. Results of the Parallel Hybrid Evolutionary Algorithm for Solving LA40 Instance

Cluster	nodo0x.ciicap.edu.mx	Nodo0x.itver.edu.mx
Node	0,1,2,3,4,5,6,7,8,9	0,1,2
Time (sec)	12618	15413
Makespan	1234	1243

150 seconds according to the number of generations. Table 4 presents the results obtained for the algorithm execution for each processor in the MiniGrid. The average time of execution in the nodes (cores) of each cluster and the best makespan is presented after finishing 6 generations. If it were possible to increase the number of processors, then the population size could be increased, without a dramatic increase in time. Increasing the number of processors would allow the algorithm to explore a larger portion of the solution space and find a better solution in less time. This shows the importance of having a larger GRID that includes computational resources for high performance applications, such as the algorithm presented here. Table 2 indicates that the number of restarts for SA for each individual was 20. According to the restart number, in the worst case, the average execution time of SA in each processor for each generation was 4,567 seconds vs. 25 seconds for the worst latency registered in the MiniGrid. Based on this observation, it can be determined that latency is not significantly important because the time processors were in use was greater than communication time between clusters (95.5% vs. 0.5%).

Table 4. Results of the parallel hybrid evolutionary Algorithm for solving YN1 instance

Cluster	nodo0x.ciicap.edu.mx	Nodo0x.itver.edu.mx
Node	0,1,2,3,4,5,6,7,8,9	0,1,2
Time (sec)	19175	25617
Makespan	933	935

6 Conclusions

It can be concluded that the use of a Grid platform for this type of problem is highly recommendable for solving complex problems in a short time. The time needed to search and find solutions diminishes, as the number of available processors in the Grid increases and there is little communication between processors. The latency varies in the MiniGrid, the time tends to decline in the evenings, so it is recommendable to work in the afternoons when implementing algorithms that use long communication times.

In order to reduce communication between processors, it is necessary to define the structure of the algorithm in a way that allows it to work with a small latency in the MiniGrid. The latency does not considerably affect the parallelization because the large tasks of the algorithm mean the processors work for extended periods in order to complete execution of the program. Thus, there are minimal communications. The algorithm requires communication only to send and receive the sub-populations, and this prevents the latency from negatively influencing the execution of the algorithm.

To increase efficiency of the algorithm in the grid environment, communication between Grid nodes should be minimized. The algorithm can be performed using all resources of the Grid presented in this work.

References

[1] Roy, Sussman: Les problemes d'ordonnancement avec contraintes disjonctives, Note D.S. no 9 bis, SEMA, Paris, France (December 1964)

[2] Papadimitriou, C.H., Steiglitz, K.: Combinatorial optimization: algorithms and complexity, p. 496. Prentice Hall Inc., USA (1982) ISBN 0-13-152462-3

[3] Díaz, A., Glover, F., Ghaziri, H.M., et al.: Optimización Heurística y Redes Neuronales, Madrid, Paraninfo (1996)

[4] Glover, F.: Future paths for integer programming and links to artificial intelligence. Computers and Operations Research 13, 533–549 (1986)

[5] Grosan, C., Abraham, A.: Hybrid Evolutionary Algorithms. In: Ajith, Ishibuchi, Hisao (eds.) XVI, 404, illus, Hardcover. Studies in Computational Intelligence, vol. 75, p. 207 (2007) ISBN: 978-3-540-73296-9

[6] Cruz-Chávez, M.A., Frausto-Solís, J.: Simulated Annealing with Restart to Job Shop Scheduling Problem Using Upper Bounds. In: Rutkowski, L., Siekmann, J.H., Tadeusiewicz, R., Zadeh, L.A. (eds.) ICAISC 2004. LNCS (LNAI), vol. 3070, pp. 860–865. Springer, Heidelberg (2004) ISSN: 0302-9743

[7] Cantu-Paz, E.: A Survey of Parallel Genetic Algorithms, Technical Report IlliGAL 97003, University of Illinois at Urbana-Champaign (1997)

[8] Beasley, J.E.: OR-Library: Distributing test problems by electronic mail. Journal of the Operational Research Society 41(11), 1069–1072 (1990); Last update 2003

[9] Sloan, J.D.: Network Troubleshooting Tools, p. 364. O'Reilly, Sebastopol (2001) ISBN 10: 0-596-00186-X

[10] Michel, M., Devaney, J.E.: A Generalized Approach for Transferring Data-Types with Arbitrary Communication Libraries. In: Proceedings of the Seventh International Conference on Parallel and Distributed Systems: Workshops, ICPADS, July 04-07, vol. 83. IEEE Computer Society, Washington (2000)

[11] Ganglia Monitoring System, Monitoring clusters and Grids since the year (2000), `http://ganglia.info/` (September 2009)

[12] Zalzala, P.J., Flemming. Zalsala, A.M.S. (Ali M.S.) (eds.): Genetic algorithms in engineering systems /Edited by A.M.S. Institution of Electrical Engineers, London (1997)

[13] Al Jadaan, O., Rajamani, L., Rao, C.R.: Improved Selection Operator for GA. Journal of Theoretical and Applied Information Technology, 269–277 (2008) ISSN 1992-8645

Dynamic Increasing the Capacity of Transmission Line Based on the Kylin Operating System

Wei Li, Zhiwei Feng, Jing Zhou, Kehe Wu, and Jing Teng

Control and Computer Engineering School,
North China Electric Power University, Beijing, China
liwei@ncepu.edu.cn

Abstract. Transmission of electric power has traditionally been limited by the conductor thermal capacity statically defined of a transmission line. However, based on real-time measurement and analysis of environmental data and transmission line characteristics, the capacity of transmission line can be greatly increased. Consequently, a monitoring system is designed and implemented to dynamic increase the transmission capacity. The system is executed on the Kylin operating system to promote its application in the electric power system. Based on the J2EE^{TM} platform, the monitoring system realizes a series of functions, such as real-time data collection, calculating and analysis, automatic storage of relevant parameters, transmission capacity prediction, etc. The monitoring system provides a reliable hardware and software platform for supervising and scheduling the grid. The effectiveness of the system is evaluated by its successful application in a grid data supervision system.

Keywords: Transmission line, Transmission capacity, J2EE^{TM} platform, Kylin operating system.

1 Introduction

With the sustained growth of economy and electricity load in recent years, traditional transmission of electric power, which has been limited by the rated conductor thermal capacity, cannot satisfy the power demands now. However, the construction of new transmission systems necessitates enormous investments in terms of time and capital[1, 2].Therefore, it is significant to improve the transmission capacity based on the existing transmission lines. In fact, the transmission capacity not only depended on the conductor temperature, but also related to the weather conditions, namely the ambient temperature, the sunshine intensity, and the wind velocity, etc. According to the current technical specification[3], the rated ampacity of a transmission line is defined based on the maximum allowed conductor temperature of 70°C, while under the "worst case" of predetermined climatic conditions - a quite high ambient temperature 40°C, a fairly low wind speed 0.5 m/s, and an extremely strong sunshine intensity 1000 W/m^2. This definition guarantees safe operation during the lifetime of transmission line.

T.-h. Kim, A. Stoica, and R.-S. Chang (Eds.): SUComS 2010, CCIS 78, pp. 10–16, 2010.

However, the rated ampacity is a very conservative theoretical value due to the extreme climatic preconditions. In fact, the practical ampacity could be much higher than the rated ampacity. In another word, a potential ampacity exists between the predetermined ampacity and the practical allowed one. Therefore, we propose a dynamic capacity-increasing approach to make full use of the potential ampacity, while guaranteeing the operation security of the transmission line.

In view of the current operation code of transmission line, and the technical specification of real-time supervision of electrical power system, a monitoring system is designed and implemented to dynamic increase the capacity of transmission line. The monitoring system is a Web application platform based on the Java™ 2 Enterprise Edition (J2EE™) distributed multi-layer application framework, using the Kylin operating system (OS) to enhance information security. The Kylin OS is a landmark achievement of the National High-Tech Research and Development Program of China (863 Program), which eliminates the monopoly of foreign OS and provides a reliable protection for Chinese information systems. The heterogeneity of the J2EE™ platform ensures the feasibility of running different operating systems on the server end and the client end, respectively. For example, the server end uses the Kylin OS for the reason of information security, whereas the client ends can still employ their original Windows® or other OSs. In conclusion, the monitoring system provides a reliable hardware and software platform for real-time supervising and scheduling the electrical grid.

In the following sections, we will first elaborate the principle of dynamic increasing the transmission capacity. Then the main functions and the architecture of the system is introduced. After that, the implementation of the system in the real world is evaluated in the section 3. Finally, we conclude and give some prospectives of the work.

2 Dynamic Transmission Capacity-Increase

Given the real-time data collected by the monitoring system, including the meteorological parameters, the conductor temperature, the current ampacity, etc, the theoretical maximum ampacity and the potential one are calculated. The calculation is based on the mathematical model defined in the power industry standard DL / T 5092-1999 "technical code for 110~500kV overhead transmission line" [3]. The formula is shown as follows:

$$I = \sqrt{\frac{9.92(T-t)(vD)^{0.485} + \pi\varepsilon DS[(T+273)^4 + (t+273)^4 - \alpha J_s D]}{R}} \tag{1}$$

where I denotes the ampacity of the transmission line, with the unit of ampere. T is the surface temperature of the line, with the unit of °C. Similarly, t is the ambient temperature, °C. v is the wind speed that is perpendicular to the line, m/s. D is the cable diameter, m. S is the Stefan - Boltzmann constant, whose value is 5.67×10^{-8} W/m^2. ε is the surface radiation coefficient of transmission line, which takes the value in the interval [0.23, 0.43]for bright new line; Whereas for old lines or the ones painted with black preservative, ε is in the interval [0.9, 0.95]. α is the surface endothermic coefficient of line. Similarly, the value of it is [0.35, 0.46] for bright new lines, and [0.9, 0.95] for old ones or lines painted with black preservative. J_s is the sunshine intensity,which takes the value of 1000W/m^2 in shiny days, and when the sunlight is perpendicular to the line. R is the ac resistance of the line, Ω/m. As have been mentioned above, the rated ampacity of a transmission line is traditionally defined using fixed values for the parameters in the equation (1). However, in the proposed monitoring system,

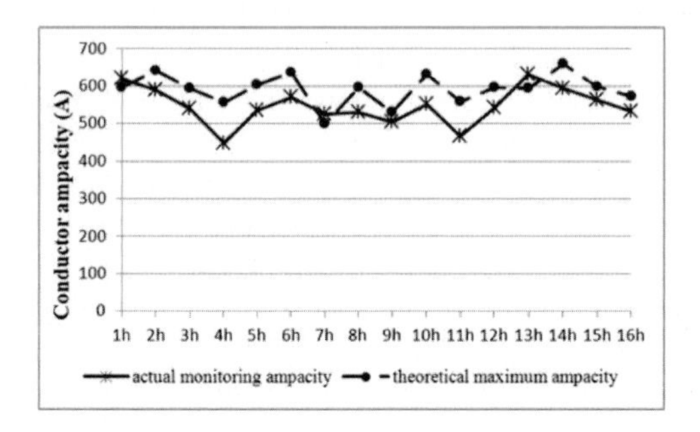

Fig. 1. Ampacity comparison diagram

we use the real-time data in the place of the fixed values to calculate the maximum allowed ampacity, which is therefore much more flexible and adaptable to the practical situation. In addition, the actual ampacity of the transmission line is real-time supervised by the system. As shown in Fig. 1, the theoretical maximum allowed ampacity is denoted by the dashed line, and the actual monitoring ampacity is demonstrated by the solid line, where their temporal values corresponding to each time point are denoted by dot and cross, respectively. The difference between the theoretical maximum ampacity and the actual monitoring one is thus the potential ampacity. Therefore, users can realize how much the transmission ampacity can be improved, while guaranteeing safe operation of the transmission line. According to our calculation, if the transmission line runs at the theoretical ampacity, we can improve the transmission capacity to more than 4 million kW $\cdot$ h in a year, which consists of 10% of its total original capacity.

3 System Design

3.1 System Architecture

The proposed monitoring system provides a platform to dynamically increase capacity, which consists of two parts, the hardware and the software. The hardware part is composed of a small meteorological station, and the sensors that are responsible of monitoring the temperature, the tension and the sag of the transmission line. The software part is based on the Brower/Server (B/S) structure, which adopts the popular J2EE^{TM} platform to ensure the heterogeneity of the system.

Hardware And OS Platform. The sensors are directly mounted on the transmission lines, which continuously detect and transfer the monitoring data to the remote data server. Therefore, the meteorological data, and the temperature, tension and sag of transmission lines around the monitoring points are collected in real time.

Since the security requirement of the electrical grid data is quite high, we choose the domestic Kylin OS as the operation platform for the monitoring software. It is also one of the innovation points of the proposed system, which realizes the application of the Kylin OS in power systems and guarantees the system information security as well.

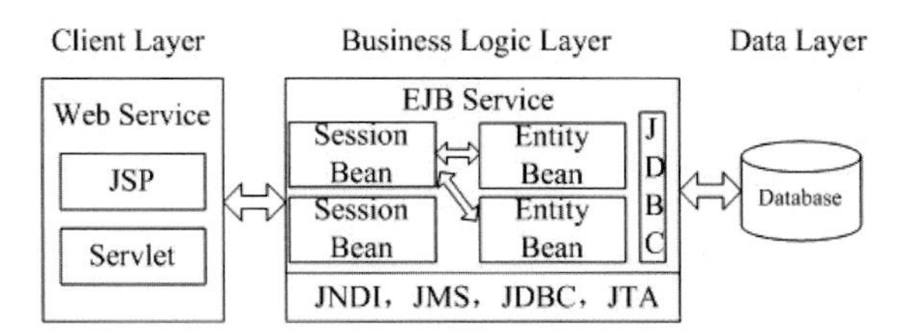

Fig. 2. Software architecture diagram

Software Architecture Design. The software part of the system uses the J2EE^{TM} platform, which adopts a reasonable and popular three-layer architecture. As shown in Fig. 2, the system is divided into the client layer, the logic layer and the data layer. The data layer adopts the Oracle® 10g Database, which mainly stores the real-time monitoring data and their upper and lower bounds. It is also responsible of transferring data to the other two layers, such as the real-time calculations of ampacity, the historical data and the related data tables, etc. Besides providing the raw data, the data layer stores the course data submitted or processed by the logic and the client layer as well.

The logic layer packages the process control logic and the operation codes of the system, which is in favor of the reusability and the scalability of the system. In this layer, the Enterprise JavaBeans^{TM} (EJB^{TM}) component is the core of the system, which realizes the system operation logic. It receives the query from the client layer, then retrieves the corresponding original data from the database and processes it, finally the processed data is sent back to the users. Furthermore, the EJB^{TM} container provides complex system services, such as transaction, life cycle, state management, multi-threading, and resource storage pool etc. Therefore, it is not necessary for the developers to program these basic but tedious system services. On the contrary, they can be much more concentrated on the operation logic. As a result, we have realized and packaged several specific function modules for the monitoring system in order to dynamically increase the transmission ampacity.

The client layer mainly consists of Java Server Page(JSP)pages, which is the interface between the system and users. In the client layer, users can inspect the functions completed by the logic layer. They can also interactive with the logic layer by input various parameters. For example, users can choose distinct monitoring station and different monitoring time; then, the logic layer generates the graphic data corresponding to these inputs; finally, the query result is returned back to the client layer, and displayed on the current page.

3.2 System Fuction

With respect to the system function, the proposed monitoring system is divided into seven modules: data collection and storage; conductor parameter management; dynamic capacity calculation; real-time display of the meteorological data, the conductor temperature and the ampacity; false alarms; historical data analysis; and statistical report. The function modules of the system are shown in Fig. 3.

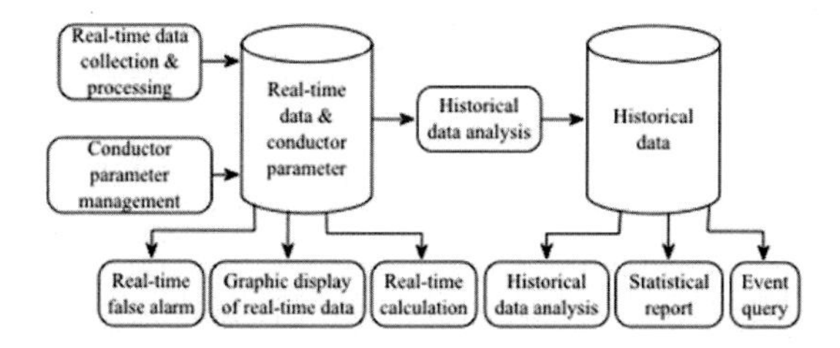

Fig. 3. Function module diagram of the system

Data Collection And Storage. There are two types of system monitoring data: real-time data and historical data. The database for real-time data stores only the daily data, which accelerates its data retrieval as a result. It also provides a caching mechanism for the data analysis and the historical data storage. The storage of the historical data prepares the source for any potential query. In fact, the database server and Web server of the system provide real-time data recording. In addition, the using of the advanced data storage strategy dynamically stores the collected data according to time stamp[5].

Conductor Parameter Management. The conductor parameters and the information of the transmission lines and the towers are managed centralized in the system, in order to provide users with rapid, accurate responses. In addition, the conductor parameter management module achieves relevant configuration of the conductor.

Real-time Data Display And Analysis. Specifically, the real-time data we discussed here include the meteorological data and the operation state of the transmission line, in which the meteorological data consists of the wind speed, the wind direction, the ambient temperature, the environment humidity and other data; the operation state of the line is composed of its temperature, sag, tension etc. The data display module uses several different forms, such as tables, charts, curves etc. to directly illustrate the monitoring result to the end users on Web pages.

The data analysis module is undoubtedly one of the key functions of the system, which provides real-time data query and analysis services to users. The results of query and analysis can be saved to Word, Excel or plain text format. By comparing, analyzing and researching the historical data, we can explore the transition principle or tendency of the transmission ampacity. Therefore, reasonable estimates and predictions on the ampacity of transmission line can be made. It consists not only the foundation of reliable and economic operation of the power system, but also plays an important part in the technical support for the power system.

4 System Implementation

4.1 Configuration

We employ the J2EE™ platform as the environment of system development. The Apache Tomcat and the JBoss® work as the web server and the application server, respectively. The Oracle® 10g acts as the database server of the data layer, and the JDBC driver of Microsoft® is used as the interface of database-driven. The Eclipse 3.2 is chosen as the development tools.

4.2 Operation Results

Real-time Monitoring The information needs to be real-time monitored including the SCADA current, the ambient temperature, the conductor temperature, the environment humidity, the wind speed and the wind direction, as shown in Fig. 4.

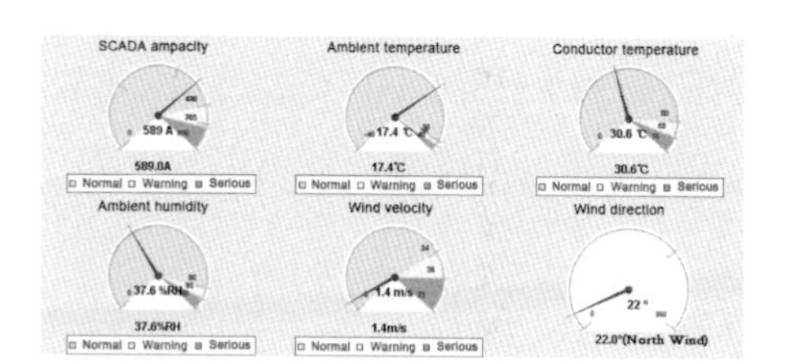

Fig. 4. Real-time monitoring diagram

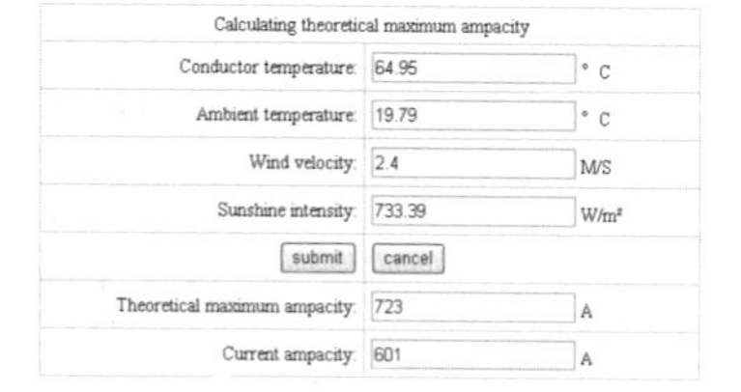

Fig. 5. Real-time monitoring diagram

Theoretial Ampacity Calculation. In fact, the system also allows users to calculate the theoretical ampacity by themselves, using the mathematical model described in the section 2. As shown in Fig. 5, users can manually submit the required parameters to obtain the theoretical maximum ampacity, and then compare it with the current ampacity.

5 Conclusion

By accurately monitoring the conductor temperature and the meteorological data in real time, the proposed system calculates the potential ampacity. Therefore, it enables the system operator to maximize the utilization of the transmission line, while ensuring safe operation. The benefits of dynamic capacity increasing consists of improving the reliability and the safety of the power system, reducing the capital expenditure, increasing the resources efficiency, etc. Based on the J2EE^{TM} multi-layer distributed architecture, the system is developed with strong information security, flexibility, cross-platform portability, scalability and maintainability. It has been successfully applied to a grid data supervision system, and has made great contribution for the development of the grid. The proposed system will help promoting the informatization, marketization and intelligentization of the power industry in the long run. However, the aging degree of the transmission line caused by maximizing its transmission capabilities is still under study.

References

1. Hongsheng, Y., Dawei, G., Weizhong, H., et al.: Feasibility Study on Increasing Conductor Allowable Temperature and Engineering Practice. Electrical Equipment 25(9), 1–7 (2004)
2. Zhiyin, Q., Qiping, Z.: Improving Transmission Ability of 500kV Lines in East China Power Grid. Electrical Equipment 10(6), 8–12 (2003)
3. Peng, Y., Xinyan, F.: Increasing transmission line capacity by DTCR model. East China Electric Power 33(3), 11–14 (2005)
4. DL/T 5092-1999P: Technical code for designing 110 500kV overhead transmission line
5. Kehe, W., Li, X., Jing, Z.: The research and implementation of a monitoring system of transmission lines to dynamic increasing capacity. The Informationization Annual Meeting of China's Power Industry (2009)
6. Fei Si Technology R & D Center: Developing Detail Explain JSP Application. Electronic Industry Press, Bingjing (2002)
7. Huoming, L.: EJB3.0 Entry Classical Arcobat. Tsinghua University Press, Beijing (2008)

Projective Illumination Technique in Unprepared Environments for Augmented Reality Applications

Giovanni Cagalaban and Seoksoo Kim*

Dept. of Multimedia, Hannam Univ., 133 Ojeong-dong, Daedeok-gu, Daejeon, Korea
gcagalaban@yahoo.com, sskim0123@naver.com

Abstract. Most augmented reality (AR) applications in prepared environments implement illumination mechanisms between real and synthetic objects to achieve best results. This approach is beneficial to tracking technologies since additional visual cues provide seamless real-synthetic world integration. This research focuses on providing a projective illumination technique to aid augmented reality tracking technologies that work in unprepared environments where users are not allowed to modify the real environment, such as in outdoor applications. Here, we address the specific aspects of the common illumination problems such as efficient update of illumination for moving objects and camera calibration, rendering, and modeling of the real scene. Our framework aims to lead AR applications in unprepared environments with projective illumination regardless of the movement of real objects, lights and cameras.

Keywords: Augmented Reality, Illumination Technique, Unprepared Environment.

1 Introduction

Augmented reality (AR) applications superimpose virtual 3D objects on a real world scene in real-time that change users' point of view. In an ideal world, they appear to user as if the virtual 3D objects actually coexist in the real environment [1]. Implementing AR not only enables users to perceive a synthesized information space, but it also allows them to naturally interact with the synthesized information using frameworks adapted from real-world experiences.

The use of computer vision algorithms in AR applications supports the detection, extraction and identification of markers in the real scene. Currently, AR applications are successful in prepared environments where they produce best results in achieving pixel-accurate registration in real time [2]. These environments enable system designers to have full control over the objects in the environment and can modify it as required. However, many potential AR applications have not been fully explored in unprepared environments due to inaccurate tracking. These include applications for drivers operating vehicles, soldiers in the field, and hikers in the jungle which could improve navigation, tracking, situational awareness, and information selection and retrieval. Several AR systems rely upon placing special markers at known locations in

* Corresponding author.

T.-h. Kim, A. Stoica, and R.-S. Chang (Eds.): SUComS 2010, CCIS 78, pp. 17–23, 2010.

the environment such as in [3]. However, this approach is not practical in most outdoor applications since one can not accurately pre-measure all objects in the environment. The inability to control the environment also restricts the choice of tracking technologies. Many trackers require placing active emitters in the environment to provide illumination mechanisms for tracking objects.

Many problems are still largely unresolved in tracking of arbitrary environments and conditions such as indoors, outdoors, and locations where the user wants to go. More so, illumination techniques in unprepared environments are difficult since the range of operating conditions is greater than in prepared environments. Lighting conditions, weather, and temperature are all factors to consider in unprepared environments. For instance, the display may not be bright enough to see on a sunny day. Visual landmarks that a video tracking system relies upon may vary in appearance under different lighting conditions or may not be visible at all at night. Additionally, the system designer cannot control the environment. It may not be possible to modify the environment.

This paper develops a projective illumination framework to aid tracking technologies for augmented reality applications. This framework is based on infrared filtering (IF) aimed to work in unprepared environments. The illumination technique addresses illumination updates, rendering, and camera calibration. Additionally, this study focuses on hybrid tracking technology that combines multiple sensors in ways that compensate for the weaknesses of each individual component. In particular, this research concentrates on the problems related to registration and calibration for real-time systems.

The rest of this article is organized as follows. Section 2 discusses related works. In section 3, we present the system requirements of the ubiquitous healthcare system. Section 4 follows with the ubiquitous healthcare system with network mobility support. Section 5 presents the implementation of the prototypical testbed network and lastly, section 6 concludes with summary and future work.

2 Related Works

Most of the AR systems have been applied to indoors and in prepared environments. Few AR systems operate in unprepared environments where the user cannot modify or control the real world. The first known system implemented in unprepared environment is the Touring Machine of Columbia [4] which uses commercially available no-source orientation sensors combined with a differential GPS.

Little attention has been given to the problems of the interaction of illumination between the real and synthetic scenes. Pioneering work in this domain has been performed by Fournier et al. [5]. This has shown how the computation of common illumination between the real and synthetic scene results in a greatly improved graphical environment with which the user can interact. The use of real video images eliminates the need to model complex environments in great detail, and provides a realistic image to the user naturally. In what concerns illumination, the introduction of virtual objects in a real scene becomes much more natural and convincing when light exchanges between real and synthetic objects are present in the composite images presented to the user. The problem of representing the dynamic range of real-world lighting conditions has been addressed by [6]. By capturing images at different levels of exposure, the response function of the imaging system may be recovered, and the images combined into a single high-dynamic range photograph. This algorithms use expensive global illumination techniques to illuminate the synthetic objects. With the exception of [7], this technique is not applicable when any form of interaction is required. Most of these algorithms for AR systems were applied in prepared environments.

3 Hybrid Tracking Algorithm

Various technical challenges are facing various applications of AR systems. One of the key challenges is the accurate tracking that measures the position and orientation of the observer's location in space. Without accurate tracking which positions virtual objects in their correct location and time, the illusion that they coexist with real objects are not possible. Central to the function of a tracking system is the exact alignment of virtual information with the objects in the real world that the user is seeing. This requires that the exact viewing position and viewing direction of the user are known.

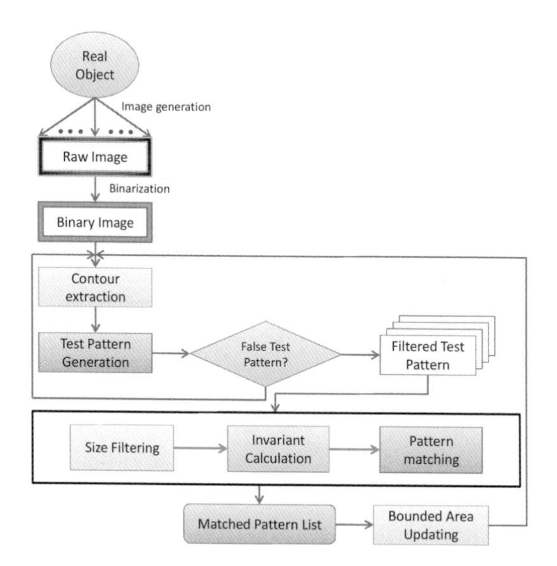

Fig. 1. Hybrid tracking algorithm

This study first presents a hybrid tracking algorithm to aid in the process of creation, identification, and tracking of patterns to be used in unprepared environments. This type of environment is defined by the properties of IF light and retro-reflexive materials used to accentuate the markers from the rest of the scene. The algorithm is shown in Fig. 1.

This algorithm was developed to work in an unprepared environment to avoid generation of false markers that would limit its performance in real-time. With the aid of IF pass filter and illumination, it discards the occurrence of false markers and enhances its performance. This algorithm performs invariant calculation to support AR applications. This can be useful in applications for adding or visualizing information over previously marked objects, where the main problem is the necessity of large patterns in the scene. The invariant feature in the hybrid tracking application can be used to generate a specific format serving as a filter to discard several candidate groups of 4 test patterns that do not fit the format. Additionally, generation of a bounded area around the pattern position in the current frame is a technique to reduce computational cost. Once the pattern is found and validated, the system creates a bounding area, used as a simple way to predict and restrict the area where a well recognized pattern may appear.

4 Implementation Design

Here, we present the proposed setup of the hybrid tracking system. Infrared emitting diodes with optimized line densities served as illumination sources for the tracking of objects. A camera and a laptop is used for position tracking and to record video data, respectively. Global positioning system (GPS) receiver unit can provide measurement of the position of any point on the Earth. Other sensors such as rate gyroscopes, compass, and tilt sensors can predict motion and provide orientation. Position as well as orientation tracking is needed. Orientation tracking is much more critical than position tracking as a small rotation of the head will have a larger visual impact than a small movement to the left or right. The system also composes of a lightweight head mounted display that offers the viewing of real and virtual objects. Fig. 2 shows the flow of data in the system.

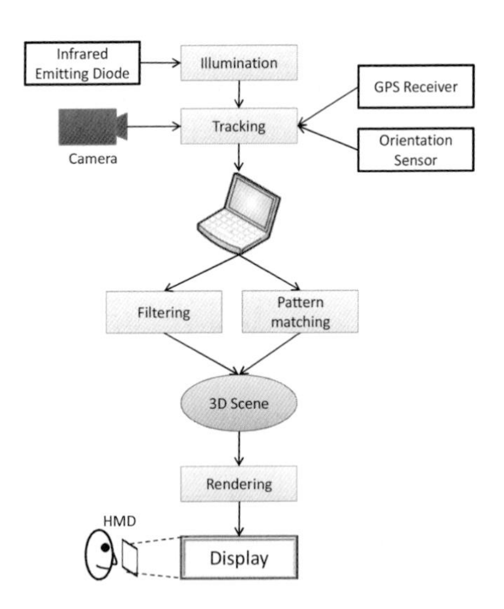

Fig. 2. Dataflow diagram

5 Camera Calibration

One of the requirement of the system is the computation of the intrinsic parameters of the camera such as focal length and aspect ratio using an image of a calibration pattern [8].We then took one image of a non-planar calibration pattern, for instance, a real object with visible features of known geometry. With minimal interaction an estimate of the camera parameters is computed. This estimate is then refined by maximizing over the camera parameters the sum of the magnitudes of the image gradient at the projections of a number of model points. The output of the process is an PRQTS matrix which is decomposed as the product of a matrix of intrinsic parameters and a SRQUS displacement matrix.

6 Projective Illumination and Image Acquisition

This section deals with the image acquisition and illumination techniques to correctly illuminate synthetic objects. This study performs illumination at a single position in the scene and assumes that this is a good approximation of the actual light illuminating each synthetic object. The image acquisition process begins with the capture of the background image. We employ a standard calibration algorithm to determine the position and orientation of the camera [9]. Then we construct a high-dynamic range omni-directional image that represents the illumination in the scene which is achieved in a similar fashion to [10], whereby multiple images of a shiny metallic sphere are captured at different exposure levels. These images are then combined into a single high-dynamic range image.

Once an omni-directional radiance-map has been captured, it must be manipulated into a form suitable for rendering. As each synthetic object is drawn, we use hardware-accelerated sphere-mapping to approximate the appearance of the object under the captured illumination conditions.

In [11] it was proposed that an omni-directional radiance-map could be pre-integrated with a bi-directional reflectance distribution function (BRDF) to generate a sphere-map that stores the outgoing radiance for each surface normal direction. This pre-integration which needs to be performed for each different BRDF, is too slow whenever an application requires the interactive manipulation of materials.

We compute a diffuse irradiance map, as well as specular maps for varying values of the surface roughness parameter. At run-time, we combine these basis maps according to the particular reflection model coefficients associated with each material, and generate the appropriate sphere-map. The final step of sphere-map generation is the application of the camera response function to translate the radiance-map into a texture map that can be utilized by the graphic device.

7 Case Study

This section presents one of the earliest algorithms for compositing synthetic objects into background images as proposed by [12]. Their solution involved estimating the location of the sun and levels of ambient light to illuminate synthetic architectural models rendered onto background photographs. An example is shown in Fig.2 [12] which shows the illumination of augmented objects being matched with the surroundings.

Fig. 3. The first image depicts the background scene while the second shows the augmented scene where the illumination of the augmented objects is matched to their surroundings

8 Conclusion

In this paper we have presented a framework for dealing with the problem of illumination between real and synthetic objects and light sources in unprepared environments. Much remains to be done to continue developing system that work accurately in arbitrary, unprepared environments. Our framework will hopefully lead to AR applications with efficient update of illumination without restrictions on the movement of real or synthetic objects, lights, and cameras. This research aimed to have easier modeling and calibration, faster illumination updates and rapid display of AR scenes.

For future work, we will focus on removing the restrictions one by one, to achieve projective illumination technique to other scenarios such as moving camera, moving lights, and moving real objects. We will also investigate the issues regarding pre-recorded video sequences, before taking the plunge into real-time acquisition.

Acknowledgements

This paper has been supported by the 2010 Preliminary Technical Founders Project Fund in the Small and Medium Business Administration.

References

1. Azuma, R.: A survey of augmented reality. Presence: Teleoperators and Virtual Environments 6, 335–385 (1997)
2. State, A., Hirota, G., Chen, D., Garrett, B., Livingston, M.: Superior augmented reality registration by integrating landmark tracking and magnetic tracking. In: Proceedings of SIGGRAPH 1996, pp. 429–438 (1996)

3. Neumann, U., Cho, Y.: A self-tracking augmented reality system. In: Proceedings of ACM Virtual Reality Software and Technology, pp. 109–115 (1996)
4. Feiner, S., MacIntyre, B., Hollerer, T.: A touring machine: prototyping 3D mobile augmented reality systems for exploring the urban environment. In: Proceedings of First International Symposium on Wearable Computers, pp. 74–81 (1997)
5. Fournier, A., Gunawan, A.S., Romanzin, C.: Common illumination between real and computer generated scenes. In: Proceedings of Graphics Interface 1993, pp. 254–262. Morgan Kaufmann, San Francisco (1993)
6. Debevec, P.E., Malik, J.: Recovering high dynamic range radiance maps from photographs. In: Proceedings of SIGGRAPH 1997, pp. 369–378 (1997)
7. Drettakis, G., Robert, L., Bougnoux, S.: Interactive common illumination for computer augmented reality. In: Eurographics Rendering Workshop 1997, pp. 45–56 (1997)
8. Robert, L.: Camera calibration without feature extraction. Computer Vision, Graphics, and Image Processing 63(2), 314–325 (1995)
9. Tsai, R.Y.: A versatile camera calibration technique for high accuracy machine vision metrology using off-the-shelf TV cameras and lenses. IEEE Journal of Robotics and Automation 3(4), 323–344 (1987)
10. Debevec, P.: Rendering synthetic objects into real scenes: Bridging traditional and image-based graphics with global illumination and high dynamic range photography. In: Proceedings of SIGGRAPH 1998, pp. 189–198 (1998)
11. Heidrich, W., Seidel, H.P.: Realistic, hardware-accelerated shading and lighting. In: Proceedings of SIGGRAPH 1999, pp. 171–178 (1999)
12. Nakamae, E., Harada, K., Ishizaki, T., Nishita, T.: A montage method: The overlaying of the computer generated images onto a background photograph. In: Computer Graphics Proceedings of SIGGRAPH 1986, vol. 20(4), pp. 207–214 (1986)

The Design of Modular Web-Based Collaboration

Ploypailin Intapong[1], Sittapong Settapat[3],
Boonserm Kaewkamnerdpong[2], and Tiranee Achalakul[1]

[1] Department of Computer Engineering
[2] Biological Engineering Program
King Mongkut's University of Technology Thonburi, Thailand
[3] Graduate School of Engineering, Shibaura Institute of Technology, Japan

Abstract. Online collaborative systems are popular communication channels as the systems allow people from various disciplines to interact and collaborate with ease. The systems provide communication tools and services that can be integrated on the web; consequently, the systems are more convenient to use and easier to install. Nevertheless, most of the currently available systems are designed according to some specific requirements and cannot be straightforwardly integrated into various applications. This paper provides the design of a new collaborative platform, which is component-based and re-configurable. The platform is called the Modular Web-based Collaboration (MWC). MWC shares the same concept as computer supported collaborative work (CSCW) and computer-supported collaborative learning (CSCL), but it provides configurable tools for online collaboration. Each tool module can be integrated into users' web applications freely and easily. This makes collaborative system flexible, adaptable and suitable for online collaboration.

Keywords: CSCW / CSCL / Web-based collaboration / Autism community.

1 Introduction

Collaboration is an activity in which two or more people or organizations interact and work together. Many collaborative systems have been implemented to support comfortable communication from different locations in virtual spaces. They provide workspace similar to the real world environment for interaction among people in virtual environment. People can use virtual tools which simulate communication activities that occur in real-world workspace. This makes collaborative systems be a popular communication channel for groups of people or organizations.

Through observing the literature, we found that there are two methods for using collaborative system. Most collaborative systems need users to install collaborative system beforehand. On the other hand, collaborative systems can also be used via a web browser of a central server with a few additional installations or even no installation required on client's side. Nevertheless, both methods provide virtual tools as a set or package for communication in virtual space.

The implementation of most existing collaborative systems combines overall virtual tools and collaborative functions together. This makes systems more complex

T.-h. Kim, A. Stoica, and R.-S. Chang (Eds.): SUComS 2010, CCIS 78, pp. 24–33, 2010.

and difficult to install; users have to handle the complexity of all functions even though they use only some. Moreover, when some parts of system fail, the main system and other parts also fail. Consequently, the system stability is decreased. In addition, combining everything together makes the systems inflexible to be applied to other works because developers cannot freely mix and match suitable virtual tools to their desired proposes. As a result, developers may need to construct new systems to cover and support all different desired requirements even though all systems hold the same basic functions. For example, collaborative systems designed for meeting focus on features that support communication whereas those designed for distance learning focus on features for presentation and explanation; both of them, however, provide same basic collaborative functions. To do so, it requires skills in programming and computer system technology which may be too difficult for normal users.

To support different requirements and to be integrated into various applications without implementing new systems, functions or tools of collaborative systems should be implemented so that they are independent from each other. Each tool can be set as a module, which can be reconfigured to suit different proposes. Users can choose modules to be integrated into their application freely and easily without implementation. This enables flexible, adaptable online collaboration.

In this paper, we propose Modular Web-based Collaboration (MWC) to provide such flexibility and adaptability to online collaboration. MWC uses the same concept as Computer Supported Cooperative Work (CSCW) and Computer-Supported Collaborative Learning (CSCL) which provide a virtual space for people to interact and cooperate with each other in some purposes via computers and networks. However, MWC provides communication tools and services for community websites or social networks that require collaboration which are called host applications. The service selection is not limited. Host applications can manage virtual tools independently and can integrate MWC services into their applications easily. MWC can support multiple host applications in the same time. When communication occurs, messages are sent/received between host applications and MWC server. Once MWC services are integrated into a host community website, end users can use collaborative system in their own environment. The interaction using MWC tools and services makes end users feel as all the communication occurs in their own community and they have privacy without having to share collaboration with other groups.

To establish the motivation through existing collaborative systems in the literature, the concept of collaboration and their related works to MWC are discussed in section 2. The design framework of MWC is described in section 3. To demonstrate the usability and flexibility of MWC, a scenario for using MWC on autism community space is presented in section 4. The conclusion of this study is included in section 5.

2 Collaboration Concept and Related Works

2.1 Web-Based Collaboration

With the rapid growth of Internet applications, many researchers in CSCW [1] and CSCL [2] have focused on the attempts to design system and architecture for distance learning in collaborative virtual environment. In this section, we present the design and implementation of existing web-based collaboration to support various objectives.

A web-based CSCW framework designed by Wang et al [3] is a web-based collaborative workspace using Java3D. This framework allows users to share workspace for design reviewing, production monitoring, remotely controlling and trouble-shooting using Java 3D model. It reduces network traffic and increases the flexibility of remote monitoring. It enables web-based synchronous collaboration with interactive control and quick response. It shows a high potential for web-based real-time distributed applications. Another CSCW framework proposed by Su et al [4] uses internet techniques including client-server and web-services to create an online collaborative design and manufacture; apart from internet techniques, web-enabled environment for collaboration, CAD/CAM, RELSP and distributed product design are employed in [4]. This framework allows users to share AutoCAD design and communicate online through speaking, writing messages on the board and seeing each other on the screen during the collaborative design process.

A research in CSCL designed by Poonam and Bhirud [5] is an interactive web-based system for learning image processing. The developed system consists of four modules including dynamic website, web application with interactive contents, quizzes and assignments. This web-based training integrates java technology platform into the web application based on JSP web application framework. It uses java applets as well as HTML contents or XML for course contents which are essentially run from a web based learning system. Interactive learning object uses MATLAB engine for image processing purposes. This system is suitable for replacing the traditional homework assignments. Other CSCL presented in [6] is the virtual math teams' project (VMT). VMT software was designed for group of 2 to 10 students to discuss mathematics in real time via chat box and whiteboard. This framework integrates synchronous and asynchronous communication together. While the users are communicating by chatting and writing on whiteboards, the systems automatically create a wiki in the portal with the same information as in the chat environment.

2.2 Basic Collaborative System's Function

By reviewing existing collaboration systems [7-9], we can classify basic functionalities for collaboration into three groups: communication, presentation, sharing functions. The most simple communication function is messaging. All exist collaborative systems have chat tools for real-time messaging. In addition, some systems also provide communication in terms of voice and video. The presentation function of collaborative systems can support image, audio and video presentations so that explanation can be done more clearly and understandably; whiteboard is an example of presentation tools. The sharing function supports the exchange of information such as files, computer screens and applications. To meet specific needs, additional virtual tools can be developed; for example, in collaboration for meeting a recording function may also be included. In addition, functionalities for collaboration also include user and room managements which support user registration and creating virtual room.

In term of utilization, current collaborative systems can be categorized into two types. On one hand, users need to install the system on the client's side. For some systems, the installation can be complicated. This makes them difficult to use. However, most of them are non-commercial. On the other hand, users can use collaboration systems via provided central servers. Hence, the installation on the client's side is reduced. Users do not need to have programming knowledge. In other words, general users can operate such systems without further implementation. Nevertheless, users need to pay for the service that lacks uniqueness as the user interface of system is fixed; users cannot modify the interface to meet their requirements. Moreover, all virtual tools are usually combined into a set or package which is not flexible to use. Users cannot choose some virtual tools as desired but have to use all virtual tools for different purposes (for example, online collaboration for meetings, selling, training and other online events) included in the same package [9]. In the case that some parts fail, it will affect other parts even the main system.

One example of existing collaborative system is an open-source collaboration system called Openmeetings [8]. This system can support 25 users per meeting room and sustain several meeting rooms at the same time. It has two kinds of meeting rooms including video conference room and audio conference room; both types of meetings can be recorded for further utilization. However, openmeetings combines everything together. This can make the system unstable and difficult to install. Other commercial collaborative system is WebEX [9]. This collaboration provides several sets of virtual tools for supporting various requirements; those tool sets, however, contain different combinations of basic collaborative functions to serve pre-defined purposes. These existing collaborative systems are not flexible to be applied to other works.

With the popularity of using online collaboration, it is beneficial to improve collaborative functions and systems so that online collaboration can be effectively used. In this research, we propose a new design framework that is flexible for different tasks; a tool set can be integrated into any web application easily without having to implement a new system to include another tool on the application.

3 The Design Framework

3.1 Conceptual Design

Modular Web-based Collaboration (MWC) is a platform which can reconfigure virtual environment for flexible usage as desired without implementing a new system. MWC designs the development virtual tools or functions independently which can be called as modules. Each module in MWC platform can be integrated into any host applications that want to add virtual tools as parts of them.

MWC provides collaboration tools in virtual space for host application. Host application's clients can use collaboration tools with their host application directly without sharing virtual space with others. Host applications can integrate MWC tools and MWC services into their applications easily without the need of writing program all over again. Firstly, host applications must register to verify their identity before using MWC tools and services via MWC website. Next, they can freely select communication tools that they want. In other words, all tools are independent so host applications can mix and match those tools suite to their applications. After registration and tool selection, host applications receive Code ID for authentication and get their selected virtual tool as an object as well as a script for calling. In the case that host applications select more than one virtual tool, host applications will receive the same number of objects as the number of selected tools; these tools are not combined together. Host applications can put these objects in their web server and insert the scripts in the source code of web page that displays collaborative system. MWC provides communication tools and services for host application including chat box, audio/video conferencing, whiteboard and desktop sharing. The other functions such as user and room management must be provided by host application. After MWC installation is done, host application's clients are allowed to interact with their community in customized virtual environment through their host application directly.

With the proposed concept, the benefits of MWC platform include the following:

- *Privacy*: MWC allows users to use collaborative functions within the host application's environment. This gives users a sense of identity. They do not need to share collaborative systems with people outside the community and also have privacy within their own community.
- *Flexibility*: The MWC-based system can be designed to suit the usage of the community. When using the system within the community, users can learn only collaborative functions required for their community. Hence, it is easy to use and suitable for general users with no sophisticated knowledge on technology.

However, as communication tools are designed to be independent of each other, it requires time for installing each of the selected tools. In addition, cross interaction between two or more tools is restricted; for example, users are not allowed to watch a video in chat box. As experienced in current collaborative systems, the number of users using MWC is limited by the performance of the hardware. Nevertheless, server farm and load balancing can be introduced to increase the capacity of the systems.

3.2 Architecture Design

As MWC provides communication tools and services to be integrated into host applications so that users feel as if they use these tools through their applications, the interaction enabling such characteristic involves clients, host applications and MWC. Clients are subscribers of host applications and MWC. They directly contact with host applications and indirectly communicate with MWC. Host applications are both providers and subscribers. They use services from MWC to provide services for their clients. MWC is the providers of all communication tools and services. The architecture overview of MWC platform is illustrated in Fig 1.

The architecture shown in Fig 1 can be referred as client-server architecture, which is software architecture describing communication between client and server. In this case, clients interact with host applications (C-H) and host applications register with MWC (H-MWC). Before host applications can provide MWC services to clients, H-MWC interaction is required; host applications must subscribe with MWC for service registration. Developer registers via MWC web interface and selects communication tools and services. At C-H interaction, users access website and interact with their host interface. When a client communicates with other clients on the same host applications, communication sessions between client application and host servers are initiated. Upon incoming requests, host servers act as an agency forwarding communication messages to MWC. Then, MWC responds and sends the message back to the host application which forwards to clients via web interface. Note that web interface and workflow design on each host application can be different.

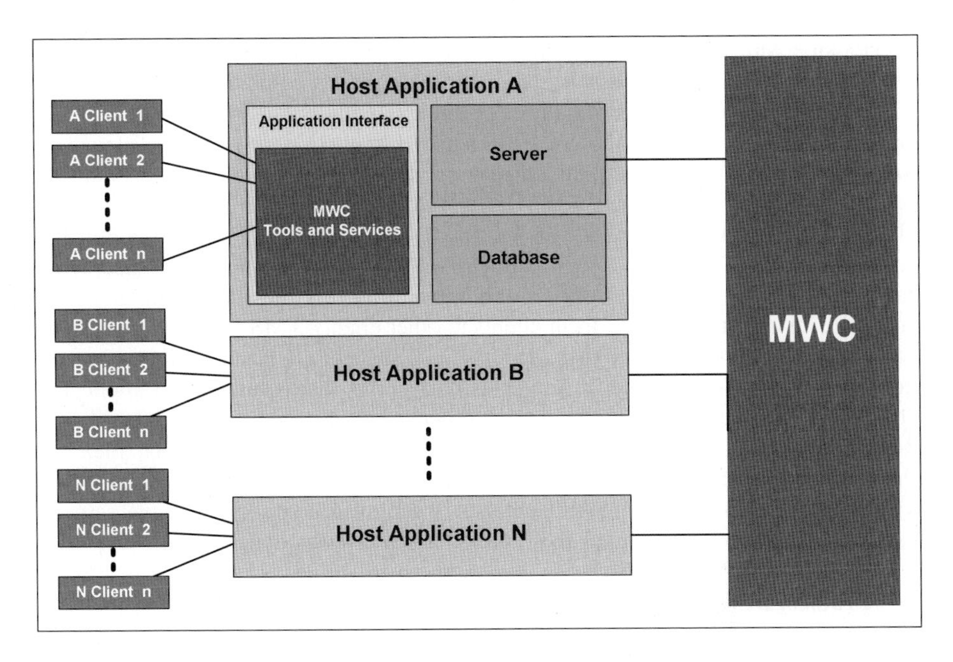

Fig. 1. Architecture overview of MWC platform

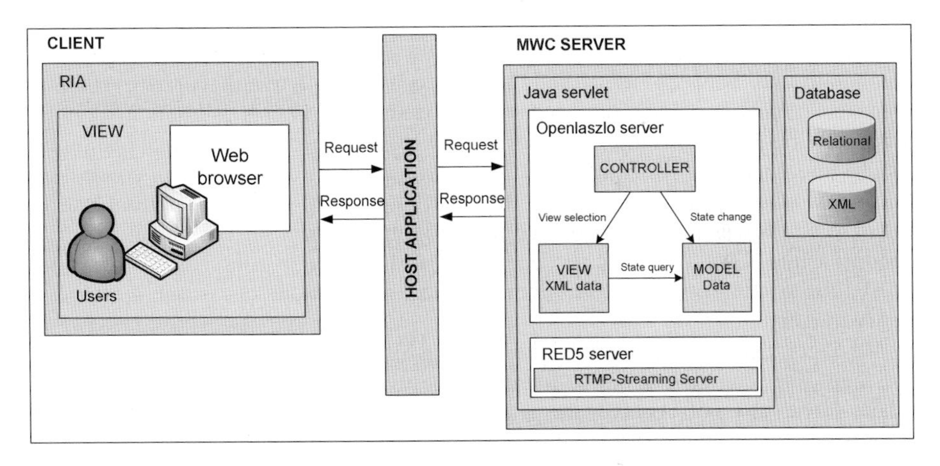

Fig. 2. Model-View-Control (MVC) architecture for MWC platform

MWC platform focusing on clients using MWC communication tools and services via agency (host application) can be elucidated in Model-View-Control (MVC) architecture presented in Fig 2. MVC architecture isolates the domain into three entities: the model (managing information), view (presenting the model) and controller (handling responses). The MWC communication tools are displayed in host application's interface. When users interact with web interface, the controller handles the event and converts into appropriate reaction. Then, the controller notifies the model of reaction and passes to view to display in web interface. This occurs in client's side with RIA technology. If it is unable to form model of reaction at client side, model will send the request to server and, then, server responds.

To provide some useful information for implementing collaborative systems based on MWC platform, we observe different technologies providing video/audio streaming used for implementing current collaborative systems and conclude the potentials as follows: *Adobe Flash* is a multimedia platform which already has libraries to support media streaming, so it helps reduce time for programming. It is positioned as a tool for RIAs which is in currently the top three platforms (apart from *Java Fx* and *Microsoft Siverlight*). Flash server transfers message communication, video stream and audio stream from clients to other clients. FMS (Adobe Flash Media Server) is the most mature one that supports mass scale of applications; it is, however, expensive. WMS (Wowza Media Server) is flash server that is suitable for application of primary and media enterprises for which cost and stability are most concerned. Red5 is an optimizing open source flash server written in Java. Adobe Flex and Openlaszlo are platforms for developing RIAs based on Adobe Flash. Adobe Flex is written in action script, whereas Openlazlo is deployed as traditional Java servlets. Laszlo applications are compiled and returned to the browser dynamically.

3.3 Protocol Design

As MWC supports multi host applications, it is essential that MWC knows from where the message was sent and to where it sends the response. The detailed protocol of MWC and supportive technology are described and illustrated in Fig 3. After host applications have already registered and integrated communication tools, communication between users and MWC occurs when a collaborative room is created. The collaborative room, a channel of one or more clients, is implicitly created when the first client joins it. First client in each channel becomes a channel operator.

The initial message for communication is sent to MWC server to open connection. It consists of three main parts including code ID, channel ID and user ID. Code ID refers to host application. Chanel ID refers to group of users such as channel code or channel name. User ID refers to client such as IP client or username. When other users join existing collaborative room, the host application sends user message to notify the server.

Messages used for communication among users consist of user ID and chat message. User can send a one-to-one message to communicate with another client; other clients do not receive the message. For one-to-one communication, the information specifying the receiving client must be added into a message. In one-to-many communication, the message is sent to all clients in channel.

If MWC server cannot find the code ID, MWC must ignore the initial message of that communication. MWC server opens connection with host application only when code ID is correct. Then, MWC server performs connection thread with initial communication information. MWC server adds users to connection thread when they join the collaborative room. It removes users from connection thread when users leave collaborative room and also closes the connection when the last user leaves the room.

When MWC server receives a message, it knows the user ID from the message and refers to corresponding thread connection. It broadcasts to all clients (or sends to single user) the chat message in the thread connection. The messages are sent according to queue as FIFO (First In First Out).

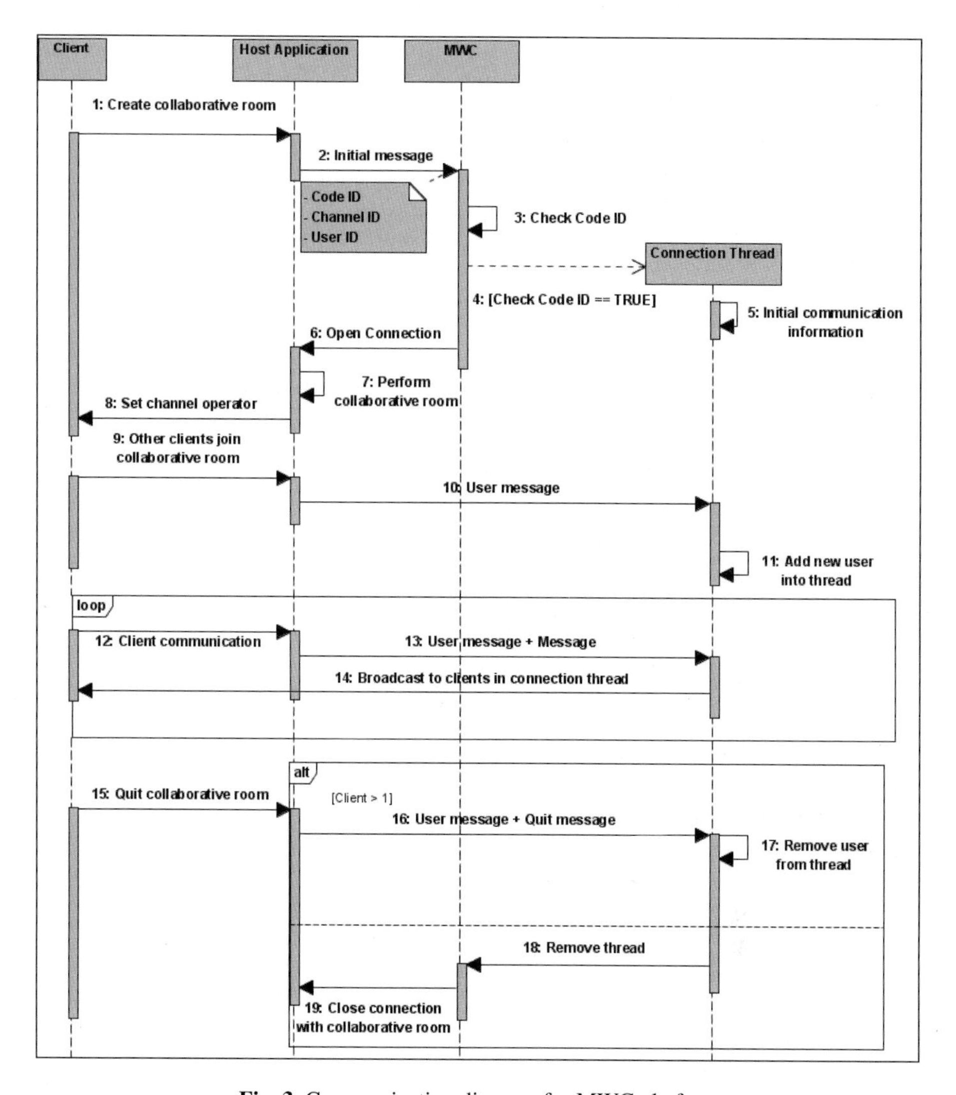

Fig. 3. Communication diagram for MWC platform

4 Scenario: Autism Community Space Based on MWC

Autism is a disorder concerning abnormalities of brains. It requires specialists to diagnose; however, there are only a few in Thailand. The learning of this medical specialty requires clinical study, but most students are doctors in local hospitals outside the cities; it makes this specialized training almost impossible. Interaction in autism community can bring about benefits of sharing information. Nevertheless, in some cultures having child with abnormalities can be seen as embarrassment to families; hence, it is better to have privacy in using online collaborative systems in their own community. Apart from that, the flexibility of collaborative systems which is easy enough for general users to operate is appreciated as communities' members may not be so skillful in computer technology to utilize an advanced system to its maximum advantage. Therefore, integrating MWC tools in community's host applications seems suitable for autism community. To demonstrate the utilization of MWC platform, a sample scenario of online collaboration in autism community in Thailand is discussed. Such online collaboration can be beneficial as a distance learning tool for local doctors and a consultation tool for teachers and parents.

Sample Scenario for Using Autism Community Space Based on MWC:

John, the director of autism community in Thailand, wants to create community website which has a virtual space for communicating and sharing information from different locations. He wants this to be easy to use; even someone who has no IT knowledge should be able to use this collaborative tools on the website. Moreover, he wants no links to other websites in the community website.

Bob, a web administrator, is assigned for this task. He creates autism community website with need collaboration tools so he accesses to MWC website and registers. After registration, he is allowed to select communications tools such as chat tools, audio/video conference tools that he can customize to suitable his purpose. Then, he receives code ID for authentication with MWC communication objects and scripts for installation in autism community website. He puts the objects in the community's web server and inserts scripts in each collaboration room design which has different virtual environments such as conferencing room, presentation room and chat room.

Alice, a pediatric doctor in local hospital, wants to consult with **Sara** who is an experienced specialist of autism center in Bangkok about clinical study of her autism patients, but it is inconvenient and too expensive to travel to Bangkok. She reserves a private conferencing room and invites Sara through autism community website. She uses video conferencing tool to communicate with Sara, uses video playing tool to show video of her patients, and also uses file sharing tool to share treatment history.

Clare, a parent of an autism patient, has doubts on her child's behavior. Sara suggests her to join autism community website. Clare creates a public chat room on child behavior topic. She shares information with other parents and seeks advices for effective treatments. Sometimes, Sara joins a chat room as an audience to observe discussions. Sara always follows information sharing in autism community website so that she could gain interesting points for further study. When finding useful information, she reserves a presentation room for training on those topics. Other members can be audiences. Sara controls audio conferencing tool for receiving questions and uses whiteboard tool to assist her explanation for better understanding.

5 Conclusion

In this paper, we propose a design of a new collaboration platform which is highly flexible and can be applied to serve different proposes without having to implement a new system for each additional specific requirement. MWC provides configurable tools for communication in virtual space which can be integrated into community own appropriate virtual environment easily. In sum, MWC platform provides privacy within the community and flexibility to collaborative systems which enables easy-to-use online collaboration. MWC can be effectively applied to fulfill different usages for autism community as demonstrated in section 4. Members can communicate and share information from different locations through autism community website easily without advanced IT knowledge. New virtual tools or versions of existing virtual tools can be easily added. The benefit of MWC can also be extended to other communities.

Acknowledgement. This research is a part of the project supported by The National Science and Technology Development Agency of Thailand (P-09-00346).

References

1. Kevin, L.: Computer-Supported Cooperative Work. In: Encyclopedia of Library and Information Science, pp. 666–677 (2003)
2. David, W., Katherine, A.: Why all CSL is CL: Distributed Mind and the Future of Computer Supported Collaborative Learning. In: Proceedings of the 2005 conference on Computer support for collaborative learning: learning 2005: the next 10 years!, Taipei, Taiwan, pp. 592–601 (2005)
3. Wang, L., Wong, B., Shen, W., Lang, S.: A Web-based Collaborative Workspace Using Java 3D. In: The sixth International Conference on Computer Supported Cooperative Work in Design, London, Ont., pp. 77–82 (2001)
4. Su, D., Li, J., Xiong, Y., Zheng, Y.: Application of Internet Techniques into Online Collaborative Design and Manufacture. In: Proceedings of the Ninth International Conference on Computer Supported Cooperative Work in Design, vol. 2, pp. 655–660 (2005)
5. Poonam, S.T., Bhirud, S.G.: Interactive Web Based Learning: Image Processing. In: International Conference on Application of Information and Communication Technologies, AICT 2009, pp. 1–4 (2009)
6. Gerry, S., Murat Perit, C.: Integrating Synchronous and Asynchronous Support for Group Cognition in Online Collaborative Learning. In: Proceedings of the 8th International Conference on International Conference for the Learning Science, Utrecht, The Netherlands, vol. 2, pp. 351–358 (2008)
7. Wenhua, H., et al.: Computer Supported Cooperative Work (CSCW) for Telemedicine. In: Proceedings of the 2007 11th International Conference on Computer Supported Cooperative Work in Design, pp. 1063–1065 (2007)
8. Openmeetings, `http://code.google.com/p/openmeetings` [February 20, 2010]
9. WebEx, `http://www.webex.com` [February 20, 2010]

Design of a Forecasting Service System for Monitoring of Vulnerabilities of Sensor Networks

Jae-gu Song[1,*], Jong hyun Kim[2], Dong il Seo[2], and Seoksoo Kim[3]

[1,3] Dept. of Multimedia, Hannam Univ., 133 Ojeong-dong, Daedeok-gu, Daejeon, Korea
[2] Electronics and Telecommunications Research Institute, Daejeon, Korea
bhas9@paran.com, jhk@etri.re.kr, bluesea@etri.re.kr, sskim0123@naver.com

Abstract. This study aims to reduce security vulnerabilities of sensor networks which transmit data in an open environment by developing a forecasting service system. The system is to remove or monitor causes of breach incidents in advance. To that end, this research first examines general security vulnerabilities of sensor networks and analyzes characteristics of existing forecasting systems. Then, 5 steps of a forecasting service system are proposed in order to improve security responses.

Keywords: Monitors Vulnerabilities, Sensor Network, Forecasting System, Security.

1 Introduction

The number of security breach is increasing every year and so is the material loss. Although the amount of damage caused by malignant codes has not been measured, the quantity is enormous, and such damage occurs mainly through common services (e.g. game hacking, messenger and voice phishing)[1]. Moreover, previous cyber attacks are now exploiting wireless or sensor networks, calling for more research efforts on new security measures. In particular, the damage arising from a zero-day attack is rapidly increasing, but forecasting or warning systems are still significantly lacking [2, 3]. Forecasting systems are especially useful in case solutions can not be provided immediately for worm/viruses or other vulnerabilities playing havoc with the society. Similar to a weather forecasting system, a cyber weather forecasting system predicts cyber threats or attacks in order to help computer users prepare for them in advance [4, 5].

In this study, a cyber security forecasting system is developed for a sensor network, which is vulnerable to an open environment. A sensor network is composed of a great number of sensors and has a limited computing capacity, which makes difficult security program installation. Also, in sensor networks, the topology constantly changes due to frequent insert and removal of sensor nodes. Considering such frequent changes, therefore, this study designs a security forecasting system and

* Corresponding author.

T.-h. Kim, A. Stoica, and R.-S. Chang (Eds.): SUComS 2010, CCIS 78, pp. 34–38, 2010.

carries out basic research for the application. Chapter 2 will explore the characteristics of forecasting systems, security forecasting systems, and sensor networks. Chapter 3 will design a management service system for prediction. Lastly, Chapter 4 will discuss improvements made in sensor networks security as well as follow-up studies.

2 Basic Research

2.1 Security Vulnerabilities of Sensor Networks

Sensor networks are randomly arranged in a limited area in order to sense objects and transmit the data to a base station. Therefore, numerous nodes increase the danger of data loss and data redundancy. Due to this nature, sensor networks are more vulnerable to security threats than general networks. Particularly, since a sense network is placed in an open environment, it becomes extremely difficult to prevent the leaking of data. As a result, the integrity of data can be damaged when malicious nodes constantly transmit data or data are easily modified, as well as causing excessive load on other nodes. In particular, because there is no special management for the nodes already arranged in WSN, malicious users may easily destroy, capture, or compromise nodes [6].

2.2 Research on Security Forecasting Systems

Most data management systems in modern societies have made various efforts to effectively respond to a multitude of breach incidents [7, 8] and, especially, governments and companies developing security products are equipped with a system to warn against security threats through announcements. Nevertheless, most hacking/phishing incidents, worm/viruses, and vulnerabilities are reported by consumers and announcements/alarms are issued based on such reports. Such a system can offer responses only after an incident occurs, unable to provide timely information to other users. Fig. 1 shows how existing systems respond to breach incidents.

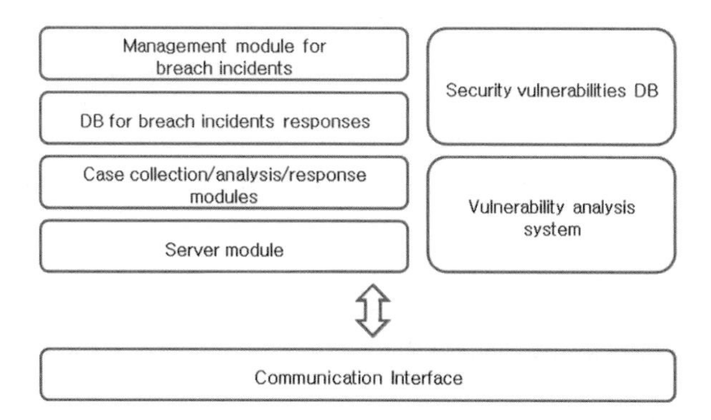

Fig. 1. Existing systems responding to breach incidents

2.3 Previous Studies on Security Forecasting Systems

For more systematic security forecasting services, existing systems need to be analyzed first. One of the most common forecasting systems found everywhere in the world is a weather forecasting system. Although it could vary according to nations, the basic structure of the system is very similar. A general weather forecasting system has the following seven steps [9].

① Observation
② Analysis
③ Model guidance and interpretation
④ Coordination within the national centers
⑤ Product generation
⑥ Product dissemination
⑦ Coordination with customers

The 7 steps above include basic elements needed for prediction such as observation and analysis, which are also required by security forecasting systems.

Also, weather forecasting systems have visible signs (e.g. clear, slightly cloudy, very cloudy, rainy, snowy, sleet, thunder, lightning, and so on) for efficient information delivering. These are similar to the 5 steps used in security announcements such as normal, attention, caution, warning, and serious, which help users figure out circumstances more clearly.

Forecasting cycles are also a critical element in the system, and it is important to detect changes at the earliest possible time for effective responses. Frequent changes detected by weather forecasting systems are expected to occur in security forecasting also. Therefore, real-time monitoring and analysis should be possible for advance announcements.

3 Forecasting Service Systems for Sensor Networks

Forecasting services for sensor networks are largely divided into prevention, detection/analysis, responses, and restoration. Here, the prevention step is to restrain breach incidents from occurring by removing or monitoring causes for such incidents in advance. Based on this step, the system can be alert to and respond to a breach incident immediately and improves the abilities to detect security threats, raising the overall quality of the system.

This study proposes the following forecasting services for sensor networks.

① Monitoring detection methods and trends of attacks
② Analyzing breach cases and affected areas
③ Producing data
④ Tracing the patterns of attacks
⑤ Providing prediction information

Figure 2 shows these processes.

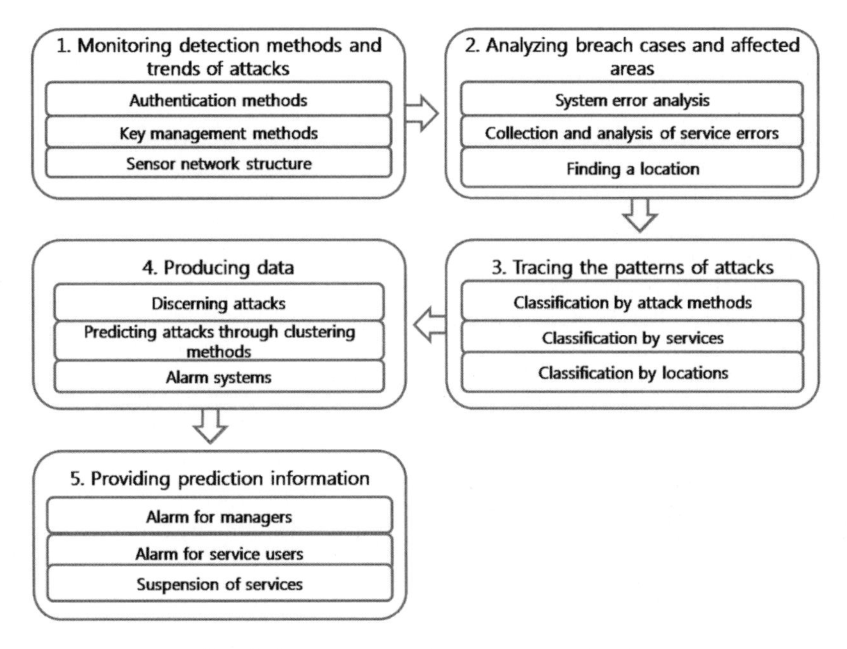

Fig. 2. Forecasting services for sensor networks

Step 1 (Monitoring detection methods and trends of attacks) is to analyze methods of attacks highly likely to cause significant damage to systems. In this stage, the false data injection attack method, which reports false data using an encryption key of a damaged node, is employed [10]. In this way, clear detection is possible through analysis of an attack signature. The system concludes that a node has been damaged if the same nodes require the same data or frequent authentication during a certain period or if they continuously attempt to transmit the same, unexpected data.

For more precise detection, the node of a damaged sensor is traced and data confirmed by using a clustering method. To that end, randomly selected data are transmitted for a specific period of time to measure average accuracy and, basing on this, the frequency of false data occurrence is verified.

Step 2 (Analysis of breach cases and affected areas) classifies breach cases according to system characteristics, locations, and time after analyzing the cases and monitoring the results. Then, Step 3 produces data based on the result while Step 4 traces affected nodes, which keep changing over time, and analyzes the overall conditions. In this step, the system discerns relations of data sets and creates clusters. This study employs EM(Expectation Maximization), a soft clustering method, for this process [11]. In this way, the system can analyze patterns of attacks and forgery or movements of nodes, as well as pinpointing a location where such attacks frequently occur and predicting patterns of future attacks.

In the final step, users are provided with the analyzed data. This involves forecasting cycles and methods of information sharing. In particular, a sensor node manager is notified of a danger and users are encouraged not to transmit important data due to unsafe network conditions. If a worst case, services are suspended first and, then, alarms are issued so as to minimize possible damage.

4 Conclusion and Follow-Up Studies

This study proposes a service system forecasting and notifying breach incidents such as false data or nodes in sensor networks. To that end, general security vulnerabilities in sensor networks were analyzed first. Also, characteristics of an existing forecasting system were examined in order to apply a forecasting system to a wireless sensor network. The forecasting service system suggested in this study can provide basic data for effectively forecasting common threats made to sensor networks. Follow-up studies can focus on analysis of system accuracy and efficiency by simulating false data attacks under the condition of location-based key pre-distribution and by implementing a forecasting system based on the results.

Acknowledgement

This paper has been supported by the Software R&D program of KEIT. [2010-10035257, Development of global collaborative integrated security control system].

References

1. Traynor, P., McDaniel, P., Porta, T.L.: On attack causality in internet-connected cellular networks. In: Proceedings of 16th USENIX Security Symposium on USENIX Security Symposium, Boston (2007)
2. Information Security.: Serious and Widespread weaknesses Persist at Federal Agencies. GAO/AMID-00-295 (2000)
3. AhnLab Security Emergency response Center, `http://download.ahnlab.com/asecReport/ASEC_Report_200902.pdf`
4. Yun, J., Paek, S.-H., Lee, D., Yoon, H.: Traffic Volume Forecasting Model for Cyber Weather Forecasting. In: Hybrid Information Technology ICHIT 2006, Cheju Island, pp. 274–276 (2006)
5. Yurcik, W., Korzyk, A., Loomis, D.: Predicting Internet Attacks: On Developing An Effective Measurement Methodology. In: 18th Annual International Communications Forecasting Conference, ICFC 2002 (2000)
6. Zhang, W., Cao, G.: Group Rekeying for Filtering False Data in Sensor Networks: A Predistribution and Local Collaboration-Based Approach. In: Proc. INFOCOM, pp. 503–514 (2005)
7. Scarfone, K., Grance, T., Masone, K.: Computer Security Incident Handling Guide, Special Publication 800-61, National Institute of Standards and Technology, U.S. Department of Commerce (2006)
8. Scarfone, K., Mell, P.: Guide to Intrusion Detection and Prevention Systems (IDPS), Special Publication 800-94, National Institute of Standards and Technology, U.S. Department of Commerce (2006)
9. Forecast process, `http://www.nhc.noaa.gov/HAW2/english/forecast_process.shtml`
10. Zhu, S., Setia, S., Jajodia, S., Ning, P.: An Interleaved Hop-by-Hop Authentication Scheme for Filtering of Injected False Data in Sensor Networks. In: Proc. S&P, pp. 259–271 (2004)
11. Borman, S.: An The Expectation Maximization Algorithm A short tutorial (2009)

A Study on M2M-based System for Hygienic Meteorology Service

Jae-gu Song[1,*], Jae Young Ahn[2], and Seoksoo Kim[3]

[1,3] Dept. of Multimedia, Hannam Univ., 133 Ojeong-dong, Daedeok-gu, Daejeon, Korea
[2] ETRI Standards Research Center, 161 Gajeong-dong, Yuseong-gu, Daejeon, Korea
bhas9@paran.com, ahnjy@etri.re.kr, sskim0123@naver.com

Abstract. M2M proposes a standardized communications technology between network and devices. This study has designed an M2M-based system to smoothly deliver information between devices which were required to provide hygienic meteorology services. Especially, an efficient plan for service provision has been studied, by classifying the types of information at each stage of user, EM, SM, HSM and SPM.

Keywords: M2M, Hygienic Meteorology, USN, Network.

1 Introduction

Recently, different studies are under way to minimize the damages by identifying man-made disasters in advance, including fire, explosion, distress, car accident as well as natural disasters such as flood, storm, heavy snowfall and earthquake. With the advance of IT, various countermeasures have been developed and introduced. Among them, M2M (Machine to Machine) is a study to deliver preliminary data in order to prepare for the contingencies, where a sensor connected between things exchanges every kinds of information including safety.

Existing sensor-based services are the approaches in physical concept, whilst in M2M, even a thing is positively used as a means of communication such as sensor node. This provides basic technology in order to realize ubiquitous computing environment by delivering information person to person, thing to thing and thing to person[1,2].

Such a M2M study focuses on the technical approach and the protocol research including platform mobility, traceability, information compatibility and discriminability, etc. The reason is that these are very important areas to ensure smooth and safe information exchange in M2M communication through protocol standardization. However, to provide an efficient system is very difficult because each service delivers data of different characters. It causes various vulnerabilities and could include unnecessary security factors[3]. Therefore, M2M service areas which satisfy the nature of each service need to be researched and the system of the areas should be designed, based on the standardized message protocol.

[*] Corresponding author.

T.-h. Kim, A. Stoica, and R.-S. Chang (Eds.): SUComS 2010, CCIS 78, pp. 39–45, 2010.

This research focused on the research of healthcare service in an M2M environment. Especially, through the hygienic meteorology closely related to preventive medicine, the research has been conducted based on the things (PDA, Smartphone) carried by users, a server that provides information and a sensor that collects environmental information.

2 Related Work

M2M is often described as the completion of wireless internet environment. That is, M2M itself is not a new concept, but containing a combined meaning applicable in more stable and different environments, by means of existing mobile communication networks. Therefore, in order to understand this, a research is needed on a service provision system in existing wireless communication environment and sensor network environment.

2.1 The Trend of M2M Service Introduction

Recently, as various M2M introduction services have appeared, the most prominent fields of research among them include remote monitoring of logistics, fleet management, machinery and facilities, technology to handle existing manual metering of the machine time of construction equipment, electricity consumption by automatic metering, and remote data process, etc. Moreover, M2M is perceived as the most important field of research for security, telemetry, healthcare and product/longevity tracking. M2M has been given attention because it is expanding its territory to the markets of various industries[4].

As M2M applications have been minimized, the trend of different service application influences even manufacturing itself. Moreover, as developers and system integrators have tried to develop a stable and flexible application which could be easily maintained with self-assessment function available, the introduction of more efficient service is being accelerated. For an effective use of this application, lowering data transmission cost of the system needs to be designed.

2.2 Wireless Data Communication Service Use Patterns

The wireless data communication can deliver directing and searching particulars promptly and correctly, through transmission of data such as letter and number, etc., and efficiently use limited frequencies with short message communication including group notification and dispatch, etc. Moreover, it can provide two-way communication and re-confirmation of transmission details since messages received can be stored. The wireless data communication service patterns and service use patterns are as follows[5];

2.3 Wireless Sensor Network

The sensor network consists of nodes arranged inside or nearby an area to be monitored. The wireless sensor network has been preferentially applied in the fields

Table 1. Wireless Data Communication Service Use Patterns

Maintenance & manufacturing industries	
Work Order	Immediate maintenances and adjustment of operating rate of the manufacturer's production line
Provision of Location Information	Efficient order and management of the places to be visited
Database Use	Tracking of the parts inventory / urgent procurement and order of the parts for maintenance
Message Communication & Electric Mail	Efficient management of work orders and receipts
Remote data collection	Remote control & maintenance
Insurance, sales industries	
Work Order	Delivery of business guidelines, discussion about a timely visit and direction of business activities
Provision of Location Information	Provision of information on the location to be visited
Database Use	Use of information on customers, price and ancillary service
Message Communication & Electric Mail	Efficient management of work orders and receipts
Private security business	
Work Order	On-site countermeasures for emergency security notice & instruction of special checkup points
Provision of Location Information	On-site instruction of security & patrol areal information
Database Use	Acquisition of the information required for various security activities
Message Communication & Electric Mail	Efficient management of work orders and receipts
Remote data collection	Remote control & maintenance
Public utilities (police, fire station, gas and electricity, etc.)	
Work Order	On-site countermeasures for emergency security notice & instruction of special checkup points
Provision of Location Information	Notifying situation & scale information of the location
Database Use	On-site acquisition of public activity data
Message Communication & Electric Mail	Efficient management of work orders and receipts
Remote data collection	Remote control & maintenance

including mechanical failure monitoring, security of home/building and automatic air-conditioning, and is being gradually expanded to the safety of public facilities, vehicle, agriculture and medical service, etc[6].

There are next generation applications requiring more intellectual system, such as weather, regional monitoring and ecosystem, and a research of analyzing the information sensed through situation awareness is recently under way. The service can be introduced in various fields of everyday life, out of limited areas[7].

2.4 Hygienic Meteorology

The officially defined hygienic meteorology is broadly classified into six viewpoints;

① Research of meteorotropic diseases and investigation of meteorological conditions

② Analysis on meteorological phenomena and causes of climate for endemic diseases caused by regional uniqueness

③ Research of the causes of seasonal diseases which are frequently found for the season in everyday life

④ Analysis of the influence of weather condition in working environment such as air conditioning, humidity and temperature on humans, with the object of improving work efficiency and public health hygiene

⑤ Investigation into meteorological influence on specific environments including mountains, coast and polar regions, etc.

⑥ Research of the relationship between clothes, residential environment and meteorology

The above classification shows how broad the meaning of hygienic meteorology is. The definition of hygienic meteorology defined up to now is newly established according to applications of research, objectives of use, and most of the definitions are included in six viewpoints. The hygienic meteorology, since it had been researched by Drude in 1930's, has been also called biometeorology, and recently various applied technologies are being researched through combination with meteorology, medicine and IT, etc. As basic information on the trend of researches is publicly available and exposed to the most of people with development of IT technology and introduction of mobile devices, the introduction of various services became possible through processing and application of information. Currently, various services are being developed, by means of the introduction of wireless sensor technology from web-based information service and the analysis of user information.

While various researches have been carried out on hygienic meteorology, the methods of approaching to the researches can be broadly classified into following three types;

① Environmental research (Environmental improvement & countermeasures)

② Research on Human Biometeorology

③ Research of service plans through combination with IT technology

The first research of environment itself, focusing on the researches depending on regional characters, has studied meteorological phenomena for regional uniqueness, climate analysis, analysis of air pollution level, character of environmental climate considering work environment, analysis of urban climate pollution and corrective measures, etc.

These kinds of researches does not end up with environmental analysis itself, but is ultimately connected to researches on human being and livestock. That is, it serves as a major preliminary study for human hygienic meteorology and is also used as a fundamental research data. Most of these researches are created by professional researches of meteorologists and geologists.

The second human biometeorology, as a research field that examines various influences of environment on human life, presents plans to improve environment by analyzing the causes [8]. According to Guidelines on Biometeorology and Air Quality Forecasts publicized by World Meteorological Organization in 2004, the human biometeorology is defined as an integration and a perception of information between human health, weather and climate (regional feature), for the purpose of providing a national public climate service.

"Poaceae pollen in Galicia (N.W. Spain): characterisation and recent trends in atmospheric pollen season", out of the currently published researches show what influence the pollen has on human life, by mentioning the relationship between seasonal germination of pollen and allergic diseases in relevant areas[9].

On the basis of environmental researches, the human hygienic meteorology is expanding to different researches related to human health.

The last research field, the service plans research through combination with IT technology, presents concrete plans to realize the aforementioned environment and human hygienic meteorology researches.

3 System for M2M-based Hygienic Meteorology Service

This study focuses on the service plans through combination with IT technologies, out of the research approach methods in hygienic meteorology. This can be realized in an environment where individuals can carry Smartphone and use wireless network as they like, and M2M is required as a core technology to communicate in this environment.

The structure of M2M service for the hygienic meteorology service proposed in this study is shown in Fig1.

The structure of this service consists of users and service provider server, and the detailed composition is as follows;

① User: A user is provided with hygienic meteorology information for his/her situation, on the basis of medical information related to the climate.

② EM(End Machine): As a device that discloses information to an end user, it provides analyzed information to home server and service provider, and transmits feedback information replied by the user.

③ SM(Sensor Machine): Collects various information of the user environment and delivers this to the home server.

④ HSM(Home server Machine): Manages user information, sensed environmental information and private medical information, collects climate information from web and creates hygienic meteorology information to be provided to the end user. Besides it maintains information and processes error information by communicating with server provider.

⑤ SPM(Service Provider Machine): Is provided with system operating information from HSM, analyses if the information is normal, and solves any problem occurred.

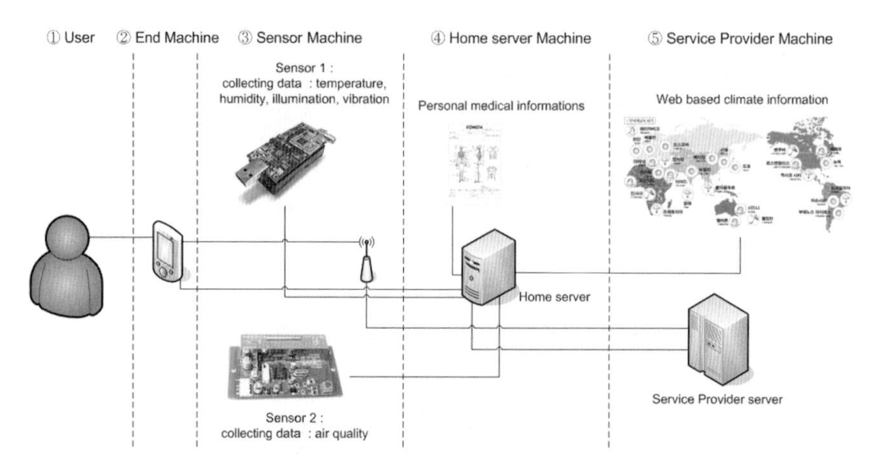

Fig. 1. M2M-based Hygienic Meteorology Service

The object of this service structure is to collect and analyze environmental, climate and private medical history information up to EM where end users identify the information, and to provide services in a safe and efficient manner through communication between the devices.

The communication between M2Ms is generated from the communication between EM, SM, HSM and HSM.

Then, the importance should be recognized by means of the detailed information. To apply hygienic meteorology service through M2M, the types of wireless communication service and use patterns should first be defined. The hygienic meteorology-based wireless communication service defined in this research is as following table;

Table 2. Hygienic meteorology Service Use Patterns

Hygienic meteorology	
Work order	Delivers local measures and specialties for environment-based health maintenance
Provision of location information	Provides information of location and environment (temperature, humidity and illumination) in a visiting area
Database use	Uses existing hygienic meteorology-related data of preventive medicine level, and uses information of the influence of climatic change on national health
Message communication & electric mail	Delivers information to individual mobile terminals
Remote data collection	Information sharing between climate information collecting server and mobile device's information provision service

As in the above table, the information generated in the communication between M2M includes daily information. However, since HSM deals with the medical history of an individual, the medical history information should be secured.

As a result, the hygienic meteorology service can be smoothly provided by applying security protocols to HSM area and general open protocols to the rest of the areas.

4 Conclusion

M2M is considered as a service technology focusing on mobility, traceability and message, rather than a physical approach from a service-centered view of the traditional sensor network. Therefore, an optimized information exchange method for each service is needed. The optimization can be realized without using unnecessary resources by clearly identifying the character of abstract information.

This research suggested the hygienic meteorology, a system structure required in specific service environments. Besides, under consideration of communication service environments, the efficient information delivery environment was presented. Applied business models of the more complex healthcare M2M fields are anticipated to be researched and developed, based on these suggestions.

Acknowledgement

This research was supported by the ICT Standardization program of MKE(The Ministry of Knowledge Economy) for the research project of ETRI.

References

1. Bachvarova, E.: Wireless M2M System Architecture For Data Acquisition And Control. In: Second Scientific Conference with International Participation Space, Ecology, Nanotechnology, Safety (2006)
2. Singer, A.: Internet Protocols Ease Development Cost and Time for M2M Communication (2006)
3. Park, Y.M., Koo, B.H., Mok, J.D.: Overview of the Wireless Communications Standardization and Standardization Efforts of Wireless LAN. ET Trends 11(1) (2007)
4. Lee, C.H., Hong, H.K.: A Study of the Message Protocols Technologies in M2M Platforms. The Journal of Korea Information and Communications Society 35(1), 53–61 (2007)
5. A Study on the Radio communications Standardization, ETRI (1995)
6. Wireless Sensors and Integrated Wireless Sensor Networks (Technical Insights), Frost & Sullivan (2002)
7. Chi, C., Hatler, M.: Wirelss Sensor Netwoks: From Dust to Reality, ON World (2002)
8. Kusch, W., Fong, H.Y., Jendritzky, G., Jacobsen, I.: Guidelines on biometeorology and air quality forecasts. World Meteorological Organization (2004)
9. Jato, V., Rodríguez-Rajo, F.J., Seijo, M.C., Aira, M.J.: Poaceae pollen in Galicia (N.W. Spain): characterisation and recent trends in atmospheric pollen season. International Journal of Biometeorology 53(4), 333–344 (2009)

Knowledge Integration and Use-Case Analysis for a Customized Drug-Drug Interaction CDS Service

Hye Jin Kam, Man Young Park, Woojae Kim,
Duk Yong Yoon, Eun Kyoung Ahn, and Rae Woong Park

Dept. of Biomedical Informatics, Ajou University School of Medicine, Suwon, Korea
{hjkam,pmy10042,locos,uni0731,justcong,veritas}@ajou.ac.kr

Abstract. Clinical decision support systems (CDSSs) are thought to reduce adverse drug events (ADEs) by monitoring drug-drug interactions(DDIs). However, clinically improper or excessive alerts can result in high alert overrides. A tailored CDS service, which is appropriate for clinicians and their ordering situations, is required to increase alert acceptance. In this study, we conducted a 12-week pilot project adopting a tailed CDSS at an emergency department. The new CDSS was conducted via a stepwise integration of additional new rules. The alert status with changes in acceptance rate was analyzed. The most frequent DDI alerts were related to prescriptions of anti-inflammatory drugs. The percentages of alert overrides for each stage were 98.0%, 96.0%, 96.9%, and 98.1%, respectively. 91.5% of overridden alerts were related to discharge medications. To reduce the potential hazards of ADEs, the development of an effective customized DDI CDSS is required, via in-depth analysis on alert patterns and overridden reasons.

Keywords: Drug-drug Interaction (DDI), Clinical Decision Support System (CDSS), CDSS Integration, alert overrides.

1 Introduction

Clinical decision support systems (CDSSs) can reduce potential medication errors by monitoring drug-drug interactions (DDI) or drug duplications at the time of prescription [1-3]. In actuality, however, approximately 35-94% of CDSSs' DDI alerts on absolutely contraindicated and interacting drug-pairs were overridden by physicians [4]. Clinically inappropriate or excessive numbers of alerts can cause 'alert fatigue' for the clinician; this can render clinicians insensitive to alerts, and can degrade their confidence in the alerts, the alerting system and CDS as a whole [5]. In order to develop an efficient/effective CDSS, analyses of alert patterns and relevant reasons for alert overrides are required; thus, methods to improve the process and knowledgebase should be searched by use-case analyses of DDI CDSS in an actual clinical situation. In this study, we conducted a 12-week pilot CDSS project in which a tailed CDSS was adopted at the emergency department of a tertiary teaching hospital in Korea. The new CDSS was conducted via a stepwise integration of new DDI rules, and alert status with changes in acceptance rate was analyzed in an effort to find ways to apply an effective CDSS.

T.-h. Kim, A. Stoica, and R.-S. Chang (Eds.): SUComS 2010, CCIS 78, pp. 46–51, 2010.

2 Methods

2.1 Heterogeneous Knowledge Integration

The subject hospital had previously adapted the national mandatory Drug Utilization Report (DUR) Knowledge for DDI CDSS. However, this was a minimal rule set for regulatory purposes, which was limited to absolute contraindications and was not sufficient for clinically meaningful presentations of DDIs. Therefore, a new set of knowledge [6] that encompasses all clinically relevant DDIs was introduced in order to strengthen the CDS function on DDIs. To perform an integrated DDI CDSS, an exclusive non-redundant knowledge set had to be prepared by joining pre-existing DDIs on contraindication with a newly introduced one on absolute contraindications and major interactions. Furthermore, as DUR knowledge is irregularly updated, an automatic merging plan/process should be designed, as shown in Fig. 1[7].

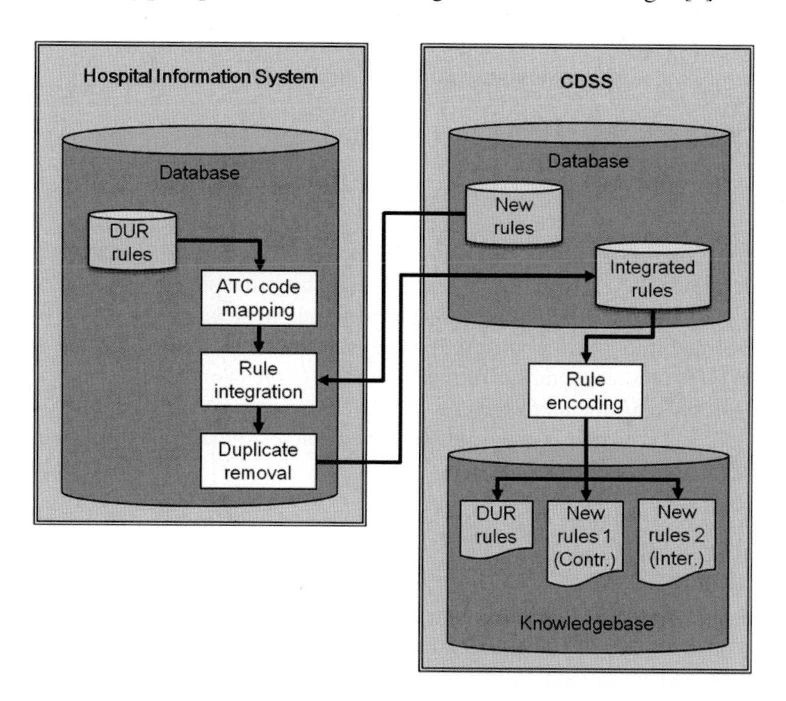

Fig. 1. A flow chart of the knowledge integration

2.2 Stepwise Knowledge Expansion

With the integrated knowledge, four sequential (3 weeks each) pilot tests were conducted. From stages 1 to 3, the amount of knowledge increased gradually in the order of DUR contraindications only, plus additional new contraindications, plus major interactions. At stage 4, all settings were identical to those in stage 1; additionally, the alerts generated by the new knowledge were not provided to the

Table 1. 12-week pilot project with a stepwise knowledge expansion

Stage	Knowledge set			Period
	DUR Contraindications	New Contraindications	New Major interactions	
1	387	-	-	Dec 10, 2009 – Dec 30, 2009
2	389	101(22)	-	Dec 31, 2009 - Jan 20, 2010
3	404	101(22)	608(592)	Jan 21, 2010 – Feb 10, 2010
4	404	101(22)	608(592)	Feb 11, 2010 – Mar 03, 2010

clinicians while logging them internally. In Table 1, the numbers of rules in each knowledgebase are the average number of rules during each test period: the numbers in parentheses indicate the numbers of rules excluding duplicates with DUR.

2.3 Use-Case Analysis

We conducted a pilot project at the emergency department (ED) of a tertiary teaching hospital. ED was selected for the new hybrid CDSS due to the following reasons: (1) prescriptions are urgently performed, (2) joint treatments are performed among several departments, (3) it has characteristics of both hospitalization and outpatient protocols, and (4) there are greater concerns about medical errors owing to lack of patient information. Information on (1) DDI alerts that occurred, (2) the causes of drug orders, (3) alert acceptances or rejections and (4) reasons for alert overrides were analyzed not only for the analysis on usability and acceptance rate on DDI CDSS and its alerts, but also for that of alert patterns and changes in acceptance rates according to the knowledge expansion.

3 Results

During the test period, the total number of prescribed patients was 684,834 (whole hospital), and 80,887 (ED, 11.8% of the whole hospital). The percentages of prescriptions and prescribed patients with DDI alerts were 0.16% and 0.36%, among the total patient records. For ED, they were 0.89% and 1.57%: this corresponds to 5.6-fold and 4.4-fold increases relative to the results from the whole set of patients. The average number of alerts for a patient with DDI alerts was similar, at 1.49 (whole) and 1.47 (ED); the average numbers of alerts over the whole patient set were 0.005 (whole) and 0.023 (ED), which is 4.6-fold higher than the number for the whole hospital.

Because the knowledge was increased gradually, the rankings for frequent alerts were rather varied according to stage: the top 10 highly ranked alerts are provided in Table 2. Despite the short application period, seven of the top 10 frequent alerts were triggered by the newly introduced rules. Most of the DDI alerts (61.8%) were related to NSAIDs. With comparisons among the prescribed clinicians' specialties, the types and ranking of frequent alerts were shifted according to the specialty, as shown in Table 3.

Table 2. Top 10 highly ranked alerts

Ingredient 1	Ingredient 2	# of alert occurred	Rate (%)
Propionic acid derivatives	Ketorolac	1,103	59.0
Acetylsalicyclic acid	Clopidogrel	187	10.1
Acetylsalicyclic acid	Heparin	146	7.8
Furosemide	Digoxin	44	2.4
Nimesulide	Ketorolac	30	1.6
Ciprofloxacin	Insulin (human)	29	1.6
Isoniazid	Rifampicin	23	1.2
Naproxen	Ketorolac	22	1.2
Spironolactone	Digoxin	18	1.0
Acetylsalicyclic acid	Ibuprofen	16	0.9

Table 3. Top 3 highly ranked alerts for top 5 highly ranked departments

Department	# of whole alerts	Ingredient 1	Ingredient 2	# of alert
Emergency	794	Propionic acid derivatives	Ketorolac	733
		Nimesulide	Ketorolac	9
		Celecoxib	Ketorolac	7
Circulatory internal medicine	153	Acetylsalicyclic acid	Heparin	45
		Acetylsalicyclic acid	Clopidogrel	31
		Losartan	Spironolactone	15
Orthopedics	38	Propionic acid derivatives	Ketorolac	22
		Furosemide	Digoxin	6
		Celecoxib	Ketorolac	4
Neurology	36	Nimesulide	Ketorolac	13
		Ketorolac	Pentoxifylline	7
		Propionic acid derivatives	Ketorolac	5
Respiratory internal medicine	30	Levofloxacin	Insulin (human)	5
		Acetylsalicyclic acid	Clopidogrel	5
		Furosemide	Digoxin	3

Although the amount of overridden alerts also varied broadly according to stage, the percentages of alert overrides for each stage were 98.0%, 96.0%, 96.9% and 98.1%, respectively (shown in Fig. 2). Owing to these high overridden rates, it was difficult to see the effects of the stepwise knowledge expansion. 91.5% of the overall overridden alerts (96.0%-98.1% of the whole alerts) were related to discharge medications.

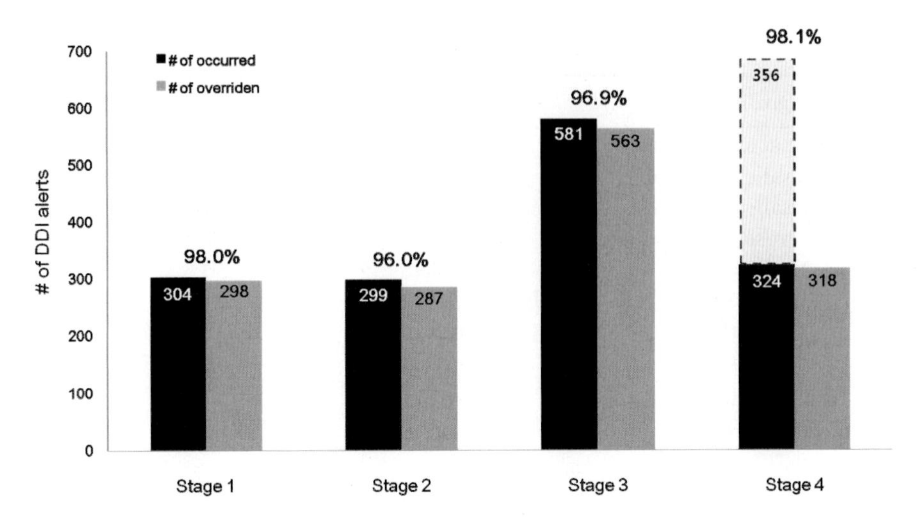

Fig. 2. The percentages of alert overrides for each stage. Dotted part of alerts (356) in stage 4 represents those that were generated by the new knowledge and were not provided to the clinicians while logging them internally: they were excluded from the calculation of percentages of alert overrides at stage 4.

4 Discussion

The applied DDI alerts on absolute contraindication and major interaction were nationally mandated or were mentioned in domestic or international references with strong interactions. Taking this into consideration, the average alert overrides (97.8%) were too high, even though the orders might have progressed with clinicians' proper judgments, from the perspective of potential risks of adverse drug events (ADEs). In order to reduce the potential hazards associated with ADEs, a great deal of education and efforts are required. Moreover, the development of an effective customized DDI CDSS requires in-depth analyses of both alert patterns and reasons for overrides.

Acknowledgments. This study was supported by a grant of the Korea Health 21 R&D Project, Ministry of Health & Welfare in Republic of Korea (A050909).

References

1. Garg, A.X., Adhikari, N.K., McDonald, H., Rosas-Arellano, M.P., Devereaux, P.J., Beyene, J., et al.: Effects of computerized clinical decision support systems on practitioner performance and patient outcomes: a systematic review. JAMA 293(10), 1223–1238 (2005)
2. Bates, D.W., Kuperman, G.J., Wang, S., Gandhi, T., Kittler, A., Volk, L., et al.: Ten commandments for effective clinical decision support: making the practice of evidence-based medicine a reality. J. Am. Med. Inform. Assoc. 10(6), 523–530 (2003)

3. Kaushal, R., Shojania, K.G., Bates, D.W.: Effects of computerized physician order entry and clinical decision support systems on medication safety: a systematic review. Arch. Intern. Med. 163(12), 1409–1416 (2003)
4. van der Sijs, H., Mulder, A., van Gelder, T., Aarts, J., Berg, M., Vulto, A.: Drug safety alert generation and overriding in a large Dutch university medical centre. Pharmacoepidemiol. Drug Saf. 18(10), 941–947 (2009)
5. Sittig, D.F., Wright, A., Simonaitis, L., Carpenter, J.D., Allen, G.O., Doebbeling, B.N., et al.: The state of the art in clinical knowledge management: an inventory of tools and techniques. Int. J. Med. Inform. 79(1), 44–57 (2010)
6. Lee, S.: Information System of National Drug Standards and Knowledgebase. Health and Medical Technololgy R&D Program, Project no. D060002, Ministry of Health and Welfare, Korea (2006)
7. Kam, H.J., Park, M.K., Kim, J.A., Cho, I., Kim, Y., Park, R.W.: An Integrated Architecture for a Customized CDSS. In: CJKMI 2009 International Symposium, p. 41 (2009)

A Study on Markerless AR-Based Infant Education System Using CBIR

Ji-hoon Lim and Seoksoo Kim*

Dept. of Multimedia, Hannam Univ., 133 Ojeong-dong, Daedeok-gu, Daejeon, Korea
siz1004@paran.com, sskim0123@naver.com

Abstract. Block play is widely known to be effective to help a child develop emotionally and physically based on learning by a sense of sight and touch. But block play can not expect to have learning effects through a sense of hearing. Therefore, in this study, such limitations are overcome by a method that recognizes an object made up of blocks, not a marker-based method generally used for an AR environment, a matching technology enabling an object to be perceived in every direction, and a technology combining images of the real world with 2D/3D images/pictures/sounds of a similar object. Also, an education system for children aged 3~5 is designed to implement markerless AR with the CBIR method.

Keywords: Markerless, AR, Infant, CBIR, SIFT, Education.

1 Introduction

Recently, an increasing number of people prefer smart phones that can provide more various contents. And one of the outstanding contents provided by smart phones is AR (augmented reality). Previously, a user had to be equipped with a number of devices including an HMD display device (glasses attached to a helmet). However, smart phones offer an optimum environment for AR including a camera, display devices, sensors, 3G, Bluetooth, Wi-Pi, and so on. Now, there are active research efforts on application of AR in various areas.

AR(augmented reality) refers to a virtual world that provides a sense of reality similar to that of the real world by combining computer graphics (objects in a virtual world) with images in the real world and through user interaction [1]. Simply put, with this technology users in the real world can view information supplemented by computer graphics.

Playing with blocks allows children to freely express what they see and experience in daily life, helping them understand the real world. In addition, they can use imagination to create a virtual world of their own and learn social space, role models, and a mode of living. However, the playing enables learning through a sense of touch and sight, not a sense of hearing.

Therefore, in this study, such limitations are overcome by a method that children can express image data such as colors, shapes, and texture based on a specific object

* Corresponding author.

T.-h. Kim, A. Stoica, and R.-S. Chang (Eds.): SUComS 2010, CCIS 78, pp. 52–58, 2010.

expressed by blocks(not marker-based methods found in many other studies), a matching technology enabling an object to be perceived in every direction, and a technology combining images of the real world with 2D/3D images/pictures/sounds of a similar object. Also, an education system is designed to maximize educational effects.

2 Related Works

2.1 CBIR

TBIR(Text-Based Image Retrieval) used by most portals retrieve images based on bibliographic information or notes [2]. But such information is provided subjectively by an image producer and, thus, even the same images can have different retrieval information. Moreover, an image producer shall provide bibliographic information or notes for every image. However, CBIR (Content-Based Image Retrieval) method searches for images based on colors, texture, or shapes of an image, providing consistency. CBIR-based systems include QBIC, MARS, WebSEEK, Photobook, and VisualSEEK [3].

In this study, CBIR is used to implement existing AR technology without markers for recognition. The blocks children play with have similar texture, which is not considered therefore in this study, and colors/shapes of the blocks are used in order to extract objects of images and to design effective education systems.

2.2 Color Histograms

In CBIR, colors of images express contents through a color histogram, and a color histogram uses color models similar to human vision (e.g. YIQ, YUV, YCbCr, and HIS) instead of the previous RGB color model as shown in Formula 1.

A color of an image is independent from sizes and direction, easy to be extracted, but very sensitive to brightness affected by illumination or outside light.

$$r = \frac{R}{(R+G+B)} \qquad g = \frac{G}{(R+G+B)} \qquad b = \frac{B}{(R+G+B)} \tag{1}$$

Hence, as depicted in Formula 2, in extracting a specific object by colors of images, the RGB model is converted into the HIS color model. The HIS model, composed of colors, chromas, and brightness, is strong against changes in brightness and not sensitive to outside light when used for image retrieval [4].

$$\begin{aligned} i &= \frac{1}{3}(r+g+b) \\ s &= 1 - \frac{3}{(r+g+b)}[\min(r,g,b)] \\ h &= \cos^{-1}\left[\frac{\frac{1}{2}[(r-g)+(r-b)}{\sqrt{(r-g)^2+(r-b)(g-b)}}\right] \\ &\text{if}, b > g \text{ then}, h = 360^\circ - h \end{aligned} \tag{2}$$

2.3 Expression through Texture

Texture refers to the way a material (e.g. wood, bricks, grain, cloth, and so on) feels such as roughness or softness, expressing data of surface structure or relations with surrounding pixels. Analysis methods are the Fourier transform, the wavelet transform, and EHD(Edge Histogram Descriptor) of MPEG-7, which provide data of frequency and multi-resolution images.

2.4 Expression through Shapes

Extraction methods using shapes of images can be divided into extraction of edges and that of key points limited to a specific area. The first method makes possible quick and easy data extraction but, in this case, even the same object could be seen differently according to angles and sizes. On the other hand, the second method has relatively slow processing speed but provides better accuracy. In this study, both of the two methods are used for more effective image retrieval. To do so, the Canny Edge method will extract edges of an image while SIFT(Scale Invariant Feature Transform) will extract key points of a specific area.

3 Processing of Input Images

3.1 Extraction of Edges by the Canny Edge

The Canny Edge algorithm [5] is the most optimized edge extraction method among general extraction operators. Comparing with surrounding pixels located in a direction of the biggest brightness change, if a gradient of a pixel is greater than that of neighboring pixels, the algorithm extracts the pixel as an edge. In this study, the Canny Edge method is used in order to make key points, extracted by the SIFT method, more robust as follows.

1. Gray scale transforming of input images
2. Removal of image noise through the Gaussian filter
3. Calculation of x/y axis gradient and intensity of each pixel
4. Removal of unnecessary edges by adjusting critical values

3.2 Extraction of Descriptors by the SIFT Method

SIFT(Scale-Invariant Feature Transform), developed by David G. Lowe, is strong against changes of sizes and rotations. The algorithm matches an requested image with those in a database so as to measure the level of similarity [6]. This technology is being used mostly for digital image processing. There are largely two steps for this image processing.

1. Extraction of key points
2. Extraction of descriptors of selected key points

After edges are extracted by the Canny Edge method, the SIFT method produces image pyramids having different Gaussian distribution values in Image I (x, y) as shown in Formula 3 in order to extract key points. G (x, y, σ) is a Gaussian distribution that has σ dispersion.

$$L(x, y, \sigma) = G(x, y, \sigma) * I(x, y) \tag{3}$$

DoG(Difference of Gaussian) image is obtained from a difference between two images, L(x, y, σ) and L(x, y, kσ), which were applied by Gaussian functions having different dispersion values (σ, kσ) as depicted in Formula 4. The obtained DoG image, D (x, y, σ), can be expressed as Formula 4.

$$\begin{aligned} D(x, y, \sigma) &= (G(x, y, k\sigma) - G(x, y, \sigma)) * I(x, y) \\ &= L(x, y, k\sigma) - L(x, y, \sigma) \end{aligned} \tag{4}$$

Next, possible images are extracted from the obtained DoG images, and key points are selected by using maximum/minimum values. The extracted key points are not affected by size changes, and an image scope can be found through dispersion values of an image. Also, the descriptor of the extracted key point is calculated by a 3D histogram composed of 128 bins, based on x/y axis locations or rotation values.

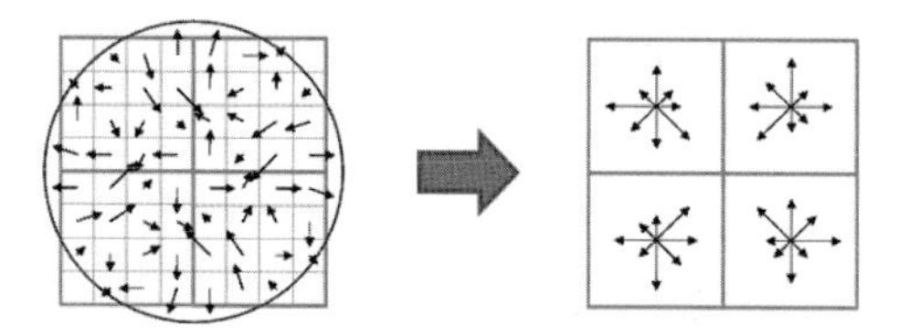

Fig. 1. Scope of an image and a histogram

The left image in Figure 1 shows the scope surrounding an extracted image and the right one, a histogram. The descriptor of a key point is composed of gradient vectors (size and direction) and calculated by the following formula.

$$0\ 000\text{Ø}0000\ 00000 = \begin{bmatrix} L(x+1,y) - L(x-1,y) \\ L(x,y+1) - L(x,y-1) \end{bmatrix} \tag{5}$$

$$m(x,y) = \sqrt{(L(x+1,y) - L(x-1,y))^2 + (L(x,y+1) - L(x,y-1))^2} \tag{6}$$

$$\theta(x,y) = \tan^{-1}((L(x,y+1) - L(x,y-1)) / (L(x+1,y) - L(x-1,y))) \tag{7}$$

3.3 Simplification of a Processed Area by the Hough Transform

The Hough Transform is a method to quickly retrieve simple shapes such as straight lines or circles from gray-scale transform images [7]. In this study, this method is

 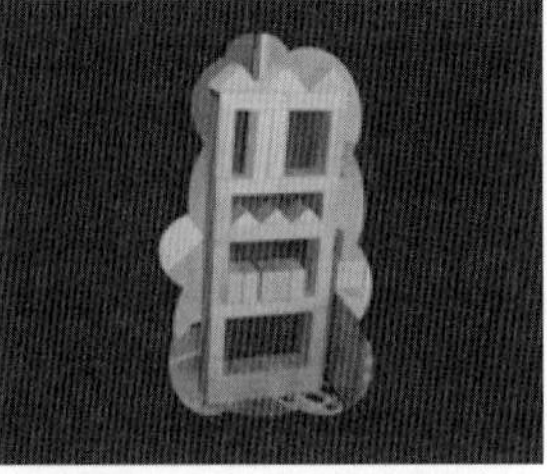

Fig. 2. Simplified processed area

used in order to simplify a processed area for key point extraction through circles treated by the Hough Transform.

As shown in Figure 2, more time is consumed for the image scope in the left picture, for the entire area should be calculated. Thus, as in the right picture, only the scope where an object belongs is captured for quick calculation.

4 AR Education System

4.1 IES System Design

The IES system, an AR education system proposed in this study, has the following structure. First, the system requires a camera receiving an input image.

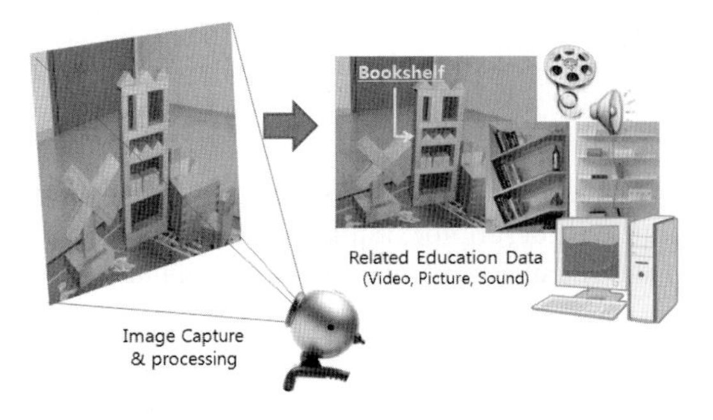

Fig. 3. Structure of the IES system

The image input by a camera is processed according to the sequence depicted in Figure 4. Then, finally, educational materials related with the object in the image is expressed through AR images, pictures, music, and so on, on the screen of a computer, a smart phone, or a PDA.

4.2 System Scenario

Children may play with blocks by creating their own object using imagination or make an object by following given instructions as if origami. For the second method,

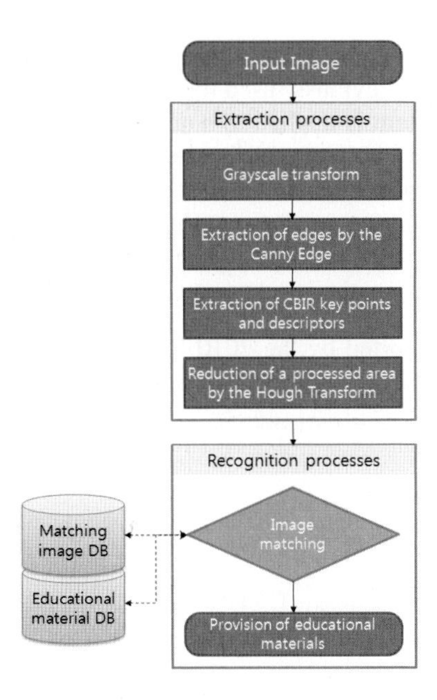

Fig. 4. Process of the IES system

instructions are provided by different levels (from low to high). Here, it is supposed that educational materials or images related with such instructions are already in the DB of the system. The scenario of the proposed system is as follows.

1. A child makes an object with blocks.
2. A parent uses a camera to input an image of the object so as to show educational materials related with the object.
3. The image input by the camera is used to retrieve related educational materials shown on the screen of a smart phone.
4. The parent shows to the child images, pictures, or music related with the object for more effective education.
5. If the parent fails to find materials related with the object, he may capture other images of the object from various directions/angles to find and save in the system more educational materials.
6. The images and materials saved are used in the future for detection and recognition of the object.

5 Conclusion

Block play is effective to help a child develop emotionally and physically but it is limited to visual learning, lacking learning through a sense of hearing. Furthermore, most education programs are available through extra charge. Likewise, although parents want to provide more educational opportunities to a child, educational resources are quite limited due to additional costs.

Therefore, this study uses CBIR to design the IES system, an AR education system, without a marker, previously used for AR technology.

The proposed IES system is applied to a single object, not a multitude of objects. Follow-up studies will provide a method to extract a number of objects with reduced time needed for image processing.

Acknowledgement

This paper has been supported by the 2010 Preliminary Technical Founders Project fund in the Small and Medium Business Administration.

References

1. Azuma, R.: A Survey of Augmented Reality. Teleoperators and Virtual Environments 6(4), 355–385 (1997)
2. Saykol, E., Sinop, A.K., Gudukbay, U., Ulusoy, O.: Content-Based Retrieval of Historical Ottoman Documents Stored as Textual Image. IEEE Tran. on images processing 13(3), 314–325 (2004)
3. Remco, Veltkamp, C., Mirela Tanase: Content-Based Image Retrieval Systems: A Survey. Technical Report UU-CS-2000-34 (2000)
4. Hsu, W., Chua, S.T., Pung, H.H.: An integrated color-spatial approach to content-based image retrieval. In: Proceedings of the third ACM international conference on Multimedia, pp. 305–313 (1995)
5. Canny, J.: A Computational Approach to Edge Detection. IEEE transactions on pattern analysis and machine intelligence 8, 679–714 (1986)
6. David, G.: Distinctive Image Features from Scale-Invariant Keypoints. International Journal of Computer Vision 60(2), 91–110 (2004)
7. Kimme, C., Ballard, D.H., Skansky, J.: Finding circles by an array of accumulators. Communication of the Association for Computing Machinery 18, 120–122 (1975)

A Study on Home Network User Authentication Using Token-Based OTP

Jung-Oh Park[1], Moon-Seog Jun[1], and Sang-Geun Kim[2]

[1] Dept. of Computer Science, Soongsil University
[2] Division of Computer Engineering, Sungkyul University
{jop07,mjun}@ssu.ac.kr, sgkim@sungkyul.edu

Abstract. The system proposed in this thesis offers authority or control over an approach to licensed users for diverse devices used in Home Network, preventing unlicensed users from inappropriate approach. In relation to communications of security certification for each device, reduction in added resources and high security can be both met through a security token generated using OTP, which is reasonably applicable to low-efficiency instruments that compose a home network.

Keywords: Home Network, OTP, SSL.

1 Introduction

In modern society, a natural connection of real to cyber spaces, combined with rapid advance in IT industry, has made home network service appear and develop besides our workplace setting. Among other IT technologies, home network service industry has been an on-going issue in this current as a prime mover for national development and new change, also with a great potential in its development ahead. With spread of home network service especially in diverse forms recently, however, scope in target for cyber attack has also enlarged, throwing an element of anxiety over our society socially and economically. This state of affairs necessitates user authentication that prevents occurrence of invasion incidents and exposure of user information in home network service.

To enable an outside client to control home network with a mobile terminal like PDA, this thesis focused on user authentication and approach control among security elements of home network. We propose a method of home network user authentication by direct access to home server from outside home, using OTP-based authentication, not via authentication server of home network service provider that was left out of consideration from the mechanical criterion for home server-oriented home network user authentication for group (TTASKO-120030) by Telecommunications Technology Association (TTA) in Korea.

This authentication uses X509 v3- based authentication for certification, controlling devices by dividing user group on its extension area, and for devices with restricted approach, it controls approach by adding ACL (Access Control List). Such a division into user with restricted approach and its manager can present approaches for each device and protect it safely from outside attack.

T.-h. Kim, A. Stoica, and R.-S. Chang (Eds.): SUComS 2010, CCIS 78, pp. 59–64, 2010.

2 Related Studies

2.1 Home Network

HNIT (Home Networking & IT) under CEA (Consumer Electronics Association), the U.S., defines home network as "a coupling together of home appliances and electronic systems for remote approach control possible." That is, through home network, each product must connect each other to share mutual service, while the user must be able to remote-control scattered instruments or use the service provided by each instrument. Setting in which such home network service has been in application is called digital home. In 2003, when the Ministry of Information and Communication designated this industry as one of next-generation growth engines for Korea, the term of digital home was first used. Digital home is the concept that unites home networking technology and information electronics embodied with this technology, suggesting that ubiquitous environment has been applied to general homes.

Home network is, as shown in Fig. 1, binding instruments at home into one network to make them capable of communication and connecting these to outside internet network to allow controlling consumer appliance from at/outside home, regardless of the user's position.

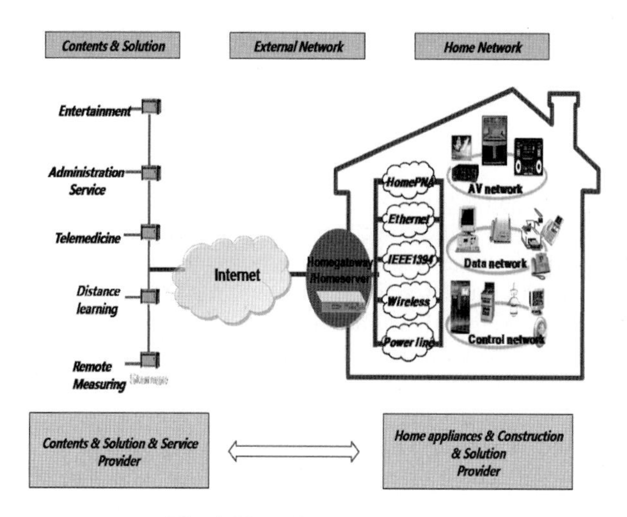

Fig. 1. Home Netowork system

It is also through home network that makes available outside service from home, such as remote medical care, remote instruction, etc. from outside contents.

2.2 OTP (One-Time Password)

OTP is the system of generating a password available for only one time, thus it authenticates user by using different passwords for each time. OTP, a typical method of double-element user authentication and basically devised on the basis of cryptographic idea, is the system of high security and convenience to use. Since it uses another password for each time, it is, unlike the system using the fixed password, proof against attack by reusing password, and since it uses cryptographic algorithm, it is also proof against prediction of a possible password in use for next time from the one currently used. Hence it is safe. If you enter a user password and input value for generating one-time password into the OPT program stored in OPT token or user PC, the system generates one-time password using cryptographic algorithm. Here, only by entering different values for each time for input values, does a one-time password come into being, and depending on what kind of value is entered for this input value, it is classified into diverse OPT methods.

3 A Proposed System

3.1 Configuration

User authentication proposed in this thesis for home network system authenticates user through OTP-based open-key authentication of cryptographic algorithm, in which home system is controlled with personal device by receiving personal authentication. This also presents a system, rather than passing through the authentication server of a service provider that must be included in existing home network setting, in which the user is able to authenticate in person from outside with a client device and home server.

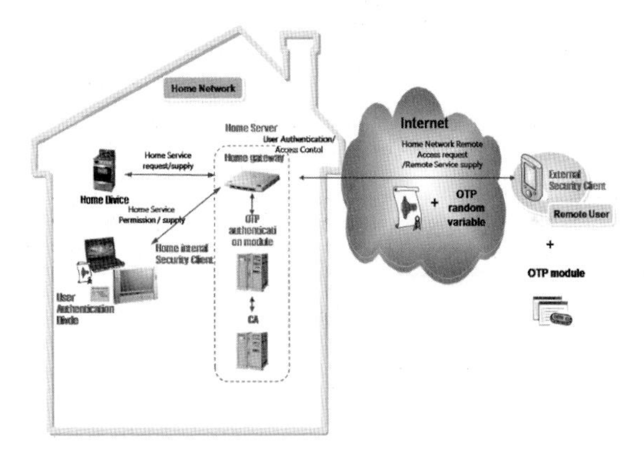

Fig. 2. Total User Authentication system proposed

In configuration of the system, as shown in Fig. 2, the user from outside, without his passage through internet network to home network of service provider, carries out approach and user authentication in person. This proposal also includes, in approaching homer server, a method of authenticating the user using OTP-based authentication and a method of controlling approach to home devices.

For user authentication, we propose a method of using OTP random variables and authentication, generated by OTP module mounted in the client device of user outside, which is a safer way to authenticate a user than existing ones. Home server is composed of the client device for outside user, home gateway for communicating between home devices within home network, and CA, OTP authentication module for user authentication.

Home device approach control, when client controls home device from outside, gives a list of approach controls differently composed for each user and provides multi-approach control for device already in use by a user so that other users cannot use it, using OTP random variables generated by each user. Manager can perform management, according to the grading of users, in separation into devices approachable and unapproachable, while a client can request addition to/deletion from a list of approach control from the manager and when the request is met, the client receives a reissue of authentication from home server.

3.2 Process of User Authentication

For process of user authentication, as shown in Fig. 3, user asks access to home network from outside through SSL (Secure Sockets Layer) with OTP random variables generated by an authentication issued for a client and OTP module mutually motivated.

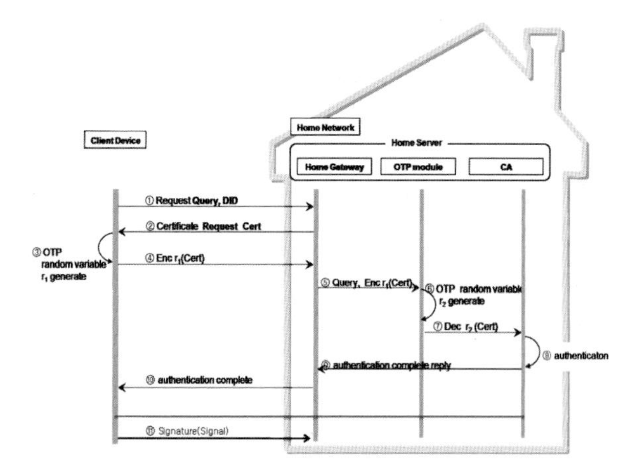

Fig. 3. Process of user authentication

Once request for user authentication has been transmitted, home server requests the information on user's authentication from the user and the user generates the value for r, an OTP random variable generated by home server and motivated OTP module, and transmits the value for r coding the authentication information with personal key in symmetric-key algorithm. OTP authentication module at home server also decodes the coded authentication information with value for r, OTP random variable generated, examines the decoded authentication information from CA. If the authentication information has been verified, it transmits the response of authentication verification to the user through home gateway, ending the authentication of the user. Then, client device can control a home device on home network.

3.3 Home Device Approach Control

When the process of user authentication is done, approaches to each device are made through home gateway, as shown in Fig. 4, When an outside client approaches home device, authority to approach home device is provided depending on the user, while home server controls approaches to home device by analyzing whether it is approachable or not through the authentication of an outside client.

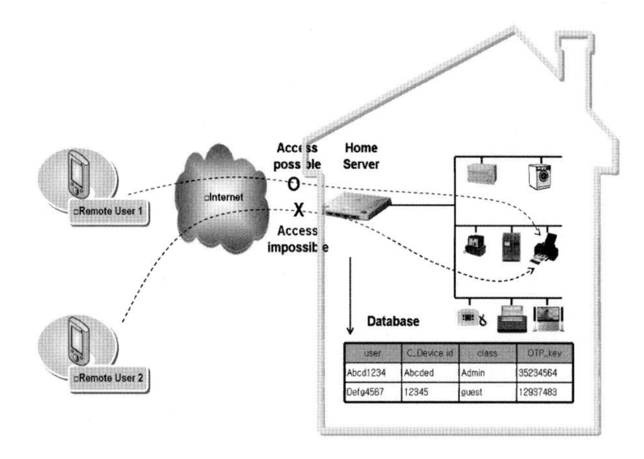

Fig. 4. Divice access control

① $Enc_{h(r\|Key)}(DIDNum, Control\ Command), DID$

- Conjoin the value for r, a random variable generated from OTP module, and one's key, and then hash.

- Transmit the value for hash, DIDnum, an X509-based attribute, coded value for Control Command message and the client's own DID to home server.

② Home server decodes these using the client's DID value to verify DIDnum and Control Command, and then notifies.

5 Conclusion

To prevent and reduce the security risks increasing together with advance in home network technology, continued concern and active study are needed on an on-going basis. Positive and continued responses to such problems are also needed from the actual spot of home network.

Above all, it requires technique for a safe user authentication in order to prevent occurrence of invasion incidents in home network service and exposure of user information. This thesis proposes a safer system of authentication in which the outside user can perform a safe user authentication with home server on the home network directly, without going through the authentication server of home network service provider.

This system has the authentication for one's client device directly issued from home server from offline, protects the authentication information safely by coding it with random variable generated by OTP module motivated between client device and home server. Besides, it authenticates user and controls device approaches using the issued authentication and the value of OPT random variable. In the protocol proposed, since data is always transmitted in encryption, when some illegal equipment has no idea of a client's personal key and random number value-r, there is no risk of exposing data information. It also has an accompanied effect of reducing communication overhead by omitting the prior movement of having to transmit from server to client by mutually motivating random number value-r generated from the existing home server.

In the course of user authentication, it encodes using home server and one-time random number value-r generated from the client's OTP to provide an authentication, safer against attacks such as snipping. It is difficult to infer the personal key and an authentication, and since it transmits the message for controlling home device by recoding the hashed value, it is impossible to infer the content of the message even if the message is interrupted halfway.

Further research task must include a study of controlling diverse home devices, a study on the method of approaching and controlling home device wireless using mobile instruments in low capacity for arithmetic operation, and a study on an even safer security protocol by applying the safe security protocol in use for existing fixed lines and this proposed method.

References

1. Kim, S.-h., Kang, C.-b., Jang, H.-j., Kim, S.-w.: A Secure Control of Home Network on a PDA. Korea Information Science Society 29 (October 2002)
2. Young-Gu, L.: Desing and Implementation of Security Protocol for Home-Network based on SOAP. Soongsil Universty (2006)
3. Rahman, M., Bhattacharya, P.: Remote Access And Networked Appliance Control Using Biometrics Features. IEEE Transactions on Consumer Electronics 49(2) (May 2003)
4. Tsai, P.-L., Lei, C.-L., Wang, W.-Y.: A Remote Control Scheme for Ubiqui- tous Personal Computing. In: International Conference on Networking, Sensing & Control (March 2004)
5. Callaway, E., Hester, L., Gorday, P.: Home networking with IEEE 802.15.4: a developing standard for low-rate wireless personal area networks. IEEE Communications Magazine 40(08) (2002)
6. Schulzrinne, H., Wu, X., Sidiroglou, S.: Ubiquitous Computing in Home Networks. IEEE Comm. Mag. (October 2003)
7. Jo, H., Youn, H.: A Secure User Authentication Protocol Based on One-Time-Password for Home Network. In: Gervasi, O., Gavrilova, M.L., Kumar, V., Laganá, A., Lee, H.P., Mun, Y., Taniar, D., Tan, C.J.K. (eds.) ICCSA 2005. LNCS, vol. 3480, pp. 519–528. Springer, Heidelberg (2005)

Web-Based Media Contents Editor for UCC Websites

Seoksoo Kim

Dept. of Multimedia, Hannam Univ., 133 Ojeong-dong, Daedeok-gu, Daejeon, Korea
sskim0123@naver.com

Abstract. The purpose of this research is to "design web-based media contents editor for establishing UCC(User Created Contents)-based websites." The web-based editor features user-oriented interfaces and increased convenience, significantly different from previous off-line editors. It allows users to edit media contents online and can be effectively used for online promotion activities of enterprises and organizations. In addition to development of the editor, the research aims to support the entry of enterprises and public agencies to the online market by combining the technology with various UCC items.

Keywords: Media Editor, Web Solution, UCC, H.264/AVC.

1 Introduction

Today, UCC strongly influences not only the Internet but various digital platforms(mobile, IPTV, etc.). The role of prosumers is not limited to a particular area but involves all aspects of the society[1]. More recently, there has been a sharp increase in the number of users who create media contents. The biggest difference between Web 2.0 and Web 1.0 is active participation and sharing among the Internet users as well as increased openness. In Web 2.0, values are created by participation of users. That is, media contents are created by common users while the distribution is done by the Internet, cutting down costs of contents development and distribution.

UCC increases competitiveness of a company through participation of many users and is being used as a means of buzz marketing through its explosive speed of spread[2]. However, Web 2.0 including UCC media contents also affects management of an enterprise, creating a gap between those who make good use of Web 2.0 and those who fail to do so. This was well demonstrated by UCC contents used for the presidential election in the US.

Therefore, this research aims to develop a web-based media contents editor to improve enterprises' competitiveness and their technology in the area. The editor is expected to help enterprises effectively express users' ideas and take an advantageous position in the area of a web-based media contents editor.

2 Related Work

The UCC media contents service system is largely divided into three parts: a user authentication system, a media contents encoding system, and a security/storage

T.-h. Kim, A. Stoica, and R.-S. Chang (Eds.): SUComS 2010, CCIS 78, pp. 65–71, 2010.

system. First, the user authentication system is comprised of a web server and an authentication server while the media contents encoding system consists of an encoding engine, an uploading server (which uploads media contents on a server), and a DB server(for management). And the security/storage system requires a DRM server, a streaming server(for transmitting), and a file/backup server(for storage). Figure 1 shows the structure of the system for UCC media contents service.

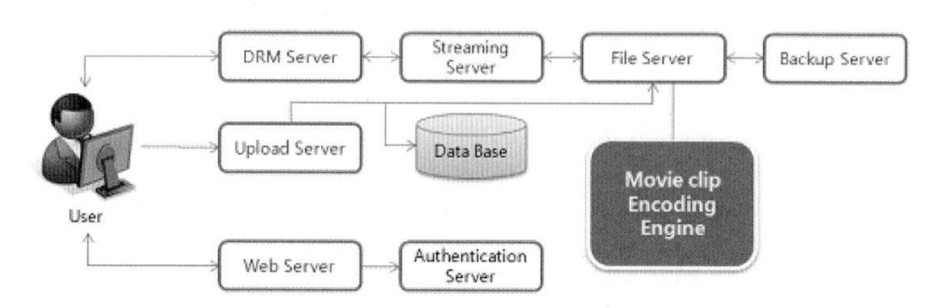

Fig. 1. Structure of UCC Media Contents Service System

UCC media contents service requires technologies related with encoding, contents transmission, and copyright. There are mainly two types of formats for UCC media contents service, which are FLV(Flash Video) and WMV(Windows Media Video). Both formats support streaming and can be easily employed by general users. MOV is offered by most UCC websites in other countries but not in Korea.

And there are largely three ways of transmitting contents: downloading(downloading of contents), progressive downloading(playing while downloading), and streaming(real-time playing of transmitted data). The first method, downloading, transmits contents to users through HTTP or FTP server with high-quality graphics and sound. However, it allows users to possess contents permanently and create excessive network traffic or downloading time if the file size is huge. Most web hosting services limit downloading of WMV, WMA, MP3, and MPEG files due to traffic overload. The second method, progressive downloading, is weak in file protection. Thus, streaming, which allows only a certain amount of data to be played instantly and other data downloaded continuously, is used for most services[3].

Lastly, the copyright for UCC is the right to exclusively use a work of study or art such as DRM(Digital Rights Management) and CCL(Creative Commons License)[4].

An editing technology should be able to offer increased convenience to users and satisfy strong desire for creativity of contents producers. Although users today make good use of developed technology of UCC media contents service, there is still not a sufficient number of user-created films, for it is too difficult for general users to employ professional editing programs and upload the media contents.

These days, some entertainment portals, which offer media contents directly created by users, are becoming increasingly popular. And more individual blogs have features to upload media contents.

Some of UCC websites that offer the feature to edit media contents include Diodeo, Mgoon, Naver, Yahoo (in Korea) as well as Videoegg and Jumpcut (overseas). Videoegg, Naver, and Yahoo allow users to cut out an unnecessary part when uploading media contents while Mgoon and Jumpcut provide more various features (insert of a text/image).

Therefore, although there is much room for development, the technology of web-based media contents editor, the most desired function among users, is expected to allow enterprises to improve their competitiveness or take an advantageous position in the industry.

3 Design of a Web-Based Media Contents Editor

3.1 Non-linear Video Editor System

For real-time file editing in a multi-media file system, video array strategies of Hermes are employed in this study. Hermes defines an extent as a group of consecutive blocks of 1~2Mb and manages data based on this unit. Also, 8 extents are grouped together in order to increase continuity of data when extents are allocated or removed. In addition, when an allocated data area is made up of consecutive extents, an array is formed by grouping for more effective management[5]. As a result, an extracted array becomes a key factor for a video edit system.

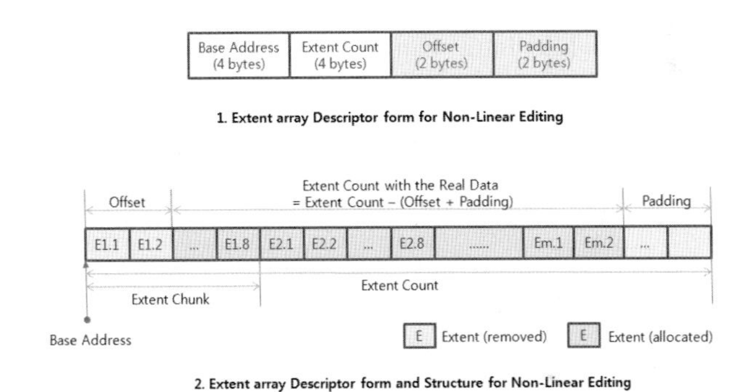

Fig. 2. Structure of a Non-linear Video Editor

3.2 Block-Level Real-Time Edit System

In order to support block-level editing with a minimum change, a descriptor should be restructured and implemented while the size of a previous array descriptor in the Hermes file system maintained. Also, a block-level edit algorithm should be applied, allowing users to change arrays and sizes.

3.3 High-Speed Extraction of Thumbnail Videos in H.264/AVC

A thumbnail image refers to a major image that helps to quickly view contents extracted by decoding a part or the entire of a video. The thumbnail image is used to provide user interfaces. A thumbnail image is extracted by down-sampling (decoding the entire of compressed bit strings)[6], by using features of compressed bit strings, or by partial decoding[7]. In this research, H.264-based SVC (Scalable Video Coding) is used, which makes possible integer down-sampling. This technology is a key factor in offering extracted information of media contents to users.

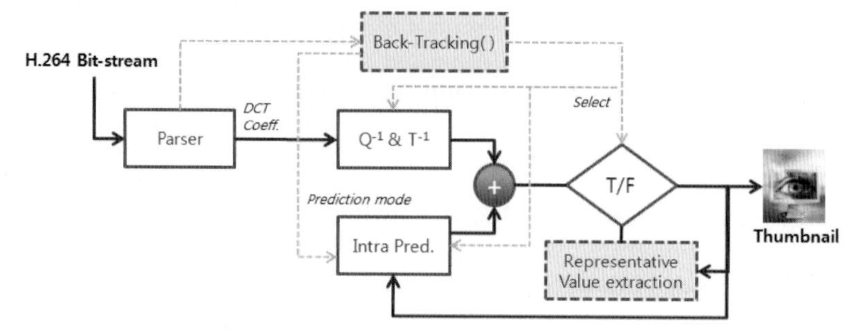

Fig. 3. H.264-based SVC (Scalable Video Coding)

3.4 High-Speed MPEG Transcoding Method

In this research, an integrated encoder, one of the most popular transcoding methods, is employed[8]. The main purpose is to correctly implement intra prediction in a pixel region, which requires the greatest amount of calculation. In H.264/AVC, regarding the luminance component of MB(Macroblock), Intra 16x16 is used for MB with uniform pixel changes while Intra 4x4(prediction changes according to each 4x4 block) is used for MB with un-uniform pixel changes.

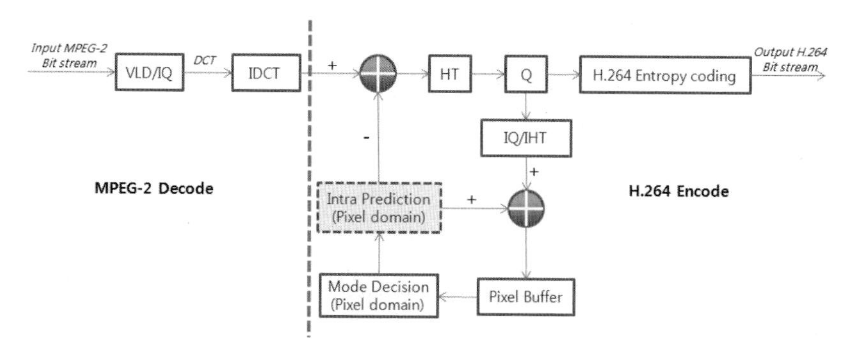

Fig. 4. High-Speed MPEG Transcoding Method

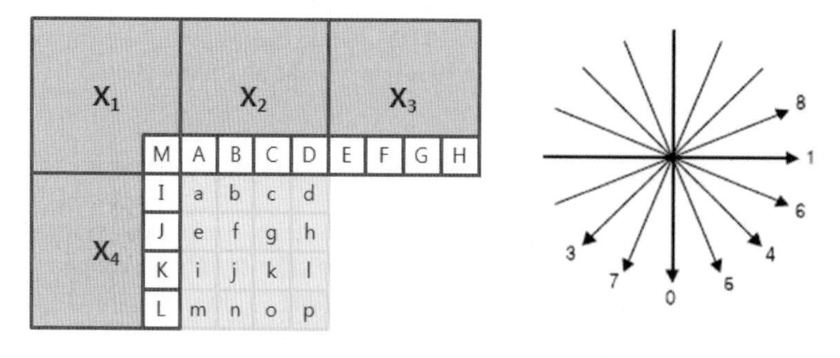

Fig. 5. Reference Pixel for Intra Prediction of 4x4 Blocks and 8 Prediction Directions

3.5 Efficient Extraction of Reduced Images in H.264/AVC Bitstream

The intra prediction methods in H264/AVC are largely divided into 16×16, 8×8, and 4x4 according to the size of blocks. There are 9 prediction modes for 4×4/8×8 sub-block but 4 modes for 16×16 blocks and 8x8 chroma blocks[9]. A reference pixel for intra prediction of 4x4 blocks and 8 prediction directions are as follows.

Intra prediction is expressed as multiplication of a matrix through prime factorization. Although each mode requires a different formula, a general formula is used in this research for intra prediction. Also, a horizontal filter and a vertical filter are defined in order to utilize frequency features of intra prediction, which, in turn, is used for an altered intra prediction formula.

$$y_{pred}^{m} = \left(\sum_{n=1}^{3} \sum_{i=1}^{4} s_i x_n c_{n,i}^{m} \right) + \left(\sum_{i=1}^{4} c_{4,i}^{m} x_4 s_i^{T} \right) \tag{1}$$

The formula above is a generalized formula for intra prediction, in which m refers to intra prediction mode, xn means an n-time neighboring block. Also, c refers to an invariable matrix multiplied to a reference block (which is a factorized matrix) while si means a shift matrix defined by the following formula.

$$s_i = (a_{ij})_{4\times4},\ a_{ij} = \begin{cases} 1, \text{ if } j=4 \text{ and } i=\alpha\ (1 \le \alpha \le 4) \\ 0, \text{ else} \end{cases} \tag{2}$$

4 Web Server System for Video Editing Services

Of the programs offering web services, Apache can be easily installed in every operation system. IIS(Internet Information Service) is applicable only in Windows and, although other web servers can be installed in all operation systems, they are mainly for private uses while Apache provides compatibility for various operation systems[10]. Even in economic terms, it is hard to find a web server superior to

Fig. 6. Structure of the Web System Based on Apache

Apache. It offers all features provided by other commercial web servers and has significantly improved its functions through the shift from Prefork to Worker. In addition, the config. of Apache is easier and more simple than that of IIS or iPlanet. As a freeware program, Apache has a wide range of data on config setup and upgrades. In this research, therefore, the web server system is designed based on Apache.

It is still not easy for many people to create UCC by using previous commercial tools. Each UCC tool has different methods of use, and contents cannot be used before a video is completed. In particular, the greatest inconvenience of previous UCC tools is that editing is not possible during playing. Therefore, the direct show technology is applied in this research in order to solve the issue as well as buffer problems that may arise during saving of a video. It also helps to easily control input/output of images.

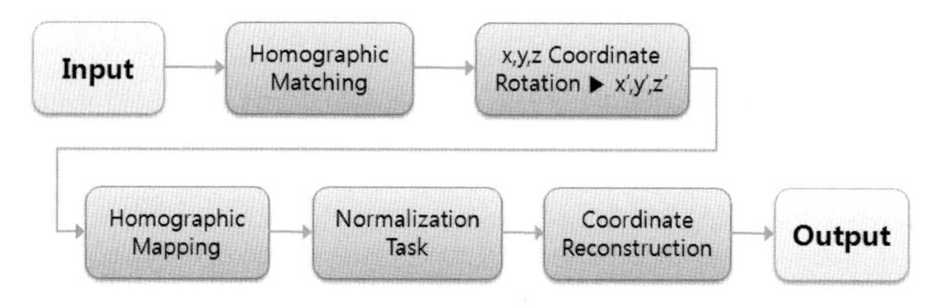

Fig. 7. Design of Direct Show Technology

5 Conclusion

The model presented in this research greatly increases users' convenience by combining a new user-based interface with a web-based media contents editor, aiming to analyze tendencies of consumers in the fast-changing UCC environment and apply the results to the system for improved efficiency. The user interface applied to the system will be easy and simple so that users of all ages can edit a video on the web to create UCC.

Acknowledgement

This paper has been supported by the 2010 Hannam University Research Fund.

References

1. Baker, S.: Portals to Paperless Prosperity. Journal of Business Strategy 20(6), 32–35 (1999)
2. Economics Review, `http://www.bookrail.co.kr`
3. Kuacharoen, P., Monney, V.J., Madisetti, V.K.: Software Streaming via Block Streaming. In: 2003 Conference, IEEE and ACM SIGDA, Munich, Germany, pp. 912–917 (2003)

4. Choi, M.J., Ji, S.W.: Broadcast Program Copyright and Video UCC in the Digital Age. Broadcast and Communication 9(2), 67–93 (2008)
5. Won, Y., Park, J., Ma, S.: HERMES: File System Support for Multimedia Streaming in Information Home Appliance. In: Shafazand, H., Tjoa, A.M. (eds.) EurAsia-ICT 2002. LNCS, vol. 2510, pp. 172–179. Springer, Heidelberg (2002)
6. Cohen, H.A.: Retrieval and Browsing of Images Using Image Thumbnails. Journal of Visual Comm. and Image Representation 8(2), 226–234 (1997)
7. Dugad, E., Ahuja, M.: A Fast Scheme for Image Size Change in the Compressed Domain. IEEE Trans., CSVT 11(4), 461–474 (2001)
8. Kalva, H., Petljanski, B.: Exploiting the directional features in MPEG-2 for H.264 Intra transcoding. IEEE Trans., on Consumer Electronics 52(2), 706–711 (2006)
9. Wiegand, T., Sullivan, G.J., Bjntegaard, G., Luthra, A.: Overview of the H.264/AVC video coding standard. IEEE Trans., Circuits Syst. Video Technol., 560–576 (2003)
10. Looking Beyond the Rhetoric, http://www.serverwatch.com

A Study on AR 3D Objects Shading Method Using Electronic Compass Sensor

Sungmo Jung and Seoksoo Kim*

Dept. of Multimedia, Hannam Univ., 133 Ojeong-dong, Daedeok-gu, Daejeon, Korea
sungmoj@gmail.com, sskim0123@naver.com

Abstract. More effective communications can be offered to users by applying NPR (Non-Photorealistic Rendering) methods to 3D graphics. Thus, there has been much research on how to apply NPR to mobile contents. However, previous studies only propose cartoon rendering for pre-treatment with no consideration for directions of light in the surrounding environment. In this study, therefore, ECS(Electronic Compass Sensor) is applied to AR 3D objects shading in order to define directions of light as per time slots for assimilation with the surrounding environment.

Keywords: Augmented Reality, 3D objects, Shading method, ECS, Mobile rendering.

1 Introduction

Studies on mobile devices have been driven by the rapid development of the wireless Internet. In particular, appearance of mobile phones supporting 3D graphics such as a PDA, smart phone, or DMB phone, has speeded up research on mobile 3D graphics[1]. A mobile environment has more limitations in hardware functions compared to a desktop environment, which makes difficult to offer various services. Therefore, much research is being done in order to implement graphics of a desktop environment through a mobile device.

A lot of research efforts are focused on NPR (Non-Photorealistic Rendering)[2], for the method uses a small amount of resources than previous graphic techniques in a mobile environment and makes possible more smooth communications with users.

In this study, a mobile device employs AR 3D objects shading method, which utilizes an electronic compass sensor, based on 3D objects using the NPR method in order to set directions of light according to time slots. The method using ECS (Electronic Compass Sensor) can measure the amount and direction of light through CV(Computer Vision) technology. Due to hardware limitations of a mobile device, however, 3D objects are shaded according to the number of shaded objects and directions defined by users rather than using CV-based measurement which requires a huge amount of calculation. In this way, a mobile device can lower the amount of calculation by simply using defined values.

---

* Corresponding author.

T.-h. Kim, A. Stoica, and R.-S. Chang (Eds.): SUComS 2010, CCIS 78, pp. 72–76, 2010.

2 Related Studies

2.1 NPR Method

During the past decade, various algorithms were studied in the area of NPR(Non-Photorealistic Rendering)[3]. While each technique creates a different style of images, Decaudin developed a cartoon rendering method, in which a shade is made by only two or three colors[4]. Here, an area is divided by n-L values. Meanwhile, Michael suggested Graphtal Textures expressing a complex object such as grass or a tree as a simple object[5]. This method perceives objects by simplifying them. However, the purpose of this study is to realize richer images using more various colors and, thus, 3D volume rendering method using gradations (suggested by Gooch)[2] is introduced in this study and real-time rendering is made possible.

2.2 Silhouette Edge Detection

One of silhouette edge detection methods[6] is a hidden line calculation algorithm which calculates visibility of silhouette edges. This algorithm makes possible silhouette edge rendering with no need to calculate by searching the entire screen and can improve frame rates. There have been studies on silhouette edge rendering through removal of hidden lines[7] but most of them failed to implement real-time rendering due to batch processes. Currently, a real-time silhouette edge rendering method has been presented through a static polyhedral model with proximal data[8] and also another method using a depth buffer and two polygon sets in order to calculate silhouette edges in real time on the screen from a given point.

3 AR 3D Objects Shading Method Using ECS

3.1 User Defined Shading

When users define gradations of discontinuous shading, number of colors, and levels of transparency, the amount of calculation required for automatic measurement (by computer vision) can be significantly reduced. Since the method does not need repeated rendering pipelines, it can be effectively applied to a mobile device.

Moreover, a backlight calculation algorithm solves the problem of blending and cartoon rendering methods which use various paths (repeated rendering pipelines of discontinuous shading) by enabling more various expressions.

The following is objects using user-defined shading (9 directions).

Figure 1 shows that 8 objects from (b) to (i) have been shaded in the (a) direction (here, (a) is a basic object shaded based on the median point).

For the defined objects, augmented time is defined per each time slot. And an ECS(Electronic Compass Sensor) is used in order to redefine an augmented object according to directions. In this way, real-time object shading, which assimilates with the surrounding environment, can be implemented.

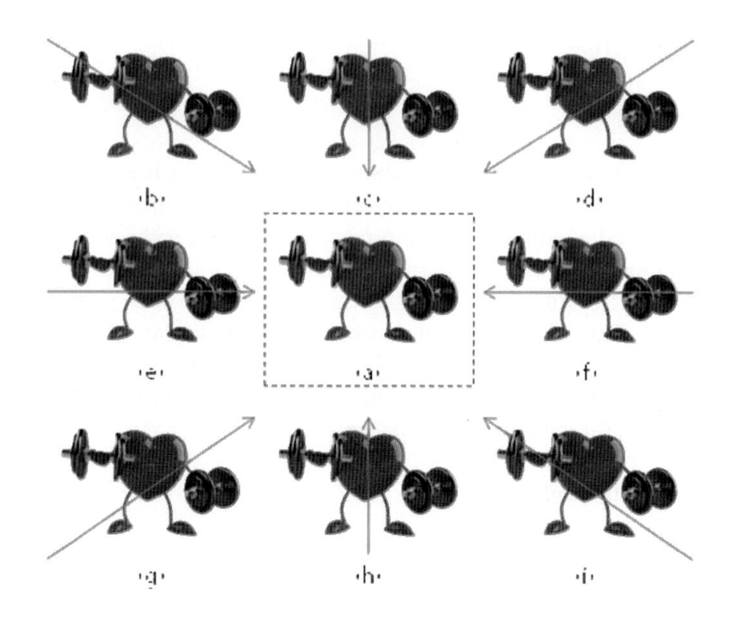

Fig. 1. Objects using user-defined shading (9 directions)

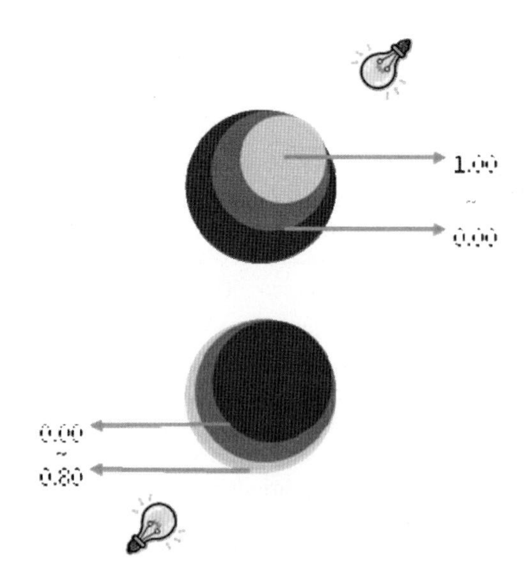

Fig. 2. Calculation of two Sources of light

3.2 Backlight Calculation Algorithm

As is the previous calculation, coordinates of texture are obtained by calculating the normal value of the object vertex ((a) of Fig.1) and Lambert light. The source of light is calculated according to Formula (1).

$$MAX\,(V \cdot L, 0)\,. \tag{1}$$

After calculating the source of light, the source of backlight is assumed for the part where the light cannot reach(≤0.00) as depicted in Figure 2 so as to calculate the intensity of light within a shade.

Formula(2) shows how to convert the two sources of light into one coordinates in order to be applied to the object. Here, V_i refers to the intensity of light of the vertex, V_{di}, the coordinates of a dark shade map, and V_{li}, those of a light shade map.

$$V_{li} = (1 - V_i) \qquad 0.5 + V_i\,,\ \ V_{di} = V_i \qquad . \tag{2}$$

The value obtained from the two sources of light is converted into 0.00~1.00 using Formula(2), and all texture maps are made into a single map. Figure 3 shows the integrated texture map for final application.

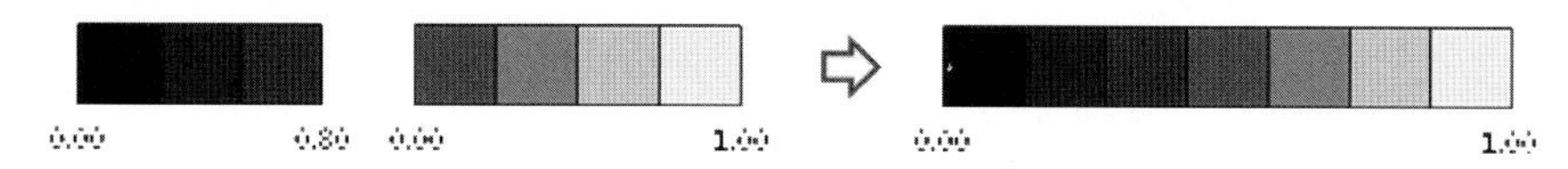

Fig. 3. Integration of the dark and light shade maps

4 Conclusion

Although there is increasing demand for the qualitative improvement and diversification of mobile contents along with the popularity of mobile devices and development of the wireless Internet, mobile devices cannot implement 3D graphic technologies used in the desktop environment due to their hardware limitations.

Hence, this study proposes a method of applying ECS (Electronic Compass Sensor) to AR 3D objects shading in order to define directions of light according to time slots for assimilation with the surrounding environment. This method requires no additional rendering but simply uses objects already rendered by user definitions. In this way, mobile devices can overcome their hardware limitations and create gradation of shades based on user preferences (Previously, mobile devices could offer only a monotonous screen due to application of solid colors and an inability to use light).

In the near future, further development of mobile hardware will make possible real-time shading using CV light measurement.

Acknowledgement

This paper has been supported by the 2010 Preliminary Technical Founders Project fund in the Small and Medium Business Administration.

References

1. Lake, A., Marshall, C., Harris, M., Blackstein, M.: Stylized Rendering Techniques For Scalable Real-Time 3D Animation. In: SIGGRAPH (2000)
2. Gooch, A., Gooch, B., Shirley, P., Cohen, E.: A Non-Photorealistic Lighting Model For Automatic Technical Illustration. In: SIGGRAPH (1998)

3. Markosian, et al.: Real-Time Nonphotorealistic Rendering. In: SIGRAPH (1997)
4. Decaudin, P.: Cartoon-Looking Rendering of 3D-Scenes, Research Unit INRIA Rocquencourt (1996)
5. Kowalski, M.A., Markosian, L., Northrup, J.D., Bourdev, L., Barzel, R., Holden, L.S., Hughes, J.F.: Art-Based Rendering of Fur, Grass, and Trees. In: SIGGRAPH (1999)
6. Smith Jr., T.G., Marks, W.B., Lange, G.D., Sheriff Jr., W.H., Neale, E.A.: Edge detection in images using Marr-Hildreth filtering techniques. Journal of Neuroscience Methods 26(1) (1988)
7. Kettner, L., Welzl, E.: Contour Edge Analysis for Polyhedron Projections. In: Strasser, W., Klein, R., Rau, R. (eds.) Geometric Modeling: Theory and Practice. Springer, Heidelberg (1997)
8. Saito, T., Takahashi, T.: Comprehensible Rendering of 3-D Shapes. In: Baskett, F. (ed.) Computer Graphics, SIGGRAPH 1990 Proceedings, vol. 24 (1990)

A New Approach for Semantic Web Matching

Kamran Zamanifar[1], Golsa Heidary[2],
Naser Nematbakhsh[1], and Farhad Mardukhi[1]

[1] Dept. of Computer Science, University of Isfahan, Isfahan, Iran
{zamanifar,nemat,mardukhi}@eng.ui.ac.ir
[2] Young Researchers Club, Computer Engineering Department,
Islamic Azad University, Najafabad Branch, Iran
golsa.heidary@gmail.com

Abstract. In this work we propose a new approach for semantic web matching to improve the performance of Web Service replacement. Because in automatic systems we should ensure the self-healing, self-configuration, self-optimization and self-management, all services should be always available and if one of them crashes, it should be replaced with the most similar one. Candidate services are advertised in Universal Description, Discovery and Integration (UDDI) all in Web Ontology Language (OWL). By the help of bipartite graph, we did the matching between the crashed service and a Candidate one. Then we chose the best service, which had the maximum rate of matching. In fact we compare two services` functionalities and capabilities to see how much they match. We found that the best way for matching two web services, is comparing the functionalities of them.

Keywords: Semantic web; matching algorithm; UDDI; OWL.

1 Introduction

Semantic web is a well defined form of the web in which computer agents are able to use information on the web in the same way as human beings do [7]. In other words, semantic of information is well defined in the semantic web to make automatic knowledge extraction possible. However, semantic web suffers from distributed and heterogeneous information.

We use OWL[1] to drive the semantic and syntactic aspects of a service. In systems which availability is one of the most important factors of quality, when a service crashes, it should be replaced by the most similar service. So a repository of web services by the name of UDDI[2] should be available for choosing a service among the candidate services.

In general, to find a similar service we can measure different features and capabilities of two services. The main features are:

---

[1] Web Ontology Language.
[2] Universal Description, Discovery and Integration.

T.-h. Kim, A. Stoica, and R.-S. Chang (Eds.): SUComS 2010, CCIS 78, pp. 77–85, 2010.

- Comparison of two services` functionality(capability)
- Comparison of quality of services (QOS)
- Comparison of two services` policies

Capability of a service refers to its functions, but policy focuses on non-functional aspects like availability, response time, reusability… which are considered at design time. One Service may have the same policy or quality as crashed service, but will not be a good choice for substitution. But if the capabilities of two services are similar, with high probability we can say that the two services can be substitute with each other. That's why we compare the functionality of crashed web service with other candidate services.

A web service provider can advertise its web through UDDI. A provider must do its work in two steps: first, he should publish his services which mean that the service should be advertised in a registry. Second, a user`s required service must be available in UDDI, by comparing the service`s capability with advertised one.

Our proposed Semantic Web matching has been done in two phases. In the first phase we compute the matching rate of two web services using by making two individual bipartite graphs. One for input matching and another for output matching. We select the most similar web service in the second phase we choose the service which has the most similarity with the crashed service. The architecture of our work is shown in "fig. 1".

In the remainder of this paper, first in Section 2, we talk about related works, in brief. Then in section 3 we define the main idea of matching semantic Web services, which is computing the matching rate by the help of bipartite graph, and after that we choose the best one .in last part of this section we analyze the time complexity of our algorithm. For better explanation of our algorithm, section 4 describes a simple scenario of matching web services. The discussion about power and weakness of our work in comparison with other works, have become in Section 5. Finally in section 6 the conclusion of our work is given.

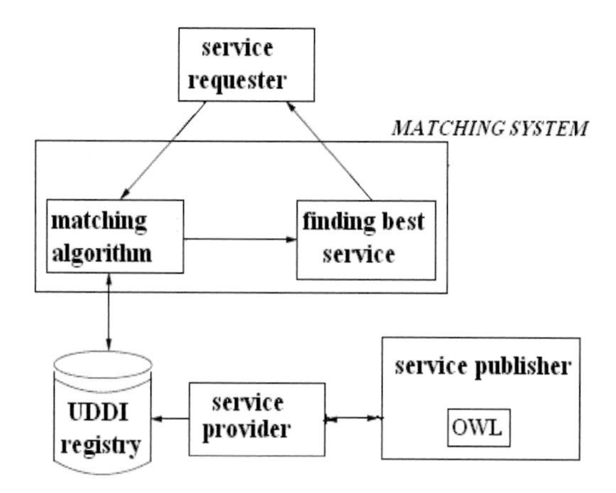

Fig .1. Shows the architecture of our matching model

2 Related Work

Web services are software building blocks that are in hand through the internet in a standard way using SOAP messaging. The most important properties of web services are reusability and the ability of composition. They allow clients to access data from remote providers without extracting it from HTML web pages or using proprietary protocols. Web services are used in several and different areas such as address validation, trading applications, product search and ordering, automatic systems, and so on. Clients can select from over 100 services listed on providers. Choosing the best service among them which satisfies our need, is as important as difficult. The first choice that comes into mind is Web Service Description Language (WSDL). In theory, semantic information in WSDL files was supposed to solve this problem, because WSDL is a way to know what a service does and how. But in practice, is not enough, because currently WSDL files don't have enough semantic information to decide substitutability or ability of compose. There is a need for automatic techniques to obtain more semantic information.

Many works have addressed ontology matching in the context of ontology design [5],[6],[7]. These works do not deal with explicit notions of similarity. However, many of them [3] have powerful features that allow for smart user interaction, or expressive rule languages for specifying mappings.

In [10] they use ws-policy for matching web services. We know that comparing functionalities will give a better result for semantic web matching.

In [9] semantic Web Service Description Ontology which is called SWSD and QOS, are used to do matching, which are not as good as functionality matching. In this work also, comparing service functionalities are considered in brief.

The similarity measured in [12], is based on statistics, but our work is precise, not on probability.

References [2],[11] have discussed about functional matching of semantic web services, but because the main idea of these works is about composition of web services, matching had not discussed very well.

3 Finding Most Similar Service

We have a repository of services that are all described in the same ontology language, OWL. So we can easily compare the capability and functionality of them. Each service does a special work by the means of some functions. These functions have inputs and outputs. Therefore, in order to compare two web services` functionalities, we must compare inputs and outputs. Now we describe the two phases of matching.

3.1 Computing the Matching Rate

The first phase is computing the similarity rate of two services. If the crashed service is C and the advertised service is ADV the inputs of each one is shown by C_{in} and

ADV_{in} and the outputs are shown by C_{out} and ADV_{out}. So we compare C_{in} with ADV_{in} and C_{out} with ADV_{out}. Four results will be achieved that the algorithm is as follows:

```
Case (Cout, ADVout):
        If Cout= ADVout then
                 return exact
        If Cout,subclass of ADVout then
                 return exact
        If ADVout subsumes Cout then
                 return plugin
        If Cout subsumes ADVout then
                 return subsumes
        Otherwise
                 return fail
```

Between all services in the repository and crashed service, this comparison should be done. An Equivalent algorithm also used for inputs.

3.1.1 Using Bipartite Graph

We do the matching by the help of bipartite graph. A Bipartite Graph is a graph $G = (V,E)$ in which the vertex set can be partitioned into two disjoint sets , $V = V_0 + V_1$, such that every edge e in E has one vertex in V_0 and another in V_1. The matching is complete if and only if, all vertices in V_0 are matched. It means that all vertices in V_0, as well as V_1, should have an edge.

Let C_{out} and ADV_{out} be the set of output concepts in C and ADV respectively. These constitute the two vertex sets of our bipartite graph. Construct graph $G=(V_0 + V_1, E)$, where, $V_0 = C_{out}$ and $V_1 = ADV_{out}$. Consider two concepts a in V_0 and b in V_1 It means that a is one of the output parameters of C and b is one of the output parameters of ADV. Let R be the result of CASE (in our algorithm, which can be Exact=E, Plugin=P, Subsume=S, Fail) between concepts a and b. It is obvious that $E > P > S > F$. We define an edge (a, b) in the graph and label this edge as R. Therefore if matching is complete (all vertices have at least one edge), now we compute the whole matching rate for these two services. In "fig. 2" we have an example of a bipartite graph which has the complete matching.

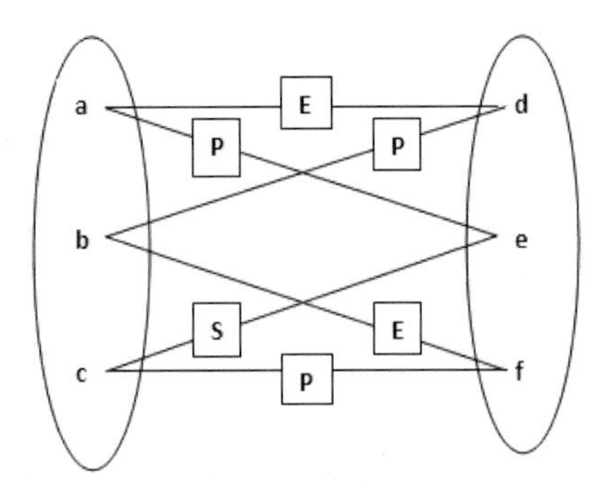

Fig. 2. Shows an example of bipartite graph of output concepts

Now we should choose the best sub graph. So for each vertex in V_0 we should choose the edge which is labled maximum, in a way that each vertex has only one edge. Two subgraphes of “fig. 2” are shown in “fig. 3” and “fig. 4”. To compute the weight of the graph ,we select the less degree of edges .

For example the weight of graph G1 in “fig. 3”, is S and the weight of graph G2 in “fig. 4” is P. The result of matching two services is the graph which has the higher weight (in our example graph G2 that has the weight P). We do all these works for inputs, too.

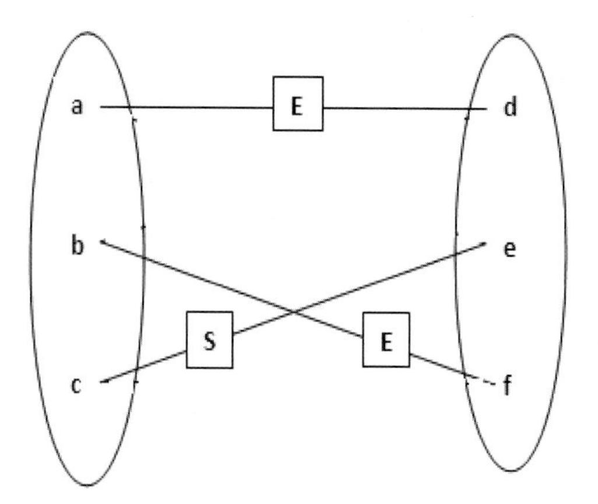

Fig. 3. Shows the matching subgraph G1

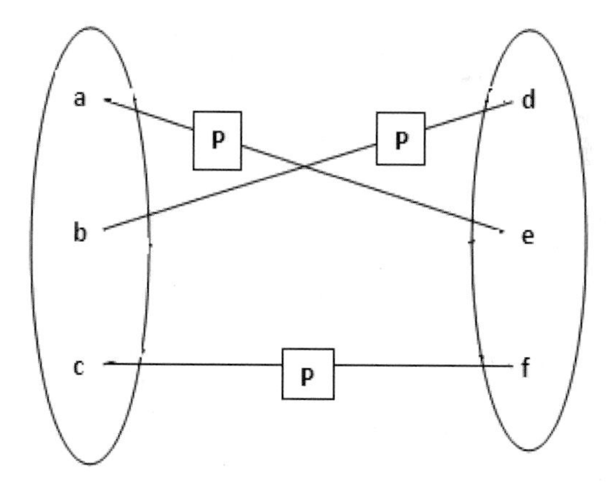

Fig. 4. Shows the matching subgraph G2

3.2 Our Matching Algorithm

If the result of matching two services` outputs is OUTSIM and the result of matching two services inputs is INSIM, the whole result of matching two services is “result” that obtains by the following algorithm . The input of our algorithm is C which is the crashed service. The outputs of our algorithm are "bestsrv" and "best result" which are the best service and the matching rate, respectively.

```
FIND MATCH (C , bestsrv, best result );
bestsrv = first service;
best result = F;

for all services in repository  (ADV)  do
    for all output parameters of  C do
          Case ( Cout , ADVout  ) ;
    make a bipartite graph for outputs ;
    OUTSIM = minimum edge in bipartite graph;

    for all input parameters of C do
          Case ( Cin , ADVin  );
    make a bipartite graph for inputs ;
    INSIM= minimum edge in bipartite graph ;

    reselt = E ;
    if  (OUTSIM=F or INSIM=F ) then
          reselt = F
                else if  (OUTSIM=S or INSIM=S ) then
                      reselt = S
                      else if  (OUTSIM=P or INSIM=P )
                then
                            reselt = P ;

if result > best result then
    best result = result ;
    bestsrv = ADV ;
if best result =E then quit ;
```

3.3 Complexity Analysis

In computing the complexity of semantic web service matching, a lot of factors have Interference such as number of services in the repository, number of input parameters and number of output parameters. In equal situation with other algorithms, we don`t consider the number of input and output parameters. So for computing time complexity of an algorithm, just the number of advertised services is important. Now if N is the number of advertised services in the repository, we choose the first service as the best one. After computing each advertised service`s similarity, If this one`s similarity rate is higher than the best one, then this service is chosen as the best and so on.

Whenever it finds the service by the similarity rate E, the work is finishing. Therefore the complexity of our algorithm is of O (N).

4 An Example

For better explanation of our algorithm, this section describes a simple scenario of matching web services.

If service C with these inputs and outputs crashes:
Inputs: (officer ID, company name)
Outputs: (name, address, phone number)

This service takes the name of a company and one of its officer`s ID. Now we want to find a service for substituting it. For example, one of the services in UDDI is ADV that has these inputs and outputs:

Inputs: (customer name, member ID)
Output: (name, mobile number, add)

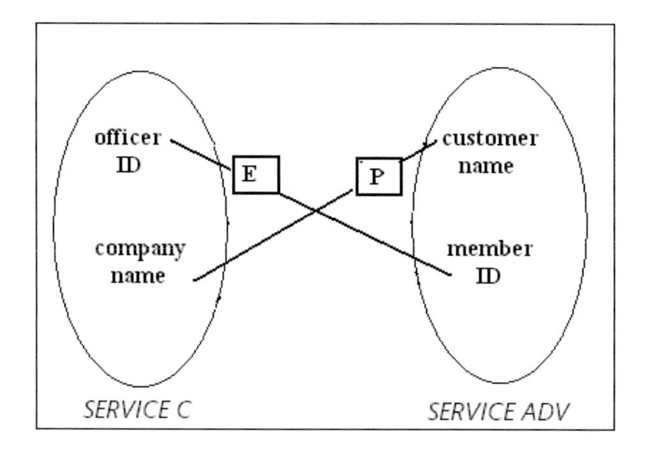

Fig. 5. Shows a bipartite graph for inputs

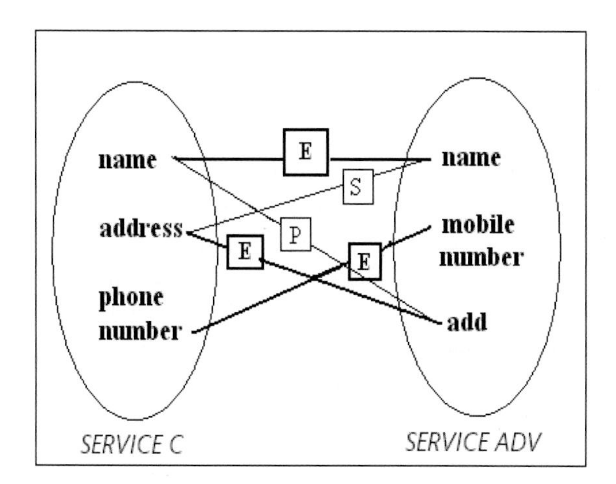

Fig. 6. Shows a bipartite graph for outputs

The bipartite graph for inputs is shown in “fig. 5”. According to our ontology, the INSIM will be P.

The bipartite graph for outputs is shown in “fig. 6”. According to our ontology, the OUTSIM will be E.

According to these two graphs, the result is P.

5 Discussion

For comparing two services, many languages are used such as : WSDL, OWL, OIL[3], DAML[4], and etc. We used the service repository (UDDI) to do the semantic matching of web services. The complexity of our algorithm is at most O(N). It means in most cases with less than N computation we can get the similar service and among the other algorithms, it is one of the most precise algorithms. Always it considers the minimum similarity between two services. Therefore if it concludes the EXACT similarity, really has found the best one. Some of other works compute the average similarity or some others say two services are exact similar if one of the input or output has the result of exact. It is obvious that sometimes our algorithm may not find a match for the crashed web service, because we don`t sacrifice the accuracy. Some other algorithms have a lot of computation to find a numeric degree of matching. Although they compute the similarity exactly, but we think that our algorithm does the same work without need to do a lot of computation.

Some services have preconditions and effects. In this work we didn`t mention these two factors for simplicity, but in future work we surely consider them.

6 Conclusion and Future Work

In this paper we have identified the problem of semantic web matching and proposed a new algorithm to solve the problem. By the help of UDDI registry, which advertises services that all are described in OWL language, we could compare the functionality of a crashed service with others. We conclude that functionality matching gives better result than QOS or policy matching. We also proposed a new architecture for our work that shows the simplicity and accuracy of our work.

The bipartite graph played the main role of matching two semantic web services in our work. At the end We had computed the time complexity of our proposed algorithm (O(N)) which is minimum among other algorithms.

Some of web services have preconditions to do a work and then have some effects. By considering these two factors, we can obtain a better result, certainly.

Our future work is focused on improving the efficiency and accuracy of this algorithm by considering preconditions and effects of a service.

---

[3] Ontology Inference Layer.

[4] DARPA Agent Markup Language.

References

1. Okutan, A.C., Kesim Cicekli, B.N.: A Monolithic Approach to Automated Composition of Semantic Web Services With The Event Calculus. knowledge-based systems 23, 440–454 (2010)
2. Abramowicz, W., Haniewicz, K., Kaczmarek, M., Zyskowski, D.: Architecture for Web Services Filtering and Clustering. In: 2nd IEEE international conference on internet and web applications and services (2007)
3. Segev, A.: Circular Context-Based Semantic Matching to Identify Web Service Composition. In: CSSSIA, Beijing, China (2008)
4. Yue, A.K., Liu, A.W., Wangb, X.: Zhou. A., Li, J.:Discovering Semantic Associations Among Web Services Based on the Qualitative Probabilistic Network. Expert systems with applications 36, 9082–9094 (2009)
5. Ivan, H., Akkiraju, R., Goodwin, R.: Learning Ontologies to Improve the Quality of Automatic Web Service Matching. In: IEEE international conference on web services (2007)
6. Doan, A., Madhavan, J., Domingos, P., Halevy, A.: Learning to Map Between Ontologies on the Semantic Web. Honolulu, Hawaii, USA (2002)
7. Ontology Matching Approaches in Semantic Web: a Survey. department of computing science, University of Alberta, Cdmonton, Canada
8. Nacer Talantikite, H., Aissani, D., Boudjlida, N.: Semantic Annotations for Web Services Discovery and Composition. Computer standards & interfaces 31, 1108–1117 (2009)
9. Fan, J.: Semantics-Based Web Service Matching Model. In: IEEE international conference on industrial informatics, pp. 323–328 (2006)
10. Verma, K., Akkiraju, R., Goodwin, R.: Semantic Matching of Web Service Policies
11. Chan Oh, S., Woon Yoo, J., Kil, H., Lee, D., Kumara, S.R.T.: Semantic Web-Service Discovery and Composition Using Flexible Parameter Matching. In: The 9th IEEE international conference on e-commerce technology and the 4th international conference on enterprise computing, e-commerce and e-services (2007)
12. Ben Mokhtar, S., Kaul, A., Georgantas, N., Issarny, V.: Towards Efficient Matching of Semantic Web Service Capabilities. In: International workshop on web services modeling and testing, pp. 137–152 (2006)
13. `http://www.w3.org/`
14. `http://semanticweb.org/wiki/Main_Page`
15. `http://en.wikipedia.org/wiki/Bipartite_graph`

Security Enhancement for Authentication of Nodes in MANET by Checking the CRL Status of Servers

Azeem Irshad, Wajahat Noshairwan, Muhammad Shafiq,
Shahzada Khurram, Ehtsham Irshad, and Muhammad Usman

University Institute of Information Technology
Pir Mehr Ali Shah, Arid Agriculture University
Rawalpindi, Pakistan
{irshadazeem2,wajahat.noshairwan,shafiq.pu,
schahzada,ehtshamirshad,manilasani}@gmail.com

Abstract. MANET security is becoming a challenge for researchers with the time. The lack of infrastructure gives rise to authentication problems in these networks. Most of the TTP and non-TTP based schemes seem to be impractical for being adopted in MANETs. A hybrid key-management scheme addressed these issues effectively by pre-assigned logins on offline basis and issuing certificates on its basis using 4G services. However, the scheme did not taken into account the CRL status of servers; if it is embedded the nodes need to check frequently the server's CRL status for authenticating any node and place external messages outside MANET which leads to overheads. We have tried to reduce them by introducing an online MANET Authority responsible for issuing certificates by considering the CRL status of servers, renewing them and key verification within MANET that has greatly reduced the external messages.

Keywords: Authentication, MANET Authority, CRL, TTP, 4G, Mobile Ad hoc Network.

1 Introduction

Mobile Ad hoc Networks (MANETs) are infrastructure-less networks comprising mobile nodes and are vulnerable to attacks for lack of any specific boundary and random entry of nodes in the network. Authentication is the hallmark of security and failure to achieving this so far is a stumbling block in the way of securing MANET. At small scale the authentication can be managed by the nodes through handshaking [6], but at larger scale it becomes complex and demands the involvement of TTP [1]. Some of the schemes are either based on self-organization in MANETs without TTP [2] where the identity is resolved by nodes themselves and some are based on absolute TTP [12], while a hybrid form of these schemes can also be used [1].

Our research work is based on the optimization of a scheme known as Tseng model [1] that gets the nodes authenticated in MANET by the use of 4th generation (4G) technology [10] and [11], a future technology that supports in communicating different platforms in a transparent manner. The Tseng model allows the

T.-h. Kim, A. Stoica, and R.-S. Chang (Eds.): SUComS 2010, CCIS 78, pp. 86–95, 2010.

authentication and distribution of certificates to nodes through the support of 4G technologies. The Tseng model did not take into account the CRL status of servers. The Tseng model shows further overheads if this feature is embedded in the scheme, since, the nodes need to check frequently the server's CRL status for authenticating a node and place external messages outside MANET. If a server finds its ID in the CA's CRL directory any time it renders all the certificates of nodes invalid in the MANET. The nodes ask their servers to find the CRL status of a corresponding node's server. The communicating nodes can be from same and different CA domains. In the worst case if nodes need to establish sessions with the nodes from different servers each time, the overhead grows even more. The Tseng model, not fulfilling the requirement of CRL for the nodes to be known before authentication, can be regarded as less secure and costly for overheads when the nodes from different servers try to communicate and verify from servers with the added feature of security.

We have tried to optimize the scheme by introducing an online MANET Certificate Authority in the network. A certificate is provided to each node by MCA after testing the CRL status of each node's server. It reduces verification visits to the server frequently to a large extent for a MANET relatively larger in size and hence less overheads enhances the efficiency of the MANET.

The paper is organized as follows: In section 2, an overview of previous schemes is presented. In section 3, we present the proposed model with certificate distribution and different communication scenarios. In section 4, we compare Tseng and proposed models and give simulation analysis while in section 6, we concluded our findings.

2 Related Work

A lot of work has been done on security problems regarding MANETS so far. We now take a brief overview of some of the related previous papers.

In threshold cryptographic scheme [3], the authority of CA is distributed among many t+1 network nodes, called servers, to minimize the chance of a single CA being compromised. All the nodes' certificates are divided into *n* shares and distributed to server nodes before network formation. If a node requires other node's public key, it requests to server nodes which generate their partial signatures individually and send to combiner to form a signature and present to the asking node. In MANET it is a cumbersome process that may cost more than a MANET's formation objective.

A similar scheme [5] is an improvement over [3] on the basis of availability. Here, the CA is a fully distributed and any $t+1$ number of nodes in MANET could behave as server nodes for issuance and verification of public keys for the nodes. Despite the advantage of availability, the scheme looses on the side of robustness with the higher values of t. The selection of t should be trade-off between both of the parameters.

In KAMAN [7], multiple Kerberos servers are responsible for distributed authentication in MANET. The servers are boot-strapped with keys shared with the client nodes. The users rely upon servers for acquiring tickets after authentication to communicate with other users which is a bottleneck for its implementation in MANETs and the servers are not trusted as there is no TTP involved initially.

In self-organized MANETS [2], the nodes rely on themselves for all routing, authentication and mobility management. The nodes issue certificates to their trustees for bringing them into MANET which are verified on the basis of repositories maintained by the nodes. Though, the scheme is self-organized but has the overheads of maintaining repositories which consumes the memory and bandwidth. Secondly, the originator blindly trusts any other node for making a new entry in the MANET.

A scheme [1] based on PKI implementation, resolves identity of nodes in MANET with the help of 4G services. The server distributes certificates to nodes through a special node using 4G services. The scheme successfully embeds TTP with MANET and getting nodes authenticated. However, it shows external message overheads when nodes from different servers communicate and verify the server's CRL status frequently. The scheme can be further optimized by reducing the overheads.

One more scheme [12] is based on certificate distribution to nodes before network formation by a trusted third party. The drawback remains with the condition of certificate issuance by TTP before network formation to all the nodes in MANET.

Some more work in this regard can be viewed in [8], [9] and [10] references.

3 Proposed Model

In Tseng model [1], the overhead tends to grow with higher proportions, as more and more nodes from different servers interact and establish sessions. If the nodes communicate recurrently, they can verify one another without server by storing CRL status. In the worst case, the communication of a node with nodes of a different server for each new session leads to external message overheads. We have tried to overcome weaknesses in Tseng model by lowering number of external messages for interacting nodes from different servers. Our scheme is based on the following assumptions.

3.1 Assumptions

1. A MANET Certificate Authority (MCA) is introduced as an independent entity authenticated by CA. The MCA has both, one homogeneous card for inter-nodes communication, and other heterogeneous card for accessing the 4G services.

2. A GN is a valid user of some server in the internet that generates its own public and private key pair.

3. There is only one MCA active in the MANET at one time, which may hand the charge over to a passive MCA in MANET any time due to any reason.

Abbreviations**:** **MID**: MCA ID, **SID**: Server ID, **NID**: GN ID, **PKNID**: Public key of GN, **EPKS**: Encryption through public key of Server, **RNID**: Random number taken by GN, **PWNID**: Password of GN, **h**: hash, $\mathbf{Cert_{MCA}}$: MCA Certificate issued by CA, **SignPRM**: Signature through private key of MCA, $\mathbf{Cert_{S}}$: Server Certificate issued by CA, **SignPRS**: Server's Signature, **PKM**: Public key of MCA, $\mathbf{Cert_{AbyS1}}$: Certificate issued by Server1 to A, **EV:** Entity Verification, **SRT:** Server Restricted life Time, **EP:** Evaluation Point, **CRL:** Certificate Revocation List, **TTP:** Trusted Third Party

3.2 System Model

In existing scheme [1], we have introduced an online MCA which establishes a secure channel with servers like special nodes in Tseng model. The nodes access servers on internet through MCA and the provided logins are basis of verifiable identities for getting certificates. All GNs generate their private and public keys through built-in PKI techniques. The authorities sign public keys for issuing certificates. The procedure of issuing certificates is defined in the following section.

Certificate Issuance.
A node having a login, that wants to become part of the MANET, sends its parameters to MCA as shown in Fig. 1. MCA sends these parameters to server along with CA certificate and its own signature, as shown in Fig. 2. In Fig. 3, the server verifies MCA certificate through CA's public key and the node's identity by decrypting parameters through its private key and public key of MCA. It generates hash value by taking hash on node password, decrypted random number, node's id and public key which is matched with the received hash. Then the server generates a certificate and sends along with its own certificate as shown in Fig. 4. MCA generates a certificate by signing node's public key, Nounce, and expiry time for the lesser time period than the SRT.

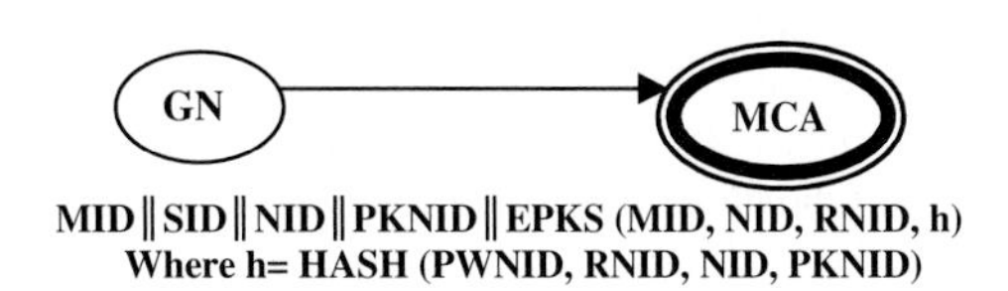

Fig. 1. Certificate Request

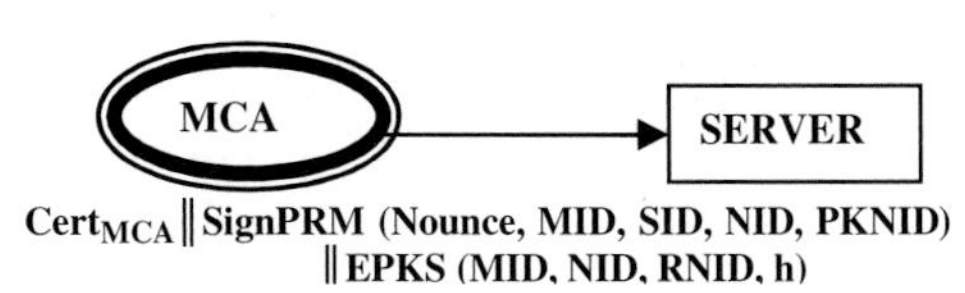

Fig. 2. Verification from Server

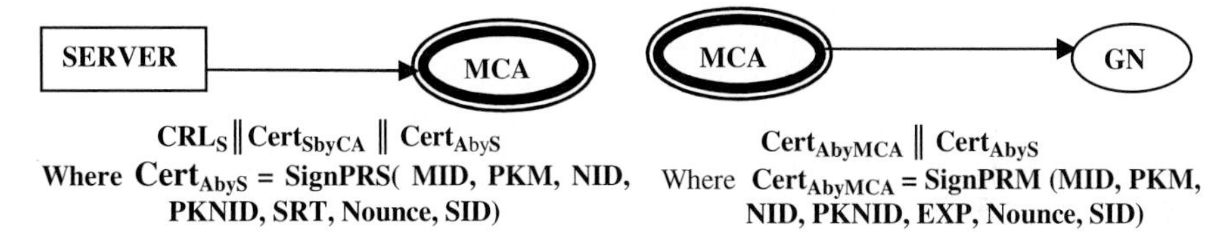

Fig. 3. Verification Acknowledgement

Fig. 4. Certificate Issued to GN

Whichever is lower of both server's CA issued certificate time and server's CRL time period, will be the certificate expiry time of node. A node accesses the public key of MCA through server's signed certificate which serves as a proof for MCA and GNs in authenticating one another. In Tseng model the scenario for different servers bears the overhead cost of entity verification. In proposed scheme the nodes in different servers scenario, establish sessions being under the MCA authority and the cost for finding server's CRL status and verification diminishes almost to zero as there is no external message cost for EV. The security is enhanced by taking into account the CRL status. MCA checks the CRL status of its member nodes' servers each time on certificate expiry to reissue certificates for validating authenticity.

Communication Scenarios and Overheads. In the following the different communication scenarios for Tseng and proposed models are explained.

New Scheme, Same Servers (NSSS) and Old Scheme, Same Servers (OSSS).

In Fig. 5, part (a) and (b) the node B can verify itself the identity of A as it knows the public keys and CRL status of its server. One drawback of OSSS is removed in NSSS as CR is done within MANET as compared to CRS. In NSSS, as authentication of a node is done within the MANET like OSSS so there is not much difference in the scenarios of both models. Now, if original MCA (OMCA) moves out of the MANET, OMCA assigns a proxy certificate to a new MCA (NMCA) after verification. The GN gets a certificate from NMCA on its certificate expiry. The two nodes should be carrying the certificate from same MCA at any instant for communication.

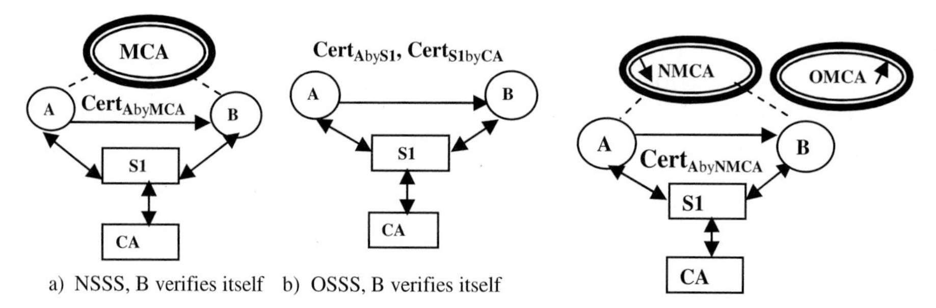

Fig. 5. Comparison of NSSS and OSSS

Fig. 6. NMCA replaces OMCA in NSSS

New Scheme, Different Servers (NSDS) and Old Scheme, Different Servers (OSDS).

The NSDS scenario overcomes the overhead in OSDS through MCA introduction. In OSDS, a node verifies the identity of other node by its server leading to overhead. In proposed scheme when nodes belonging to different servers come under MCA, the EV is performed by nodes within the MANET as the public key of MCA is known to all nodes. Our scheme do not incur cost for EV and overhead is reduced which leads to efficiency for the MANET as shown in Fig. 7. If OMCA moves out, the nodes switch to the NMCA as shown in Fig. 8. The nodes may regenerate certificates before the certificate expiry in case of urgent need for making contact to a node that has switched to NMCA. OMCA provides a list of node IDs to NMCA while moving out. The NMCA issues certificates to the nodes after verification of those IDs.

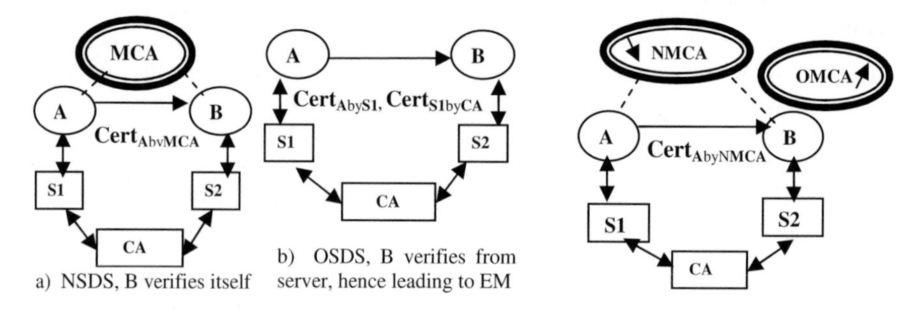

Fig. 7. Comparison of NSDS and OSDS

Fig. 8. NMCA replaces OMCA in NSDS

4 Comparison with Simulation Results

The purpose of this section is to draw the comparison of both schemes on the basis of calculations and proved results.

4.1 Major Differences in Tseng and Proposed Models

The differences in both schemes are based on the number of external messages for Certificate Renewals and Entity Verifications as described under:

First Time Certificate Issuance (FCI) and Certificate Renewals in both models.

The FCI in both models takes two external messages as shown in the Fig. 9 part (c). The CR from MCA (CRM) in proposed model relies on internal messages, while in Tseng model CR from Server (CRS) relies on external messages which are an overhead cost as shown in Fig. 9 part (a) & (b).

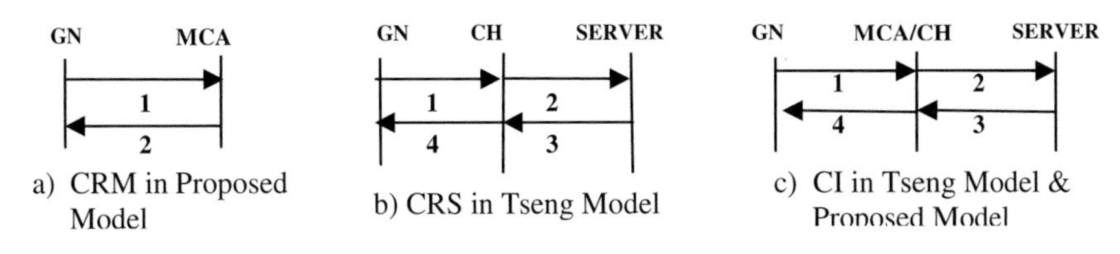

Fig. 9. Certificate Renewal and CI in Tseng and Proposed Models

Authentication Cost in Terms of External Messages.

The authentication cost varies with the scenarios of both schemes. In OSSS and NSSS, the EV cost is limited to internal messages. In OSDS, the nodes place external messages to servers as an overhead cost. In NSDS, the external messages cost is eliminated by introducing MCA in the MANET. We briefly show the exchange of messages for Tseng model in Fig. 10. The further details in this regard can be accessed from [1]. The table 1 shows the comparison of both models in terms of external messages for CRs and EVs. The NSSS and NSDS scenarios contain only internal messages without EV cost. The OSDS scenario bears the external message cost for EV.

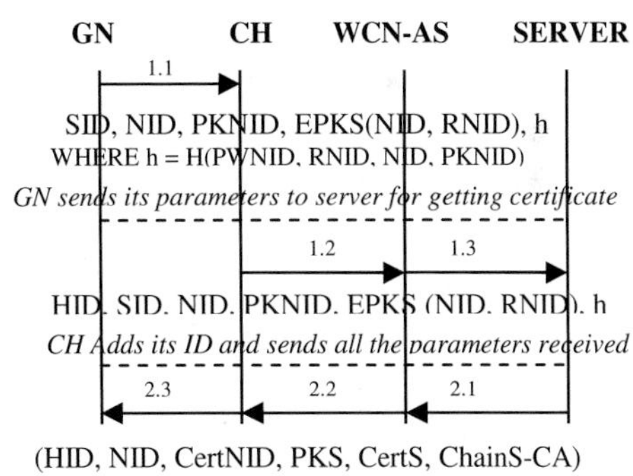

Fig. 10. Exchange of Messages in Tseng Model

Table 1. Comparison of EV cost for scenarios of both schemes

Activities / Scenario	Internal Messages		External Messages	
	CRM	CH	CRS	EV
OSSS	-	2	2	-
OSDS	-	2	2	2
NSSS	2	-	-	-
NSDS	2	-	-	-

Our simulation results are supported with the following case study. Assume a node N, in OSDS, authenticates the nodes by placing messages to its server. If node N establishes 4 sessions on average in an EP then the number of messages for CR and EV up to 3 EPs amounts to as following: (The EP session is taken equivalent to SRT).

OSDS ➔ **EV:** 12 * 2 = **24** NSDS ➔ **EV: 0**

CR: 16 * 2 = **32** **CR:** 4 * 2 = **8**

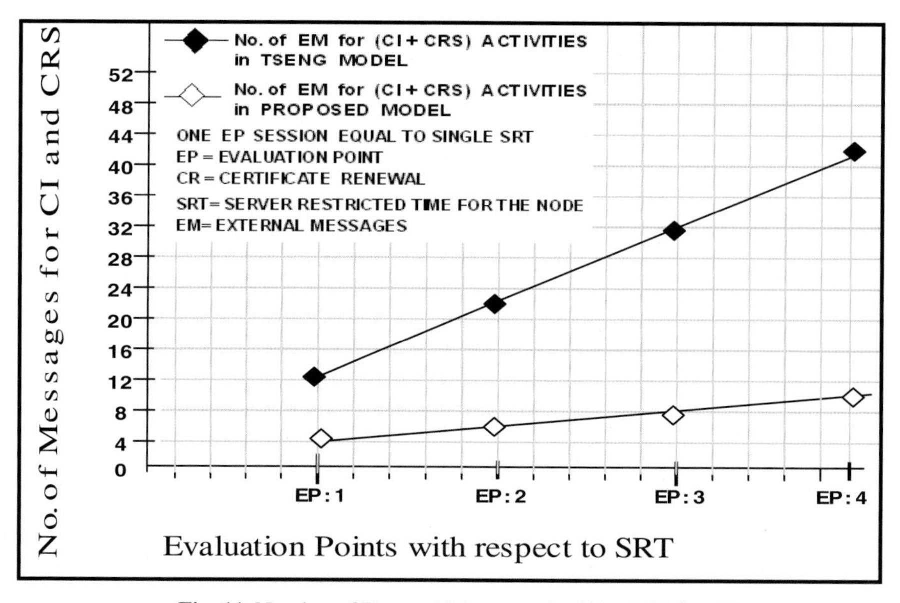

Fig. 11. Number of External Messages for CI and CRS at EPs

The number of external messages for 3 EP sessions is 56 for OSDS. In NSDS scenario there is only cost for CR as 8 external messages. We generate a function for calculating CR messages as [(E+1)*2] for proposed model and [(5E+1)*2] for Tseng model. E is the number of EP sessions. EV cost is calculated in OSDS by multiplying the number of EVs in all EP sessions with 2. The proposed scheme bears no cost in NSDS for EV activity.

In Fig. 11, the difference for number of messages based on CI and CRS in both models are shown. The curves for Tseng model rises sharply while the curves for the proposed model rises slowly which is an indicator of efficiency of proposed scheme.

In Fig. 12, the time line for certificate renewal from MCA has been drawn.

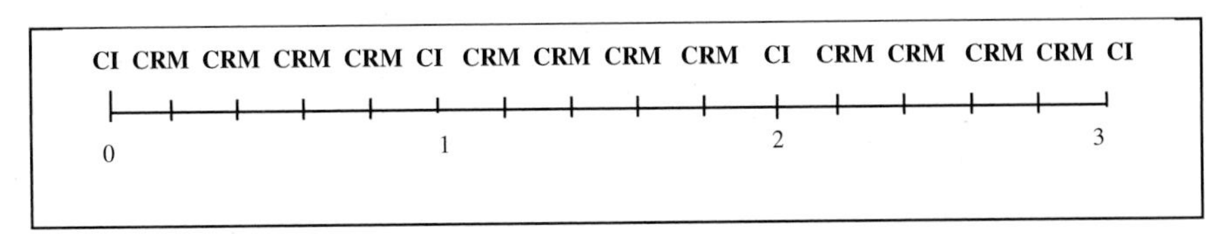

Fig. 12. Time line showing CRM for Proposed scheme

In Fig. 13, the time line for certificate renewal from server has been drawn.

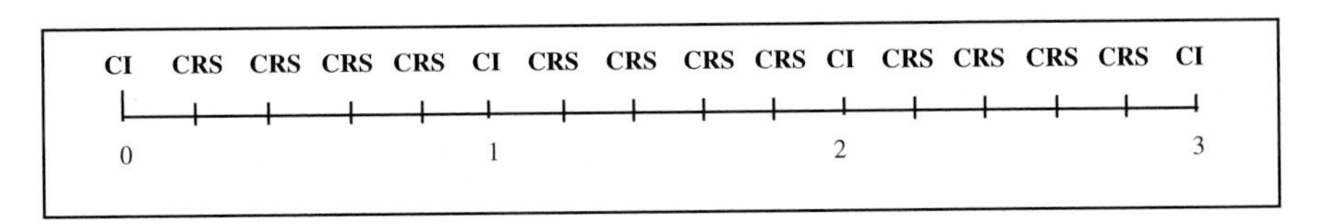

Fig. 13. Time line showing CRS for Tseng Model

The above figures support the analysis and simulation results and help us come to the conclusion that the proposed model reduces the overheads with enhanced security features of certificate revocation status.

5 Conclusion

In this paper, we have tried to overcome the weaknesses in Tseng model. This model does not take the CRL status of servers into account which leads to lack of security on the part of nodes and their servers. When this feature is embedded in Tseng model it shows no more optimal results and comes with external message overheads. In proposed scheme the nodes authenticate other nodes' servers within the MANET leaving the hassle of finding CRL status to an online authority, which helps saving the external messages to a large extent as evident by the simulation analysis. Secondly, the certificate renewal becomes more efficient and is performed within MANET without resorting to server. Our scheme can be regarded as the extension of previous scheme with improved features.

References

1. Tseng, Y.M.: A heterogeneous-network aided public-key management scheme for MANETS. Int. J. Net. Mgmt. 17, 3–15 (2006)
2. Capkun, S., Buttyan, L., Hubaux, J.P.: Self-Organized Public-Key Management for Mobile Ad Hoc Networks. IEEE Transactions on Mobile Computing 2(1), 52–64 (2003)

3. Zhou, L., Haas, Z.J.: Securing Ad Hoc Networks. IEEE Net. J. 13(6), 24–30 (1999)
4. Brandt, I., rd, D., Landrock, P., Pedersen, T.: Zero-Knowledge Authentication Scheme with Secret Key Exchange. Journal of Cryptology (1998)
5. Kong, J., Zerfos, P., Luo, H., Lu, S., Zhang, L.: Providing robust and ubiquitous security support for mobile ad hoc networks. In: IEEE (ICNP 2001), pp. 251–260 (November 2001)
6. Stajano, F., Anderson, R.J.: The resurrecting duckling: Security issues for ad-hoc wireless networks. In: Malcolm, J.A., Christianson, B., Crispo, B., Roe, M. (eds.) Security Protocols 1999. LNCS, vol. 1796, p. 204. Springer, Heidelberg (2000)
7. Pirzada, A., Mc Donald, C.: Kerberos Assisted Authentication in Mobile Ad-hoc Networks. In: The 27th Australasian Computer Science Conference (2004)
8. Weimerskirch, A., Thonet, G.: A Distributed Light-Weight Authentication Model for Ad-hoc Networks. In: Kim, K.-c. (ed.) ICISC 2001. LNCS, vol. 2288, pp. 6–7. Springer, Heidelberg (2001)
9. Zhu, S., Xu, S., Setia, S., Jajodia, S.: Establishing pair wise keys for secure communication in ad hoc networks: a probabilistic approach. In: The 11th IEEE International Conference on Network Protocols (2003)
10. Schulzrinne, H., Wu, X., Sidiroglou, S.: Ubiquitous Computing in Home Networks. IEEE Commun. Mag., 128–135 (November 2003)
11. Hui, S.Y., Yeung, K.H.: Challenges in the Migration to 4G Mobile Systems. IEEE Commun. Mag., 54–59 (December 2003)
12. Varadharajan, V., Shankaran, R., Hitchens, M.: Security for cluster based ad hoc networks. Computer Communications 27(5), 488–501 (2004)

Azeem Irshad has been doing Ph.D after completing MS-CS from UIIT, PMAS Arid Agriculture University Rawalpindi. and is currently working as a Research Associate for the research department of UIIT. His research interests include MANET security, merger of Ad hoc networks with other 4G technology platforms, multimedia over IP and resolving the emerging wireless issues.

Wajahat Noshairwan is currently working as a Assistant Professor of Wireless Networks in UIIT, PMAS, Arid Agriculture University Rawalpindi, Pakistan. He provides professional consultancy in cutting edge aspects of Wireless Security. He has the extensive knowledge on networks related disciplines. He is a gold medalist at MAJU University in MS-CS programme. His research interests include MANET security, mobility, vertical handoff for 4G Networks, latest issues in multimedia over IP.

Muhammad Shafiq has been doing Ph.D after completing MS-CS from UIIT, PMAS Arid Agriculture University Rawalpindi. He has also done MIT from University of the Punjab. He is currently serving as an Assistant Professor in Raees-ul-Ahrar College in computer science department. His research interests include issues in Ad hoc Networks, Telecom Networks Management and Wireless Security.

Ehtsham Irshad. The author has done MS-CS (Networks) from UIIT, PMAS, Arid Agriculture University, Rawalpindi, Pakistan. He has completed BS-CS from AIOU, Islamabad. He has been serving as Networks Administrator in International Islamic University, Islamabad. The area of research is wireless networks, mobility and security in Ad hoc networks and seamless vertical handoff among heterogeneous networks.

Shahzada Khurram has done MS-CS (Networks) from UIIT, PMAS, AAUR. He is currently working as a Networks Administrator in the UIIT, PMAS, Arid Agriculture University Rawalpindi. He has been organizing LINUX, CCNA and CCNP courses under TEVTA, Govt. of Pakistan. His research interests include WIFI security, multimedia over IP and issues in the recent developments on (IEEE 802.11n) standards for wireless.

Muhammad Usman has done MS-CS (Networks) from UIIT, PMAS Arid Agriculture University Rawalpindi and is currently working as a Research Associate for the research department of UIIT. He has recently contributed for seamless handover techniques for GPRS and WLAN platforms. His research interests include resolving the wireless handover problems and ongoing issues in multimedia over IP.

Performance Evaluation Analysis of Group Mobility in Mobile Ad Hoc Networks

Ehtsham Irshad, Wajahat Noshairwan, Muhammad Shafiq, Shahzada Khurram, Azeem Irshad, and Muhammad Usman

University Institute of Information Technology
Pir Mehr Ali Shah, Arid Agriculture University
Rawalpindi, Pakistan
{ehtshamirshad,wajahat.noshairwan,shafiq.pu, schahzada,irshadazeem2,manilasani}@gmail.com

Abstract. Mobility of nodes is an important issue in mobile adhoc networks (MANET). Nodes in MANET move from one network to another individually and in the form of group. In single node mobility scheme every node performs registration individually in new MANET whereas in group mobility scheme only one node in a group i.e group representative (GR) performs registration on behalf of all other nodes in the group and is assigned Care of Address (CoA). Internet protocol (IP) of all other nodes in the group remains same. Our simulated results prove that group mobility scheme reduces number of messages and consumes less time for registration of nodes as compared to single node mobility scheme. Thus network load is reduced in group mobility scheme. This research paper evaluates the performance of group mobility with single node mobility scheme. Test bed for this evaluation is based on Network Simulator 2 (NS-2) environment.

Keywords: Mobile Adhoc network, Group Registration, Group Representative, Care of Address.

1 Introduction

The application of mobile adhoc network is increasing in the modern age. It is a network that is developed on the run without any prior infrastructure [1]. Every node in the network acts as a router or relay station to forward data to the designated node. In mobile adhoc network (MANET) nodes are mobile and constantly changes its location from one MANET to another.

MANET is developed without any prior infrastructure so security is a big issue in it [2]. Nodes in MANET roam from one network to another. In roaming between different networks the main aim is to receive data in its new location without any delay and disruption. For this purpose mobile IP functionality is used to allow roaming of nodes [3]. Mobile IP allows a node to acquire a virtual IP address called Care of Address (CoA) in the new network.

T.-h. Kim, A. Stoica, and R.-S. Chang (Eds.): SUComS 2010, CCIS 78, pp. 96–103, 2010.

In mobile IP scenario [4] when a mobile node (MN) is in its original location, that network is known as home network. When it moves to a new network that network acts as a foreign network. When a mobile node moves from its home network to foreign network, it sends registration request to foreign agent. The foreign agent sends registration request to home network of the mobile node. The home agent sends registration reply to the foreign agent. The foreign agent send registration reply to mobile node and it is assigned a Care of Address in the foreign network.

A group registration with group mobility scheme is proposed [6]. In this scheme nodes moves in the form of a group. Groups are established on the basis of similar interests like army soldiers can form infantry, artillery or armored group. Groups can be identified as business or educational groups. There is a fixed node in the network which is the overall controller of MANET. This fixed gateway keeps the mobility location of all nodes in the network. Every group has a group representative (GR) which is controller of the group. When nodes move in the form of a group, this GR performs registration on behalf of all other nodes in the group. Only GR is assigned an individual CoA. IP of all other nodes in the group remains same.

The rest of this paper is organized as; section 2 describes the related work. The mathematical function is described in section 3. In section 4 simulation results are described. Section 5 concludes the contribution of this paper. At the end of this paper references are given.

2 Related Work

There are various schemes for mobility in MANET. Mobile IP routers are used to support mobility [7]. This router has two interfaces. One interface is connected to MANET and other is connected to outside world. A scheme is proposed by Zhao [8] for adhoc network connection to the internet. In this scheme dynamic gateways are used that uses mobile IP functionality. Mobile nodes in adhoc network use these dynamic gateways to connect to the internet. The dynamic gateway architecture has several advantages, like it eliminates the need of fixed gateways, reduces system complexity, improves reliability and lowers the cost.

Mobile IP functionality is used in mobile adhoc network [5]. All the nodes in the network are mobile except one node that is fixed. This fixed node acts as a gateway. This concept allows single node mobility. When a node moves, it asks for registration request and it is assigned a Care of Address (CoA). All the nodes are assigned an individual CoA in the new network. A scheme explained internet based mobile ad hoc networking [9]. Each node in a mobile adhoc network (MANET) logically consists of a router with possibly multiple IP addressable hosts. End devices in MANET are mobile.

3 Mathematical Analysis

In this section we have performed mathematical analysis of group mobility scheme with single node mobility scheme [6].

The mathematical function shows the relationship of number of nodes with number of messages for single node mobility scheme.

$$f(n) = m * n \quad \text{for all } n \geq 1$$

m = no. of messages which is constant i.e 5

n =no. of nodes

If n=1 $f(1) = 5 * 1 = 5$

If n=2 $f(2) = 5 * 2 = 10$

If n=3 $f(3) = 5 * 3 = 15$

If n=4 $f(4) = 5 * 4 = 20$

If n=5 $f(5) = 5 * 5 = 25$

The relationship of number of nodes with number of messages in group mobility scheme is

$$f(n) = m \quad \text{for all } n \geq 1$$

If n =1 $f(1) = 5$

If n =2 $f(2) = 5$

If n =3 $f(3) = 5$

If n =4 $f(4) = 5$

If n =5 $f(5) = 5$

In single node mobility scheme as no. of nodes increases in the MANET, no of messages increases. In group registration concept as the number of nodes increases, the no. of messages to register node in the new group remains same because in this scheme only one node i.e GR performs registration on behalf of all other nodes in the group.

Table 1. Comparison of messages between single & group mobility scheme

SCHEMES	**NUMBER OF NODES**				
	1	2	3	4	5
	NUMBER OF MESSAGES				
SINGLE NODE MOBILITY SCHEME	05	10	15	20	25
GROUP MOBILITY SCHEME	05	05	05	05	05

In Fig. 1 the number of messages is shown at horizontal axis while the vertical axis shows the number of nodes. In single mobility scheme the number of messages tends to grow with the increasing number of nodes. While in group mobility scheme the number of messages remains constant with the increasing number of nodes as compared to single node mobility scheme.

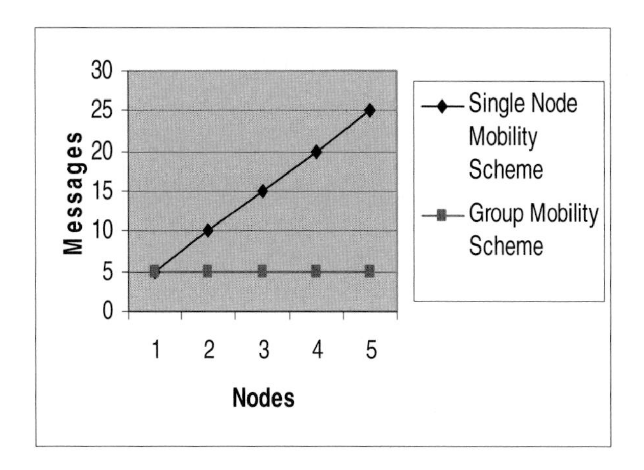

Fig. 1. Comparison of message consumption for registration between single and group mobility schemes

4 Simulations

In this section we have discussed the simulations environment, simulation detail and its results. We have performed our simulation results on NS-2.

4.1 Message Size

The architecture consists of 5 mobile nodes in a group, 5 in other group and one node that is fixed in group mobility scheme. In this scheme when a group moves, GR of group sends 5 messages to register itself in new group. Each message is of 128 bits in length. So total size of messages for registration of GR is 640 bits in length.

In single node mobility scheme there are 5 mobile nodes in a MANET and 5 nodes in other MANET. One node in each MANET is fixed that acts as a gateway. In this scheme every node sends individual registration request message. The message size is 128 bits in length and one node sends 5 messages for registration so it takes 640 bits for registration of one node, so total size of message for registration of 5 nodes is 3200 bits length.

4.2 Registration Time

In group registration scheme when a group moves, GR of group only send registration request message. This registration process of whole group is completed in 4.30 ms. In this scheme only GR sends registration request as it does registration on behalf of all other nodes in the group. So registration of whole group is completed in this time.

In single node mobility scheme when 1st node moves to a new MANET, it completes registration process in 4.30 ms. Second node completes its registration process in 8.60 ms. Third node completes this process in 12.90 ms. Fourth node completes this process in 17.20 ms. Fifth node completes this process in 21.50 ms.

Table 2. Registration Time for Nodes

SCHEMES	NUMBER OF NODES				
	1	2	3	4	5
	TIME FOR REGISTRATION (ms)				
SINGLE NODE MOBILITY SCHEME	4.3	8.6	12.9	17.2	21.5
GROUP MOBILITY SCHEME	4.3	4.3	4.3	4.3	4.3

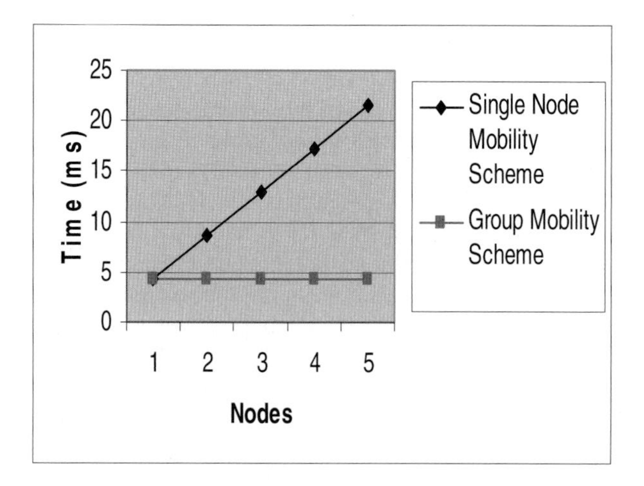

Fig. 2. Registration time for nodes

In Fig. 2 the time in milliseconds has been shown at horizontal axis while the vertical axis shows the number of nodes. In single mobility scheme the registration time tends to grow with the increasing number of nodes. While in group mobility scheme the registration time remains constant with the increasing number of nodes as compared to single node mobility scheme.

4.3 Network Load

Network load is reduced in group mobility scheme as compared to single mobility scheme. In group mobility scheme network load for registration of a group is 0.625 Kbps. The network load in single node mobility scheme for registration of nodes in MANET is 3.125 Kbps.

Table 3. Network load for Single &Group mobility scheme

NETWORK LOAD	Kbps
SINGLE NODE MOBILITY SCHEME	3.125
GROUP MOBILITY SCHEME	0.625

5 Conclusion

In this research paper we have simulated the results of group mobility and single node mobility schemes in MANET. We have discussed the simulation scenarios, mathematical functions and simulation results which validate the efficiency of group mobility scheme. The validated result shows that group mobility scheme not only reduces the number of messages, it also reduces the network load and time for registration of nodes.

References

1. Breed: Wireless Ad hoc Networks Basic Concepts. High Frequency Electronics, 44–47 (2007)
2. Burg, A.: Ad hoc Network Specific Attacks, Seminar Ad hoc Networking Concepts. Applications and Security, 12 (2003)
3. Anonymous: Mobility and Mobile IP Introduction, White paper, Doc no. IPU-000 Rev C IPunplugged, p. 17 (2003)
4. Tseng, Y., Shen, C., Chen, W.: Integrating Mobile IP with Ad hoc Networks. IEEE Computer, 48–55 (2003)
5. Tseng, Y., Shen, C., Chen, W.: Mobile IP and Ad hoc Networks: An Integration and Implementation Experience. IEEE Computer 36(5), 48–55 (2003)
6. Irshad, E., Noshairwain, W., Usman, M., Irshad, A., Gilani, M.: Group Mobility in Mobile Ad hoc Networks. In: WWW/Internet, IADIS Germany, October 13-15 (2008)

7. Kock, B., Schmidt, J.: Dynamic Mobile IP Routers in Adhoc Networks. In: International Workshop on Wireless Ad hoc Networks (IWWAN) OULU Finland (2004)
8. Zhao, J., Wang, L., Kim, Y., Jiang, Y., Yang, X.: Secure Dynamic Gateway to Internet Connectivity for Ad hoc networks. In: NIST, pp. 305–319 (1989)
9. Corson, M., Macker, J., Cirincione, G.: Internet Based Mobile Ad hoc Networking. IEEE Internet Computing, 63–70 (1999)

Ehtsham Irshad. The author has done MS-CS (Networks) from UIIT, PMAS, Arid Agriculture University, Rawalpindi, Pakistan. He has completed BS-CS from AIOU, Islamabad. He has been serving as Networks Administrator in International Islamic University, Islamabad. The area of research is wireless networks, mobility and security in Ad hoc networks and seamless vertical handoff among heterogeneous networks.

Wajahat Noshairwan is currently working as a Assistant Professor of Wireless Networks in UIIT, PMAS, Arid Agriculture University Rawalpindi, Pakistan. He provides professional consultancy in cutting edge aspects of Wireless Security. He has the extensive knowledge on networks related disciplines. He is a gold medalist at MAJU University in MS-CS programme. His research interests include MANET security, mobility, vertical handoff for 4G Networks, latest issues in multimedia over IP.

Muhammad Shafiq has been doing Ph.D after completing MS-CS from UIIT, PMAS Arid Agriculture University Rawalpindi. He has also done MIT from University of the Punjab. He is currently serving as an Assistant Professor in Raees-ul-Ahrar College in computer science department. His research interests include issues in Ad hoc Networks, Telecom Networks Management and Wireless Security.

Azeem Irshad has been doing Ph.D after completing MS-CS from UIIT, PMAS Arid Agriculture University Rawalpindi. and is currently working as a Research Associate for the research department of UIIT. His research interests include MANET security, merger of Ad hoc networks with other 4G technology platforms, multimedia over IP and resolving the emerging wireless issues.

Muhammad Usman has done MS-CS (Networks) from UIIT, PMAS Arid Agriculture University Rawalpindi and is currently working as a Research Associate for the research department of UIIT. He has recently contributed for seamless handover techniques for GPRS and WLAN platforms. His research interests include resolving the wireless handover problems and ongoing issues in multimedia over IP.

Shahzada Khurram has done MS-CS (Networks) from UIIT, PMAS, AAUR. He is currently working as a Networks Administrator in the UIIT, PMAS, Arid Agriculture University Rawalpindi. He has been organizing LINUX, CCNA and CCNP courses under TEVTA, Govt. of Pakistan. His research interests include WIFI security, multimedia over IP and issues in the recent developments on (IEEE 802.11n) standards for wireless.

Procedure of Partitioning Data Into Number of Data Sets or Data Group – A Review

Tai-hoon Kim*

Multimedia Engineering Department, Hannam University,
Daejeon, Korea
taihoonn@hnu.kr

Abstract. The goal of clustering is to decompose a dataset into similar groups based on a objective function. Some already well established clustering algorithms are there for data clustering. Objective of these data clustering algorithms are to divide the data points of the feature space into a number of groups (or classes) so that a predefined set of criteria are satisfied. The article considers the comparative study about the effectiveness and efficiency of traditional data clustering algorithms. For evaluating the performance of the clustering algorithms, Minkowski score is used here for different data sets.

Keywords: Clustering metric, Euclidean Distance, K-Means Clustering, Genetic Algorithm, Fuzzy C Means Clustering.

1 Introduction

Clustering involves dividing a set of data points into non-overlapping groups, or clusters, of points, where points in a cluster are "more similar" to one another than to points in other clusters. The term "more similar," when applied to clustered points, usually means closer by some measure of proximity. When a dataset is clustered, every point is assigned to some cluster, and every cluster can be characterized by a single reference point, usually an average of the points in the cluster. Any particular division of all points in a dataset into clusters is called a partitioning [2].

Iterative refinement clustering algorithms (e.g. K-Means, EM) converge to one of numerous local minima. It is known that they are especially sensitive to initial conditions [3].

A number of researchers have proposed genetic algorithms for clustering [1, 5, 6]. The basic idea is to simulate the evolution process of nature and evolve solutions from one generation to the next.

K-Means (KM) is considered one of the major algorithms widely used in clustering. However, it still has some problems, and one of them is in its initialization step where it is normally done randomly. It is well known that KM might converge to a local optimum, and its result depends on the initialization process, which randomly generates the initial clustering. It is sensitive to the initially selected points, and so it

* Corresponding author.

T.-h. Kim, A. Stoica, and R.-S. Chang (Eds.): SUComS 2010, CCIS 78, pp. 104–115, 2010.

does not always produce the same output. This algorithm does not guarantee to find the global optimum, although it will always terminate [1,7].

Image segmentation is the decomposition of a gray level or color image into homogeneous regions [8]. In image segmentation, cluster analysis is used to detect borders of objects in an image.

Fuzzy C-means Clustering algorithm (FCM) is a method that is frequently used in pattern recognition. It has the advantage of giving good modeling results in many cases, although, it is not capable of specifying the number of clusters by itself [13]. Clustering algorithms aim at modeling fuzzy (i.e., ambiguous) unlabeled patterns efficiently [14]. A number of methods have been proposed for clustering microarray data. Hierarchical clustering, self-organizing maps, K-means, and fuzzy c-means have all been successful in particular applications.

This paper is organized as follows. Clustering is surveyed in section II. Performance Metric is discussed in section III. In section IV a comparative study or result analysis is shown. Finally, we conclude in Section V.

2 Previous Work

We review three of the most popular widely used clustering techniques in this section. The algorithms sated here are taken from the listed references.

These are:

a. K-means Clustering Algorithm,
b. Clustering Using Genetic Algorithm (for fixed number of cluster,k)
c. Fuzzy C-means Clustering Algorithm,

Clustering as a fundamental pattern recognition problem can be characterized by the following design steps [9]:

- Data Representation: What data types represent the objects in the best way to stress relations between the objects, e.g., similarity?
- Modeling: How can we formally characterize interesting and relevant cluster structures in data sets?
- Optimization: How can we efficiently search for cluster structures?
- Validation: How can we validate selected or learned structures?

Clustering in N-dimensional Euclidean space R^N is the process of partitioning a given set of n points into a number, say K, of groups (or, clusters) based on some similarity / dissimilarity metric [1].Let the set of n points $\{x_1,x_2,\ldots\ldots,x_n\}$ be represented by the set S and the K clusters be represented by $C_1,C_2,\ldots\ldots\ldots,C_K$. and corresponding Cluster centers are $Z=\{ z_1,z_2, \ldots\ldots, z_K \}$ Then

$$C_i \neq \emptyset \; for \; i = 1,2,\ldots\ldots\ldots\ldots K$$

$$C_i \cap C_j = \emptyset \; for \; i = 1,2,\ldots\ldots\ldots K, \qquad j = 1,2,\ldots\ldots\ldots\ldots K \; and \; i \neq j$$

$$\text{And } \cup_{i=1}^{K} C_i = S$$

The clustering metric that has been adopted is the sum of the Euclidean distances of the points from their respective cluster centers. Mathematically, the clustering metric M for the K clusters C_1, C_2,……, C_K is given by

$$\boldsymbol{M}(\boldsymbol{C_1}, \boldsymbol{C_2}, \dots\dots\dots\dots\dots \boldsymbol{C_K}) = \sum_{i=1}^{k} \sum_{x_j \in C_i} |\, \boldsymbol{x_i} - \boldsymbol{z_i} \,| \tag{1}$$

Where $|\, \boldsymbol{x_i} - \boldsymbol{z_i} \,|$ is a chosen distance measure between a data point x_i and the cluster centre z_i , is an indicator of the distance of the n data points from their respective cluster centers.

2.1 K-Means Clustering Algorithm

K-means clustering is an algorithm to classify or to group data based on attributes/features into **K** number of group. The basic steps of K-means Clustering Algorithm [1], are shown below.

Input:
A data set X of n points in d dimensional space in to K Clusters.
$X : \{x_1, x_2, \dots\dots\dots\dots, x_n\}$
where $x_i = [x_{i1}, x_{i2}, \dots, x_{id}]$ for i=1,2,…,n.
Z : Set of Cluster Centers
$Z = \{z_1, z_2, \dots\dots\dots, z_K\}$
Where $z_i = [z_{i1}, z_{i2}, \dots, z_{id}]$ for i = 1,2,……..,K
Output:
n_i = number of points in cluster $i=1,2,\dots,K$.

Objective function:
This algorithm aims at minimizing an *objective function/clustering Metric*, in this case a squared error function i.e the clustering metric described in equation (1).

The algorithm is composed of the following steps:

Step 1. Choose K initial cluster centers $z_1, z_2, \dots\dots z_k$ randomly from the n points $\{x_1, x_2, \dots\dots x_n\}$

Step 2. Assign point x_i , i =1,2,3,……n to cluster $C_j, j \in \{1,2,3, \dots\dots K\}$ iff

$$|x_i - z_j\,| < |\, x_i - z_p\,|\; and\; j \neq p$$

Ties are resolved arbitrarily.

Step 3. Compute New Cluster centers $z_1^*, z_2^*, \dots\dots z_k^*$ as follows

$$z_i^* = \frac{1}{n_i} \sum_{x_j \in C_i} x_j, \; i = 1,2,3, \dots\dots K$$

Where n_i is the number of elements belonging to cluster C_i

Step 4. If $z_i^* = z_i, i = 1,2,3 \dots K$. Then terminate otherwise continue from step 2

In case the process does not terminate at Step 4 normally, then it is executed for a maximum fixed number of iterations.

Example:

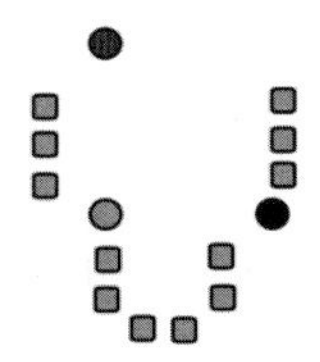

Fig. 1. K initial cluster center(in this case k=3) are randomly selected from the data set

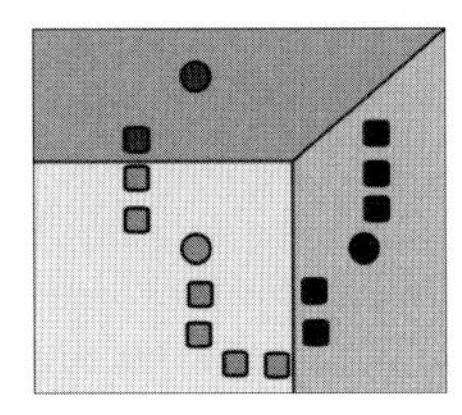

Fig. 2. K clusters are created by associating every observation with the nearest mean

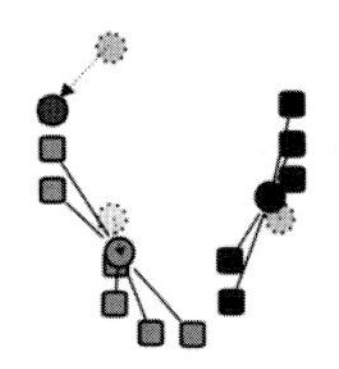

Fig. 3. The centroid of each of the K clusters becomes the new means of corresponding clusters

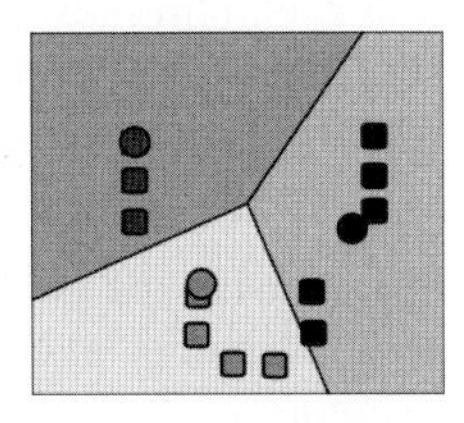

Fig. 4. After few iteration Clustered Data and the Updated Cluster Center

2.2 Clustering Using Genetic Algorithm:

The searching capability of GAs has been used in this article for the purpose of appropriately determining a fixed number K of cluster centers in R^N; thereby suitably clustering the set of n unlabelled points. For this case *objective function / clustering Metric* in equation (1) , is used.

The basic steps of GAs, which are also followed in the GA-clustering algorithm[5,1], are shown below.

Begin
1. Generation , t = 0
2. Initial Population Creation P(t)
3. Compute Fitness P(t)
4. t = t+1
5. If Termination Condition achieved go to Step 10
6. Select P(t) from P(t-1)
7. Crossover P(t)
8. Mutate P(t)
9. Go to Step 3
10. Output Best and Stop

End

a) String Representation:

Each string is a sequence of real numbers representing the *K* cluster centres. For an *N*-dimensional space, the length of a chromosome is *N***K* words, where the first *N* positions (or, genes) represent the *N* dimensions of the first cluster centre, the next *N* positions represent those of the second cluster centre, and so on. As an illustration let us consider the following example.

Example:

Let *N*=2 and *K*=3, i.e., the space is two dimensional and the number of clusters being considered is three. Then the chromosome

51.6 72.3 18.3 15.7 29.1 32.2

represents the three cluster centers (51.6 72.3), (18.3 15.7) and (29.1 32.2) . Note that each real number in the chromosome is an indivisible gene.

b) Population Initialization:

The *K* cluster centres encoded in each chromosome are initialized to *K* randomly chosen points from the data set. This process is repeated for each of the *P* chromosomes in the population, where *P* is the size of the population.

c) Fitness computation:

The fitness computation process consists of two phases. In the first phase, the clusters are formed according to the centres encoded in the chromosome under consideration. This is done by assigning each point x_i, i=1, 2,……,*n*, to one of the clusters C_j with centre z_j such that

$$|x_i - z_j| < |x_i - z_p| \; p = 1,2,\ldots\ldots,K \text{ and } j \neq p$$

All ties are resolved arbitrarily. After the clustering is done, the cluster centres encoded in the chromosome are replaced by the mean points of the respective clusters. In other words, for cluster C_i, the new centre z_i* is computed as

$$z_i^* = \frac{1}{n_i} \sum_{x_j \in C_i} x_j, \qquad i = 1,2,3,\ldots\ldots.K$$

These z_i* s now replace the previous z_i s in the chromosome. As an illustration, let us consider the following example.

Example:

The first cluster centre in the chromosome considered in Example 1 is (51.6, 72.3).With (51.6, 72.3) as centre, let the resulting cluster contain two more points, viz., (50.0, 70.0) and (52.0, 74.0) besides itself i.e., (51.6, 72.3). Hence the newly computed cluster centre becomes ((50.0+52.0+51.6)/3, (70.0+74.0+72.3)/3) = (51.2, 72.1). The new cluster centre (51.2, 72.1) now replaces the previous value of (51.6, 72.3).

Subsequently, the clustering metric M is computed as follows:

$$\boldsymbol{M} = \sum_{i=1}^{K} \boldsymbol{M}_i$$

$$\boldsymbol{M}_i = \sum_{X_j \in C_i} |\boldsymbol{X}_j - \boldsymbol{Z}_i|$$

The fitness function is defined as $f = 1/M$, so that maximization of the fitness function leads to minimization of M.

d) Selection:

The selection process selects chromosomes from the mating pool directed by the survival of the fittest concept of natural genetic systems.

e) Crossover

Crossover is a probabilistic process that exchanges information between two parent chromosomes for generating two child chromosomes. Here single point crossover with a fixed crossover probability of μ_C is used. For chromosomes of length *l*, a random integer, called the crossover point, is generated in the range (1, l-1). The portions of the chromosomes lying to the right of the crossover point are exchanged to produce two offspring.

f) Mutation:

Each chromosome undergoes mutation with a fixed probability k_m. For binary representation of chromosomes, a bit position (or gene) is mutated by simply flipping its value. Since we are considering floating point representation in this article, we use the following mutation. A number $\boldsymbol{\sigma}$ in the range [0, 1] is generated with uniform distribution. If the value at a gene position is *v*, after mutation it becomes

$$\boldsymbol{v} \mp 2 * \boldsymbol{\sigma} * \boldsymbol{v}, \boldsymbol{v} \neq 0$$

$$\boldsymbol{v} \mp 2 * \boldsymbol{\sigma}, \boldsymbol{v} = 0$$

The '+' or '-' sign occurs with equal probability.

g) Termination Criterion:

Here the processes of fitness computation, selection, crossover, and mutation are executed for a maximum number of iterations.

2.3 Fuzzy C-Means Clustering Algorithm

Fuzzy C-means (FCM) is a method of clustering which allows one piece of data to belong to two or more clusters. The method are available in the literature [13]. It is based on minimization of the following objective function:

$$J_m = \sum_{i=1}^{N}\sum_{j=1}^{K} u_{ij}^{m} | x_i - z_j |^2 \ , 1 < m < \propto$$

where m is any real number greater than 1, u_{ij} is the degree of membership of x_i in the cluster j, x_i is the i th of d-dimensional measured data, z_j is the d-dimension center of the cluster, and | * | is any norm expressing the similarity between any measured data and the center.

Fuzzy partitioning is carried out through an iterative optimization of the objective function shown above, with the update of membership u_{ij} and the cluster centers z_j by:

$$u_{ij} = \frac{1}{\sum_{k=1}^{K}\left(\frac{|x_i - z_j|}{|x_i - z_k|}\right)^{\frac{2}{m-1}}}$$

$$z_j = \frac{\sum_{i=1}^{N} u_{ij}^{m} x_i}{\sum_{i=1}^{N} u_{ij}^{m}}$$

This iteration will stop when

$$max_{ij}\{| u_{ij}^{(k+1)} - u_{ij}^{(k)} | < \varepsilon\}$$

where ε is a termination criterion between 0 and 1, whereas k are the iteration steps. This procedure converges to a local minimum or a saddle point of J_m.

The algorithm is composed of the following steps:

Step 1. Initialize U=[u_{ij}] matrix, $U^{(0)}$

Step 2. At k-step: calculate the centers vectors $Z^{(k)}=[z_j]$ with $U^{(k)}$

$$z_j = \frac{\sum_{i=1}^{N} u_{ij}^{m} x_i}{\sum_{i=1}^{N} u_{ij}^{m}}$$

Step 3. Update $U^{(k)}$, $U^{(k+1)}$

$$u_{ij} = \frac{1}{\sum_{k=1}^{K}\left(\frac{|x_i - z_j|}{|x_i - z_k|}\right)^{\frac{2}{m-1}}}$$

Step 4. If $\| U^{(k+1)} - U^{(k)} \| < \varepsilon$ then STOP; otherwise return to step 2.

Examples:

In fuzzy clustering, each cluster is a fuzzy set of all the patterns.

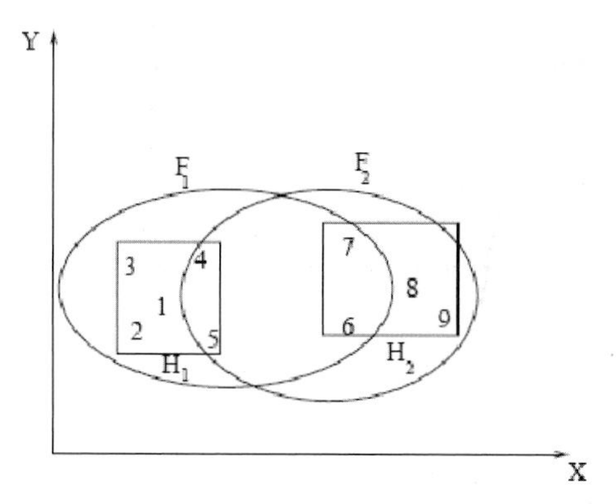

Fig. 5. Fuzzy Cluster

The rectangles enclose two "hard" clusters in the data:

H_1 = {1,2,3,4,5} and H_2 = {6,7,8,9}. A fuzzy clustering algorithm might produce the two fuzzy clusters *F*1 and *F*2 depicted by ellipses. The patterns will have membership values in [0,1] for each cluster. Details shown in Fig.5

For example, fuzzy cluster *F*1 could be compactly described as

·Σ1,0.9),(2,0.8),(3,0.7),(4,0.6),(5,0.55),(6,0.2),(7,0.2),(8,0.0), Σ9,0.0)∝

and F_2 could be described as

·Σ1,0.0), (2,0.0), (3,0.0), (4,0.1), (5,0.15), (6,0.4), (7,0.35), (8,1.0), (9,0.9)}

3 Performance Metric

3.1 Performance Metric

For evaluating the performance of the clustering algo-rithms, Minkowski score [12,10], are used for different data sets.

a) Minkowski score:

Minkowski score (MS) is define as follows : A clustering solution for a set of n elements can be represented by an n × n matrix C, where Ci,j = 1 if point i and j are in the same cluster according to the solution, and Ci,j = 0 otherwise. The Minkowski score of a clustering result C with reference to T , the matrix corresponding to the true clustering, is defined as

$$MS(T,C) = \frac{|T - C|}{|T|}$$

Where

$$|T| = \sqrt{\sum_i \sum_j T_{i,j}}$$

The Minkowski score is the normalized distance between the two matrices. Lower Minkowski score implies better clustering solution, and a perfect solution will have a score zero.

4 Results

The experimental results comparing the above discussed algorithms for three artificial data sets [1, 17] (*Data 1*, *Data 2*, *Data 3*) and two real-life data sets (Soybean, Balance-Scale) respectively. These are first described below:

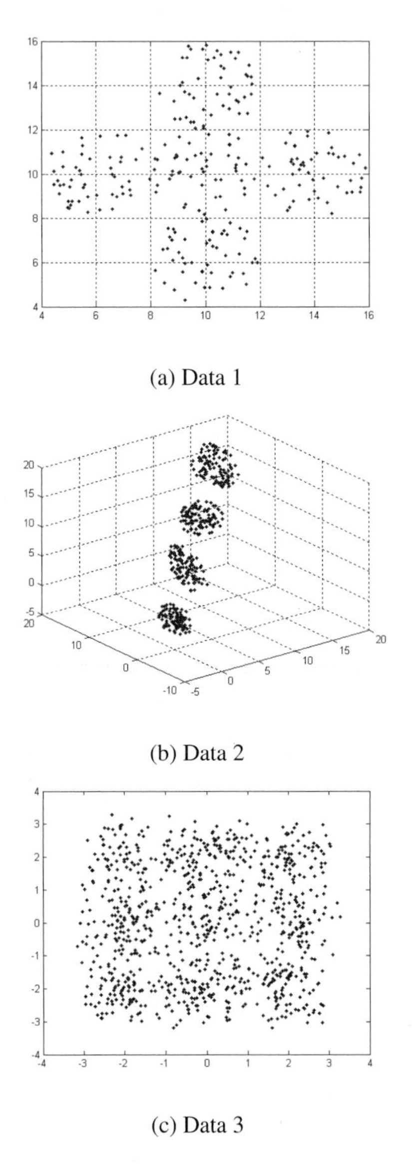

(a) Data 1

(b) Data 2

(c) Data 3

Fig. 6. Artificial Data Set (a)Data 1 (b) Data 2 (c) Data 3

4.1 Artificial Data Sets

Data 1: This is a two-dimensional data set where the number of clusters is five. It has 250 points. The value of K is chosen to be 5 for this data set. Shown in fig.6 (a)

Data 2: This is a nonoverlapping three-dimensional data set where the number of clusters is four. It has 400 data points. The data set are shown in fig.6 (b)

Data 3: This is an overlapping two-dimensional triangular distribution of data points having nine classes where all the classes are assumed to have equal a priori probabilities (1/9). It has 900 data points. Shown in fig.6 (c)

4.2 Real life Data Sets

Soybean: The Soybean data set contains 47 data points on diseases in soybeans. Each data point has 35 categorical attributes and is classified as one of the four diseases, i.e., number of clusters in the data set is 4.
Balance-Scale: This is a weight and distance Database. The Balance-Scale data set contains 625 data points. Each data point has 4 categorical attributes. Number of clusters in the data set is 3. The Information about the attribute of this data set are Left-Weight, Left-Distance, Right-Weight and Right-Distance. Attributes are given in numerical from such as 1 to 5.

The real life data sets mentioned above were obtained from the UCI Machine Learning Repository,

website: http://www.ics.uci.edu/~mlearn/MLRepository.html

The stated algorithms discussed in the section II has been applied to well known data sets. Column 2 of table I represent the number of clusters for different data sets. Column 3 of Table 1 represents the number of data points respectively for the data set. The Remaining columns represent the Minkowski score respectively. Each algorithm has been run for 20 times. The average of Minkowski score (described in section III) has been re-ported.

Performance and comparison for the above discussed algorithm in section II are shown in the following TABLE:I using Minkowski score(MS).:

Lower Minkowski score implies better clustering solution, and a perfect solution will have a score zero.

Table 1. Experiments on different data sets

Data Set	No of Cluster	No of Data Points	K-Means	GA Clustering ()	Fuzzy C Means
Data Set 1	2	250	0.4243	0.4210	0.4404
Data Set 2	4	400	0	0	0
Data Set 3	9	900	0.5454	0.5314	0.5314
Soybean	4	47	0.8087	0.8060	0.7763
Balance-Scale	3	625	0.9783	0.9690	0.9667

5 Conclusions and Future Directions

The performance of the discussed algorithms have been demonstrated using Minkowski score shown in Table 1, for different data sets. The effectiveness of the combination of genetic algorithms and cluster analysis are very much useful for solving the initialization problem in K-means.

Acknowledgement. This paper has been supported by the 2010 Hannam University Research Fund.

References

1. Maulik, U., Bandyopadhyay, S.: Genetic Algorithm-based Clustering Technique. Pattern Recognition Society, 1455–1465 (2000)
2. Faber, V.: Clustering And the Continuous K-means Algorithm. Los Alamos Science (22) (1994)
3. Fayyad, U.M., Reina, C., Bradley, P.S.: Initialization of Iterative Refinement Clustering Algorithms. In: 4th International Conference on Knowledge Discovery & Data Mining (KDD 1998), pp. 194–198 (1998)
4. Bradley, P.S., Fayyad, U.M.: Refining Initial Points for K-Means Clustering. In: Shavlik, J. (ed.) International Conference on Machine Learning (ICML 1998), pp. 91–99 (1998)
5. Krishna, K., Murty, M.: Genetic K-Means Algorithm. IEEE Transactions on Systems, Man and Cybernetics- Part B: Cybernetics 29(3), 433–439 (1999)
6. Dehuri, S., Ghosh, A., Mall, R.: Genetic Algorithms for Multi-Criterion Classification and clustering in Data Mining. International Journal of Computing and Information Sciences 4(3), 143–154 (2006)
7. Al-Shboul, B., Myaeng, S.-H.: Initializing k-Means Using Genetic Algorithm. World Academy of Science, Engineering and Technology 54 (2009)
8. Comaniciu, D., Meer, P.: Mean Shift: A Robust Approach Toward Feature Space Analysis. IEEE Transactions on Pattern Analysis and Machine Intelligence 24(5) (May 2002)
9. Buhmann, J.: Data clustering and learning. In: The Handbook of Brain Theory and Neural Networks, pp. 308–312. The MIT Press, Cambridge (2003)
10. Saha, I., Mukhopadhyay, A.: Improved Crisp and Fuzzy Clustering Techniques for Categorical Data. IAENG International Journal of Computer Science (2008) ISSN: 1819-9224 (online version); 1819-656X, print version
11. Maulik, U., Mukhopadhyay, A., Bandyopad-hyay, S.: Effcient clustering with multi-class point identification. Journal of 3D Images 20, 35–40 (2006)
12. Maulik, U., Bandyopadhyay, S.: Performance Evaluation of Some Clustering Algorithms and Validity Indices. IEEE Transactions On Pattern Analysis And Machine Intelligence 24(12) (December 2002)
13. Alata, M., Molhim, M., Ramini, A.: Optimizing of Fuzzy C-Means Clustering Algorithm Using GA. World Academy of Science, Engineering and Technology (39) (2008)
14. Baraldi, A., Blonda, P.: A Survey of Fuzzy Clustering Algorithms for Pattern Recognition—Part I. IEEE Transactions On Systems, Man, And Cybernetics—Part B: Cybernetics 29(6) (December 1999)

15. Uslan, V., Bucak, Ð.Ö.: Microarray Image Segmentation Using Clustering Methods. Mathematical and Computational Applications 15(2), 240–247 (2010)
16. Kim, S.Y., Choi, T.M.: Fuzzy Types Clustering for Microarray Data. World Academy of Science, Engineering and Technology (4) (2005)
17. Bandyopadhyay, S., Pal, S.K.: Classification and Learning Using Genetic Algorithms: Applications in Bioinformatics and Web Intelligence. Springer, Heidelberg (2007)
18. Liu1, H., Li, J., Chapman, M.A.: Automated Road Extraction from Satellite Imagery Using Hybrid Genetic Algorithms and Cluster Analysis. Journal of Environmental Informatics 1(2), 40–47 (2003)

Processing of Handwritten Signature Image for Authentication

Tai-hoon Kim[*]

Multimedia Engineering Department, Hannam University,
Daejeon, Korea
taihoonn@hnu.kr

Abstract. Since last few years, Handwritten Signature Authentication is a classical research work area in the line of Computer Science. Various new techniques of Image Analysis also attracting the Computer Scientists as well. Firstly, Pixel clustering is used to transform the signature image into bi-color image. Then secondly, instead of considering the whole image, only signature area is extracted. Thirdly, by using Image scaling technique the signature image resized along the coordinate directions. As different techniques are used to subsample (image after transformation) which will be discussed in turn. Fourthly, a different technique is used for thinning to reduce the threshold output of an edge detector algorithm is used to lines of a single pixel thickness. In this paper we propose the above mentioned series of techniques as the preprocessing analysis part of Handwritten Signature Recognition.

Keywords: Skeletonization, Scaling, ITA (Image Thinning Algorithm), ROI (Region of Interest).

Subject index: CO18 Signal Processing, CO00 Others.

1 Introduction

Handwritten Signature Recognition is a generalized way of authenticity. However, it is easy to copy, signature of one person may vary in different times, and it is still more common and widely recognized technique for authentication. There are 2 (Two) approaches available for Handwritten Signature Recognition: a) On-Line and b) Off-Line. This Research is highlighted on the static features of a Handwritten Signature, which can be considered as Off-Line Approach [7].

The Images of the Handwritten Signature is taken repeatedly considering as the "Signature of a person may vary widely time to time", which are special type of objects. From these sample Signatures an average will be taken and stored for authentication in future. At this point the type of errors likes to reduce the chance of rejection of genuine Signatures and improve forgery resistance. Incorporating those two aspects – acceptance of the variance and the requirement for exactness of certain features in one system is a very difficult task and still there is no perfect solution. The

[*] Corresponding author.

T.-h. Kim, A. Stoica, and R.-S. Chang (Eds.): SUComS 2010, CCIS 78, pp. 116–123, 2010.

techniques developed so far, is that, to extract Morphological Features from Handwritten Signature Image and by analyzing that Image decisions can be taken [6].

2 Previous Works

Image can be represented morphologically, using Image Dilation and Erodes. The fundamental operations associated with an object are the standard set operations union, intersection, and complement plus translation, and Boolean Convolution [1].

Binary Image Morphology is taken into account based on behavior of Binary Images, Erosion and Dilation, consideration of foreground and background of images, Blur, Effect of addition of noise, translationally invariant methods for pattern matching [2].

Morphological Filtering of image, a theory introduced in 1988 in the context of mathematical morphology. Research on lattice framework. The emphasis is put on the lattices of numerical functions in digital and continuous spaces and Morphological Filters [3].

The usefulness of the hit-miss transform (HMT) and related transforms for pattern matching in document image application is examined. HMT is sensitive to the types of noise found in scanned images, including both boundary and random noise, a simple extension, the Blur HMT, is relatively robust. The noise immunity of the Blur HMT derives from its ability to treat both types of noise together, and to remove them by appropriate dilations [4].

The adaptation is achieved using a tradeoff parameter in the form of a nonlinear function of the local saturation. To evaluate the performance of the proposed algorithm, a deigned psychophysical experiment is used to derive a metric denoted as the average value for the psychophysical evaluation in percent (APE%). Results of implementing the proposed APE show that an APE=73 to 96% can be achieved for basic morphological operators, i.e., dilation, erosion, opening, and closing. APE value depends on the size and shape of the structuring element as well as on the image details. The proposed algorithm has also been extended to other morphological operators, such as image smoothing (noise suppression), top hat, gradient, and Laplacian operators. In the case of a smoothing operation, an average peak signal-to-noise ratio (PSNR)=31 to 37 dB is achieved at various structuring elements and applied noise variances, while good results are achieved with the proposed top-hat operators [5].

Whenever an image is digitized, i.e., converted from one form to another some form of degradation occurs at output [6]. There is no image processing system which can produce an ideal image. Image enhancement is the improvement of the appearance of the image.

Enhancement can be done via, contrast intensification, smoothing and edge sharpening. Algorithm for spatial domain and frequency domain techniques are used widely. Spatial domain is dealt with neighborhood of single pixel and frequency domain dealt with global filters (masks) [6].

Alessandro Zimmer, Lee Luan Ling, 2003, proposed a new hybrid handwritten signature verification system, where the on-line reference data acquired through a digitizing tablet serves as the basis for the segmentation process of the corresponding scanned off-line data. Local foci of attention over the image were determined through a self-adjustable learning process in order to pinpoint the feature extraction process.

Both local and global primitives were processed and the decision about the authenticity of the specimen defined through similarity measurements. The global performance of the system was measured using two different classifiers [7].

A method for the automatic verification of handwritten signatures was described by Ramanujan S. Kashi, William Turin, and Winston L. Nelson in 1996. The method based on global and local features that summarize aspects of signature shape and dynamics of signature production. They compared with their previously proposed method and shown the improvement of current version [8].

3 Our Work

We propose following four algorithms to achieve our goal. Those algorithms are:

3.1 Transform Gray Signature Image to Bi-color Signature Image

Input: Gray scale Signature Image

Output : Bi-Color Signature Image

a. Open Gray scale Signature Image in Read Mode
b. Read the Pixel
c. Check the Pixel intensity value: if the value is less than 255 (gray value for white color) Then convert it to 0 Else no modification in the Pixel value
d. Rewrite the Pixel with changed intensity value
e. If not 'end of file' Then go to Step-b.
f. Close image file

3.2 Extracting Region of Interest (ROI)

Input: Bi-Color Signature Image (Output of 4.1 Algorithm)

Output : Image only with Signature Region

a. Open Image1 (Bi-Color Signature Image) File in Input Mode
b. Open Image2 File in Output Mode
c. Declare an Integer 2D Matrix of [n x m], where, n and m are width and height of Image1
d. Get RGB Value[i, j] of Image1 and store it to Matrix[i, j] position
e. GotoStep-4 until end of Image1 File Matrix [n, m] is generated with RGB Weight of Image1.
f. Identify First row where First Black RGB Color is occurred in Matrix[n, m], i.e., p
g. Identify First column where First Black RGB Color is occurred in Matrix[n, m], i.e., q
h. Here, Matrix[p, q] is the starting position of Signature Region of Image1
i. Identify Last row where Last Black RGB Color is occurred in Matrix[n, m], i.e., x

j. Identify Last column where Last Black RGB Color is occurred in Matrix[n, m], i.e., y
k. Here, Matrix[x, y] is the end position of Signature Region of Image1
l. Get RGB Values of the Matrix.....[p, q] to [x, y] Position and Write into Image2 File

3.3 Scaling

Considering the resultant bi-color signature image from the algorithm mentioned in 4.2. Mathematics behind the scaling we used and tested randomly as given below....

a. Input image is loaded via Toolkit and Media-Tracker.
b. Four (4) arguments contain the maximum size of the Image to be created. The actual size of the Image will be computed from that maximum size and the actual size of the image (all sizes are given as pixels). The code will scale the Input Image correctly.
c. If the two arguments for the maximum Image size are both 100 and the image that was loaded is 400 times 200 pixels large, we want the image to be 100 times 50 pixels large, not 100 times 100, because the original image is twice as wide as it is high. A 100 times 100 pixel image would contain a very skewed version of the original image.
d. Now that we have determined the size of the image we create a BufferedImage of that size, named iImage. We have taken another object for that new image and call its drawImage method to draw the original image on that new image. The call to drawImage does the actual scaling.
e. The rendering and bilinear interpolation can be used (performance will slowdown) and speed more important. For nicer results (at least in some cases) we have used INTERPOLATION BICUBIC instead of INTERPOLATION BILINEAR.
f. In order to save the scaled-down image to a file, we have created a buffered FileOutputStream with the second argument as name and initialize the necessary objects. The quality argument from the command line is converted from the interval 0 to 100 to the interval 0.0f to 1.0f, because that's what the codec expects (I mostly used 0.75f). The higher that quality number is, the better the resulting image quality, but also the larger the resulting file.

3.4 Image Thinning Algorithm (ITA)

Input: Resultant Signature Image from 4.3

Algorithm

Output : Thinned Signature Image

a. Take the surrounding pixels of foreground.
b. Foreground points must have at least a single background neighbor.
c. Reject points that with more than one foreground neighbor.
d. Continue Steps [b to d] until locally disconnect (divided into 2 parts) region with Pixel iterate until convergence.

Implemented pseudocode:

```
BufferedImage bi = ImageIO.read (new File("Signature_Image"));
int[][] matrix = new int[bi.getWidth()][bi.getHeight()];
for(int i=0; i<bi.getWidth();++i)
{
  for(int j=0;j<bi.getHeight();++j)
  {
          matrix[i][j] = bi.getRGB(i,j);
  }
}

int rows = bi.getWidth(), columns = bi.getHeight();
for(int i = 0; i < rows; ++i)
{
  for(int j = 0; j < columns; ++j)
  {
                if((i==0) || (j==0) || (i==(rows-1)) || (j==(columns-1)))
                matrix[i][j] = 0;
  }
}
for(int r = 1; r < rows-1; r++)
{
  for(int c = 1; c < columns-1; c++)
  {
          if ((matrix[r][c] == 1) && (matrix[r][c+1] == 1)
          && (matrix[r+1][c] == 1) && (matrix[r+1][c+1] == 1))
             matrix[r][c] = 0;
  }
}
for(int r = 1; r < rows-1; r++)
{
  for(int c = 1; c < columns-1; c++)
  {
          if ((matrix[r][c] == 1) && ((matrix[r][c-1] == 0)
          && (matrix[r-1][c] == 0) && (matrix[r][c+1] == 0)
          && (matrix[r+1][c] == 0)))
                  matrix[r][c] = 0;
  }
}
```

4 Result

Extensive testing has been done with a Signature Database composed of 131 users (individuals) Signatures, each user with 24 Handwritten Signatures and 10-trained forgery Signatures [Handwriting Databases: http://www.gpds.ulpgc.es/download/].

One such testing result is taken and shown here in this paper. Fig. 1 shows a user signature in original 256-color image and different ink color is used, but, other than black.

Bi-Color clustered image is the output image shown in Fig. 2, and this is the output of our "Transform Gray Signature Image to Bi-Color Signature Image" algorithm. The one more advantage of this algorithm is that it is very much useful for noise reduction; this can be achieved by tuning the threshold value during conversion to Bi-Color Image. The resultant image is passed through our "Extracting Region of Interest (ROI)" algorithm; the effect is shown in Fig. 3.

Fig. 1. 256 BMP Image (Gray) [246 x 146], Before Pixel Clustering

Fig. 2. Bi-Color Image [246 x 146], after Pixel Clustering

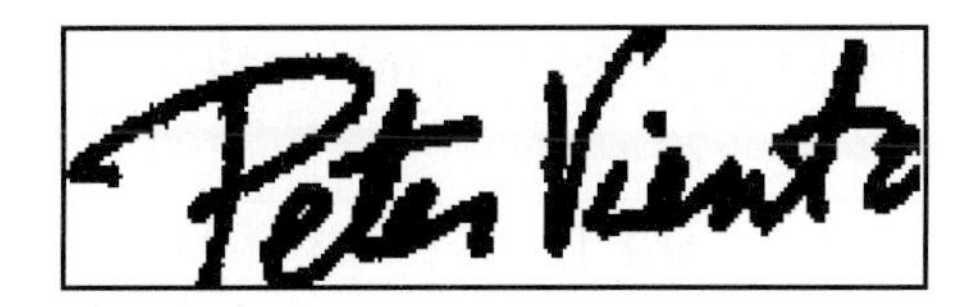

Fig. 3. Bi-Color Image (ROI) [223 x 73]

Fig. 4. Bi-Color Scaled Image [150 x 70]

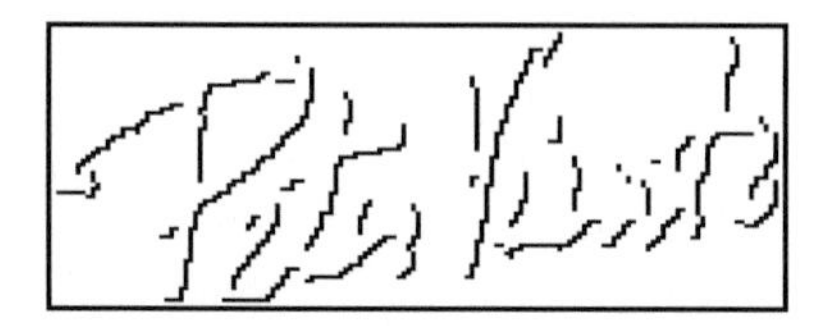

Fig. 5. Thinned Signature Image [150 x 70]

Then the Resultant Signature Image is passed through our Scaling algorithm and the result is shown in Fig. 4. Then the image is passed through our ITA algorithm, where we have used our own technique (Fig. 5), but, other than some widely used masking algorithms.

Table 1. Time Taking for Gray to Bi-Color and ROI

Test No. #	Image size [n x m]	Millisec.	Sec.
1	246 x 146	234	0.234
2	422 x 229	283	0.283
3	670 x 352	310	0.310

Algorithms are implemented in jdk1.6 with Java Advance Imaging (JAI 1.2.2). Hardware configuration is Pentium Dual CPU with 1.8 GHz Processor with 1 GB RAM. Fedora 9 is used as Operating System. Table 1 shows the various experimental times with different image sizes. Programs are giving results in polynomial times. Times taken by first 2 algorithms are shown in Table 1.

Table 2 shows the times taken by Scaling and ITA algorithms. Both the sets of result in Table 1 and Table 2 proves that size of images do not affect much in the final output.

Table 2. Time Taking for Scaling and ITA

Test No. #	Image size [n x m]	Millisec.	Sec.
1	223 x 73	193	0.193
2	337 x 113	232	0.232
3	535 x 252	281	0.281

5 Conclusion

This paper is emphasized on various Image Processing algorithms. These are preprocessing part of Signature Images required for future authentication and recognition. In this paper we propose four different algorithms very newly. Initially, gray scale signature image has been taken and gradually through four different algorithms final output is extracted. The result derived finally, has been outlined in term of times consumed. We are also working towards the authentication and recognition of signature and will be detailed in future publications.

Acknowledgement. This paper has been supported by the 2010 Hannam University Research Fund.

References

1. Young, T., Gerbrands, J.J., van Vliet, L.J.: Morphology-based Operations, `http://www.ph.tn.tudelft.nl/Courses/FIP/noframes/fip-Morpholo.html` (last visited on May 16, 2009)
2. Bloomberg, D.S., Vincent, L.: Pattern Matching using the Blur Hit-Miss Transform. Journal of Electronic Imaging 9(2), 140–150 (2000)
3. Serra, J., Vincent, L.: An Overview of Morphological Filtering. Circuits, Systems and Signal Processing 11(1), 47–108 (1992)
4. Bloomberg, D.S., Vincent, L.: Blur Hit-Miss Transform and its Use in Document Image Pattern Detection. In: Proceedings of SPIE, Document Recognition II, San Jose, CA, March 30, vol. 2422, pp. 278–292 (1995)
5. Al-Otum, H.M., Uraikat, M.T.: Color image morphology using an adaptive saturation-based technique. Optical Engineering, Journal of SPIE 43(06), 1280–1292 (2004)
6. Dutta Majumder, D., Chanda, B.: Digital Image Processing and Analysis. Prentice-Hall of India Pvt. Ltd., Englewood Cliffs (2006)
7. Zimmer, A., Ling, L.L.: A Hybrid On/Off Line Handwritten Signature Verification System. In: International Conference on Document Analysis and Recognition, Edinburgh, Scotland, August 3-6, vol. 1, pp. 424–428 (2003)
8. Kashi, R.S., Turin, W., Nelson, W.L.: On-line handwritten signature verification using stroke direction coding. In: Optical Engineering, SPIE Digital Library, vol. 35, pp. 2526–2533 (1996)

Access Requirement Analysis of E-Governance Systems

Tai-hoon Kim[*]

Multimedia Engineering Department, Hannam University,
Daejeon, Korea
taihoonn@hnu.kr

Abstract. The strategic and contemporary importance of e-governance has been recognized across the world. In India too, various ministries of Govt. of India and State Governments have taken e-governance initiatives to provide e-services to citizens and the business they serve. To achieve the mission objectives, and make such e-governance initiatives successful it would be necessary to improve the trust and confidence of the stakeholders. It is assumed that the delivery of government services will share the same public network information that is being used in the community at large. In particular, the Internet will be the principal means by which public access to government and government services will be achieved. To provide the security measures main aim is to identify user's access requirement for the stakeholders and then according to the models of Nath's approach. Based on this analysis, the Govt. can also make standards of security based on the e-governance models. Thus there will be less human errors and bias. This analysis leads to the security architecture of the specific G2C application.

Keywords: G2C, Broadcasting model, Critical flow model, Comparative analysis model, E-advocacy model, Interactive model.

1 Introduction

Using Information Communication Technology, e-governance is approaching the citizens to provide them web-based services at lower cost and higher efficiency. Among the various components of e-governance, using G2C the citizen can directly interact with Govt. Citizens will be reluctant to use the web based services offered by the government, due to their poor skill, lack of confidence, security and privacy concerns [Backus 2001]. Many of the citizens of the developing countries like India are illiterate. Also, presently there is a lack of inexpensive and easy-to-use security infrastructure in G2C applications. Moreover, to support all levels of the society, there must be a proper access control technology according to the models.

In our previous paper [5], G2C applications are classified based on the Egov models. The information flow follows the work flow in these models. The information

[*] Corresponding author.

T.-h. Kim, A. Stoica, and R.-S. Chang (Eds.): SUComS 2010, CCIS 78, pp. 124–130, 2010.

provided to the citizens is to be reliable, but the information channels are almost always public and insecure. Moreover, there is a necessity for maintaining a secure environment for hosting the govt's data.

This paper looks at the access permissions of the users of G2C applications from a model driven approach. The users and their roles have been identified. Based on this analysis, access requirements of the roles have been identified. This analysis leads to the security architecture of the specific G2C application.

2 Previous Work

Based on different classes of information, their sources and frequency of updation and exchange, various models of E-governance projects can be evolved. The National E-governance Action Plan of the Government of India [NeGP] can act as a model for such projects. Other sources of information include and DigitalGovernance.org web page of Mr. Vikas Nath [Nath, 2005]. In the latter, Nath has classified the models into the following categories:

- Broadcasting model
- Critical flow model
- Comparative analysis model
- E-advocacy model
- Interactive model

2.1 Broadcasting Model

The model is based on broadcasting or dissemination of useful governance information which already exists in the public domain into the wider public domain through the use of ICT and convergent media. The utility of this model is that a more informed citizenry is better able to benefit from governance related services that are available for them.

2.2 Critical Flow Model

The model is based on broadcasting or dissemination information of 'critical' value (which by its very nature will not be disclosed by those involved with bad governance practices) to targeted audience using ICT and convergent media. Targeted audience may include media, opposition parties, judicial bench, independent investigators or the wider public domain itself.

2.3 Comparative Analysis Model

Comparative Knowledge Model is one of the least-used but a highly significant model for developing country which is now gradually gaining acceptance. The model can be used for empowering people by matching cases of bad governance with those of good governance, and then analyzing the different aspects of bad governance and its impact on the people.

2.4 E-Advocacy Model

E-Advocacy / Mobilization and Lobbying Model is one of the most frequently used Digital Governance model and has often come to the aid of the global civil society to impact on global decision-making processes. The strength of this model is in its diversity of the virtual community, and the ideas, expertise and resources accumulated through this virtual form of networking.

2.5 Interactive Model

Interactive-Service model is a consolidation of the earlier presented digital governance models and opens up avenues for direct participation of individuals in the governance processes. Fundamentally, ICT have the potential to bring in every individual in a digital network and enable interactive (two-way) flow of information among them.

3 Our Works and Analysis

Access to protected information must be restricted to people who are authorized to access the information. The computer programs, and in many cases the computers that process the information, must also be authorized. This requires that mechanisms be in place to control the access to protected information. The sophistication of the access control mechanisms should be in parity with the value of the information being protected- the more sensitive or valuable the information the stronger the control mechanisms need to be.

In this section, the roles of the stakeholders as the user will be identified according to the models. The stakeholders of egovernance systems are Government officials, LUB/LSG, private sector, NGO/civil society, organizations and citizens. The users can be classified as Database managers, Developers, Implementation officers, Information officers, Chief information security officers, officers and employees, Agents, Public.

We are considering here that at the client side only the information are being accessed with the help of some security checking. All access permissions according to the model are described below:

3.1 Broadcasting Model

Broadcasting / Wider Disseminating Model

Public Domain ⟶ Wider Public Domain

In broadcasting model, generally data are for public use. So read permission must be granted to all the stakeholders. Other permissions for the stakeholders are described in the following table:

Users	Read	Write	Delete	Execute	Modify	Append
DBM	Y	Y	Y	Y	Y	Y
Dev	Y	Y	N	Y	N	Y
IO	Y	N	N	Y	N	Y
Info	Y	Y	Y	Y	Y	Y
Ciso	Y	N	N	Y	N	Y
Emp	Y	N	N	Y	N	N
A	Y	N	N	Y	N	N
P	Y	Y	N	Y	N	Y

3.2 Critical Flow Model

Critical Flow Model

Critical Domain ⟶ Targeted / Wider Domain

In critical flow model, data must reach to the targeted domain not to all. All access requirements for the stakeholders are described below:

Users	Read	Write	Delete	Execute	Modify	Append
DBM	Y	Y	Y	Y	Y	Y
Dev	Y	Y	N	Y	N	Y
IO	Y	N	N	Y	N	Y
Info	Y	Y	N	Y	N	Y
Ciso	Y	N	N	Y	N	Y
Emp	Y	N	N	Y	N	N
A	Y	N	N	Y	N	N
P	Y	N	N	Y	N	Y

3.3 Comparative Analysis Model

Comparative Analysis Model

Private / Public Domain + Public / Private Domain

⟶ Wider Public Domain

In comparative analysis model, the analysis is done based on old records. All access requirements for the stakeholders are described below:

Users	Read	Write	Delete	Execute	Modify	Append
DBM	Y	Y	Y	Y	Y	Y
Dev	Y	Y	N	Y	N	Y
IO	Y	N	N	Y	N	Y
Info	Y	Y	N	Y	N	Y
Ciso	Y	N	N	Y	N	Y
Emp	Y	N	N	Y	N	N
A	Y	N	N	Y	N	N
P	Y	N	N	Y	N	Y

3.4 E-Advocacy Model

Mobilisation and Lobbying Model

Networking Networks for Concerted Action

E-advocacy model has come to the aid of the global civil society to impact on global decision making process. All access requirements for the stakeholders are described below:

Users	Read	Write	Delete	Execute	Modify	Append
DBM	Y	Y	Y	Y	Y	Y
Dev	Y	Y	Y	Y	Y	Y
IO	Y	N	N	Y	N	Y
Info	Y	Y	N	Y	N	Y
Ciso	Y	N	N	Y	N	Y
Emp	Y	N	N	Y	N	N
A	Y	N	N	Y	N	N
P	Y	Y	Y	Y	Y	Y

3.5 Interactive Model

Service Delivery Model

Citizen ⇄ Government

In interactive model, information flows in two ways. All access requirements for the stakeholders are described below:

Users	Read	Write	Delete	Execute	Modify	Append
DBM	Y	Y	Y	Y	Y	Y
Dev	Y	Y	Y	Y	Y	Y
IO	Y	N	N	Y	N	Y
Info	Y	Y	Y	Y	Y	Y
Ciso	Y	N	N	Y	N	Y
Emp	Y	N	N	Y	N	N
A	Y	N	N	Y	N	N
P	Y	Y	N	Y	N	Y

DBM → Database Manager
Dev → Developers
IO → Implementation Officers
Info → Information Officers
Ciso → Chief Information security officer
Emp → Officers and Employees
A → Agents
P → Public

The above tables provide the broad access control mechanisms for different stakeholders for different models. The government can also make standards of security based on the egovernance models. Thus there will be less human errors and bias.

4 Conclusion

In summary, this paper presents a methodology to formulate the access control mechanisms for different stakeholders of different G2C applications in a model driven manner. The methodology can be used for protecting all G2C applications from unauthorized user s.

This paper contains a general study but not from the point of individual application. Here model based access controls have been designed. Further work in this direction is the development of control directories as per the accepted ISO 27001 standards.

Acknowledgement. This paper has been supported by the 2010 Hannam University Research Fund.

References

1. [Ammal 2007] Ammal, Anantalakshmi, R.: E-Governance Application Life Cycle Management - Issues and Solutions. egovasia (2007)

2. Backus, M.: E-governance in Developing Countries., International Institute of Communication & Development (IICD), Research Brief No. 1 (March 2001), http://www.ftpiicd.org/files/research/reports/report3.pdf
3. Mazumdar, C., Kaushik, A.K., Banerjee, P.: On Information Security Issues in E-Governance: Developing Country Views. CSDMS journal, July 6 (2009)
4. Nath, V.: Digital Governance Initiative, http://www.DigitalGovernance.org
5. Kaushik, A.K., Mazumdar, C., Bhattacharjee, J., Saha, S.: Model Driven Security Analysis of Egovernance Systems. in eIndia (2008), November edition, http://www.egovonline.net/Resource/eindia08-full-paper-for-abstract-173.pdf

Cognitive Informatics in Medical Image Semantic Content Understanding

Marek R. Ogiela[1] and Lidia Ogiela[2]

AGH University of Science and Technology
[1] Institute of Automatics
[2] Faculty of Management
al. Mickiewicza 30, PL-30-059 Krakow, Poland
{mogiela,logiela}@agh.edu.pl

Abstract. This publication presents the idea of cognitive categorisation systems as used in the cognitive informatics field using the example of image analysis systems – UBIAS. Cognitive categorisation systems execute semantically-oriented data analysis processes, i.e. analyse meaning based on the semantic contents of analysed data sets. Cognitive data analysis processes lead to the in-depth interpretation and understanding of various types of data. This publication presents a selected class of cognitive categorisation systems called the UBIAS (Understanding Based Image Analysis Systems), also referred to as image analysis systems.

Keywords: Cognitive informatics, cognitive processes, cognitive categorization systems, UBIAS systems (Understanding Based Image Analysis Systems).

1 Introduction

Understanding analysed data is extremely complex because in purely human cognitive processes it comprises many different, extremely important thought processes, including:

- perception – sensory perception, the process of perceiving (differentiating, recognising, perceptual categorisation, orientation),
- attention – unintentional attention, free attention,
- memory – declarative (overt), sensory, short-term, long-term (episodic, semantic, contextual), implicit memory,
- thinking,
- calculating,
- abstracting,
- forming concepts,
- taking decisions,
- planning, predicting,
- problem solving,
- understanding.

T.-h. Kim, A. Stoica, and R.-S. Chang (Eds.): SUComS 2010, CCIS 78, pp. 131–138, 2010.

Human cognitive processes leading to interpreting and describing complex data form the foundation for designing computer data analysis systems. The thought processes taking place in the human brain are used as the basis for designing a computer system which attempts to imitate natural processes and thus aims at the automatic, computer analysis of data. Cognitive data analysis systems are designed based on cognitive resonance, which has been described in [7-11, 13, 14] and is presented in Figure 1.

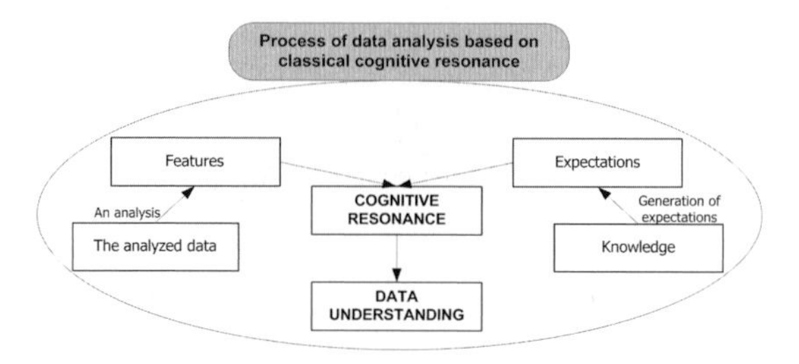

Fig. 1. Cognitive resonance in the process of data analysis and understanding

Traditional data analysis processes based on cognitive resonance are extended in this publication to include stages at which the system learns using the knowledge collected in its knowledge bases and by analysing situations it does not understand. If the system encounters a situation it does not understand, i.e. one undefined in its knowledge base, it cannot correctly classify it and match the pattern. In this situation the system enters a state of surprise and incomprehension of data, which it can, however, use to supplement the knowledge base with new cases of pattern classification and data understanding, as a result of which it becomes necessary to add new, undefined examples to the expert knowledge base. This situation is presented in Figure 2.

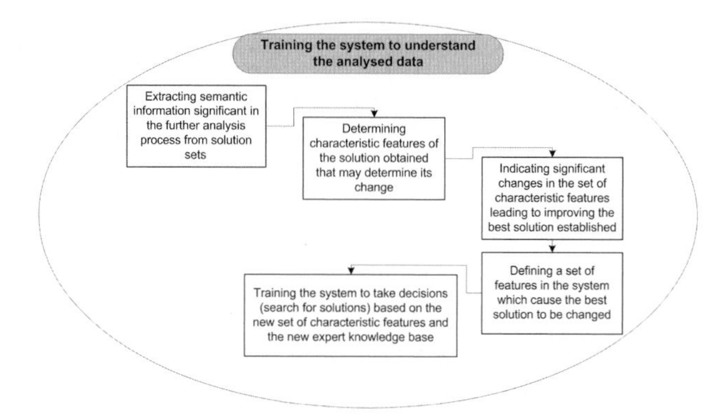

Fig. 2. Training the system to understand the analysed data

Supplementing data analysis with stages at which the system learns new solutions means that the cognitive resonance must be repeated in the in-depth data analysis process, and if the learning process is multiplied, then cognitive resonance must be repeated more than once. Incorporating new system learning solutions in the data analysis process means that data analysis processes are much more complex (Fig. 3.).

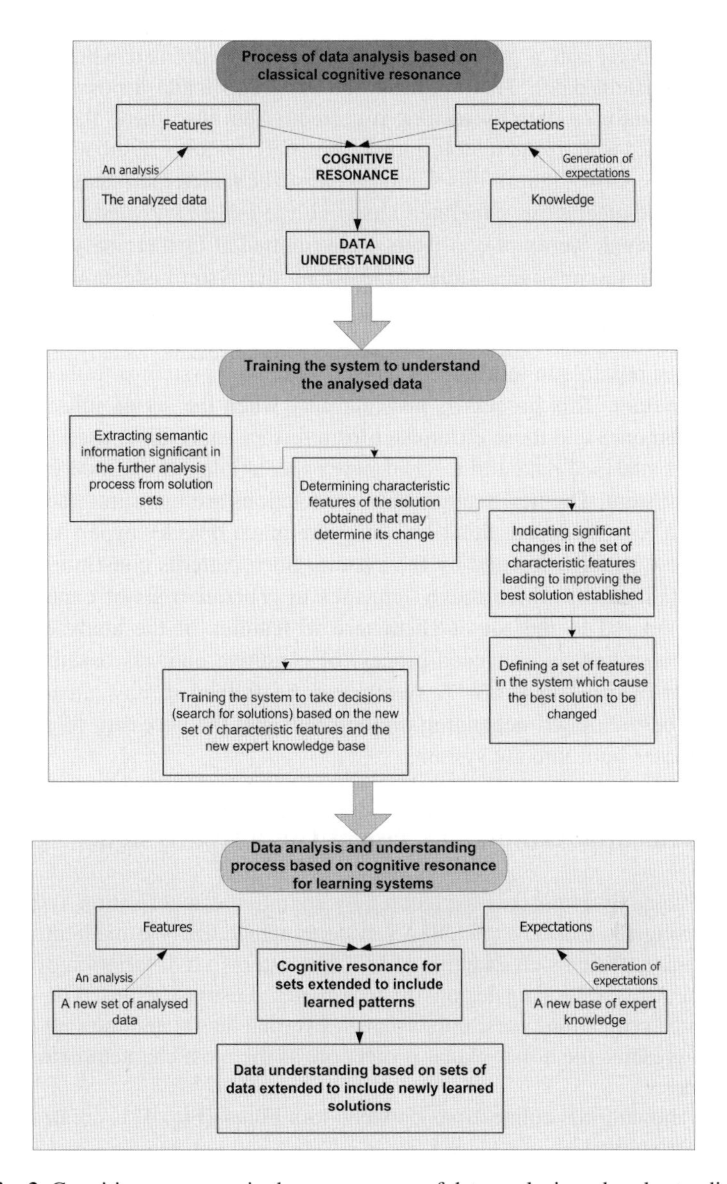

Fig. 3. Cognitive resonance in the new process of data analysis and understanding

In the cognitive data analysis process, the process whereby the system learns new solutions which may impact the decision-making solution obtained is of key importance. So far, in the cognitive categorisation processes, the understanding of analysed data was based on the classical cognitive analysis process, whereby connections were indentified between pairs of consistent expectations of the system acquired from expert knowledge bases and characteristic features extracted from analysed datasets, and this led to cognitive resonance during which the above connections were determined as consistent or inconsistent. Only pairs that were completely consistent were selected for further analysis and a group of solutions could be defined which included the identified consistent pairs. This definition of the group made it possible not only to recognise the analysed data by naming it correctly, but also made the understanding of data complete, which lead to determining semantic features of the analysed data.

Research has shown that in this solution, pairs of characteristic features of the analysed data and expectations generated based on the expert knowledge base collected in the system which were not consistent were omitted at further analysis stages. This made it possible to envisage a situation in which the system encounters a solution it does not know and which is not defined at all in its bases. The question is, is it possible to recognise this type of a situation? Yes, the solution proposed in this publication shows that it is possible to introduce a stage at which the system is trained in solutions new to the system. This process is possible only when the set of solutions obtained (both optimum ones and those eliminated from further analysis) is used to create a set of features of analysed data and a set of new expectations not defined in the original bases of the system. The new features and expectations are input into the system base in which data is re-analysed, this time using the much broader expert knowledge set containing new patterns learned by the system. Such patterns constitute an extended expert knowledge base which the system uses to generate a set of expectations, and these are compared to the set of characteristic features of the analysed data. This process thus becomes an enhanced process of cognitive analysis based on cognitive resonance for learning systems. A system can be trained in any situation and this training can be multiplied depending on the needs and the necessity of extending the knowledge bases built into the system.

2 UBIAS as an Example of Class of Cognitive Systems

Semantic data analysis processes executed by cognitive categorisation systems can be presented using the example of UBIAS systems which analyse medical image data. This publication presents an example of UBIAS systems - a description of the process of analyzing x-rays showing lesions in foot bones. Lesions of this type can be presented in three projection types: dorso-planar, external lateral and internal lateral. This publication presents the dorso-planar projection, as the most characteristic of the three possible types.

A graph showing the connections between foot bones (Fig. 4) is created for the selected projection type.

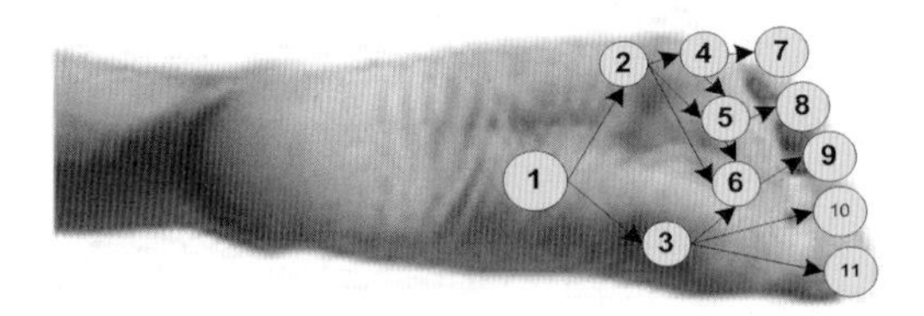

Fig. 4. A graph describing the foot bone skeleton in the dorsoplanar projection

Topographic relationships were introduced for the thus defined, spanned graph describing the foot bone skeleton in the dorsoplanar projection. These relationships describe the location of particular structures in relation to one another, as well as the possible pathological changes within the foot (Fig. 5).

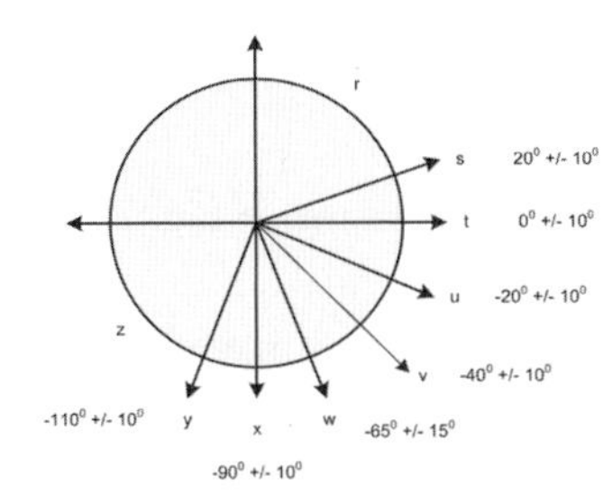

Fig. 5. A relation graph for the dorsoplanar projection of the foot

The introduction of such spatial relationships (Fig. 5) and the representation in the form of a graph spanned on the skeleton of foot bones were used to define the graph proper, in which all the adjacent foot bones were labeled as appropriate for the analysed dorsoplanar projection (Fig. 6).

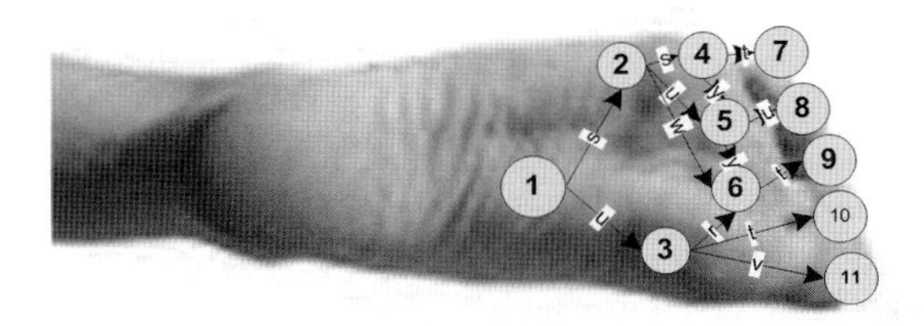

Fig. 6. A graph with numbers of adjacent bones marked based on the relation graph for the dorsoplanar foot projection

This graph shows bones that are already numbered and which have been assigned labels in line with searching the graph across (bfs/wfs-wide first serach). Such a representation creates a description of foot bones using the so-called IE graph. This is an ordered and oriented graph for which the syntactic analysis will start from the distinguished apex number 1 (Fig. 6).

For the purposes of the analysis conducted, a formal definition of the graph grammar was introduced, which takes into account the developed linguistic description of correct connections between foot bones:

$$G_{dors} = (N, \Sigma, \Gamma, ST, P)$$

where:

- The set of non-terminal labels of apexes:

N = {ST, CALCANEUS, OS NAVICULARE, OS CUBOIDEUM, OS CUNEIFORME MEDIALE, OS CUNEIFORME INTERMEDIUM, OS CUNEIFORME LATERALE, M1, M2, M3, M4, M5}

- The set of terminal labels of apexes:

  Σ = {s, t, u, v, w, x, y, c, on, oc, ocm, oci, ocl, m1, m2, m3, m4, m5},
  Γ – the graph shown in Fig. 6.,
  The start symbol S = ST,
  P – a finite set of productions shown in Fig. 7.

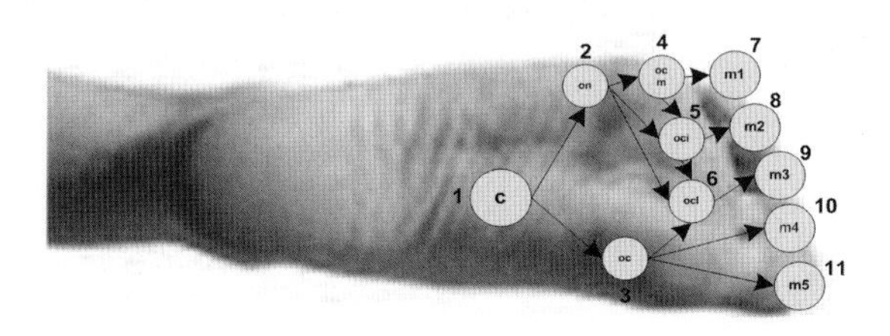

Fig. 7. A set of productions defining the interrelations between particular elements of the structure of foot bones for the dorsoplanar projection

Our analysis of image-type data understanding was aimed at an in-depth understanding of the images analysed, in this case also of specific lesions. Figure 8 shows the possibilities for describing various disease cases by expanding the set of linguistic rules to include additional grammatical rules.

The presented examples of the cognitive analysis and interpretation of data, describing the lesions appearing in foot bones, show possible cases, namely: fractures and deformations foot.

Semantic data analysis systems which present images of lesions found in foot bones are used to semantically understand the analysed medical images. This is because they present not just the recognition of the lesion, but also describe it, define directions in which necessary preventive treatment should go and project changes that may occur in the future.

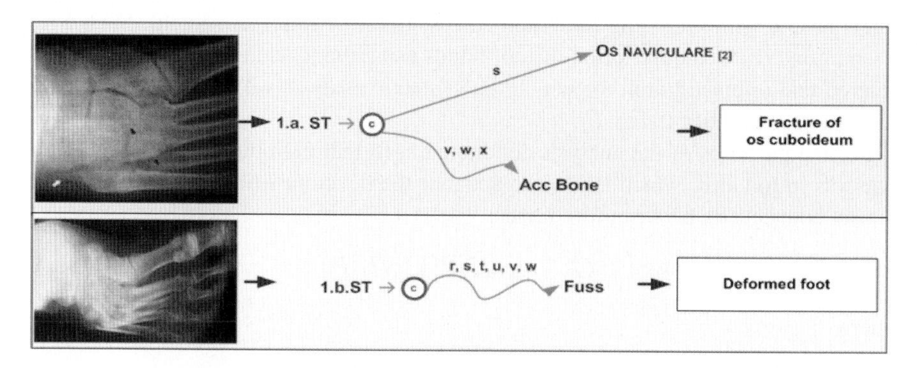

Fig. 8. Examples of using the automatic understanding of foot bone lesions detected by the system in the dorsoplanar projection

3 Conclusion

Cognitive data analysis systems are employed for executing processes of analysing, interpreting and reasoning using expert knowledge bases deployed in this type of systems. Data is analysed based on its semantic contents, which makes it possible to conduct meaning-based data analysis. This analysis supports not only recognising the analysed data, but also understanding it and reasoning based on the decisions taken by the system. Cognitive categorisation systems enhanced with learning stages represent a new class of intelligent cognitive information systems. Their distinguishing characteristic is that their knowledge bases are constantly extended to include new interpretational patterns of data analysis.

Acknowledgement. This work has been supported by the Ministry of Science and Higher Education, Republic of Poland, under project number N N516 196537.

References

1. Meystel, A.M., Albus, J.S.: Intelligent Systems – Architecture, Design, and Control. John Wiley & Sons, Inc., Chichester (2002)
2. Bechtel, W., Abrahamsen, A., Graham, G.: The live of cognitive science. In: Bechtel, W., Graham, G. (eds.) A Companion of Cognitive Science, pp. 1–104. Blackwell Publishers, UK (1998)
3. Branquinho, J. (ed.): The Foundations of Cognitive Science. Clarendon Press, Oxford (2001)
4. Cohen, H., Lefebvre, C. (eds.): Handbook of Categorization in Cognitive Science, The Netherlands. Elsevier, Amsterdam (2005)
5. Duda, R.O., Hart, P.E., Stork, D.G.: Pattern Classification, 2nd edn. A Wiley-Interscience Publication John Wiley & Sons, Inc. (2001)
6. Minsky, M.: The Society of Mind. Simon & Schuster, New York (1987)
7. Ogiela, L.: Modelling of Cognitive Processes for Computer Image Interpretation. In: Al-Dabass, D., Nagar, A., Tawfik, H., Abraham, A., Zobel, R. (eds.) EMS 2008 European Modelling Symposium, Second UKSIM European Symposium on Computer Modeling and Simulation, Liverpool, United Kingdom, September 8-10, pp. 209–213 (2008)
8. Ogiela, L., Ogiela, M.R.: Cognitive Techniques in Visual Data Interpretation. Studies in Computational Intelligence, vol. 228. Springer, Heidelberg (2009)
9. Ogiela, L., Tadeusiewicz, R., Ogiela, M.R.: Cognitive techniques in medical information systems. Computers In Biology and Medicine 38, 502–507 (2008)
10. Ogiela, M.R., Tadeusiewicz, R.: Modern Computational Intelligence Methods for the Interpretation of Medical Images. Springer, Heidelberg (2008)
11. Ogiela, M.R., Tadeusiewicz, R., Ogiela, L.: Image languages in intelligent radiological palm diagnostics. Pattern Recognition 39, 2157–2165 (2006)
12. Rutkowski, L.: Computational Intelligence, Methods and Techniques. Springer, Heidelberg (2008)
13. Tadeusiewicz, R., Ogiela, L.: Selected Cognitive Categorization Systems. In: Rutkowski, L., Tadeusiewicz, R., Zadeh, L.A., Zurada, J.M. (eds.) ICAISC 2008. LNCS (LNAI), vol. 5097, pp. 1127–1136. Springer, Heidelberg (2008)

14. Tadeusiewicz, R., Ogiela, L., Ogiela, M.R.: The automatic understanding approach to systems analysis and design. International Journal of Information Management 28, 38–48 (2008)
15. Wang, Y.: The Theoretical Framework and Cognitive Process of Learning. In: Proc. 6th International Conference on Cognitive Informatics (ICCI 2007), pp. 470–479. IEEE CS Press, Lake Tahoe (2008)
16. Wilson, R.A., Keil, F.C.: The MIT Encyclopedia of the Cognitive Sciences. MIT Press, Cambridge (2001)

An Attack on Wavelet Tree Shuffling Encryption Schemes

Samuel Assegie, Paul Salama, and Brian King

Purdue School of Engineering & Tech.
Indiana University - Purdue University Indianapolis
{psalama,briking}@iupui.edu

Abstract. With the ubiquity of the internet and advances in technology, especially digital consumer electronics, demand for online multimedia services is ever increasing. While it's possible to achieve a great reduction in bandwidth utilization of multimedia data such as image and video through compression, security still remains a great concern. Traditional cryptographic algorithms/systems for data security are often not fast enough to process the vast amounts of data generated by the multimedia applications to meet the realtime constraints. Selective encryption is a new scheme for multimedia content protection. It involves encrypting only a portion of the data to reduce computational complexity(the amount of data to encrypt)while preserving a sufficient level of security. To achieve this, many selective encryption schemes are presented in different literatures. One of them is Wavelet Tree Shuffling. In this paper we assess the security of a wavelet tree shuffling encryption scheme.

1 Introduction

Internet multimedia applications has become extremely popular. Valuable multimedia content such as digital images and video, are vulnerable to unauthorized access while in storage as well as during a transmission over a network. Streaming of secure images/real-time video in the presence of constraints, such as bandwidth, delay, computational complexity and channel reliability is one of the most challenging problems. For example, a 512×512 color image at 24 bits/pixel would require 6.3Mbits. While the bandwidth issue can be resolved using compression, securing multimedia data still remains a big challenge, especially in light of the diversity of devices (in terms of resource availability) that will transmit and receive the content.

Traditional image and video content protection schemes are fully layered, the whole content is first compressed, and then the compressed stream is encrypted using a standard cryptographic technique (such as TDES, AES, ...)[16]. However, the requirement for high transmission rate with limited bandwidth makes this traditional technique inadequate. In the fully layered scheme compression and encryption are two different(disjoint) processes. The multimedia content is processed as a classical text assuming that all the bits in the plaintext are equally important. But with constrained resources (in real-time networking, high

T.-h. Kim, A. Stoica, and R.-S. Chang (Eds.): SUComS 2010, CCIS 78, pp. 139–148, 2010.

definition delivery, low power, low memory and computational capability) this scheme is inefficient. Thus techniques for securing multimedia data requiring less complexity and less adverse effect on the compression without compromising the security of the data is required. One such technique is to use selective encryption [3,18]. Selective encryption is an encryption technique based on combining the encryption and compression process, that will reduce computational complexity as well as bandwidth utilization, by encrypting only the "essential parts of the image". In Section 3, we briefly discuss some selective encryption techniques.

This work focuses on the analysis of a selective encryption technique that is based on the permutation (shuffling) of wavelet trees, a technique that was suggested by Kwon, et. al. [4] to be used as the sole base of providing privacy. Here we demonstrate that as the sole cryptographic primitive it is weak.

2 Wavelet Based Compression Techniques

Over the past several years, the wavelet transform has gained widespread acceptance in image compression research. Since there is no need to divide the image into macro blocks (no need to block the input image), wavelet based coding at higher compression avoids blocking artifacts. Wavelet transform can decompose a signal in to different subbands. There are many compression techniques that use the wavelet transform, including JPEG-2000, EZW [14] and SPIHT [12]. We now briefly discuss EZW and SPIHT.

In 1993, Shapiro [14] presented an algorithm for entropy encoding called *Embedded Zerotree Wavelet (EZW) algorithm*. After applying the wavelet transform, the coefficients can be represented using trees because of the subsampling that is performed in the transform. A coefficient in a low subband can be thought of as having four descendants in the next higher subband. The four descendants each have four descendants in the next higher subband and we see a quad-tree structure emerges and every root has four leafs. A zerotree is a quad-tree of which all nodes are equal to or smaller than the root. The zero tree structure is based on the hypothesis that if a wavelet coefficient at a coarse scale is insignificant with respect to a given threshold T, then all wavelet coefficients of the same orientation in the same spatial location at finer scales are likely to be insignificant with respect to T. The idea is to define a tree of zero symbols that start at a root that is also zero and label as end-of-block. Many insignificant coefficients at higher subbands (finer resolution) can be discarded since the tree grows as powers of four.

The EZW algorithm encodes the obtained tree structure. The resulting output is such that the bits that are generated in order of importance, yielding a fully embedded code. The main advantage of this encoding technique is the encoder can terminate the encoding at any point, thereby allowing one to achieve a target bit rate (i.e. rate scalable). Similarly, the decoder can also stop decoding at any point resulting an image that would have been produced at the rate of the truncated bit stream. Since EZW generate bits in order of importance, those bits that affect the perceived quality of the decompressed image/video most can be placed at the beginning of the data stream; since the entire stream depends on those bits they can be a good candidate for selective encryption.

In 1996, Said and Pearlman [12] introduced a computationally simple compression technique (algorithm) called *SPIHT (Set Partitioning In Hierarchical Trees)* that is based on the wavelet transform. SPIHT uses set partitioning and significance testing on hierarchical structures of transformed images to extend/improve on the work of Shapiro [14]. SPIHT is also a good candidate to be used in a selective encryption scheme.

3 Prior Work on Selective Encryption

Selective encryption has been suggested and adopted as a basic idea for encryption of digital images and videos, aiming to achieve a better trade off between the encryption load and the security level. Selective encryption is a method of selectively concealing portions of a compressed multimedia bitstream while leaving the remaining portions of the stream unchanged.

There are a number of selective encryption techniques. Here we briefly discuss only a few schemes. For more details and a more thorough discussion we suggest the reader look at [18,6].

In 2002, Podesser, Schmidt and Uhl [10] applied the following technique. They proposed a selective bitplane encryption using AES. They conducted a series of experiments on 8-bit grayscale images, and observed the following: (1) encrypting only the MSB is not secure; a replacement attack is possible, (2) encrypting the first two MSBs gives hard visual degradation, and (3) encrypting three bitplanes gives very hard visual degradation.

Zeng and Lei [20] proposed a selective encryption scheme in the frequency domain (wavelet domain). The general scheme consists of selective scrambling of coefficients by using different primitives (selective bit scrambling, block shuffling, and/or rotation). The input video frames are transformed using wavelet transform and each subband represents selected spatial frequency information of the input video frame. The authors propose two ways to scramble the coefficients. In their first suggestion, they observed that some bits of the transform coefficients have high entropy and can thus be encrypted without greatly affecting compressibility. In their second suggestion, the authors observed that shuffling the arrangement of coefficients in a transform coefficient map can provide effective security without destroying compressibility, as long as the shuffling does not destroy the low-entropy aspects of the map relied upon by the bitstream coder. To increase security, the authors suggested block shuffling. Each subband is divided into a number of blocks of equal size (the size of the block can vary for different subbands) and within each subband, blocks of coefficients will be shuffled according to a shuffling table generated using a key.

Kwon, Lee, Kim, Jin, and Ko [4] described a scheme which involves shuffling of spatial orientation trees(SOT) to secure multimedia data. The authors mentioned the deficiency of traditional block shuffling technique and proposed *wavelet tree shuffling* as an alternate security mechanism as part of the security architecture for multimedia digital rights management. The authors proposed a 4-level wavelet transform. According to Shapiro's[14] algorithm, this will result in 13 sub-bands, and the wavelet coefficients are grouped according to wavelet trees.

In 2005, Salama and King [13] proposed a joint encryption-compression technique (Selective Encryption) for securing multimedia data based on EZW. Their approach is selectively encrypting those bits for which the entire bit stream depends. Through a serious of experiments/simulations, the authors found that encrypting the leading 256 bits of a 512×512 image will provide sufficient security. In their scheme, first the image will be transformed using discrete wavelet transform, apply EZW, and then entropy encoded before it is encrypted using the proposed Selective Encryption technique. The authors developed a security analysis of the proposed joint compression-encryption technique, and demonstrated an attack called *Database Attack*. In this attack, an adversary(unintended receiver) can intercept the encrypted signal and attempt to replace the encrypted portion of the data stream by another portion that he/she would generate. For the attack to be successful, the interceptor would need a selective database (small enough for computations to be feasible and large enough to include all possible target images), that contains at least one of the possible images that can be transmitted. The attacker then performs a brute force attack by encoding all images in the database and comparing the unprotected part of the stream with the corresponding part of the compressed images from the database. If there is a match, the attacker can replace one stream with another. As a countermeasure to this attack, the authors propose *Randomly shuffling the SOT prior to encoding*, shuffling the SOT (spatial orientation trees) after wavelet transform. This will frustrate brute force database attack without affecting the compression performance.

In 2006, Wu and Mao [8] proposed a shuffling technique as part of their selective encryption architecture. The authors use the MPEG-4 fine granularity scalability (FGS) functionality provided by the MPEG-4 streaming video profile [5] to illustrate their concept and approach. A video is first encoded into two layers, a base layer that provides a basic quality level at a low bit rate and an enhancement layer that provides successive refinement. The enhancement layer is encoded bitplane by bitplane from the most significant bitplane to the least significant one to achieve fine granularity scalability. The authors propose an intra bit plan shuffling on each bit plane of n-bits according to a set of cryptographically secure shuffle tables and using a run-EOP approach. In addition to bit-plane shuffling, the authors also proposed randomly flipping the sign bit s_i of each coefficient according to a pseudo-random bit b_i from a one-time pad, i.e., the sign remains the same when $b_i = 0$ and changes when $b_i = 1$.

4 A Generic Framework of a Wavelet Tree Shuffling Encryption Scheme

Here we outline the construction of an encryption scheme which is based on the use of permuting the trees which are produced by the wavelet transform. This scheme was suggested by Kwon et. al. in [4].

Suppose the image I is of size $M \times N$. Then I can be represented as

$$I = \begin{bmatrix} m_{0,0} & \cdots & m_{0,N-1} \\ \vdots & \ddots & \vdots \\ m_{M-1,0} & \cdots & m_{M-1,N-1} \end{bmatrix}_{M \times N} \tag{1}$$

where $m_{i,j}$ is the i, j pixel of I.

For a level L of wavelet decomposition the number of SOT's (spatial orientation tree) is

$$T = \frac{M \cdot N}{2^{2L}} \tag{2}$$

If $M = 2^d$ then the maximum level of decomposition will be d. Thus, if an image I of size 512×512 is decomposed using 4 levels of decomposition then there will be 1024 trees. Since there are 1024! permutations of the trees, this would require a key of at least 1024 bits. A symmetric cryptosystem which uses a key of size 1024 should provide security for well over 50 years [1,2]. However such a scheme does not possess such security.

Encryption:
In this procedure, the coefficient matrix I of size $M \times N$ is shuffled using a permutation (determined by symmetric key K) to form a corresponding image C. This is achieved by applying a permutation (shuffling) to the SOT's that were created during the wavelet transform. More formally let $\mathcal{WT}$ denote the 2D discrete wavelet transform and PERM_K denote the permutation that shuffles the wavelet trees. Then the ciphertext C is generated as follows:

First the wavelet transform is applied to I,

$$\mathcal{WT}(I) = (\mathcal{T}_1, \mathcal{T}_2, \ldots, \mathcal{T}_T)$$

(here $\mathcal{T}_i$ denotes the ith tree produced by the wavelet transform). Then given key K, the trees $(\mathcal{T}_1, \mathcal{T}_2, \ldots, \mathcal{T}_T)$ are then permuted as

$$C = \text{PERM}_K(\mathcal{T}_1, \mathcal{T}_2, \ldots, \mathcal{T}_T).$$

This process is illustrated in Fig. 1.

Decryption:
Given the ciphertext C, the image I is then reconstructed as follows. First the inverse of the permutation (that was induced by key K) is applied. Thus

$$(\mathcal{T}_1, \mathcal{T}_2, \ldots, \mathcal{T}_T) = \text{PERM}_K^{-1}(C).$$

Then the $\mathcal{IWT}$ (inverse wavelet transform) is applied. The result is

$$I = \mathcal{IWT}(\mathcal{T}_1, \mathcal{T}_2, \ldots, \mathcal{T}_T).$$

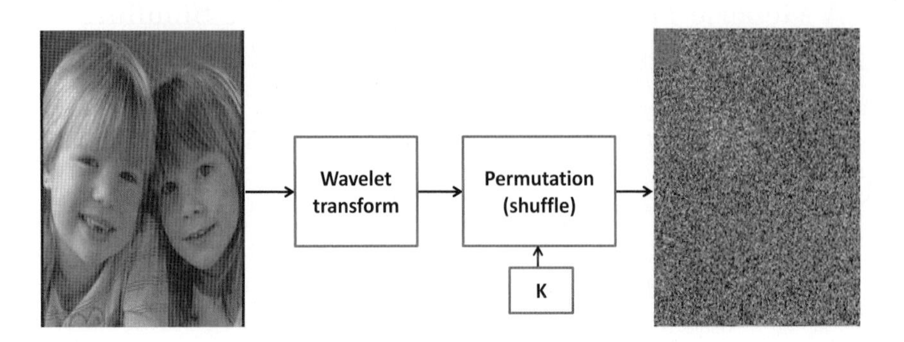

Fig. 1. Encrypting by shuffling wavelet trees

5 Attacking a Wavelet Tree Shuffling Encryption Scheme

In this work, we are analyzing shuffling of the wavlet trees (SOT's) as if it were the only mechanism used for encryption [4].

Norcern and Uhl [9] also discussed the insecurity of a system (in terms of compression performance) that was based on randomly permuting wavelet-subbands incorporated in the JPEG2000 or the SPIHT coder proposed by Uehara et. al. in [17]. Their work differs from ours in the sense that they were attacking a scheme that was randomly permuting the coefficients of wavelet subbands, while our attack is on encryption schemes that are shuffling the tree structure created after the wavelet transform.

The basis of our attack will be a chosen plaintext attack, in particular a lunch time attack We will assume that the attacker has temporary possession of the encryption machine and can feed the machine selected plaintext and will receive the corresponding ciphertext. >From this lunch time attack the adversary will be able to determine the permutation (encryption) key.

First consider an image of size $\mathbf{M} \times \mathbf{N}$ and suppose it is transformed in to wavelet coefficients using the L-level discrete wavelet transform(DWT). With an L-level decomposition, we have $3L + 1$ frequency bands. In Fig 4, when $L = 4$, the lowest frequency subband is located in the top left(i.e., the LL4 subband, and the highest frequency subband is at the bottom right.(i.e., the HH1 subband). The relation between this frequency bands can be seen as a parent-child relationship [14]. Thus, for an image of size $M \times N$ with L-level of decomposition, in total we will have $\frac{M}{2^L} \cdot \frac{N}{2^L}$ trees. After constructing wavelet trees, a secret key K is used to randomly shuffle the trees.

Clearly if an attacker can guess the size of the image and the number of levels of decomposition, then shuffling of the wavelet trees (SOT's) is vulnerable to the lunch time attack. A simple scenario of a lunch time attack: a legitimate user s away from the desk (computer) without locking their computer/machine, the attackers can use a chosen plaintext attack. A chosen plaintext attack can be launched by any party (adversary) who can access the machine while the user is absent. For the attack to be successful, the adversary needs to have access for the encryption machine. Once the adversary has access to the encryption machine, he/she can choose a series of chosen plaintexts and feed them to the encryption machine as shown in Fig. 5. Thus if I_j denotes a chosen plaintext, and if we denote the wavelet tree shuffling encryption scheme (as illustrated in Fig 1) by $E(I_j, K)$ where K is the key, then the adversary will generate chosen plaintexts

$$E(I_1, K), E(I_2, K), \ldots, E(I_r, K).$$

The adversary can use these to determine the image size and the level of decomposition. Thus we will assume that the adversary knows these parameters.

In [4], Kwon et. al. proposed the shuffling of wavelet trees of a wavelet coefficient which undergoes 4-levels of wavelet decomposition as part of their security architecture for digital rights management. According to Equation 2, for an image of size 512×512 which undergoes 4-levels of wavelet decomposition, there will be 1024 trees. Any shuffling of these trees using a randomized shuffling key(shuffling matrix) should provide the security of 1024 bit key size.

Though Kwon et. al. [4] were using a level $L = 4$ of decomposition for an image of size 512×512, we will provide an analysis/experiments based on image size of 256×256. Only a slight modification of our experiments would have to be constructed to attack an image of size 512×512. The following table demonstrates the relationship between the level of decomposition and the number of trees.

Table 1. Relationship between no. of decomp. and no. of trees for an image of size 256×256

No. of decomp.	No. of trees
3	1024
4	256
5	64
6	8
7	4

Assume for the moment that the level of decomposition was $L = 5$. In Fig 2, we provide the original chosen plaintext image. This image was selected, an image consisting of a series of integers from 1 to 64, where each integer is in white placed within a small back box, in order to clearly determine the permutation key and the levels of decomposition. Using the encryption machine, and applying the shuffle to this plaintext we get the ciphertext in Fig 3, for a level of decomposition $L = 3$, we get the ciphertext in Fig 4, for a level of decomposition $L = 4$, and we get the ciphertext in Fig 5, for a level of decomposition $L = 5$. Observe the clarity of the image Fig 5, the digits are clearly visible thus revealing that we have determined the level of decomposition. Once the adversary possesses Fig 5, they know the level of decomposition as we well as the permutation, hence they know the key, and so they have broken the cipher.

Fig. 2. Original Image

Fig. 3. Shuffled Image where $L = 3$

So due to the nature of wavelet trees if an adversary would have guessed the correct size of the image and the number of level of decomposition, he/she can get a ciphertext that will leak information about the randomized key used to shuffle the wavelet trees (SOT's). The adversary needs to encrypt only $\log_2(V)$ times for each carefully chosen plaintext images to find the key, where V is $max(M, N)$ for an image of size $M \times N$. This is because the maximum number of possible levels of decomposition is bounded by the logarithm of the dimensions as observed in Section 4. An image of size 256×256 has at most 8 levels of decomposition, as illustrated in the above figures, by the choosing an appropriate plaintext, and viewing the potential ciphertexts from the vantage of different levels of wavelet decomposition (such as $L = 3$, $L = 4$, $L = 5$), the correct level and key can be determined. That is, the image is a carefully chosen image, determined by the adversary and the resulting ciphertext (Fig. 3, 4, 5) will be analyzed, and if the adversary can easily deduce how the wavelet trees are shuffled then they have found the key. For example, for an image of size 256×256 which has gone through 4 levels of wavelet decomposition we need to run the encryption machine for the chosen plaintext at most 8 times and as it can be easily be seen how the trees are shuffled by looking at Fig 5. The series of chosen plaintext (image) used by the adversary will be limited since the possible number of decomposition are limited (small numbers) to distort the image visually. For example, for an image of size 512×512 a wavelet transform of the image with 8 level of decomposition will have only 4 wavelet trees and the key size to shuffle these trees also be 4, and will not distort the image. Thus, the adversary can easily guess the key used to shuffle the wavelet trees.

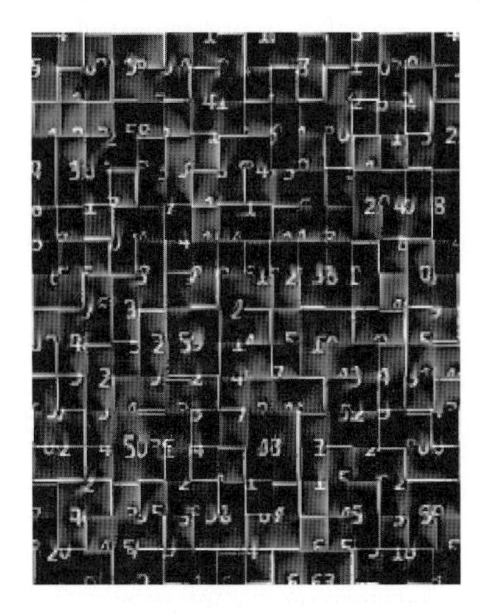

Fig. 4. Shuffled Image where $L = 4$

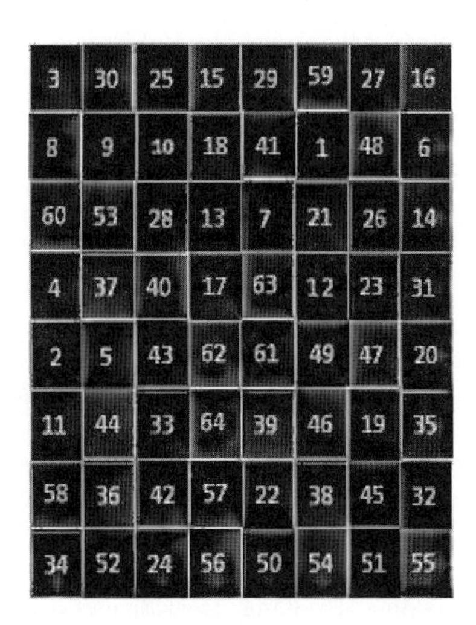

Fig. 5. Shuffled Image where $L = 5$

6 Conclusion

We have discussed the wavelet tree shuffling encryption scheme [4]. We have showed that the shuffling of the wavelet trees, when used as a sole security mechanism, is insecure against a chosen plaintext attack. However despite its weaknesses, shuffling of wavelet trees can add security when other cryptographic primitives are used. For example, in [13] the authors use wavelet tree shuffling as a supplementary security mechanism to strengthen the security of multimedia data (to protect against the *database attack*).

References

1. Recommendation for Key Management, Special Publication 800-57 Part 1, NIST, 03/2007
2. Keylength - Cryptographic Key Length Recommendation, `http://www.keylength.com`
3. Furht, B., Kirovski, D. (eds.): Multimedia Security Handbook. CRC Press, Boca Raton (2004)
4. Kwon, G., Lee, T., Kim, K., Jin, J., Ko, S.: Multimedia digital right management using selective scrambling for mobile handset. In: Hao, Y., Liu, J., Wang, Y.-P., Cheung, Y.-m., Yin, H., Jiao, L., Ma, J., Jiao, Y.-C. (eds.) CIS 2005. LNCS (LNAI), vol. 3802, pp. 1098–1103. Springer, Heidelberg (2005)
5. Li, W.: Overview of Fine Granularity Scalability in MPEG-4 Video Standard. IEEE Trans. on Circuits & Systems for Video Technology 11(3), 301–317 (2001)
6. Massoudi, A., Lefebvre, F., De Vleeschouwer, C., Macq, B., Quisquater, J.: Overview on selective encryption of image and video: challenges and perspectives. EURASIP J. Inf. Security, 1–18 (2008)

7. Naor, M., Yung, M.: Public-key cryptosystems provably secure against chosen ciphertext attacks. In: Proceedings of the 22nd Annual Symposium on Theory of Computing. ACM, New York (1990)
8. Mao, Y., Wu, M.: A joint signal processing and cryptographic approach to multimedia encryption. IEEE Trans. Image Processing 15(7), 2061–2075 (2006)
9. Norcen, R., Uhl, A.: Encryption of wavelet-coded imagery using random permutations. In: Proceedings of the IEEE International Conference on Image Processing (ICIPŠ 2004), Singapore. IEEE Signal Processing Society (October 2004)
10. Podesser, M., Schmidt, H., Uhl, A.: Selective bitplane encryption for secure transmission of image data in mobile environments. In: Proceedings of the 5th Nordic Signal Processing Symposium, NORSIG 2002 (2002)
11. Pommer, A., Uhl, A.: Selective encryption of wavelet-packet encoded image data: efficiency and security. Multimedia Systems 9(3), 279–287 (2003)
12. Said, A., Pearlman, W.A.: A new, fast and efficient image codec based on set partitioning in hierarchical trees. IEEE Trans. Circuits and Systems for Video Technology 6(3), 243–250 (1996)
13. Salama, P., King, B.: Efficient secure image transmission: compression integrated with encryption. In: Proc. SPIE, vol. SPIE-5681, pp. 47–58 (2005)
14. Shapiro, M.: Embedded Image Coding using Zerotrees of Wavelet Coefficients. IEEE Trans. Signal Processing 41, 3445–3462 (1993)
15. Shamir, A.: How to share a secret. Comm. of ACM 22(11), 612–613 (1979)
16. Stinson, D.: Cryptography: theory and practice, 2nd edn. CRC Press, Boca Raton (2002)
17. Uehara, T., Safavi-Naini, R., Ogunbona, P.: Securing wavelet compression with random permutations. In: Proceedings of the 2000 IEEE Pacific Rim Conference on Multimedia, Sydney, pp. 332–335. IEEE Signal Processing Society (December 2000)
18. Uhl, A., Pommer, A.: Image and Video Encryption: From Digital Rights Management to Secured Personal Communication. Springer, Boston (2005)
19. Valens, C.: As appeared in (1999), `http://pagesperso-orange.fr/polyvalens/clemens/ezw/ezw.html`
20. Zeng, W., Lei, S.: Efficient frequency domain selective scrambling of digital video. IEEE Transactions on Multimedia 5(1), 118–129 (2003)

Context Aware Systems, Methods and Trends in Smart Home Technology

Rosslin John Robles and Tai-hoon Kim*

Multimedia Engineering Department, Hannam University,
Daejeon, Korea
rosslin_john@yahoo.com, taihoonn@hnu.kr

Abstract. Context aware applications respond and adapt to changes in the computing environment. It is the concept of leveraging information about the end user to improve the quality of the interaction. New technologies in context-enriched services will use location, presence, social attributes, and other environmental information to anticipate an end user's immediate needs, offering more-sophisticated, situation-aware and usable functions. Smart homes connect all the devices and appliances in your home so they can communicate with each other and with you. Context-awareness can be applied to Smart Home technology. In this paper, we discuss the context-aware tools for development of Smart Home Systems.

Keywords: Context-aware computing, smart home, Ubiquitous, Automation.

1 Introduction

Context awareness originated as a term from ubiquitous computing or as so-called pervasive computing which sought to deal with linking changes in the environment with computer systems, which are otherwise static. [1] Research in the development of context aware applications has gained attention since the early 1990s.

Smart Home is the integration of technology and services through home networking for a better quality of living. Smart homes or buildings are usually a new one that is equipped with special structured wiring to enable occupants to remotely control or program an array of automated home electronic devices by entering a single command.

Context awareness plays a big role in developing and maintaining a Smart Home. On the following parts of this paper, we discuss Context Awareness, Smart Home, and the Context aware tools used in Smart Home Development.

2 Context Awareness

Devices may have information about the circumstances under which they are able to operate and based on rules, or an intelligent stimulus, react accordingly. Context

* Corresponding author.

T.-h. Kim, A. Stoica, and R.-S. Chang (Eds.): SUComS 2010, CCIS 78, pp. 149–158, 2010.

aware devices may also try to make assumptions about the user's current situation. [1] Context awareness refers to the idea that computers can both sense, and react based on their environment.

While the people from the field of computer science has initially perceived the context as a matter of user location, in the last few years this notion has been considered not simply as a state, but part of a process in which users are involved; thus, sophisticated and general context models have been proposed, tailor the set of application-relevant data, to support context-aware applications which use them to adapt interfaces, increase the precision of information retrieval, make the user interaction implicit, discover services, or build smart environments.

A context aware mobile phone may know that it is currently in the meeting room, and that the user has sat down. The phone may conclude that the user is currently in a meeting and reject any unimportant calls. [2]

Context aware systems are deals with the acquisition of context, the abstraction and understanding of context, and application behavior based on the recognized context. As the user's activity and location are crucial for many applications, context awareness has been focused more deeply in the research fields of activity recognition and location awareness.

Context-aware computing refers to a general class of mobile systems that can sense their physical environment, like their context of use, and adapt their behavior accordingly. Such systems are a component of a ubiquitous computing or pervasive computing environment. Three important aspects of context are: where you are; who you are with; and what resources are nearby.

Although location is a primary capability, location-aware does not necessarily capture things of interest that are mobile or changing. Context-aware in contrast is used more generally to include nearby people, devices, lighting, noise level, network availability, and even the social situation; e.g., whether you are with your family or a friend from school. [3]

3 Smart Home Technology

Smart Home is the term commonly used to define a residence that uses a Home Controller to integrate the residence's various home automation systems. Integrating the home systems allows them to communicate with one another through the home controller, thereby enabling single button and voice control of the various home systems simultaneously, in preprogrammed scenarios or operating modes. [4]

Research and development in the Home Automation field is expanding rapidly as electronic technologies converge. The home network encompasses security, communications, entertainment, information systems, and convenience. [5]

Powerline Carrier Systems (PCS) is a technology which is used to send coded signals along a home's existing electric wiring to programmable switches, or outlets. These signals convey commands that correspond to "addresses" or locations of specific devices, and that control how and when those devices operate. A PCS transmitter, for instance, can send a signal along a home's wiring, and a receiver plugged into any electric outlet in the home could receive that signal and operate the appliance to which it is attached. [5]

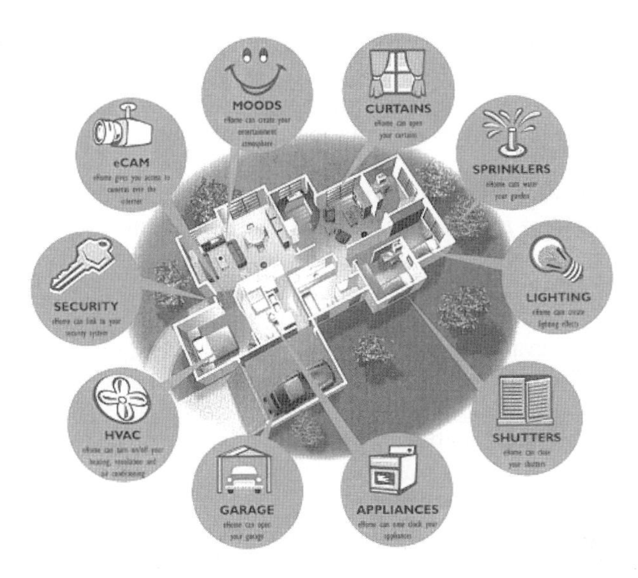

Fig. 1. Smart Home Technology Automation

For PCS X10 is a common protocol, it is a signaling technique for remotely controlling any device plugged into an electrical power line. X10 signals, which involve short radio frequency (RF) bursts that represent digital information, enable communication between transmitters and receivers. In Europe, technology to equip homes with smart devices centers on development of the European Installation Bus, or Instabus. This embedded control protocol for digital communication between smart devices consists of a two-wire bus line that is installed along with normal electrical wiring. The Instabus line links all appliances to a decentralized communication system and functions like a telephone line over which appliances can be controlled.

3.1 Setting Up a Smart Home

Technologies such asZ-Wave, X10, Insteon and ZigBee just provide the technology for smart home communication. Manufacturers have made alliances with these systems to create the products that use the technology. Here are some examples of smart home products and their functions:

- Cameras will track your home's exterior even if it's pitch-black outside.
- Plug your tabletop lamp into a dimmer instead of the wall socket, and you can brighten and dim at the push of a button.
- A video door phone provides more than a doorbell -- you get a picture of who's at the door.
- Motion sensors will send an alert when there's motion around your house, and they can even tell the difference between pets and burglars.
- Door handles can open with scanned fingerprints or a four-digit code, eliminating the need to fumble for house keys.

- Audio systems distribute the music from your stereo to any room with connected speakers.
- Channel modulators take any video signal -- from a security camera to your favorite television station -- and make it viewable on every television in the house.
- Remote controls, keypads and tabletop controllers are the means of activating the smart home applications. Devices also come with built-in web servers that allow you to access their information online.The Remote keypad will send a message to your lamp.

These products are available at home improvement stores, electronics stores, from technicians or o-nline. Before buying, check to see what technology is associated with the product. Products using the same technology should work together despite different manufacturers, but joining up an X10 and a Z-Wave product requires a bridging device.

When designing a smart home, you can do as much or as little home automation as you want. You could begin with a lighting starter kit and add on security devices later. If you want to start with a bigger system, it's a good idea to design carefully how the home will work, particularly if rewiring or renovation will be required. In addition, you'll want to place strategically the nodes of the wireless networks so that they have a good routing range. The cost of a smart home varies depending on how smart the home is. One builder estimates that his clients spend between $10,000 and $250,000 for sophisticated systems. If you build the smart home gradually, starting with a basic lighting system, it might only be a few hundred dollars. A more sophisticated system will be tens of thousands of dollars, and elements of home theater systems raise the cost of a system about 50 percent. [6] Technologies related to Smart Home will is discussed in Section 4.

3.2 Benefits of Smart Home

Obviously, Smart homes have the ability to make life easier and more convenient. Home networking can also provide peace of mind. Wherever you are, the smart home will alert you to what's going on, and security systems can be built to provide an immense amount of help in an emergency. For example, not only would a resident be woken with notification of a fire alarm, the smart home would also unlock doors, dial the fire department and light the path to safety.

Smart homes also provide some energy efficiency savings. Because systems like Z-Wave and ZigBee put some devices at a reduced level of functionality, they can go to "sleep" and wake up when commands are given. Electric bills go down when lights are automatically turned off when a person leaves the room, and rooms can be heated or cooled based on who's there at any given moment. One smart homeowner boasted her heating bill was about one-third less than a same-sized normal home. Some devices can track how much energy each appliance is using and command it to use less.

Smart home technology promises tremendous benefits for an elderly person living alone. Smart homes could notify the resident when it was time to take medicine, alert the hospital if the resident fell and track how much the resident was eating. If the elderly person was a little forgetful, the smart home would perform tasks such as shutting off the water before a tub overflowed or turning off the oven if the cook had wandered away. It also allows adult children who might live elsewhere to participate in the care of their aging parent. To those with disabilities or a limited range of movement, Easy-to-control automated systems would provide similar benefits.

3.2.1 Security

The electricity that flows into a home can be adjusted automatically to different voltages for different appliances and devices. In addition, unlike conventional electric systems within a house, the electrical system in a smart home provides energy only to outlets that have appliances plugged in and turned on. An intelligent home controller also monitors the circuits inside the house and disconnects power in case of short circuit. Sensors in a house chip capable of detecting water leaks and gas as well as the first signs of smoke and alert residents inside.

3.2.2 Savings

Because a smart home may regulate the use of public utilities for the better and most effective owners and rental residents can realize significant savings on utility bills. Utilities can be adjusted to take advantage of lower off peak rates.

3.2.3 Environmental Benefits

Smart homes are also often green homes. Due to the improved efficiency of public services in a smart home, water consumption, electricity and gas are limited – preserving natural resources and fossil fuels. Smart homes also include innovations such as solar panels to further decrease the need for conventional fossil fuels for energy needs. Green roofs, gardens and rain lasting improve the aesthetics of the home loans in addition to environmental benefits.

4 Smart Home Software and Technology

All the devices and appliances are receivers, and the means of controlling the system, such as remote controls or keypads, are transmitters. If you want to turn off a lamp in another room, the transmitter will issue a message in numerical code that includes the following:

- An alert to the system that it's issuing a command,
- An identifying unit number for the device that should receive the command
- A code that contains the actual command, such as "turn off."

All of this is designed to happen in less than a second, but X10 does have some limitations. Communicating over electrical lines is not always reliable because the lines get "noisy" from powering other devices. An X10 device could interpret electronic interference as a command and react, or it might not receive the command at all. While X10 devices are still around, other technologies have emerged to compete for your home networking dollar.Instead of going through the power lines, some systems use radio waves to communicate, which is also how WiFi and cell phone signals operate. However, home automation networks don't need all the juice of a WiFi network because automation commands are short messages. The two most prominent radio networks in home automation are ZigBee and Z-Wave.

Fig. 2. The dots represent devices that could be connected to your smart home network

4.1 Z-Wave

Z-Wave uses a Source Routing Algorithm to determine the fastest route for messages. Each Z-Wave device is embedded with a code, and when the device is plugged into the system, the network controller recognizes the code, determines its location and adds it to the network. When a command comes through, the controller uses the algorithm to determine how the message should be sent. Because this routing can take up a lot of memory on a network, Z-Wave has developed a hierarchy between devices: Some controllers initiate messages, and some are "slaves," which means they can only carry and respond to messages.

4.2 ZigBee

ZigBee's name illustrates the mesh networking concept because messages from the transmitter zigzag like bees, looking for the best path to the receiver. While Z-Wave uses a proprietary technology for operating its system, ZigBee's platform is based on the standard set by the Institute for Electrical and Electronics Engineers (IEEE) for wireless personal networks. This means any company can build a ZigBee-compatible product without paying licensing fees for the technology behind it, which may eventually give ZigBee an advantage in the marketplace. Like Z-Wave, ZigBee has fully functional devices and reduced function devices.

4.3 Insteon

Insteon (commonly written INSTEON) is a system for connecting lighting switches and loads without extra wiring, similar to the X10 standard, designed specifically to address the inherent limitations in the X10 standard but also to incorporate backward compatibility with X10.

Using a wireless network provides more flexibility for placing devices, but like electrical lines, they might have interference. Insteon offers a way for your home network to communicate over both electrical wires and radio waves, making it a dual mesh network. If the message isn't getting through on one platform, it will try the other. Instead of routing the message, an Insteon device will broadcast the message, and all devices pick up the message and broadcast it until the command is performed. The devices act like peers, as opposed to one serving as an instigator and another as a receptor. This means that the more Insteon devices that are installed on a network, the stronger the message will be. [6]

5 Context Aware Systems, Methods and Current Trends in Smart Home Technology

Context awareness plays a big role in developing and maintaining a Smart Home. The following are researches and development of tools and technologies utilizing Context Awareness in developing a Smart Home.

5.1 Context Awareness for Pervasive Home Environments

CAMUS-MS is a framework which collects contexts from SA and uses the contextual information. In addition, it supports a variety of functions for context-aware application development. In particular, CAMUS-MS controls all information about user context including user preference used in contents recommendations and environmental contexts, then sends events by context's changes to applications and helps applications to perform suitable actions for the context. There is another point that is important for the CAMUS-MS: it offers a service framework that can connect to the basic service agent of a robot controller and a variety of software, such as voice recognition, image recognition, and motion detection. In the next section, we describe the detailed components of CAMUS-MS. [7]

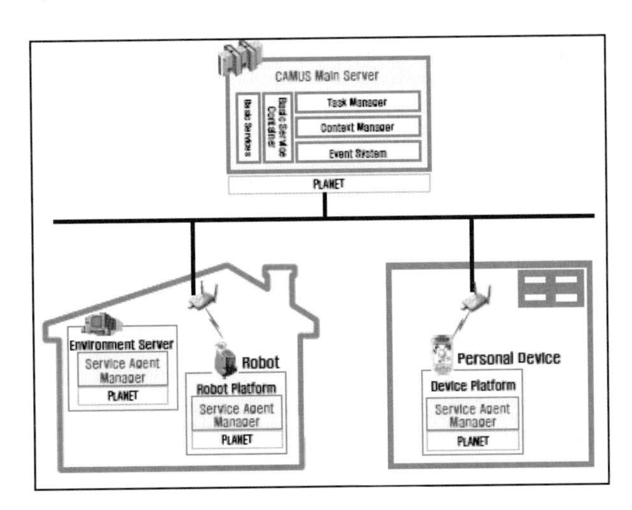

Fig. 3. CAMUS System Architecture

SAM is a program that manages and controls SAs within the environment. To do this, SAM is installed within various environments in a location, obtains information from a variety of sensors in the environment, sends the information to CAMUS-MS, receives instructions from CAMUS-MS, and controls the SA in the environment. Therefore, SAM can be installed in any location, such as rooms or offices, and also in a robot platform or PDA.[7]

The following figure shows the system architecture of CAMUS. CAMUS is composed of four parts: CAMUS-MS (CAMUS Main Server), SAM (Service Agent Manager), SA (Service Agent) and Planet.

SA executes the functions of legacy applications and sensors installed on physical places through communication with SAM and CAMUS-MS. SAs are the software module interfaces of devices and applications that interact with CAMUS-MS. Because SA exposures attribute and action with interfaces that are accessed by CAMUS-MS. For example, suppose that the SA, which provides the information of the user's location, has interfaces o send the id and physical location for the user and the RFID sensor exists in the environment. [7]

5.2 Positioning Method for Applications in Smart Home

The next figure shows the proposed smart home system architecture. In this system, several devices called the beacons were pre-installed in the interesting area. The mobile users are required to equip with a mobile device called the badge which periodically broadcasts echo requests to collect the positioning signals from the beacons and then forward the positioning signals to a nearby router. The positioning signal of a beacon contains the ID and the signal strength of the beacon. [8]

Whenever a router received the positioning signals from a badge, it forwards the positioning signals to its coordinator directly or through the help of other routers. In Zigbee network, the coordinator is used to coordinate the operation of all the routers in the same network and collect packets from the routers.

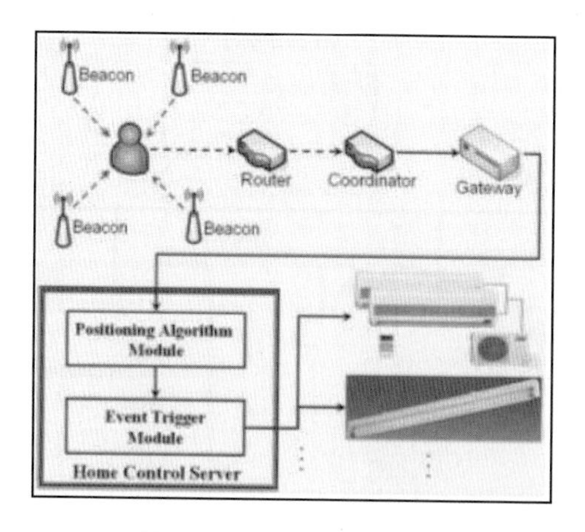

Fig. 4. System Architechture

The packets are then forwarded to the home control server through the gateway which equips with the Zigbee communication module and the Ethernet communication module. The positioning module in the home control server is used to estimate the positions of the badges according to the positioning signals from the badges. The estimate positions of the badges are then forwarded to the event trigger module. The event trigger module maintains numerous rules which control the household appliances. If the estimate position of a badge satisfies certain rules, the corresponding events will be triggered and some commands will be sent to corresponding household appliances to change their states. [8]

5.3 Context Aware Gateway

A home inhabitant could answer the incoming call by choosing his/her preferred SIP UA devices and post the outgoing presence information on an "as needed" basis (based on pre-defined context aware service criteria such as entity-to-entity relationships and current situations). An effective way to do this is to deploy an intelligent gateway to select/control communication sessions between home networks and the Internet. SIP Context Aware Gateway (SIP-CAG), serving as a SIP-based home application secretary, provides tremendous flexibility and capability in adapting to diverse environment situations. [9]

SIP-CAG supports the SIP-based context aware applications based on the inhabitant's motivations, behaviors and needs. The next figure shows the proposed SIP-CAG architecture that comprises the following major components: SIP Server and Presence Server, Stateful Inspection Layer (SIL), Content Filter (CF), SIP Application NAT (SAN), Feature Selector (FS), Policy of Context (PoC) and SIP-CAG Management Agent (SMA). Components are collaborated together to perform the context management. In the following paragraphs, we describe the function of each component in more detail.[9]

6 Conclusion

In developing and maintaining a Smart Home, Context awareness plays a big role. We discuss the context aware tools which can be implemented in Smart home. Although Smart Home brings a lot of advantages, there are also challenges that comes with it. One of the challenges of installing a smart home system is balancing the complexity of the system against the usability of the system.

References

1. Wikimedia - Context awareness, `http://en.wikipedia.org/wiki/Context_awareness` (accessed: May 2010)
2. Schmidt, A., Aidoo, K.A., Takaluoma, A., Tuomela, U., Van Laerhoven, K., Van de Velde, W.: Advanced Interaction in Context (PDF). In: Gellersen, H.-W. (ed.) HUC 1999. LNCS, vol. 1707, pp. 89–101. Springer, Heidelberg (1999), `http://www.teco.edu/~albrecht/publication/huc99/advanced_interaction_context.pdf` (accessed: June 2010)

3. Wikimedia - Context-aware pervasive systems, http://en.wikipedia.org/wiki/Context-aware_pervasive_systems (accessed: May 2010)
4. SmartHomeUSA.com, What is a Smart Home, http://www.smarthomeusa.com/info/smarthome/ (accessed: May 2010)
5. Redriksson, V.: What is a Smart Home or Building (2005), http://searchcio-midmarket.techtarget.com/sDefinition/,,0sid183_gci540859,00.html# (accessed: May 2010)
6. Edmonds, M.: How Smart Homes Work - Setting Up a Smart Home, http://home.howstuffworks.com/home-improvement/energy-efficiency/smart-home1.htm (accessed: May 2010)
7. Moon, A., Kim, M., Kim, H., Lee, K.-W., Kim, H.: Development of CAMUS based Context-Awareness for Pervasive Home Environments. International Journal of Smart Home 1(1) (January 2007)
8. Jin, M.-H., Fu, C.-J., Yu, C.-H., Lai, H.-R., Feng, M.-W.: I3BM Zigbee Positioning Method for Smart Home Applications. International Journal of Smart Home 2(2) (April 2008)
9. Cheng, B.-C., Chen, H., Tseng, R.-Y.: A Context Aware Gateway for SIP-based Services in Ubiquitous Smart Homes. International Journal of Smart Home 2(1) (January 2008)

Data Hiding a Key Management for Interoperable Urban Services

Maricel O. Balitanas and Taihoon Kim*

Hannam University, Department of Multimedia Engineering,
Postfach, 306 791
jhe_c1756@yahoo.com, taihoonn@empal.com

Abstract. Availability of a reliable urban services data is the key component for an industrialized area. Urban settings are challenging places for experimentation and deployment and along with its complexity insecurity also contributes a bigger challenge. To address such issues this paper has defined the implementation issues in integrating Geospatial services data and web services technologies and proposed a methodology of securing the systems. The proposition presented earlier is a symmetric encryption which is to share the common key for doing both encryption and decryption secretly, and periodically.

Keywords: Urban Services, Data Hiding, Cryptography.

1 Introduction

Urban computing is an integration of computing, sensing and actuation technologies into everyday Urban Settings and lifestyle. Such setting include in streets, pubs, squares, shops, cafes and buses. Urban settings are challenging places for experimentation and deployment, however, they remain little explored as pervasive environments for largely practical reasons. One reason is the complexity in terms of ownership; another is denseness as to what and who will participate in an application or system. People constantly enter and leave urban spaces, occupying it with high variable and changing usage pattern within a day. [1]

Despite the complexities in urban computing, yet it has already gained a widespread of interest among researchers. Almost half the world's population roughly lives in Urban Environment where PDA's and laptops are commonly own by most people and the phenomenal mobile phone which has capabilities beyond simple voice call. Connectivity is extensive, mobile phones are now equipped with Bluetooth for short-range communication, in addition to long-range cellular data connections. Wi-Fi networks are also commonplace and technology is put to more interesting use than one might at first think.

In this paper we have consider the previous research by [2] as the scenario and proposed a key management for data hiding in interoperable urban services. Chapter 1 on this paper provides an introduction and background. It will be followed by Chapter 2 which will tackle the Interoperable service for accessing Urban Service data.

* Corresponding author.

T.-h. Kim, A. Stoica, and R.-S. Chang (Eds.): SUComS 2010, CCIS 78, pp. 159–166, 2010.

Chapter 3 will be the Data type handling, chapter 4 will show the data hiding technique and lastly, the conclusion of this paper.

2 Interoperable Service for Accessing Urban Service Data

In any city, availability of a reliable urban services data is the key component for it to be considered a successful city management. Traditional public administration's inability to meet citizen's need affectively led a search for new methods that grant more efficient, effective and reliable provision of services to citizens. Because there are many users with distinct computing platforms and preferences, urban services data must be accessible using interoperable solutions. [2][3]

There are four types of users in the GI and IT communities. Table 1 shows the classification of users, the functionalities they require, their communication modes and the types of access they require in order to fulfill their responsibilities. All departments are mandated to comply on the set of predefined regulations for developing and implementing software systems and infrastructure such as using homogeneous platforms. Therefore there is significant need for standard-base interoperable services for accessing and processing USD for all categories of users who make up the GI and IT communities.

Table 1. Urban Services data and Platform

Users	Connectivity and Platform
Municipality Services Department	Homogeneous computing platform (Windows-based applications programming languages)
Other Municipality Organization	Municipality Internal Wide Network of fiber optic
Public and Private Sector and Academia	World Wide Web
Citizens	Homogeneous computing platform (Wide range of operating system applications, programming language and Web browsers)

3 Data Type Handling

Web Services use SOAP messages to communicate with each other. Open Geospatial Consortium services use a specific request and response formats. All OGC services request data using URL-encoded parameters in the HTTP request on in XML messages. [3] Both methods can be converted to SOAP messages using a simple wrapper solution. Problem arises in creating response messages when raw binary data must be retrieved from OGC service. Addressing such concern in data type handling issue, four approaches can be followed:

3.1 Translator Web Service

The translator Web Service approach is a simple Web service which receives the client's request and translates it into the equivalent URL-encoded string for OGC services. The URL-encoded string is then sent back to the client in the SOAP response from the Web Service. The client must then use this URL-encoded string to retrieve the data directly from the OGC service. In such case OGC service ask the server engine to generate binary data using the request parameters and then sends this back to the client. [4]

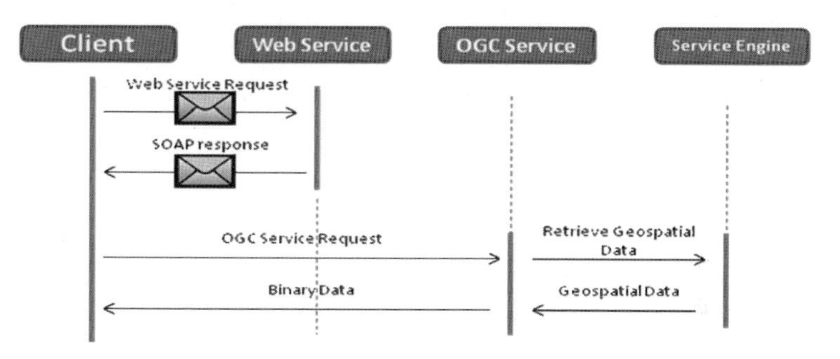

Fig. 1. Translator Web Service Approach

3.2 Physical Web Service

SOAP request for a Web service is translated by the Web service and sent to the OGC service. The OGC service asks the server engine component to generate a binary file and then stores the binary file on a permanent location on the server. [4]After, the physical address of the binary file is dispatched to the Web service as the response from the OGC service. Lastly, the client retrieves and stores the binary file using its physical address. As shown in the next figure.

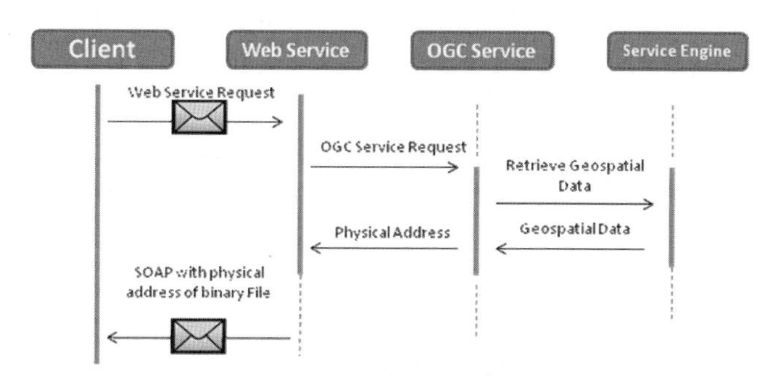

Fig. 2. Retrieving file using physical address

3.3 Wrapper Web Service

Next approach is a Web service completely wraps around an OGC service. The response from the OGC service is retrieved by the Wrapper Web Service and then resent to the client using Web service-specific encoding. Using a Wrapper Web Service makes all communications between the client and Web service Web service based.[5] In this case, wrapping the OGC service with a Web service requires that binary data are included in the Web service's response as shown in the next figure.

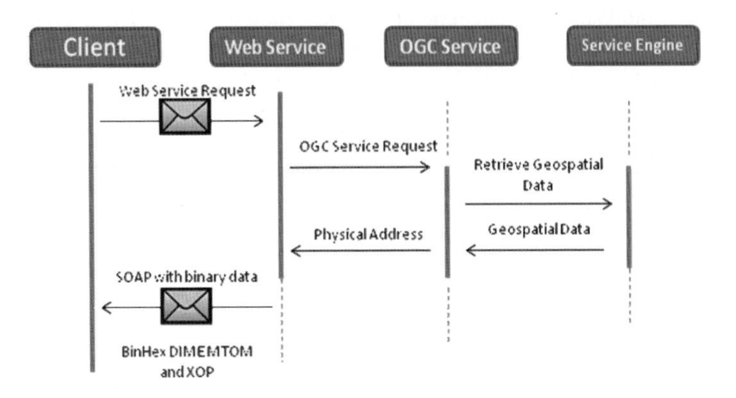

Fig. 3. Web Service Technique in BinHex, DIME and MTOM

Binary data can be transmitted using SOAP response. It can be encoded in BinHex encoding. Using BinHex, binary data can be encoded to ASCII character and then included in any XML document, like SOAP messages. In this case, the Wrapper Web Service encode data retrieved from OGC service and returns the Web service's response to the client. The client must then decode the BinHex data before using it. A more efficient approach is to use Direct Internet Message Encapsulation to send a large amount of binary data. DIME is a message When BinHex encodings are used, the Web service and client processing become complicated and time-consuming when attachments are very large. It has been proven that BinHex encoded data expand by a factor of 2 [6] Dime protocol describes a mechanism for sending one or more attachment externally to the SOAP message. By keeping the attachment and the SOAP message separate, the processing load can decrease dramatically. Further, given the nature of parsing a SOAP message, he entire message would need to be read into memory before any internally included data could be decoded and used; for large binary files this would require large amounts of memory on both the client and Web service sides. On the other hand, an XML parser can read the SOAP header to locate the SOAP message or specific attachments without loading the whole message, thus providing better performance and scalability. MTOMis used to optimize the transmission of a SOAP message by selectively encoding portions of the message while still presenting an XML infoset to the SOAP application. In other words MTOM is utilized to optimize the transmission of large binary messages while keeping them in accordance with other Web services extensions. MTOM is able to transmit binary data as a raw bytes, saving encoding/decoding time and resulting in smaller messages. MTOM thus provides a better approach than DIME. When DIME is used, the data are external to the SOAP messages.

3.4 Common Backend

The OGC service provides OGC specific encodings as responses (like GML and binary image data), while the Web service provides XML-based and Web service compliant responses (Fig. 4). In this case, the Web service can provide a response to the Web service client with multiple options, such as including the physical address of the binary file, encoding binary data with Base64 or BinHex, using optimized techniques like XOP and MTOM or even delivering any other specific formats like KML and GeoRSS.

Web browsers, along with more limited clients, such as cell phones or other mobile devices), it is important to recognize that—in many cases—the Presentation will be physically separated from the UI logic. Business logic is the combination of validation, edits, database lookups, policies, and algorithmic transformations that constitute an enterprise's way of doing business. More simply, the Business Logic layer includes all business rules, data validation, manipulation and security for the application. It is important to recognize that a typical application will use Business Logic in a couple of different ways. Most applications have some user interactions, such as allowing the user to draw a new shape (and providing snapping capabilities) or displaying a form in which the user views or enters attribute data of a feature into the system (where the data are validated as soon as the user enters them) or edits existing data.

Moreover, most applications have some non-interactive processes, such as calculating statistical parameters (which need whole data to be analyzed), posting a newly-added feature to a database or comparing two versions of the same dataset. Ideally, if the Business Logic layer is used in a rich and interactive way, it can be physically deployed on the client workstation (for desktop applications) or on the Web server (for Web applications) to provide the high levels of interactivity that users desire. To support non-interactive processes, on the other hand, the Business Logic layer often needs to be deployed on an application server, or as close to the database server as possible. The Data Access layer interacts with the Data Management layer to retrieve, insert, update, and remove information. The Data Access layer does not actually manage or store the data; it merely provides an interface between the Business Logic and the database. Logically defining Data Access as a separate layer enforces a separation between the Business Logic and any interaction with a database (or any other data source). This separation provides the flexibility to choose later whether to run the Data Access code on the same machine as the Business Logic, or on a separate machine thus providing the flexibility to deploy software application as a standalone or network-based application or even providing high scalability using clustering and Web farm concepts. It also makes it much easier to change data sources without affecting the whole application. This is important because it enables the use of third party services that provide geospatial data, instead of using centralized or distributed geospatial data servers within an organization (an approach that is encouraged in the GI community by concepts such as spatial data sharing, spatial data warehouses and spatial data infrastructures). The Data Access layer is typically implemented as a set of classes, with each class containing methods that are called by the Business Logic layer to retrieve, insert, update, or delete data. Finally, the Data Storage and Management layer handles the physical creation, retrieval, update, and deletion of data. This is different from the Data Access layer, which requests the creation, retrieval, update, and deletion of data. The Data Management layer actually implements these operations within the context of a database or a set of files.

4 Data Hiding

Encryption is the most common and powerful approach to secure the system. There are two fundamental alternatives for the location of encryption gear or device: Link encryption and end-to-end encryption. There are so many cryptographic methods already developed for encryption and decryption of information. Those cryptographic algorithms can briefly be categorized into two different groups: symmetric and asymmetric encryption approaches. Symmetric encryption algorithm is characterized by the fact that the decryption key is identical to the encryption key. Symmetric encryption, also referred to as conventional encryption or single-key encryption, was the only type in use prior to the development of public key encryption in the 1970s[7]

The key should be exchanged in advance between sender and receiver in a secure manner and kept secret [8], which is seen in Fig. 4

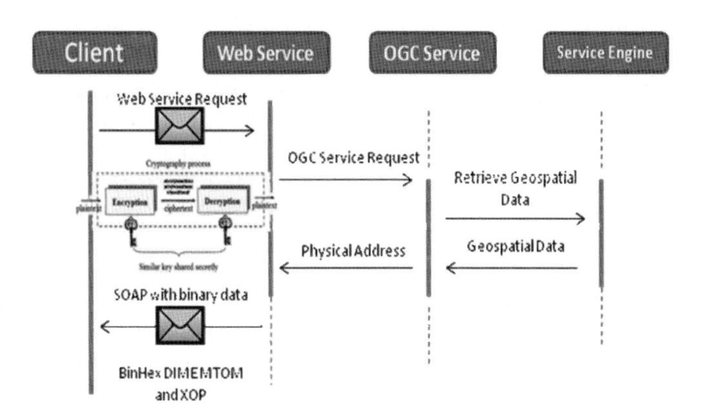

Fig. 4. Symmetric Encryption

In the symmetric encryption, two parties involved in communication should share the same key and that key should be protected from the accesses by others [7]. There are several ways for this key distribution. Here two kinds of key distribution are presented. The first one is that the communication initiator makes the key and sends it to the responder. This method is called a decentralized key distribution, which is portrayed in Fig. 5. There is no key distribution center in this method. Initiator A requests B to send the session key by r and B responds to A with the key encrypted with the master key already shared with A by s. And then A confirms the key distribution process by t. The second one is that the third party makes the key and distributes it to the initiator or both of them, which is called a centralized key distribution in Fig. 6. The flow her is notated with solid lines and Arabic numbers indicate the first method in which the key is distributed only to the communication initiator A, while the one with dotted line and alphabets stands for the second method in which the key is distributed to both parties related to communication. In this case, we call for the KDC (Key Distribution Center) to carry out the key management. In the Internet environment, it is better to enable the KDC to manage the key distribution process since there are lots of hosts in the network and thereby more communication combinations.

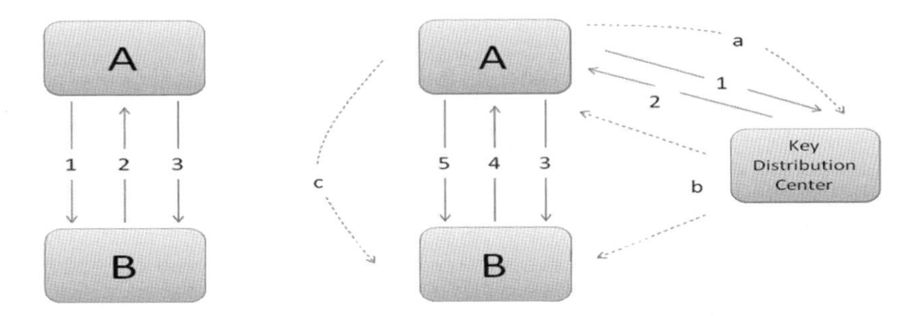

Fig. 5. Decentralized key distribution

Fig. 6. Centralized key Distribution

4.1 Determination of Key Distribution Period

There is an on-going conflict between system performance and security strength. In this paper, system performance is deliberated as a function of time delay in communication when the encryption process is added to the communication between the client and the Services. Also, the reinforcement of security enhances the network performance by reducing the risk of potential attacks. Viewed in this term, there should be balance in relations with two factors though there is the difficulty of adjusting two factors on account of different dimension. We aim to unify the units of two values which are required to be commensurable. Moreover, we need to define the QoS function composed of network traffic part and security part, which represents the overall quality of service in communication.

$$QoS = PI + SI \tag{1}$$

PI and SI designate performance index and security index respectively. PI is computed based on the time delay caused by the encryption process and the value of PI is distributed between 1 and 0. There may be a couple of reasons for communication errors. However we simply owe it to the delay by encryption and decryption processes. Suppose that there is a functional relationship between the communication delay and the key distribution period. As the key distribution period gets shorter, the communication delay will accordingly be longer since the frequent key changing increases the network traffic therein.

In Figure 7, α is inversely proportional to t_d or T/ td, where T is the period of communication in the Network systems. To normalize the value, it becomes $(T-\alpha)/T$. When α reaches T, the performance value converges to 0, which means the communication failure. Wheen it is assumed that the functional relationship between

α and T_d is provided in a form of time $\alpha = k / t_d$ where k is a constant, the performance index related to the time delay incurred by the encryption process is implemented.[9]

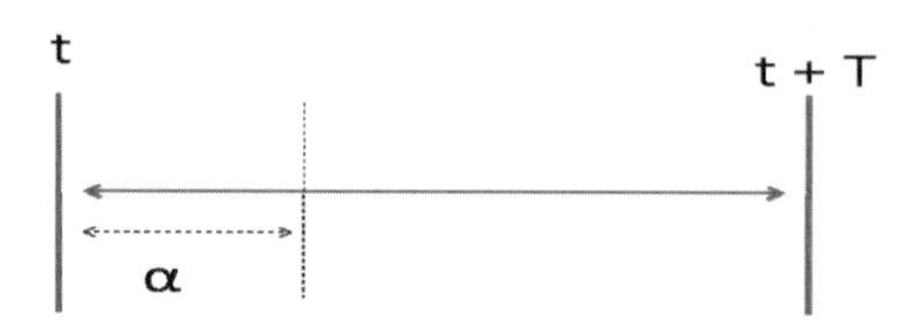

Fig. 7. Delay tie by encryption process

5 Conclusion

Availability of reliable and secure urban services data is a key component of a successful city management. To provide an interoperable solution should be integrated efficiently. This paper has defined the implementation issues in integrating Geospatial services data and Web services technologies and proposed a methodology of securing the systems. In conclusion, the proposition presented earlier is a hot issue in the symmetric encryption which is to share the common key for doing both encryption and decryption secretly, and periodically. Given that this key is used for a long time, it increase the risk to be cracked and takes it for granted to change the key frequently enough. Moreover, from our point of view is the decision on the optimal key distribution period based on the QoS function for maximizing the network performance and security within a permissible level.

References

1. Hadar, E., Perreira, M.: IEEE International Conference on Web Services, ICWS 2007 (2007)
2. Erl, T.: Service-oriented architecture: Concepts, technology, and design, p. 792. Prentice Hall PTR, Englewood Cliffs (2005)
3. Gailey, J.H.: Understanding Web services specifications and the WSE, p. 272. Microsoft Press, Washington (2004)
4. Newcomer, E., Lomow, G.: Understanding SOA with Web services. Addison Wesley, Inc., Maryland (2005)
5. Rammer, I.: Advanced.NET Remoting, 2nd edn., p. 608. Apress Publishing, California (2005)
6. Box, Bosoworth, Gudgin, Nottingham, & Orchard (2003); Lawrence (2004)
7. Stallings, W.: Cryptography and network security: principles and practices, 4th edn. Prentice Hall, New Jersey (2006)
8. Dzung, D., Naedele, M., Von Hoff, T.P., Crevatin, M.: Security for industrial communication systems. Proc. IEEE – Special Issue Indust. Commun. System 93(6), 1152–1177 (2005)
9. Kokai, Y., Masuda, F., Horiike, S., Sekine, Y.: Recent development in open systems for EMS/SCADA. Int. J. Electric Power Energy Syst. 20(2), 111–123 (1998)

Machine-Type-Communication (MTC) Device Grouping Algorithm for Congestion Avoidance of MTC Oriented LTE Network

Kwang-Ryul Jung[1], Aesoon Park[1], and Sungwon Lee[2]

[1] Electronic and Telecommunications Research Institute, Korea
krjung@etri.re.kr
[2] Department of Computer Engineering, Kyung Hee University, Korea
drsungwon@khu.ac.kr

Abstract. Machine-Type-Communication (MTC) is a new paradigm in mobile wireless network domains such as Mobile WiMAX (IEEE 802.16e) and 3GPP LTE (3rd Generation Partnership Project Long Term Evolution). We explain the background for MTC environments, and its key issues. Then we focus on the uplink traffic aggressiveness characteristics in major applications such as Smart Grid. Then, we propose a new congestion avoidance algorithm to reduce the congestion of the uplink intensive applications.

1 Introduction

In these days, mobile wireless communication is widely used in human communications such as voice call, messaging, and web browsing. However, these kinds of services and technologies are matured and new kinds of services and technologies are requested. Among these requests, MTC is the hottest issue in the standardization and industry areas.

MTC means the communication by machines. It is expected that among total 6 billion people in the world, 2.7 billion people will use mobile phone at the year 2020. However, 50 billion machines are expected to use wireless/wireline communication technologies at the year 2020. Already, millions machine devices use cellular technologies in telematics, security, automatic meter reading, payment, and vending machines[1][2].

- Telematics: Tracking and tracing of vehicles. Comprehensive solution consisting of 2G/3G cellular technologies, Global Positioning System (GPS) and Radio Frequency Identification (RFID). Status of individual packets such as temperature, unloading time/place are continuously sent to the back office
- Security: Pan-European service (eCall) based on a single European Emergency number – 112. Audio/Video call together with Minimum Set of Data (MSD). Basis for additional vehicle data concepts such Intelligent Transport System (ITS) [3]

T.-h. Kim, A. Stoica, and R.-S. Chang (Eds.): SUComS 2010, CCIS 78, pp. 167–178, 2010.

- Smart Metering: Automated two way communication between utility meters and utility company's backend IT infrastructure. Remote data transmission and remote control.
- Payment: "Pay as you drive" service which establishes information such as travel time and distance, recording of accident data, transmission of data to insurer, and tracking of stolen vehicles.
- Vending Machines: Vending machine with an integrated cellular module enables a smart service. Vending machine can be remotely monitored via a cellular connection to actively check for fill level or a possible malfunctioning.

According to these trends, mobile broadband technologies can't enable whole these things by alone. Heterogeneous network technologies such as 4G mobile broadband technologies (Mobile WiMAX, LTE, etc), 2G/3G cellular technologies, WLAN/WPAN technologies, sensor technologies should be integrated to take advantage of individual technology strength and efficiency.

In this paper, we will explain the state of the art of the machine communication in mobile broadband communication domain, and its service examples and requirements in chapter.2. Then, we explain the congestion issues in chapter.3. The proposed solution for congestion is explained in chapter.4. Performance of the proposed method will be explained in chapter.5. Finally, we conclude this paper in chapter.6.

2 The State of the Art of MTC

Smart Grid and Automated Metering are the most driving forces for MTC environments.

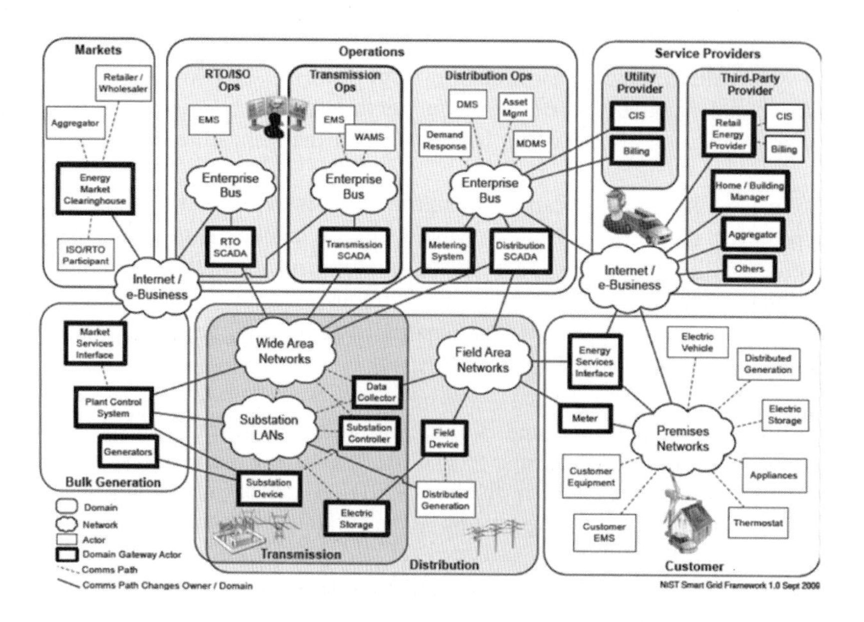

Fig. 1. NIST Smart Grid Framework

2.1 Smart Grid and Automated Metering Activities

Smart Grid is a hottest issue in the communication area in these days. Smart Grid has several definitions. However, Smart Grid can be conceptually defined as the intelligent power supply network with two-way wireless and/or wireline communication technologies to increase the productivity and efficiency.

National Institute of Science and Technology (NIST) takes in charge of the national policy of Smart Grid in US. And, NIST considers mobile broadband technologies such as 2G/3G/4G as key enablers for Smart Grid networks as in Fig.1 [4][5][6].

Electric Power Research Institute (EPRI) also considers mobile broadband technologies such as Mobile WiMAX, GPRS and LTE as key enablers for automated metering infrastructure (AMI) as in Fig.2 [7].

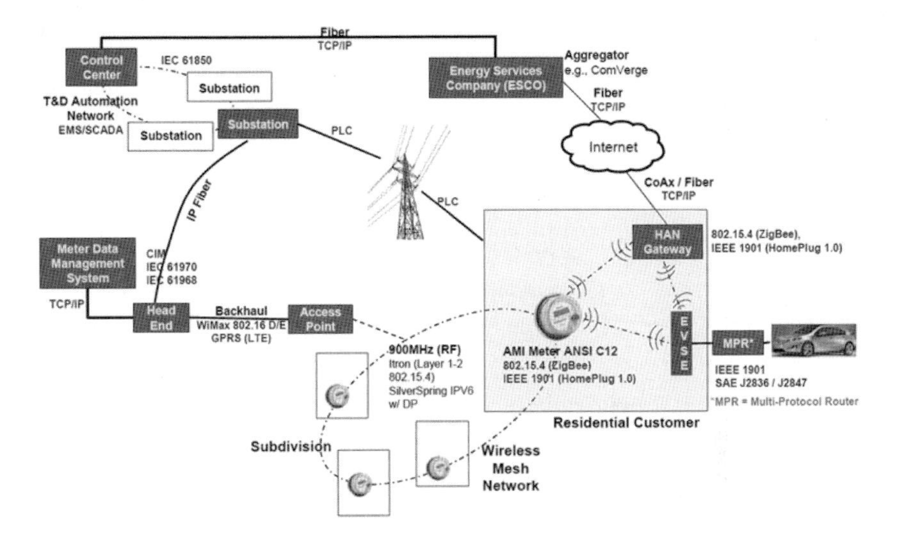

Fig. 2. EPRI AMI Architecture

2.2 ITU and 3GPP Standardization Activities

International Telecommunication Union (ITU) and 3GPP are already pushed by conventional Machine-to-Machine (M2M) and MTC commercialization. According to the Netherlands report, in the near future, it is very likely that the use of M2M communication will grow significantly. This growth will most likely take place in various different applications that will require M2M communication, and it is expected that many of such applications will be large in scale. For example, an application such as smart energy metering or eCall , which might become widely used in the near future, could lead to around six million connections for each of these applications in a country like the Netherlands[8][9][10][11][12].

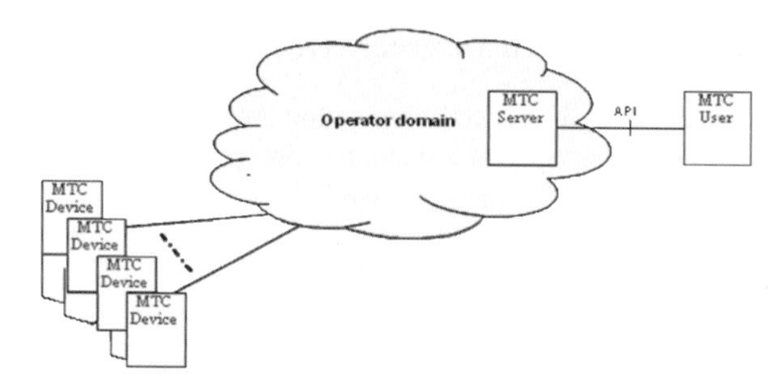

(a) MTC Device & Operator MTC Server Scenario

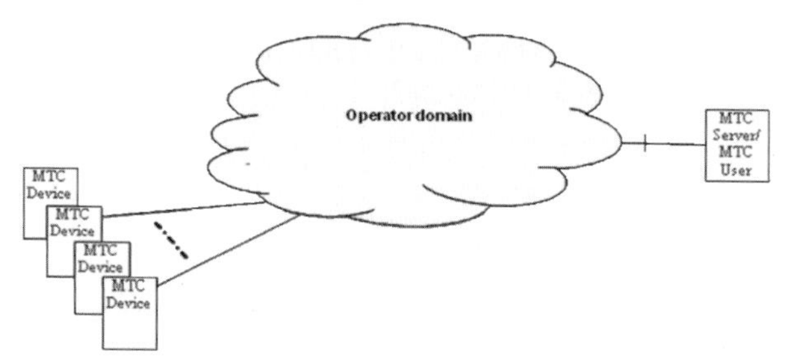

(b) MTC Device & ISP MTC Server Scenario

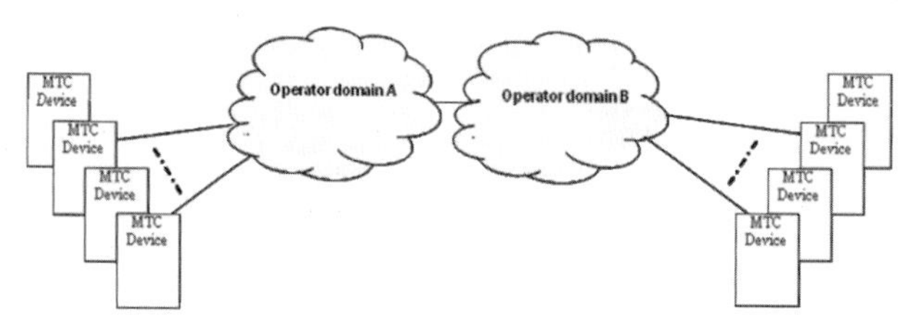

(c) MTC Device Only Scenario

Fig. 3. MTC Communication Scenarios

A. MTC Network Architecture

3GPP checked several feasibility points over years and published the 3GPP TR 33.812 as the results of the feasibility study on the security aspects of remote provisioning and change of subscription for M2M equipment. Also, in Release 10, 3GPP published TS 22.368 and TR 23.888. Several MTC service scenarios are considered and defined in these documents.

Actually, 3GPP considers three scenarios for MTC communications. First scenario is depicted in (a) of Fig.3. It shows the communication scenario with MTC devices communicating with MTC server. MTC server is located in the operator domain. Second scenarios is depicted in (b) of Fig.3. It shows the communication scenario with MTC devices communicating with MTC server. MTC server is located outside of the operator domain. Third scenario is depicted in (c) of Fig.3. MTC devices communicate directly with each other without intermediate MTC server. But, Third scenario is not considered in 3GPP specification. Thus, MTC devices should communicate with MTC server at this time in 3GPP standardization movement.

The end-to-end application, between the MTC device and the MTC server, uses services provided by the 3GPP system. The 3GPP system provides transport and communication services including 3GPP bearer services, and IP multimedia subsystem (IMS), and short messaging service optimized for the machine type communication. MTC device connects to the 3GPP network. MTC device communicates with a MTC server or other MTC devices using the 3GPP bearer services. The MTC server is an entity which connects to the 3GPP network and thus communicates with MTC devices. MTC server may be an entity outside of the operator domain or inside an operator domain.

Table 1. MTC Service Examples

Service Area	MTC applications
Security	Surveillance systems Backup for landline Control of physical access Car/driver security
Tracking & Tracing	Fleet Management Order Management Pay as you drive Asset Tracking Navigation Traffic information Road tolling Road traffic optimisation/steering
Payment	Point of sales Vending machines Gaming machines
Health	Monitoring vital signs Supporting the aged or handicapped Web Access Telemedicine points Remote diagnostics
Remote Maintenance/ Control	Sensors Lighting Pumps Valves Elevator control Vending machine control Vehicle diagnostics
Metering	Power Gas Water Heating Grid control Industrial metering
Consumer Devices	Digital photo frame Digital camera eBook

Table 2. MTC Service Characteristics

Characteristics	Description
Low Mobility	Do not move, move infrequently, Move within a certain region
Time Controlled	Tolerate to send or receive data only during defined time interval
Time Tolerant	Delay data transfer
Packet Switched (PS) Only	Only require packet switched services
Small Data Transmissions	Send or receive small amounts of data
Mobile Originated Only	Only utilize mobile originated communications
Infrequent Mobile Terminated	Mainly utilize mobile originated communications
MTC Monitoring	Monitoring MTC device related events
Priority Alarm Message (PAM)	Issue a priority alarm in the event of theft or other immediate attention
Secure Connection	Require a secure connection between the MTC server and services
Location Specific Trigger	Trigger MTC devices which are known by the MTC application in certain area
Network Provided Destination for Uplink Data	All data from an MTC device to be directed to a network provided destination
Infrequent Transmission	Send or receive data infrequently
Group Based MTC Features	Group based collection of MTS features
Group Based Policing	Group based management of MTS policies
Group Based Addressing	Group based addressing of MTS identification

B. MTC Services Examples

Expected MTC services by 3GPP are summarized in Table.1. Smart Grid service can be mapped to the Metering service. Most services are already conceptually explained in chapter.1. Especially, Consumer Electronic (CE) devices are also considered in MTC communication area.

C. MTC Service Characteristics and Requirements

MTC services show different characteristics over conventional human oriented service, and its major characteristics are summarized in Table.2.

Some MTC devices are fixed or low mobility. For example, most Smart Meters at home don't move as located at each home.

Most scenarios assume broadcast or multicast transfer from the MTC server to MTC devices which means, in most cases, the same information is transferred from the MTC server to all MTC devices. Thus, "group based MTC device management and downlink communication" may be the characteristics for most MTC scenarios. However, MTC devices transfer own information to the MTC server independently through uplink communication.

According to uplink communication, the mobile originated information transfer from the MTC device to the MTC server may be the general case for MTC environments, and it is considered as the most of unicast traffic in the network between MTC devices and MTC server. Also, MTC devices are considered as packet only devices with small data transmission such as the metered information per month.

3 Network Congestion Issues

MTC Network congestion can be happed in areas: radio network, signaling network and core network[12].

Congestion is caused by "the Single Consumer and the Multiple Generator" network characteristics, where multiple MTC devices generate information and transfer to the single MTC server at the same time.

3.1 Radio Network Congestion

Mass concurrent data transmission takes place in some MTC applications. For example, all sensors can transmit the data almost simultaneously likes bridge monitoring with a mass sensors, hydrology monitoring during the heavy rain, building monitoring for intruder detection. Thus, the optimization for a mass simultaneous transmission is required to reduce the radio network congestion.

3.2 Core Network Congestion

Despite of the avoidance of radio network congestion, congestion over core network, MTC server, and core network links can be caused by the simultaneous transmission from the large group of MTC devices as in Fig.4.

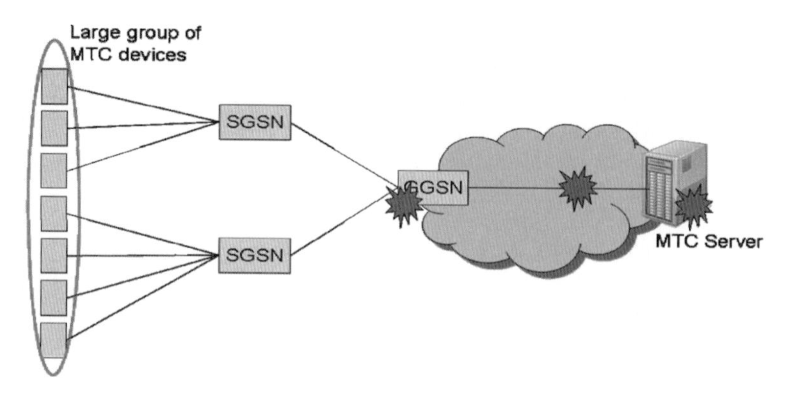

Fig. 4. Core Network Congestion Issues

3.3 Signaling Network Congestion

Signaling network congestion is triggered when large machines try signaling simultaneously. Simultaneous attach, activation, modification, and deactivation cause signaling overload over signaling network elements such as SGSN, HSS, MME and other signaling devices as in Fig.5.

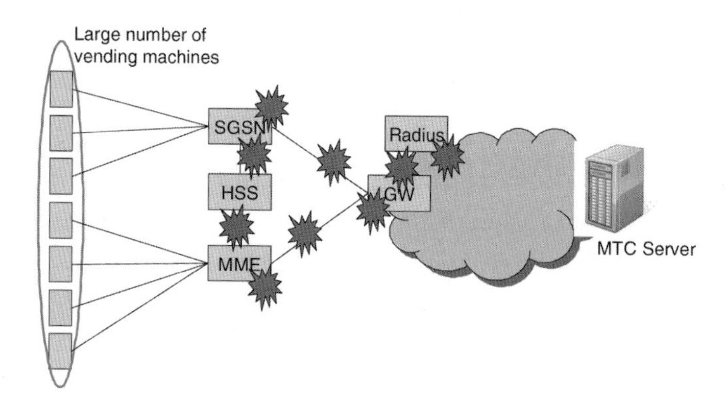

Fig. 5. Signaling Network Congestion Issues

4 Proposed Congestion Avoidance Method

We propose the co-operative manipulation and report method for aggressive MTC sensor device environments. In the proposed method, MTC devices are metering devices, and transfer metered information to the MTC server periodically.

4.1 Basic Assumptions

We target to the traditional MTC service scenarios where multiple MTC devices and a single MTC server are considered. And the MTC server is located outside of the operator domain.

MTC devices are some type of reporting devices such as smart meters, sensor devices, and vital sign detectors. These devices are located at dense areas.

MTC devices support multiple communication technologies both mobile broadband technology (such as Mobile WiMAX, LTE), and local area networking technology. Local area technology can be wireless personal area network (WPAN), and other coming technologies such as power line communications (PLC).

In this paper, we assume that MTC device supports both LTE and Bluetooth (or PLC).

4.2 Group Construction and Leader Selection

MTC devices at dense area construct the managed group by using Bluetooth (or PLC) technology as in Fig.6. The managed group can be constructed by single hop limited or multi hop constrained principles. For the grouping, simple protocol was proposed to find each other, and make a group within a local communication area. Also, the group leader selection algorithm is designed based on communication capability, communication link quality, storage status and battery status of each node. Group leader takes in charge of the mobile broadband communication, and other group member can transfer and receive information to/from the MTC server through the group leader.

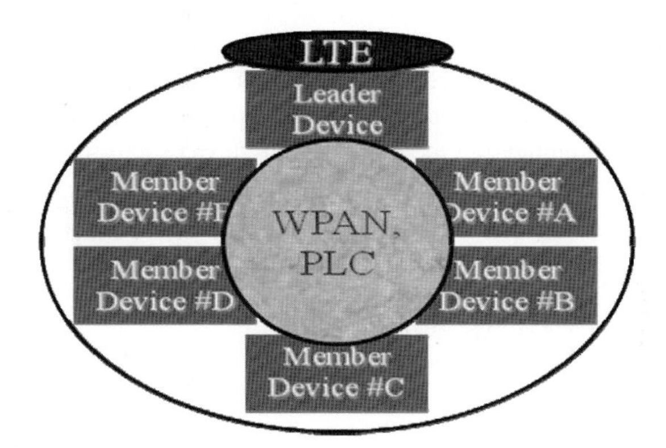

Fig. 6. Node Grouping and Group Leader

4.3 Congestion Avoidance via Cooperative Transfer

MTC server and MTC devices communicate via LTE communication link through the group leader. By using a LTE communication link, MTC server can gather information and manage the MTC device group. Thus, MTC server should manage the table which describes the relationship between a group leader and its group members.

Most important role of the group leader is an aggregation of the uplink message transfer as in Fig.7. In aggregation process, MTC devices transfer uplink information to group leader, and a group leader does not transfer the received information immediately to MTC server. However, a group leader accumulates the received uplink traffic to the MTC server at its own internal buffer, and periodically merges the messages to efficient format for mobile wireless link.

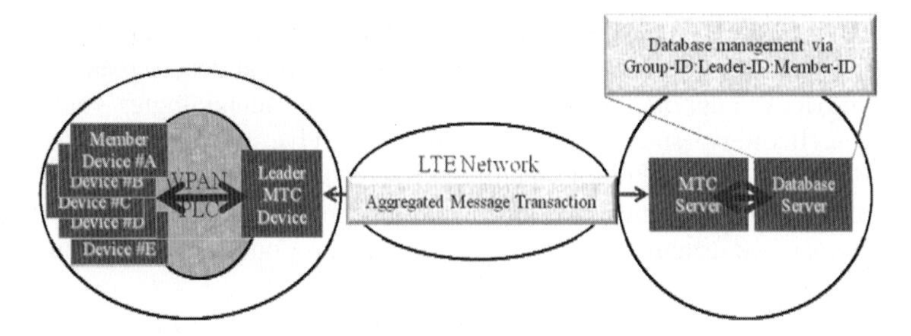

Fig. 7. Aggregated Uplink Transfer to MTC Server

Not only for uplink, but also downlink can be more efficient than conventional 1:1 communication between MTC devices and MTC server. When MTC server wants to transmit the same information to multiple MTC devices, MTC server just send a single message to a group leader, then group leader distributes the information for group members. Thus, downlink traffic can be more efficient for MTC environments.

5 Performance Evaluation

We evaluate and analyze the performance of the proposed method over LTE and Bluetooth based MTC device/server type scenarios. Especially for evaluation, we use the OMNeT++ network simulator. We compare the performance with the conventional 1:1 communication method between devices and server.

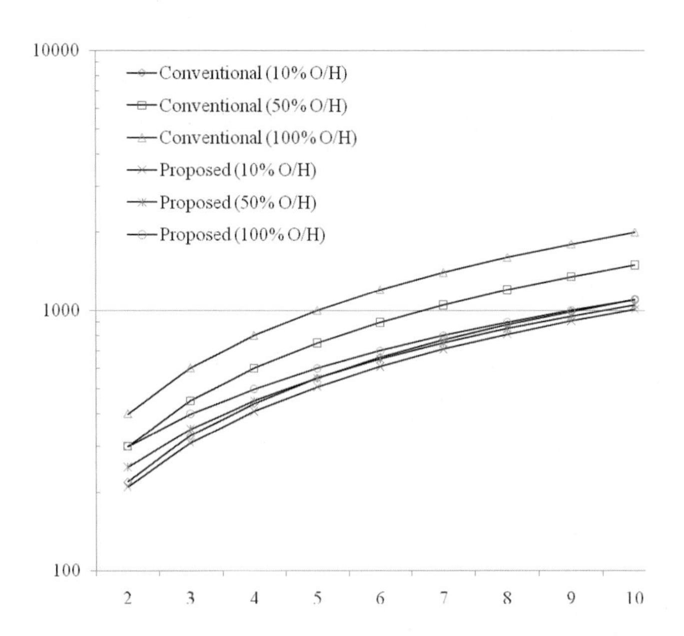

Fig. 8. Traffic Volume over Uplink Connection

Performance metrics are considered as mobile wireless communication link utilization, and power saving effect of MTC devices including a group leader and group members. Performance results show that the proposed method shows enhanced power saving effect with efficient link utilization. However, as we can easily expect, delay performance is increased due to the group leader buffering and processing delay.

Average traffic volume over uplink connection is described in Fig.8. It is calculated as the summation of the total uplink transmission per transaction. X-axis means the number of MTC devices. Conventional approach means all MTC device transfer independently through LTE link. In the performance evaluation we vary the overhead information relative to traffic size. As the results show that the proposed algorithm efficiently uses the LTE link. Also, the proposed method shows better performance when the overhead size is restively large in transactional services.

6 Conclusion

M2M or MTC is the most important issue in mobile broadband network such as Mobile WiMAX and LTE as a key enabling technology for the coming decades. However, its service requirements are not clearly researched at these days, and analyzed characteristics show that the fundamental evolution or change is required for conventional 3G/4G technologies. Especially, when we consider MTC device and MTC server scenarios as in 3GPP, where intensive uplink traffic can be easily expected, possible congestion points are not only limited to radio network but it includes core and signaling networks.

In this paper, we analyze the state of the art of the machine type communication in mobile broadband communication domain, especially for 3GPP standardization issues. Then, we analyze and summarize MTC service examples and requirements. Based on MTC service examples, we clarify the congestion issues in radio network, core network and signaling network. Then, we propose the co-operative manipulation and report method for aggressive MTC sensor device environments. The proposed method reduces the network load at uplink link by using a grouping method at dense area local network. Group leader is selected based on parameter metrics, and it connects group members with MTC server.

Based on performance evaluation, the proposed method shows better performance over conventional approach.

Acknowledgements. This work was supported by the IT R&D program of MKE/KEIT. [KI001484, Research on service platform for the next generation mobile communication]

References

1. Allied Business Intelligent, Cellular M2M Connectivity Service Providers (2010)
2. ABI, The Cellular M2M Module Market (2007)
3. Safety Forum, Recommendation of the DG eCall, for the introduction of the pan-European eCall (March 15, 2006)

4. NIST, NIST Framework and Roadmap for Smart Grid Interoperability Standards Release 1.0 (Draft) (September 2009)
5. ATIS (by Dr. George W. Arnold), Response for General Comments NIST Framework and Roadmap for Smart Grid Interoperability Standards Release 1.0 (Draft) in Docket No.0909291327-91328-01 (November 9, 2009)
6. ATIS, Response to the Federal Communications Commission's (Commission) above-referenced Public Notice released (September 4, 2009) (October 2, 2009)
7. EPRI, Electric Transportation Update (June 30, 2009)
8. The Netherlands, Machine-to-Machine consequence for number resources E.164 and E.212, ITU Telecommunication Standardization Sector, Study Group 2 – Contribution 63 (November 2009)
9. 3GPP TR 33.812, Feasibility Study on the Security Aspects of Remote Provisioning and Change of Subscription for M2M Equipment (December 2009)
10. 3GPP TR 22.868, Study on Facilitating Machine to Machine (December 2009)
11. 3GPP TR 23.888, System Improvements for Machine-Type Communications (March 2010)
12. 3GPP TS 22.368, Service Requirements for Machine-Type Communications (April 2010)

An Architecture for the Emotion-Based Ubiquitous Services in Wearable Computing Environment

Haesung Lee and Joonhee Kwon

Department of Computer Science, Kyonggi University
San 94-6, Yiui-dong, Yeongtong-ku, Suwon-si, Gyeonggi-do, Korea
{seastar0202,kwonjh}@kyonggi.ac.kr

Abstract. The exhibition of informatics services does not always attend to the context that involves the user's physiologically emotional signals. Wearable computing services with user's emotional information could be extremely effective in various applications. In this paper, with the aim of developing more effective human-centric services, we propose a novel methodological architecture. Proposed architecture employs the convergence of wearable computing technique, human's emotional information and context tagging technique to develop more realistic and robust human-centric service.

Keywords: emotion based service, wearable computing, ubiquitous computing.

1 Introduction

Today, many techniques such as context-aware computing and wireless network computing significantly enhance the functionality of ubiquitous computing services and applications, and enrich the way they interact with users and resources in the environment. Many people in ubiquitous computing environment think that the device should be intelligent enough to understand our needs in order to answer our dynamically changed needs. For instance, if we are in an unknown place and we are hungry we want that our smart phone help us in the process of choosing a restaurant according to our gastronomic preferences. For doing this it is necessary to give the devices the ability to interact with the environment and at the same time the ability to search for information. These features are related with context-aware computing. However, there are no realistic and robust human-centric services so far.

With the improvement of human's life, the advent of the mobile era makes human-centric service technologies to be more needed. For development of those technologies, human's emotion is focused from many research field such as HCI(human-computer interaction), wireless advertising and mobile contents service[1]. Also, together with the remarkable development of wireless sensor network, wearable computing is being important technique for human-centric service in ubiquitous computing environment [3].

The objective of this paper is to propose a novel methodological architecture that employs the convergence of wearable technique, human's emotional context and tagging technique of web in order to develop more realistic and robust human-centric service.

T.-h. Kim, A. Stoica, and R.-S. Chang (Eds.): SUComS 2010, CCIS 78, pp. 179–187, 2010.

2 Related Works

2.1 Previous Works of Wearable Computing Environment

With the development of wireless computing and the miniaturization of electrical component, wearable computing systems have drawn a lot of attention from the numerous and yearly increasing corresponding research as one of the key techniques in ubiquitous computing environment. Because of increasing need to personalized services, there has been a need to wearable computing system in order to provide more personalized services to a person [4]. To address this demand, a variety of system prototypes and commercial products such as a smart cloth and an armband have been produced in the course of recent years, which aim at providing real-time feedback information about one's physical condition [5][6].

Especially, for efficient healthcare monitoring system, wearable computing techniques are mainly focused from most researchers of bio informatics domains. Many medical informatics researches like [2] and [3] propose WHMS (wearable health-monitoring systems) which constitute a new means to address the issues of managing and monitoring patients, and persons with their physical condition information.

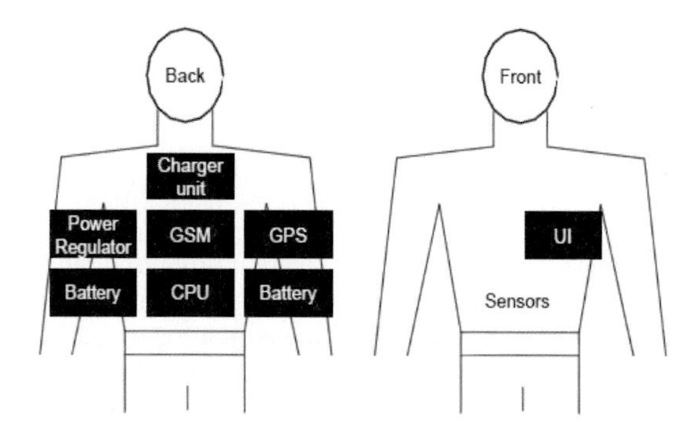

Fig. 1. The Architecture of the smart cloth prototype

Beside medical informatics domain, wearable systems can be integrated into cloths [5]. Smart clothing is made from fabrics that are wireless and washable that integrate computing fibers and materials into the integrity of the fabrics. Reima-Tutta Corporation has introduced the world's first large scale smart clothing prototype, which is intended for the survival in the arctic environment [7]. Figure 1 shows the architecture of the prototype of Reima-Tutta Corporation. The addition of sensors to a cloth allows for the wearable's behavior to be related to the wearer's current activity or situation.

Most applications in ubiquitous computing environment have to provide personalized services to each user without restrictions of the time and locations. However, majority of the existing systems such like what we introduce above have limits to provide human-centric information services because they not consider any ways to provide proper contents to a user who potentially wants to find those contents.

2.2 The Use of Human's Emotion Information

Up to the present, the categories of basic emotion for human being have been completely defined with psychological theory or psychological experiments [8][9]. An emotion is characterized as a set of reactions that a human being has when facing several situations. These reactions, which vary from individual to individual, are influenced by each person's personality by the way a person observes the world around him as well as by his emotional state at the time when those situations take place. Emotions have an important influence on the life of the human being, influencing many aspects of biological functioning, psychological functioning and social behavior. Therefore, if we want to develop more real and credible personalized service applications, the use of user's emotional information is a much effective approach. Several studies were performed aiming recognizing and simulating emotion in computer system. [10][11][12]. Rosalind Picard studies the use of emotions in informatics systems from the recognition, representation and simulation to the research that involves emotions in human-machine interactions [13].

Despite the increasing need to human centric services, however, there are much few applications which provide various human centric services in these days. For this, we propose the more useful architecture for various human centric services in wearable computing environments.

The use of user's physiological signals for the development of efficiently personalized service system is very powerful approach to capture user's current emotional status. Therefore, capturing user's current emotional status, we can not only provide more personalized service each person but consider more various application scenarios.

3 Wearable System for Emotion Based Service

For useful emotion based service, wearable system has to be needed. A wearable system in which various sensors are embedded senses physiological data, determines user's emotional status and triggers different service applications based on emotion.

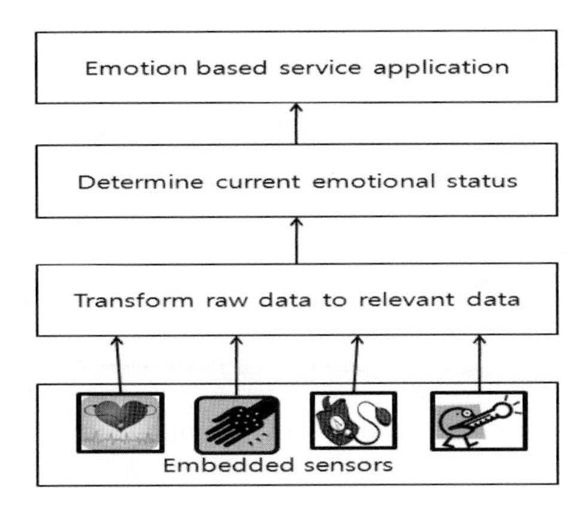

Fig. 2. Four layered wearable system

In this chapter, we define a wearable system which is composed of four layers. Figure 2 represents four layers of wearable system. The basement of Figure 2 represents the sensor part which consists of various sensors. Those Sensors collect different types of information by sensing physical signals from a user. The third layer in Figure 2 transforms collected raw sensor data to the valid data. In the second layer, user's emotional status is defined based on valid sensor data which is sent from the third layer. In the top layer, defined user's emotional status information is used in various application services.

4 Emotion Tagged Content Repository

Tagging is not only an individual process of categorization, but implicitly it is also a social process of indexing. In IR (information retrieval) a tag is means of tag querying.

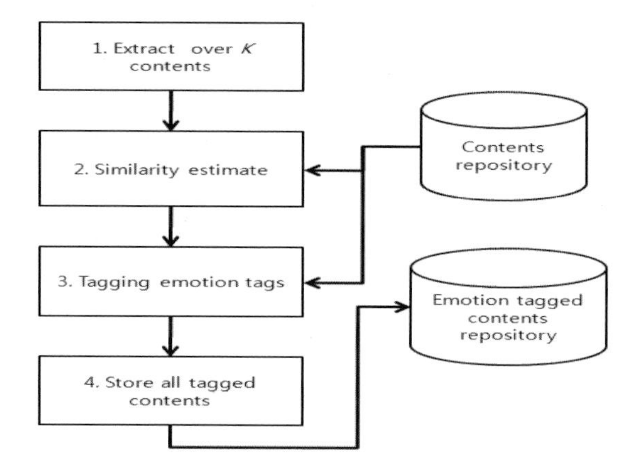

Fig. 3. The Construction process of Emotion Tagged contents repository

In this paper, we use tagging technique for tagging emotion tags on each content. But, our proposed tagging process is more automatic than typical things. Figure 3 shows our construction process of emotion tagged contents repository.

The first step in the process is extracting over k contents representing k different basic emotions. For example, if we use the categories of six basic emotions; *sadness, anger, happiness, surprise, fears and timid,* we can extract over six contents representing these six different emotions. The fact that extracting represented contents is done manually by some experts could take advantage of a better reflection of the human perception.

The next step is computing k similarities between the set of extracted k contents and each one of all contents stored in contents repository. In the example described above, we can compute six values the similarity for each of all contents. In other words, we define emotional feature of content by considering similarity between each content and k contents representing basic k emotions.

In the third step, content is tagged with k emotion tags in which emotional information and the value of similarity are described. If the tag T describing sadness has the largest similarity among other tags, the content tagged with T can be considered as an appropriate content for human's sadness status. Figure 4 visualizes the example of similarity estimation and emotion tagging in our process.

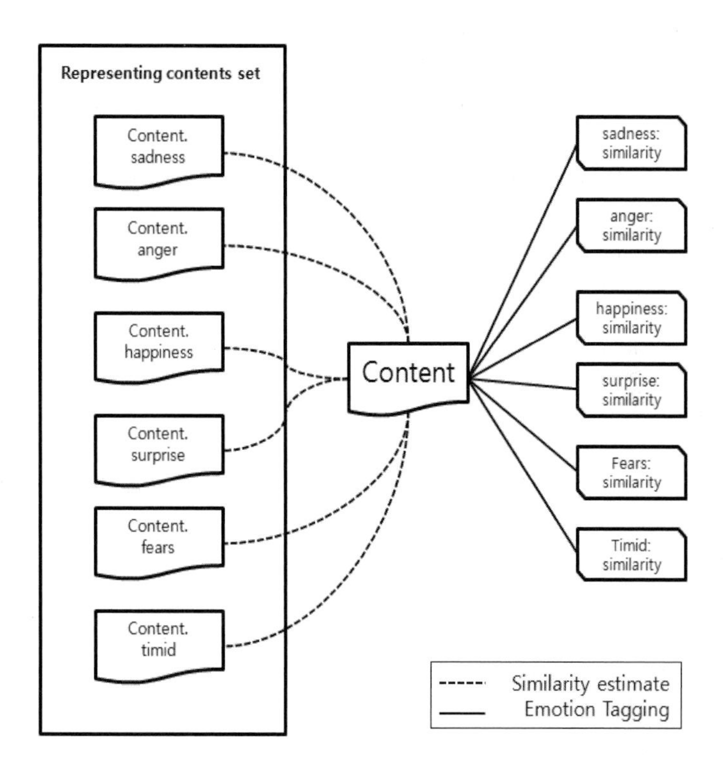

Fig. 4. Visualization of Similarity estimate and Emotion tagging

Finally, all tagged contents are stored into emotion tagged contents repository. Emotion tagged contents repository is accessed at the time when proper contents are searched by emotion tag querying based on user's current emotional status.

5 Providing Emotion Based Contents

In this chapter, we propose an approach to providing proper personalized contents to a user based user's emotional status. To find appropriate contents, we use emotion tagged contents repository proposed in preceding chapter 4 and emotion information returned from a wearable system.

Figure 5 depicts the process of emotion based contents service. The process is composed of two sub processes, the search process and scoring & ranking process. And, in the process, we use two types of the query. One is user's input query and the other is automatically generated query by a wearable system.

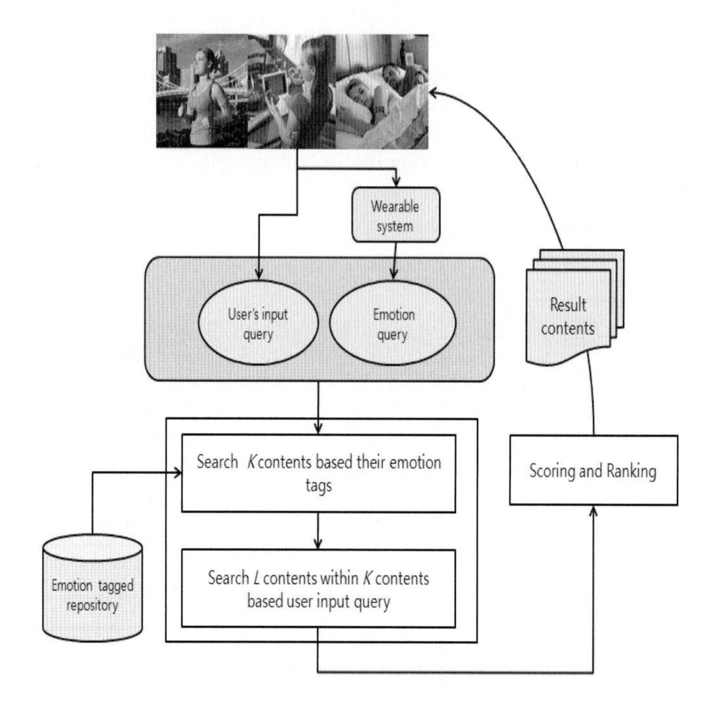

Fig. 5. The process of emotion based contents service

The query inputted by a user presents user's need to particular contents. If a user wants to see a movie not to listen to music, the user oneself input a term *movie* as a query. The emotion query is transferred from wearable system described in chapter 3. The two types of the query are used to search proper contents to user's need and user's emotional status.

The search process depicted in figure 5 consists of two levels. Firstly, k contents are extracted from emotion tagged contents repository through exact matching the emotion query with emotion tags which are tagged on contents within emotion tagged contents repository. Then, in the second level, l contents are extracted from k contents $(l \leq k)$ by checking whether each of k contents is related to user's input query or not. After search process, finally extracted contents are ordered through the process of scoring and ranking by their relevance with user's needs and user's emotional status.

Therefore, the top content or top k contents among ordered contents are provided to each user.

With our proposed approach, the follow two scenarios can be considered:

- *Scenario 1.* User A is in sadness emotional status because he wrangled with his girl friend. And he wants to see a movie. In this case, using our proposed approach, the application can provide a movie which is to be brightly happy ending not horror or sad ending.
- *Scenario 2.* User B fails to get to sleep as her emotional status are much fears and surprised. So, user B wants to listen to music while she lies down on the bed. In this case, the application using our approach can provide comfortable music proper to insomnia not rock or dance music.

6 The Architecture for Emotion Based Service

The development of wireless computing and the miniaturization of electrical components have accelerated the production of different mobile devices such as smart phones, wrest computing devices, and net books. Because of the advance of wireless computing and those various types of computing devices, it is possible that human's needs to various personalized services much increase.

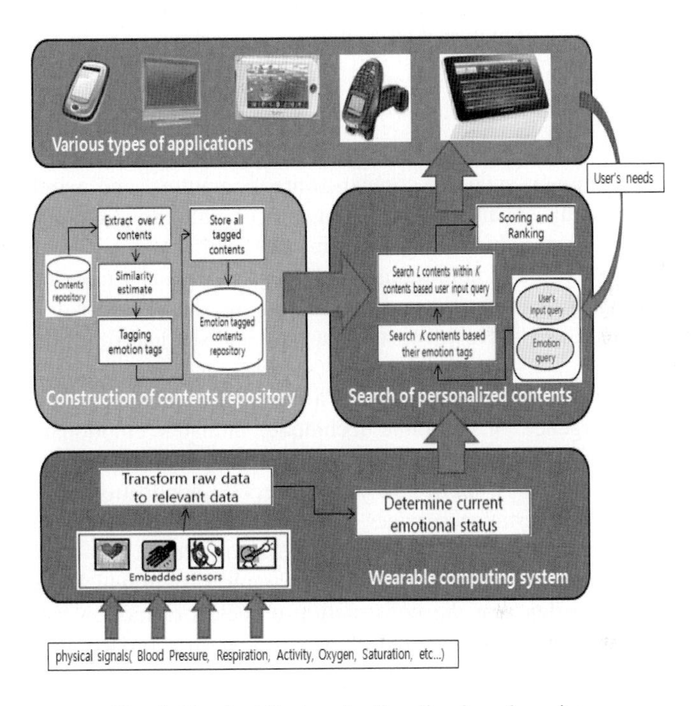

Fig. 6. The Architecture for Emotion based services

In this chapter, we proposed an architecture for various emotion based services. Our proposed architecture is applied in various service application domains such as multimedia services, advertising services and smart home services. Therefore, using our architecture, the various types of applications are considered. Figure 6 depicts our proposed architecture. The proposed architecture consists of four parts, each of which is previously described in chapter 3, 4, 5.

Key concepts of proposed architecture are:

- Wearable computing system described in chapter 3 can sense physiological data, determine emotional status and execute a service based human's emotion.
- Emotion tagged contents repository described in chapter 4 is used in the process of searching contents proper user's needs and user's emotional status. To construct emotion tagged contents repository, we propose emotion tagging process in chapter 4.

- In the process of searching contents based on emotion described in chapter 5, two types of queries are used; *user's input query, automatically generated emotion query*, in order to provide more realistic and personalized services. And the process of searching contents consists of two phrases. The first phrase is extracting contents based on an emotion query. Then, in second phrase, the final set of contents is extracted based on user's input query.
- Before providing emotionally personalized contents to a user, the ultimately extracted contents are ordered by a scoring and ranking function in order to choose only one or high ranked contents.

7 Conclusions

Today, many techniques such as context-aware computing and wireless network computing significantly enhance the functionality of ubiquitous computing services and applications, and enrich the way they interact with users and resources in the environment. Many people in ubiquitous computing environment think that the device should be intelligent enough to understand our needs in order to answer our dynamically changed needs. However, there are no realistic and robust human-centric services so far.

The objective of this paper is to propose a novel methodological architecture that employs the convergence of wearable technique, human's emotional context and tagging technique of web to develop more realistic and robust human-centric service. Our proposed architecture is applied in various service application domains such as multimedia services, advertising services and smart home services. Therefore, using our architecture, the various types of applications are considered.

For our proposed architecture, we define a wearable system which is composed of four layers. Also, we describe the construction process of emotion tagged contents repository using tagging technique. Finally, we propose an approach in order to prove proper personalized contents to a user based on user's emotional status.

Acknowledgment

This work was supported by the Gyonggi Regional Research Center (GRRC) and Contents Convergence Software (CCS) research center.

References

1. Gatzoulis, L., Iakovidis, I.: Wearable and portable health systems. IEEE Eng. Med. Biol. Mag. 26(5), 51–56 (2007); Clerk Maxwell, J.: A Treatise on Electricity and Magnetism, 3rd edn., vol. 2, pp. 68–73. Clarendon, Oxford (1892)
2. Lymperis, A., Dittmar, A.: Advanced wearable health systems and applications, research and development efforts in the european union. IEEE Eng. Med. Biol. Mag. 26(3), 29–33 (2007)
3. Bonato, P.: Advances in wearable technology and applications in physical medicine and rehabilitation. J. NeuroEng. Rehabil. 2, 2 (2005)

4. Bonato, P.: Wearable sensors/systems and their impact on biomedical engineering. IEEE Eng. Med. Biol. Mag. 22(3), 18–20 (2001)
5. Farringdon, J., Moore, A.J., Tilbury, N., Church, J., Biemoon, P.D.: Wearable Sensor Badge and Sensor Jacket for Context Awareness. In: Proceddings of the 3rd IEEE International Symposium on Wearable Computers, p. 107 (1999)
6. Jakicic, J.M., Marcus, M., Gallagher, K.I., Randall, C., Thomas, E., Goss, F.L., Robertson, R.J.: Evaluation of the SenseWear Pro Armband^{TM} to Assess Energy Expenditure during Exercise. Med. Sci. Sports Exerc. 36(5), 897–904 (2004)
7. Rantanen, J., et al.: Smart Clothing for the Arctic Environment. In: Proc. Of the Fourth International Symposium on Wearable Computers, Altranta, GA, October 16-17, pp. 15–23. IEEE, Los Alamitos (2000)
8. Ortony, A., Turner, T.J.: What's basic about basic emotions? Psychological Review 97, 315–331 (1990)
9. Plutchik, R.: The Multifactor-Analytic Theory of Emotion. The Journal of Psychology 50, 153–171 (1960)
10. Gratch, J., Marsella, S.: Evaluation a computational model of emtion. Journal of Autonomous Agents and Multiagent Systems 11(1), 23–43 (2006)
11. Hudicka, E.: Depth of Feelings: Alternatives for Modeling Affect in User Models, pp. 13–18 (2006)
12. Picard, R.: Affective Computing. M.I.T Media Laboratory Perceptual Computing Section Technical Report, V. 321 (November 26, 1995), `http://vismod.media.mit.edu/tech-reports/TR-321.pdf` [September 21, 2008]
13. Parada, R., Pavia, A.: Teaming up humans with autonomous syntehtic characters. Artifical Intelligence 173(1), 80–103 (2009)

Adaptive Data Dissemination Protocol for Wireless Sensor Networks*

Byoung-Dai Lee

The Department of Computer Science, Kyonggi University, Suwon, Korea
blee@kgu.ac.kr

Abstract. In terms of data delivery required by the application, WSN can be categorized as continuous or event-driven. The former type provides an accurate snapshot of relevant attributes with high energy consumption, whereas the latter type is more energy efficient and yet less accurate because only sudden and drastic changes in the value of a sensed attribute trigger data delivery. In this paper, we propose a hybrid data delivery protocol that adaptively switches between these two data dissemination schemes. That is, during the period when the condition referring to an event is met, sensor nodes continuously send data to an observer. However, when the condition is no longer valid, the protocol behaves as event-driven, thus eliminating unnecessary data transmission. As such, the proposed protocol is able to enable an accurate analysis of the environment being monitored with moderate consumption of valuable resources.

Keywords: Data Dissemination, Energy Efficiency, WSN.

1 Introduction

Recently, advances in wireless communications technology and microelectromechanical systems (MEMS) have enabled the development of low-cost, low-power, network-enabled, multi-functional microsensors. These sensor nodes are capable of monitoring various physical conditions, such as temperature, ultrasonic waves, soil composition, and object motion, performing simple computations, and communicating among peers or directly to an external base station. A typical wireless sensor network (WSN) consists of hundreds to thousands of these nodes deployed over an area and integrated to collaborate through a wireless network. Due to its ease of deployment, reliability, scalability, flexibility, and self-organization, the existing and potential applications of WSNs span a wide spectrum in various domains, in which environment and technical requirements may greatly differ. Examples of representative WSN applications include military applications ([3][8]), habitat monitoring applications ([1][4][13]), environmental observation and forecasting systems ([2][16][17]), scientific exploration ([10]), home and office intelligence ([9][20]), and medical care ([5][18]).

* This work was supported by the GRRC program of Gyeonggi province [GRRC Kyeonggi 2010-B01: Developing and Industrializing Core-Technologies of the Smart-Space Convergence Framework].

T.-h. Kim, A. Stoica, and R.-S. Chang (Eds.): SUComS 2010, CCIS 78, pp. 188–195, 2010.

In terms of data delivery required by the application, WSN can be classified as continuous, when sensor nodes collect data and send them to an observer continuously throughout time, and as event-driven, when sensor nodes react immediately to sudden and drastic changes in the value of a sensed attribute due to the occurrence of a certain event [7]. As communication energy is a major contributor to the total energy consumption and is determined by the total amount of communication and transmission distance, the event-driven data dissemination scheme may lead to less energy consumption and thus, prolong the network lifetime. On the other hand, as the amount of received data determines the level of accuracy, the continuous data dissemination scheme is more suitable when higher accuracy is demanded. Therefore, it is essential to make a trade-off between the level of accuracy desired and resource consumption.

In this paper, we propose a hybrid data dissemination system that dynamically switches between these two data dissemination schemes. Basically, the proposed system behaves as event-driven, and, therefore, an event triggers data delivery by sensor nodes. However, from the point when an event occurs to the point when the event becomes no longer valid, sensor nodes continuously send data to an observer, so more accurate analysis becomes possible. The novel aspect of our approach is that only a subset of sensor nodes, deployed in the areas where the event occurs and the areas in close proximity to them, take participation in continuous data dissemination. In particular, this capability is helpful in accurate analysis as data from relevant areas can be collected proactively without the intervention of an observer. Overall, the proposed system enables accurate analysis of the environment being monitored with moderate consumption of valuable resources.

We envision that many real-life WSN applications will benefit from the proposed hybrid system. For instance, consider a forest fire detection system. When a fire is detected, by continuously receiving the raw temperature data from not only the burning area, but also the "neighboring" areas, where no fire is yet detected, an observer is able to accurately analyze where the fire originates, forecast where it is potentially heading, and plan where the fire extinguishing operation should be focused.

The paper is organized as follows: Section 2 summarizes related work and Section 3 describes the sensor network model. In section 4, we explain the proposed system in detail. Finally, we conclude in section 5.

2 Related Work

Significant research has been conducted to develop energy efficient processing techniques that minimize power requirements across all levels of the protocol stacks in WSNs. In this paper, however, we explore only those efforts that deal with sensor network protocols.

LEACH [11] introduces adaptive clustering protocol that allows for a randomized rotation of the cluster head's role after a given interval to make energy dissipation in the sensor nodes uniform. This protocol is appropriate when constant monitoring by the sensor nodes is needed. The sensor network model of our approach is based on the hierarchic clustering model of LEACH.

TEEN [14] is a representative example of reactive network protocols in WSN. It uses a hard threshold, which is the threshold value of the sensed attribute, and a soft threshold, which is a small change in the value of the sensed attribute that triggers the sensor node to transmit data. Thus, the hard threshold tries to reduce the number of transmissions by allowing sensor nodes to transmit data only when the sensed attribute is in the range of interest. The soft threshold, on the other hand, further reduces the number of transmissions that might have otherwise occurred when little or no change occurs in the sensed attribute.

APTEEN [15] is a hybrid protocol that changes the periodicity or threshold values used in the TEEN protocol according to user needs and type of application. Along with the threshold values described, APTEEN parameterizes the maximum time period between two successive reports sent by a sensor node so that a sensor node is forced to transmit sensed data if it has not sent data for a long period of time. Thus, APTEEN combines both proactive and reactive policies. Both TEEN and APTEEN have many similarities with our approach. For instance, sensor nodes implementing TEEN or APTEEN protocols are engaged in data dissemination only when a certain event occurs. However, the difference lies in the application model. That is, their work is best suited for time critical applications such as intrusion detection, whereas ours can be used for applications that require fast and accurate analysis or diagnosis of the environment when an event of interest occurs.

The SPIN family of protocols [12] uses data negotiation and resource-adaptive algorithms. For this purpose, sensor nodes use three types of messages to advertise new data, to request the data, and to send the actual data, respectively. These protocols suppress duplicate information and prevent redundant data from being sent to neighbors by conducting a series of negotiation messages before real data transmission begins.

SINA [19] selects the most appropriate data distribution and collection method based on the nature of queries and current network status. In particular, individual sensor nodes make autonomous decisions about whether they should participate in the information gathering process based on a given response probability. Furthermore, sensor nodes are able to defer sending data for some period of time. These methods maximize the quality of response in terms of their number and responsiveness while minimizing network resource consumption.

Although both SPIN family protocols and SINA selectively include the sensor nodes for data dissemination, as with our approach, the application models considered are different. However, parts of their work, such as resource-adaptive algorithms, can also be applied in our approach.

3 Sensor Network Model

The sensor network model that we consider is the multilevel hierarchical clustering-based network [6]. It consists of a set of sensor nodes and a base station (BS), through which an observer can interact with the sensor network. In this model, sensor nodes form clusters based on, for example, proximity, and a cluster head for each cluster is elected according to a set of rules. Similarly, cluster heads at the same level can also form clusters and thus a hierarchy of clusters forms a tree. Consequently, when a

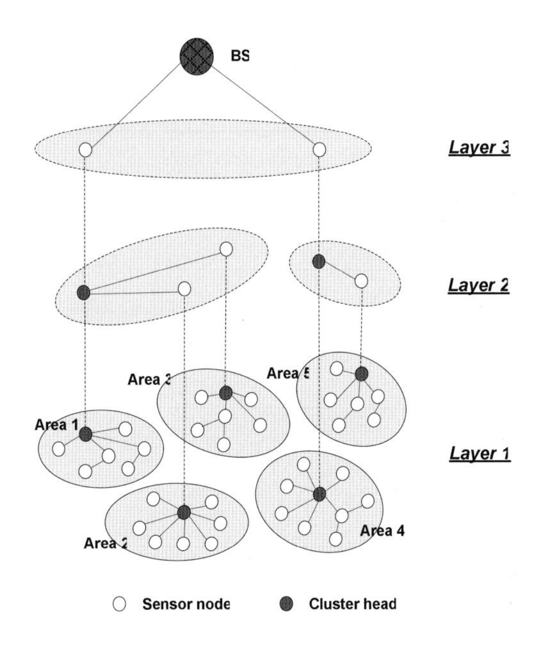

Fig. 1. An example of three-level hierarchical clustering

sensor node wishes to report data, it must forward data to its cluster head, which, in turn, forwards the data to its own cluster head until the data reach the BS (see Fig. 1). Control information, on the other hand, can be flooded in the reverse direction, from cluster heads to cluster members.

In our model, the cluster heads at each layer periodically run algorithms to determine the data dissemination scheme to use for the next time interval. The parameters required by the decision and the number of member nodes are assumed to be known to the cluster heads after clusters are formed.

Although the clustering-based approach can greatly contribute overall system scalability, lifetime, and energy efficiency, care must be taken when organizing clusters. For instance, as the communication depends on the cluster heads, the energy depletion of cluster heads is faster than that of other sensor nodes. Furthermore, a balance between cluster size and cluster diameter is typically desired because these parameters control the load on the cluster heads, and the cost of communication between the cluster head and the cluster members. As various approaches have been proposed in the literature, we do not consider the cluster organization problem. We believe that this assumption does not significantly impact the effectiveness of our approach because the proposed system runs after cluster formation is completed.

4 Adaptive Data Dissemination Protocol

The integral parts of our approach are that 1) it switches between the event-driven data dissemination scheme and the continuous data dissemination scheme at the occurrence of events of interest, and 2) sensor nodes in close proximity to those nodes

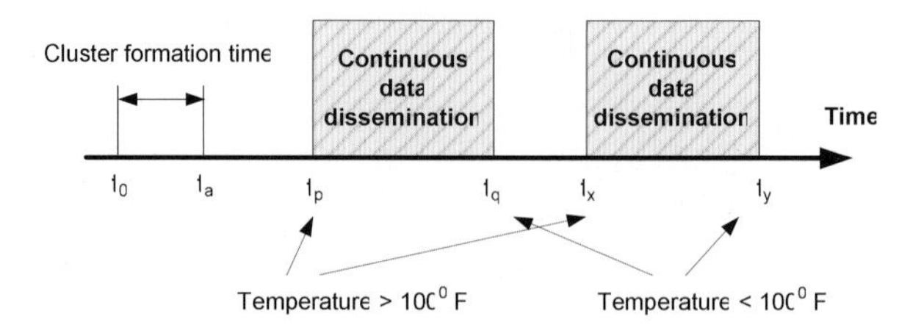

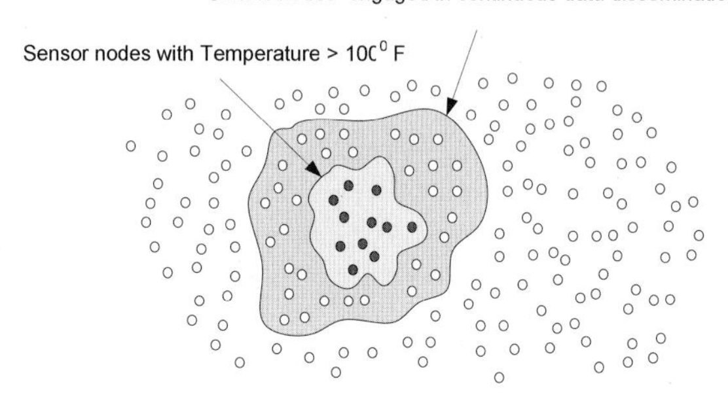

Fig. 2. Operational features in terms of the time and the space domains

detecting the events are also engaged in continuous data dissemination (see Fig. 2). In order to fulfill these requirements, it must be decided in a timely manner when to switch between the two data dissemination schemes, and which sensor nodes to involve in the continuous data dissemination. Depending on the answers to these questions, several policies can exist. In this paper, we present two algorithms, called the *PED (Parameter-based Event Detection)* algorithm and the *PAD (Parameter-based Area Detection)* algorithm.

4.1 PED Algorithm

The PED algorithm, shown in Fig. 3, is given a threshold value and two pairs of parameters controlling the level of aggressiveness for changes in data dissemination schemes. When the cluster members are the sensor nodes that actually sense the physical conditions, the threshold value determines what percentage of sensor nodes within the same cluster detects the event. Otherwise, the percentage of lower level clusters reporting in a continuous manner is used for the threshold value. Each pair of parameters, (P_{start}, Q_{start}) and (P_{stop}, Q_{stop}), is used to determine when to start or stop the continuous data dissemination, respectively. P_{start} denotes the maximum number of time intervals with an increasing slope of the threshold variable, and Q_{start} denotes the maximum number of time intervals that the threshold is exceeded, regardless of the slope. Similarly, when P_{stop} and Q_{stop} are used, a decreasing slope of the threshold variable and the time intervals that do not exceed the threshold variable are applied, respectively.

```
PED(α, Pstart, Qstart, Pstop, Qstop)
{
/*
Scurr: the threshold variable over current time window
Sprev: the threshold variable over previous time window
α: the threshold value
p,q: counter variable corresponding to Pstart/Pstop and Qstart/Qstop
*/
  if (current mode is event-driven)   {
    if (Scurr ≥ Sprev) {
       if (Scurr ≥ α)  p++;
       else p = 0;
      q++;
    }
    else
       p = q = 0;

    if ((p ≥ Pstart) or (q ≥ Qstart)
      switch to "continuous" mode
  }
  else   {
    if (Scurr ≤ α) {
      if (Scurr ≤ Sprev) p++;
      else p = 0;
      q++;
    }
    else
       p = q = 0;

    if ((p ≥ Pstop) or (q ≥ Qstop)
      switch to "event-driven" mode
  }
}
```

Fig. 3. PED algorithm

Use of P is to enable a rapid response to a change but to prevent transient response. If the test on P fails, the test on Q is applied, which is more forgiving (e.g., the requirement that the slope is monotonically increasing or decreasing is relaxed). Smaller values of P and Q lead to more aggressive action and each cluster header is free to select these parameters differently. When a cluster head determines to change the current data dissemination scheme, it then reports its decision to its cluster head so that it can make a decision whether other clusters at the lower level must change the data dissemination scheme currently employed.

Fig. 4 illustrates PED algorithm with (P_{start} = 3, Q_{start} = 5) and (P_{stop} = 3, Q_{stop} = 5). At t_7 and t_{16}, the algorithm starts and stops the continuous data dissemination based on Q_{start} on P_{stop}, respectively. Note that failure of tests on both P_{start} and Q_{start} does not affect the normal behavior of the sensor networks. That is, the data capturing an event are still propagated all the way to the BS.

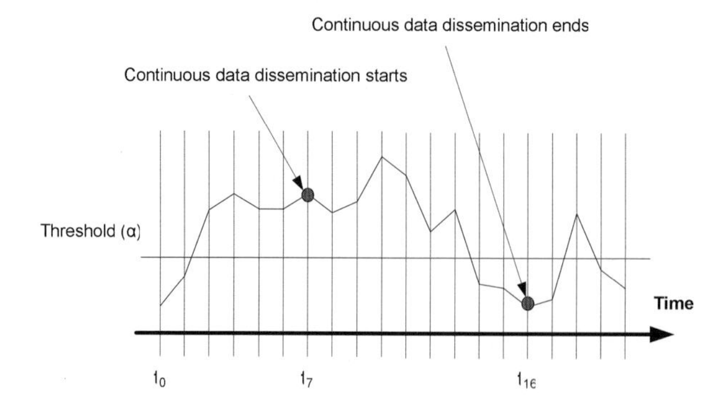

Fig. 4. An illustration of PED algorithm

4.2 PAD Algorithm

Once cluster heads decided to switch the data dissemination scheme, the next decision to make is which sensor nodes must be affected by that decision. In our approach, the PAD algorithm deals with the cluster members as a whole. In other words, when the data dissemination scheme is selected by a cluster head, then all of the sensor nodes within the cluster will be directed to report data in that manner.

Our simple approach is based on the argument that the closer a sensor node is located to the sensor nodes that detect an event, the more likely it is that the sensor node is relevant to that event in the near future, as clusters are typically formed according to proximity. For instance, the current temperature measured by sensor nodes located close to the fire, say, may not be high enough to trigger an event, but would be much higher than those of nodes tens of miles away from the fire. Therefore, the possibility that those sensor nodes soon trigger the event becomes high.

However, we acknowledge that this approach may not work very well in some cases. For instance, if sensor nodes detecting an event are located at the border between clusters, those nodes in other clusters can be included only when enough clusters at the same level have initiated continuous data dissemination. One of our future projects is to develop algorithms to address this issue.

5 Conclusion and Future Work

In this paper, we propose a hybrid data dissemination system that dynamically switches between a continuous data dissemination scheme and an event-driven data dissemination scheme. The benefit of the system is that it enables fast and accurate analysis of the environment being monitored with moderate consumption of energy.

The future work lies in several areas. First, we plan to investigate the validity of our approach using simulation. In addition, as described previously, the PAD algorithm we presented has some limitations. Therefore, one of our future research objectives is to develop algorithms to address such limitations.

References

1. Akyildiz, I.F., Su, W., Sankarasubramaniam, Y., Cayirci, E.: Wireless sensor network: a survey. Computer Networks 38(4), 393–422 (2002)
2. Automated Local Evaluation in Real-Time (ALERT), `http://www.alertsystems.org`
3. Biagioni, E., Bridges, K.: The application of remote sensor technology to assist the recovery of rare and endangered species. Special Issues on Distributed Sensor Networks for the International Journal of High Performance Computing Applications 16(3) (August 2002)
4. Cerpa, A., Elson, J., Estrin, D., Girod, L., Hamilton, M., Zhao, J.: Habitat monitoring: application driver for wireless communications technology. In: Proceedings of 1st ACM SIGCOMM Workshop on Data Communications, pp. 20–41 (2001)
5. Easton, M.: Using space technology to fight malaria. Queen's Gazette 12 (April 2003)
6. Estrin, D., Govindan, R., Heidemann, J.: Embedding the Internet. ACM Communications 43(5) (May 2000)
7. Ilayas, M., Mahgoub, I. (eds.): Handbook of Sensor Networks: Compact Wireless and Wired Sensing Systems. CRC Press, Boca Raton (2005)
8. Goldsmith, A.J., Wicker, S.B.: Design challenges for energy-constrained ad hoc wireless networks. IEEE Wireless Communications, 8–27 (2002)
9. Herring, C., Kaplan, S.: Component-based software system for smart environments. IEEE Personal Communications, 60–61 (2000)
10. Howard, A., Mataric, M., Sukhatme, G.S.: Mobile sensor network deployment using potential fields: a distributed, scalable, solution to the area coverage problem. In: Proceedings of 6th International Symposiums on Distributed Autonomous Robotics Systems, pp. 299–308 (June 2002)
11. Heinzelman, W., Chandrakasan, A., Balakrishnam, H.: Energy-efficient communication protocol for wireless microsensor networks. In: Proceedings of Hawaiian International Conference on Systems Science (January 2000)
12. Kulik, J., Heinzelman, W.R., Balakrishnan, H.: Negotiation-based protocols for disseminating information in wireless sensor networks. Wireless Networks 8, 169–185 (2002)
13. Mainwaring, A., Polastre, J., Szewczyk, R., Culler, D., Anderson, J.: Wireless sensor network for habitat monitoring. In: Proceedings of 2002 ACM International Workshop on Wireless Sensor Networks and Applications (September 2002)
14. Manjeshwar, A., Agarwal, D.P.: TEEN: a routing protocol for enhanced efficiency in wireless sensor networks. In: Proceedings of International Workshop on Parallel Distributed Computing (April 2001)
15. Manjeshwar, A., Agarwal, D.P.: APTEEN: a hybrid protocol for efficient routing and comprehensive information retrieval in wireless sensor networks. In: Proceeding of International Symposiums on Parallel Distributed Processing (2002)
16. National Interagency Fire Center, `http://www.nifc.gov`
17. Rabaey, J.M., Ammer, M.J., Silva, J., Patel, D., Roundy, S.: PicoRadio supports ad hoc ultra-low power wireless networking. IEEE Computer Magazine 33(7), 42–48 (2000)
18. Schwiebert, L., Gupta, S., Weinmann, J.: Research challenges in wireless networks of biomedical sensors. Mobile Computing and Networking, 151–165 (2001)
19. Srisathapornphat, C., Jaikaeo, C., Shen, C.: Sensor information networking architecture. In: Proceedings of International Workshop on Parallel Processing, pp. 23–30 (August 2000)
20. Srivastava, M.B., Muntz, R., Potkonjak, M.: Smart kindergarten: sensor-based wireless networks for smart developmental problem-solving environments. Mobile Computing and Applications, 132–138 (2001)

Scalable Multicast Protocols for Overlapped Groups in Broker-Based Sensor Networks

Chayoung Kim[1] and Jinho Ahn[2]

[1] CCSRC in GRRC, in Kyonggi University
[2] Dept. of Computer Science, in Kyonggi University
Suwon-si Gyeonggi-do, Korea
{kimcha0,jhahn}@kgu.ac.kr

Abstract. In sensor networks, there are lots of overlapped multicast groups because of many subscribers, associated with their potentially varying specific interests, querying every event to sensors/publishers. And gossip based communication protocols are promising as one of potential solutions providing scalability in P(Publish)/ S(Subscribe) paradigm in sensor networks. Moreover, despite the importance of both guaranteeing message delivery order and supporting overlapped multicast groups in sensor or P2P networks, there exist little research works on development of gossip-based protocols to satisfy all these requirements. In this paper, we present two versions of causally ordered delivery guaranteeing protocols for overlapped multicast groups. The one is based on sensor-broker as delegates and the other is based on local views and delegates representing subscriber subgroups. In the sensor-broker based protocol, sensor-broker might lead to make overlapped multicast networks organized by subscriber's interests. The message delivery order has been guaranteed consistently and all multicast messages are delivered to overlapped subscribers using gossip based protocols by sensor-broker. Therefore, these features of the sensor-broker based protocol might be significantly scalable rather than those of the protocols by hierarchical membership list of dedicated groups like traditional committee protocols. And the subscriber-delegate based protocol is much stronger rather than fully decentralized protocols guaranteeing causally ordered delivery based on only local views because the message delivery order has been guaranteed consistently by all corresponding members of the groups including delegates. Therefore, this feature of the subscriber-delegate protocol is a hybrid approach improving the inherent scalability of multicast nature by gossip-based technique in all communications.

Keywords: Sensor network, Group communication, Overlapped multicast groups, Scalability, Reliability.

1 Introduction

A new data dissemination paradigm for such sensor networks is different from mobile ad-hoc networks in a method of designing data propagation and aggregation generated by various and lots of sensor nodes[5]. There are several researches based on the P (publish)/S (subscribe) paradigm in the area of sensor network communications to

T.-h. Kim, A. Stoica, and R.-S. Chang (Eds.): SUComS 2010, CCIS 78, pp. 196–205, 2010.

address the problem of querying sensors from mobile nodes in order to minimize the number of result sent packets[5],[7]. In P/S paradigm systems, a query node periodically runs an algorithm to identify the sensors it wishes to track and "subscribe" to these sensors of their interest, and the sensors periodically "publish"[5],[7]. So, the intermediate sensors in the networks, along the reverse path of interest propagation might aggregate the query results by combining reports from several sensors[5]. An important feature of that interest and data propagation and aggregation are determined by localized sensors interactions[5]. And performing local computations to reduce data before transmission can obtain orders of magnitude energy savings[5],[7]. Recently, gossip-based protocols seem more appealing in many P/S systems because they are more scalable than traditional reliable broadcast[1] and network-level protocols deriving from IP Multicast for many of the various applications requiring reliable dissemination of events[4],[6]. Gossip-based protocols have turned out to be adequate for large scale settings by achieving a "high degree of reliability" and strong message delivery ordering guarantees offered by deterministic approaches[2],[4],[6]. The seminal probabilistic broadcast (pbcast) algorithm of Birman et. al.[2] is originally described as a broadcast presented in global view based system and Eugster's algorithm(lpbcast) [4] is implemented for P/S systems as a broadcast. These previously developed gossip-based protocols implicitly assume that all processes in a group are interested in all events[2]. And, such a flooding technique is not adequate when many events are only of interest for lower than half the processes in the overall groups[3]. PMCAST [3] deals with the case of multicasting events only to subsets of the processes in a large group by relying on a specific orchestration of process as a superimposition of spanning trees. The delegates of this protocol[3] yet are themselves not interested in the same topics as subscribers have. But, when multicasting an event, PMCAST[3] follows the underlying tree, by gossiping depth-wise, starting at the root and the interested subscribers receive their event messages[3]. And these messages must be delivered to the overlapped multicast groups, while causally ordered delivery properties are guaranteed on gossip-based protocols. An atomic broadcast on gossip-based protocols(atomic probabilistic broadcast) is implemented in Birman et. al.[2] and Eugster et. al.[3] for P/S systems. But, these protocols are performed by hierarchical membership protocols[3] for each delegate group or the totally ordered delivery properties are maintained by global member views[2]. These features are likely to be highly overloaded on each member and not scalable. Also, there is no causal order guaranteeing multicast protocol supporting overlapped multicast groups, useful for many distributed applications such as video-conferencing, multi-party games and private chat rooms, based on publisher(sensor-broker) in the previously developed protocols. In this paper, we present two versions of causal order guaranteeing multicast protocols. The one is sensor-broker(publisher-broker) based protocol that the sensor-broker as delegates aggregate the information of results, periodically gossip about the messages of them and guarantee causally ordered delivery of the messages in the face of transient member population. And the other is subscriber-delegate based protocol that all messages including node join/leave and causally ordered delivery information are communicated by the gossip-style dissemination based on partially randomized local views with a part of delegate selected to represent that subgroups[3], leading to a form of compound spanning tree by subgroup select/merge-procedure.

2 Two Types of Protocols

2.1 Sensor-Broker Based Protocol

Basic Idea. In sensor-broker based protocol, some sensors that are designed as brokers might lead to make overlay networks and query nodes subscribe to all the topics that match their interest. The mapping of subscribers and brokers is entirely driven by the query application. Recently, much research has been devoted to designing broker selection methods that best suits application needs[6],[7]. Each sensor can provide information periodically with some of its brokers by peer-sampling services[6], assumed to be implemented in the systems and the brokers might aggregate the query results by combining reports from several sensors. Query nodes subscribe to the corresponding sensor-broker networks of their interest and receive the results of the queries. The sensor-broker periodically gossip about the messages of the results[7] and guarantee causally ordered delivery of the messages with aggregating the information of the results in subscribers-overlapped networks. In this protocol, because every broker knows every other brokers, a VT(vector time) for each broker, p_i is a vector of length n, where n = the number of broker members. And this protocol leads us to extend single VT to multiple VTs because a broker belongs to several interest groups[1]. Sensor-broker maintain VT for each group and attach all the VTs to every message that they multicast. If all brokers representing a sensor grid may be stale, all members of brokers are not changed because of periodical peer-sampling services[6], designed to suit application needs. But, if all suscribers in a grid lose their interests in information published on a particular topic, the brokers representing the grid send leave messages to the corresponding overlay network groups and then all of their group members are updated by leaving brokers. Also, epoch protocol[1] is assumed to be implemented for the member leave processing. And the digest information of vector and membership list are sent or received periodically using IP-Mulitcast or, if IP multicast is not available, using a randomized dissemination protocol and the subsequent gossip-based anti-entropy protocols[2]. Therefore, this protocol might make up transient faults of sensor-broker with the peer-sampling services[6] and deal with brokers' leave by membership management. The processing of temporal faults is different from that of brokers' leave. So, these features of this protocol might result in its very low membership management cost compared with the cost incurred by maintaining hierarchical member list for delegate groups in the previous protocols[2].

An Example and Algorithm Description.

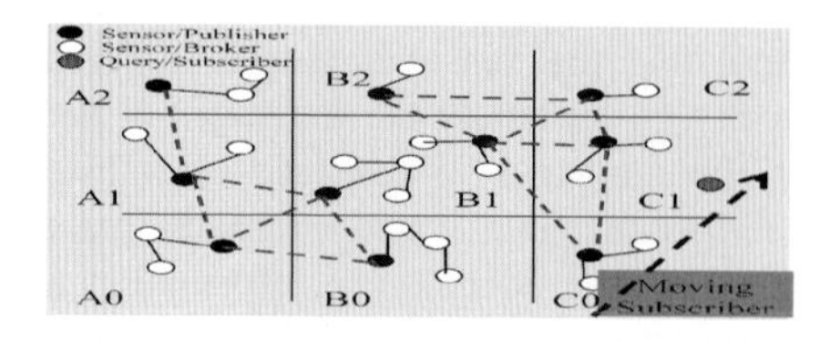

Fig. 1. Publisher vs. Subscribers in a Sensor Network

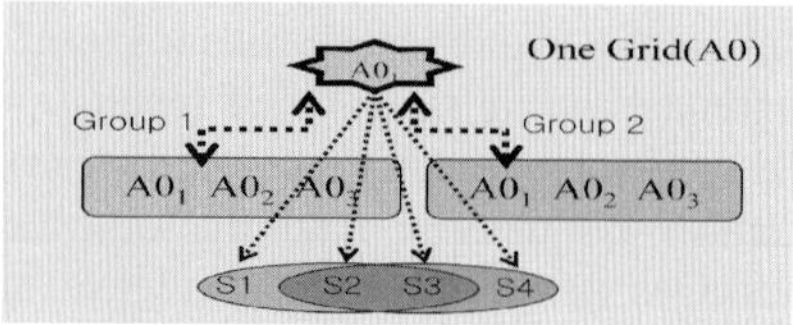

Fig. 2. Sensor-Broker A1 covering a Grid

In figure 1, there is a two-dimensional area of interest(AoI), which the sensor-broker publish messages to a particular topic, while query nodes subscribe to all the topics that match their interests. In this sensor-broker based protocol, we present an causal order guaranteeing multicast protocol supporting overlapped subscriber groups and useful for many distributed applications such as video-conferencing, multi-party games and private chat rooms requiring causally ordered delivery of messages like sensor-broker based networks in figure 1. We can see that a sensor-broker $A0_1$ in a grid A0 of a sensor network like one of figure 1 publishes desired messages of query results to all query nodes, {s1, s2, s3} and {s2, s3, s4} subscribing to their topics, Group1 and Group2 respectively, using gossip-style disseminations, in figure 2. And there are overlapped multicast members={s2, s3} subscribing to the topics of Group1 and Group2.

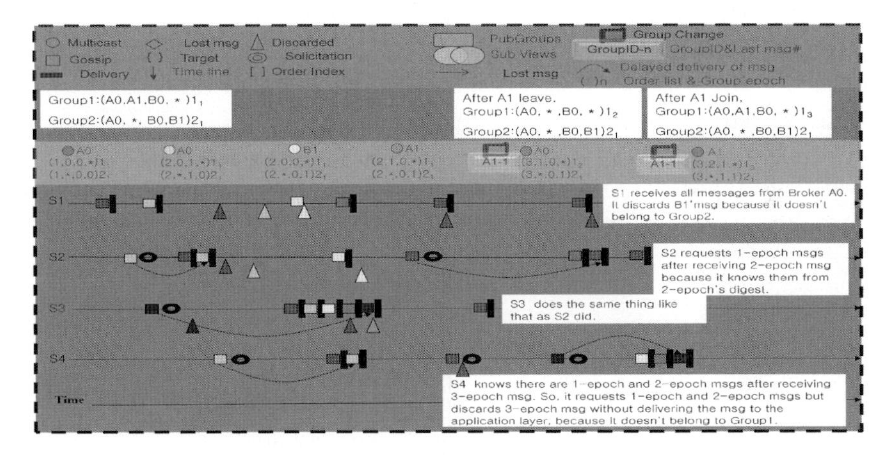

Fig. 3. An example of message deliveries from brokers to subscribers

Figure 3 shows that sensor-broker gossip about multicast messages piggybacked with all VT clocks for all of interesting groups to guarantee causally ordered delivery of messages in a sensor network like one of figure 2. In this cases, there are undesired message are sent to a subscriber, forcing it to discard them. In the figure 3, there are VT clocks $((0,0,0,*)1_1, (0,*,0,0)2_1)$ for each group, Group1 and Group2, to depict each VTg_e as a vector of length n, with a subscript epoch variable, $_e$ for covering the cases of a process leave and join and a special entry * for each process that is not a member of Group$g$. For each message generated by a member, each $VTg_e(p_i)[i]$ is incremented by 1. So, if a member A0 generates a multicast message, then $VT1_1$ and $VT2_1$ is $((1,0,0,*)1_1$ and $(1,*,0,0)2_1)$ respectively. And when a member A1 leaves and joins Group1 again, VT1 for Group1 is from $(*,*,0,*)1_2$ to $(*,0,0,*)1_3$ because subscript epoch is changed from 2 to 3. Figures 2 and 3 show an example of sensor-broker based protocol with causal ordering VT clocks in what order is A0->A0->B1->A1->A0->A1. This example in figure 3 illustrates Group1={A0,A1,B0}, Group2={A0,B0,B1}, Subscribers={S1,S2,S3,S4} and maximum number of Gossip Rounds = 2. The overlapped subscriber members={S2,S3} receive all messages from Group1 and Group2, S1 receives messages only from Group1 and S4 receives messages only from Group2. In this case that subscribers know what messages should be delivered according to causal ordering VT clocks piggybacked by multicast messages and discard them after comparing their causal ordering VT clocks and validating their receipt of predecessor messages.

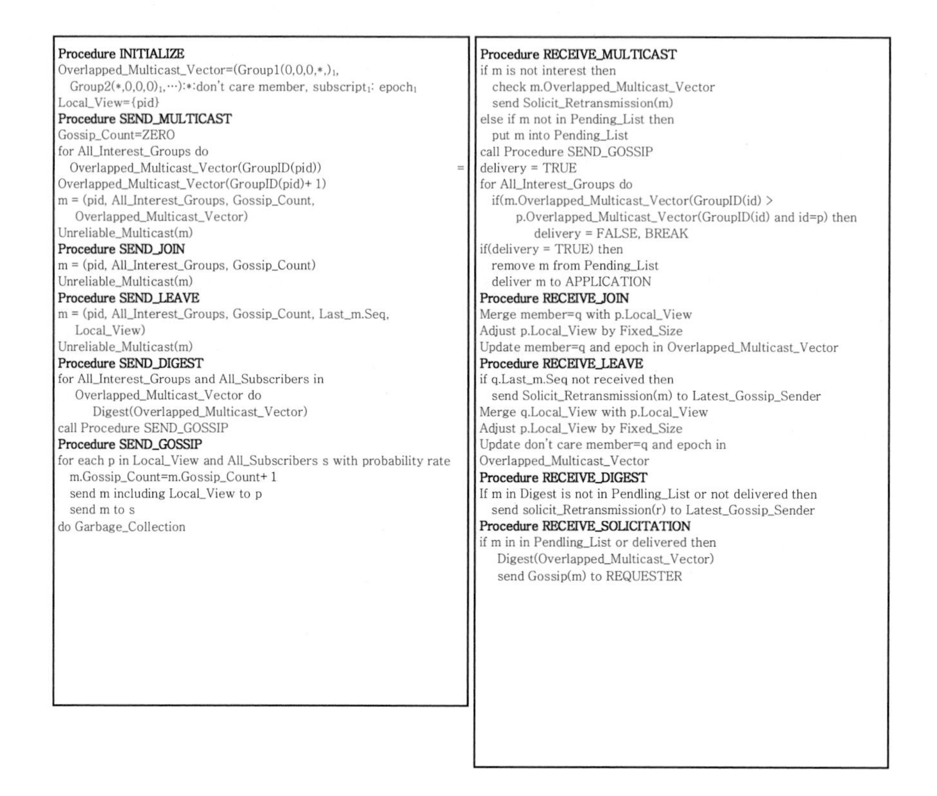

Fig. 4. Algorithm Description(*Cont'd*)

Fig. 5. Algorithm Description

2.2 Subscriber-Delegate Based Protocol

Basic Idea. In subscriber-delegate based protocol, node join and leave messages are disseminated by gossiping. And it is guaranteed that the last message generated by a leave member has been already delivered before changing group membership using the flush protocol[1] performed by delegate members[3]. In the previous developed protocols of distributed systems, there is a consistency on the message delivery order using the flush protocol[1] performed by delegates detecting a failed process. So, in this protocol, a message delivery order consistency is guaranteed by relying on PMCAST delegates[3]. The upper level node of such a subgroup is merged with the one of a set of neighbor subgroups. The select/merge-procedure is performed recursively, leading to a form of compound spanning tree like in figure 6[3],[4]. To represent the interests of all processes, the interests of the respective processes must be regrouped by reducing the complexity of the interests like figure 7[3]. In figure 7, we can see that the member p2 in Group1={p1, p2, p4} can receive all messages belonging to only Group2={p1, p3, p4} because it is a member of delegates={p2, p3}. Also, we know that the message delivery order consistency must be guaranteed for overlapped members subscribing Group1 and Group2, p1 and p4 in figure 7. Multicast messages piggybacked with causal ordering graph[8] on the basis of each overlapped group ID are sent and received based on gossiping for causally ordered delivery according to a set of overlapped groups. And all members deliver the gossip messages to the application layer after comparing their causal ordering graph and validating their receipt of predecessor messages.

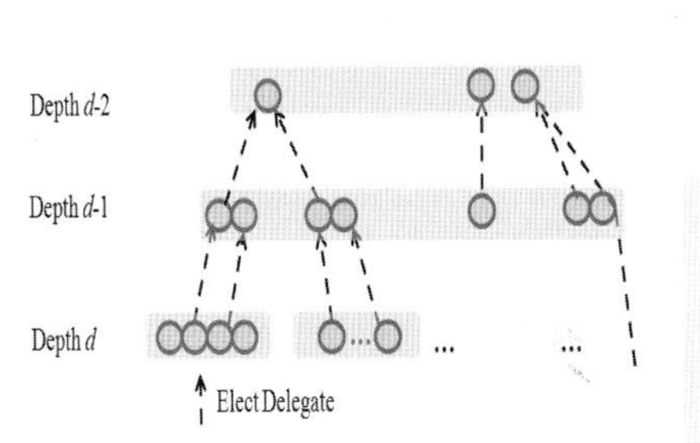

Fig. 6. Membership of Delegates selected from Subscribers

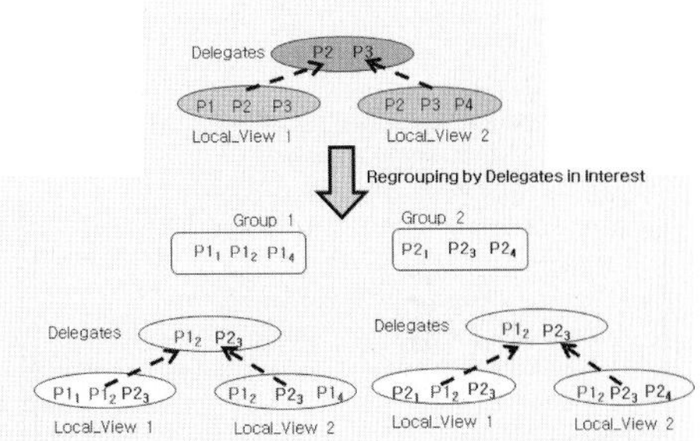

Fig. 7. Processes regrouping by Delegates in their interests

One member detecting a failed member of the group sends a message about the last message sent by the failed member and the failed member itself to the subscriber-delegate providing message ordering guarantees using traditional reliable communicating protocols. And subscriber-delegate perform flush protocol[1] for holding newly generated message until all members update each group membership. All members receiving FLUSH messages send an ACK message including the newly updated last message sent by the failed member to subscriber-delegate. And subscriber-delegate receiving ACK message from all members update the latest version of the last message and gossip about newly changed each group membership based on randomized local views. After maximum gossip round, all members start the membership change of the overlapped multicast groups by their own members. So, the proposed subscriber-delegate protocol is a hybrid approach because it is a conservative solution in case that delegates perform flush protocol and members send ACK messages[1] as same as traditional one, but improves the inherent scalability of multicast nature by gossip-based technique in updating group membership list with fully decentralized methods.

An Example and Algorithm Description. Figures 10 shows an example of the subscriber-delegate based protocol with a failed process p3 and lost messages. This example illustrates Group1={p1, p2, p4}, Group2={p1, p3, p4}, Local-View1={p1,p2,p3}, Local-View2={p2,p3,p4}, Gossip-Target = 2, and maximum number of gossip rounds = 2. The first sending process p1 multicasts m1 to gossip-targets, {p2, p3} selected from Local-View1={p1,p2,p3}. But, this message sent to p2 is lost in transit. On receiving m1 of Group1, p3 discards m1 because it doesn't belong to Group1 but gossips about m1 to gossip-targets, {p2, p4} selected from Local-View2={p2,p3,p4}, because it is one of delegates. Process p4 of Group1, it delivers m1 to the application layer after comparing causal ordering graph and gossips about m1 to {p2} except {p3}, the recent gossip-sender. But m1 sent to p2 is lost in transit. Process p3 of Group2 generates m3 and sends it with summary of predecessor, m1 as a message of Group1 and gossips about m3 to gossip-targets, {p2, p4} selected from Local-View2={p2,p3,p4}. After sending the message, p3 fails and m3 sent to p4 is lost in transit.

```
Procedure RECEIVE_MULTICAST
_m_is_Interest
if m in Group_ViewID_ContextGraph then
   m.gossip_count + ONE
else if m not in Group_ViewID_Pending_List then
   put m into Group_ViewID_Pending_List
for each p in Partial_ViewID with probability rate
   send Gossip(m) to p
     delivery = TRUE
for all item r=[m.Group_ViewID_Order] as ancestors do
   if r not in Group_ViewID_ContextGraph up to R-level then
        send solicit_Retransmission(r) to Latest_Gossip_Sender
        delivery = FALSE
if(delivery = TRUE) then
   remove m from Group_ViewID_Pending_List
   for all item s=[m.id, m.Group_ViewID_mSeq] as leaves
      in Group_ViewID_ContextGraph do
         attach m to every s
   deliver m to APPLICATION
_m_is_Not_Interest
call Procedure SEND_GOSSIP
Procedure RECEIVE_LEAVE_RELIABLE
_Process_P
if m=p'Last_q_mSeq not same in FLUSH then
  change m.Last_q_mSeq
  Send Reliable_MULTICAST(ACK(m)) to Delegates_D
     in Partial_View
_Delegates_D
m = (q_id, Group_ViewID, Last_q_mSeq, gossip_count)
Send Reliable_MULTICAST(FLUSH(m)) to Process_P
   in Partial_View
After receiving ACK from all members,  Update Last_MSG and
  Group_ViewID
call Procedure SEND_GOSSIP
```

Fig. 8. Algorithm Description(*Cont'd*)

```
Procedure INITIALIZE
id = p, Group_ViewID={p}, Group_ViewID_Order ={ },
   Partial_ViewID={p}
Procedure RECEIVE_LEAVE
Update Group_ViewID and Partial_ViewID After MGR
call Procedure SEND_GOSSIP
Procedure RECEIVE_JOIN
Update Group_ViewID and Partial_ViewID After MGR
call Procedure SEND_GOSSIP
Procedure SEND_MULTICAST
Group_ViewID_mSeq = Group_ViewID_mSeq+ 1
gossip_count=ZERO
Group_ViewID_Order={ }
for all item s=[m.id, m.Group_ViewID_mSeq] as leaves
   in Group_ViewID_ContextGraph do
        attach m to every s
for all item s=[m.id, m.Group_ViewID_mSeq] as ancestors
   upto R-level in Group_ViewID_ContextGraph do
     put s into Group_ViewID_Order as m's predecessors
     m = (id, Group_ViewID_mSeq, gossip_count,
          Group_ViewID_Order)
Unreliable_Multicast(m)
Procedure SEND_JOIN
m = (q_id, Group_ViewID, gossip_count)
call Procedure SEND_GOSSIP
Procedure SEND_LEAVE_RELIABLE
m = (q_id, Group_ViewID, Last_q_mSeq, gossip_count)
Send Reliable_Multicast(LEAVE(m)) to Delegates_D
   in  Partial_View
Procedure SEND_DIGEST
for all m in Group_ViewID_ContextGraph do
     m.gossip_count = m.gossip_count + ONE
Digest(Group_ViewID_ContextGraph))
call Procedure SEND_GOSSIP
Procedure SEND_GOSSIP
for each p in Partial_ViewID with probability rate
   send Gossip_MSG to p
do Garbage_Collection
```

Fig. 9. Algorithm Description

Thus, each local view is changed. The one becomes {p1, p2} and the other becomes {p2, p4}. Process p4, the first process to find out about failed p3, generates and sends a reliable message verifying the failed process p3 to the one of delegates, {p2} selected from its local view={p2,p4} without the last message information about p3, because p4 didn't receive m3 generated by p3. Although p2 does not belong to Group2, it can have the information of a leaving member and its last messages because it is shared by Local_View1= {p1, p2} and Local_View2= {p2, p4}. One of delegates, p2 updates the last message as m3, selects {p1} from Local_View1 and {p4} from Local_View2, respectively, and send the leave member information to the members of Group2, {p1, p4} using flush protocol. On receiving FLUSH message and the last message, m3, p1 and p4 send ACK to the delegate member={p2}. The delegate member={p2} receiving ACK messages from the members of Group2={p1,p4} updates the latest version of the last message as m3 and gossips about new group membership list as {p1,p4}. Process p1 and p4 receiving the new membership list deliver m3 to the application layer after comparing causal ordering graph and after maximum gossip round, they start changing the membership list of Group2 as {p1, p4} by themselves. Also, we can see that like the cases in figures 8, the last message m3 is eventually delivered to non-faulty members of Group2 by p2, even though p2 does not belong to Group2, p3 fails and m3 is lost in transit. The one of delegate member={p2} can receive an information about leave member from other members using reliable group communication, update new group membership and gossips about it to other members of Group2 with reasonable reliability and consistency guarantees because p2 is shared in the two local views.

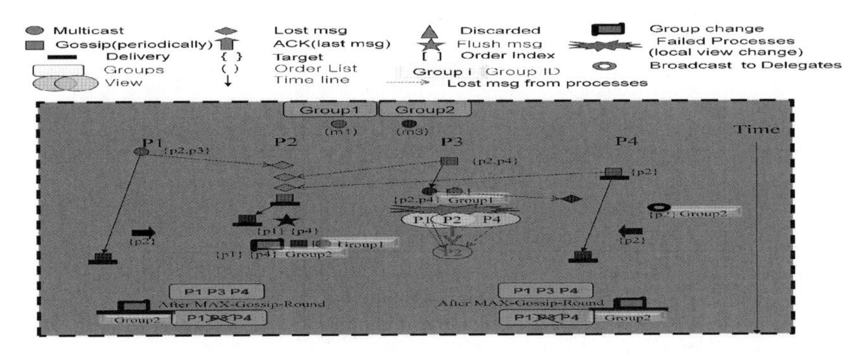

Fig. 10. An example of a failed member and lost messages

3 Performance Comparison

In this section, we compare average throughput of our sensor-broker based protocol with that of a previous hierarchical membership protocol based on traditional reliable delegates [1] and reliability of our subscriber-delegate based protocol with that of a fully decentralized membership protocol based on a partially randomized local views[4]. In this comparisons, we rely on a set of parameters referred to Bimodal Multicast[2] and LPBCast[4]. And we assume that processes gossip in synchronous rounds, gossip period is constant and identical for each process and maximum gossip round is logN. The probability of network message loss is a predefined 0.1% and the probability of process crash during a run is a predefined 0.1% using UDP/IP. The group size of each sub-figure is 32(2), 64(4), 128(8) and 256(16).

Figure 11 shows the average throughput as a function of perturb rate for various group sizes. The x-axis is the group size (the number of overlapped groups) and the y-axis is the number of messages processed in the perturb rate, (a)20%, (b)30%, (c)40%

and (d)50%. In the four sub-figures from 11(a) to 11(d), the average throughput of causally ordered delivery protocol based on sensor-broker is not a rapid change than that of the protocol based on traditional reliable delegates. Especially, the two protocols are compared to each other in terms of scalability by showing how the number of messages required for maintaining membership list in perturbed networks with processes join and leave. Our sensor-broker based protocol is more scalable because the brokers are selected by peer-sampling services[6],[7] and all messages including join and leave are gossiped by them. And in the four sub-figures from 12(a) to 12(d), the reliability of causally ordered delivery protocol based on fully decentralized views is more slightly sur-linear than that of our protocol based on subscriber-delegate. These results imply that our proposed protocol is not seriously affected by becoming larger of the group size and the gossip-targets size. Also, we can see that the message complexity becomes much bigger as the gossip-target size increase. The message complexity of our proposed protocol doesn't increase significantly because of highly reasonable reliability guaranteed based on subscriber-delegate. However, that of the protocol based on fully decentralized views increases much more in a sur-linear way as the gossip-target size is larger because of more frequently disseminated messages, especially, at high perturb rates.

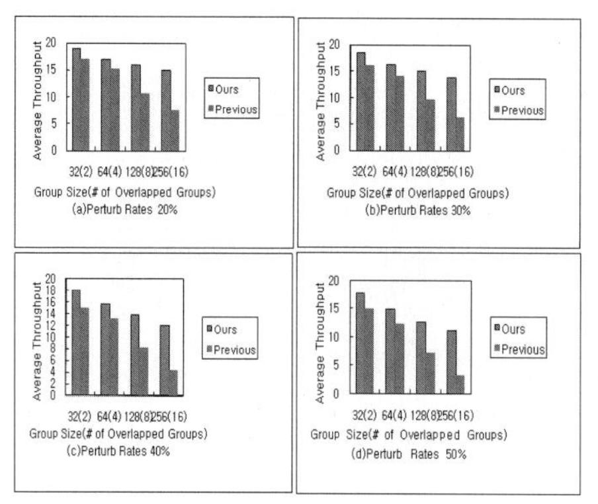

Fig. 11. Average Throughput by Perturb Rates

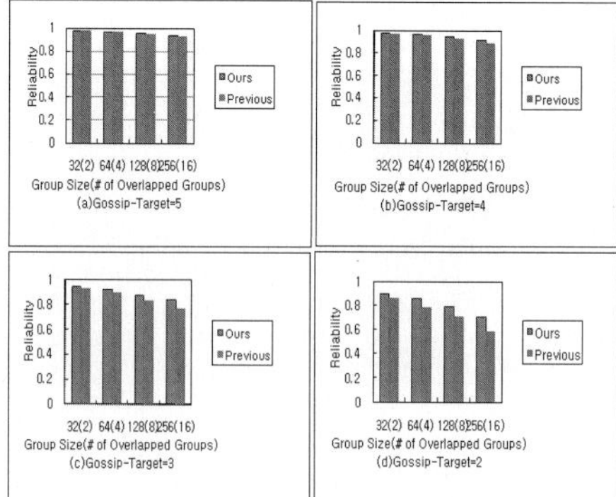

Fig. 12. Reliability by Gossip-Targets

4 Conclusion

In this paper, we present two versions of causal order guaranteeing multicast protocol. The one is sensor-broker(publisher-broker) based protocol that the brokers as delegates periodically gossip about the messages in overlapped multicast groups and guarantee causally ordered delivery of the messages. And the other is subscriber-delegate based protocol that all messages including node join/leave and causally ordered message delivery information are communicated by the gossip-style dissemination based on partially randomized local views with a part of delegates, selected to represent each group interesting to their subscribers. In sensor-broker based protocol, some brokers lead to make overlay networks and query nodes subscribe to the corresponding brokers networks of their interest and receive the results of the queries, aggregated messages of the information. And the information of vector by each interesting group piggybacked on every multicast message for causally ordered delivery are sent or received periodically using gossip-based protocols by sensor-broker. So, these features of this protocol might be that causally ordered delivery properties by sensor-broker are the same as those properties by delegates[1] but result in its very low communication cost compared with the cost incurred by hierarchical member list based on global membership[1] because of gossip-style disseminations. In subscriber-delegate based protocol, on the other hand, one member detecting a failed process sends a message about its failure and lastly sent message to the subscriber-delegate by traditional reliable communication protocols for node join and leave processing. And subscriber-delegate perform flush protocol[1] for holding newly generated messages until all members update each group membership. The subscriber-delegate receiving ACK messages from all members update the latest version of the last message and gossip about newly changed group membership based on randomized local views. After the maximum gossip round, all members start the membership change by themselves. Multicast messages including a causal ordering graph[8] on the basis of each overlapped group ID are sent and received based on gossiping for causally ordered delivery. So, the proposed subscriber-delegate protocol is a hybrid approach because it is a conservative solution in case that delegates perform flush protocol and members send ACK messages[1] as same as the traditional one, but improves the inherent scalability of multicast nature by gossip-based technique in updating group membership list with fully decentralized methods on local views including a part of subscriber-delegate. And for future works, we clearly show the pros and cons according to applications by comparing two different protocols, the sensor-broker based one and the subscriber-delegate based one.

Acknowledgments. This work was supported by Gyeonggido Regional Research Center program grant (References Developing and Industrializing Core Technologies of the Smart Space Convergence Framework).

References

1. Birman, K., Schiper, A., Stephenson, P.: Lightweight Causal and Atomic Group Multicast. ACM Trans. Comput. Syst. 9(3), 272–314 (1991)
2. Birman, K., Hayden, M., Ozkasap, O., Xiao, Z., Budiu, M., Minsky, Y.: Bimodal Multicast. ACM Trans. Comput. Syst. 17(2), 41–88 (1999)
3. Eugster, P., Guerraoui, R.: Probabilistic Multicast. In: the 2002 International Conference on Dependable Systems and Networks (DSN 2002), pp. 313–324. IEEE Computer Society Press, Vienna (2002)
4. Eugster, P., Guerraoui, R., Handurukande, S., Kouznetsov, P., Kermarrec, A.-M.: Lightweight probabilistic broadcast. ACM Trans. Comput. Syst. 21(4), 341–374 (2003)
5. Intanagonwiwat, C., Govindan, R., Estrin, D.: Directed diffusion: A scalable and robust communication paradigm for sensor networks. In: The Sixth Annual International Conference on Mobile Computing and Networking (MobiCOM 2000), pp. 56–67. ACM SIGMOBILE, Boston (2000)
6. Jelasity, M., Voulgaris, S., Guerraoui, R., Kermarrec, A.-M., Steen, M.: Gossip-based Peer Sampling. ACM Trans. Comput. Syst. 25(3), 1–36 (2007)
7. Pleisch, S., Birman, K.: SENSTRAC:Scalable Querying of SENSor Networks from Mobile Platforms Using TRACking-Style Queries. Int. J. of Sensor Networks 3, 266–280 (2008)
8. Peterson, L., Buchholzand, N., Schlichting, R.: Preserving and using context information interprocess communication. ACM Trans. Comput. Syst. 7(3), 217–246 (1989)

On Reducing the Impact of Exceptional Conditions on Museum Sightseeing Crowdedness Control Mechanisms

Yoondeuk Seo and Jinho Ahn

Department of Computer Science University of Kyonggi,
Suwon-si, Gyeonggi-do, Korea
{seoyd,jhahn}@kgu.ac.kr

Abstract. As ubiquitous computing technology convergences into many industrial domains, most museums want to apply this technique to their domain. However, most ubiquitous museums merely provide the simplest service giving visitors only static information of artifact. To resolve limitations of the existing ubiquitous museums, we had proposed Visitor Preference based Museum Viewing Search Algorithm that provides the best path for visitors to reflect their preferences. However, since the Visitor Preference based Museum Viewing Search Algorithm did not consider abnormal situations that occur while visitors look at the exhibit. So when abnormal situations occur, the exhibition may cause congestion problems that may make visitors feel very uncomfortable. In this paper, we propose an efficient congestion control algorithm to solve these problems. This algorithm automatically re-finds proper alternative paths for avoiding congestion resulting from the abnormal ones occurring during the museum viewing. The proposed algorithm improves comfortable museum viewing services by preventing congestion in advance when exceptional conditions occur. For the experiment of the proposed algorithm, we show that the algorithm can provide the best path without congestion exhibition.

Keywords: Distributed systems, Ubiquitous computing, RFID, Greedy, Preference-based.

1 Introduction

'Ubiquitous' is originated from a Latin word, and means 'anytime, anywhere' or 'everywhere' etc. The term was known to the general public, when Mark Weiser of Xerox Company introduced the concept of the 'Ubiquitous Computing' in 1988. Ubiquitous computing is the method of enhancing computer use by making many computers available throughout the physical environment, but making them effectively invisible to the user [1]. A ubiquitous network society is a society where it is possible to seamlessly connect "anytime, anywhere, by anything and anyone", and to exchange a wide range of information by means of accessible, affordable and user friendly devices and services [2].

The significantly increasing interest in the potential of ubiquitous computing technology leads to enormous related research and development activities being

T.-h. Kim, A. Stoica, and R.-S. Chang (Eds.): SUComS 2010, CCIS 78, pp. 206–212, 2010.

ongoing actively. Also currently in Korea, ubiquitous computing business is a kind of popular variety unfolding in public and private sectors. Although this trend advances related technology such as high performance, short-range wireless communication technologies, RFID, USN, and home network, initially, each technology independently is being applied and developed. However, all of these different technologies are being converged and then development of much more various ubiquitous services and technologies through this convergence is considerably accelerating.

Beyond the simple concept formulation phase, ubiquitous devices with development of ubiquitous computing technology are increasingly embodied in many areas such as u-City [3], u-Health [4], u-traffic [5], u-education, and u-distribution / logistics [6]. As ubiquitous computing technology converges into many industrial domains, this promising gear allows museum tour culture to be dramatically changing in terms of convenience and information richness. Using this technology, the existing museums may provide not only the ancient artifact's text-formed information but also many different kinds of information such as sound or media [7-9].

However, the most ubiquitous museums merely provide the simplest services that give visitors only static information about artifacts without fully utilizing smarter and high level of ubiquitous computing technology. In addition, the most ubiquitous museums provide several uniform paths to every visitor, not properly reflecting his or her accommodation and requirements.

To resolve limitations of the existing ubiquitous museums, we had proposed Visitor Preference based Museum Viewing Search Algorithm that provides the best path for visitors to reflect their preferences. However, since Visitor Preference based Museum Viewing Search Algorithm did not consider abnormal situations that may occur while visitors look at the exhibits. When abnormal situations occur such as resting somewhere privately, finding new interests and entering the exhibition that is not assumed to be in visitor's selected path, the exhibition may be highly congested, making the visitors feel a big discomfort resulting from the crowded exhibition.

In this paper, we propose an efficient congestion control algorithm to solve these problems. This algorithm automatically re-finds proper alternative paths for avoiding congestion resulting from the abnormal ones occurring during the museum viewing. The proposed algorithm may significantly improve the level of comfort of museum viewing services by preventing congestion in advance when abnormal situations occur. When abnormal situations occur during the museum viewing, the proposed algorithm can detect these conditions and visitors continue doing their comfortable museum viewing by getting their proper new paths through our automatically re-finding path selection methods.

2 Related Work

The National Museum of Korea services PDA image guidance system [10]. PDA image guidance system is a mobile service. PDA image guidance system is a new concept museum guidance system which provides visitor`s current position and information about optimized, but uniform viewing movement as well as simply information about exhibits to visitors through mobile device (PDA). When visitors who rent PDAs stand in front of artifacts and exhibits, the PDA image guidance system provides information about artifacts and exhibits by video and audio through information interchange infrared sensors of mobile devices and infrared generator installed artifacts.

However, if you wear a PDA around your neck during 1-2 hours, your neck will be tired because of considerable size and weight. In addition, since font size is very small, visitors may feel uncomfortable to use.

The Louvre museum provides a multimedia guide service through PDA [11]. The multimedia guide service provides information about exhibits by museum staffs in more than seven languages such as French, English, German, Spanish, Italian, Japanese, Korean, etc.

To use the multimedia guide service, visitors enter artifact's ID on the right or left side of the corresponding artifact. Then visitors may hear information about artifacts. The multimedia guide service provides a variety of information about their authors and copyright information as well as information of exhibits. Also visitors freely enjoy the exhibits by selecting their favorite tour courses according to theme, duration, and the difficulty.

However, the multimedia guide service causes the problem that the exhibitions are seriously congested or are crowded by visitors in popular and famous exhibitions only by providing several uniform courses.

The National Science Museum operates u-museum services using Mobile RFID technology [12, 13]. Visitors can utilize U-museum services by inputting their basic information, name, telephone number, e-mail, etc, after visitors has leased dongles in mobile RFID terminal rental at the museum entrance. Visitors place their phones close to exhibits and can obtain a variety of multimedia information though the exhibits or their web servers. Visitors can hear the narration about description of exhibits by earphones and see description of exhibits using mobile phones.

U-museum services have many advantages that visitors provide comfortable service by their mobile phones. But u-museum services are being limited only in some forms of the provided static services.

The Seoul Museum of History provides the u-exhibit guidance system [14]. The u-exhibit guidance system guides a variety of facilities as well as provides information about exhibits. The u-exhibit guidance system is linked to the museum website, providing richer museum information with its visitors. The u-exhibit guidance system provides automatically description of exhibits and artifacts in a variety of forms, flash video, photos, text, voice, etc, and the exhibition guides without visitor`s special operation using ubiquitous sensors. In addition, the u-exhibit guidance system can provide detailed and professional information about exhibits through touch screen, so visitors are easy to acquire the desired information.

However, although a variety of ubiquitous technologies being applicable for enabling visitors to much more easily access and be immersed in the ubiquitous museum according to their interests, but the u-exhibit guidance system only provides uniform and static types of information. In addition, PMP is too big and heavy, so visitors experience inconvenience.

3 Congestion Control Algorithm

If visitors can take a break during the museum viewing, it makes time differentiation between algorithm providing entering time and visitor's actual entering time for the exhibition which causes congestion of the exhibition. And this congestion may occur in any exhibition by visitor's unexpected behaviors like: they could find new interests on their way, be confused about the exhibition, and enter the wrong room.

To solve this problem, we attach a sensor on entrance of each room in order to monitor that visitors are coming in and out of the room [15]. If the unexpected behavior occurs, it causes huge differentiation between the time given by Visitor Preference based Museum Viewing Search Algorithm and actual visitor's entering time. We analyze and then provide alternative, but proper paths for the corresponding visitors. By this process, we enable visitors to continue their viewing comfortably on the fly regardless of their unexpected behaviors.

And also, if a visitor enters the exhibition that is not in the recommended path, our system will notify visitors of this situation and set the exhibition to the next starting point for them in order to re-find their proper path. After that, we provide the re-selected path to the visitor, which makes him or her enjoy the rest of their viewing without feeling inconvenience.

And if visitors find some new interests, they can request re-finding their appropriate paths considering their interests.

Figure 1 shows how our system may handle abnormal situations appropriately during visitor's museum viewing.

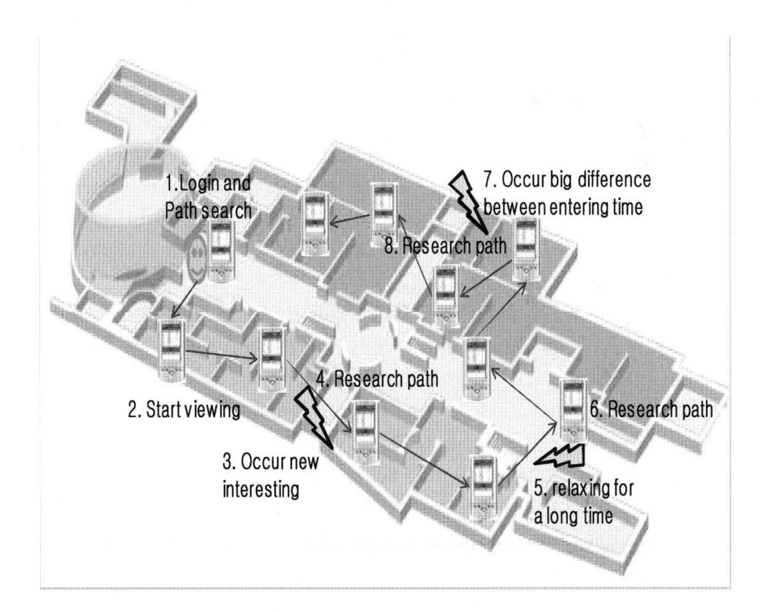

Fig. 1. An example scenario of abnormal situation

4 Experiments

Experiments are performed to evaluate how effectively the proposed congestion control algorithm can help visitors do their comfortable viewing even in abnormal situations. Our performance evaluation environment is as follows.

Table 1. Performance Evaluation Environment

Classification	Description
Operating System	Windows XP Service Pack3
Implementation Language	C#
CPU	Intel® Core™ 2 Duo CPU E8300 @ 2.83GHz
Memory	2GB

For this, experimental conditions are set as follows: Experiments were carried out by emulating the situation that visitors should enter into the museum within at least every five minutes. In our experiment, one minute of visitor's actual viewing is translated into one second in simulation time. Under the assumption mention earlier, visitors entering the museum are randomly generated up to 100 persons. It is assumed that each exhibition is held only at one room. In here, the maximum number of visitors each exhibition may accommodate without making them feel discomfort resulting from congestion is set up to 10 persons. The number of exhibitions is set to 43.

The initial experimental procedure for each visitor entering the museum is following. First, he or she randomly selects from at least 5 to up to 43 exhibitions he or she wants to see. Then, his or her phone sends the list of rooms selected to our system server. Next, it gets back visitor's appropriate path from the server after the server has analyzed with the selection information through our congestion control algorithm. In order to find out the degree of congestion that occurs during the museum viewing, the number of visitors in each exhibition is checked every 10 minutes intervals from the time of the first entrance into the exhibition to the time of the last exit out of the exhibition after all visitors have finished their viewing with or without our congestion control algorithm.

Figure 2 shows the effectiveness of our congestion control algorithm applied in an abnormal situation. In this figure, the x-axis represents the room number for each exhibition and the y-axis, the highest number of visitors entering the exhibition measured every ten minutes. When our algorithm isn't applied, a lot of exhibition rooms are seriously congested, having one and half times more than the affordable number of visitors recommended for each room. However, with our proposed congestion control algorithm, the number of rooms having each more than ten persons decreased dramatically. In this case, the reason a few places have more than 10 persons even after executing our algorithm, is that, among all the exhibition rooms, some are very interesting spots and some others, unpopular ones. So, from these results, we can see that our congestion control algorithm are very effective when unexpected behaviors occur to provide high quality of comfort in museum viewing for visitors.

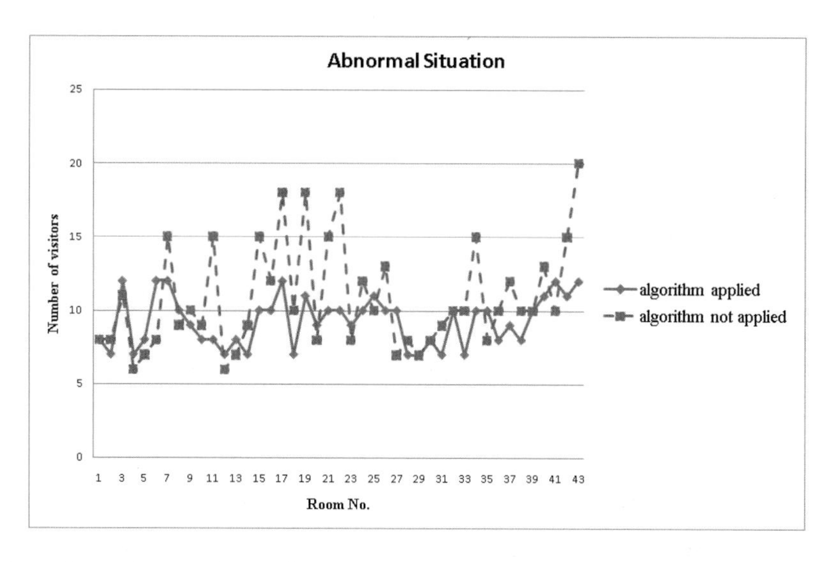

Fig. 2. The highest number of visitors entering each exhibition measured every ten minutes

5 Conclusion

As ubiquitous computing technology converges into many industrial domains, this promising gear allows museum tour culture to be dramatically changing in terms of convenience and information richness.

However, most ubiquitous museums merely provide the simplest forms of services giving visitors only uniform and static information of artifact, not fully utilizing smarter and high level of ubiquitous computing technology. In addition, the convenience of visitors is not reflected properly to their viewing. It allows them to follow several uniform paths, which makes them feel a big discomfort resulting from crowded exhibition.

To resolve limitations of the existing ubiquitous museums, we had proposed Visitor Preference based Museum Viewing Search Algorithm that provides the best path for visitors to reflect their preferences. However, since the Visitor Preference based Museum Viewing Search Algorithm did not consider their unexpected behaviors like finding new interests during their viewing, relaxing suddenly and privately during viewing, etc.,.

In this paper, we proposed an efficient congestion control algorithm. The algorithm helps visitors to have a more comfortable viewing by preventing congestion that could be caused by their unexpected behaviors

Acknowledgments. This work is supported by CCSRC of Gyeonggi-do Regional Research Center Project(Project Title: Developing and Industrializing Core-Technologies of the Smart-Space Convergence Framework.).

References

1. Weiser, M.: Some computer science issue in ubiquitous computing. Communications of the ACM 36(7), 75–84 (2003)
2. Yau, S., et al.: Reconfigurable Context-Sensitive Middleware for Pervasive Computing. IEEE Pervasive Computing 1(3), 33–40 (2002)
3. Lee, S.W., Choi, S.M., Ku., J.E.: Regulatory Issues on u-City Revitalization. Electronics and Telecommunications Research Institute 24(2), 77–83 (2009)
4. Lee, M., Han, D., Jung, S., Cho, C.: A Platform for Personalized Mobile u-Health Application Development. Korean Institute of Information Scientists and Engineers 35(2), 154–158 (2008)
5. Lee, S., Choi, Y.-W.: A Study on the Development of u-Traffic User Services. Journal of the Korean Society of Civil Engineers 27(1), 43–51 (2007)
6. Lee, M.Y., Kim, M.J.: Trend of Event-Driven Service Technology. Electronics and Telecommunications Research Institute 21(5), 61–68 (2006)
7. Saha, D., Mukherjee, A.: Pervasive Computing: A Paradigm for the 21st Century. IEEE Computer 36(3), 25–31 (2003)
8. Hightower, J., Borriello, G.: Location Systems for Ubiquitous Computing, vol. 34(8), pp. 57–66. IEEE Computer Society Press, Los Alamitos (2001)
9. Hung, N.Q., Ngoc, N.C., Hung, L.X., Lei, S., Lee, S.: A Survey on Middleware for Context-Awareness in Ubiquitous Computing Environments. Korean Information Processing Society Review (2003) ISSN1226-9182
10. The National Museum of Korea, `http://www.museum.go.kr`
11. The Louvre museum, `http://www.louvre.fr`
12. The National Science Museum, `http://www.science.go.kr`
13. Hwang, D.-R.: A Study on the Functional Change and Application Scheme of U-library/museum. Journal of the Korean Society for Library and Information Science 41(4), 181–199 (2007)
14. The Seoul Museum of History, `http://museum.seoul.kr`
15. Kawahara, Y., Minami, M., Morikawa, H., Aoyama, T.: Design and Implementation of a Sensor Network Node for Ubiquitous Computing Environment. In: IEEE VTC2003-Fall. IEEE Press, Orland (2003)

A Communication Architecture for Monitoring and Diagnosing Distribution Systems*

Yujin Lim[1], Sanghyun Ahn[2], and Jaesung Park[3]

[1] Department of Information Media, University of Suwon,
2-2 San, Wau-ri, Bongdam-eup, Hwaseong-si, Gyeonggi-do, 445-743, Korea
[2] School of Computer Science, University of Seoul,
90 Jeonnong-dong, Dongdaemun-gu, Seoul, 130-743, Korea
[3] Department of Internet Information Engineering, University of Suwon,
2-2 San, Wau-ri, Bongdam-eup, Hwaseong-si, Gyeonggi-do, 445-743, Korea
{yujin,jaesungpark}@suwon.ac.kr
ahn@uos.ac.kr

Abstract. In power distribution system, a sink of each wireless sensor network is required to deliver monitoring data or fault diagnosis to a gateway for ultimate delivery to a monitoring center located in an external network. In this paper, we propose an efficient path management mechanism between the sink and the gateway. We have proved that our proposed mechanism performs much better than existing routing mechanisms in terms of the service time and the data loss rate.

1 Introduction

In this paper, we consider the utilization of wireless sensor networks for data transfer needs of condition monitoring, maintenance and remote diagnosis for power distribution system. Data delivery mechanism is proposed to deliver monitoring data and fault diagnosis from large scale distant transformer to remote monitoring center.

Typically, the wireless sensor network is a multi-hop wireless network whose main purpose is to deliver sensed data collected from multiple sensors to one or more data collecting devices (or sinks) [1]. Wireless sensor networks have a wide range of applications. Especially, wireless sensor networks are useful in continuously acquiring information in inaccessible or perilous areas for some time duration. In this case, sensed data collected at a sink have to be delivered to a remote monitoring center through an intermediate system, such as a gateway (GW), connected to an external network like the global Internet. In general, sensors are static and densely deployed within an area, so they can be easily connected to a sink. However, unlike

* This research was supported by Basic Science Research Program through the National Research Foundation of Korea (NRF) funded by the Ministry of Education, Science and Technology (No.2010-0017251).

This research was supported by the MKE (The Ministry of Knowledge Economy), Korea, under the HNRC (Home Network Research Center) –ITRC(Information Technology Research Center) support program supervised by the NIPA(National IT Industry Promotion Agency (NIPA-2010- C1090 - 1011 – 0010).

T.-h. Kim, A. Stoica, and R.-S. Chang (Eds.): SUComS 2010, CCIS 78, pp. 213–221, 2010.

static and densely deployed sensors, sinks can be static or mobile and are relatively sparsely deployed, so providing the data delivery service between sinks and the GW may be difficult without any other devices or mechanisms. Thus, an additional relay network may have to be deployed for the support of the data delivery between sinks and the GW. The relay network is constructed with relay points (RPs) and provides the data delivery between sinks and the GW. RPs are supposed not to generate or consume data, but to just forward data from sinks to the GW. In this paper, we assume RPs are placed in a grid pattern and the performance of the relay network is not subject to the varying location and the number of sinks.

One of the issues to be considered on the relay network is the efficient path management. In the relay network, the destination of monitoring data from a number of transformers (i.e., sensors) is one single point, the GW. Therefore, we propose a mechanism that can improve the service time and the service quality of the relay network through the efficient management of paths between sinks and the GW within the large-scale power distribution system.

2 Related Work

Since data measured at a sensor is transmitted to the sink, sensors near to the sink are congested with network traffic. Thus, the network lifetime is reduced because the sensors near to the sink unduly consume more energy than others. In order to overcome this problem, the mobile sink mechanisms have been proposed. Because a mobile sink collects data from sensors while moving within a network, the moving path and the mobility pattern of the sink can affect the overall network performance. In the case when the moving path of the sink can be known in advance, the data delivery path for data collection can be optimized. The optimal trajectory of the sink requires the location information of sensors for the sink to collect sensed data in a timely and efficient manner. However, it is difficult to determine the optimal trajectory of the sink in advance since sensors are usually randomly deployed. On the other hand, when the moving path of the sink cannot be known a priori, an additional mechanism for the successful data delivery to the sink from sensors is in need. TTDD [2], one of the most representative mobile sink mechanisms, prevents repetitive flooding of the frequently changing location information of the sink to the entire network.

3 Tree-based Path Management Mechanism

The tree-based path management mechanism proposed in this paper consists of tree construction, tree maintenance and directional flooding procedures. By the tree maintenance procedure, a failed tree branch resulted from an RP failure is restored. And, the directional flooding is used for the delivery of the first data message from a sink to the GW in order to reduce the data delivery delay.

When a mobile sink moves in or a static sink is powered on in the service area of an RP, the RP constructs a tree branch from itself to the GW for the delivery of data from the sink to the GW. If the RP already has a branch to the GW, it uses the existing branch. While a new tree branch is constructed, if it meets an existing tree branch, the new branch is merged with the existing one. If an RP detects the failure of its upstream RP, it starts the local repair procedure to guarantee the network connectivity. In our tree-based path management mechanism, because only the PROBE message is broadcasted by the GW at the network initialization stage and all other tree management procedure is carried out in a distributed fashion, the management overhead imposed on the GW is ignorable.

3.1 Tree Branch Construction

Cost factors in constructing the least cost path from an RP to the GW can be the distance or hop from the RP to the GW, the remaining energy of intermediate RPs or the number of sinks within the service area of the RP. From the PROBE message broadcasted by the GW, each RP can infer its path cost to the GW. This path cost is used in determining the serving RP of a sink and branch RPs on the tree branch from a serving RP to the GW. The detailed procedure of the tree branch construction is depicted in Fig. 1.

1) Serving RP Selection
The tree branch construction is initiated at an RP when the RP receives a data message from a sink for the first time. Since a wireless data transmission is broadcast in nature, data from a sink can be received by more than one RPs. Thus, an RP (i.e., serving RP) that is in charge of delivering data from the sink to the GW has to be decided. The serving RP of a sink is selected in the following distributed manner.

Each RP receiving data from a sink sets its timer value in proportion to the path cost from itself to the GW. If the path cost from an RP to the GW is k, the timer of the RP is set to a random value within [k-1, k) in order to reduce the probability of RPs with the same cost selecting the same timer value. The RP having not received any ADVERTISEMENT messages from its neighboring RPs for the timer interval declares itself as the serving RP by sending an ADVERTISEMENT message to its neighboring RPs. The RP having received an ADVERTISEMENT message before its timer expires assumes that another RP has been selected as the serving RP and cancels its timer and finishes the serving RP selection procedure.

2) Branch RP Determination
After the serving RP of a sink is selected, the serving RP continues the tree branch construction by sending a CONSTRUCT message to its neighboring RPs. For the selection of RPs on the tree branch from the serving RP to the GW, the serving RP includes the path cost from itself to the GW in the CONSTRUCT message.

Upon receiving the CONSTRUCT message, an RP compares the path cost in the message with the path cost from itself to the GW. If its path cost is less than that in the CONSTRUCT message, the RP becomes a candidate branch RP which can be a branch RP of the tree branch. Since there can exist more than one candidate branch RPs, one of the candidate branch RPs is chosen as an upstream branch RP (in short, upstream RP) of the serving RP by using a procedure similar to the serving RP selection procedure. That is, a candidate branch RP sets up its timer value proportional to the path cost from itself to the GW. If a candidate RP has not received any SUPRESS messages from its neighboring RPs during the timer interval, it becomes a branch RP. A new branch RP continues the branch RP selection procedure by sending a CONSTRUCT message with the path cost from itself to the GW. This process is repeated until when a CONSTRUCT message is delivered to the GW or an existing tree branch is met. The downstream branch RP (in short, downstream RP) having received a CONSTRUCT message from its upstream branch RP sends a SUPRESS message to make other candidate branch RPs terminate the branch RP selection procedure.

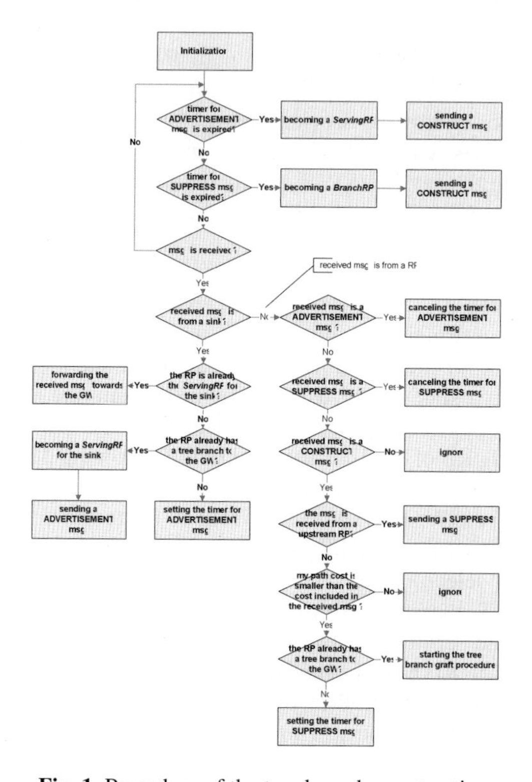

Fig. 1. Procedure of the tree branch construction

3.2 Tree Branch Graft

When an RP looking for its upstream RP encounters a branch RP, it can use the branch RP as its upstream RP and this will reduce the control overhead related to the tree branch construction and speed up the process. We call this the tree branch graft whose procedure is pretty similar to the branch RP construction except for the fact that the tree branch graft occurs when an existing branch RP rather than the GW is met.

Fig. 2 shows an example of the tree branch graft. The tree branch for sink1 has been already made with RP1, RP2, RP3 and RP4. If we assume that RP5 is the serving RP for sink2, RP5 sends a CONSTRUCT message to establish a tree branch for sink2. Since the message is forwarded via the one-hop broadcast, RP2 which already belongs to the tree branch for sink1 receives the CONSTRUCT message. If the path cost in the message is larger than that of RP2, RP2 sends an AFFILIATE message so that RP5 sets RP2 as its upstream RP in the tree branch for sink2. On the receipt of an AFFILIATE message, RP5 sends a SUPPRESS message to prevent other RPs from becoming the upstream RP. RP2 does not send a CONSTRUCT message and the tree branch construction for sink2 is completed.

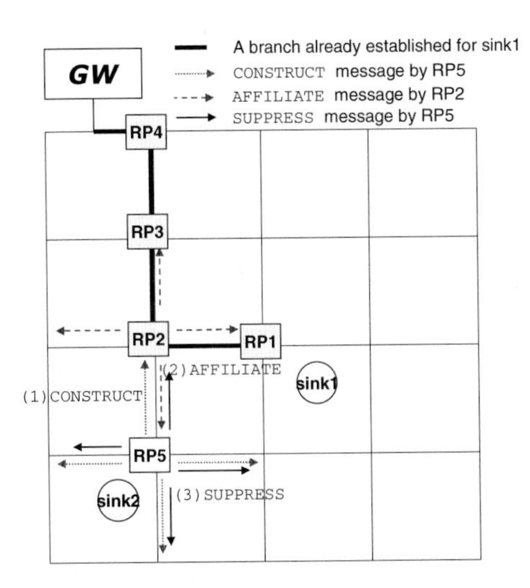

Fig. 2. An example of the tree branch graft

3.3 Tree Maintenance

Each branch RP maintains the soft state on its upstream RP and downstream RPs. Each serving RP maintains the information on its upstream RP and the sinks that it is serving. Each branch RP maintains the soft state on its downstream RPs by sending HELLO messages periodically. If a downstream RP does not receive any HELLO messages from its upstream RP for some time interval, it removes the information about the upstream RP. After that, the RP sends a CONSTRUCT message to initiate the tree branch reconstruction by performing the above mentioned tree branch construction and tree branch graft procedures.

When a branch RP does not receive data for a time interval (Data timer expiration), it assumes that it does not have sinks to serve through this tree branch and, then, releases all the states it maintains for the tree branch. This makes unnecessary tree branches pruned from the tree.

3.4 Directional Flooding

Before the tree branch construction is completed, the first data message from a sink is delivered to the GW via the directional flooding by the serving RP of the sink. This initial directional flooding is designed to facilitate the delivery of the first data message and to reduce the overhead of the original flooding by limiting the transmission of the message only to the direction closer to the GW. The RP having received the first data message from a sink includes its path cost in the data message and sends the message to its neighboring RPs via the one-hop flooding. Only the neighboring RP whose path cost is less than that in the message forwards the message to its neighboring RPs. Since the directional flooding is applied only to the first message, the effect of the directional flooding on the overall network lifetime is marginal.

4 Performance Evaluation

In this section, we compare the performance of the proposed tree-based mechanism with that of DSDV [3] and AODV [4] in terms of the following performance factors by carrying out the NS-2 [5] based simulations:

- **Control overhead** the value computed by dividing the number of control messages required for the path setup and maintenance by the service time of the relay network.
- **Service time of the relay network** the time duration from when the network is initialized to when the network is unable to provide a path from a sink to the GW.
- **Data delivery delay** the time needed to forward data from a sink to the GW.
- **Data loss rate** the number of lost data messages per second.

For the construction of the relay network, 11 × 11 RPs are placed in a grid pattern with the grid cell size being 24m × 24m. IEEE 802.11 is used for MAC protocol. We have used the shadowing model in [6] to model radio propagation environment inside of an office building. A sink sends a data message for every 2 seconds, and the number of sinks is set to 15% of the number of RPs (i.e., 20 sinks). Also, mobile sinks follow the random waypoint model [7], and channels for the communication between RPs are separated from those for the communication between RPs and sinks.

Fig. 3 ~ 6 show the control overhead, the service time of the relay network, the average delay of data delivery and the data loss rate when the maximum speed of mobile sinks is varied from 1km/h to 10km/h. The number in each legend of the figures represents the percentage of mobile sinks. For example, Tree-20% means the tree-based mechanism with mobile sinks being 20% of sinks. All of the simulation results are normalized by the results of the proposed tree-based mechanism with the percentage and the maximum speed of mobile sinks being 20% and 1km/h, respectively.

Fig. 3 shows the control overhead for path setup with varying the mobility speed of mobile sinks. In the tree-based mechanism and AODV, as the speed of mobile sinks or the number of mobile sinks increases, the possibility of mobile sinks moving to neighboring grid cells gets higher and, as a result, the number of path setup requests increases. Due to this, the control overhead increases twice on average as the sink mobility increases, and 1.2 times on average as the number of mobile sinks increases. On the other hand, DSDV is not affected by the sink mobility speed or the number of mobile sinks because DSDV periodically sends path update messages without regard to path setup requests. Therefore, we can see that, on average, the control overhead of AODV and that of DSDV is almost 20 and 40 times larger than that of the tree-based mechanism, respectively.

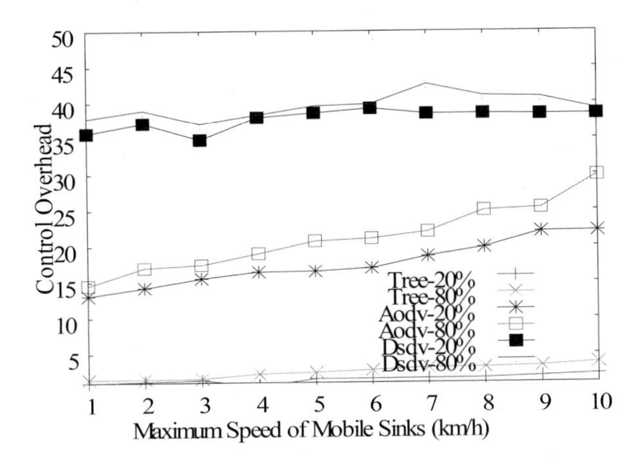

Fig. 3. Sink mobility speed *vs.* control overhead

Fig. 4 is the graph showing the service time of the relay network with various sink mobility speeds. The service time of the relay network can be a good measure for us to expect the maintenance cost of the relay network. The most affecting factor to the service time of the relay network is the path setup control overhead. As more control messages are generated by RPs, RPs consume more energy and result in failures. As mentioned in the paragraph related to Fig. 3, the service time of the relay network operating with flooding based DSDV or AODV is significantly less than that operating with the tree-based mechanism. For instance, even with the ratio of mobile sinks being 80% and the sink mobility speed being 10km/h, the service time of using the tree-based mechanism is at least 4 times larger than that of using DSDV.

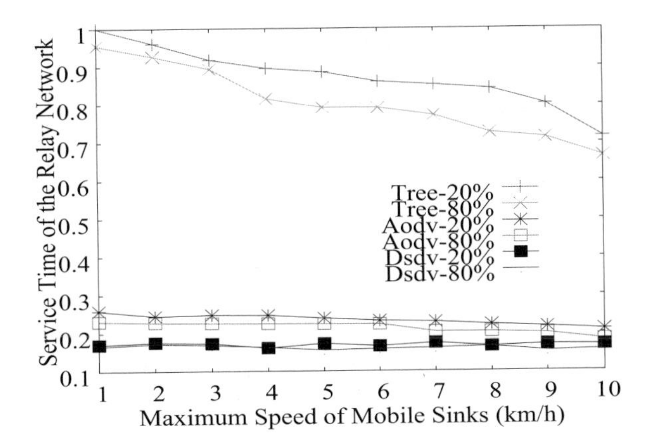

Fig. 4. Sink mobility speed *vs.* service time

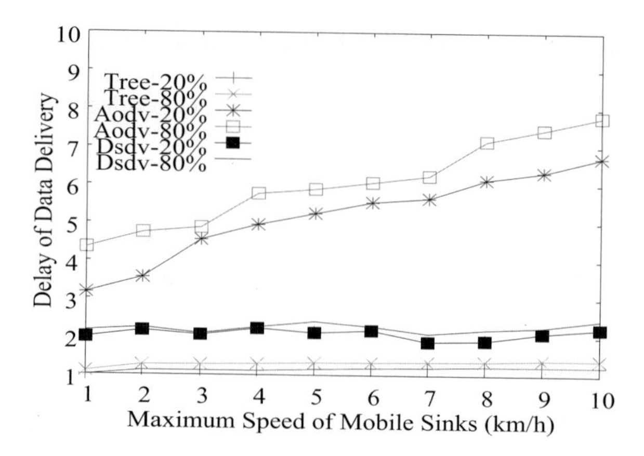

Fig. 5. Sink mobility speed *vs.* data delivery delay

Fig. 5 shows the data delivery delay. In AODV, since the path setup procedure is initiated when an RP receives data for the first time from a sink, the data delivery is delayed until the path from the RP to the GW is completely established. Thus, AODV produces the largest delay which is almost 5 times larger than that of the tree-based mechanism. On the other hand, since DSDV establishes all the paths from RPs to the GW regardless of data reception from sinks, data can be delivered from an RP without waiting for a path to the GW to be built up. As a result, DSDV gives much less data delivery delay than AODV, but DSDV requires longer initial path setup time because of flooding based path setup for all possible paths from all RPs to the GW at the network initialization stage.

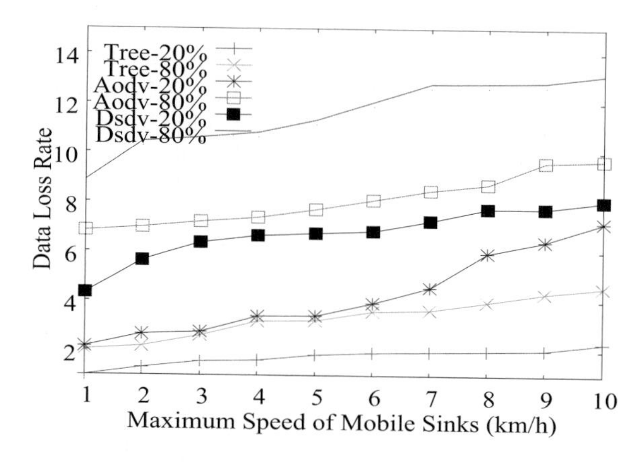

Fig. 6. Sink mobility speed *vs.* data loss rate

Fig. 6 shows the data loss rate for various sink mobility speeds. The data loss rate is affected by path recovery since data are buffered and removed from the buffer if the buffer overflows or the buffering time exceeds some pre-specified time (in our simulations, 30 seconds) during path recovery. In addition to path recovery, the sink mobility speed and the number of mobile sinks affect the data loss rate. As the sink mobility speed or the number of mobile sinks increases, the possibility of a mobile sink moving to another grid cell during path recovery gets higher. Thus, the data loss rate increases twice, on average, as the sink mobility speed or the number of mobile sinks increases. DSDV using flooding for path recovery experiences more data losses than AODV and the tree-based mechanism which restore failed paths by exchanging recovery related messages locally. The average data loss rate of DSDV is 6 times larger than that of the tree-based mechanism. Since AODV floods path recovery related control messages in almost 1/3 of the network, the data loss rate of AODV is almost 4 times larger than that of the tree-based mechanism which exchanges path recovery messages only with neighboring RPs.

5 Conclusion

In power distribution system, monitoring data and fault diagnosis collected at a sink have to be delivered via the GW to a remote monitoring center located in an external network like the global Internet. In this case, the deployment of an additional relay network was considered for the support of the network connectivity between sinks and the GW. In this paper, we proposed an effective path management mechanism for the relay network to provide the network connectivity between sinks and the GW and to improve the service time and the service quality of the relay network. From simulation results, we showed that our proposed mechanism outperforms the existing mechanisms in terms of the control overhead, the service time of the relay network, the data delivery delay and the data loss rate.

References

1. Romer, K., Mattern, F.: The Design Space of Wireless Sensor Networks. IEEE Wireless Communications Magazine 11(6), 54–61 (2004)
2. Ye, F., Luo, H., Cheng, J., Lu, S., Zhang, L.: A Two-Tier Data Dissemination Model for Large-Scale Wireless Sensor Networks. In: Proc. Annual International Conference on Mobile Computing and Networking (Mobicom), pp. 148–159. ACM Press, New York (2002)
3. Perkins, C., Bhagwat, P.: Highly Dynamic Destination-Sequenced Distance-Vector Routing (DSDV) for Mobile Computers. ACM SIGCOMM Computer Communication Review 24(4), 234–244 (1994)
4. Perkins, C., Belding-Royer, E., Das, S.: Ad Hoc On-Demand Distance Vector (AODV) Routing. IETF RFC 3561 (2003)
5. The network simulator, ns-2, http://www.isi.edu/nsnam/ns/
6. Rappaport, T.S.: Wireless Communications, principles and practice. Prentice Hall, New Jersey (1996)
7. Resta, G., Santi, P.: An Analysis of the Node Spatial Distribution of the Random Waypoint Mobility Model for Ad Hoc Networks. In: Proc. ACM Workshop On Principles Of Mobile Computing (POMC), pp. 44–50 (2002)

Network Infrastructure for Electric Vehicle Charging*

Yujin Lim[1], Jaesung Park[2], and Sanghyun Ahn[3]

[1] Department of Information Media, University of Suwon,
2-2 San, Wau-ri, Bongdam-eup, Hwaseong-si, Gyeonggi-do, 445-743, Korea
[2] Department of Internet Information Engineering, University of Suwon,
2-2 San, Wau-ri, Bongdam-eup, Hwaseong-si, Gyeonggi-do, 445-743, Korea
[3] School of Computer Science, University of Seoul,
90 Jeonnong-dong, Dongdaemun-gu, Seoul, 130-743, Korea
{yujin,jaesungpark}@suwon.ac.kr
ahn@uos.ac.kr

Abstract. Controlled charging of electric vehicles can take care of fluctuating electricity supply. In this paper, we design network infrastructure to collect and deliver data of charging data of electric vehicles to remote monitoring center. In our network infrastructure, we analyze and compare the existing routing mechanisms for multi-hop wireless networks from aspect of the control overhead for the path establishment.

1 Introduction

Electric Vehicle (EV) has attracted great attention in recent years due to its ultralow emissions, high fuel economy. Controlled charging of EVs can take care of fluctuating electricity supply. To control the charging of EVs, the design of network infrastructure is needed to collect and deliver charging data of EVs to remote monitoring center. Fig. 1 shows basic structure of EV charging. A charging station needs to interact with the remote monitoring center beyond simple collecting charging data of EVs. To do this, an additional relay network may have to be deployed for the support of the network connectivity between sinks in charging stations and a gateway (GW) in remote monitoring center.

Fig. 2 depicts our network infrastructure for EV charging consisting of 3 tiers. The lowest tier, the charging station tier, is composed of more than one charging stations each of which collects data and delivers from EV to the mobile or static sink of the station. In order to deliver data collected at sinks to the GW connected to the remote monitoring center, the relay network tier is constructed with relay points (RPs) and provides the network connectivity between sinks and the GW. RPs are supposed not

* This research was supported by Basic Science Research Program through the National Research Foundation of Korea (NRF) funded by the Ministry of Education, Science and Technology (No.2010-0017251).

This research was supported by the MKE (The Ministry of Knowledge Economy), Korea, under the HNRC (Home Network Research Center) –ITRC(Information Technology Research Center) support program supervised by the NIPA(National IT Industry Promotion Agency (NIPA-2010- C1090 - 1011 – 0010).

T.-h. Kim, A. Stoica, and R.-S. Chang (Eds.): SUComS 2010, CCIS 78, pp. 222–229, 2010.

to generate or consume data, but to just forward data from sinks to the GW. Typically, for monitoring applications requiring the relay network, such as a military zone, disaster area, or underwater, it is not expected to have the support of wired networks or power lines. Therefore, in this paper, we assume the multi-hop wireless relay network composed of static and battery-powered RPs.

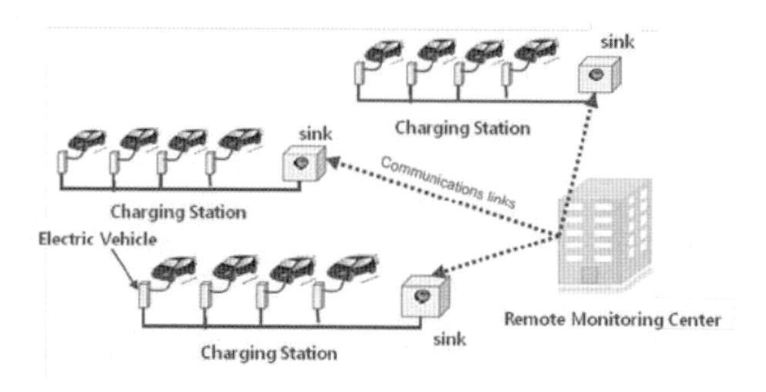

Fig. 1. Basic structure of EV charging

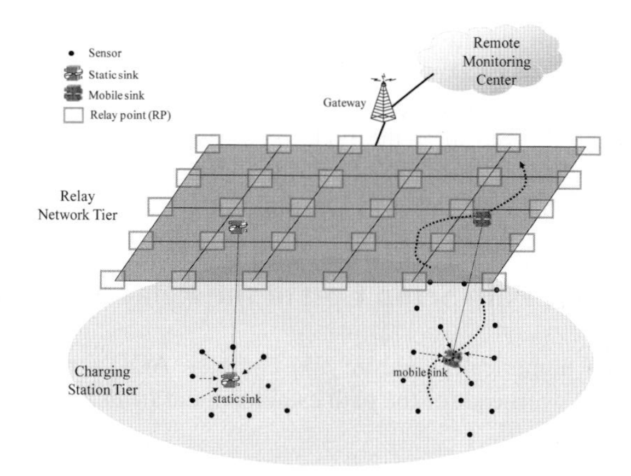

Fig. 2. Network infrastructure for EV charging

One of the issues to be considered on the relay network is the optimal placement of RPs. In our network infrastructure, since the number of sinks changes dynamically and the trajectory of mobile sinks cannot be predicted accurately, the optimal placement of RPs are not possible. Thus, in this paper, we place RPs in a grid pattern so that the performance of the relay network is not subject to the varying location and the number of sinks.

Another issue to be considered is the efficient path management. In the relay network, the destination of data from a number of charging stations is one single point, the GW. The traditional mobile ad hoc network (MANET) routing protocols [1] such as AODV [2] and DSDV [3] can be considered for path management between sinks and the GW in the multi-hop wireless relay network. While the MANET routing protocols are designed to deal with dynamic changes of the network topology, the topology of the relay network is relatively static. Moreover, an RP does not need to setup and maintain routes for all other RPs since the final destination of all data in the relay network is the GW. So, if a MANET routing protocol is used in the relay network, it may impose unnecessary overhead. Thus, in order to validate the performance of the existing routing mechanisms for multi-hop wireless networks, we numerically analyze the path setup cost from a sink to the GW.

2 Related Work

In AODV, a data source first requests a network to set up a path to a destination whenever it has data to send. If an intermediate node between the source and the destination does not have a path record, it creates a routing entry to the destination and forwards the request. The path record at an intermediate node is deleted if the record has not been used for its lifetime. Since a path is created when it is needed, the network does not have to maintain unnecessary path records.

In case of DSDV, each node in a network broadcasts its routing table at a predefined route update interval. Accordingly, each node can maintain path information to all nodes in the network. On the contrary to AODV, DSDV does not induce path setup delay because a path already exists whenever a source wants to send a data. However, the issue with DSDV is that the network convergence time to a stable state becomes longer with the increase of the network size because the routing table of a node is stabilized after the node receives the routing tables from all the other nodes in the network. Since the size of the routing table also increases according to the network size, DSDV may cause a scalability issue.

3 Analysis on the Path Setup Cost of the Relay Network

In this section, we numerically analyze and compare AODV with DSDV from the aspect of the control overhead for the path establishment from an RP to the GW.

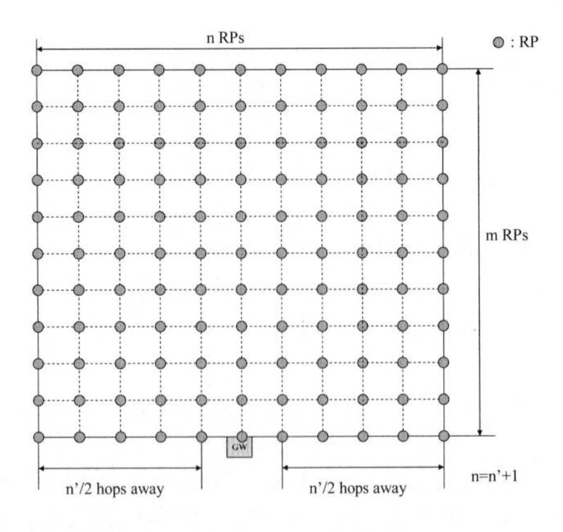

Fig. 3. A relay network in network infrastructure for EV charging

3.1 Relay Network Model

For the comparison of path management mechanisms, we consider a relay network with $nRP \times mRP$ in a grid pattern as shown in Fig. 3. We assume that one GW is located at the center of the lower line. If we assume that the data transmission range of an RP is the inner area of the circle with radius r and one RP is located at each grid point, the distance between two RPs at two farthest corners of a grid cell is $\sqrt{2}r$ and each RP has 4 neighboring RPs. We assume that the communication channel between a sink and an RP is different from that between RPs and, at the network setup stage, channels are assigned via a channel allocation mechanism so that channel interference can be minimized [4][5].

3.2 Analysis on the Path Setup Cost

In this subsection, we analyze the performance of AODV and DSDV in terms of the average path cost from an RP to the GW. Even though the size of each control message may be different, for the simplicity of the analysis, we assume that the transmission cost (e.g., determined by the message transmission power) of a control message is C_t and the receiving cost (e.g., determined by the message receiving power) of a control message is C_r. We also assume that a hop count is used as the path cost and no error occurs during the message transmission and reception.

In AODV, the data transmission path from an RP to the GW is established once a serving RP (an RP that is in charge of delivering data from the sink to the GW) of a sink broadcasts an RREQ message and the GW unicasts an RREP message to the serving RP as a response. Since the RREQ message is broadcast, all RPs in the relay network receive one RREQ message from each of its neighboring RPs and sends one RREQ messages to its neighboring RPs.

Since the relay network is composed of $nRP \times mRP$, the control message transmission cost of AODV becomes:

$$\begin{aligned} C_{AODV}^{RREQ} &= (n-2)(m-2)(C_t+4C_r)+2[(n-2)+(m-2)](C_t+3C_r)+4(C_t+2C_r) \\ &= nmC_t+2(2nm-n-m)C_r \end{aligned} \quad (1)$$

A branch RP (an RP on the tree branch from a serving RP to the GW) receives an RREP message from its upstream RP (a branch RP closer to the GW of a tree) and unicasts an RREP message to its downstream RP (a branch RP farther from the GW of a tree). Thus, if the number of hops from a serving RP to the GW is a , the cost of RREP message transmission is:

$$C_{AODV}^{RREP} = (C_t + 2C_r)a + (2a+4)C_r = aC_t + 4(a+1)C_r \quad (2)$$

From (1) and (2), the path setup cost from a serving RP to the GW using AODV becomes:

$$C_{AODV} = C_{AODV}^{RREQ} + C_{AODV}^{RREP} \quad (3)$$

In DSDV, each RP in the relay network sends a DSDV path update message periodically for every T_u to maintain the network topology information. Since a path update message generated by an RP is flooded through the network, each RP receives one DSDV path update message from each of its neighboring RPs. Hence, in the relay network with $nRP \times mRP$, the cost incurred by an update message from an RP becomes:

$$C_{DSDV,RP} = [(n-2)(m-2)(C_t + 4C_r)] \\ + 2[(n-2)+(m-2)](C_t + 3C_r) + 4(C_t + 2C_r) \tag{4}$$

The first term in (4) is the cost for RPs inside the grid network and the second term for RPs at the boundary of the grid network and the third for RPs at the corners of the grid network.

The control overhead of AODV is determined by the number of hops from a serving RP to the GW. In Fig. 3, we assume $n = n'+1$, n is an odd number, and $m \geq n'/2$, for the simplicity of the analysis. Then, the total number of RPs within the $nRP \times mRP$ grid network is $A = mn - 1$. Therefore, the probability $p(a)$ of the number of hops from a serving RP to the GW being a is:

$$p(a) = \begin{cases} (2a+1)/A & , \quad 1 \leq a \leq n'/2 \\ (n'+1)/A & , \quad n'/2 < a \leq m-1 \\ 2(m+n'/2-a)/A, & m \leq a \leq n'/2+m-1 \end{cases} \tag{5}$$

Now, by (3), the average values of C_{AODV} is determined by (6):

$$E[C_{AODV}] = \overline{C}_{AODV} = \sum_{k=1}^{m+n'/2-1} [C_{AODV}]p(k) \tag{6}$$

3.3 Path Setup Cost Determined by the Sink Mobility and Other System Parameters

In this subsection, we analyze the control message transmission cost of path management mechanisms for the relay network. For this, we consider the network topology change time, the sink density and the sink mobility. When a sink moves out of the service range of a serving RP and contacts a new RP, if the new serving RP already has a path to the GW, the existing path can be used for the newly coming sink. However, if the new serving RP does not have a path to the GW, the RP has to establish a path to the GW for the sink. Thus, the sink mobility and the sink density within the network affect the control message transmission cost of the path setup in the relay network.

For the analysis, we assume that, after a sink joins the relay network, the sink moves through the service range of RPs. If we let the dwell time of the sink in the service range of the ith serving RP be X_i, we can depict the sink mobility model for the time duration τ during which the network topology is kept unchanged as shown in Fig. 4. The relay network topology can be changed due to physical or logical reasons. If an RP is removed because of power depletion or malfunction or an RP is added, the network topology can be changed physically. Also, the logical topology can be changed if path cost such as link bandwidth, transmission delay varies between RPs.

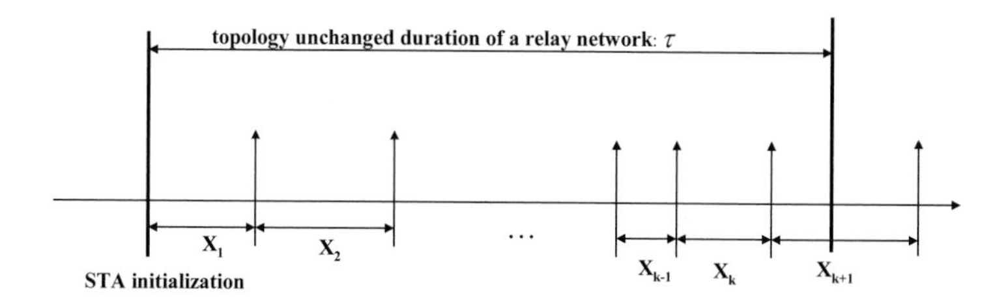

Fig. 4. Sink mobility model

If we assume that X_i's in $\{X_i : i = 1,2,...\}$ are independent and share the same average μ_X, variance σ_X^2 and probability density function $f_X(x)$, the probability that a sink changes its serving RP k times for τ is:

$$f_H(k \mid \tau) = \Pr(X_1 + X_2 + \cdots + X_k \le \tau < X_1 + X_2 + \cdots + X_{k+1}) \tag{7}$$

If we denote $R_k = X_1 + X_2 + \cdots + X_k$, the cumulative distribution function (CDF) $F_{R_k}(r)$ of R_k is obtained as the following by the central limit theorem:

$$\lim_{k \to \infty} F_{R_k}(r_k) = \Phi\left(\frac{r_k - k\mu_X}{\sqrt{k\sigma_X^2}}\right) \tag{8}$$

Here, $\Phi(x)$ is the CDF of the standard normal distribution function. Thus, (7) becomes:

$$f_H(k \mid \tau) = \Phi\left(\frac{\tau - k\mu_X}{\sqrt{k\sigma_X^2}}\right)(1 - \Phi\left(\frac{\tau - (k+1)\mu_X}{\sqrt{(k+1)\sigma_X^2}}\right)) \tag{9}$$

If we assume that N_s static sinks are uniformly distributed within t the $nRP \times mRP$ network, the probability of static sinks existing in the service range of RPs becomes:

$$p_s = \frac{N_s}{nm} \tag{10}$$

The probability of a mobile sink is located in the service range of an RP is $p_r = 1/(nm)$. If we denote N_m as the number of mobile sinks, then the probability that there is no mobile sink in an RP when a mobile sink moves into the service range of the RP is:

$$p_m = (1 - p_r)^{N_m}. \tag{11}$$

A new serving RP of a mobile sink establishes a path to the GW only when it does not serve any static or mobile sinks. Since the position of a mobile sink is determined regardless of the position of static sinks, if all sinks are assumed to always have data to send, by (10) and (11), the probability that a new serving RP of a sink establishes a path to the GW is:

$$p_{path} = (1 - p_s) p_m \tag{12}$$

Thus, the probability of a mobile sink requesting path setup i times for k serving RP changes for τ becomes:

$$p(i \mid k, \tau) = {}_k C_i \, p_{path}^i (1 - p_{path})^{k-i}. \tag{13}$$

From (9) and (13), the probability of a mobile sink requesting path setup i times for τ becomes:

$$p(i \mid \tau) = \sum_{k=0}^{\infty} p(i \mid k, \tau) \cdot f_H(k \mid \tau). \tag{14}$$

Now, with assuming that a mobile sink always has data to send, from (6) and (14), the control message transmission cost for path setup of a mobile sink for τ in AODV is:

$$C_{AODV}(\tau) = \sum_{i=0}^{\infty} i \cdot \overline{C}_{AODV} \cdot p(i \mid \tau). \tag{15}$$

In DSDV, since RPs are not synchronized in time, they do not generate path update messages simultaneously even if they send path update messages for every T_u time interval. For τ, each RP can send up to $j = \lfloor \tau / T_u \rfloor$ path update messages before a network topology change. If we assume that the number of RPs sending path update messages for the time duration ($\tau - jT_u$) is a random variable X and X follows the Poisson distribution, the average control message transmission cost of DSDV for τ in the $nRP \times mRP$ network is:

$$C_{DSDV}(\tau) = (mn-1) j C_{DSDV,RP} + \sum_{i=1}^{mn-1} i \cdot C_{DSDV,RP} (\tau - jT_u)^i \frac{e^{-(\tau - jT_u)}}{i!} \tag{16}$$

Fig. 5 shows the average path setup cost per second for the 101 × 101 grid topology for various τ s. The dwell time of a sink in a grid cell is assumed to follow the exponential distribution with the average of 30 seconds. As the frequency of topology changes increases, the number of path setup requests increases. Hence, as τ increases, the path setup cost of AODV increases since it is affected by the dwell time of mobile sinks in grid cells. On the other hand, DSDV is not affected by τ, but by T_u. In case of DSDV, as the path update period increases, the path setup cost decreases significantly, but the network service quality like the data delivery delay or the data loss rate may significantly deteriorate due to inadaptability to network topology changes.

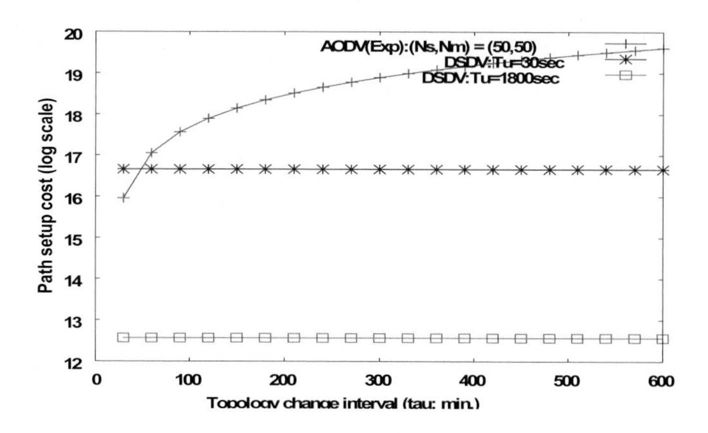

Fig. 5. Topology change interval vs. path setup cost

4 Conclusion

Controlled charging of electric vehicles can take care of fluctuating electricity supply. In this paper, we design network infrastructure to collect and deliver data of electric vehicles charging data to remote monitoring center. To do this, we employ relay network for the support of the network connectivity between sinks and a gateway. In our network infrastructure, we analyze and compare the existing routing mechanisms for multi-hop wireless networks from aspect of the control overhead for the path establishment.

References

1. Al-Karaki, J.N., Kamal, A.E.: Routing techniques in wireless sensor networks: a survey. IEEE Wireless Communications 11(6), 6–28 (2004)
2. Perkins, C., Belding-Royer, E., Das, S.: Ad Hoc On-Demand Distance Vector (AODV) Routing. IETF RFC 3561 (2003)
3. Perkins, C., Bhagwat, P.: Highly Dynamic Destination-Sequenced Distance-Vector Routing (DSDV) for Mobile Computers. ACM SIGCOMM Computer Communication Review 24(4), 234–244 (1994)
4. Raniwala, A., Chiueh, T.: Architecture and Algorithms for an IEEE 802.11-Based Multi-Channel Wireless Mesh Network. In: Proc. IEEE INFOCOM, pp. 2223–2234 (2005)
5. Alicherry, M., Bhatia, R., Li, L.E.: Joint Channel Assignment and Routing for Throughput Optimization in Multiradio Wireless Mesh Networks. IEEE Journal on Selected Areas in Communications 24(11), 1960–1971 (2006)

Prediction of Personal Power Consumption Using the Moving Average Technique

Jongwoo Kim[1], Sanggil Kang[1], and Hak-Man Kim[2]

[1] Computer Science and Information Engineering
Inha University, Incheon, Korea
[2] Dept. of Electrical Engineering
University of Incheon, Incheon, Korea
skytolove@inha.edu, sgkang@inha.ac.kr, hmkim@incheon.ac.kr

Abstract. In this paper we introduce a new prediction method of personal power consumption using the moving average technique. Unlike typical methods, our method considers the trend of consumer's statistical power consumption changes for estimating the statistical future power consumption. In the simulation section, we verify that the performance of our method is better than that of typical method.

Keywords: prediction of power consumption; moving average; conditional probability.

1 Introduction

In past years, power energy paradigm is the unidirectional service that energy providing service provides generated electrical power to consumers according to weekly, monthly, or yearly based demographical and regional energy consumption information. Recently, service paradigm of power enterprise has been rapidly changing from panning of electric energy consumption to marketing exploration. To accomplish the new power service mechanism, we need fast prediction of power consumption for each consumer [1]. Various power load forecasting methods have been presented. Alves et al. [2], Charytoniuk et al. [3], and Darbellay et al. [4] presented the short term electricity demand prediction using neural network. Cottet et al. [5] proposed a load forecasting problem using Bayesian modeling. Those services have built power utilization management information systems based on online transaction processing. So the systems should have huge amount of data accumulation and computation load. And various applications using the data are restricted to big consumers not individuals, to our best knowledge.

In this paper we propose a personal power prediction system using moving average technique which is commonly utilized in forecasting applications. For that, we first collect personal power consumption history during predetermined time interval. Based on the history, we compute the statistical power consumption prediction with evidence given. In the procedure of the prediction, we consider the variation of prediction differences during some time interval, which are represented in time series

T.-h. Kim, A. Stoica, and R.-S. Chang (Eds.): SUComS 2010, CCIS 78, pp. 230–234, 2010.

[6]. In other words, we estimate one-step future personal power prediction at present by adding the mean variation of the past statistical power consumption to the current statistical power consumption.

The remainder of this paper is organized as follows. Section I describes our system architecture. Section II derives our algorithm for predicting consumer's future power consumption using the moving average technique. In Section III, we show the simulation results using artificially generated data. We then conclude our paper in Section IV

2 Prediction of Power Consumption Using Mat

In this section, we derive our algorithm for power consumption prediction from the statistical probability of random variables of input features. Fig. 1. is an acyclic graphical expression of the causal relationship between expected power consumption at time t denoted as $C(t)$ and its input features denoted X_is such as $\{C_k(t\text{-}D)$, Week($X_1(t)$), Season($X_2(t)$), Time interval($X_3(t)$)$\}$. Here, $C_k(t\text{-}D)$ is the average power consumption history during D past time from time t. Let's consider each input feature as a random variable.

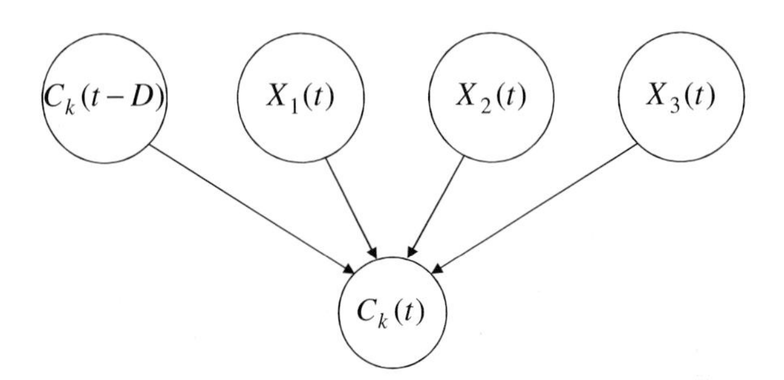

Fig. 1. Acyclic graphical expression of the causal relationship of variables

Each variable X_i can have γ_i possible attributes and can be expressed as $\mathrm{X_i} = \lfloor \mathrm{x_{i1}}\ \mathrm{x_{i2}} \cdots \mathrm{x_{ij}} \cdots \mathrm{x_{i\gamma_i}} \rfloor$ where x_{ij} indicates the jth attribute in X_i. The statistical power consumption at t denoted as $p_{\mathrm{C_k}}(\mathrm{t})$ of a state $C_k(\mathrm{t})$ in a variable $C(t)$ given evidences for the input features can be calculated with the conditional probability as below;

$$p_{C_k}(t) \equiv prob(C_k(t)|E) = n_{C_{t-D,t}} / N_{C_{t-D,t}} \tag{1}$$

where $N_{C_{t-D,t}}$ and $n_{C_{t-D,t}}$ are the total number of power consumption data collected during $t\text{-}D$ to t and the sample number of C_k under input feature evidence E , respectively. E is a metric of attributes of input features, i.e., $E=\{x_{12}, x_{23}, x_{32}\}$ in Fig. 1.

For the statistical prediction models [7, 8], the goal is to obtain the optimal estimate of the conditional probabilities for the future state. As shown in the equation (1), the typical method does not consider the trend of personal energy consumption change so the estimated conditional probability is not suitable for representing the future energy consumption, which usually changes in time. In order to compensate the drawback, we apply the moving average technique (MAT) [9] to update the conditional probability.

Equation (1) needs to be modified in a time series. The modification starts with expressing the conditional probabilities in terms of time t. As shown in Fig. 2, the conditional probability of C_k at time t is computed from the conditional probabilities of C_k collected during time duration D (window size) between t-D and t with given evidence (E).

The change of conditional probabilities for $\mathrm{C_k}$ between two consecutive time steps can be obtained by calculating the probability variation (PV), denoted as $\Delta \mathrm{p}_{\mathrm{C_k}}(\mathrm{t}+1)$ as seen in Equation (2).

$$\Delta p_{C_k}(t+1) = p_{C_k}(t+1) - p_{C_k}(t) \tag{2}$$

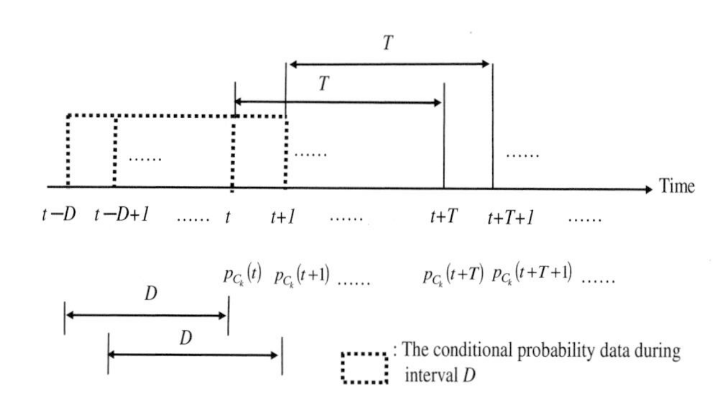

Fig. 2. Schematic representation of the probability update

A set of the successive PVs can be obtained during a predetermined period D from t to $t+T$ such as

$$\Delta p_{C_k}(t, t+T) = \left[\Delta p_{C_k}(t+1), \Delta p_{C_k}(t+2), \cdots, \Delta p_{C_k}(t+T)\right] \tag{3}$$

The set $\Delta \mathrm{p}_{\mathrm{C_k}}(t, t+T)$ is the history of consumer's power consumption changes during a time period T from a time t so we use it for predicting the future power consumption for the attribute C_k at time $t+T+1$. If the PVs are normally distributed or similarly normally distributed, the mean of them can be considered as the maximally likely occurring variation for the one-step-ahead future by the concept of maximum likelihood estimation. Thus, the prediction, denoted as $\hat{p}_{C_k}(t+T+1)$, of the conditional probabilities $p_{C_k}(t+T+1)$ for time $t+T+1$ can be formulated by adding the mean of the PVs, denoted as $\Delta\bar{p}_{C_k}(t, t+T)$, to the probability $p_{C_k}(t+T)$ as shown in Equation (4).

$$\hat{p}_{C_k}(t+T+1) = p_{C_k}(t+T) + \Delta\bar{p}_{C_k}(t, t+T) \tag{4}$$

where $\Delta\bar{p}_{C_k}(t, t+T) = (1/T)\sum_{i=1}^{D} \Delta p_{C_k}(t+i)$. From Equation (4), the performance of our method depends on $\Delta\bar{p}_{C_k}(t, t+T)$. The larger the value of T, the more accurate the estimated conditional probability can be.

3 Experimental Results

Our proposed method is performed using randomly generated data as follows:

$$C = \begin{cases} 0, 0\sim5\text{kW} \\ 1, 6\sim10\text{kW} \\ 2, 11\sim15\text{kW} \\ 3, \text{Over } 16\text{kW} \end{cases} \qquad X_1 = \begin{cases} 0, \; x_{11}=\text{Weekday} \\ 1, \; x_{12}=\text{Weekend} \end{cases}$$

$$X_2 = \begin{cases} 0, x_{21}=\text{Spring} \\ 1, x_{22}=\text{Summer} \\ 2, x_{23}=\text{Fall} \\ 3, x_{24}=\text{ Winter} \end{cases} \qquad X_3 = \begin{cases} 0, \; x_{31}=\text{A.M.} \\ 1, \; x_{32}=\text{P. M.} \\ 2, \; x_{33}=\text{Night} \end{cases}$$

The states in C, X_1, X_2 and X_3 can be considered the evidences (E) for estimating the conditional probabilities of the power consumption.

As varying the values of D and T, the performance of our method is compared with that of the original method as shown in Equation (4), by calculating the mean error between the estimated probabilities and the true probabilities for 10,000 consumers. For each viewer, the error can be calculated as shown in Equation (5).

$$Error = (1/(N \cdot K)) \sum_{k=1}^{K} \left| p_{C_k}(t+T+k) - \hat{p}_{C_k}(t+T+k) \right| \tag{5}$$

where K is the number of unit for the average power consumption.

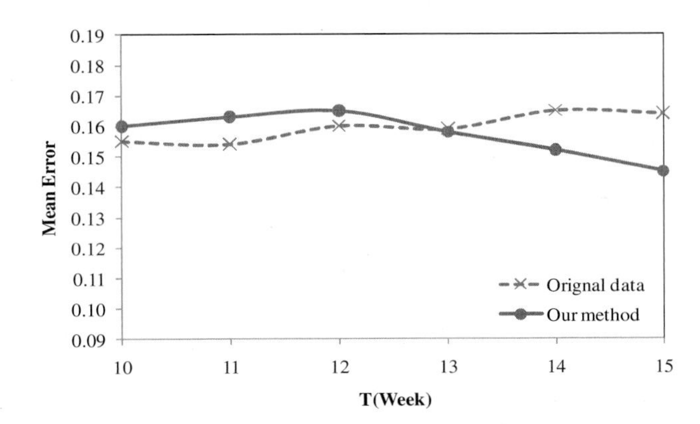

Fig. 3. Performance comparison of our method and the typical method as varying the value of T and D

As shown in Fig. 3, the error of both our method and the original data depends on the values of D because the reliance of $p_{C_k}(t)$ in Equation (2) places on the population of $N_i(t)$ which depends on the window size D. That is, the larger size of $N_i(t)$, the more representative $p_{C_k}(t)$ is, if a consumer's power consumption does not change drastically. The value of T also gives the effect on yielding the optimal performance of our method because the mean value of the PVs depends on the value of T as seen in the equations (3) and (4). From Fig. 1, for D=10, the performances of our method is superior to those of the original data as the value of T is larger at 13 week. However, the initialization of our method gets delayed as the values of T and D increase.

4 Conclusion

In this paper we introduce a new prediction method of personal power consumption using the moving average technique. From the experimental results, our method is valid for the prediction of short-term performance has shown. Our method outperforms that the values of T is large enough to reflect a consumer's power consumption changes for predicting the future power consumption. However, we determined the optimal values of the parameters from the exhaustive empirical experience using randomly generated data. The randomly generated data might not be enough for the exhaustive experiment. Therefore, the real data of power consumption will need to collect. Also, we need to do further study for developing an automatic algorithm to determine the optimal values of parameters T and D for the each power consumption.

References

1. Bergey, P.K., Hoskote, M.: A decision support system for the electrical power districting problem. Decision Support Systems 36(1) (September 2003)
2. Alves da Silva, A.P., Ferreira, V.H., Velasquez, R.M.G.: Input space to neural network based load forecasting. International Journal of Forecasting 24, 616–629 (2008)
3. Charytoniuk, W., Chen, M.S.: Very short-term load forecasting using artificial neural networks. IEEE Transactions on Power Systems 15, 263–268 (2000)
4. Darbellay, G.A., Slama, M.: Forecasting the short-term demand for electricity – Do neural networks stand a better chance? International Journal of Forecasting 16, 71–83 (2000)
5. Cottet, R., Smith, M.: Bayesian modeling and forecasting of intraday electricity load. Journal of the American Statistical Association 98, 839–849 (2003)
6. Zakeri, I., Adolph, A.L., Puyau, M.R., Vohra, F.A., Butte, N.F.: Application of cross-sectional time series modeling for the prediction of energy expenditure from heart rate and accelerometry. Journal of Applied Physiology 104, 1665–1673 (2008)
7. Akaike, H.: A new look at the statistical model identification. IEEE Transactions on Automatic Control 19, 716–723 (1974)
8. Buntine, W.: A guide to the literature on learning probabilistic networks from data. IEEE Transactions on Knowledge and Data Engineering 8, 195–210 (1996)
9. Box, G.E.P., Jenkins, G.: Time Series Analysis, Forecasting and Control. Holden-Day (1990)

Personalized Energy Portal Service Using Consumers' Profile Information

Jongwoo Kim[1], Juwan Kim[1], Sanggil Kang[1], Hak-Man Kim[2], and Young-Kuk Kim[3]

[1] Computer Science and Information Engineering
Inha University, Incheon, Korea
[2] Dept. Electrical Engineering
University of Incheon, Incheon, Korea
[3] Dept. Computer Engineering
Chungnam National University
skytolove@inha.edu, kenstein@inhaian.net, sgkang@inha.ac.kr,
hmkim@incheon.ac.kr ykim@cnu.ac.kr

Abstract. We propose a personalized energy portal service system using a collaborative filtering technique, which provides a personalized energy information in an energy portal site according to consumer's energy interest tendency. The proposed system analyzes the registered consumers' actions on the site, such as "clicking" energy and "downloading" on the information related personal energy interest. According to the different actions, we provide a weight for calculating the consumers' tendency of energy consumption. However, the information is uniformly provided to the individual consumers. In order to avoid the problem, we customize the order of energy information according to whether there is any mismatching of profiles among registered consumers and target consumers.

Keywords: personalized, energy portal, collaborative filtering, smart energy.

1 Introduction

Energy portal service offers customers real-time information on their energy use and induces them to save energy and reduce its related cost. Under the system, customers can check the rates of the previous meter-checking date, and can obtain the information of consumer's energy-consuming pattern during some period, according to the state-run energy firm. Most of the general energy portal service provides uniformly problems-reliability, high prices, and reliance on imports-to individual consumers without consumers' discretion. Also, consumers spend much time to look up the flood of the unwanted energy information. To solve the problems, many researchers [1][2][3] developed the personalized energy portal service.

In this paper, we propose a personalized energy portal service system using a collaborative filtering technique, which provides personalized energy information according to consumers' tendency. The proposed system analyzes the registered consumers' actions on the site, such as "clicking" energy and "downloading" on the information related personal energy interest. According to the different actions, we

T.-h. Kim, A. Stoica, and R.-S. Chang (Eds.): SUComS 2010, CCIS 78, pp. 235–241, 2010.

provide a weight for calculating the consumers' tendency of energy consumption. However, the information is uniformly provided to individual consumers when they search with same keywords. In order to avoid the problem, we customize the order of the energy information according to whether there is any mismatching of profiles among registered consumers and target consumers or not.

The remainder of this paper consists as follows. Chapter 2 briefly explains the overall architecture of our system. Chapter 3 explains our proposed personalization technique. In Chapter 4, we show the simulated results of our system. Finally chapter 5 will conclude

2 Overall System Architecture

Figure 1 shows the overall architecture of the proposed energy portal service system which collect consumer's profile information through two-communication network, including metering. The architecture is composed of three modules such as Log Analyzer (LA), Personalization Inference (PI), and Dynamic HTML Generation Machine (DHGM). The LA module collects the registered consumers' clicking frequency of energy consumption from their log information. If a target consumer requests a query, the PI module infers the personalized information by our algorithm explained in the following section, which utilizes the obtained consumers' clicking information, the downloading information stored in the DB, and consumers' profile information stored in the Profile DB. The DHGM module shows the inferred personalized information to the target consumer through Displayed Information.

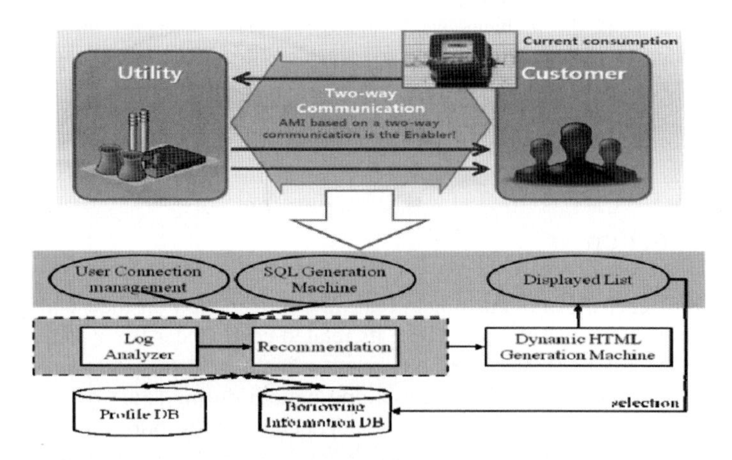

Fig. 1. The overall architecture of the proposed energy portal system

3 Proposed Energy Portal System

For energy retained in energy portal service such as reliability, high prices, and reliance, etc, the tendency of energy consumption can be expressed by the actions taken by consumers. There are two types of actions on the site, such as "clicking" energy and "downloading" on the information related personal energy interest. In conventional method, the tendency of energy is expressed with the frequency of clicking for the energy by consumers during a predetermined period, as seen in Equation (1).

$$\mathrm{Pre}f_i = \sum_{k=1}^{K} c_{k,e_i} \tag{1}$$

where $\mathrm{Pre}f_i$ is the tendency of energy e_i, c_{k,e_i} is the frequency of clicking energy e_i by consumer k and K is the total number of consumers registered in the energy portal service system. Equation (1) is the tendency obtained using the action of clicking only. However, the action "downloading" is usually stronger index for estimating the tendency than the action "clicking." In order to take into the consideration, we provide a weight w to the action "downloading" in calculating the frequency of the action of clicking in Equation (1).

$$\mathrm{Pre}f_i = \sum_{k=1}^{K} (c_{k,e_i} + w \cdot b_{k,e_i}) \tag{2}$$

where b_{k,e_i} is the frequency of downloading energy e_i by consumer k and $w > 1$. As seen in Equation (2), the accuracy of the tendency depends on the value of w. As the frequency of downloading an energy increases, the tendency for the energy linearly increases with slop w. Based on the value of the tendency, the order of energy in the information as the result of a target consumer's query using one or more than one keyword is determined. However, the information is uniformly provided to the individual consumers when they search with same keywords. In order to avoid the problem, we customize the order of the information according to consumers' profile.

In general, the action information of consumers with the same profile as a target consumer is more useful for predicting the target consumer's usage behavior than that of consumers with different profile. For example, let a target consumer's profile information be ut= (Student, Housewife, Employee and Owner-operator). Also, there is two consumers' (Consumer 1 and Consumer 2) action information for utility bills in the home page of tax office. Consumer 1 with profile information $u1$= (Student, Housewife and People on none incomes) saved the utility bills once. Consumer 2 with profile information $u2$ = (Employee, Owner-operator and People on incomes) discounted the utility bills once. In this case, even though the action of downloading is more effective on computing the tendency of the utility bills than the action of clicking, the action of clicking is more reliable because the profile information of Consumer 1 is identical with the target consumer. In order to consider this problem for computing the tendency, we modify the frequency of the actions by providing a penalty according to the degree of mismatching profiles between the consumers in the database and the target consumer.

$$\mathrm{Pre}f_i = \sum_{k=1}^{K} P_k (c_{k,e_i} + w \cdot b_{k,e_i}) \tag{3}$$

where P_k is the penalty for the mismatching and $P_k \leq 1$. If there is no mismatching between consumer k and the target consumer then $P_k=1$.

By using Equation (3) for computing the tendency of energy consumption, the personalized search information can be provided by according to different target consumers. Also, as seen in Equation (3), the accuracy of estimating the tendency of energy consumption depends on the values of the variables w and P_k. In the experimental section, we show the optimal values of those variables from the empirical experience.

4 Experiment

In order to investigate the effectiveness of our recommendation method, we develop a dummy website for personalized energy portal service. We implemented our system using the JAVA web server in the Window NT environment. In the server, we used the JSDK which is Java servlet developer kit 1.4 to run our personalized energy portal service program. MS SQL server 2000 was used as the relational database. Also, JDBC (Java Database Connectivity) was used in order to connect database with servlet. Figure 2 is a dummy website for personalized energy portal service which shows consumer's personalized energy information in the main page.

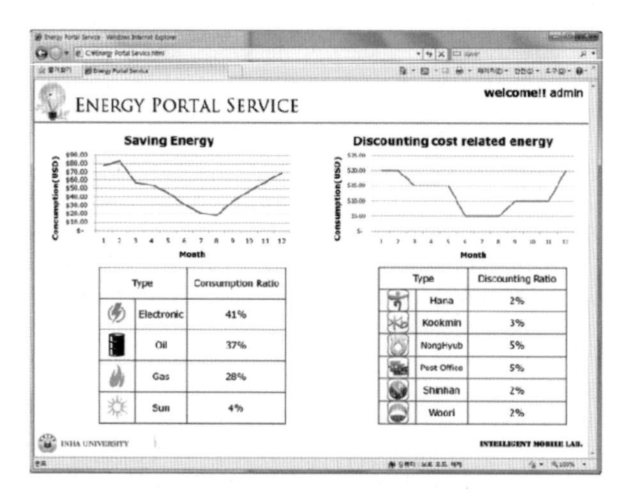

Fig. 2. An example of the personalized energy portal service

Also, we made a sample energy portal service website with artificially generated profile information of clicking and downloading for 100 persons. Out of the persons, 80 persons are used for training our system, the rest is used for test. For the weight of downloading action in Equation (3), we chose w=4. Also, the penalty of mismatching between the kth registered consumer and a target consumer is set to P_k=0.6. We evaluate our recommendation method with other collaborative filtering (CF) [4] based recommendation method in terms of recommendation accuracy. The accuracy is measured with the percentage of 'clicking' or 'downloading' from the recommended information. We compare the performance of our method with CF based recommendation technique that uses Pearson Correlation Coefficient [5] and Cosine Similarity. We also compared the accuracy of our method with a recent work by Lee et al. [6] where they propose a new method, named as Inner Consumption Measure, for recommendation. Figure 3 shows the performance comparison of our method with Inner Consumption Measure, Cosine Similarity, and Pearson Correlation Coefficient method in terms of accuracy according to the purchase rate form the recommended consumptions. The accuracy was compared for top 5, 10, 15, and 20 recommend information.

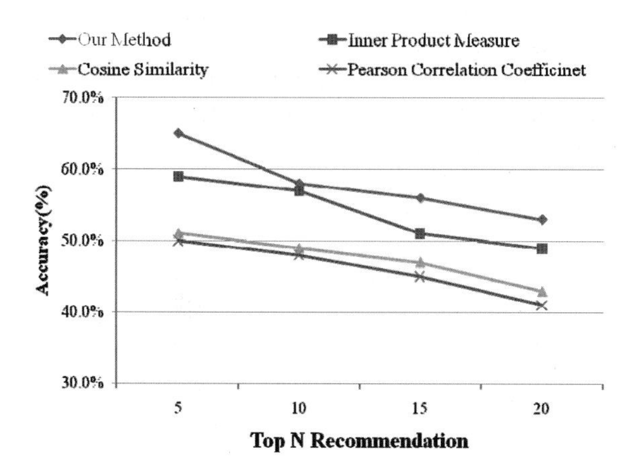

Fig. 3. The accuracy of our method and other existing methods for recommendation

As seen in Figure 3, our recommendation method shows the higher accuracy than other existing methods for recommendation. It means that customers are more frequent to spend on the consumption from the recommended consumption provided by our method. Our method achieves the best accuracy when the number of recommended consumption is 5 and achieves almost similar accuracy as Inner Consumption Measure when the number of recommended consumptions increases. However, our method shows remarkable accuracy over other methods when the number of recommended consumption is 20 or more than that. For instance, when the number of recommended consumption is 20 our method shows 7% batter accuracy than Inner Consumption Measure. Also our method shows 12.05% and 10.3% batter accuracy than Pearson Correlation Coefficient and Cosine Similarity method, respectively. This is because of our approach of grouping where we introduce the age and gender to find the customers with similar tendency.

Also we measure the accuracy of our method for different age and gender group and finally we observed the three test customers purchasing pattern and we evaluate the recommendation in terms of accuracy

5 Conclusion

In this paper, we proposed the personalized energy portal service system by considering energy portal service consumers' actions such as "clicking" energy and "downloading" on energy information and their profile information. From the experimental section, it is shown that our system can give accuracy to energy portal service consumers, compared to the existing energy portal service system.

However, the weight of downloading action and the penalties of mismatching used in the experiment were obtained from the exhaustive empirical experience. We need to do further study for developing an automatic algorithm in determining the weight and the penalties for various situations.

Acknowledgments. This research was supported by the MKE(The Ministry of Knowledge Economy), Korea, under the ITRC(Information Technology Research Center) support program supervised by the NIPA(National IT Industry Promotion Agency). (grant number NIPA-2010-C1090-1031-0002)

References

1. Bollacker, K.D., Lawrence, S., Giles, C.L.: A System for Automatic Personalized Tracking of Scientific Literature on the Web. In: Proc. ACM Conference on Digital Libraries, pp. 105–113 (1999)
2. Lee, W.P., Yang, T.H.: Personalizing Information Appliances: A Multi-agent Framework for TV Program Recommendations. Expert Systems with Applications 25(3), 195–210 (2003)
3. Kamba, T., Bharat, K., Albers, M.C.: An Interactive Personalized Newspaper on the Web. In: Proc. International World Wide Web Conference, pp. 159–170 (1995)
4. Goldberg, D., Nichols, D., Oki, B., Terry, D.: Using collaborative filtering to weave an information tapestry. Communications of the ACM 35(12), 61–70 (1992)
5. Benesty, J., Chen, J., Huang, Y., Cohen, I.: Pearson Correlation Coefficient. Springer Topics in Signal Processing 2, 1–4 (2009)
6. Lee, T.Q., Park, Y., Park, Y.: A Similarity Measure for Collaborative Filtering with Implicit Feedback. In: Huang, D.-S., Heutte, L., Loog, M. (eds.) ICIC 2007. LNCS (LNAI), vol. 4682, pp. 385–397. Springer, Heidelberg (2007)
7. Froehlich, J., Everitt, K., Fogarty, J., Patel, S., Landay, J.: Sensing opportunities for personalized feedback technology to reduce consumption. In: Proc. CHI Workshop on Defining the Role of HCI in the Challenge of Sustainability (2009)
8. Tomain, J.P.: Energy portal service Path: How Willie Nelson Saved the Planet. Cumberland Law Review 8 (2006)
9. Rabaey, J., Arens, E., Federspiel, C., Gadgil, A., Messerschmitt, D., Nazaroff, W., Pister, K., Varaiya, P., Oren, S.: Energy portal service distribution and consumption: Information technology as an enabling force. Technical report, Center for Information Technology Research in the Interest of Society, CITRIS (2001)
10. Rabaey, J., Arens, E., Federspiel, C., Gadgil, A., Messerschmitt, D., Nazaroff, W., Pister, K., Varaiya, P., Oren, S.: Smart energy distribution and consumption: Information technology as an enabling force. Technical report, Center for Information Technology Research in the Interest of Society, CITRIS (2001)

11. Kim, K., Seo, S., Sung, J., Seo, H., Oh, S., Kwak, H.: Optimal Planning of Energy portal service System and its Applications. The Korean Society of Mechanical Engineers, 3452–3457 (2007)
12. Chris, M., Stadler, M., Siddiqui, A., Aki, H.: Control of Carbon Emissions in Zero-Net-Energy Buildings by Optimal Technology Investments in Energy portal service Systems and Demand-Side-Management. In: 32nd IAEE International Conference, Energy, Economy, Environment: The Global View, June 21-24 (2009)
13. Chan, M., Viard, T., Caillavet, M.L., Campo, E.: Smart Technology for the Elderly and Disabled Consumers at Home, 2nd TIDE Congress. In: The European Context for Assistive Technology, pp. 393–396 (April 1995)
14. Mazza, P.: The Energy portal service Network: Electrical Power for the 21st Century. Climate Solutions (2002)
15. Nadel, S., Geller, H.: Energy portal service Policies: Clicking Money and Reducing Pollutant Emissions Through Greater Energy Efficiency. American Council for an Energy-Efficient Economy, Washington, DC (2001)
16. Ahmed, S., Ko, P., Kim, J., Kang, S.: An Enhanced Recommendation Technique for Personalized e-commerce Portal. In: IEEE/IITA (2008)
17. Heck, W.: Energy portal service meter will not be compulsory (April 2009), `http://www.nrc.nl/international/article2207260.ece/Smart_energy_meter_will_not_be_compulsory`
18. Park, W., Kim, W., Kang, S., Lee, H., Kim, Y.: Personalized Digital E-library Service Using Users' Profile Information. In: Gonzalo, J., Thanos, C., Verdejo, M.F., Carrasco, R.C. (eds.) ECDL 2006. LNCS, vol. 4172, pp. 528–531. Springer, Heidelberg (2006)

Reliable Power Quality Data Delivery Mechanism Using Neural Network in Wireless Sensor Network*

Yujin Lim[1], Hak-Man Kim[2], and Sanggil Kang[3]

[1] Department of Information Media, University of Suwon,
2-2 San, Wau-ri, Bongdam-eup, Hwaseong-si, Gyeonggi-do, 445-743, Korea
yujin@suwon.ac.kr
[2] Department of Electrical Engineering, University of Incheon,,
12-1 Songdo-dong, Yeonsu-gu, Incheon, 406-772, Korea
hmkim@incheon.ac.kr
[3] Department of Computer Science and Information Engineering, Inha University,
253 Yonghyun-dong, Nam-gu, Incheon 402-751, Korea
sgkang@inha.ac.kr

Abstract. Power grids deal with the business of generation, transmission, and distribution of electric power. Current systems monitor basic electrical quantities such as voltage and current from major pole transformers with their temperature. We improve the current systems in order to gather and deliver the information of power qualities such as harmonics, voltage sags, and voltage swells. In the system, data delivery is not guaranteed for the case that a node is lost or the network is congested, because the system has the in-line and multi-hop architecture. In this paper, we propose a reliable data delivery mechanism by modeling an optimal data delivery function by employing the neural network concept.

Keywords: neural network, sensor network, cost function, data delivery mechanism, power quality.

1 Introduction

Power grids involve generation, transmission and distribution of electric power. The electrical distribution system delivers electric power through feeders and pole transformers from distribution substations to end users such as houses, office buildings, and factories. Power quality is any power problem manifested in voltage, current, or frequency deviations, that results in failure or malfunctioning of the customer equipments [1].

In general, current systems monitor basic electrical quantities such as voltage and current from major pole transformers with their temperature. For evaluating current status of power quality, finding places occurring power quality problems, and planning measures, we need more extra information of power quality. We improve the current systems in order to gather and deliver the information of power qualities such as harmonics, voltage sags, and voltage swells.

---

* This work was supported by Inha University Research Grant.

T.-h. Kim, A. Stoica, and R.-S. Chang (Eds.): SUComS 2010, CCIS 78, pp. 242–249, 2010.

For expanding the power quality monitoring system, there are various issues such as measurements, controls, databases, and communications. In order to design the communication network, wireless multi-hop communication paradigm is often employed to construct in electrical distribution system (EDS) to reduce the deployment and management cost. Many studies have been paid an attention on building EDS using wireless sensor networks (WSNs) [2, 3]. The reason of using WSN is its efficiency in monitoring of numerous computing and sensing devices distributed within a large-scale environment.

WSN for the power quality monitoring system delivers power quality information generated by pole transformers to remote monitoring center in the residential division. Usually, the power quality information is periodically measured, gathered, and transmitted towards the monitoring center. Once power quality measured at a pole transformer is out of a normal range, an alarming message with detailed contents is promptly sent in the event based manner. In the system, data delivery is not guaranteed for the case that a node is lost or the network is congested, because the system has the in-line and multi-hop architecture.

To solve the problem, we propose a reliable data delivery mechanism by modeling an optimal data delivery function. The performance of the function lies in determining the optimal coefficients in the function with considering of the wireless propagation environment or the topological environment around the node. To do that, we employ the neural network (NN) concept [4].

The remainder of this paper is structured as follows. Section II describes our system architecture. Section II explains our data delivery mechanism. Following this, we verify the designed system by NS-2 simulations in Section 4. Finally, Section 5 summarizes our results, discusses our future plans, and offers conclusions.

2 System Architecture

The EDS has tens of thousands of pole transformers ranging widely over hundreds of square kilometers. A monitoring center in a residential division of a city is a data collecting point which gathers the power quality information from scattered pole transformers deployed over the city. The distribution network for EDS consists of three subsystems as shown in Fig. 1; a collection subsystem, relay subsystem, and monitoring subsystem. The collection subsystem is composed of several distribution substations (Hereafter, the term 'substation' is exchangeable of 'distribution substation'). Each substation is connected to several feeders. Each feeder collects the power quality data from hundreds of pole transformers and delivers them to the substation. Since pole transformers have been deployed sparsely at distance of hundred meters. WSN by using IEEE 802.11b standard [5] is employed to construct the collection subsystem in order to reduce the deployment and management cost. The relay subsystem is responsible for delivering the data gathered by the substations to the monitoring subsystem via wired infrastructure due to the long distance between the relay subsystem and the monitoring subsystem. The monitoring center in the monitoring subsystem processes the power quality data to recognize current status of situations and takes appropriate actions on the assessed situation [6]. Since substations in the relay subsystem are connected to the monitoring center through a high-speed wired network, the communication between them is highly reliable. Thus the problem of data delivery in EDS is the same as the data delivery problem at the collection subsystem.

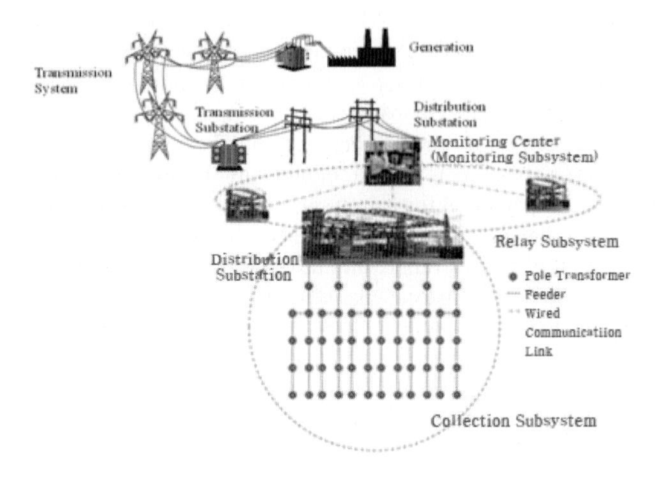

Fig. 1. Network infrastructure for EV charging

3 Data Delivery Mechanism

3.1 Data Forwarding Mechanism

In EDS, all pole transformers (Hereafter, we will use the term 'node') can be data sources, while the monitoring center alone is a data sink. In addition, the network topology in EDS is stationary. We design a reliable data forwarding protocol for the collection subsystem.

Since the packet loss probability in wireless multi-hop communication environment increases with the number of hops [7], we choose the Hop Distance (HD) from the node to the substation as one of the metrics for path management. Besides, it is well known that packet loss is due to either collision or weak signal [8]. By exchanging HELLO messages among nodes, each node measures Received Signal Strength (RSS) and HELLO Message Reception Ratio (HMRR) of its neighbor nodes. HMRR represents the ratio of the number of HELLO message received from a neighbor node to the number of the Hello message sent by the node. HMRR reflects the impact of channel contention from neighbor nodes. Finally, in order to reflect the degree of congestion of a node, Queue Length (QL) of each node is also employed as one of cost factors and QL is included and delivered in HELLO message.

At the network initialization stage, the substation floods a PROBE message over the entire network so that each node in the network can infer the minimum number of hops from the substation to itself. Thereafter, the substation floods a PROBE message periodically so that nodes can update their hop distance from the substation. The path cost is used in constructing the path between the substation and one node.

A node periodically sends a HELLO message including its QL to its neighboring node as its heartbeat. When a node receives a HELLO message, the node updates the soft state on the node having sent. If a node or the wireless link to the node fails, any HELLO messages from the node are not arrived for a given amount of time. Thus, the soft state on the node is released. The node detecting the node failure tries to repair the broken path by sending a REPAIR message to its neighboring nodes via the one-hop flooding. Once a neighboring node receives the REPAIR message, it responds with a REPAIR_ACK message having its path cost. Then, the node having sent the REPAIR message receives the REPAIR_ACK message(s) and it selects the node having the least path cost as its next node towards the monitoring center. To select the next-hop node, it is important to determine an optimal link cost function.

Once the data forwarding path is constructed, the power quality data is delivered to the monitoring center through the path. Whenever a node has data to send, periodically or in the event-based manner, the node transmits the data to its next node. This forwarding process continues until the monitoring center receives the power quality information.

3.2 Modeling Cost Function by Employing NN Concept

The link cost function depends on the input features based on the characteristics of wireless propagation, channel contention, and topological environment surrounding a node such as HMRR denoted as x_1, QL denoted as x_2, RSS denoted as x_3, and HD denoted as x_4. Also, the link cost function can be characterized as a nonlinear function of a weighted sum of the inputs as seen in (1). We employ the log function as a nonlinear function because of its promising characteristic. In general, each weight value is determined by the importance of the corresponding input.

$$Cost_i = \log\left(\sum_{i=1}^{4} x_i \cdot w_i\right) \tag{1}$$

where $Cost_i$ is the link cost of the ith neighbor node out of N neighbor nodes and w_i is the connection weight corresponding to input x_i. (1) can be represented in a neural network structure as seen in Fig. 2.

Now, let's discuss about the connectivity of the inputs in the network. As seen in Fig. 2, all inputs are fully connected to the function. It is just like black-box style connection which is commonly used in neural network (NN). However, we intuitively know that some inputs are highly correlated to generate the output of the cost function. For instance, if HMRR(x_1) is high then QL(x_2) and RSS(x_3) are high because they are correlated among them. To take this into the consideration, we connect the inputs in the coupled and uncouple connection style according to whether inputs are correlated or uncorrelated as seen in Fig. 3. For instance, the connection of x_1, x_2, and x_3 is coupled and that of x_4 is uncoupled in the network. From Fig. 3, the cost function (1) is modified as (2).

$$Cost_i = \log\left(\sum_{i=1}^{3} x_i \cdot w_i\right) + \log(x_4 \cdot w_4) \tag{2}$$

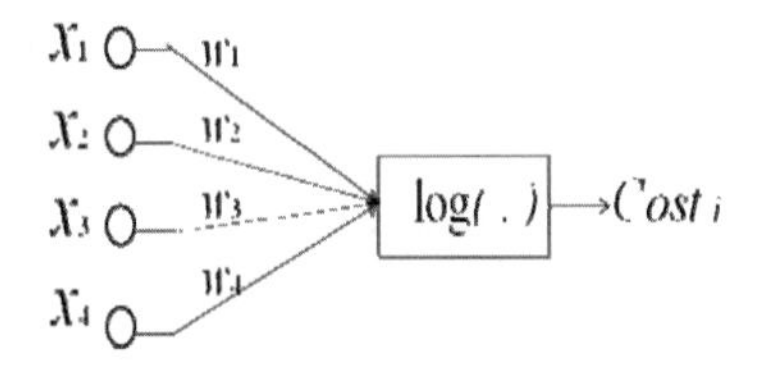

Fig. 2. Link cost function represented in NN

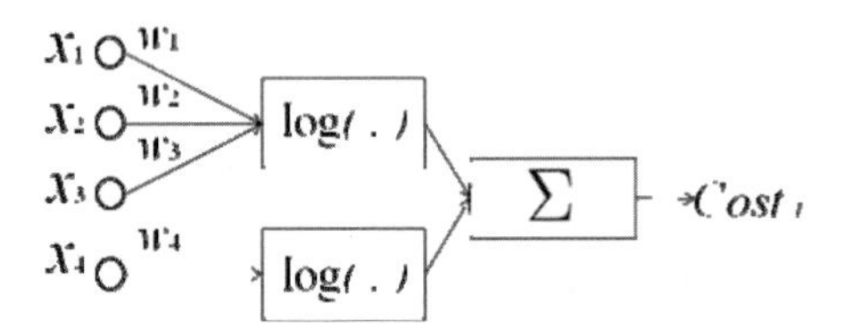

Fig. 3. Link cost function represented in our NN

The optimal performance of the cost function depends on the proper weight vector which can be obtained by training the NN to reach to the maximum of the packet transmission success ratio. To find an optimal weight vector, we imitate the training process [4] of finding the optimal weights in an NN. Each weight in an NN means the importance of each input for producing the link cost. Thus, each weight can be obtained by weight sensitivity with respect to the packet transmission success ratio. Based on the training process with the rationale, the NN is trained with the following steps:

i. Set to 1s' to each weight as an initial weight vector (called old weight vector for a convenience).

ii. Calculate $Cost_i$, i=1,2,…,N, using (2).

iii. Select the neighbor node having max link cost and forward data packets to the selected node, denoted as (3).

$$Cost_{\max} = \max\{\cos t_1, \cos t_2, ..., \cos t_N\} \tag{3}$$

iv. Calculate the packet transmission success ratio ($SR_{max.Costi}$) which is the ratio of the number of packets transmitted successfully to the node with max. link cost to the total number of packets sent to the node.

v. Update one weight w_k by adding very small amount δ (learning ratio) to the weight at one time as in (4).

$$w_k \leftarrow w_1 + \delta \tag{4}$$

vi. Repeat the steps from *ii* to *iv*, using the updated weight in *v*. If there occurs any improvement on the packet transmission success ratio at step *vi*, then we replace the old weight to the updated weight.

vii. Repeat the steps from *ii* to *vi* by varying the learning ratio δ such as δ+Δ until the packet transmission success ratio converges.

During the training process, it is challenging to determine an optimal learning ratio. If we choose too large value of δ, it causes high convergence speed but it has high possibility of missing the optimal weight values. Too small value of δ is vice versa of too large value of δ where convergence speed is too small but low possibility of missing the optimal weight values. We determine the learning ratio from exhaustive empirical experiment in the following subsection.

4 Performance Evaluation

To validate the performance of our data delivery mechanism in the collection subsystem, we compare the performances of our method with those of Fully Connected NN, (FCNN) and the conventional method. In the conventional method, only one input out of the four inputs is used in computing the cost function, i.e., f(x_1), f(x_2), f(x_3), and f(x_4). For the construction of the single-hop collection subsystem, 20 nodes are randomly placed in the 500m x 500m area. From preliminary experimental results, the optimal learning ratio (δ=0.3) is derived to maximize the packet transmission success ratio. Besides, the convergence speed is not issued in the experiments because the NNs are trained within about 5-10 seconds.

We experiment our mechanism using the NS-2 simulator. We use the log-normal model to model radio propagation environment. A node sends a HELLO message for every 100 milliseconds. IEEE 802.11 standard is used as the MAC layer. The transmission range of a node is 250m, and the total simulation time is 360 sec.

To analyze the effect of collision on packet transmission success ratio, we vary the probability of packet collision, using Gaussian distribution with zero mean and standard deviation (σ_1) as seen in Fig. 4. Methods using the NNs are more robust than the conventional method, irrespective of the degree of packet collision. Method using our method improves the performance compared to the FCNN and the conventional method, by about 22% and 45% respectively.

Fig. 5 is the packet transmission success ratios obtained by varying the QL, using Gaussian distribution with zero mean and standard deviation (σ_2). It shows the effect of network congestion on the packet transmission success ratio. The results indicate that method using our method delivers more packets than the FCNN and the conventional methods, by about 17% and 51%.

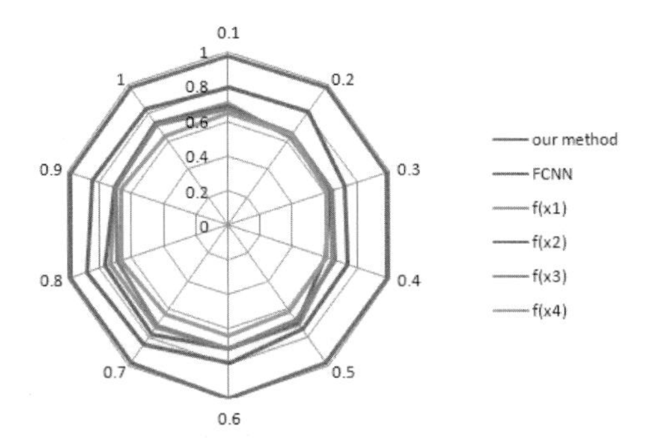

Fig. 4. Packet transmission success ratio with varying HMRR using Gaussian distribution with $N(0, \sigma_1)$

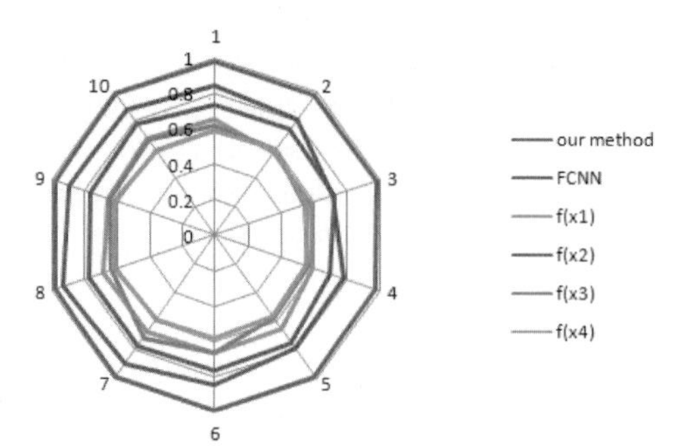

Fig. 5. Packet transmission success ratio with varying QL using Gaussian distribution with N(0, σ_2)

Fig. 6 is the packet transmission success ratios obtained by varying the RSS in the shadowing propagation model. For varying, we add the log normal random fading with zero mean and standard deviation (σ_3). From the figure, we can see that method using our method is more robust despite of dynamic random fading and also improve the packet transmission success ratio (about 23% and 43%), compared to the FCNN and the conventional method.

From the above experimental results, we can conclude that the data delivery mechanism using our method improves the packet transmission success ratio without the burden of large overhead occurred during training our method.

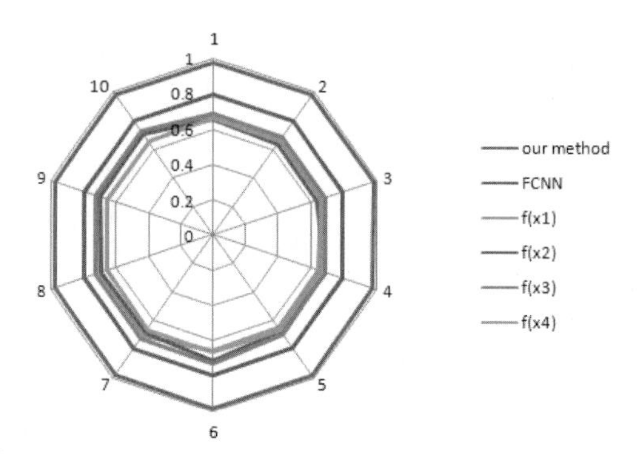

Fig. 6. Packet transmission success ratio with varying RSS by adding log normal random fading with N(0, σ_3)

5 Conclusion

In this paper we propose a reliable data delivery mechanism for power quality monitoring system, by modeling an optimal data delivery function. In our method, we extracted four input features to consider the environments surrounding a node such as HMRR, QL, RSS, and HD. We showed the feasibility of our method from comparison of our method with the FCNN and the conventional method. From the comparison, we can conclude that the performance of our method is better than those of the conventional methods with perspective of the packet transmission success ratio.

References

1. Santoso, S., Beaty, H.W., Dugan, R.C., McGranaghan, M.F.: Electrical Power Systems Quality. McGraw-Hill, New York (1996)
2. Al-Karaki, J.N., Kamal, A.E.: Routing Techniques in Wireless Sensor Networks: A Survey. IEEE Wireless Communications 11(6), 6–28 (2004)
3. Niculescu, D.: Communication Paradigms for Sensor Networks. IEEE Communications Magazine 43(3), 116–122 (2005)
4. Haykin, S.: Neural Networks: A Comprehensive Foundation. Prentice-Hall, New Jersey (1998)
5. IEEE Std. 802.11. Wireless LAN Medium Access Control (MAC) and Physical Layer (PHY) Specification: Higher Speed Physical Layer (PHY) Extension in the 2.4GHz Band (1999)
6. Anagnostopoulos, C., Hadjiefthymiades, S.: Enhancing Situation-Aware Systems through Imprecise Reasoning. IEEE Trans. on Mobile Computing 7(10), 1153–1168 (2008)
7. Toumpis, S., Goldsmith, A.J.: Capacity Regions for Wireless Ad Hoc Networks. IEEE Trans. on Wireless Communications 2(4), 736–748 (2003)
8. Rayanchu, S., Mishra, A., Agrawal, D., Saha, S., Banerjee, S.: Diagnosing Wireless Packet Losses in 802.11: Separating Collision from Weak Signal. In: Proc. IEEE Conf. on Computer Communications (INFOCOM), pp. 735–743. IEEE Press, Los Alamitos (2008)

A New Challenge of Microgrid Operation

Hak-Man Kim[1] and Tetsuo Kinoshita[2]

[1] Dept. of Electrical Engineering, Univ. of Incheon, Korea
hmkim@incheon.ac.kr
[2] Graduated School of Information Science, Tohoku Univ., Japan
kino@riec.tohoku.ac.jp

Abstract. The microgrid is a small-scale power system, which is composed of distributed generators, distributed storage systems, and loads. Currently, the microgrid has been studied in many countries because of being an eco-friend power system, supplying good power quality, and including renewable energy sources such as solar power and wind power. It is anticipated that the microgrid will be introduced to distribution power systems in the near future. According to operational and geographical conditions, the microgrid is operated by two modes: the grid-connected mode and the islanded mode. In this paper, backgrounds of the microgrid are described. Microgrid operation is classified in detail. A mathematical model based on linear programming and an example of a multiagent system for microgrid operation are introduced.

Keywords: microgrid, microgrid operation, grid-connected mode, islanded mode, agent-based microgrid operation.

1 Introduction

Recently, interests of renewable energy have been increasing because of issues related to energy and environment. In relation, many projects for research, development and demonstration of microgrids have been undergoing [1, 2]. To introduce microgrids to power grids, many technologies are still required. Microgrid operation is one of important technologies.

Basically, microgrids are operated in the grid-connected mode in the normal condition and are operated in the islanded mode in a geographically isolated place from a power grid or in abnormal conditions such as fault occurrence. Furthermore, there are some differences according to operation environments such as one-owner microgrid operation, authorized microgrid operation, and competitive operation under an energy market.

Recently, agent-based microgrid operation has been studied [3-5]. The approach is based on the characteristics of the intelligent agent: reactivity, pro-activeness, and social ability [6]. For this reason, microgrid operation can be a good application of information science and computer science engineers. However, it is not easy for them to understand the microgrid and its operation.

Therefore, this paper introduces the concept of the microgrid and its operation modes, explains the characteristics of microgrid operation according to above-mentioned

T.-h. Kim, A. Stoica, and R.-S. Chang (Eds.): SUComS 2010, CCIS 78, pp. 250–260, 2010.

operation environments. To help understanding of microgrid operation, we illustrate a mathematical model based on linear programming (LP) and its examples. In addition, we introduce the concept of agent-based microgrid operation and an example for building multiagent system for microgrid operation. Finally, we suggest challenging areas of microgrid operation.

2 Microgrid Operation

2.1 Microgrid

The microgrid is a small-scale power system composed of distributed generation systems (DGs), such as solar power, wind power, and fuel cells, distributed storage systems (DSs), and loads as shown in Fig. 1 [5]. The microgrid provides electricity and/or heat to customers such as residential buildings, commercial buildings, public offices and industrial compounds, as shown in Fig. 1. In Fig. 1, CHP means combined heat and power and PCC is an abbreviation for the point of common coupling [7].

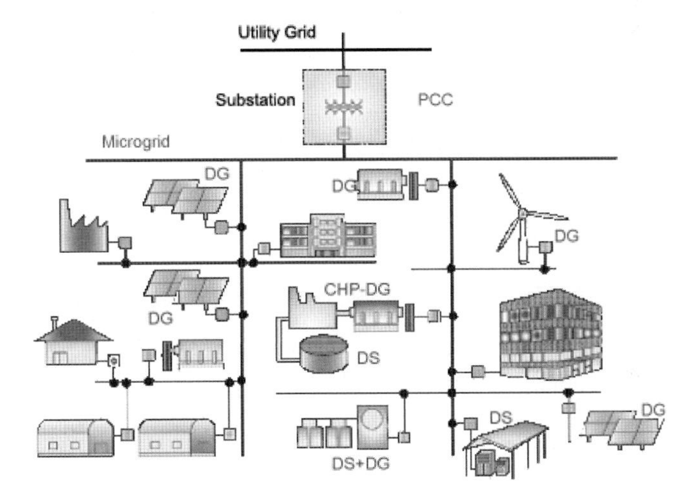

Fig. 1. Typical configuration of microgrids [7]

2.2 Operation Procedure of Microgrids

Since it is difficult to find some literatures to explain the detailed procedure of real microgrid operation, it is assumed that microgrids are operated by two steps: planning and implementation in some papers, as shown in Fig. 2 [4, 5]. The Microgrid Operation and Control Center (MGOCC) should establish an operation plan for the next interval and should implement the operation plan established during the previous interval. The interval period depends on operation rules.

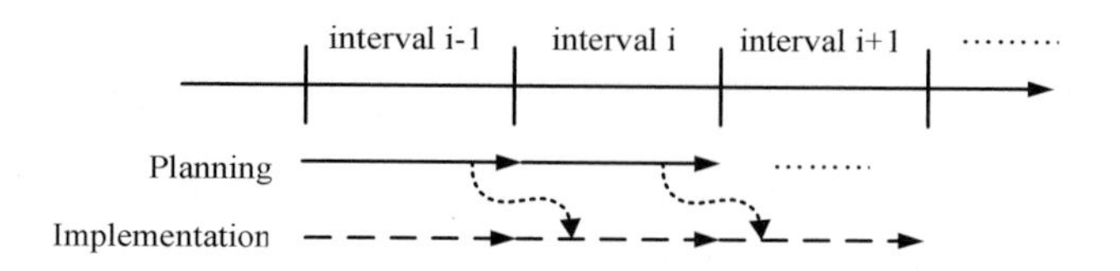

Fig. 2. Microgrid operation procedure [4, 5]

2.3 Operation Modes of Microgrids

Maintaining specific frequency, such as 50 Hz or 60 Hz, is related to a balance between power supply and power demand and therefore is an important requirement for microgrid operation. More precisely, frequency goes up in the case of supply surplus but goes down in the case of supply shortage.

Microgrids are operated by two modes according to system conditions and geographical environments: the grid-connected mode and the islanded mode. On normal conditions, microgrids exchange power with an interconnected power grid to meet a balance whenever supply shortage or supply excess in microgrids occurs. This operation mode is the grid-connected mode. Because of abnormal conditions such as fault occurrence at the interconnected power grid or a geographical environment such as locating at a small island, microgrid can be isolated from a power grid and therefore should operate without interconnection with any power grid. This operation mode is the islanded mode. In the islanded mode, the following actions are performed to meet a power balance;

- The decrease in generation and the charge action of DS in the case of power supply surplus
- Load-shedding, which is intentional load reduction, and the discharge action of DS in the case of power supply shortage.

Especially, load-shedding should be performed as minimally as possible from the viewpoint of supply reliability. Fig. 3 shows operation procedures for an islanded microgrid.

2.4 Operation Environments Relating to Ownership or Authority

There are some differences between microgrid operation according to ownership or authority and such as one-owner microgrid operation, authorized microgrid operation, and competitive operation under an energy market. Microgrids are classified into two operation environments: centralized operation, distributed operation, and semi-distributed operation as follows.

2.4.1 Centralized Operation

Centralized operation is that the MGOCC unilaterally establish operation plans and implements the plans to minimize operation costs and to maximize the operation profits. Here, participants such as DGs, DSs, and loads take part in the operation

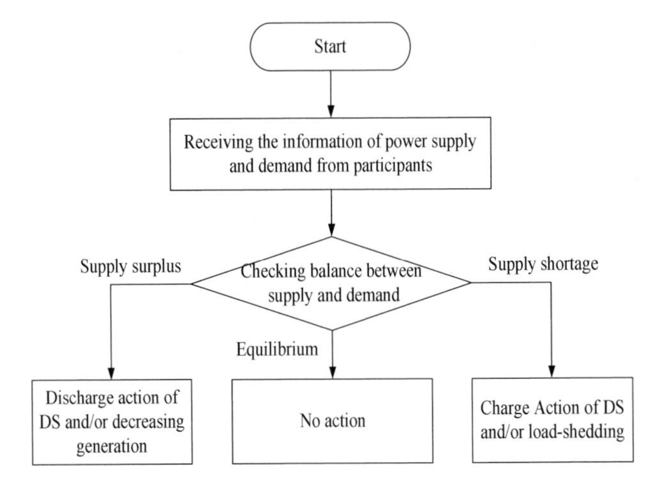

Fig. 3. Operation procedures for an islanded microgrid

procedure with passive actions such as informing their information to the MGOCC and receiving instructions from the MGOCC. This kind of operation is common in the case of one-owner operation or fully authorized operation from whole participants.

In centralized operation, big advantage is to maximize profits of a microgrid but there are restrictions relating to autonomous operation of participants.

2.4.2 Distributed Operation

This operation is generally performed on competitive energy market environments. Here, the MGOCC takes part in management of autonomous participation of participants like bidding. This kind of operation is effective to guarantee autonomous rights of participants. In distributed operation, autonomous participant of participants is guaranteed but to maximize total profits is not guaranteed.

2.4.3 Mixed Distributed Operation

This operation is basically based on distribution operation but the MGOCC has some operational authority relating to critical decision such as load-shedding to maximize total profits.

2.5 Human Operation and Agent-Based Operation

Recently, multiagent system technologies for microgrid operation have been studied [3-5]. Agents have the following capabilities: reactivity, pro-activeness, social ability, and so on [6]. Agents are able to sense the external environment, effectively make a decision based on design purpose against the environment and act according to the decision by their capabilities. Furthermore, it is well-known that agent-based approach is effective in distributed systems or problems and is related to the following merits [4];

- Economy of scale related to high costs required for hiring human operators
- Privacy problems occurred by hiring human operators for houses

- Fast processing time of decision-making
- Flexibility and adaptability based on inherent characteristics of agents.

On the other hand, human operation is based on operator's experience and/or mathematical optimization algorithms.
In reality, it is anticipated that the number of human operators for microgrid operation will be minimized because of above-mentioned restrictions

3 LP-based Optimal Operation

3.1 Model

We establish a simple model based on LP, which is a well-known optimization technique for linear systems, for grid-connected operation to understand the concept of microgrid operation. This operation is suitable to centralized operation mentioned in Sec. 2.4. Especially, a LP-based model is composed of a linear objective function and linear constraints. Fig. 4 shows simplified microgrid configuration [5]. In this paper, we don't consider models of DSs for simplicity. The LP-based model is as follow [4].

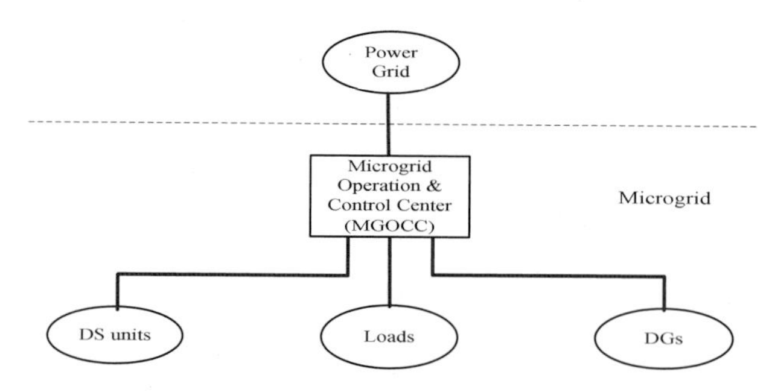

Fig. 4. Microgrid configuration [5]

In grid-connected operation, a microgrid exchanges power with an interconnected power grid on normal conditions to meet power balance. An objective function is modeled like (1) to minimize generation costs of DGs and purchase costs of power from the power grid, and to maximize profits from selling power to the power grid. In (1), i is the identifier of DGs, $Cost_i$ is a generation cost of DG_i, PR_B is a buying price of power from the power grid, PR_S is a selling price of power to the power grid, P_{Buy} is power purchased from the power grid, and P_{Sell} is power selling to the power grid.

$$Min \ \sum_i Cost_i \cdot DG_i + PR_B \cdot P_{Buy} - PR_S \cdot P_{Sell} \tag{1}$$

As a constraint, total power produced by DGs is equal to the sum of power of DGs supplied to loads in the microgrid and power of DGs sold to the power grid. In (2), DG_{io} is power supplied from DG_i for the demand of Microgrid and DG_{is} is power sold from DG_i to the power grid.

$$DG_{i0} + DG_{is} = DG_i \tag{2}$$

The following equation shows each DG produce power within its capacity. Here, $DGcap_i$ is production capacity of DG_i.

$$DG_i \leq DGcap_i \tag{3}$$

Microgrid loads are supplied from DGs and the power grid. Here, D_{MG} is total loads of the microgrid.

$$\sum_i DG_{i0} + P_{Buy} = D_{MG} \tag{4}$$

In addition, the following equations are used. Equation (6) means power purchased from the power grid is not limited. M means a big number.

$$P_{Sell} = \sum_i DG_{is} \tag{5}$$

$$P_{Buy} \leq M$$

3.2 Experiment

Fig. 5 shows a test microgrid for testing the LP-based model. Table 1 shows transaction prices [4].

Table 1. Transaction prices

Experiment	P_S of the next interval	P_B of the next interval
Experiment 1	50¢/ kWh	90¢/kWh
Experiment 2	30¢/ kW h	70¢/kWh
Experiment 3	80¢/ kWh	110¢/kWh

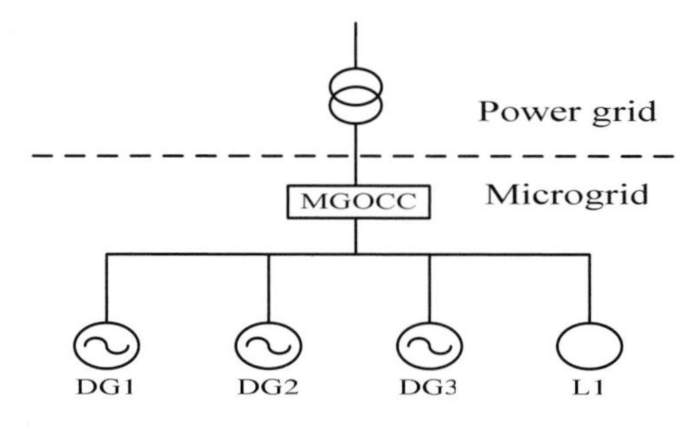

Fig. 5. Test microgrid [4]

Production costs of DGs and a forecasted load for the next interval are as follows:

- DG1 = production cost: 60¢/kWh, capacity: 10kWh
- DG2 = production cost: 75¢/kWh, capacity: 25kWh
- DG3 = production cost: 20¢/kWh, capacity: 5kWh
- L1 = 35kWh

Table 2 shows results of three experiments.

Table 2. Expriment results

Variables	Experiment 1	Experiment 2	Experiment 3
DG_{10}	10 kWh	10 kWh	10 kWh
DG_{1s}	-	-	-
DG_{20}	20 kWh	-	25kWh
DG_{2s}	-	-	5 kWh
DG_{30}	5 kWh	5 kWh	-
DG_{3s}	-	-	5 kWh
P_{Sell}	-	-	5 kWh
P_{Buy}	-	20 kWh	-

4 Agent-based Microgrid Operation

4.1 Design

In this section, we introduce a concept of agent-based microgrid operation and a simple design for building a multiagent system in grid-connected mode. Agents take charge of the MGOCC, DGs, DSs, and loads for microgrid operation. Fig. 6 shows the conceptual figure of agent-based microgrid.

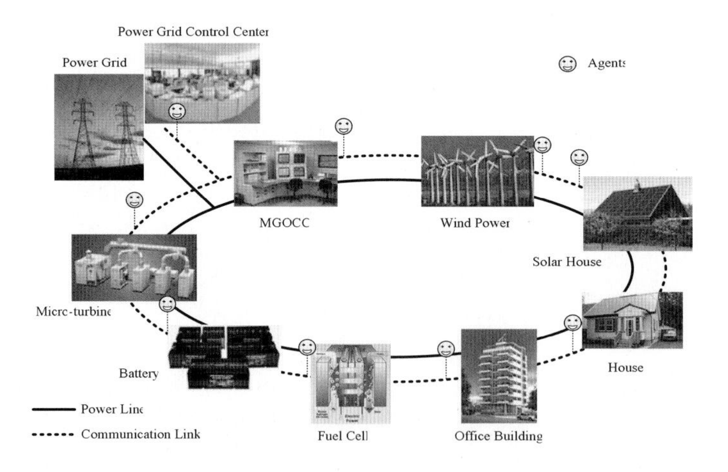

Fig. 6. Agent-based microgrid [4]

The design procedure of a multiagent system is as follows [8];

- Step 1: Problem definition
- Step 2: System requirement analysis
- Step 3: Agent system design
- Step 4: Agent design.

To build a multiagent system for microgrid operation, the following agent set (Ag) is defined [4].

$$Ag = \{Ag_{MGOCC}, AG_{L}, AG_{DG}, AG_{DS}, Ag_{PG}\} \tag{7}$$

In (7), Ag_{MGOCC} is the MGOCC agent and has a lot of knowledge and information to solve problems. Ag_{MGOCC} takes charge of a manager agent of total operation procedures. AG_L is a set of load agents (Ag_L) and Ag_L takes charge of a load or a group of loads located at same place. AG_{DG} is a set of DG agents (Ag_{DG}) and Ag_{DG} governs a DG or a group of DGs located at same place. AG_{DS} is a set of storage device agents (Ag_{DS}) and Ag_{DS} takes charge of a DS or a group of DSs located at same place. Ag_{PG} is an agent of a power grid.

The agent communication language (ACL) and knowledge sharing among agents are required to construct a multiagent system [6, 8]. Here, we use the Contract Net Protocol (CNP) [6, 9, 10] for interactions among agents. The following is to show the main process of the CNP [10].

Step 1: A manager announces the existence of a task via a broadcast message.
Step 2: Agents evaluate the announcement and capable agents submit bids.
Step 3: The manager awards a contract to the most suitable agent among bidding agents as a contractor for the task.

The following is an example of the procedure of agent-based operation based on the CNP and on an energy market environment [4];

Step 1: Ag_{MGOCC} receives transaction prices (PR_S and PR_B) for the next interval from Ag_{PG}.
Step 2: Ag_{MGOCC} announces a new task with the transaction prices via a broadcast message to every Ag_{DG}, Ag_{L}, and Ag_{DS}.
Step 3: Ag_L and Ag_S, as consumers, inform Ag_{MGOCC} of their power demand.
Step 4: Ag_{DG} and Ag_S, as suppliers, evaluate the task using (8) and eligible agents submit bids, where P_{bid} is a bid price.

$$Bid = \begin{cases} 1 & if\ P_{bid} \leq P_B \\ 0 & other \end{cases} \tag{8}$$

Step 5: Ag_{MGOCC} selects final suppliers as contractors by a merit order algorithm, which is a classical algorithm for economic dispatch, and awards contracts.
Step 6: Surplus power unselected at a price lower than or equal to PR_S is sold to the power grid and shortage power is purchased from the power grid.
Step 7: The task is complete by submitting a report after performing the contract.

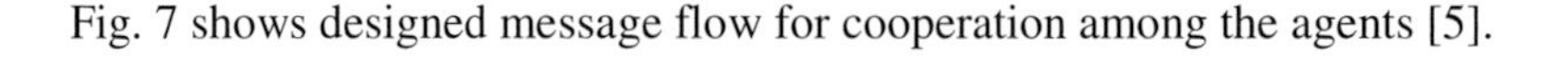

Fig. 7 shows designed message flow for cooperation among the agents [5].

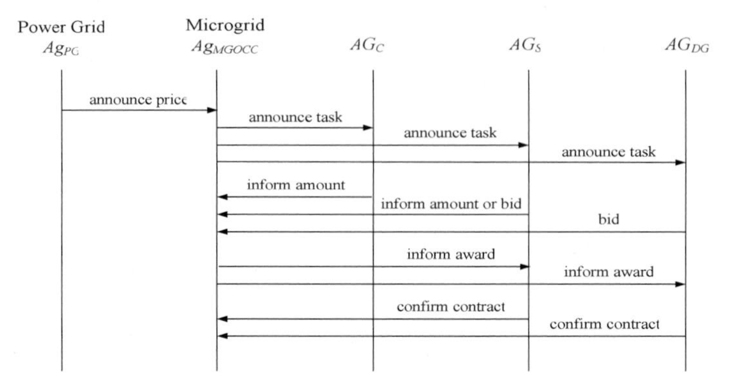

Fig. 7. Message flow among agents [5]

The multiagent system is designed by Distributed Agent System based on Hybrid Architecture (DASH) as a multiagent platform, Interactive Design Environment for Agent Designing Framework (IDEA) as a GUI-based interactive environment for the DASH platform , and Java for user defined functions [4, 11-13].

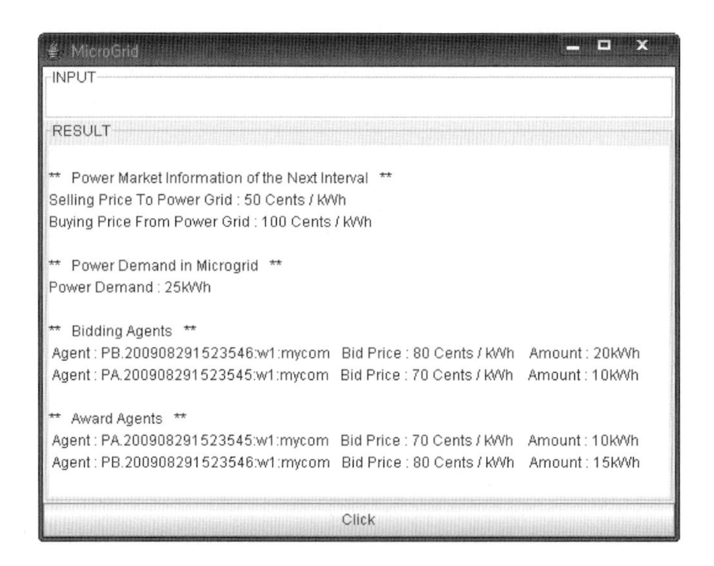

Fig. 8. A snapshot of the simulation result by IDEA [5]

4.2 Experiment

The developed multiagent system is tested with the following data;

- PR_S of the next interval: 50¢/ kWh
- PR_B of the next interval: 100¢/kWh

- Load in next interval = 25kWh
- PA = production cost: 70¢/kWh, capability: 10kWh
- PB = production cost: 80¢/kWh, capability: 20kWh

In this test, we assume DG agents bid with their power production cost. Fig. 8 shows the simulation result of the microgrid operation [5].

5 Conclusion

Microgrid operation is an important problem required for introduction of microgrids into power grids. In this paper, we dealt with backgrounds of microgrid and discussed characteristics of microgrid operation according to operation environments and with their classification. We introduced a mathematical model for microgrid operation based on LP and its some examples. We introduced a concept of agent-based microgrid operation as a challenging technology. In addition, we illustrated a design example for building a multiagent system for microgrid operation and a test example using the multiagent.

For more practical application of microgrid, there are the following interesting challenges to be solved.

- Forecasting electricity prices, loads, and power production of renewable energy sources such as wind power and solar power because changes of their output according to weather conditions
- Detailed mathematical model
- Consideration on heat or thermal energy produced from combined heat and power (CHP) sources like fuel cells in the microgrid
- Effective microgrid operation rules
- More efficient cooperation schemes among participants
- Agent's decision-making strategies

Finally, we hope new approaches from computer and information engineering societies as well as power engineers for more effective microgrid operation.

References

1. Hatziargyriou, N., Asano, H., Iravani, H.R., Marnay, C.: Microgrids. IEEE Power and Energy Magazine 5(4), 78–94 (2007)
2. Barnes, M., Kondoh, J., Asano, H., Oyarzabal, J., Ventakaramanan, G., Lasseter, R., Hatziargyriou, N., Green, T.: Real-Word MicroGrids – An Overview. In: 2007 International Conference of System of Systems Engineering, pp. 1–8 (2007)
3. Dimeas, A.L., Hatziargyriou, N.D.: Operation of a Multiagent System for Microgrid Control. IEEE Trans. on Power Systems 20(3), 1447–1455 (2005)
4. Kim, H.-M., Kinoshita, T.: A Multiagent System for Microgrid Operation in the Grid-interconnected Mode. Journal of Electrical Engineering & Technology 5(2), 246–254 (2010)

5. Kim, H.-M., Kinoshita, T.: Multiagent System for Microgrid Operation based on a Power Market Environment. In: INTELEC 2009, incheon, Korea (October 2009)
6. Wooldridge, M.: An Introduction to Multiagent Systems, 2nd edn. A John Wiley and Sons, Ltd., Chichester (2009)
7. Katiraei, F., Iravani, R., Hatziargyriou, N., Dimeas, A.: Microgrids Management-Controls and Operation Aspects of Microgrids. IEEE Power Energy 6(3), 54–65 (2008)
8. Barnes, M., Kondoh, J., Asano, H., Oyarzabal, J., Ventakaramanan, G., Lasseter, R., Hatziargyriou, N., Green, T.: Real-Word MicroGrids – An Overview. In: 2007 International Conference of System Engineering, pp. 1–8 (2007)
9. Kinoshita, T. (ed.): Building Agent-based Systems, The Institute of Electronics, Information and Communication Engineers (IEICE), Japan (2001) (in Japanese)
10. Smith, R.G.: The Contract Net Protocol: High-level Communication and Control in a Distributed Problem Solver. IEEE Trans. on Computer C-29(12) (December 1980)
11. Weiss, G. (ed.): Multiagent Systems: A Modern Approach to Distributed Artificial Intelligence. The MIT press, Cambridge (1999)
12. Kinoshita, T., Sugawara, K.: ADIPS Framework for Flexible Distributed Systems. In: Ishida, T. (ed.) PRIMA 1998. LNCS (LNAI), vol. 1599, pp. 18–32. Springer, Heidelberg (1999)
13. Uchiya, T., Maemura, T., Xiaolu, L., Kinoshita, T.: Design and Implementation of Interactive Design Environment of Agent System. In: Okuno, H.G., Ali, M. (eds.) IEA/AIE 2007. LNCS (LNAI), vol. 4570, pp. 1088–1097. Springer, Heidelberg (2007)
14. IDEA/DASH Tutorial, `http://www.ka.riec.tohoku.ac.jp/idea/index.html`

Authentication System for Electrical Charging of Electrical Vehicles in the Housing Development

Wang-Cheol Song

Department of Computer Engineering, Jeju National University,
66 Jejudaehakno, Jeju-si, Jeju, 690-756, South Korea
philo@jejunu.ac.kr

Abstract. Recently the smart grid has been a hot issue in the research area. The Electric Vehicle (EV) is the most important component in the Smart Grid, having a role of the battery component with high capacity. We have thought how to introduce the EV in the housing development, and for proper operation of the smart grid systems in the housing area the authentication system is essential for the individual houses. We propose an authentication system to discriminate an individual houses, so that the account management component can appropriately operate the electrical charging and billing in the housing estate. The proposed system has an architecture to integrate the charging system outside a house and the monitoring system inside a house.

Keywords: Smart Grid, Electric Vehicle, EV, Authentication, Home Electricity Management box, RFID tag.

1 Introduction

In a few years interest about the micro grid has been transferred to the smart grid. It integrates the information and communications technologies (ICTs) into the power grid at transmission and distribution levels to monitor and manage energy usage and maximize the efficiency, stability and reliability of the power system [1]. NIST [2] makes the standards for smart power grid in the United States, and the European Commission has issued a mandate M/441 to the three European Standardization organizations CEN [3], CENELEC [4] and ETSI [5] to define an open architecture for utility meters and services as well as interoperability of technologies and applications. Also, Distribution Automation Systems (DAS) has been defined by the IEEE as systems to enable an electric utility to monitor, coordinate, and operate distribution network components in real-time mode from remote control centers [6]. As the DAS server exchanges information with field equipments for the system operation, operations and communications are the key technologies in the distribution system.

In this paper, a management box as a DAS system and its authentication mechanism are proposed for operations of smart grid system in housing development. Fig. 1 shows the basic components in the smart grid system. As the Electric Vehicle (EV) is regarded as the key component in the smart grid, we have considered how to operate EV's electric charging in the housing development in the following points:

T.-h. Kim, A. Stoica, and R.-S. Chang (Eds.): SUComS 2010, CCIS 78, pp. 261–266, 2010.

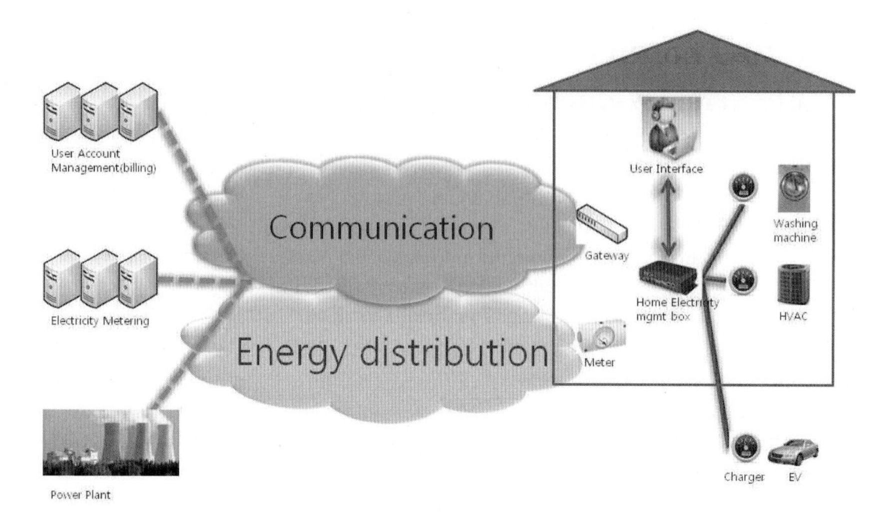

Fig. 1. Smart grid components

1. The electricity rate for charging the EV should be different from usual electricity rate. In order to control the load, the rate should depend on how slow it charges, when it charges, and so on. Especially in Korea, the progressive rate is applied to the usual living places. If the same rate is applied to electric charging of the EV, it cannot be considered to charg the EV at home.
2. The electric power company should provide separate power lines for EV charging. It is because EV charging must not affect usual power consumption of houses. So, I guess EV charging in the housing development can be developed as a new service business.
3. In the resident user's viewpoint, the electric power is preferred to be properly displayed as well as well managed. A home electricity management box can be set up at home and play a role of the control center as the DAS system. It should control electric charging of EVs and transaction of electricity as well as interact with the electric power company.
4. The EV is the key component in the smart grid, but located outside the house. Also, the charging system might be shared with other people, especially in the housing development. Hence, proper authentication should be applied to identify whose EV is charging. The authentication is required between the charger and the user as well as needed between home electricity management box and electric power companies for the billing.
5. The home electricity management box should integrate all kinds of management at home. The management may include the home electricity management, billing management, electricity trading management, electric charging management of the EVs, electricity load distribution management and so on.

Recently smart grid has been one of the hot issues in both research fields and industry fields. In a few years many research results have been published. T. Verschueren and et al. [7] proposes a common service architecture that allows houses with renewable energy generation and smart energy devices to plug into a distributed energy management system, integrated with the public power grid. In the architecture, Home Energy Management Box is proposed and plays the integral role. However, it just study only about the general aspect for the architecture and does not consider the security aspect for applying EVs in the architecture. A. Hamlyn and H. Cheung show computer network security management and authentication aspects for smart grid operations [8-9]. Their topics are written as authentication, but they explain only the access control to the power grid resources and the stability control. K. Mets and et al. [10] introduce how to adequately deal with pluggable (hybrid) electrical vehicles (PHEV) and propose Home Energy control box. Their research results show minimizing the peak load and flattening the overall load profile for electric charging of PHEV. But, it does not explain the architecture for it. K. Clement-Nyns and et al. [11] analyze how the extra electrical loads for charging the PHEV have an impact on the distribution grid in terms of power losses and voltage deviations, and propose coordinated charging to minimize the power losses and to maximize the main grid load factor. W. Kempton and et al. [12] generally speak about power grids and several aspects and explain Vehicle-to-grid power (V2G) implementation.

Based on the several research results, this paper intend to propose an architecture to introduce the EV in the housing development. Also, an authentication system is proposed to discriminate an individual houses, so that the account management component can appropriately operate the electrical charging and billing in the housing estate. Section 2 explains the Home Electricity Management Box, and the authentication mechanism is described in section 3. We conclude in section 4.

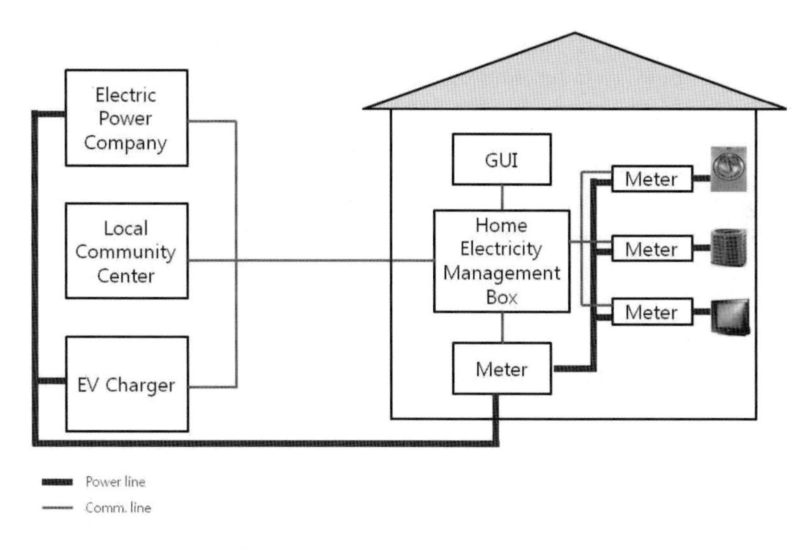

Fig. 2. The proposed architecture with HEMB

2 Home Electricity Management Box

Before the smart grid, the general users have just consumed the electricity. That is, the users do not need to control the electricity flow and what they can do their best is just to passively monitor the usage to minimize the electricity consumption. When they join the smart power grid, however, they can control the electricity flow and negotiate with the electricity power company to trade their electricity charged in their own batteries. In order to manage these kinds of activities, every home is required to have a management center with user friendly GUI. Therefore, this paper proposes a Home Electricity Management Box (HEMB) as the central component in the in-house energy management, and to reside at the consumers premises shown in Fig. 2. This HEMB is connected to the local metering devices to monitor the electricity usage and control them to connect or be circuit-broken when needed. And it can interact with electricity managing devices at the local community center and share information to maximize the efficiency of the energy usage. It can trade its own electricity charged in the EV with other people or the electricity power companies, too.

This HEMB box should be designed to have the intelligent functionality. Some people may want to control their electricity usage every day, but others may want the electricity to be controlled automatically. Then, it should do all of things such as negotiation of prices when electricity trading, load balancing when electric charging of EVs, monitoring the usage, alerting overload and so on.

HEMB has to discriminate the users to charge the EV at the charger shared in the housing development. As it is related to the billing system, HEMB is required to interact with the electricity power company. In Section 3 the authentication mechanism is proposed.

3 Authentication for EV Charging and Account Management

In order to introduce EV in the housing area, several points must be considered. Many people say EVs can be charged in the night when the electricity load is small, but if the number of EVs increases, the load must be handled separately from usual electricity power consumption and separate power lines must be newly built in the area. As one of the considering points, as the EV could be parked and charger be built outside the houses, when a EV want to charge the electricity, authentication must be performed and the charging amount must be properly billed. Therefore, we would like to propose an authentication mechanism.

The authentication is divided as two regions: between HEMB and the electricity charger for EV and between the electric power company and HEMB. For authentication between HEMB and the electric charger, the secret keys can be used. As the resident users can register its secret key to their HEMB directly, the secret key mechanism is sufficient for security. Then, we propose to use the RFID tag as the secret key. These days many vehicles change the car key to RFID tag. So, if the user registers the RFID tag as the secret key, it could be used to identify who want to charge electricity of the EV with the procedure shown in Fig. 3.

The authentication between the electric power company and HEMB can use the public key mechanism. The company can maintain the Certificate Authority (CA) to handle the keys, and a public and private key pair could be assigned to a HEMB by the CA. With the keys, MEMB can interact with the electric power company. The interaction includes billing, electricity trade, electricity load balancing and so on. All of exchanged messages can be encrypted by the public keys of the other side after authentication of HEMB as shown in Fig. 4.

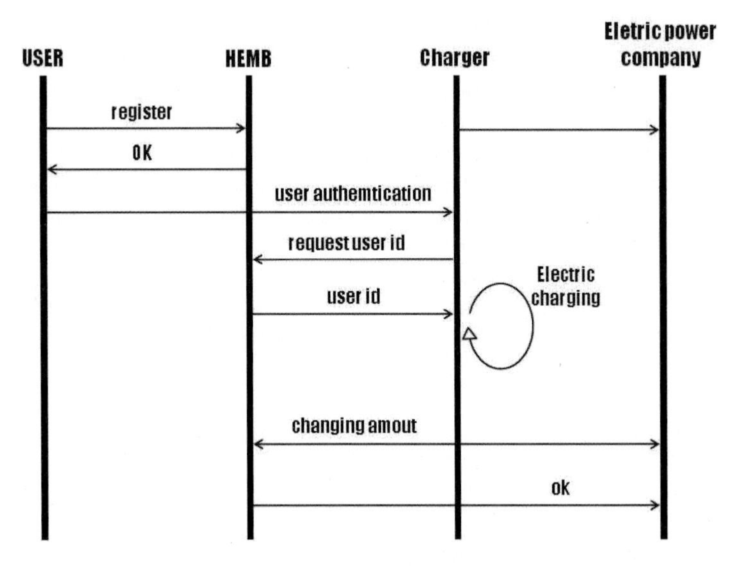

Fig. 3. Authentication of EV

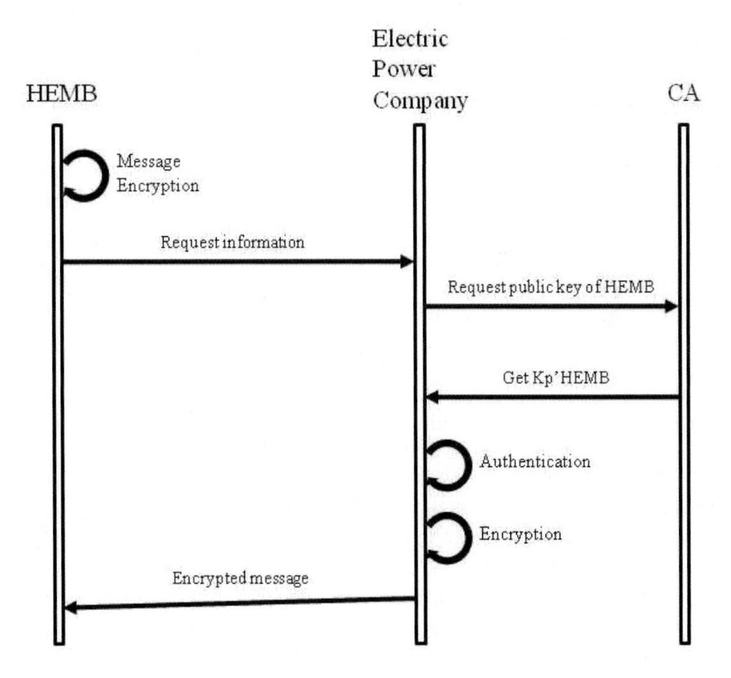

Fig. 4. Authentication between HEMB and the electric company

4 Conclusion

In this paper, we have proposed the authentication mechanism for introducing the EVs in housing development. Also, a Home Electricity Management Box is proposed as the control center in the smart grid system. As the EV is the key component in the smart power grid, we have made HEMB properly manage EV's electric charging. In order to do that, HEMB has been needed to discriminate who charges the EV's electricity. Also, through information exchanged with the electric power company HEMB can manage billing, electric trade, usage monitoring and so on. The information can be security sensitive. So, we have applied the authentication and message encryption. We think this system can be essential to introduce the smart power grid including EV in the housing development.

Acknowledgments. "This research was supported by the MKE(The Ministry of Knowledge Economy), Korea, under the ITRC(Information Technology Research Center) support program supervised by the NIPA(National IT Industry Promotion Agency)" (NIPA-2010-(C1090-1011-0009)).

References

1. Wikipedia, `http://en.wikipedia.org/wiki/Smart_grid`
2. NIST, National Institute of Standards and Technology (January 2010), `http://www.nist.gov/`
3. CEN, The European Committee for Standardization (January 2010), `http://www.cen.eu/cenorm/`
4. CENELEC, European Committee for Electrotechnical Standardization (January 2010), `http://www.cenelec.eu/`
5. ETSI, European Telecommunications Standards Institute (January 2010), `http://www.etsi.org/`
6. Bassett, D., Clinard, K., Grainger, J., Purucker, S., Ward, D.: Tutorial course: distribution automation. IEEE Tutorial Publ. 88EH0280-8-PWR (1988)
7. Verschueren, T., Haerick, W., Mets, K., Develder, C., De Turck, F., Pollet, T.: Architectures for smart end-user services in the power grid. In: 2010 IEEE/IFIP Network Operations and Management Symposium Workshops (NOMS Wksps), April 19-23, pp. 316–322 (2010)
8. Hamlyn, A., Cheung, H., Mander, T., Wang, L., Yang, C., Cheung, R.: Computer network security management and authentication of smart grids operations. In: 2008 IEEE Power and Energy Society General Meeting - Conversion and Delivery of Electrical Energy in the 21st Century, July 20-24, pp. 1–7 (2008)
9. Cheung, H., Hamlyn, A., Yang, C.: Network security authentication of power system operations. In: Canadian Conference on Electrical and Computer Engineering, CCECE 2008, May 4-7, pp. 001687–001692 (2008)
10. Mets, K., Verschueren, T., Haerick, W., Develder, C., De Turck, F.: Optimizing smart energy control strategies for plug-in hybrid electric vehicle charging. In: 2010 IEEE/IFIP Network Operations and Management Symposium Workshops (NOMS Wksps), April 19-23, pp. 293–299 (2010)
11. Clement-Nyns, K., Haesen, E., Driesen, J.: The Impact of Charging Plug-In Hybrid Electric Vehicles on a Residential Distribution Grid. IEEE Transactions on Power Systems 25(1), 371–380 (2010)
12. Kempton, W., Tomić, J.: Vehicle to Grid Implementation: from stabilizing the grid to supporting large-scale renewable energy. J. Power Sources 144(1), 280–294 (2005)

Design of a Multi-agent System for Personalized Service in the Smart Grid*

Jinhee Ko[1], In-Hye Shin[1,**], Gyung-Leen Park[1], Ho-Yong Kwak[2], and Khi-Jung Ahn[2]

[1] Department of Computer Science and Statistics
[2] Department of Computer Engineering
Jeju National University, Jeju, Korea
littletomato7942@hotmail.com,
{ihshin76,glpark,kwak,kjahn}@jejunu.ac.kr

Abstract. This paper designs a multi-agent system capable of providing personalized services in the smart grid, defining the relevant agent modules. The proposed system provides electricity consumers with personalized power purchase recommendation. Our framework consists of four agents and seven object categories. For the operation center which manages and controls the whole system, an adaptive agent, a coordination agent, and a filtering agent are defined, while a consumer agent is define for each home to collect the history of power consumption. Based on the analysis of the consumer, power market, residence, power consumption, appliance, family member, and electric vehicle objects, those agents autonomously cooperate to provide a personalized power service to each smart grid entity. In addition, adaptive learning capability further improves the recommendation quality.

1 Introduction

The Smart Grid combines the traditional power system and the information and communication technology, or ICT. It will be characterized by the bidirectional flow of electricity and information to create an automated and widely distributed energy delivery network [1]. The smart grid brings many benefits such as energy saving by real-time pricing and efficient energy management. In addition, it can accelerate the deployment of electric vehicles and renewable energy which can significantly reduce the greenhouse gas emissions [2]. After all, this technology can cope with high oil prices and global climate changes. The Smart Grid provides a new paradigm of the power industry and a next-generation power environment, having diverse forms to achieve the given system goals. It enables self-healing, consumer participation, attack resistant, providing high quality power, optimizing assets and operating efficiently.

---

* This research was supported by the MKC Korea, under the ITRC support program supervised by the NIPA (National IT Industry Promotion Agency) (NIPA-2010-(C1090-1011-0009)).
This research was also supported by the MKE (Ministry of Knowledge Economy) through the project of Region technical renovation, Korea.
** Corresponding author.

T.-h. Kim, A. Stoica, and R.-S. Chang (Eds.): SUComS 2010, CCIS 78, pp. 267–273, 2010.

The national government of the Republic of Korea defined the national roadmap for the smart grid in January, 2010, aiming at building the world's first nation-wide smart grid system by 2030. As a preliminary step for the enterprise, the national government opened the Jeju smart grid complex which covers all key technologies belonging to the smart grid. Currently, 12 consortiums consist of 171 companies are participating in the enterprise, categorized into 5 groups of the smart power network, the smart power market, the smart grid consumer, smart transportation, and smart renewable energy [4].

In the mean time, in the future, smart grid will integrate a sophisticated personalization service which is commonly used in the recommender system for tour, movies, and books. This service can also recommend a smart energy activity, namely, generation and consumption, to the individuals according to their personalities, taking into account the diverse electricity pricing policy and energy management automation system. In this regard, this paper is to propose an autonomous and active personalization service in the smart grid based on a multi-agent system, where more than one agent cooperatively solves a given complex problem.

The rest of this paper is structured as follows. Section 2 introduces the personalization and the multi-agent system, discussing related work in the smart grid. Section 3 proposes multi-agent system architecture for the personalized service in the smart grid. Finally, Section 4 summarizes and concludes this paper.

2 Related Work

2.1 Personalization and the Multi-agent System

Personalization means offering goods, services, or related information to each customer based on his personal characteristics. The basic goal of the personalized system is to offer a service that customers may want or need even though they didn't ask or choose it [5]. The methods of personalization include rule-based filtering, collaboration filtering, learning agent, and content-based filtering, while some schemes can be combined to create a hybrid method.

Rule-based filtering is the most basic and general personalization scheme, which creates a user profile by asking users on their personalities, interests, and preferences. However, it takes quite a long time to obtain a reasonable answer, as this scheme needs a lot of preference data. Collaboration filtering autonomously estimates the user's concern or interest based on the taste information obtained from the sufficient number of users. This scheme relies on the assumption that the user's interest in the past will not greatly change in the future. Its main characteristic lies in that the taste information is collected not from the specific user groups but from many others. The learning agent observes users' activities on the web, decides which subject the user is interested in, and finally provides the recommendation to the user. This scheme requires that the predefined metadata be stored in the database. Content filtering is the technique whereby content is blocked or allowed based on the analysis result of its content, rather than its source or other criteria. It is most widely used on the Internet to filter e-mail and web access [6].

It is also possible to request too much information from users. In this case, users may refuse to provide the information or sometimes answers meaningless data, creating a great amount of garbage. The multi-agent system can solve this problem, as it actively collects the user information and adaptively analyzes the information. The agent system automatically handles the work needed by users. Moreover, the multi-agent system is a conglomerate of agent systems. The cooperation of multi-agent solves complicated problems, which cannot be solved by a single agent [7].

2.2 The Smart Grid

As shown in Figure 1, the smart grid delivers electricity from power providers to consumers using the digital technology, enabling the devices to communicate with the operation system. The operation system plays a role of the broker between consumers and providers, performs monitor-and-control functions, and analyzes the collected data. Many governments, looking for ways to improve energy efficiency to overcome global warming issues, seriously take into account these electricity networks. A smart grid network will also give consumers more choices, for example, enabling them to pay different rates for difference appliances. We can also introduce different electricity rates for power generated from green technology such as wind and solar power generators. Moreover, advanced smart meters can establish a network capable of informing people of when it is cheapest to use electricity. It is estimated that household electricity bills could be cut by around 15 percents [1,4].

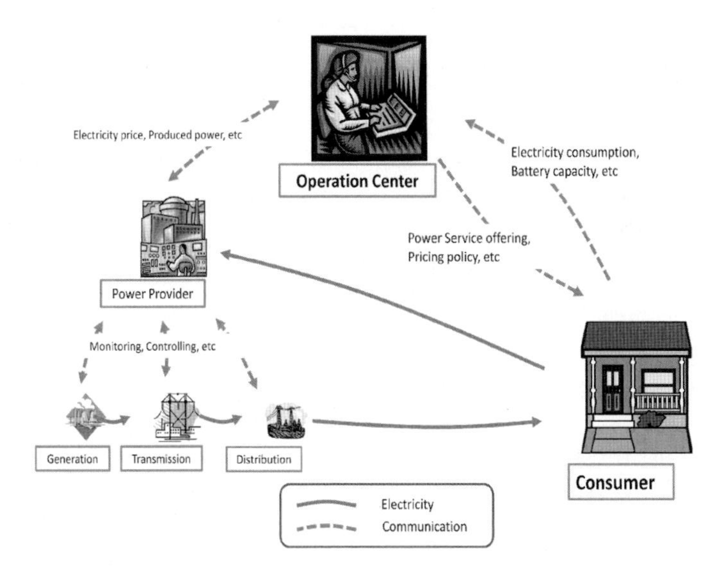

Fig. 1. Data flow in the smart grid

By way of the bidirectional interaction with consumers, the operation center can collect the diverse usage patterns of respective consumers, making it possible to provide a personalized recommendation service and an efficient power network management. For example, the peak power load can be forecasted and mitigated by the power consumption pattern analysis. For the consumer side, he can provide his profile and power consumption history in real-time, optimize the power consumption, and finally select the pricing policy advantageous to his consumption pattern.

3 The Proposed System

3.1 The Multi-agent System Structure

Our system consists of 4 agents and 7 object categories as shown in Figure 2. To begin with, the consumer agent, created for each home, collects the profile and power usage history in its database. The power communication network delivers the data monitored by the consumer agent to the operation center, which improves the power network efficiency by analyzing the power load, production, and consumption. Agents in the operation center autonomously cooperate with one another by the coordination agent.

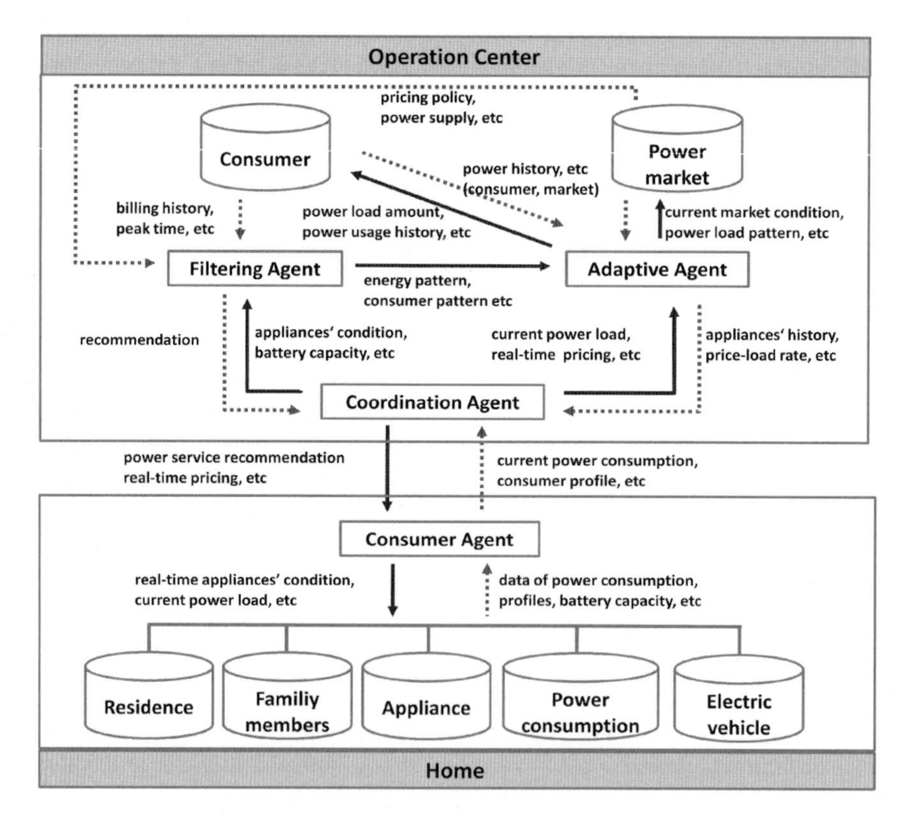

Fig. 2. The proposed multi-agent system architecture

3.2 Operation Center

3 agents and 2 object categories are defined for the operation center. First, the coordinator agent opens a communication interface for consumer agents in their own homes. Every message is distributed by this agent to the appropriate server or agent. Second, the filtering agent selects the specific data set according to the given filtering strategy. For example, for the new subscriber, the agent searches the user profile and the power consumption history out of existing subscribers having the reasonable level of similarity with the new subscriber. Based on the power consumption pattern of those similar subscribers, the filtering agent recommends an appropriate service to the new comer and guides a battery-charging schedule. As contrast, for the existing subscribers, the content-based filtering scheme is basically exploited, giving a cross recommendation using the collaborative filtering strategy which takes into account the power usage pattern of other consumers. In addition, the learning agent can enhance the recommendation quality by tuning the relevant parameters based on the feedback from the customer on the current recommendation.

The data generated at the filtering agent are transferred to the adaptive and consumer agents. The adaptive agent modifies the information in the object base according to the data received from the filtering agent. Here, the reinforcement learning mechanism, specifically, the Q-learning algorithm, keeps updating the weight for the data associated with each agent. For this purpose, this agent defines a weight evaluation function to optimize the recommendation from each agent.

3.3 Home

An agent and 5 object categories are defined for homes. The consumer agent monitors the power usage pattern in each home for the sake of responsive and accurate power control, collecting the real-time data on user profile, current appliance power consumption, and the charge rate for electric vehicles. The collected data are stored in the home database and will be transmitted to the operation center. The home agent can trigger the data analysis on the consumer and the power market in the operation center, when it discovers that a new consumption policy is necessary. The coordination agent delivers the analysis results obtained from all agents in the operation center to the consumer agent. Based on those results, the later controls power consumption of electrical appliances, recommends the personalized power service to the consumer, and so on.

3.4 Data Classification

2 and 6 object categories are defined in the database for the operation center and homes as shown in Figure 3 and Figure 4, respectively. First, the operation center object includes consumer and power market objects which mainly cover home profiles, power usage pattern, and power consumption history. Additionally, hourly, daily, and seasonal power load, current power supply and consumption status, and the like are belonging to this object category. Next, the home object includes the residence type, battery capacity, members, power features of electronic appliances, and the electric vehicle profile. This object also covers the hourly, daily, and monthly power load dynamics obtained from monitoring of the consumer agent.

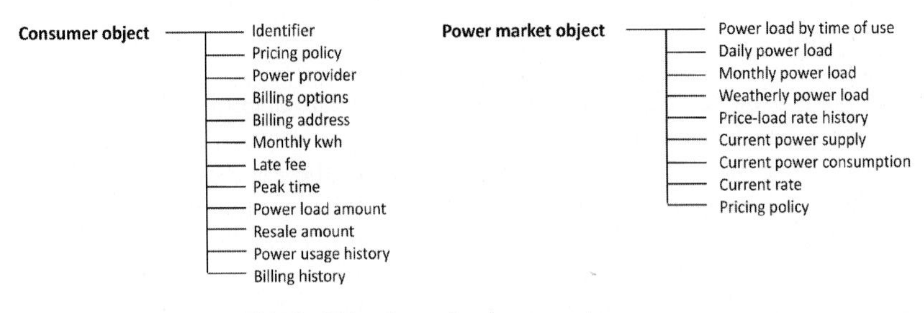

Fig. 3. Object base for the operation center

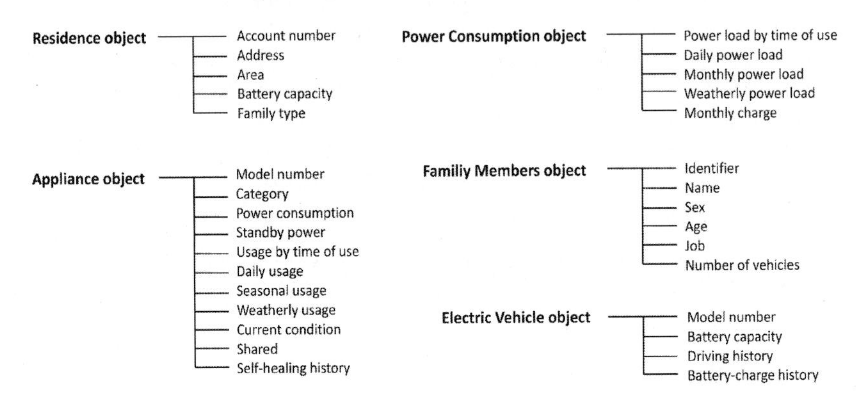

Fig. 4. Object base for home

4 Conclusion

This paper has designed a multi-agent system capable of providing personalized services in the smart grid, after defining the relevant agent modules. The proposed system can adapt to the change in the customer characteristics and in the environmental condition. In our design, the operation center includes a coordination agent, a filtering agent, and an adaptive agent along with 2 object categories. As contrast, a consumer agent and 5 object categories are defined for each of homes. Noticeably, agents monitor the power network to collect the usage pattern and other power related data. Then, they recommend an efficient and system-specific power service by the cooperative analysis on the power operation and available services. As a result, it is expected that our analysis model can provide an efficient guideline or recommendation to consumers, provides, and service developers.

As future work, we are first planning to assess our design by implementing the prototype and then develop a mobile interface with the corresponding web service.

References

1. EPRI: Report to NIST on the Smart Grid Interoperability Standards Roadmap. NIST (2009)
2. Pipattanasomporn, M., Feroze, H., Rahman, S.: Multi-agent Systems in a Distributed Smart Grid; Design and Implementation. In: 2009 IEEE PES Power Systems Conference and Exposition, pp. 1–8 (2009)

3. Son, S., Chung, B.: A Korean Smart Grid Architecture Design for a Field Test based on Power IT. In: Transmission & Distribution Conference & Exposition, Asia and Pacific, pp. 1–4 (2009)
4. Korea Smart Grid Institute, `http://www.smartgrid.or.kr`
5. Na, Y., Ko, I., Han, K.: A Design of a Recommendation System for One to one Web Marketing. The Korea Information Processing Society Transaction: Part D 11(7), 1537–1542 (2004)
6. Wikipedia, `http://www.wikipedia.org`
7. Ko, J., Hur, C., Kim, H.: A Personalized Mobile Learning System Using Multi-agent. In: Dean, M., et al. (eds.) WISE 2005 Workshops. LNCS, vol. 3807, pp. 144–151. Springer, Heidelberg (2005)
8. Lee, J., Kang, E., Park, G.: Design and Implementation of a Tour Planning System for Telematics Users. In: Gervasi, O., Gavrilova, M.L. (eds.) ICCSA 2007, Part III. LNCS, vol. 4707, pp. 179–189. Springer, Heidelberg (2007)
9. Rui, D., Deconinck, G.: Future electricity market interoperability of a multi-agent model of the Smart Grid. In: 2010 International Conference on Networking, Sensing and Control (ICNSC), pp. 625–630 (2010)
10. Bhuvaneswari, R., Srivastava, S.K., Edrington, C.S., Cartes, D.A., Subramanian, S.: Intelligent Agent Based Auction by Economic Generation Scheduling for Microgrid Operation. In: Innovative Smart Grid Technologies (ISGT), pp. 1–6 (2010)

An Efficient Scheduling Scheme on Charging Stations for Smart Transportation*

Hye-Jin Kim[1], Junghoon Lee[1], Gyung-Leen Park[1], Min-Jae Kang[2], and Mikyung Kang[3,**]

[1] Dept. of Computer Science and Statistics
[2] Dept. of Electronic Engineering
Jeju National University, 690-756, Jeju Do, Republic of Korea
[3] University of Southern California - Information Sciences Institute, VA22203, USA
{hjkim82,jhlee,glpark,minjk}@jejunu.ac.kr, mkkang@isi.edu

Abstract. This paper proposes a reservation-based scheduling scheme for the charging station to decide the service order of multiple requests, aiming at improving the satisfiability of electric vehicles. The proposed scheme makes it possible for a customer to reduce the charge cost and waiting time, while a station can extend the number of clients it can serve. A linear rank function is defined based on estimated arrival time, waiting time bound, and the amount of needed power, reducing the scheduling complexity. Receiving the requests from the clients, the power station decides the charge order by the rank function and then replies to the requesters with the waiting time and cost it can guarantee. Each requester can decide whether to charge at that station or try another station. This scheduler can evolve to integrate a new pricing policy and services, enriching the electric vehicle transport system.

1 Introduction

The Republic of Korea was nominated as world's leading nation in the smart grid technology [1]. The smart grid is the next generation power network which combines information technology with the legacy power network to optimize the energy efficiency [2]. It can also make it possible to exchange information on power generation and consumption between those parties, bringing the era of *prosumer*, which means any individual can be both consumer and producer of energy at the same time. The Korean national government opened the smart grid complex in Jeju area, pursuing 5 goals of smart power grid, smart place, smart transportation, smart renewable energy, and smart electricity service [1]. Among these, the smart transportation part installs electric charging stations along the road network and at homes to accelerate the deployment of electric vehicles [3]. More specifically, the charge station will be installed in the existing gas

* This research was supported by the MKE, Korea, under the ITRC support program supervised by the NIPA. (NIPA-2010-(C1090-1011-0009)).

** Corresponding author.

T.-h. Kim, A. Stoica, and R.-S. Chang (Eds.): SUComS 2010, CCIS 78, pp. 274–278, 2010.

stations and LPG filling stations, public institution buildings, shopping malls, and airports.

Electric vehicles are charged on any charging stations, but it takes quite a long time in stations. Moreover, the requirement on the charge is usually different vehicle by vehicle. For example, a vehicle arrives at the station at 2 PM, needs 5 KW with the unit price less than 1 USD, and can afford to wait until 3 PM. Thus, the charging station must schedule the service order for multiple vehicles to meet the requirement of as many vehicles as possible. In this regard, this paper is to parameterize the vehicle-side requirement on battery charging and propose a scheduling scheme which decides the charge order to improve the satisfiability of vehicles. The station charges the vehicles according to this order and informs a vehicle of the estimated service time. The vehicle can confirm its reservation, renegotiate with a modified requirement, or choose another station. This paper is organized as follows: After issuing the problem in Section 1, Section 2 describes the background of this paper. Section 3 explains the service scenario and proposes the rank function. Section 4 summarizes and concludes this paper with a brief introduction of future work.

2 Background

Smart transportation is one of the most important areas in the smart grid. Electric vehicles need nation-wide power charge infrastructure, possibly creating a new business model embracing diverse vehicles, charging stations, and corresponding services [4]. Based on the provided information such as price plan of each station and a personal schedule, a user can decide when to charge his car, while reselling the surplus back to the power company during the peak hours. In addition, the battery-charged power can be used as back-up power source [5], so we can expect the improvement in the power network efficiency and reliability as well as the reduction of greenhouse gas emissions. The charging station can be installed in diverse places as shown in Figure 1. Drivers can charge their vehicles at their homes, offices, public institutes, shopping malls, charging stations, and the like. Noticeably, while the car is being charged, the driver can work at his office or take shopping at the mall. In those places, many vehicles will be concentrated and they must be served according to a well-defined reservation strategy.

3 Scheduling Scheme

3.1 Service Scenario

To simplify the problem, this section first assumes that the station charges one vehicle at a time, however, this restriction can be easily eliminated. In our scenario, a driver tries to make a reservation at a charge station before it arrives at the station via the vehicular network, specifying its requirement details as shown in Figure 2. Each requirement consists of expected price, estimated arrival time,

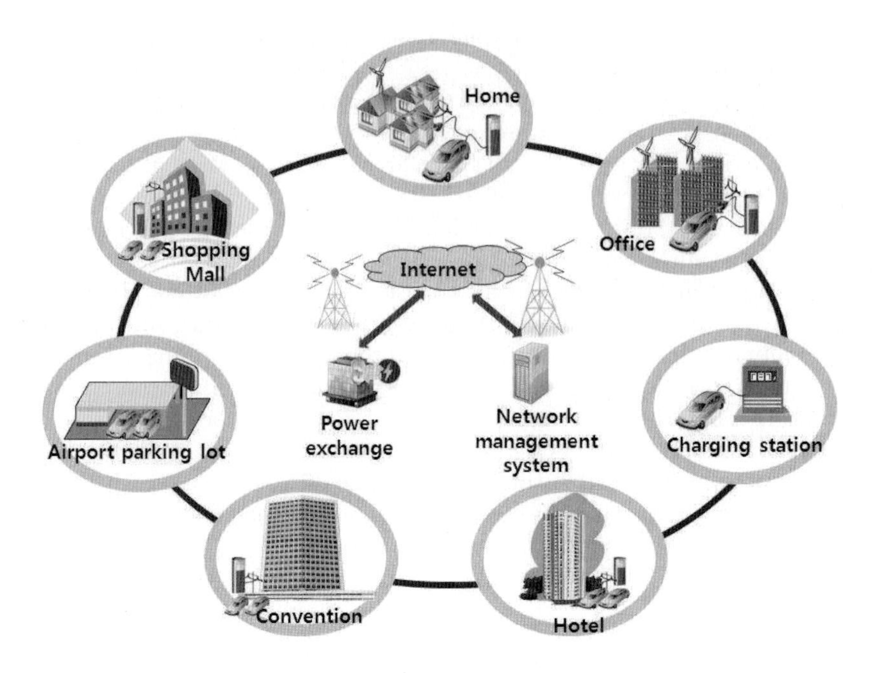

Fig. 1. Various places where many EVs can charge

tolerance bound on waiting time, minimum and maximum charge amount, and so on. Receiving the request, the scheduler calculates the rank function for the new request, reorders the request along with the existing ones, and checks whether the station can meet the requirement of the new request without violating the constraints of already admitted requests. The result is delivered back to the vehicle, and the driver can confirm the request, attempt a renegotiation, or choose another station. Here, it must be mentioned that there are several commercially available vehicular networks, for example, DSRC (Dedicated Short Range Communication) and IEEE 802.11 WLAN [6].

3.2 Rank Function

Each vehicle sends a reservation request message consist of the fields shown in Figure 3(a) via its in-vehicle telematics device and the corresponding vehicle network [7]. The scheduler processes requests one by one, namely, reorders the requests based on the rank function, estimates the service completion time, checks whether the completion time lies within the tolerance bound for all requests, and finally sends back to the requester whether the station can accept the request or not. The scheduler defines the rank function, T_v, as shown in Eq. (1).

$$T_v = ETA_v + WT_v + \frac{C_v}{r}, \tag{1}$$

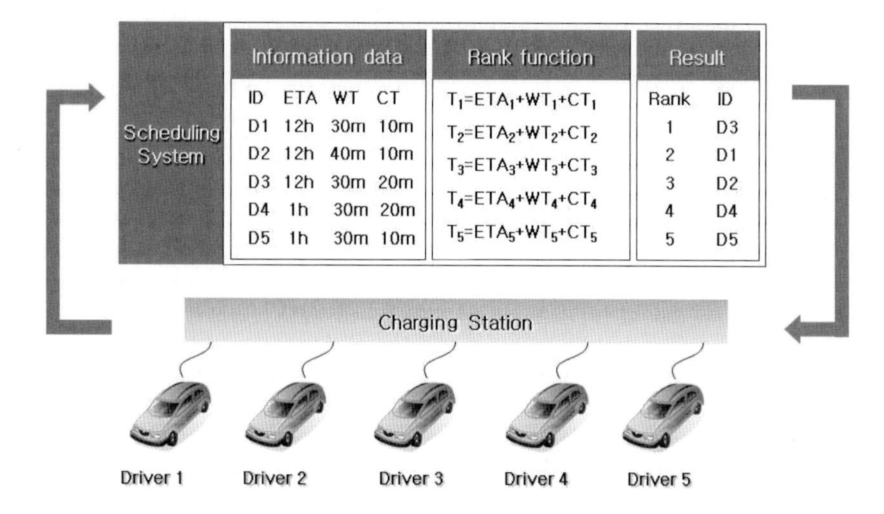

Fig. 2. Service scenario

where ETA_v denotes the estimated arrival time of vehicle v and WT_v denotes the tolerance bound on the waiting time, that is, how long v can wait until it is served. In addition, C_v is the charge amount of v, and r is the charge speed in the station, hence, CT_v, or $\frac{C_v}{r}$ means the charging time. The rank function can be executed in $O(n)$ time complexity, where n is the number of requests. It can avoid the time-consuming search space traversal that takes $O(n!)$ complexity, possibly giving the prompt reply to the vehicle so that it can renegotiate or try another station.

(a) Request specification				(b) Scheduling			
Req	ETA	VT	CT	T	Rank	Service time	Decision
A	12:00	30m	10m	12:20	2	12:30	Accept
B	12:00	40m	10m	12:30	3	12:40	Accept
C	12:00	30m	20m	12:10	1	12:20	Accept
D	12:00	40m	20m	12:20			Reject
E	13:00	30m	20m	13:20	5	13:20	Accept
F	13:00	30m	20m	13:30	6	13:30	Accept

Fig. 3. Operation of the scheduler

Figure 3 shows the sample scenario to describe how the proposed scheme works. The requests from A to F arrive at the scheduler sequentially, and each of them invokes the scheduler, respectively. Until request C, the service order decided by T_v can charge all vehicles within their tolerance bound. However, for D, the service order (C, A, D, B) cannot meet the tolerance bound requirement for B and D. As a result, the scheduler rejects D. For requests E and F, which have the later arrival time, can be served and accepted.

The proposed rank function is highly likely to admit the request having a long tolerance bound, as it can wait a relatively long time and give flexibility to the scheduler. The station prefers those requests and can possibly give a discount. In addition, the estimated arrival time can be decided by the in-vehicle navigation module by the locations of the current vehicle and the charging station. We can assume that the estimation is quite accurate. Generally, the in-vehicle computer system has sufficient computing power especially in electric vehicles, as it handles a lot of stream data to monitor and sometimes control vehicles [8]. However, if the vehicle arrives ahead of schedule, it must wait. On the contrary, if the vehicle arrives after the reserved time, its reservation is adjusted or sometimes cancelled.

4 Concluding Remarks

This paper has designed a reservation-based scheduling scheme for the charging station to decide the service order of multiple vehicles to improve the number of charging requests the station can serve. The proposed rank function takes into account the estimated arrival time, delay tolerance bound, and charging speed. The rank function decides whether a new request can be served in a linear execution time. It can also integrate additional criteria such as pricing policy, for example, which gives a discount to the request having a long tolerance bound. As future work, we are first planning to verify the efficiency of our scheme in terms of schedulability, comparing with the brute force scheme which can find the optimal solution even in unacceptable time. Next, a charging station selection algorithm is to be designed for the convenient driving of electric vehicles.

References

1. Korean Smart Grid Institute, http://www.smartgrid.or.kr/eng.htm
2. Gellings, C.W.: The Smart Grid: Enabling Energy Efficiency and Demand Response. CRC Press, Boca Raton (2009)
3. Guille, C., Gross, G.: A Conceptual Framework for the Vehicle-to-grid (V2G) Implementation. Energy Policy 37, 4379–4390 (2009)
4. Kaplan, S.M., Sissine, F.: Smart Grid: Modernizing Electric Power Transmission and Distribution; Energy Independence, Storage and Security. The Capitol.Net (2009)
5. Markel, T., Simpson, A.: Plug-in Hybrid Electric Vehicle Energy Storage System Design. In: Advanced Automotive Battery Conference (2006)
6. Society of Automotive Engineers: Dedicated Short Range Communication Message Set Dictionary. Technical Report, Standard J2735 (2006)
7. Lee, J., Park, G., Kim, H., Yang, Y., Kim, P., Kim, S.: A Telematics Service System Based on the Linux Cluster. In: Osvaldo, G., Gavrilova, M. (eds.) ICCSA 2007, Part III. LNCS, vol. 4490, pp. 660–667. Springer, Heidelberg (2007)
8. Schweppe, H., Zimmermann, A., Grill, D.: Flexible In-vehicle Stream Processing with Distributed Automotive Control Units for Engineering and Diagnosis. In: IEEE 3rd International Symposium on Industrial Embedded Systems, pp. 74–81 (2008)

Design for Run-Time Monitor on Cloud Computing*

Mikyung Kang[1], Dong-In Kang[1], Mira Yun[2], Gyung-Leen Park[3], and Junghoon Lee[3,**]

[1] University of Southern California – Information Sciences Institute, VA, USA
[2] Dept. of Computer Science, The George Washington University, Washington DC, USA
[3] Dept. of Computer Science and Statistics, Jeju National University, Jeju, South Korea
{mkkang,dkang}@isi.edu, mirayun@gwu.edu,
{glpark,jhlee}@jejunu.ac.kr

Abstract. Cloud computing is a new information technology trend that moves computing and data away from desktops and portable PCs into large data centers. The basic principle of cloud computing is to deliver applications as services over the Internet as well as infrastructure. A cloud is the type of a parallel and distributed system consisting of a collection of inter-connected and virtualized computers that are dynamically provisioned and presented as one or more unified computing resources. The large-scale distributed applications on a cloud require adaptive service-based software, which has the capability of monitoring the system status change, analyzing the monitored information, and adapting its service configuration while considering tradeoffs among multiple QoS features simultaneously. In this paper, we design Run-Time Monitor (RTM) which is a system software to monitor the application behavior at run-time, analyze the collected information, and optimize resources on cloud computing. RTM monitors application software through library instrumentation as well as underlying hardware through performance counter optimizing its computing configuration based on the analyzed data.

Keywords: Run-Time Monitor, Cloud Computing, QoS, library instrumentation, performance counter.

1 Introduction

Cloud computing is a new information technology trend that moves computing and data away from desktops and portable PCs into large data centers. The basic principle of cloud computing is to deliver applications as services over the Internet as well as infrastructure. A cloud is the type of a parallel and distributed system consisting of a collection of inter-connected and virtualized computers that are dynamically

* This research was supported by the MKE (The Ministry of Knowledge Economy), Korea, under the ITRC (Information Technology Research Center) support program supervised by the NIPA (National IT Industry Promotion Agency (NIPA-2010-C1090-1011-0009)) and OPERA Software Architecture Project.
** Corresponding author.

T.-h. Kim, A. Stoica, and R.-S. Chang (Eds.): SUComS 2010, CCIS 78, pp. 279–287, 2010.

provisioned and presented as one or more unified computing resources. The large-scale distributed applications on a cloud require adaptive service-based software, which has the capability of monitoring the system status change, analyzing the monitored information, and adapting its service configuration while considering tradeoffs among multiple QoS features simultaneously.

Recently, multi-core and many-core architectures are becoming more and more popular due to diminishing returns from traditional hardware innovations such as caching and deep pipeline architectures. With more cores in a processor, it is easier to get performance gains by parallelizing applications than traditional approaches. In addition, traditional processors consume large amounts of power to achieve high performance by using high frequencies. By using multiple cores at a lower frequency, and consequently lower voltage, multi-core architectures can achieve higher performance with lower power consumption.

There have been many multi-core processors from commercial vendors [1][2][3]. Among them, Tilera Corporation offers three processor families with the largest number of cores on a general-purpose chip available on the market [4]. Boeing has developed a processor called MAESTRO, to be used in space, based on the first Tilera processor, TILE64 [5]. The TILE64 has 64 cores on a chip. Each core has a three-instruction-wide Very Long Instruction Word (VLIW) pipeline, memory management unit, L1 and L2 cache, so each core itself is a complete processor, which can run a complete operating system like Linux (although more commonly, a single operating system instantiation is used to control multiple cores).

We are to design Run-Time Monitor (RTM) which is a system software to monitor the characteristics of applications at run-time, analyze the collected information, and optimize resources on cloud node which is consisted of multi-core processors. The rest of the paper is organized as follows. In Section 2, the system architecture is briefly described. Our proposed Run-Time Monitor is described in Section 3. Implementation result is described in Section 4, and Section 5 concludes the paper.

2 System Architecture

Eucalyptus (Elastic Utility Computing Architecture for Linking Your Programs To Useful Systems) project began from California University at Santa Barbara, and mainly was targeting at building a private open-source cloud platform [6]. Now Eucalyptus is an open-source implementation of Amazon EC2 (Elastic Compute Cloud) and compatible with most business interfaces [7][8]. Eucalyptus is an elastic computing structure that can be used to connect the user's programmers to the useful systems and it is an open-source infrastructure using clusters or workstations implementation of elastic, utility, and cloud computing. Figure 1 demonstrates the topology structure of Eucalyptus resources. In this figure, the node controller is a component running on the physical resources. On each node, all kinds of entities of virtual machines can run. Logically connected nodes form a virtual cluster, and all nodes belonging to the same virtual cluster receive a command from the cluster controller and then report to the same controller. Parallel HPC applications often need to distribute large amounts of data to all compute nodes before or during a run [9]. In a cloud, these data are typically stored in a separate storage service. Distributing data from the storage service to all compute nodes is essentially a multicast operation. This paper targets multi-core processor with a single node controller on each node. After receiving data and command, each node processes data while monitoring performance and optimizing resources. And then it returns the result to the cluster controller.

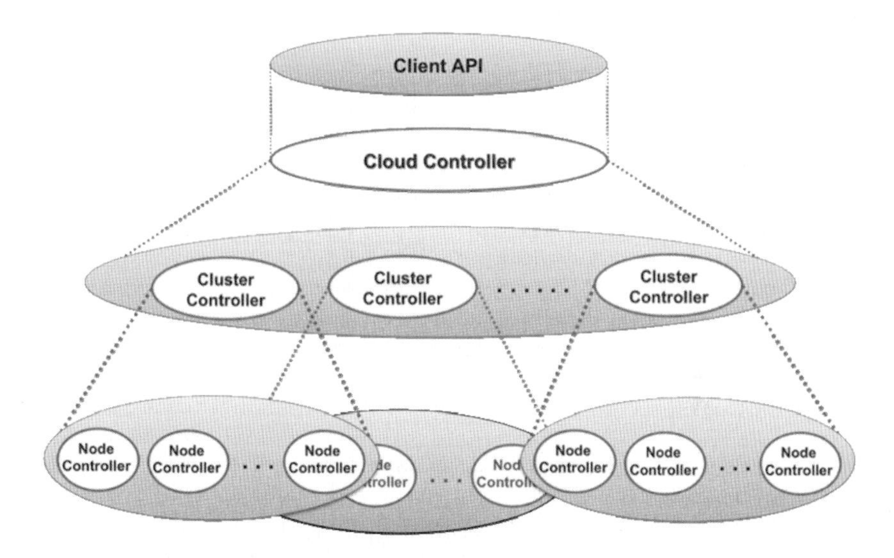

Fig. 1. The resource topology structure of Eucalyptus

3 Run-Time Monitor (RTM)

Run-time monitor is a system software to monitor the characteristics of applications at run-time. As shown in Figure 2, RTM monitors both application software and hardware. We implemented RTM through library instrumentation for software information and through Perfmon2/PAPI for hardware information. The collected software and hardware information can be used by Parallel Performance Analysis Tool and Run-Time system.

RTM implementation targets multi-core processors such as Tilera's TILE64 processor or MAESTRO processor. The implementation has been developed on top of a modified version of libraries such as MPI [10], pthread, iLib [5] and so on.

For the dynamic linking at runtime via preload, we exploited library interposition with the library instrumentation as shown in Figure 3. The interposition is a technique that allows an additional function to be automatically called whenever a library function is called. In RTM model, an interposed library layer was added so that original library modification/recompilation is not needed, no source is needed for anything, and no recompilation/redo on library version update (only on API change). So in each interposed library, needed information was collected after calling unmodified binary library.

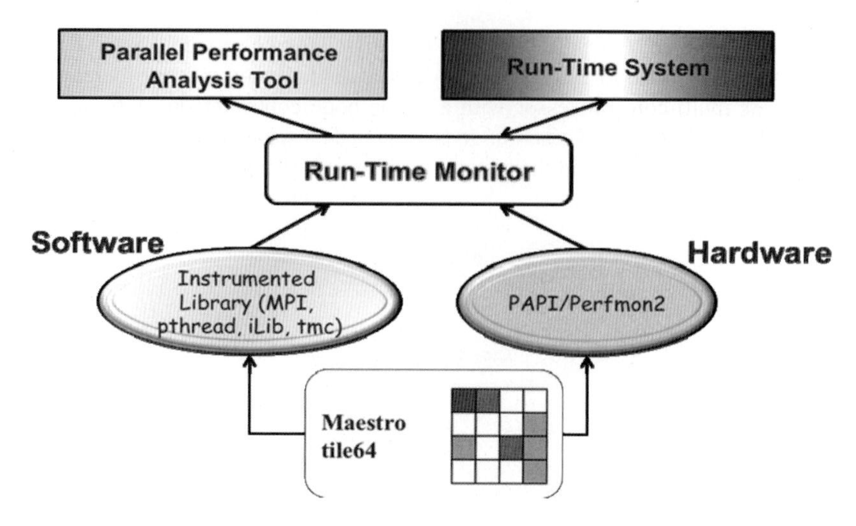

Fig. 2. Run-Time Monitor Overview

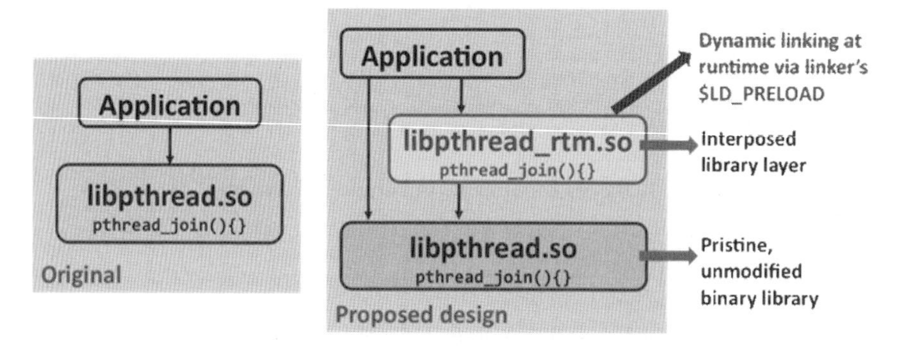

Fig. 3. Library Interposition

Let's see this sample interposed call, pthread_join, in Figure 4. In the application, pthread_join is called. Then, in the interposed library layer, unmodified original library is called and the return value was saved. The *dlsym* is a routine that gives the user direct access to the dynamic linking facilities. The *dlsym* allows a process to obtain the address of a symbol defined within a shared object previously opened by *dlopen*. If handle is *RTLD_NEXT*, the search begins with the *next* object after the object from which *dlsym* was invoked. The *pthread_join* is the symbol's name as a character string. And then other information needed for RTM was calculated, set, and saved in FIFO system. After saving necessary information, interposed library sends the information to the RTM server at the instrumentation layer. Basic library functions and inline/macro functions were instrumented for the interposition.

The RTM server and client concept is illustrated in Figure 5. After loading the RTM software and hardware servers, RTM client program begins to run. Whenever an event happens on each tile, RTM client sends the information to the RTM server

```
int main()
{ ...
  pthread_join(..);
}
```

Application

```
int pthread_join(pthread_t thread, void **value_ptr)
{ static void * (*func)();
  if(!func)
   func = (void *(*)()) dlsym(RTLD_NEXT, "pthread_join");

  fcall.start_cycle = time(NULL);
  int retval = (int) func(thread, value_ptr);
  fcall.end_cycle = time(NULL);

  fcall.tid = (long long) pthread_self(); // Other info
  // Send info to RTM-server
  write(rtm_fifo, &fcall, sizeof(fcall));
  return(retval);
}
```

Interposed library (instrumentation layer)

```
int pthread_join(pthread_t thread, void **value_ptr)
{ ...
}
```

libpthread.so (unmodified binary)

Fig. 4. Sample Interposed Call

using system FIFO. Then the RTM software server calculates the communication pattern and synchronization information providing task graph and synchronization graph XML files periodically. The RTM hardware server also collects hardware information from hardware client on each tile through Perfmon2.

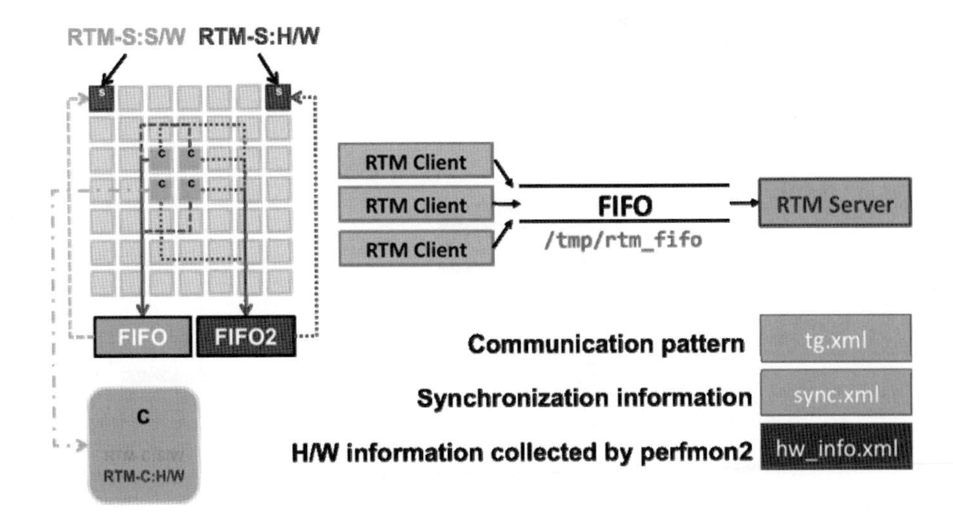

Fig. 5. RTM Server and Client

In the message passing model, source and destination rank and tile location, transferred data amount, timestamp for each event are saved whenever an event happens and then collected by monitoring tile periodically according to the pre-defined interval. Using statistical results for each (source, destination) pair/process calculated by the RTM software server, the user can know that task dependency and the load. At the hardware level, Perfmon2 counter values were collected on each tile, transferred to RTM hardware and then provided as an XML file. The performance counters such as ONE, MP_BUNDLE_RETIRED, TLB_EXC, HIT, L2_HIT, MP_DATA_CACHE_STALL, MP_INSTRUCTION_CACHE_STALL, MISS_I, MISS_D_RD, and MISS_D_WR, were used for the hardware information. This information is collected by way of multiplexing on each tile and sent to the RTM hardware server.

In the shared memory model, the event name (Barrier/Mutex/Lock/Conditional events), the number of occurrences for each event, Max/Min/Ave time of each thread/process for each event, process/pthread ID, and the address of each event group are saved whenever an event happens and then collected by monitoring tile periodically according to the pre-defined interval. Using statistical results for each process/pthread/event, the user can know the synchronization information. At the hardware level, it has no difference from the message passing model.

4 Implementation

The 64 cores are interconnected with mesh networks, while each processor executes at up to 866 MHz to achieve up to 443 billion operations per second. For supporting several libraries, we implemented MPI on the TILE64/MAESTRO based on MPI 1.2 specification. Also Perfmon2/PAPI was ported on multi-core architecture. Figure 6 depicts the RTM eclipse plug-in which can be used for analyzing the periodic hardware and software results on TILE64 or MAESTRO.

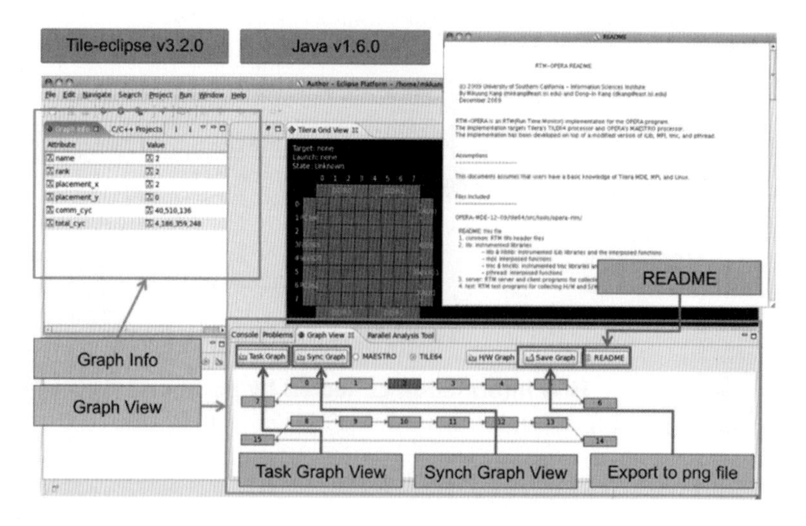

Fig. 6. RTM Graph View and Graph Info

Figure 7 shows the snapshot of the task graph for the communication pattern. As we mentioned in the previous section, the graph view and the related informaiton can be provided. We can know the event's source, destination, total count, data amount, distance, sender cycles, time, CPW (Cyles Per Word), bandwidth, and receiver cycles, time, CPW, bandwith.

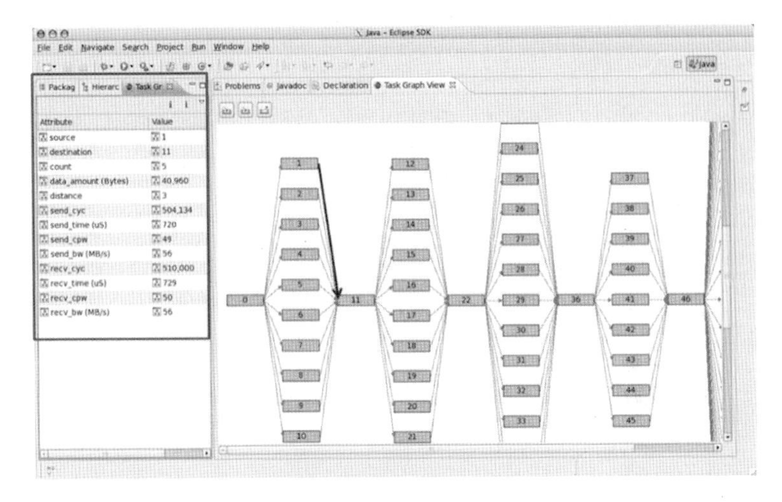

Fig. 7. RTM Task Graph

Figure 8 shows the snapshot of the sync graph for the synchronization information. The first and second images show the result when the <link to the event / event> is selected. We can know the rank, the number of occurrences, maximum/minimum/ average cycles, and the average execution time between each event.

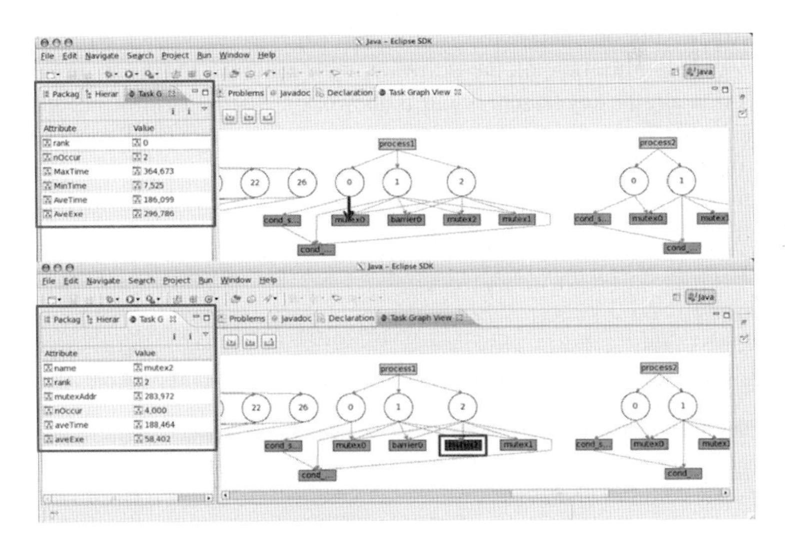

Fig. 8. RTM Sync Graph

The RTM hardware server gathers performance counter information and provides in a XML file. Using this XML file, user can know the current status which tile is running.

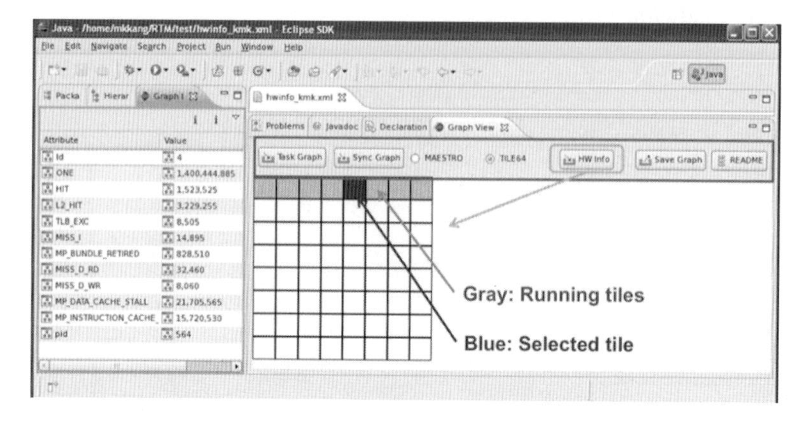

Fig. 9. RTM Hardware Info Graph

5 Conclusions

In this paper, we have designed Run-Time Monitor which is a system software to monitor the characteristics of applications at run-time, analyze the collected information, and optimize resources on cloud computing. RTM monitors both application software through library instrumentation and underlying hardware through performance counter optimizing its computing configuration based on the analyzed data. For future work, we are planning to develop a dynamic run-time self-morphing software framework on multi-core systems. It is expected to provide a framework for an automated optimization of the software with minimal overhead on multi-core systems. After all, the key feature of our framework is that 1) performance monitoring is detached from the applications to system-wide run-time manager, 2) application has a range of morphing at run-time, 3) run-time manager monitors the application's performance and morphs the application at run-time for either better performance or adapting to situation.

References

1. Feng, W., Balaji, P.: Tools and Environments for Multicore and Many-Core Architectures. IEEE Computer 42(12), 26–27 (2009)
2. Schneider, S., Yeom, J., Nikolopoulous, D.: Programming Multiprocessors with Explicitly Managed Memory Hierarchies. IEEE Computer 42(12), 28–34 (2009)
3. Multi-core processor (2009), http://en.wikipedia.org/wiki/Multi-core_processor

4. Bell, S., Edwards, B., Amann, J., Conlin, R., Joyce, K., Leung, V., MacKay, J., Reif, M., Bao, L., Brown, J., Mattina, M., Miao, C.-C., Ramey, C., Wentzlaff, D., Anderson, W., Berger, E., Fairbanks, N., Khan, D., Montenegro, F., Stickney, J., Zook, J.: TILE64 Processor: A 64-Core SoC with Mesh Interconnect. In: Proc. IEEE International Solid-State Circuits Conference (ISSCC), pp. 88–98 (2008)
5. Tilera Corporation (2010), http://www.tilera.com/
6. http://www.eweek.com/c/a/Cloud-Computing/Eucalyptus-Offers-OpenSource-VMwareBased-Cloud-Platform-100923/
7. Peng, J., Zhang, X., Lei, Z., Zhang, B., Zhang, W., Li, Q.: Comparison of Several Cloud Computing Platforms. In: International Symposium on Information Science and Engineering, pp. 23–27 (2009)
8. Hill, Z., Humphrey, M.: A Quantitative Analysis of High Performance Computing with Amazon's EC2 Infrastructure: The Death of the Local Cluster? In: 10th IEEE/ACM International Conference on Grid Computing, pp. 26–33 (2009)
9. Chiba, T., Burger, M., Kielmann, T., Matsuoka, S.: Dynamic Load-Balanced Multicast for Data-Intensive Applications on Clouds. In: 10th IEEE/ACM International Conference on Cluster, Cloud, and Grid Computing, pp. 5–14 (2010)
10. Message Passing Interface Forum, MPI: A Message Passing Interface Standard (2009), http://www.mpi-forum.org/docs/

Design of an Advertisement Scenario for Electric Vehicles Using Digital Multimedia Broadcasting*

Junghoon Lee[1], Hye-Jin Kim[1], In-Hye Shin[1,**], Jason Cho[2], Sang Joon Lee[3], and Ho-Young Kwak[3]

[1] Dept. of Computer Science and Statistics
[3] Dept. of Computer Engineering
[2] i SET Co., Ltd, Republic of Korea
Jeju National University, 690-756, Jeju Do, Republic of Korea
{jhlee,hjkim82,ihshin76,sjlee,kwak}@jejunu.ac.kr, jkboss@iset-dtv.co.kr

Abstract. This paper designs an integrative advertisement system based on digital multimedia broadcasting for the electric vehicles, which need a lot of driving information for battery efficiency and charge planning. The advertiser interface interacts with the advertisement processing system to pay the fee and have the contents endorsed. The advertisement contents are registered, monitored, encoded, and finally delivered to vehicles according to the contract via the broadcasting center. Here, this paper defines a new frame format on the data service stream and is in the process of developing and verifying the encoder and decoder modules. Our system is expected to provide the fundamentals for the development of diverse electric vehicle services.

1 Introduction

The state-of-the-art vehicle is generally equipped with a relatively high-bandwidth wireless communication interface, a display unit, and the high-capacity computing device. Electric vehicles, having more electric and electronic components such as electric motors, need to be controlled more sophisticatedly and digitally [1]. So, they embed well-defined control logic and run it mainly under the control of the in-vehicle computing device. Moreover, ever-growing wireless communication technology makes it possible for the fast moving vehicles to access the global network such as the Internet. The drivers can possibly obtain useful driving information such as a fuel-saving route, current traffic information, and charging station availability. This information can be provided in diverse ways the advertisement is one of the most promising ones.

* This research was supported in part by the MKE, Korea, under the ITRC support program supervised by the NIPA (NIPA-2010-(C1090-1011-0009)), by KIAT under the Regional Industry and Technology Development Project, and the MKE again through the project of Region technical renovation, Korea.

** Corresponding author.

T.-h. Kim, A. Stoica, and R.-S. Chang (Eds.): SUComS 2010, CCIS 78, pp. 288–291, 2010.

Digital contents consist of text, moving picture, and location information, while they must be delivered with a low price and updated in real-time. Being an instance of the digital content, the advertisement content additionally has location-dependent and time-dependent features. In addition, the created content must reach as many clients who want the content as possible. To this end, a content manager system is indispensable and this system must be able to not just tie the advertiser, content manager, and the system operator, but also interact with the available communication facility. In the mean time, DMB (Digital Multimedia Broadcasting) is a stable digital radio transmission technology for sending multimedia to mobile devices such as mobile phones and in-vehicle telematics devices [2]. Hence, DMB technology is considered to be one of the most cost-efficient and easy-to-install wireless carriers for electric vehicles. In this regard, this paper is to design an advertisement contents service system, possibly for electric vehicles, capable of integrating the DMB infrastructure.

2 Service Scenario and System Design

After being created as a form of texts, images, and moving pictures, the mobile advertisement must be delivered to multiple mobile users in a reasonable price. DMB is capable of meeting such requirements, as mobile hosts can receive the content broadcasted over a wide area even if they are moving fast. Particularly, it can cooperate with the digital map service to create a location-dependent advertisement. Moreover, the 2-tier network architecture can provide more advanced content download mechanism. Besides the basic broadcast channel, it can download the large volume data, entering the high-speed network coverage area. Those data are cached in the DMB terminal and much reduces the update time.

For the DMB-based advertisement, it is necessary to connect advertisers, content creators, communication facilities, and clients. Figure 1 shows the framework we are currently developing. This architecture consists of an advertiser part, an advertisement processing system, and a broadcasting center. An advertiser pays the advertisement fee via the Internet connection. The price can be decided by the location of vehicles and the price plan for peak, mid-peak, and off-peak interval, respectively, while it can be paid on hourly, daily, and monthly basis. After the operator endorses the advertisement content, it will be registered in the server system. This step is necessary to prevent illegal content from being displayed to the public clients including youths. Then, this content is encoded and multiplexed to the DMB signal in the broadcasting center and transmitted via the DMB network. The client terminal, receiving the contents, displays them to the client.

Figure 2 depicts the transmission system through which DMB contents are encoded, multiplexed, and then broadcasted. The data service turns into the bit stream along with the video and audio content, while the error correction code is attached to overcome the unavailability of interactive error control in the one-way transmission. For the transmission part, current terrestrial technology

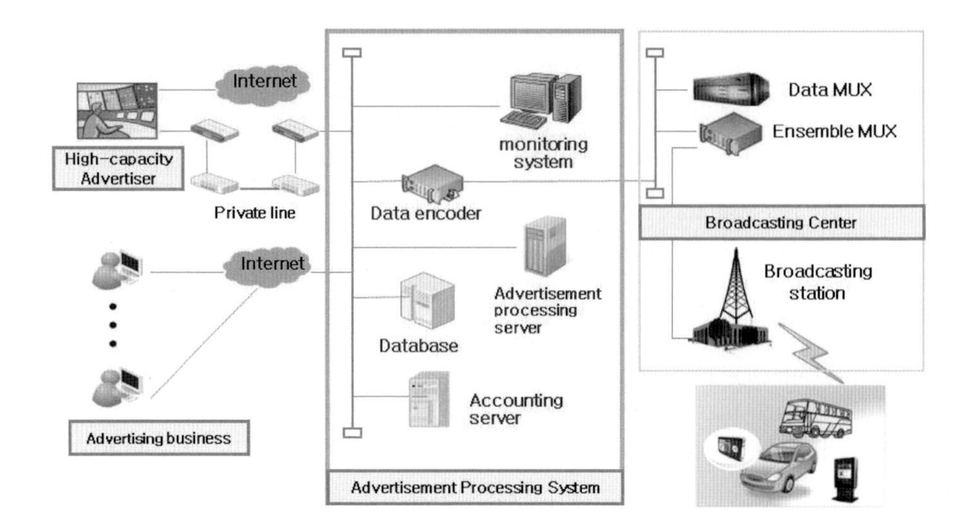

Fig. 1. Advertisement service scenario

is well-established as well as cheap in maintenance and transmission. Our design makes the advertisement be transmitted via the data service stream, as it can combine video, audio, and text more flexibly. As contrast to the video and audio streams which are owned by the major broadcasting companies, the data service stream allows to define an additional data format and attach the corresponding encoder/decoder modules. Each field is defined for the integration and presentation of the embedded contents, constituting the entire message hierarchically [3].

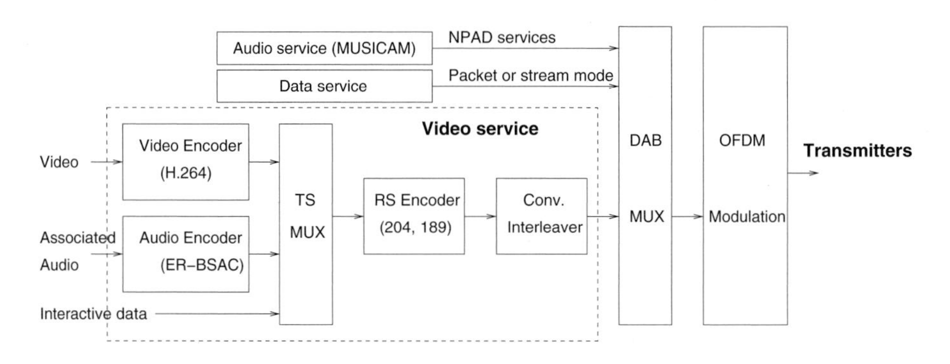

Fig. 2. Transmitter module

Figure 3 shows the simplified architectures of DMB receivers for full and convergency modules. The receiver receives the signal and displays to the user with a radio frequency tuner, a broadband demodulator, and processing circuits. The in-vehicle telematics device can integrate this component for the sophisticated electric vehicle services. For the display in the end-user side, the travel time is

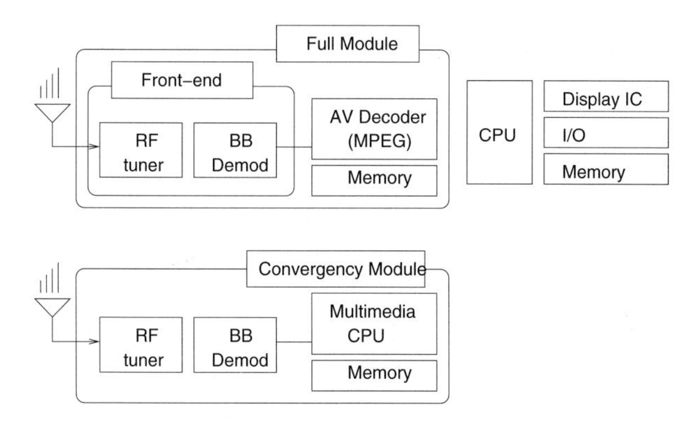

Fig. 3. Receiver module

different according to the vehicle type, travel hour and the like. For example, our previous work has analyzed the traffic pattern and travel time distribution for the taxi passengers according to the pick-up point and travel hour [4]. This analysis shows that about 40 % of travels belong to the interval between 6 and 8 minutes. In addition, the travel time can be estimated before the departure of a vehicle based on the source and destination, so the display schedule can decide the play sequence of location-dependent and time-dependent contents.

3 Concluding Remarks

This paper has designed an integrative advertisement system based on digital multimedia broadcasting for the electric vehicles, which need a lot of driving information for battery efficiency. The designed system consists of the advertiser part, the advertisement processing system, and the broadcasting center. A new hierarchical frame format is defined for the advertisement content on the data service stream for the flexible encoder/decoder implementation. Our system is expected to provide the fundamentals for the development of diverse electric vehicle services. As future work, we are planning to conduct a field test for the implemented system to verify the correctness and efficiency of our design.

References

1. Kempton, W., Dhanju, A.: Electric Vehicles with V2G: Storage for Large-Scale Wind Power. Windtech International, 18–21 (2006)
2. Manson, G., Berrani, S.: Automatic TV Broadcast Structuring. International Journal of Digital Multimedia Broadcasting (2010)
3. Paik. R., Chang, M., Kim, J., Choo, D., Jo, M., Bagharib, A., Tan, R.: On the Development of a T-DMB TPEG Traffic and Travel Service. In: Intelligent Transport System Asia Pacific Forum (2008)
4. Lee, J.: Traveling Pattern Analysis for the Design of Location-Dependent Contents Based on the Taxi Telematics System. In: International Conference on Multimedia, Information Technology and its Applications, pp. 148–151 (2008)

Development of Wireless RFID Glove for Various Applications

Changwon Lee[1], Minchul Kim[1], Jinwoo Park[1], Jeonghoon Oh[2], and Kihwan Eom[1]

[1] Department of Electronic Engineering, Dongguk University,
26, Pil-dong 3-ga, Jung-gu, Seoul, Korea
[2] Daeduk College, Daejeon, Korea
kihwanum@dongguk.edu

Abstract. Radio Frequency Identification is increasingly popular technology with many applications. The majority of applications of RFID are supply-chain management. In this paper, we proposed the development of wireless RFID Glove for various applications in real life. Proposed wireless RFID glove is composed of RFID reader of 13.56 MHz and RF wireless module. Proposed Gloves were applied to two applications. First is the interactive leaning and second is Meal aid system for blind people. The experimental results confirmed good performances.

Keywords: Wireless RFID glove, RFID Reader, RF Wireless Module, Interactive learning, Meal aid system.

1 Introduction

RFID (Radio Frequency Identification) is a fast growing field and increasingly in many applications. Its techniques exchange remotely information using radio frequency. Characteristics of radio frequency are long distance of recognition, the various mind tag of recognition and freely date of change. RFID tag is applied to miniaturization, low price, object recognition and USN environment [1-2].

RFID technique was defined the radar concept in period of second world war and the program development was begun in order to distinguish our military and the enemy air vehicle. Since the late 1960s, it is used hazardous materials monitoring, distribution, security and car differentiation.

Recently iGlove and iBracelet was developed from Intel Research Seattle group. iGlove of the glove form and iBracelet of the bracelet form were developed with the wearing style RFID systems which use Mica2Dot sensor networks. According to conduct of the user RFID read the tag which is used conduct of user and analysis of situation. In addition, Worn RFID that is able to game at the tag etc was developed [3-4].

In this paper, we propose the development of wireless RFID glove for various applications in real life. Developed the wireless RFID gloves is composed of RFID reader which read tag information, and the wireless module deliver tag information in the computer. In order to confirm the usefulness of developed wireless RFID we are implementing and two kinds of applications. First, number card game and puzzle etc

T.-h. Kim, A. Stoica, and R.-S. Chang (Eds.): SUComS 2010, CCIS 78, pp. 292–298, 2010.

is interactive learning program. Second, blind people apply meal aid system with one of the various welfare policies.

2 Proposed Wireless RFID Gloves

Proposed wireless RFID glove system is Fig. 1. PC operation signals send Base modules. Base modules with 433MHz radios deliver signal to RF modules. A RFID reader is operated by RF module delivered signal, and read the RFID tag information. That module receives the tag information is transmitted wirelessly to PC.

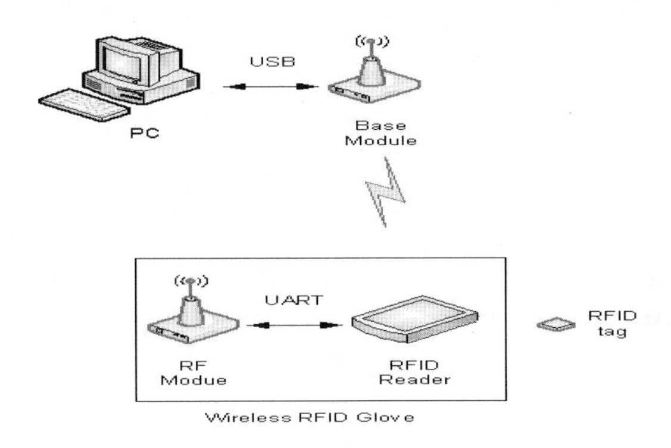

Fig. 1. Block diagram of wireless RFID glove system

Fig. 2 is the photograph of designed wireless RFID glove. A wireless RFID gloves is composed of RF wireless module, RFID leader and power supply battery.

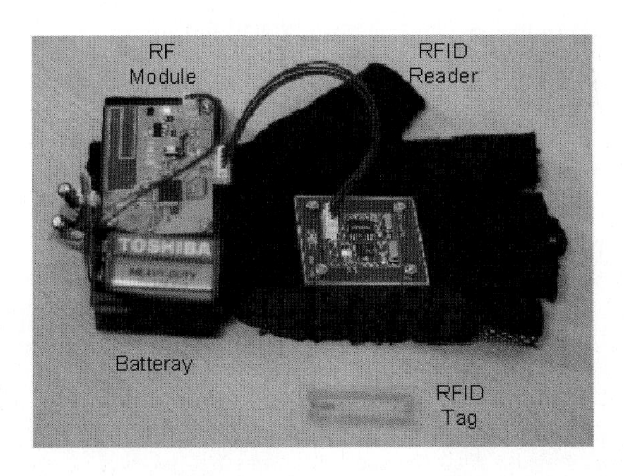

Fig. 2. Wireless RFID glove

2.1 RF Wireless Module

Base module and RF module is Fig. 3. Antenna at RF module is designed in the PCB by pattern, and it minimizes size of module.

(a) Base module

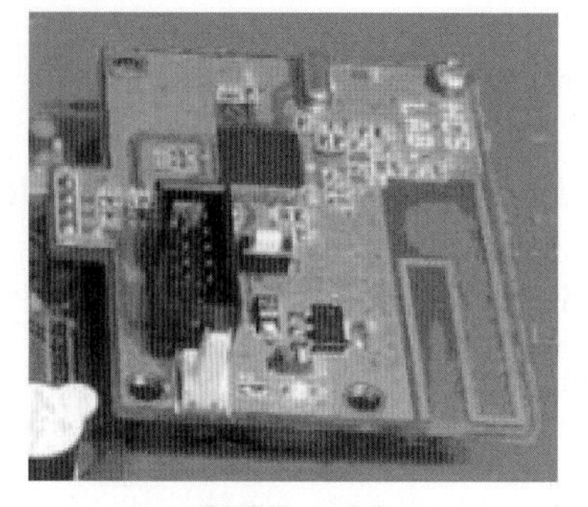

(b) RF module

Fig. 3. Base module and RF module

Table 1. Wireless module specifications

Information	Content
RF Frequency	433MHz
Modulation	FSK
RF Data Rate	38.4kbaud
RF Output Power	10dBm (10mW)
Range	interior : 50m outdoor : 140m
Microcontroller	8051 built-in
Memory	32kB flash memory
Power	3.3V

(a) Wireless RFID glove reader

(b) Set system reader

Fig. 4. The photograph of RFID reader

The CC1010 has a RF and 8051 in the single chip, and table1 shows the specifications [5].

2.2 RFID Reader

The RFID readers used 13.56 MHz, and implemented by connecting the wireless module and UART is used. An antenna is built substrate with pattern on the outside. Fig. 4 is the photograph of RFID reader and its specifications are table 2 [6].

Table 2. RFID reader specifications

Information	Content
RF Frequency	13.56MHz
RF Data Rate	26.kbps
Range	80mm
RFID reader	EM4094
Microcontroller	ATmega8
Interface	TTL UART
Power	5V

3 Two Applications

In this paper, proposed wireless RFID glove system applies to two kinds of application in order to confirm the usability.

3.1 Interactive Learning

First, wireless RFID glove apply to number game of Interactive learning.

The data structure of two-way data form between Base module and RF module is table 3. In order that various people participate in learning program at same time, each RF module has ID and between single Base module and 1: N is designed.

Table 3. Data Structure

Information	**Size (byte)**	**Content**
Destination ID	1	Receive device ID
Source ID	1	Outgoing device ID
Command	1	Operation instruction
RFID Data	3～12	Data

RFID data include RFID tag in Unique UID information, data of RFID tag memory and operation of RFID reader. Command data decide kind of data which send wireless module from Base module. Test program with wireless RFID glove is Fig. 5.

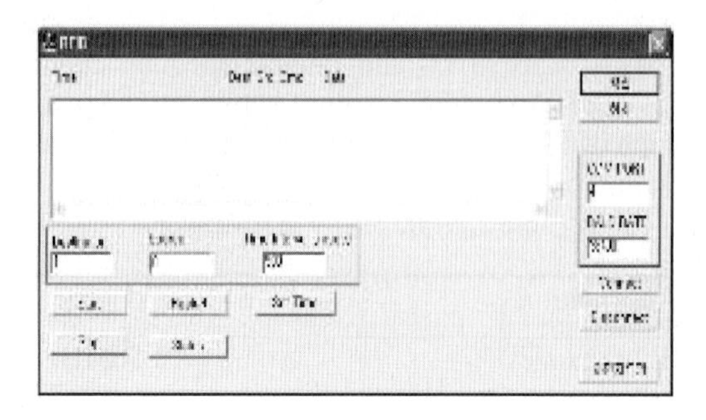

Fig. 5. Test Program

Number game attached RFID tags in the card like Fig. 6. Number, operation sign and input (enter) save RFID data memory. Multiplication table problem question at terminal and wireless RFID glove solve problem by reading the number card.

Many people participate in game method, if wearers with wireless RFID glove give question for people by speaking to number and operation sign, the others input solution.

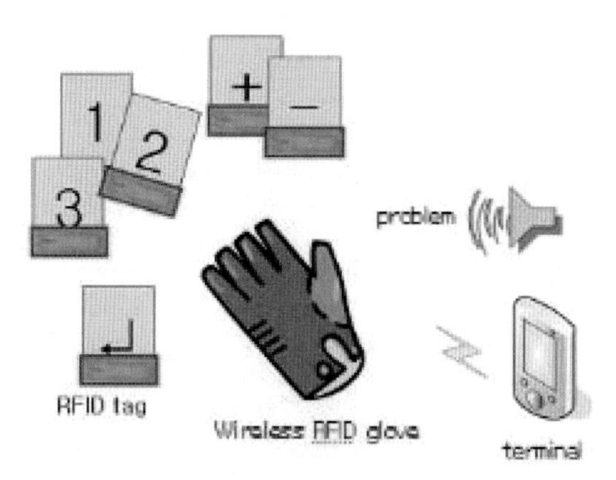

Fig. 6. Number game

Various objects information card and color card etc is used in various learning, and can be used for a variety of interactive learning.

3.2 Meal Aid System

Various welfare policies increase activity territory of blind people. But helper necessary works still remain. Especially the case of meal occupied the major portion the role of the helper. The helper to inform the location and a type of the food, blind people is able to meal [7]. In order to improve discomfort, meal aid system of blind people apply wireless RFID glove. The configuration of blind people meal aid system is Fig. 7.

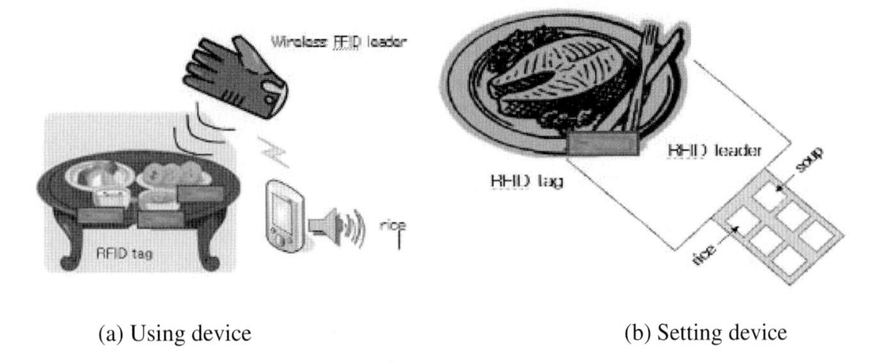

(a) Using device (b) Setting device

Fig. 7. Meal aid system for blind people

RFID tags attach in the food plate. Food set-up of kind through set device save RFID tag. Blind people wearing wireless RFID glove read RFID tag and they confirm food present location and kind. Kinds of food appear device through output sound.

Because commercial product size fixed, RFID reader consider attached location. In order that hand moving and operation is not influenced, RFID reader experiment on palm and back of hand Attachment. The result of recognition rate is shown Fig. 8.

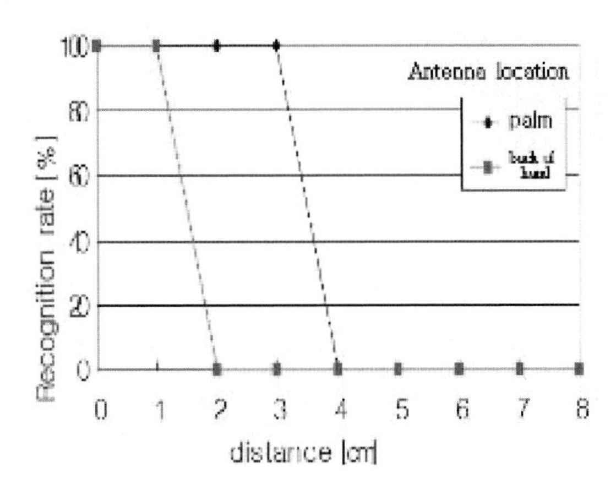

Fig. 8. The recognition rate of RFID tag and reader distance between

In the Fig.8, the RFID reader attaching in the palm has been recognized in the tags within 3cm.

4 Conclusion

In this paper, we proposed the development of wireless RFID gloves for various applications in real life. Developed the wireless RFID gloves is composed of RFID reader which read tag information, and the wireless module deliver tag information in the computer. The RFID readers used 13.56 MHz, and implemented by connecting the wireless module and UART is used. The wireless module system is composed of base module and RF module. The wireless RFID glove apply to two applications in order to confirm practicality of that. First, number card game and puzzle etc is interactive learning program. Second, blind people apply meal aid system with one of the various welfare policies. Through two experiments confirmed the usefulness of the proposed wireless RFID gloves.

Project in future increases RFID recognition distances, and miniaturization of wireless RFID glove.

Acknowledgements

This work was supported by the Hub University for Industrial Collaboration (HUNIC) grant funded by the Korea government Ministry of Knowledge Economy and Ministry of Education, Science and Technology.

References

1. Weinstein, R.: RFID: a technical overview and its application to the enterprise. IT Professional 7(3), 27–33 (2005)
2. Landt, J.: The history of RFID. IEEE Potentials 24(4), 8–11 (2005)
3. Fishkin, K.P., Philipose, M., Rea, A.: Hands-On RFID: Wireless Wearables for Detecting Use of Objects. In: Proceedings of Ninth IEEE International Symposium on Wearable Computers, pp. 38–41 (October 2005)
4. Konkel, M., Leung, V., Ullmer, B., Hu, C.: Tagaboo: A Collaborative Children's Game Based upon Wearable RFID Technology. Personal and Ubiquitous Computing 8(5), 382–384 (2005)
5. Chipcon, http://www.ti.com/lprf
6. Firmsys, http://www.firmsys.com
7. Jung, K.-Y.: Daily life training of Blind people. Korea Blind Union (2001)

Implementation of RFID Tag for Metal Surface Mount

Chong Ryol Park, Sang Won Yoon, Kyung Kwon Jung, and Ki Hwan Eom

Department of Electronic Engineering, Dongguk University,
3-26, Pil-dong, Joong-gu, Seoul, Korea
parkcr@smba.go.kr,
{sangwony,kwon,kihwanum}@dongguk.edu

Abstract. This paper described a metal mount RFID tag that works reliably on metallic surface. The proposed method is to use commercial RFID tags, Styrofoam103.7 material is attached on back side of RFID tag. Styrofoam103.7 material which has 2.5 mm thickness and 1.03 of relative permittivity was attached on back side of RFID tag. In order to verify the performance of proposed method, we evaluated the experiment on the supply chain system of electric transformers. The experimental results on supply chain of electric transformers show that the proposed tags can communicate with readers from a distance of 2 m. The results of recognition rates are comparable to commercial metallic mountable tags.

Keywords: RFID tag, metallic surface, supply chain, electric transformers.

1 Introduction

Radio frequency identification (RFID) is a technology used for object identification, which finds various applications in retail, transportation, manufacturing and supply chains. RFID comprises readers, and transponders also known as tags. Most RFID tags contain at least two parts. One is an integrated circuit for storing and processing information, modulating and demodulating a RF signal, and other specialized functions. The second is an antenna for receiving and transmitting the signal. RFID tags are generally cheaper and simpler compared to active ones. They contain no batteries and can be fully encapsulated for ruggedness and protection.

The numerous potential applications of the RFID system make ubiquitous identification possible at frequency bands of 125 KHz (LF), 13.56 MHz (HF), and 860-960 MHz (UHF). In the recently years, three key factors drove a significant increase in RFID usage: decreased cost of equipment and tags, increased performance to a reliable identification, and a stable international standard around UHF band. As the use of RFID systems increases, manufacturers are pushing toward higher operating frequencies (UHF band) for long reading range, high reading speed, capable multiple accesses, anti-collision, and small antenna size compared to the LF or HF band RFID system.

Passive UHF RFID tags are able to provide good read ranges for object identification compared with LF or HF RFID tags, and they are also seen as potentially low cost. Traditional passive RFID tags have under-performed in metal

T.-h. Kim, A. Stoica, and R.-S. Chang (Eds.): SUComS 2010, CCIS 78, pp. 299–306, 2010.

rich environments, limiting their utility. If RFID tag is attached directly to metallic object, it may work poorly. The antenna performance is seriously decreased because of the reactance variation on the antenna impedance. Metal reflects the Radio Frequency of RFID tags, and therefore the tags either need to be specially designed for metal or attached without actually touching the metal with special spacers. Many applications for supply chain and materials management require RFID tags with long read ranges, durability and reliable read-rates while mounted on a radio-interfering metal surface. Bubble wrap, Styrofoam and other neutral materials can be used to attach non-metallic tags to metal, or we can use one of our specially designed made for metal tags [1-3].

A number of commercially available planar and label-like passive UHF RFID tags have been tested against a large aluminum plate. Results from the testing have shown that as the tags are brought closer to the aluminum plate, the read range decreases. A patch antenna with an electromagnetic band gap ground plane has been used in the tag design. Yu, Kim, and Son have offered different tag designs that use patch antennas [4].

In 2008 more than a dozen new passive UHF RFID tags emerged to be specifically mounted on metal. ODIN technologies of Ashburn, Virginia, produced a benchmark which showed varying performance of metal mount tags, with the greatest read distance being just over 25 feet in real-world conditions. However, these tags are much more expensive [5].

In this paper, we propose a dielectric foam attached structure for metallic objects. The proposed method is to use commercial RFID tags, Styrofoam103.7 material is attached on back side of RFID tag. It is possible to make simple and low cost. In order to verify the performance of proposed method, we evaluate the experiment on the supply chain system of electric transformers.

The RFID and its characteristics near metallic surface are detailed in Section 2. Section 3 presents a proposed RFID tag design, which also includes the simulation results. Section 4 shows an implementation and experimental results. Finally, some conclusions are presented in Section 5.

2 Related Work

In this section, we discuss the inner workings of RFID systems. For the most part, RFID systems comprise three principal components, as shown in Figure 1.

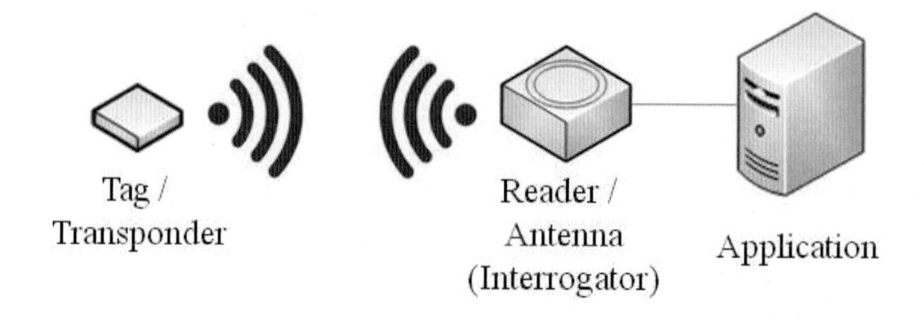

Fig. 1. RFID system structure

The first is the tag, which is affixed to the item that is to be tracked or identified within the supply chain by the RFID system. The reader, which has a number of varied responsibilities including powering the tag, identifying it, reading data from it, writing to it and communicating with a data collection application. The data collection application receives data from the reader, enters the data into a database, and provides access to the data in a number of forms that are useful to the sponsoring organization.

RFID system communicates by electromagnetic waves. When designing the RFID tag antennas mountable on metallic platforms, it is very important to understand the behavior of the electromagnetic fields near metallic surfaces since the antenna parameters (the input impedance, gain, radiation pattern, and radiation efficiency) can be seriously affected by metallic platforms. In this section, the behavior of electromagnetic fields near metallic surfaces will be considered [3].

For a boundary that lies between two media in space with medium 1 characterized by dielectric permittivity ε_1, magnetic permeability μ_1, and electric conductivity σ_1, and medium 2 characterized by ε_2, μ_2, and σ_2.

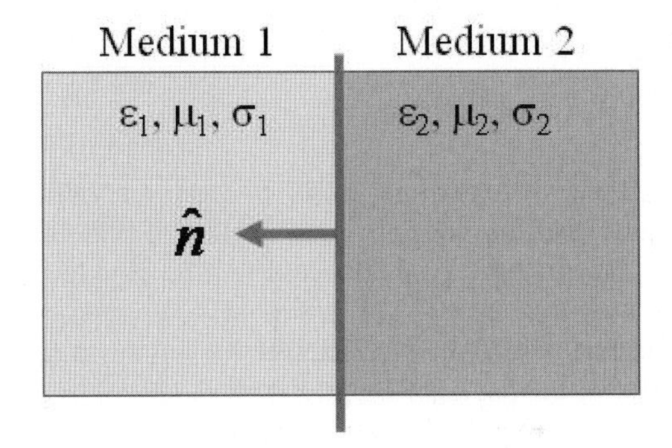

Fig. 2. Boundary between two media

The electromagnetic boundary conditions for a general case can be expressed as follow Equations.

$$\hat{n} \times (E_1 - E_2) = 0, \tag{1}$$

$$\hat{n} \cdot (D_1 - D_2) = \rho_s, \tag{2}$$

$$\hat{n} \times (H_1 - H_2) = J_s, \tag{3}$$

$$\hat{n} \cdot (B_1 - B) = 0. \tag{4}$$

Where,

$\hat{n}$ is the unit normal vector to the boundary directed from medium 2 to medium 1
E is the electric field intensity (V/m), D is the electric flux density (C/m^2)
H is the magnetic field intensity (A/m), B is the magnetic flux density (W/m^2)
ρ_s is the surface charge density (C/m), J_s is the surface current density (A/m^2).

If medium 1 is a metallic medium and we assume as a practical approximation that it is a perfect electric conductor with infinite conductivity, there will be no electric field in this medium. Consequently, $D_1=0$, $B_1=0$, and $H_1=0$. Hence, for this case, the boundary condition become

$$\hat{n} \times E_1 = 0, \tag{5}$$

$$\hat{n} \cdot D_1 = \rho_s, \tag{6}$$

$$\hat{n} \times H_1 = J_s, \tag{7}$$

$$\hat{n} \cdot B_1 = 0. \tag{8}$$

It is noticed that there are no tangential components of the electric field on a perfect electric conductor. On the other hand, there are only tangential components of the magnetic field directly next to a perfect electric conductor. Hence, not all components of electromagnetic fields are available near a perfect electric conductor.

3 Metallic RFID Tag Design

According to the theory of electromagnetic boundary conditions, there are only tangential components and no normal components of the magnetic field to the metallic surface. In addition, the magnetic field will be doubled when it is very near the metallic surface.

The RFID tag design here exploits this fact by having a gap between metallic surfaces. The structure of the tag is shown in Figure 3.

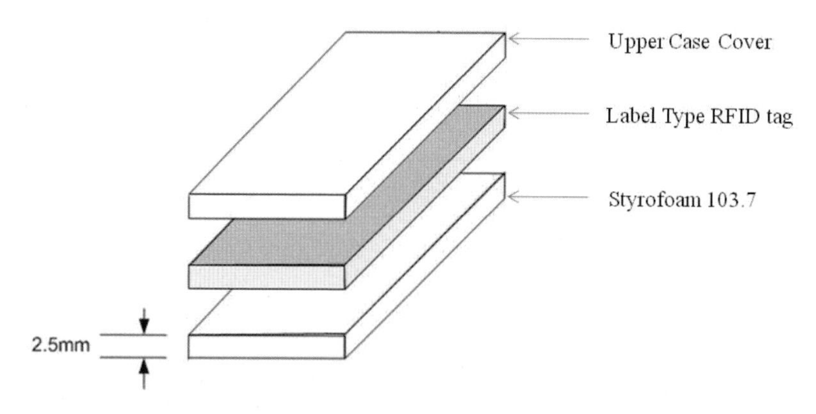

Fig. 3. Proposed RFID tag structure

The proposed method is to use commercial RFID tags, Styrofoam103.7 material is attached on back side of RFID tag. The Stryrofoam103.7's thickness is 2.5 mm and the relative permittivity is 1.03. Using the dielectric material like a Styrofoam provided to radiate electromagnetic wave on top side of RFID tag.

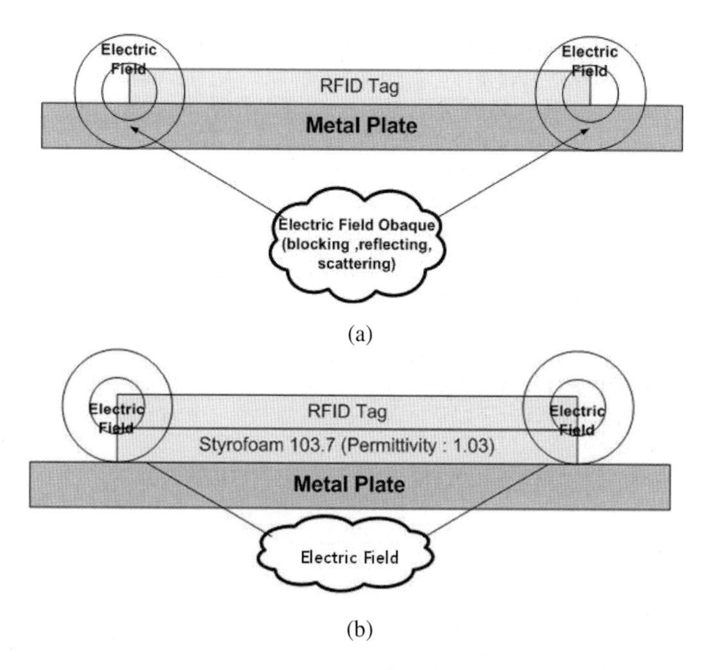

Fig. 4. Electromagnetic field

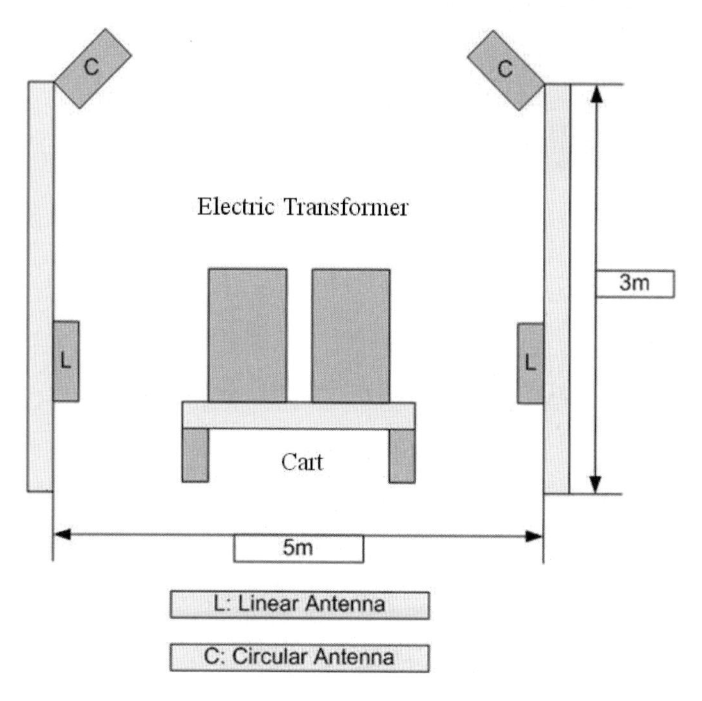

Fig. 5. Warehouse gate

4 Implementation and Experiments

In this section, we describe the performance of the proposed system and the evaluation of the supply chain of electric transformers.

The electric transformers consist of metallic surface. They pass through the warehouse gate installed 4 antennas as shown in Figure 5.

In order to improve the performance of identifying, two linear antennas mounted on the side of gate and two circular antennas mounted on the top of gate. Figure 6 is the photo of experimental set-up.

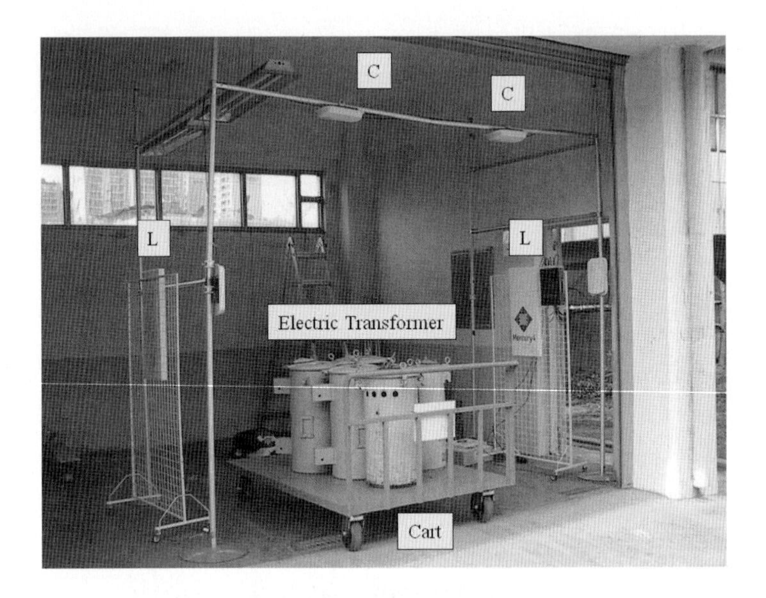

Fig. 6. Experimental set-up

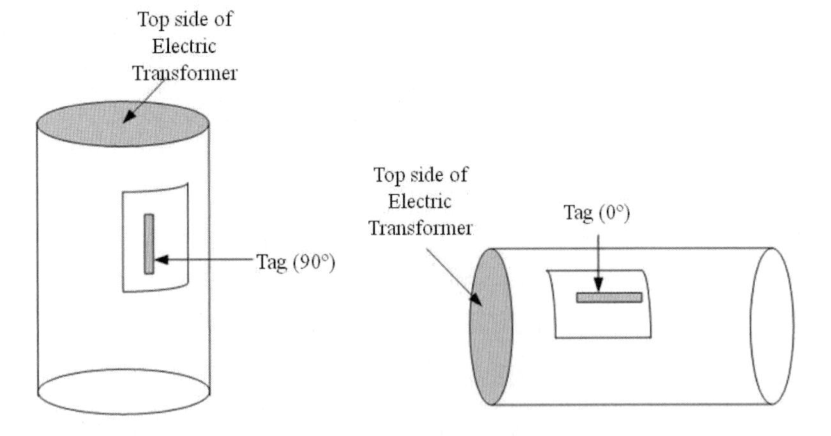

Fig. 7. Attachment direction of RFID tags

The RFID tag is attached on the surface of the electric transformer in two different directions, horizontal (0°) and vertical (90°) as shown in Figure 7.

We compared proposed RFID tag with Sontec, Prenix, Smart1, and AWID tags which are commercial metallic mountable tags, and cost about \$20 ~ \$30. We used general purpose RFID tags, ALL-9354-02 which is about \$0.5, for proposed RFID tag design.

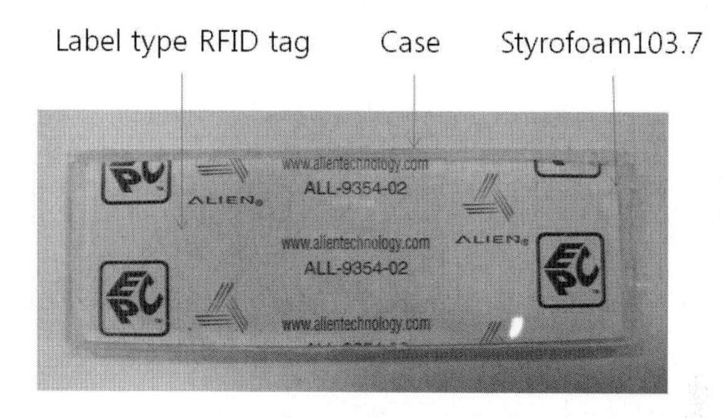

Fig. 8. Proposed RFID tag

We achieved measurement results of detection range between RFID reader and tag with varying Styrofoam103.7 from 1 mm to 5 mm as shown in Figure 9. The results showed that the RFID tag detected 2 m range with 2.5 mm thickness of Styrofoam103.7.

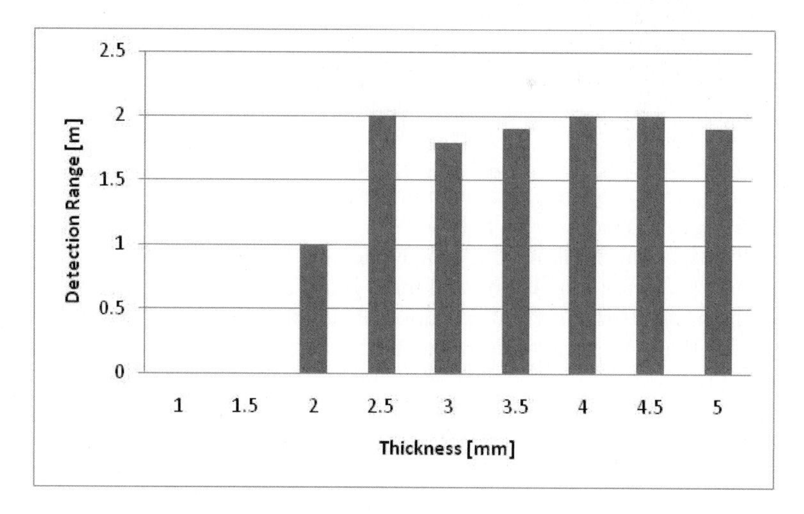

Fig. 9. Detection range vs. Styrofoam thickness

In Table 1, the experimental results demonstrate recognition ranges with respect to the tag direction. The tag recognition test is applied for constant speeds of the cart, 0.5 m/sec.

A total of 100 test iterations are executed at two tag directions. Two hundred examples are obtained as historical experimental data. The recognition success rates are 100 % of each tag includes proposed tag. The experimental results showed the effectiveness of the proposed method using low cost label type RFID tags.

Table 1. Results of detection range

Tags	Direction	Range [m]
Sontec	90 °	4.9
	90 °	6.3
Prenix	90 °	7.4
	90 °	8~
Smart1	90 °	7.5
	90 °	8~
AWID	90 °	5.6
	90 °	6.9
Proposed tag (ALR'M' + Styrofoam103.7)	90 °	3.8
	90 °	4.8

5 Conclusion

In this paper, we described a metal mount RFID tag that works reliably on metallic surface. The proposed tag consisted of commercial tag, and gap material between tag and metallic surface. Styrofoam103.7 material which has 2.5 mm thickness and 1.03 of relative permittivity was attached on back side of RFID tag. The experimental results on supply chain of electric transformers show that the proposed tags can communicate with readers from a distance of 2 m. The results of recognition rates are comparable to commercial metallic mountable tags.

This tag can be used to identify and track goods and articles in various production, supply chain, asset management scenarios or ubiquitous computing, including item-level applications. We plan to do research on different types of object and additional influence factors such as the number and height of antennas in the future.

References

1. Glover, B., Bhatt, H.: RFID Essentials. O'Reilly Media, Sebastopol (2006)
2. Finkenzeller: RFID Handbook. Wiley, Chichester (2003)
3. Ahson, S., Ilyas, M.: RFID handbook: applications, technology, security, and privacy. CRC Press, Boca Raton (2008)
4. Want, R.: An Introduction to RFID Technology. IEEE Pervasive Computing 5(1), 25–33 (2006)
5. ODIN Technologies, `http://www.odintechnologies.com`

U-Bus System Design Based on WSN for the Blind People

Trung Pham Quoc[1], Minchul Kim[1], Hyunkwan Lee[2], and Kihwan Eom[1]

[1] Electronic Engineering, Dongguk University, 26, Pil-dong 3-ga jung-gu Seoul, Korea
[2] Honam University, Eodeungno 330, Gwangsan-gu, Gwangju, Korea
kihwanum@dongguk.edu

Abstract. This paper proposed the U-bus system design based on wireless sensor network (WSN) for blind people. This system has two main parts. First part is blind people recognition. Another part is communication between a bus and bus station. Blind people recognition part is constructed simple device and system. This part decides existing or non-existing of the blind at bus station. And then if pre-process recognize blind people, the bus station will communicate the bus. We make up the announcement system about arrived bus information for the blind people using these parts. This announcement about arrived bus is very useful to blind people for taking the bus.

Keywords: WSN, Blind people, Recognition, U-Bus system, Announcement.

1 Introduction

With WSN technology, not only small applications such as smart house but also in the larger society such as environmental, military ,health and commercial applications. More and more we see the importance of WSN technology. To make human life become more convenience we should apply development technologies [1][2].

One of the main reasons we are interested in the design and desire to deploy U-bus system are: In addition to the facilities such as subway which are developed fairly complete, with sound and image messages, so that it is very well to support for people with disabilities, especially blind people. Taxis are very convenient but the cost is so expensive, this is not an advantage to blind people usually take taxi. For the bus, it is waste if blind people can not join in bus traffic. For reasons above we show a system. This system is designed for the purpose to transfer sound message of bus number which is currently parked in front of the blind. So that this system can help blind people to recognize the bus which blind people want to take without the help of others. With this system, again WSN technology is applied.

Nowadays have many systems support blind people are researched and developed. Not only simple applications such as Braille keyboard for the blind, the system protects the blind when crossing the street, but also complex systems such as car for the blind and so many other systems. Maybe the practicality of the system needs some time in the future but for the purpose of the blind are closer to society, to avoid the inferiority with society as well as research, design and deployment of this system is absolutely necessary. This system, have some main advantages: such as creating the possibility to participate in bus transportation for the blind are more convenient, Blind

T.-h. Kim, A. Stoica, and R.-S. Chang (Eds.): SUComS 2010, CCIS 78, pp. 307–315, 2010.

can go to somewhere by bus or subway, some places that subway can not come blind people can go there by bus. Weaknesses exist: some things to be upgraded and designed to more efficacy in use, with this system blind people can not search the destination which they want to go, how to recognize exactly blind people who are staying at the bus station and want to take bus. This paper is divided into five sections, in the next section discusses the technology are used to design systems, how K-mote is applied in this system. Section 3 is system propose with flowchart and block diagram of system. Section 4 is the analysis of advantages and disadvantages. Section 5 shows other issue and future researching for this system. Finally, section 6 concludes the paper.

2 Technology Construction of WSN

The U-bus system is designed to combine two elements: hardware part and software part. The hardware part use INtech's K-mote. The K-mote have CC2420 chip for RF communication so we can send or receive useful information through using this chip. Fig. 1 is a device used this paper.

Fig. 1. K-mote

And next part is software part. We use TinyOS for this design project with OS (Operating System). TinyOS is a free and open source component-based operating system and platform targeting wireless sensor networks (WSNs). It started as collaboration between the University of California, Berkeley in co-operation with Intel Research and Crossbow Technology, and has since grown to be an international consortium, the TinyOS Alliance [3]. Especially we use TinyOs -1.x version for stable development. The above version has been developed for a long time so more stabilized than TinyOs-2.x version. TinyOS is an embedded operating system written in the nesC programming language as a set of cooperating tasks and processes [3].

Fig. 2. nesC compiler process

It is a kind way of the pre-processor. The nesC is a component-based and syntax is similar to the C programming language. The nesC compiler convert source code into C program file and this file is responsible compile and link through the GCC(GNU Compiler Collection) compiler [4]. Fig. 2 is the nesC compiler process.

3 Proposed U-Bus System Propose

3.1 System Design

Fig. 3 shows this proposed system block diagram and process block diagram. This system has two important parts that are recognition of blind people and communication between bus and bus station. First blind people recognition using switch. The bus station has two areas include normal people area and blind people area, and is Fig. 4.

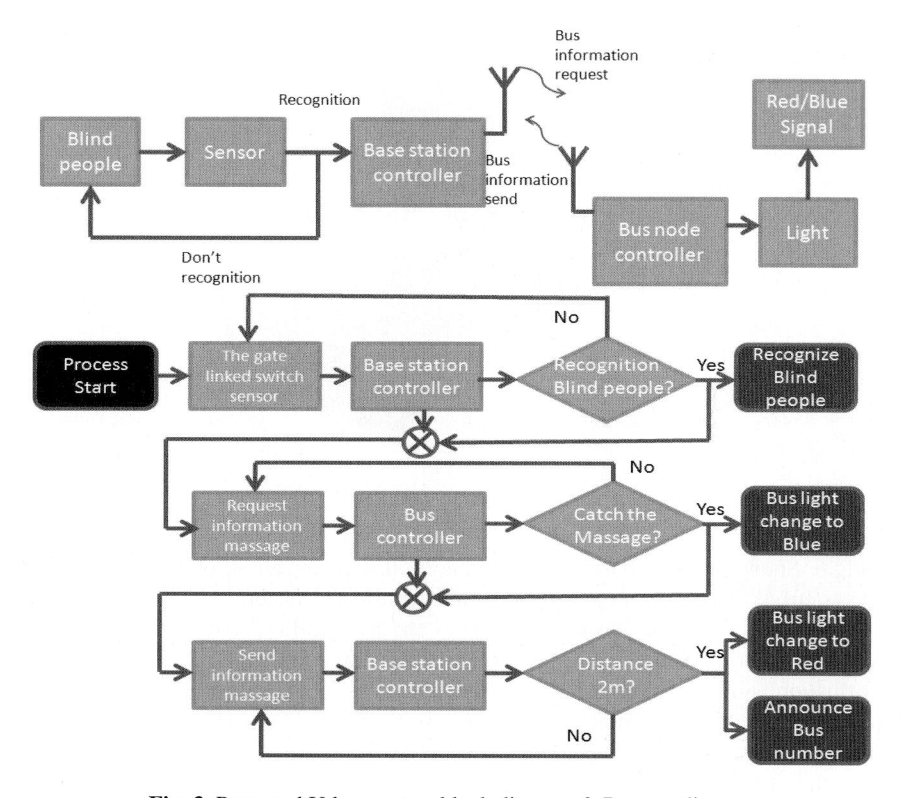

Fig. 3. Proposed U-bus system block diagram & Process diagram

Doorway of blind people area has gate consist of two stick that link switch. If person into this area through the gate, we can assume that someone exists in this area.

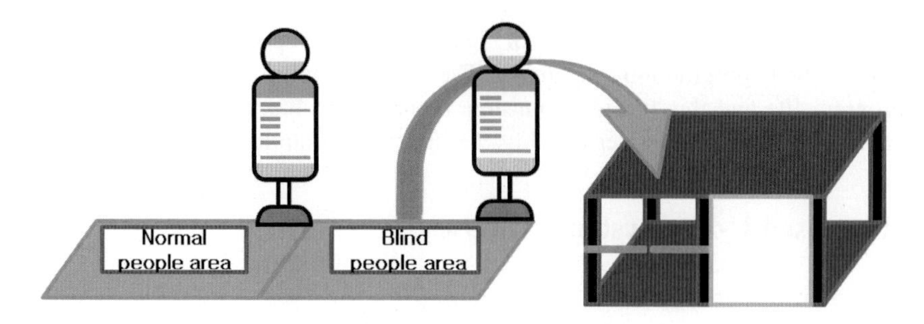

Fig. 4. Bus station Structure

When the system recognizes people as above, next bus and bus station communicate. The bus station announces exiting blind people to any bus in the RF communication area. If the bus catches this massage, it will send information about self bus number etc. Fig. 5 shows bus and bus station communication process.

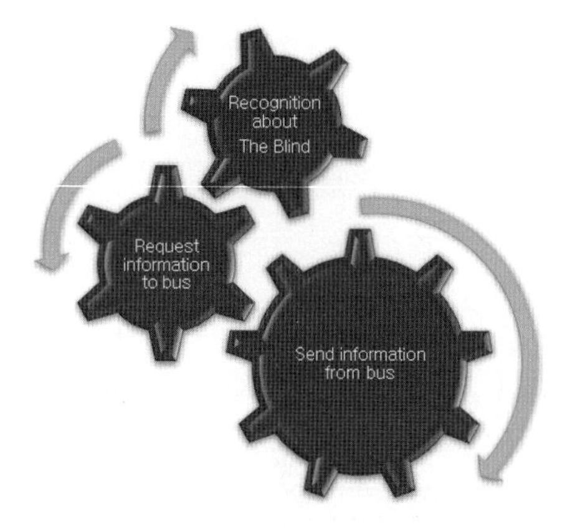

Fig. 5. Bus and bus station communication process

Fig.6. Show the flowchart of U-bus system. It explains about overall proposed U-bus system. We previously examined the behavior of the system.

In addition, bus light system announces the availability of the blind. Each one means sign that red is non-existing and blue is existing. And this system announce the information about bus number to blind people just one and distance 2m between the bus and bus station. So, after the bus station announce the number through speaker, this system reset to start line. And distance is assumed by RSSI (Received Signal Strength Indication).

3.2 Experiment

This project is necessary to estimate the distance through using RSSI. So we experiment distance follow RSSI measurement at first.

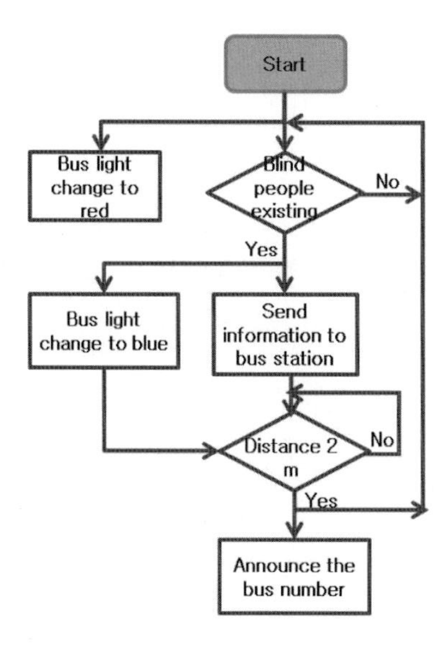

Fig. 6. Proposed U-bus system flow chart

Table 1. Distance follow RSSI measurement experiment result

Purpose	Maximum Effective Range	RSSI Magnitude	Error Distance
Request msg. range	20m	-150dbm	±2m
For Announcement	2m	-60dbm	±0.1m

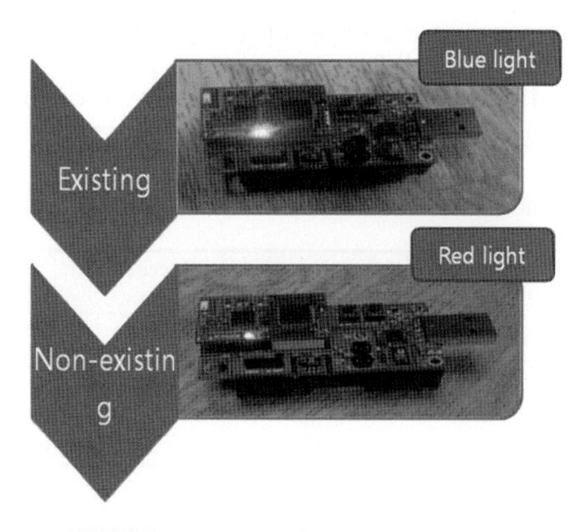

Fig. 7. Bus light change experiment result

Distance follow RSSI measurement has huge error according to surrounding environment. Therefore, assuming the exact distance is very hard. However this project is not required the exact value. So, we can use approximate value. Table1 shows the experimental value of 30 times the average data at the space without obstacles. Following this experimental result data, when the bus station request information to the bus, the bus light change in require massage range

Fig. 7. shows the result color of bus light. It change the color according to check existing of the blind and check the result of announce bus number.

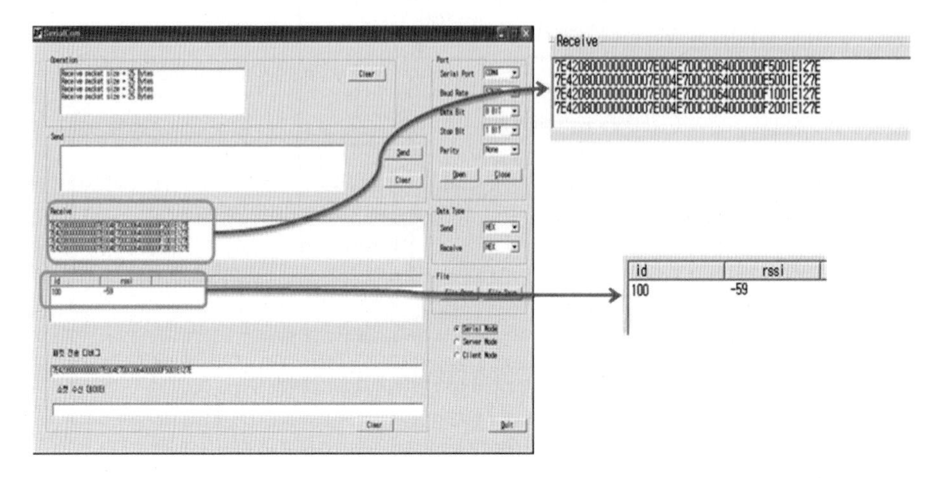

Fig. 8. MFC program for identification about bus number information

When the whole process is completed for communicate between the bus and bus station, final process works. This final process is announcement process. We make MFC cording for check this process result. Fig. 8 shows the result of it. The bus send massage 25 byte data include RSSI magnitude, tag node id and bus number etc. We can find thus data information at Receive part of this program. However, interpreting this data is very hard. So we classify data for using easy. The id part and RSSI part is that. We can understand bus number about approached bus through classified data.

Table 2. Success rate about how many bus arrived station

How many Bus	Between Buses Distance	Success rate
1 bus	-	100%
2 buses	5m	100%
2 buses	10m	100%
3 buses	5m	100%
3 buses	10m	100%

Finally, we experiment that change number of buses arriving at the station and each between the buses distance: The between buses distance not include bus length and we suppose the bus lengths 12m. Each experiment carries out 15th times. Table 2 is result about this experiment.

4 Advantage and Disadvantage

In previous section we presented the model of system proposed and principle of system. So with features of system we can clearly understand the advantages that this system can support the blind. Blind people will not worry because they could not go to places where subway can not go. How can the blind correct bus which they want to take without the help of other people around them. In addition to the blind, the elderly have been slow or other disabled people can also get the benefits of participating in the use of U-Bus system. The bus will stop in a good position and with time enough for blind people take bus. It is really convenience for disable people, especially the blind.

It is wrong if we think that U-bus which we presented above is complete. Have some disadvantage is exiting. We will fell complex to recognize the person who is staying at bus station is blind person or normal person - sometime people may wrong place. Because we used only one sensor to recognize blind people so that if the blind change them mind and move out blind place, how station can recognize. For finding information also important in this system but we did not have occasion to mention, how blind people can find destination which they will come and how station can recommend the blind what the bus number you should take. The last one, only bus which blind people want to take should stop at the blind people area.

5 Other Issue and Future Researching

With the application and implementation of this system we hope to apply this principle on many other systems as Kiosk System and other public systems. To support for the blind in particular and disabled people in general we had researching for the purpose of optimizing the function of the system.

In blind recognition part, we interested to use RFID technology which is very relevant and high efficiency when the blind can be controlled to allow bus station identifies they are or are not staying at the bus station. With this solution, build the blind place is not necessary. We will present more detail in next paper.

For the information searching part, with the use of Braille PDA Desk movies for the blind will help blind people find accurate information on intelligent vehicles need to be able to reach the desired destination. In addition to the more complex technologies such as voice recognition can also be applied aimed to optimize the system. U-bus system will give a best support for the participants to use, especially the blind.

In Fig.9. shows complete proposal of U-bus system based on WSNs for the blind. In this system we want to show some advantage features

a) PDA for the blind:
   - Integrated RFID tag
   - Use to search destination and chose bus number which the blind want to take.
   - Receive information from bus station by headphone

b) Bus station (BS):
   - Recognize the blind who want to take the bus by RFID technology

- Send bus number that the blind want to take to the bus
- Receive the bus number of the bus and sent to the blind by PDA (the blind can hear information about bus number by headphone)

c) Bus:

- Receive information from BS
- If bus have the blind want to take, bus will stop at the blind area.
- If the bus does not have the blind want to take, need not stop at the blind area.
- If the bus does not have the blind want to take, need not stop at the blind area.

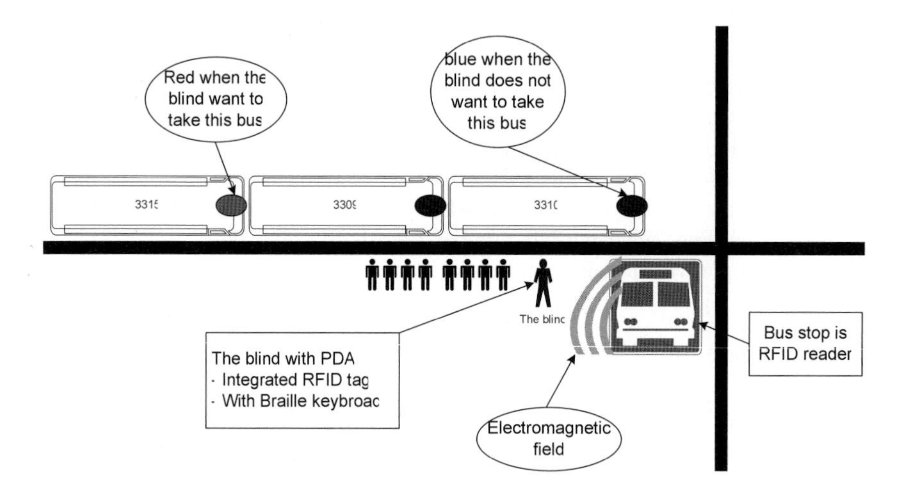

Fig. 9. Complete proposal of U-bus system based on WSNs for the blind

PRINCIPLE

The blind with PDA (Integrated RFID tag) go to Electromagnetic field of BS. The blind can search destination and register the bus number which they want to take to BS.

BS recognized the blind in the bind area and BS will announce exiting blind people and bus number that the blind want to take to any bus in the RF communication area

The bus number matching will send back BS its number when it stop at the blind area, else need not to stop at the blind area.

The blind can hear number of the bus which they want to take when this bus stop in front of the blind.

The blind can easy to take the bus.

6 Conclusion

Through this paper we hope to study the subject of applications to support disable people, especially blind people increasingly spend a lot more attention. The idea of the U-bus system was again demonstrate the broad applicability of the WSN technology. Although we realize the limitations of the system that still exit as blind people identification at the bus station, or need to add some features such as equipment necessary to search destination information for blind people etc, but this system has reduced the inconvenience of the blind people when they participate bus transportation.

This designed system constructs TinyOs and K-mote based on WSN (wireless Sensor Network). We take experiment two parts for check about performance this system. First part particularizes distance according to RSSI (Received Signal Strength Indication). Another part checks process whole system. We can expect excellence more easily to take bus by experiment result.

With the proposed a complete U- bus system in section 5 of this paper, the reference to the real difficulties of the blind when using this system, we will research more carefully about the design systems and technology used to design. So that we will design next version of this system early with more complete, as well as optimization function for the blind when they participate U-bus system. Finally, we hope principle of U-bus system can apply to other systems such as Kiosk or Automatic shop to support disable people in near future.

Acknowledgements

This work was supported by the Hub University for Industrial Collaboration (HUNIC) grant funded by the Korea government Ministry of Knowledge Economy and Ministry of Education, Science and Technology.

References

1. Nam, S.-Y., Jeong, G.-I., Kim, S.-D.: Ubiquitous Sensor Network Structure & Application, pp. 13–19. Sanghakdang (2006)
2. Korea Institute of Science and Technology: Information Sensor Network Embedded Software Technology Trend (December 2004)
3. Wikipedia, `http://en.wikipedia.org/`
4. nesC Structure, `http://nescc.sourceforge.net`
5. Kang, J.-H.: Sensor Network Open Source Project. In: Information & Communication Technology, 1st edn., vol. 18 (2004)
6. Chae, D.-H., Han, G.-H., Lim, G.-S., Ahn, S.-S.: Sensor Network Outline and Technology trend. Journal of Information Science 12th edition 22 (2004)
7. Levis, P.: TinyOS: Getting Started (March 2002)
8. Getting up and running with Tiny OS, Crossbow (2002)

Measurements and Modeling of Noise on 22.9-kV Medium-Voltage Underground Power Line for Broadband Power Line Communication

Seungjoon Lee[1], Donghwan Shin[1], Yonghwa Kim[2], Jaejo Lee[2], and Kihwan Eom[1]

[1] Department of Electronic Engineering, Dongguk University, Seoul, Korea
[2] Korea Electrotechnology Research Institute (KERI), Ansan, Korea
{acousticjoon,accsimov}@naver.com,
{yongkim,jjlee}@keri.re.kr,
kihwanum@dongguk.edu

Abstract. This paper proposed the measurements and modeling of noise on the 22.9-kV Medium-Voltage (MV) underground power distribution cable for Broadband Power Line Communication (BPLC). The proposed measurement system was composed of inductive coupler and Digital Phosphor Oscilloscope (DPO). The measurement noise data was obtained from thirty-two pad mounted transformers in the test field located in Choji area of Ansan city. After conducting analysis of noise characteristics in time and frequency domain, the noise model are presented. In order to analyze the noise in frequency domain, Power Spectral Density (PSD) was computed with empirical data using Welch's method. The modeling of the power line noise at each frequency carried out using Cumulative Probability Distribution (CPD) of the noise power. It compared with common Cumulative Distribution Functions (CDF) of Nakagami-m distribution, Gaussian distribution, Gamma distribution. In low frequency range, gamma distribution was fitted with the CPD. Nakagami-m distribution provides a good fitting to the noise CPD above 20MHz frequency range.

Keywords: Broadband Power Line Communication, Noise Modeling, Noise measurement, Medium-Voltage Power Distribution Line, Cumulative Probability Distribution.

1 Introduction

The underground power distribution cable is the most important communication media as a backbone network for the BPLC. Formerly, cable network or fiber optics is used for the purpose of backbone network because of cost-effective Power Line Communication (PLC) technology. Until now, a number of studies of PLC was performed for Low-Voltage (LV) PLC and In-House PLC. Recently, MV power line has been considered as a backbone network of whole power line network. Although, a group of studies shows in some detail information of the impedance, transmission properties and noise characteristics of the medium voltage power network, the characteristics of 22.9-kV MV underground power line is not so well known[1,7].

T.-h. Kim, A. Stoica, and R.-S. Chang (Eds.): SUComS 2010, CCIS 78, pp. 316–324, 2010.

Detailed channel parameters such as noise, frequency, signal attenuation, impedance and location are required for high performance PLC systems. Noise characteristic of the channel is play an important role to determine performance of PLC in those parameters [2]. Accurate noise model is required to realize high bit rate transmission of the BPLC. The noise model in 10-kV MV, LV and in-house was done by a lot of researchers. However, the model of the noise on 22.9kV MV power line is rarely regarded in the literature. This paper only considered noise on 22.9kV MV power line which is critical factor in performance of PLC systems for BPLC.

To model the noise of MV power line, this paper presents measurements results of the 22.9kV MV underground power line noise based on empirical noise data obtained from thirty-two pad mounted transformers in the test field of a suburb of Ansan city near Seoul, Korea. After conducting careful analysis with empirical data in time and frequency domain, modeling of the data at each frequency was also carried out for accurate noise model.

Section II presents measurement environment and system for the noise characteristics. In Section III, results of measured data in time-domain and computed PSD are discussed. The appropriate noise model at each frequency is described in Section IV. The concluding remark is given in Section V.

2 Measurements

2.1 Measurement Environment

The test field using underground 22.9kV MV power distribution line has been constructed by Korea Electrotechnology Research Institute (KERI) in Choji area located in Ansan. The measurements were conducted at thirty-two sites of 3-phase pad mounted transformers, Poonglim - T02 (Transformer number), T0 T04, T05, T06, T07, T08, T09, T18, T19, T20, T21, T22, T233, T24, T25, T26, T27, T28, T30, T31, T32, T33, T40, T42, T43, T44, T45, T46, T47, T52 and T53, located in 1B/L and 11B/L of the test field as shown in Figure 1.

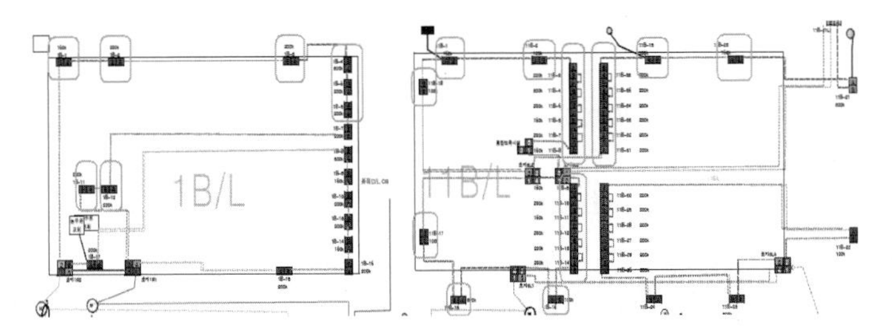

Fig. 1. Location of the 3-phase pad mounted transformers

2.2 Measurements Setup

Figure 2 describes composed noise measurement system in the test field. The data was obtained by inductive coupler and DPO of the system. The coupling unit connected with phase A of the 3-phase transformer and the oscilloscope.

In the noise measurements, we used a TEKTRONICS DPO3032 for recording data with USB sticks. A sample rate 100 mega samples per second was chosen for spectral analysis to 50MHz frequency range by Shannon sampling theorem.

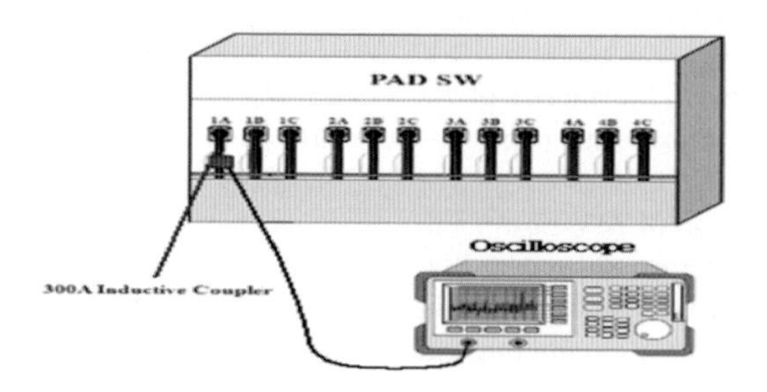

Fig. 2. Measurement system

In order to overcome harsh measurement environment and prevent high-voltage, MATTRON MTR-ICU-H58 (see Figure 3) inductive coupler was used in the system. The inductive coupler employed a principle of magnetic induction easily connected to outer of power line different from capacitive coupler. Since high current flows through the MV underground power line, a magnetic core of inductive coupler should have high permeability and saturation current characteristic. The coupler employed for the measurement system has highest permeability, lowest power loss and the magnetic properties without magnetic saturation on the subterranean power line transferred high current of 300A[3].

Fig. 3. A photograph of the inductive coupler of the system

3 Experimental Results

3.1 Characteristics of the Noise

The noise in LV power line channel can be separated into five classes, colored background noise, narrow band noise, periodic impulsive noise asynchronous to the mains frequency, periodic impulsive noise synchronous to the mains frequency and asynchronous impulsive noise. Since colored background noise and narrow band noise vary slowly over time, they can be regarded as background noise. The latter three types have rapid time-varying properties, so that they can be summarized as impulsive noise [4].

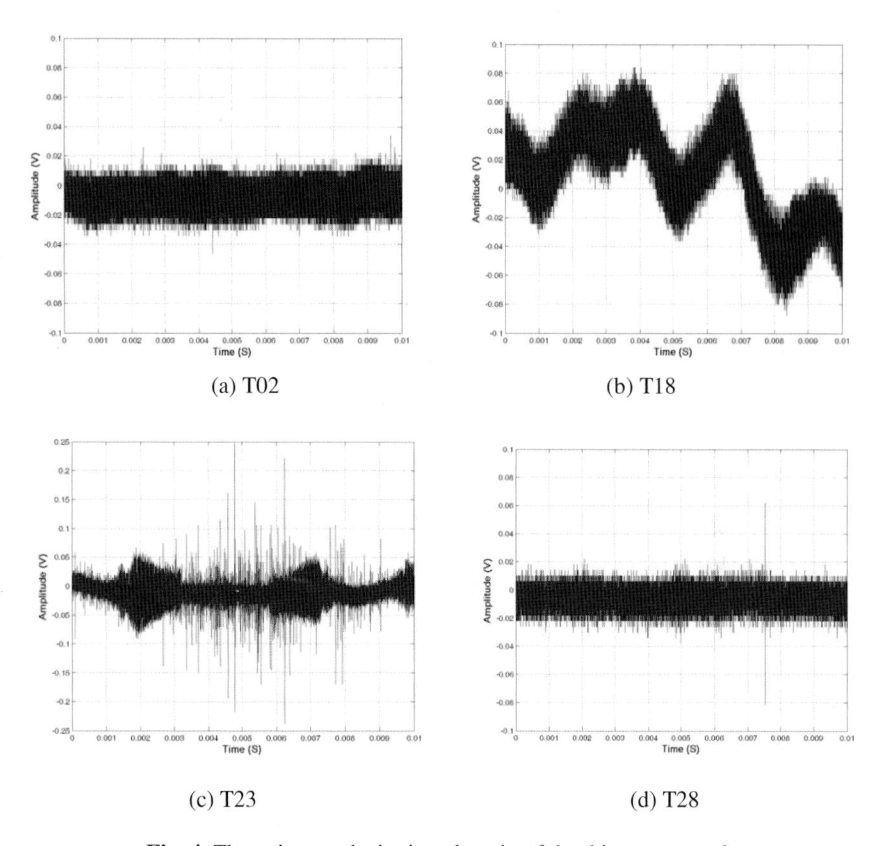

(a) T02 (b) T18

(c) T23 (d) T28

Fig. 4. The noise graphs in time domain of the thirty-two graphs

The classification of noise in LV power line can be adopted to represent characteristics of noise of MV power line. However, since wireless disturbances impact LV power line network heavily, narrowband noise amplitude of MV power line with the pad mounted transformers is less than of LV power line. Impulsive noise amplitude mainly related with appliance noise were also lower than LV power line, because MV power line was not directly connected with house appliances.

As stated above, the measurements were conducted in thirty-two pad mounted transformers in the test field. The record of each measurement has a length of 0.01s, so that 1 million samples of each measurement were obtained. The thirty-two graphs can be classified into four representative types following its line shape (see Figure 4). The results show the MV power line noise can be considered sum of the background noise and the impulsive noises.

The noise graphs of Figure 4(a) and 4(b) stationary remain over measurement period, so that they can be regard as a background noise. Figure 4(c) and 4(d) show impulsive noise comprised background noise on MV power line. Table 1 shows computed mean and variance of noise amplitude at each transformer to give noise level on the MV power line. The noise data-sets have mean value a range of -12mV ~ -1mV except for T18 which has a mean value of 10.90mV and high variance with T52.

Table 1. Mean and vriance of noise data at each transformer

Transformer	Mean(mV)	Variance(mV^2)	Transformer	Mean(mV)	Variance(mV^2)
T02	-5.87	35.5	T26	-7.25	139.6
T03	-6.06	25.0	T27	-8.73	193.2
T04	-6.28	57.9	T28	-4.87	25.6
T05	-7.48	128.4	T30	-5.93	66.3
T06	-6.18	129.6	T31	-5.38	37.3
T07	-5.37	83.5	T32	-6.48	36.0
T08	-5.22	56.1	T33	-3.77	59.4
T09	-4.34	68.2	T40	-4.58	287.4
T18	10.90	954.2	T42	-7.81	175.0
T19	-5.21	50.0	T43	-8.42	142.8
T20	-8.52	113.3	T44	-9.99	114.0
T21	-8.41	57.2	T45	-5.40	89.2
T22	-1.53	60.4	T46	-5.38	56.4
T23	-9.21	95.5	T47	-8.18	54.2
T24	-6.83	92.1	T52	-11.68	1348.8
T25	--3.43	149.1	T53	-10.42	100.8

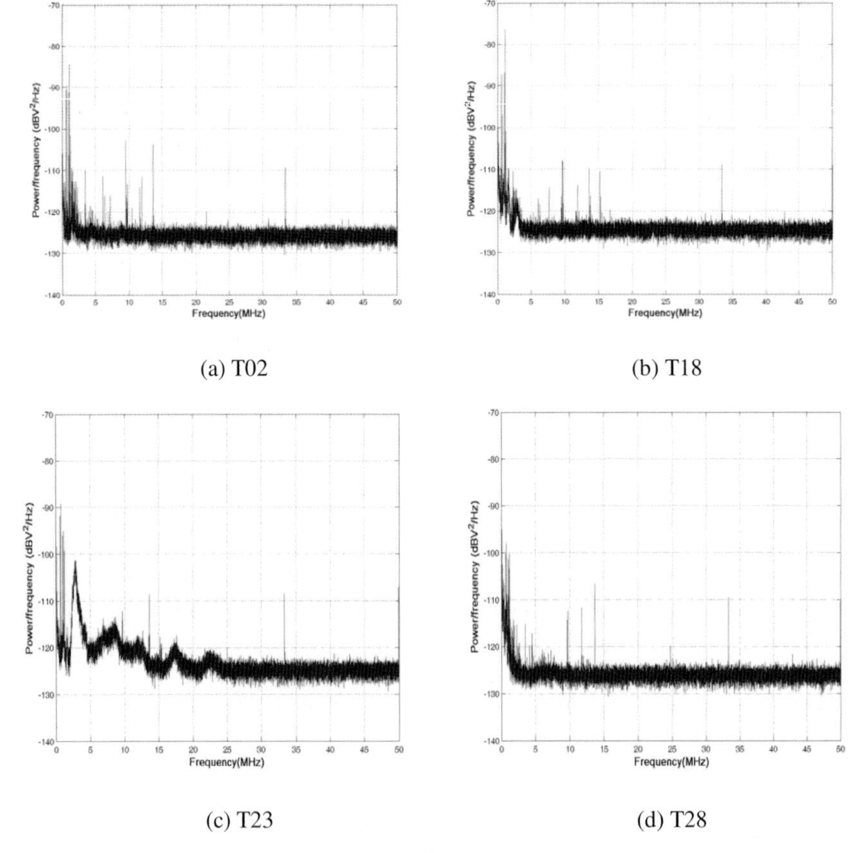

(a) T02 (b) T18

(c) T23 (d) T28

Fig. 5. Computed noise PSD of empirical data at each transformer

3.2 Spectral Analysis

In order to assess impact of the noise on communication with limited bandwidth, the spectral analysis of the noise data is a better approach [4]. Also, for more effective modeling of the noise, the PSD of the noise is considered. The spectral estimation was carried out using Welch's method from the thirty-two noise data-sets [5].

Figure 5 shows computed PSD from the noise data at the T02, T18, T23 and T28, the same as Figure 4. The PSD of background noise included narrow band disturbances is shown in Figure 5(a), 5(b) and 5(d). A noise data of the T28 (see Figure 4(d)) describes a simple impulsive noise characteristic in time domain, however, in frequency domain it can be regarded as the background noise (see Figure 5(d)). The periodic impulsive noise asynchronous to the mains frequency has a relatively high PSD value (see Figure 5(c)). From the figure 5, there are more disturbances in the MV power line network which affect low frequency range below 15 MHz such as short wave radios.

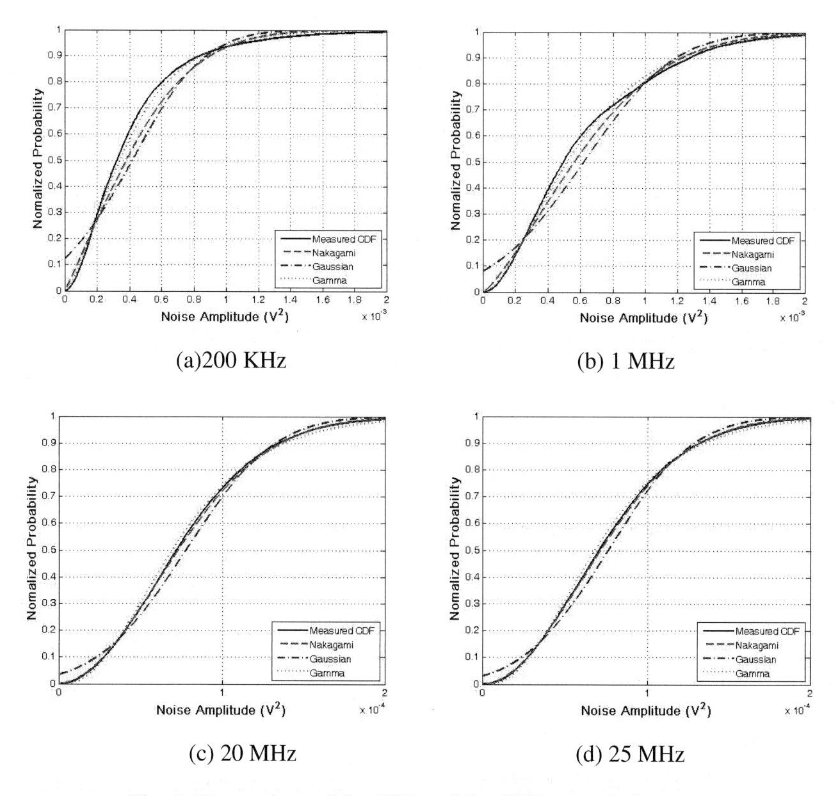

Fig. 6. Comparison of the CPD and the CDFs at each freqeuncy

4 Noise Modeling and Verification

Many articles regarding some techniques of modeling of the LV power line noise in frequency domain and time domain were presented in the literature [4,6,7,8]. For obtaining average noise PSD, the modeling method in frequency domain was used in the literature, but it does not represent any noise information comprised random behavior of the noise at each frequency [7]. To model the noise of MV power line at each frequency, the CPD was used for the noise distribution. Computed CPD from empirical data is compared with some well known CDFs, Nakagami-m distribution [6], Gaussian distribution [8] and Gamma distribution [7]. Figure 6 shows the comparison between the calculated CPD and the CDFs at each frequency. Since the CPD at each frequency was fitted into Gamma distribution below 20MHz of frequency range and Nakagami-m distribution above 20MHz range, a certain frequency of 200 kHz, 1 MHz, 20 MHz and 25 MHz was chosen as shown in Figure 5.

The Nakagami-m PDF can be written as

$$p(r|m,\Omega) = \frac{2m^m}{\Gamma(m)\Omega^m} r^{2m-1} e^{\frac{-mr^2}{\Omega}} , \qquad (1)$$

where m is defined as the ratio of moments $m = E^2[X^2]/VAR[X^2]$, Ω is the mean power of the random variable r, $\Gamma(\cdot)$ is the Gamma function. And the Gamma CDF is as the following:

$$F(r|a,b) = \frac{1}{b^a\Gamma(a)} \int_0^x t^{a-1} e^{-t/b}\, dt , \qquad (2)$$

where a is a=m and b is $b=\Omega/m$.

The computed CPD at 200 kHz and 1 MHz is becoming Gamma CDF as shown in Figure 6(a) and 6(b). The Nakagami-m distribution provides excellent fitting into the CPD at 20MHz and 25MHz (see Figure 6c and 6d). The value of m for CDFs is 0.52 and at 200 kHz, 0.63 at 1MHz, 0.82 at 20MHz and 0.98 at 25MHz, respectively, and the value of Ω is 3.02×10^{-7} at 200 kHz, 5.82×10^{-7} at 1MHz, 5.81×10^{-7} at 20MHz and 5.82×10^{-7} at 25MHz (see Figure 7).

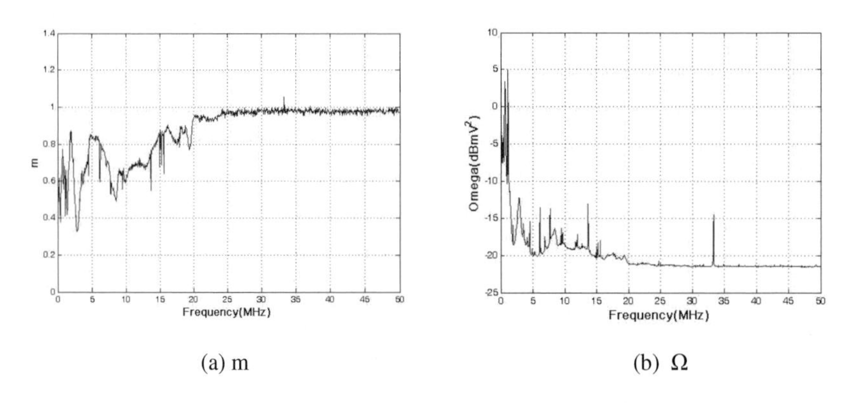

(a) m (b) Ω

Fig. 7. The result of m and Ω profiles at certain frequency

Variation at each frequency of the value of m and Ω was shown in Figure 7 (a) and (b) respectively. The Pearson's chi-squared test was used to evaluate difference between the observed CPD and the theoretical CDFs. Usually, the chi-squared value of the test-statistic can be written as

$$X^2 = \sum_{i=1}^{n} \frac{(O_i - E_i)^2}{E_i} , \qquad (3)$$

where n is the number of divided intervals, O_i is the empirical value, E_i is the theoretical value in the ith interval. To calculate the test-statistic value, data sample was divided into ten intervals. Since the CDFs have two unknown parameters, degree of freedom can be presented 7 (=10-2-1). Therefore, computed the standard value X^2 is $X^2_{1-\alpha} = X^2_{0.99} = 18.48$, when the α is α =1%. The standard value of X^2 compared to the calculated X^2 value to find goodness of fit in the CDFs and CPDs.

By performing the X^2 test, assumed the Gamma distribution and the Nakagami-m distribution was accepted for modeling in below 20MHz and above 20MHz, respectively. The value of $X^2$of the Gamma distribution is 4.57 at 200kHz and 4.81 at 1MHz. The computed value X^2 of the nakagami-m distribution is 0.06 at 20MHz and 0.09 at 25MHz. Under the standard value of $X^2_{0.99}$, the Gamma distribution and Nakagami-m distribution is accepted to the noise model at certain frequency.

5 Conclusion

This paper proposed the measurements and modeling of noise on the 22.9-kV Medium-Voltage (MV) underground power distribution cable for Broadband Power Line Communication (BPLC). The proposed measurement system was composed of inductive coupler and Digital Phosphor Oscilloscope. The measurements were carried out in the test field located in Choji area of a suburb of Ansan city near Seoul, Korea. The time domain analysis of the measured noise was conducted, also frequency domain analysis of the noise data was performed with PSD using Welch's method. Based on the noise data, the CPDs were calculated at each frequency for the model of noise. And then, common CDFs were compared with the CPD at each frequency. The Gamma distribution was fitted into the CPD below 20MHz frequency range. Above the 20MHz, Nakagami-m distribution was accepted to the noise model. It is different results of the noise model on LV power line which has a model of Nakagami-m distribution in entire PLC frequency band.

Acknowledgement

This work was supported by the Power Generation & Electricity Delivery of the Korea Institute of Energy Technology Evaluation and Planning (KETEP) grant funded by the Korea government Ministry of Knowledge Economy (No.G2020).

References

1. Lee, J.-J., Hong, C.S., Kang, J.-M., Hong, J.W.-K.: Power line communication network and management in Korea. International Journal of Network Management (IJNM) 13(6), 443–457 (2006)

2. Kim, Y.H., Song, H.H., Lee, J.H., Kim, S.C.: Wideband Channel Measurements and Modeling for In-House Power Line Communication. In: Proc of ISPLC 2002, March 27-29, pp. 189–193 (2002)
3. Kim, J.-R., Kim, H.-S., Huh, J.-S., Lee, H.-Y., Lee, J.-H., Oh, Y.-W., Byun, W.-B.: The Design and Characteristics of the Inductive Coupler Using the Nanocrystalline Materials. Journal of the Korean Magnetics Society 16(6), 300–304 (2006)
4. Zimmermann, M., Dostert, K.: An Analysis of the Broadband Noise Scenario in Powerline Networks. In: 2000 International Symposium on Powerline Communications and its Applications, Limerick, Ireland (April 2000)
5. Stoica, P., Moses, R.L.: Introduction to Spectral Analysis, pp. 52–54. Prentice-Hall, Englewood Cliffs (1997)
6. Meng, H., Guan, Y.L., Chen, S.: Modeling and analysis of noise effects on broadband power line communications. IEEE Transactions on Power Delivery 20(2) part 1, 630–637 (2005)
7. Tao, Z., Xiaoxian, Y., Baohui, Z., Xu, N.H., Xiaoqun, F., Changxin, L.: Statistical Analysis and Modeling of Noise of 10-kV Medium-Voltage Power Lines. IEEE Transactions On Power Delivery 22(3), 1433–1439 (2007)
8. Benyoucef, D.: A New Statistical Model of the Noise Power Density Spectrum for Power line Communication. In: Proceedings of the 7th International Symposium on Power-Line Communications and its Applications, Kyoto, Japan (2003)

Optimal Control Method of Electric Power Generation in Multi Level Water Dams

Yeosun Kyung[1], Joowoong Kim[1], Sungboo Jung[2], and Kihwan Eom[1]

[1] Dpartment of Electronic Engineering
Dongguk University: 26, Pil-dong 3-ga, Jung-gu, Seoul, Korea
[2] Seoil College, 49-3, Myeonmok-dong, Seoildaehak-gil-22,
Jungnang-gu, Seoul, Korea
kihwanum@dongguk.edu,
csbcsb@seoil.ac.kr

Abstract. Generally, about 67% of hydropower generation is depending on the rainy season. In other word the process of development of water resources is too limited to a particular season.

As a solution to this problem, this paper proposes an optimal control method of hydroelectric power development in multi level water dams.

The main advantage of the system is stabilization of the amount of power generate in dams. Simulation results show that the proposed optimal control method indicates development rate in Chungpyung Dam is about a 40%, 25% in Chungju Dam and 37.5% in Paldang.

Keywords: Hydroelectric Power Plant, Multi Dam, Optimal Control Method, Power Generate.

1 Introduction

Korea four distinct seasons, rainy days cannot be constant. Therefore, Korea's hydroelectric dam water is not constant, and about 67% of hydropower generation is depending on the rainy season. Since the operation ratio of power system for drainage system in Han River of Seoul is 30%, the cost for thermal power generation is saved if the hydroelectric power generation becomes higher. There is a problem that reserves the water below the current average water level to prevent from flood in July and August, flood season. There is not enough power generated in February and March, dry season due to water supply. In flood season, enough water should be reserved by means of the precise system that predicts, and prevents the flood.

In this paper, we propose the optimal control method of electric power generation in multilevel water dams for water resources problem. In order to verify the development rate of the proposed optimal control method, and to compare it with the rainy season, we perform simulation on the total control system for hydroelectric power in Han River of Seoul.

T.-h. Kim, A. Stoica, and R.-S. Chang (Eds.): SUComS 2010, CCIS 78, pp. 325–334, 2010.

2 Power System Optimal Control Method by the Regulation of Dam Water Level -Simple Model

2.1 Optimal Control of Water Level

Consider the following feedback problem

$$J(x,\tau) = E\int_{\tau}^{T}\{h[x(t) - \bar{x}]^2 + c[u(t) - \bar{u}]^2\}e^{-\rho t}dt + \bar{s}\,h[x(T) - \bar{x}]^2 e^{-\rho(T-\tau)} \quad (1)$$

$$\frac{dx(t)}{dt} = [-ax(t) - u(t) + s(t)] + \sigma\omega(t) \quad (2)$$

$$dx(t) = [-ax(t) - u(t) + s(t)]dt + \sigma d\omega(t) \quad (3)$$

where, $0 \le \tau \le t \le T$, $x(0) = x_0$. Some variables used in this problem is given by

$0 \le \tau \le t \le T$: time
$x(t)$: dam water level at time t.
$u(t)$: generation speed(amount of release) at time t.
x_0: water level at t = 0, starting point.
$\bar{x}$: target water level of x(t).
$\bar{u}$: target value for generation speed(amount of release), u(t).
h: penalty for the deviation of water level at time t, x(t), from target water level($\bar{x}$).
c: penalty for the deviation of generation speed, u(t), from the target value, $\bar{u}$.
ρ: a constant for easy calculation by the computer.
Let $T = \infty$, $(0 \le \rho < 1)$; unnecessary to solve the differential equation,
a: natural extinction rate of water level(drying rate data is given)
s(t): water level change rate by external source (+inflow – drinking water – water for agriculture: data is given).
ω(t): water level change rate by random source; 0, average of ω(t), σ variance.
E: expected value of integral.

The model above is to minimize the cost for dam water level and for dam water release during the period [0,*T*] when there is relationship of differential equation (2) between water release, inflow, and water level. The determinant variable, u(t), should be defined for dam water releasing rate at each time to minimize the total cost. If u(t) is determined, then the optimized x(t) is also solved by equation (2), and the cost function (1) becomes minimized. The optimization of a general model for the problem stated above is like shown below [1],[2],[3].

$$J(x,\tau) = E\{\int_{\tau}^{T} f^0 e^{-\rho t}dt + S(T, x_T)e^{-\rho(T-\tau)}\} \quad (4)$$

$$\frac{dx(t)}{dt} = f + \sigma\omega, x_0 = x_0 \quad (5)$$

$$-J_t = \min_u\{f^0 + J_x f + \frac{1}{2}J_{xx}\sigma^2 - \rho J\} \quad (6)$$

$$f^0 = h[x(t) - \bar{x}]^2 + c[u(t) - \bar{u}]^2 \quad (7)$$

$$f = -ax(t) - u(t) + s(t) \quad (8)$$

The algorithm to find an optimized solution is derived as follows by letting the function like below.

$$J(t,x) \equiv xSx + qx + r \equiv V(t,x)$$
$$J_t \equiv x\dot{S}x + \dot{q}x + \dot{r} \left(* \, x\dot{S}x + \dot{q}x + \dot{r} = 0 \text{ for independant } t \text{ and } x \text{ in } V(t,x)\right)$$
$$J_x \equiv 2Sx + q$$
$$J_{xx} \equiv 2S$$

The algorithm to find an optimized solution is derived as follows by letting the function. The s, q and r are sub-variables. The optimization is shown below by using partial derivatives. From last equation shown above, it should be 0 in order to satisfy x(t), water level, at any time. Therefore, we have

$$S(T) = \bar{S}he^{-\rho T}$$
$$q(T) = -2\bar{S}h\bar{x}e^{-\rho T}$$
$$r(T) = 0 \tag{9}$$

Let $0 < t < T$, then the equations is as follows: $\dot{S}(t) \neq 0, \dot{q}(t) \neq 0, \dot{r}(t) \neq 0$. S(t),q(t), and r(t) can be calculated with Euler method or Runge-kutta methods when $0 \leq t \leq T$. Then the feedback control can be obtained with the following equation if x(t) is given by

$$u^*(t) = \bar{u}(t) + \frac{1}{2c}J_x = \frac{1}{c}S(t)x(T) + \left(\bar{u}(t) + \frac{1}{2c}q(t)\right) \tag{10}$$
$$* \left(* \, u_{stable}(t) = -a\bar{x} + s(t), 0 \leq u(t) \leq u_{max} \, , u_{max} \, : \text{Maximum Capacity}\right)$$

2.2 Simulation of a Model – Hwacheon Dam-

We can review Hwacheon dam, independently. In other words, the impact of change in the amount of release from Hwacheon dam can be relieved at dams in downstream such as Chuncheon dam, Uiam dam, and Chungpyung dam.

Table 1. Amount of inflow to Hwacheon dam per month. This data is an average during 1950-1982. The unit used in this table is CMS, cubic meter per second = m^3/sec.

Month	**1**	**2**	**3**	**4**	**5**	**6**
Inflow	10.7	9.6	28.0	76.6	56.6	80.4
Cumulative	10.7	20.3	48.3	124.9	181.5	261.9
						(m^3/sec)
Month	**7**	**8**	**9**	**10**	**11**	**12**
Inflow	248.6	322.2	253.1	26.3	23.8	21.3
Cumulative	510.5	832.7	1085.8	1112.1	1135.9	1157.2
						(m^3/sec)

Average per month: 96.4 m3/sec =1157.2(m3/sec)/12

Power plant statistics of Hwacheon dam as follows :

Capacity of Facility, MW: 108 MW=27MW*4;
Capacity of Facility, MW per year: 326,000 MWh;

Turbine: capacity: 30MW; numbers: 4, total capacity: 120MW;
Max usage: 185 m3/sec = 46.25 m3/sec* 4 turbines;

Simulation conditions are can be written as

ψψ= 24; ψψψ= 0ψ1;ψ ψ ψ= 240; ψ -= ψ0 = 50;ψψ= 1; ψψ= 0ψ01; x̄ψ= ψψψ= 60; ūψ= ψψψ= 10;

u_{max} = 11; ρψ= 0ψ1; ψψ= 0ψ05;ψψ= 0; a= 0;

The result is shown in the Fig.1.

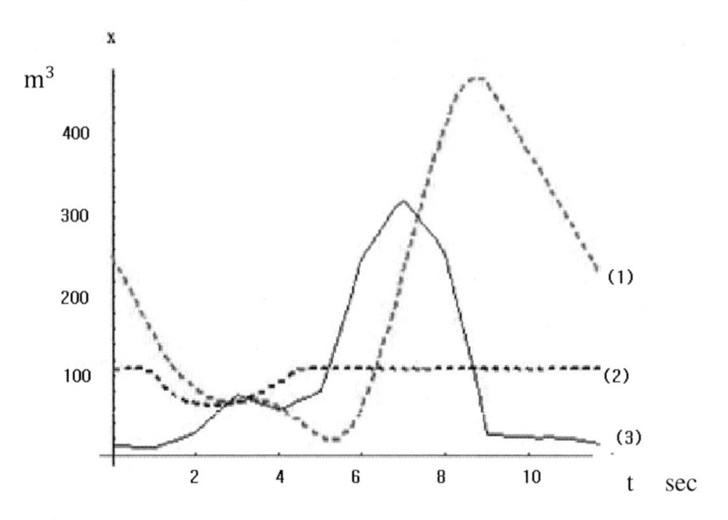

Fig. 1. Simulation of simple model in Hwacheon dam

In the figure, (1) is water level with an unit of m^3, (2) is water for generation in proposed algorithm with an unit of m^3/sec and (3) is inflow with an unit of m^3 /sec.

3 Total Optimal Control Method for Hydroelectric Power Plant in Han River

3.1 Discrete Optimization Model

The previous researches have been on similar problems to next model with cost reduction model to generate the power.

$$\min F = \sum_{t=1}^{T}\left\{\sum_j h_j\left(u_j(t) - Q_{j0}\right) + \left[P_R(t) - \sum_j h_j\left(u_j(t) - Q_{j0}\right)\right]^2\right\} \tag{11}$$

$$x(t+1) = x_j(t) - u_j(t) + s_j(t) \tag{12}$$

$$0 \le t \le T, x_j(0) = x_0$$

Definition of variables can be written as

$s_j(t)$: [+inflow – drinking water – water for agriculture] at j reservoir at time t.
$x_i(t)$: amount of water reserved in j reservoir at time t.

$u_j(t)$: output power(converted value that was used) of j hydroelectric power plant at time t.
$\overline{P}_R(t)$: expected power consumption of j at time t.
$\overline{R}_j(t), \overline{S}_j(t), \overline{Q}_j(t), \overline{P}_{Hj}(t), \overline{P}_{Si}(t)$: indicate the maximum of variables
$\underline{R}_j(t), \underline{S}_j(t), \underline{Q}_j(t), \underline{u}_{Hj}(t), \underline{v}_{Si}(t)$: indicate the minimum of variables.
$\overline{P}_R(t)$: power consumption at time t.
Explanation of cost:
$\sum_j h_j(u_j(t) - Q_{j0})$: cost for hydroelectric power generation.
$[P_R(t) - \sum_j h_j(u_j(t) - Q_{j0})]^2$: An approximate cost value for shortage of power supply or over power supply compared to power consumption.

$$P_R(t) \cong \sum_j [h_j(u_j(t) - Q_{j0})], P_{Hj}(t) \cong h_j(u_j(t) - Q_{j0}) \tag{13}$$

Each variable is located in the middle between maximum and minimum values.

$\underline{u}_j(t) \le u_j(t) \le \overline{u}_j(t)$, $\underline{s}_j(t) \le s_j(t) \le \overline{s}_j(t)$, $\underline{P}_{Hj}(t) \le P_{Hj}(t) \le \overline{P}_{Hj}(t)$, $\underline{P}_{Si}(t) \le P_{Si}(t) \le \overline{P}_{Si}(t)$.

3.2 Total Optimized Control Model

The problem above can be converted to an optimized control problem.

$$\min J = \frac{1}{2}\int_0^T \{[x(t) - \tilde{x}(t)]'Q[x(t) - \tilde{x}(t)] + [u(t) - \tilde{u}(t)]'R[u(t) - \tilde{u}(t)] + [v(t) - \tilde{v}(t)]'P[v(t) - \tilde{v}(t)]\} dt + \frac{1}{2}[x(t) - \tilde{x}(t)]'Q[x(t) - \tilde{x}(t)] \tag{14}$$

$$\frac{dx(t)}{dt} = Ax(t) + Bu(t) + Cv(t) + g(t), \ 0 \le t \le T, \ x(0) = x_0 \tag{15}$$

$$\underline{u}_j(t) \le u_j(t) \le \overline{u}_j(t), \underline{v}_j(t) \le v_j(t) \le \overline{v}_j(t), \underline{x}_j(t) \le x_j(t) \le \overline{x}_j(t), 0 \le t \le T$$

n: number of hydroelectric power plants; ′: transpose.
u(t): water release related to power generation.
v(t): drinking water + water release for flood prevention.
$u(t) = [u_1(t), \dots, u_n(t)]$; $v(t) = [v_1(t), \dots, v_n(t)]$; $x(t) = [x_1(t), \dots x_n(t)]$

This can't be solved by feedback algorithm due to inequality sign. To solve the problem, apply Augmented Lagrangian Method [4],[5]. Lagrangian function L is defined above, $\lambda(t)$ and sub variables like $\overline{z}(t)$, and z(t) are inserted. In order to solve in numeral method, effectively, multiply $\gamma/2$ to inequality sign, and relative term, and square them. The converted problem can be solved by numerical method, Gradient method. Euler method is used to solve the differential equations, and I can simplify if there is no restriction. The algorithm for optimal numerical method is used for this system. Superscript i means an ith approximate value improved, calculate $x^{i+1}(t)$ value improved in left side by using the value in the right side. The calculation period and boundary condition are shown next.

$$0 \le t \le T, \ x^i(0) = x_0, \lambda'(t) = Q[x^i(t) - \tilde{x}(t)]$$

From this, $x^i(t)$ is obtained when t=0, and $x^i(0) = x_0$ is known, and divide the period $0 \le t \le T$, into Δ_t. Calculate the value, $x^{i+1}(t+\Delta_t)$ by changing the period from t = 0, Δ_t, $2\Delta_t$, …, T. Variable, $\lambda(t)$, is the last value at t = T, $\lambda'(t) = Q[x^i(t) - \tilde{x}(t)]$ divide the period, $0 \le t \le T$, into Δ_t. Calculate the value, $\lambda^{i+1}(t-\Delta_t)$ by changing the period from t = T, T-Δ_t, T-2Δ_t, …, 0. Hydroelectric power dams in Han River are shown in Fig. 2.

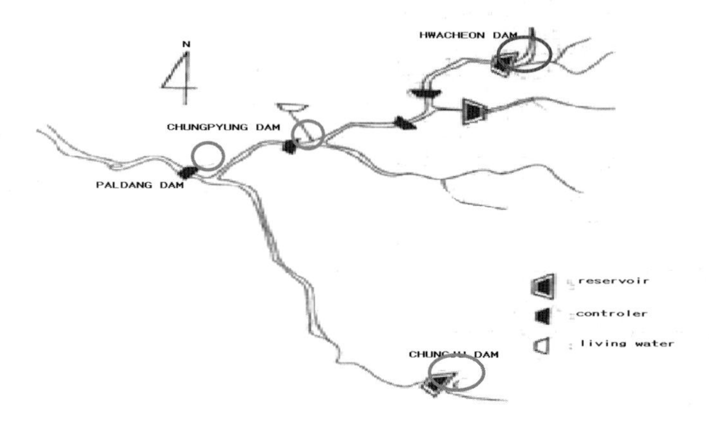

Fig. 2. Dams in Han River

Let's review to optimize all three dams in Han River: Paldang dam which is important for hydroelectric power, supplying water to Seoul, and flood control, and Chungpyung dam, and Chungjoo dam which are connected to Paldang dam. The general model of vector matrix above can be rewritten like the following:

$$\min J = \frac{1}{2}\int_0^T\{q_1[x_1(t)-\tilde{x}_1(t)]^2 + q_2[x_2(t)-\tilde{x}_2(t)]^2 + q_3[x_3(t)-\tilde{x}_3(t)]^2 + r_1[u_1(t)-\tilde{u}_1(t)]^2 + r_2[u_2(t)-\tilde{u}_2(t)]^2 + r_3[u_3(t)-\tilde{u}_3(t)]^2 + p_1[v_1(t)-\tilde{v}_1(t)]^2 + p_2[v_2(t)-\tilde{v}_2(t)]^2 + p_3[v_3(t)-\tilde{v}_3(t)]^2\}\,dt + \frac{1}{2}\bar{q}_1[x_1(T)-\tilde{x}_1(T)]^2 + \frac{1}{2}\bar{q}_2[x_2(T)-\tilde{x}_2(T)]^2 + \frac{1}{2}\bar{q}_3[x_3(T)-\tilde{x}_3(T)]^2 \tag{16}$$

$$\frac{dx_1(t)}{dt} = -a_1x_1(t) - [b_1u_1(t) + c_1v_1(t)] + g_1(t) \tag{17}$$

$$\frac{dx_2(t)}{dt} = -a_2x_2(t) - [b_2u_2(t) + c_2v_2(t)] + g_2(t) \tag{18}$$

$$\frac{dx_1(t)}{dt} = -a_3x_3(t) - [b_3u_3(t) + c_3v_3(t)] + [b_2u_2(t) + c_2v_2(t)] + [b_1u_1(t) + c_1v_1(t)] + g_3(t) \tag{19}$$

$$\underline{u}_j(t) \le u_j(t) \le \bar{u}_j(t), \underline{v}_j(t) \le v_j(t) \le \bar{v}_j(t), \underline{x}_j(t) \le x_j(t) \le \bar{x}_j(t)\,, 0 \le t \le T$$

Subscript 1 in all variables indicates the Chungpyung dam, subscript 2 is Chungjoo dam, and subscript 3 is Paldang dam. a_1, a_2, and a_3 indicate natural evaporation from the amount of water reserved in dams. g(t) is the amount of inflow from raining, and other region. u(t) is the amount of released and v(t) is drinking water and the amount of water released for flood control. The review period: $0 \leq t \leq T$, initial water reserved in each dam: $x_1(0) = x_{10}, x_2(0) = x_{20}, x_3(0) = x_{30}$. To find out the best optimum solution for 3 dam models with the algorithm to calculate the general optimization solution above, see below

- Chungpyung Dam

Table 2. Chungpyung Dam

month	Water level (EL.m)	Rainfall (mm)	Inflow (m^3/sec)	Used for Generation (m^3/sec)	Released for prevention (m^3/sec)	Fluctuation Pondage (m^3/sec)
1	50.271	1.000	2335.800	2210.300	0.000	125.500
2	50.505	35.000	2270.500	2232.900	0.000	37.600
3	50.556	61.000	2967.400	3026.300	0.000	-58.900
4	50.407	75.000	3167.500	3146.300	0.000	21.200
5	50.426	110.500	4594.500	4645.200	0.000	-50.700
6	49.937	174.500	6675.100	6813.599	0.000	-138.500
7	49.645	836.500	29398.201	9631.800	19735.301	31.101
8	49.705	283.500	15410.399	10533.599	4884.300	-7.500
9	49.945	22.500	4464.000	4268.100	0.000	195.900
10	50.571	53.500	2025.400	2045.600	0.000	-20.200
11	50.592	53.000	2198.600	2198.400	0.000	0.200
12	50.270	19.000	2914.200	2931.800	0.000	-17.600
Aver/sum	50.236	1725.000	78421.602	53683.895	24619.602	118.104

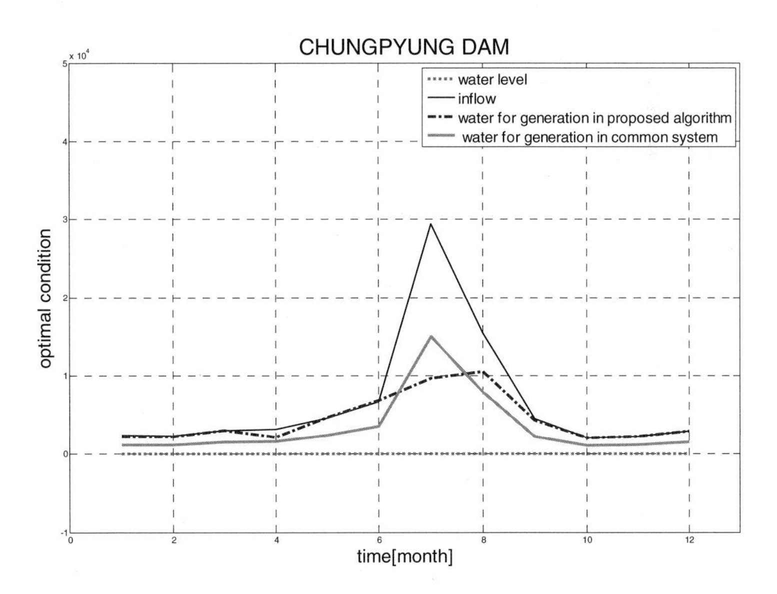

Fig. 3. Optimal solution of Chungpyung dam

- Chungjoo dam

Table 3. Chungjoo dam

month	Water level (EL.m)	Rainfall (mm)	Inflow (m^3/sec)	Used for Generation (m^3/sec)	Released for prevention (m^3/sec)	Fluctuation Pondage (m^3/sec)
1	64.642	16.300	2347.803	2353.512	0.000	-5.709
2	64.537	27.000	1896.696	1876.879	40.616	-20.799
3	64.395	43.900	2035.505	2023.125	0.000	12.380
4	64.535	26.500	1991.507	1862.260	106.519	22.728
5	64.666	97.100	2463.776	2482.174	0.000	-18.398
6	64.631	96.700	2569.069	2456.870	84.315	27.884
7	64.617	356.700	25599.471	3399.189	22205.396	-5.115
8	64.754	144.000	6241.611	3937.936	2320.850	-17.175
9	64.745	56.700	2294.223	2282.063	0.765	11.395
10	64.724	38.900	2193.484	2183.365	8.683	1.436
11	64.671	41.700	1879.880	1874.649	11.674	-6.443
12	64.650	35.400	2452.162	2461.088	0.956	-9.882
Aver/sum	64.631	980.900	53965.188	29193.107	24779.773	-7.698

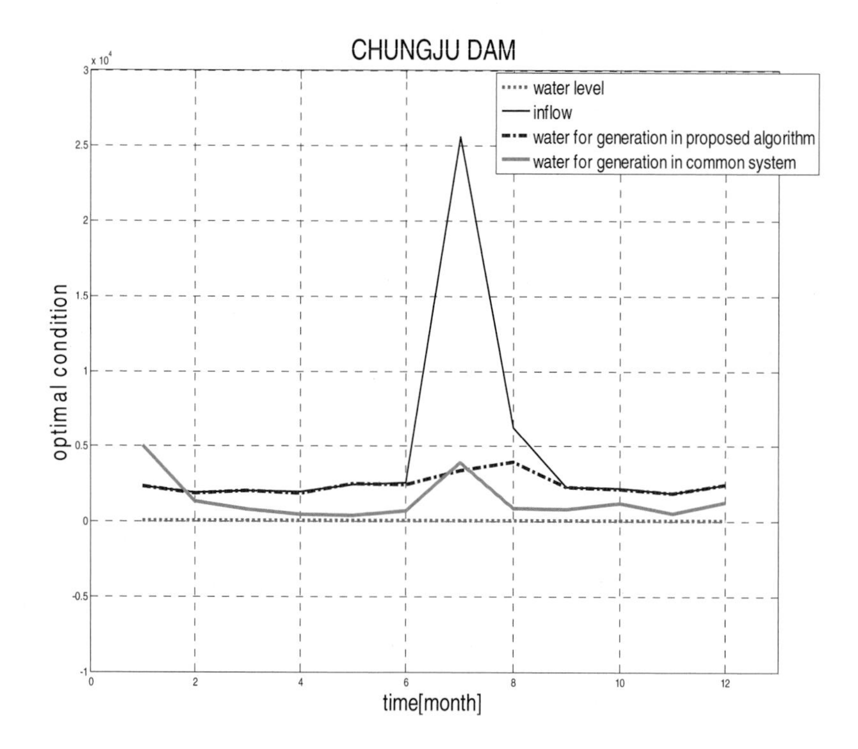

Fig. 4. Optimal solution of Chungju dam

- Paldang dam

Table 4. Optimal solution Paldang dam

month	Water level (EL.m)	Rainfall (mm)	Inflow (m^3/sec)	Used for Generation (m^3/sec)	Released for prevention (m^3/sec)	Fluctuation Pondage (m^3/sec)
1	25.165	0.500	4000.300	3912.601	0.000	87.700
2	25.216	35.500	4056.600	4085.700	0.000	-29.100
3	25.176	65.000	4976.100	4960.801	2.900	12.399
4	25.101	53.000	4865.801	4881.700	0.000	-15.899
5	25.095	110.000	7630.000	7642.401	0.000	-12.400
6	25.050	131.000	9445.700	9483.700	0.000	-38.000
7	24.887	782.000	79447.516	15363.700	64054.902	28.913
8	24.910	248.000	26799.604	17914.201	8915.000	-29.598
9	25.120	48.500	6692.400	6592.400	0.000	100.000
10	25.125	53.000	4096.600	4167.500	0.000	-70.900
11	25.170	49.500	4313.600	4288.800	0.000	24.800
12	25.114	17.500	5101.800	5068.400	0.000	33.399
Aver/sum	25.096	1593.500	161426.016	88361.898	72972.797	91.315

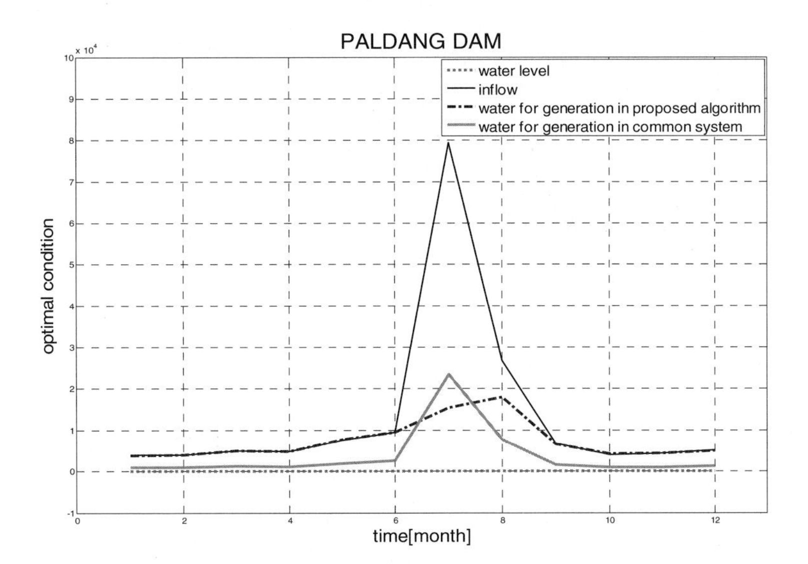

Fig. 5. Optimal solution of Paldang dam

4 Conclusion

Generally, about $\frac{2}{3}$ (67%) of hydropower generation is depend on the rainy season. In other word the process of development of water resources is too limited to a particular season. In order to develop this problem, we proposed an optimal control method of hydroelectric power development in multi level water dams. To verify the development rate of the proposed optimal control method, we simulated on the total control system for hydroelectric power in Han River of Seoul.

Simulation results show that the proposed optimal control method indicates development rate in Chungpyung Dam is about a 40%, 25% in Chungju Dam and 37.5% in Paldang. The main advantage of the proposed optimal control method is stabilization of the amount of power generate in dams.

Through the system for simple model, Hwacheon Dam, the development rate of hydropower is stabilized without big impact from inflow even in the rainy season or dry season. It would be helpful to maintain the water level in each dam not exceeding the limits, and the generate rate would be kept the constant power generation. However, this system has some negative side because of labile factor. It has not been performed that accurate measurement of each element in Han-River. The data of each element in this paper is approximate number derived by a formula. When the data of each factors correctly attained, this system can be valuable method.

References

1. Bryson, A.E., Ho, Y.C.: Applied Optimal Control. Bleisdell Publ., Waltham (1969)
2. Bryson, A.E.: Applied Linear Optimal Control. Cambridge Univ. Pr., Cambridge (2002)
3. Bensoussan, A.: Stochastic Control of Partially Observable Systems, pp. 20–25. Cambridge Univ. Pr., Cambridge (1992)
4. Evtushenko, Y.G.: Numerical Optimization Techniques. Springer, Heidelberg (1985)
5. Bertsekas, D.P.: Parallel and Distributed Computation. In: Numerical Methods. Athena Scientific, Belmont (1997)
6. Morari, M.: Mixed Logic Dynamical Model of a Hydroelectric Power Plant. ETH
7. Morari, M.: Supervisory Water Level Control for Cascaded River Power Plants

Real-Time Hand Gesture SEMG Using Spectral Estimation and LVQ for Two-Wheel Control

Mohammad `Afif B Kasno[1], Jihoon Ahn[1], Kyungkwon Jung[1], Yonggu Lee[2], and Kihwan Eom[1]

[1] Department of Electronic Engineering, Dongguk University, 26, Pil-dong 3-ga, Jung-gu, Seoul, Korea
[2] Department of Medical Instrument and Information, Hallym College, Chuncheon City, Gangwond-do, Korea
kihwanum@dongguk.edu

Abstract. In this paper, a real-time experimental of Hand Gesture SEMG using Spectral Estimation and Linear Vector Quantization for Two-Wheel Machine Control is proposed. The raw SEMG signals been captured from SEMG amplifier and the Auto Regressive (AR) Covariance returned the power spectral density (PSD) magnitude squared frequency response. Up to 4 channels of AR data will be combined and a fine tuning step by using LVQ will then incorporate for pattern classification. The database then been build and use for real-time experimental control classification. Captured data will send through serial port and Two-Wheel Machine will receive and move accordingly. The detail of the experiment and simulation conducted described here to verify the differentiation and effectiveness of combined channels PSD method SEMG pattern classification of hand gesture for real-time control.

Keywords: SEMG pattern classification, LVQ, Spectral Estimation, Human-Computer Interaction, Two-Wheel Machine Control.

1 Introduction

Since the past three decades ago, biomedical signal control has becoming more popular for its application in rehabilitation and human-computer interfaces (HCI). Surface electrode sensors use for capturing Surface electromyography (SEMG) signal which will be useful to measure the activities of the musculature system [1]. The application of SEMG varies in term of the various acquired signals from different tissue, organ, or musculature. Various approach for the use of SEMG has been presented, and such approaches are extremely valuable to physically disable persons like unvoiced speech recognition and the hands-free SEMG mouse [2].

There are many useful pattern recognition method has been proposed recently to help identify and distinguish each different hand gesture SEMG signals. Kyung Kwon Jung et. al [3] proposed the SEMG Pattern Classification using Spectral Estimation and Neural Network and 4th-order Yule-Walker algorithm is been proposed to estimates the power spectral density (PSD) of the SEMG signals. This method gives

T.-h. Kim, A. Stoica, and R.-S. Chang (Eds.): SUComS 2010, CCIS 78, pp. 335–344, 2010.

the success rate of the proposed classifier about 78 percent. Furthermore, M. A. Kasno proposed the experimental improvement of [4] by using 100th –order merged Covariance AR data and LVQ and the proposed classifier reach 99 percent of success rate.

In order to make each hand gestures well distinguish from each other is of crucial importance [5]. The hand gesture figure should be more attentive so the surface electrode sensors area which the musculature placed on could react and gives the sufficient distinguish information for each of the hand gesture. There are several of hand gesture figures been experiment like Korean Hand Gesture [3], [4], Punch Movement Hand Gesture [2], and Thumb-up Rotation Hand Gesture [5].

In this paper, the real-time Hand Gesture SEMG using Spectral Estimation and LVQ for Two-Wheel Control is been proposed. The proposed system use hand gesture 100th order merged Covariance AR and LVQ to control two-wheel vehicle simultaneously. Different of Hand Gesture approach as well as SEMG captured signal channels is been experiment and discussed.

2 SEMG Measurement System

The way SEMG-based attached to the musculature for measuring makes HCI the most practical because of conveniently and safely compare to the other neural signals [6]. Referring to [4], the SEMG Measurement Equipment was built and test with similar set of measurement. Fig. 1 shows the SEMG Measurement Equipment. It is consists of SEMG electrodes, SEMG amplifier circuitry and A/D converter for computer interfacing.

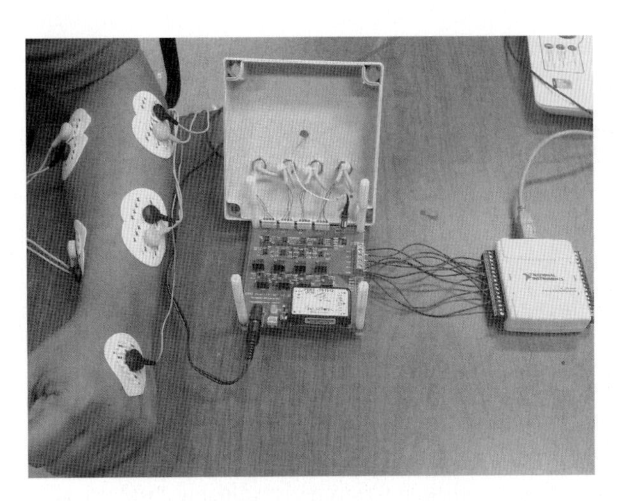

Fig. 1. SEMG measurement equipment

The SEMG amplifiers are instrumentation amplifier, namely INA2128 chip for the preamplifier and OPA2604 for body reference circuit. The ADC used is NI-DAQ USB 6009. Fig. 2 shows the block diagram of an SEMG measurement system.

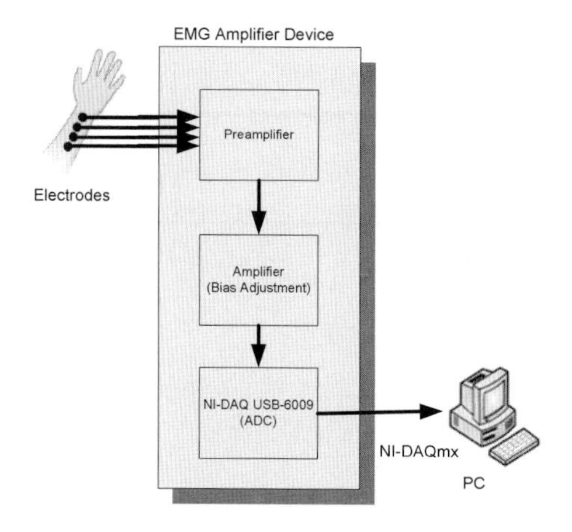

Fig. 2. Block diagram of SEMG measurement system

3 Two Wheel Vehicle System

Two-Wheel Vehicle System is made by microprocessor, DC Stepping Motor and also circuitry. In this project, Atmega128 microprocessor is been used as CPU system [7].

Sanyodenki 103H5205-0480) Stepping Motor has been used for the wheel control. The stepping motor is DC 24V-1.2A input voltage, 1.8°/step, and it been controlled by SLA7024 High-Current PWM. Fig. 3 shows the Two-Wheel vehicle system, and Table 1 shows the details about wheel movement direction.

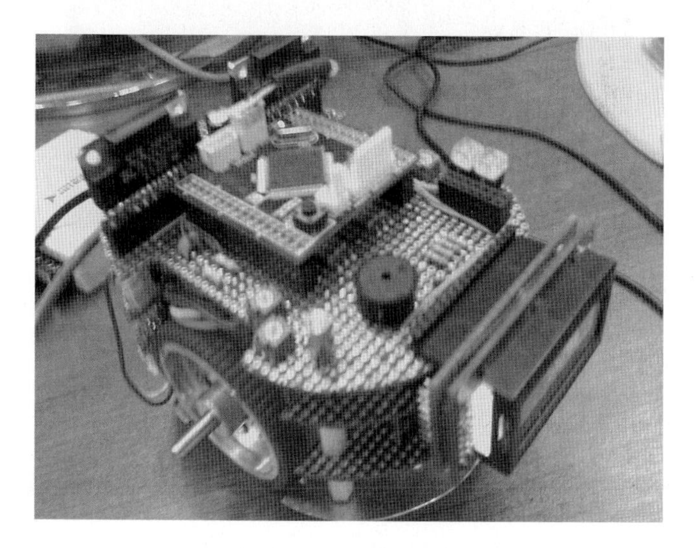

Fig. 3. Two-Wheel Vehicle System

Table 1. The details of wheel direction movement. It is consists of 2 stepping wheel and 3 different movements, forward, reverse and stop.

Action(Flag)	Wheel	
	Left	Right
Forward(w)	○	○
Reverse(x)	●	●
Right(d)	○	-
	○	○
Left (a)	-	○
	○	○
Stop(s)	-	-
○: On Forward, ●: Turn Reverse, - : Stop		

For computer interfacing, MFC(Microsoft Foundation Class) program was used to take the output result from SEMG Measurement Equipment. The MFC program also provides GUI in 3D simulation. The GUI as shown in Fig. 4.

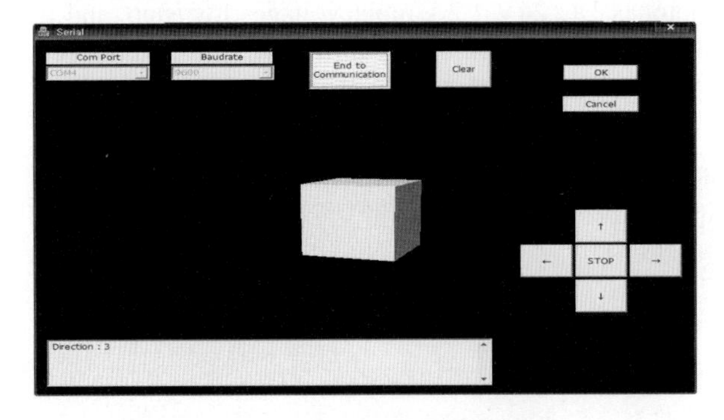

Fig. 4. MFC GUI 3D Simulation

4 Pattern Classification

4.1 Covariance Method

The Covariance method is a technique for estimating the AR parameters [8]. The Covariance equations can be written as

$$\begin{bmatrix} r(0) & r(-1) & \cdots & r(-n) \\ r(1) & r(0) & \cdots & \vdots \\ \vdots & \vdots & \ddots & r(-1) \\ r(n) & \cdots & \cdots & r(0) \end{bmatrix} \begin{bmatrix} 1 \\ a_1 \\ \vdots \\ a_n \end{bmatrix} = \begin{bmatrix} \sigma^2 \\ 0 \\ \vdots \\ 0 \end{bmatrix} \tag{1}$$

where *r(k)* is autocovariance a_n is AR parameter, σ^2 is variance. If $\{r(k)\}_{k=0}^{n}$ were known, equation (1) could be solve for $\theta = [a_1, \dots, a_n]^T$.

To explicitly stress the dependence of θ and σ^2 on the order *n*, (1) can be written as

$$R_{n+1}\begin{bmatrix}1\\ \theta_n\end{bmatrix} = \begin{bmatrix}\sigma^2\\ 0\end{bmatrix} \tag{2}$$

4.2 LVQ

LVQ network has a first competitive layer and a second linear layer. The competitive layer learns to classify input vectors. The linear layer transforms the competitive layer's classes into target classifications defined by the user. The net input of the first layer of the LVQ is written as

$$n_i^1 = -\| {}_i w^1 - p \|, \tag{3}$$

Or in vector form as

$$n^1 = -\begin{bmatrix} \| {}_1 w^1 - p \| \\ \| {}_2 w^1 - p \| \\ \vdots \\ \| {}_{S^1} w^1 - p \| \end{bmatrix}, \tag{4}$$

where n is neuron, w is weight and p is the input. The second layer of the LVQ is represented by W^2 matrix where the columns represent subclasses and rows represent classes. The row in which the 1 occurs indicates which class the appropriate subclass belongs to is written as

$$(w_{ki}^2 = 1) \Rightarrow \text{Subclass } i \text{ is a part of class } k \tag{4}$$

The classes learned by the competitive layer are referred as subclasses and classes of the linear layer as target classes [9].

5 Proposed Method

The system built is been showed in Fig. 5. The proposed of Real-Time Hand Gesture SEMG control using Spectral Estimation and LVQ for Two-Wheel method configuration is shows on Fig. 6. The simulation coding on how the real-time will operate is described with a flow chart in Fig. 7.

Fig. 5. The full system of Real-Time hand gesture SEMG

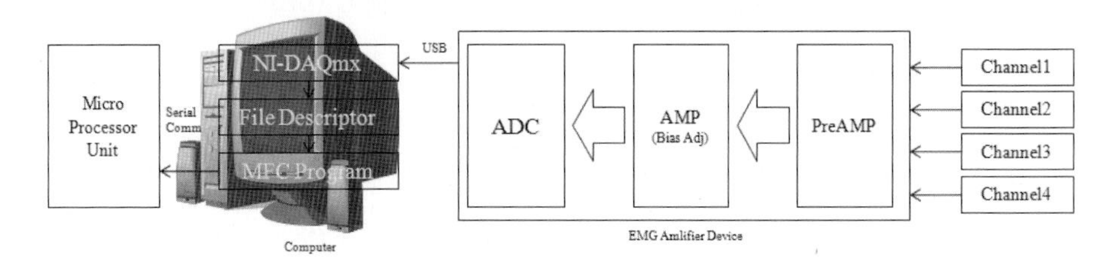

Fig. 6. Proposed Real-Time Hand Gesture SEMG control using Spectral Estimation and LVQ for Two-Wheel vehicle

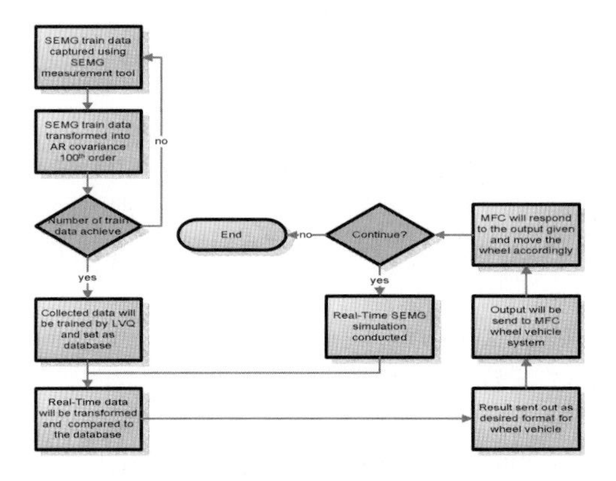

Fig. 7. Flow chart of real-time simulation coding

First, the SEMG train data will be captured using SEMG measurement equipment. The required channels are depending on the simulation needed. The data will be sampled at 10 kHz in 2 seconds. Then, the SEMG train data will transformed into AR Covariance 100th order by using spectral estimation covariance. Until the number of required data train is achieved, the simulation will keep continuing data train captured loop.

After the number of required data train is achieved, the simulation proceeds to pattern classification and the trained data will be conducted to LVQ training phase. The LVQ neural network has 4 neurons on the input layer, and 6 neurons on the output layer. The initial step size in interval location step has been set to 0.01, learning rate is set to 0.9, and the learning rate is decreased in the course of time.

The trained data will be set as database, and real-time simulation will be conducted. As the train data is sampled at 10 kHz in 2 seconds, the real-Time simulation will also be conducted in the same duration. The captured real-time data will be transformed into AR covariance 100th order and will be compared with the database. Result will be sent out as desired format for wheel vehicle system.

Result from SEMG measurement equipment will be read as input for MFC two-wheel vehicle system. The result then been compared with the setting and the MFC two-wheel will move the wheel accordingly

6 Experimental Result

The experiment conducted to showed the differentiation of each hand gestures for single, double, triple or quadruple channels setting in terms of successful recognition are described below. Table 2 shows the hand gestures for real-time simulation.

Table 2. The hand Gestures for real-time simulation. The 1st is Korean Hand Gesture referred similar to [4], Punch Hand Gesture [2] and Thumb up hand Gesture [5].

Korean Hand Gesture	*Punch Hand Gesture*	*Thumb up Hand Gesture*	*Control Direction*
ㄴ		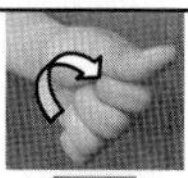	Forward (w)
ㄱ			Back (x)
 ㄷ		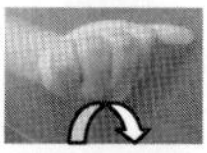	Left (d)
 ㅅ		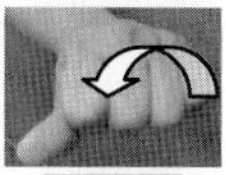	Right (a)
 ㅂ	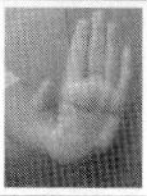	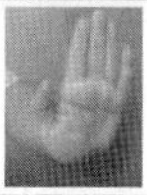	Stop (s)

The real-time experimentation was conducted by 20 samples for training and 10 samples for testing. Results for different hand gestures are showed in Table 3, 4 and 5. It is showed the differentiation of each hand gestures for single, double, triple or quadruple channels setting in terms of successful recognition.

Table 3. The Korean hand Gestures for real-time simulation. The table consists of the Korean Hand Gesture figures, Time presented for train data and real-time simulation classified time for different channels combination.

Korean Hand Gesture	*Time presented for train data*	*Time presented for testing data*	*Real-time simulation classified times*			
			Ch1	Ch1 and Ch2	Ch1, Ch2 and Ch3	Ch1, Ch2, Ch3 and Ch4
ㄴ	20	10	7 (70%)	7 (70%)	8 (80%)	8 (80%)
ㄱ	20	10	1 (10%)	2 (20%)	3 (30%)	3 (30%)
ㄷ	20	10	1 (10%)	2 (20%)	3 (30%)	3 (30%)
ㅅ	20	10	3 (30%)	3 (30%)	4 (40%)	5 (50%)
ㅂ	20	10	0 (0%)	0 (0%)	2 (20%)	3 (30%)

Table 4. The Punch Movement hand Gestures for real-time simulation. The table consists of the Punch Movement Hand Gesture figures, Time presented for train data and real-time simulation classified time for different channels combination.

Punch Movement Hand Gesture	*Time presented for train data*	*Time presented for testing data*	*Real-time simulation classified times*			
			Ch1	Ch1 and Ch2	Ch1, Ch2 and Ch3	Ch1, Ch2, Ch3 and Ch4
	20	10	5 (50%)	5 (50%)	6 (60%)	6 (60%)
	20	10	0 (0%)	0 (0%)	1 (10%)	2 (20%)
	20	10	0 (0%)	0 (0%)	3 (30%)	3 (30%)
	20	10	2 (20%)	2 (20%)	2 (20%)	2 (20%)
	0	10	0 (0%)	1 (10%)	2 (20%)	2 (20%)

The results showed that the hand gestures response to each of the SEMG electrodes channels differently. Some of the hand gestures did not response well with only one channel and need other channels in order to distinguish the hand gestures preferably. The more channels used for the simulation, the more accurate pattern recognition can be done in order to distinguish the hand gestures. During real-time experimentation, Korean Hand Gestures showed the best results with high successful rate compared to the other hand gestures. It was because of more flexion happen during contradiction of muscle movement when most of finger muscles involved compare to the other hand gestures.

Table 5. The Thumb-Up Rotation hand Gestures for real-time simulation. The table consists of the Thumb-Up Rotation Hand Gesture figures, Time presented for train data and Real-time simulation classified time for different channels combination.

Thumb-Up Rotation Hand Gesture	*Time presented for train data*	*Time presented for testing data*	*Real-time simulation classified times*			
			Ch1	Ch1 and Ch2	Ch1, Ch2 and Ch3	Ch1, Ch2, Ch3 and Ch4
	20	10	3 (30%)	4 (40%)	6 (60%)	7 (70%)
	20	10	1 (10%)	2 (20%)	3 (30%)	3 (30%)
	20	10	2 (20%)	2 (20%)	3 (30%)	3 (30%)
	20	10	1 (10%)	2 (20%)	2 (20%)	2 (20%)
	20	10	2 (20%)	2 (20%)	4 (40%)	4 (40%)

7 Conclusion

In this paper, real-time simulation of Hand Gesture SEMG using Spectral Estimation and LVQ for two-wheel machine control is proposed. Proposed method is combining of SEMG measurement equipment and MFC Two-Wheel vehicle system. The use of the Covariance Method is to estimates the power spectral density of the signal. The spectral estimate returned is the magnitude squared frequency response of AR model. The pattern classification used LVQ Neural Network and once the database is set up as training data, the real-time simulation taken place in order to give the instruction control for two-wheel MFC control system.

In order to verify the effectiveness of the proposed method, the experimental of 3 different SEMG hand gesture has been conducted. Experimental results showed the differentiation of each hand gestures for single, double, triple or quadruple channels setting in terms of successful recognition.

References

1. Chen, A., Kevin, B.: Continous Myoelectric Control for powered protheses using Hidden Markov Models. IEEE Transactions on Biomedical Engineer 52, 123–134 (2005)
2. Kim, J.-.S., Jeong, H., Son, W.: A new means of HCI: EMG-MOUSE. In: 2004 IEEE International Conference on Systems, Man an Cybernetics, October 10-13, vol. 1, pp. 100–104 (2004)
3. Jung, K.K., Kim, J.W., Lee, H.K., Chung, S.B., Eom, K.H.: EMG Pattern Classification using Spectral Estimation and Neural Network. In: SICE Annual Conference 2007, pp. 1108–1111 (2007)
4. Kasno, M.A., Jung, K.K., Eom, K.H.: Improvement of SEMG Pattern Classifier using Covariance AR and Linear Vector Quantization. In: Wish Well 2010, July 19- 21 (2010)
5. Xu, Z., Xiang, C., Wen-hui, W., Ji-hai, Y., Lantz, V., Kong-qiao, W.: Hand Gesture Recognition and Virtual Game Control Based on 3D Accelerometer and EMG Sensors. In: IUI 2009, pp. 401–405 (2009)
6. Reaz, M.B.I., Hussain, M.S., Mohd-Yasin, F.: Techniques of EMG signal analysis: detection, processing, classification and applications. Biological Procedures Online 8(1), 11–35 (2006)
7. Osborne, A.: An Introduction to Microcomputers, 2nd edn. Basic Concepts, vol. 1. Osborne-McGraw Hill, Berkely (1980)
8. Stoica, P., Moses, R.L.: Introduction to Spectral Analysis. Prentice Hall, Englewood Cliffs (1997)
9. Hagan, M.T., Demuth, H.B., Beale, M.: Neural Network Design. PWS Pusblising Company (1996)

A Self-deployment Scheme for Mobile Sensor Network with Obstacle Avoidance*

Chan-Myung Kim, Yong-hwan Kim, Hee-Sung Lim, and Youn-Hee Han**

School of Computer Science and Engineering,
Korea University of Technology and Education
{cmdr,cherish,tlshrns00,yhhan}@kut.ac.kr

Abstract. In our previous study, we showed that the coverage area can be efficiently expanded by having sensors move to the centroids of their Voronoi polygon generated using the location information of neighboring sensors. In this paper, we present an energy-efficient self-deployment scheme to utilize the attractive force generated from the centroid of a sensor's local Voronoi polygon as well as the repulsive force frequently used in self-deployment schemes using the potential field. By using the proposed scheme, mobile sensors also have the obstacle-avoidance capability. The simulation results show that our scheme can achieve a higher coverage and enables less sensor movements in shorter times in a region with obstacles than self-deployment schemes using the traditional potential field.

Keywords: Mobile sensor networks, Self-deployment, Voronoi diagram, Potential field, and Obstacle-avoidance.

1 Introduction

The sensors used in many application fields are primarily required to be well deployed in order to cover a region of interest (ROI) where interesting events might happen and need to be detected [4]. It is sometimes difficult for humans to directly deploy sensors in unexploited, hostile, or disaster areas. In these cases, it is possible to have sensors scattered randomly, for example, by using an aircraft. Using such a random deployment of sensors, however, makes it difficult to ensure the required coverage in the ROI. Therefore, if the sensors (a.k.a mobile sensors) have locomotive capability (in addition to the main functions of sensing, computation, and communication), they can intelligently move to the correct places by themselves so as to achieve the required coverage.

* This work was supported by the IT R&D program of MKE/KEIT [10035245: Study on Architecture of Future Internet to Support Mobile Environments and Network Diversity] and also supported by the Ministry of Knowledge Economy (MKE) and Korea Institute for Advancement in Technology (KIAT) through the Workforce Development Program in Strategic Technology.

** Corresponding Author

T.-h. Kim, A. Stoica, and R.-S. Chang (Eds.): SUComS 2010, CCIS 78, pp. 345–354, 2010.

The self-deployment problem of sensors in a given ROI relates to how sensor coverage can be maximized in less movements and shorter time. The most widely known self-deployment strategy is one using a *Potential Field* where sensor nodes are treated as virtual particles, subject to virtual forces [5,8,7,1,9]. These forces repel the sensor nodes from each other and from obstacles, and ensure that sensor nodes will quickly spread out to maximize the coverage area. Self-deployment schemes using the potential field do not require models of the environment (e.g., any map of obstacles in a given ROI) ahead of the deployment time, and so it is both simple and highly scalable. The *Voronoi diagram* has also been actively used as a fundamental tool for resolving the coverage problem of wireless sensor networks [10,2,3,4]. Given a set of sensors $S = \{s_1, s_2, s_3, \ldots, s_N\}$, the Voronoi diagram divides a two-dimensional plane into N convex polygons. Each polygon c_i has only one sensor s_i and consists of the set of edges that are equidistant from neighboring nodes. In our previous study [6], we proposed a new scheme to fix the coverage hole by using the local Voronoi polygon and its centroid (geometric center). When sensors are randomly deployed at the initial time, a sensor calculates the centroid of its local Voronoi polygon as the next position and moves to it.

In this paper, we propose an improved mobile sensor self-deployment scheme using a new virtual force, called the *Centroid-directed Virtual Force*. Similar to existing self-deployment schemes, our approach allows the mobile sensors in the networks to move from a high density area to a low density area, while avoiding obstacles in a given ROI. Each sensor and obstacles behave as a source of force for all other sensors. If a sensor is placed too close to other sensors or obstacles, it exerts repulsive forces on them. This ensures that the sensors are not overly clustered, which would lead to poor coverage regions in other parts of the given ROI. The distinguishing feature of the proposed scheme is the utilization of the attractive force generated from the centroid of the Voronoi polygon as well as the repulsive force. That is, the proposed scheme incorporates the coverage hole fixing capability with the spreading ability using the traditional potential field. Our simulation results show that our scheme can achieve a higher coverage and less movements in a shorter time in an ROI with obstacles than the existing scheme.

The rest of this paper is as follows. Section 2 defines the self-deployment problem addressed in this paper, describes the assumptions made to solve that problem, and provides the metrics to measure the performance of the mobile sensor self-deployment schemes. Section 3 proposes the strategy of our scheme and describes the procedure of our self-deployment scheme in detail. Section 4 provides gives a comparative performance analysis and Section 5 describes the conclusions and future works.

2 Problem Definition

Let A denote the ROI on a two-dimensional field. In a given ROI, there may or may not be obstacles that need to be avoided by mobile sensors. The field boundaries mimic wall-like obstacles. Each sensor is assumed to be equipped with a device (e.g. supersonic range finder) that can grasp the exact range and bearing of surrounding obstacles. Assume that N sensors have the communication range R_c and the sensing range R_s. Sensors within R_c of a sensor are called the sensor's neighboring nodes. To model the coverage and connectivity, we use the binary model where the quality of sensing (communication) is constant within R_s (R_c) and is zero outside the sensing (communication) range. The sensors are capable of omni-directional motion and move in rounds of variable distance. Each sensor is also assumed to know their location information by applying an arbitrary localization method and is able to determine the location of neighboring nodes by communication (e.g., periodic beacon messages).

Based on the above assumptions, the problem to be solved in this paper is stated as follows: *Given N mobile sensors with sensing range R_s and radio communication range R_c and given the ROI with an arbitrary initial sensor distribution, how should they deploy themselves so that they maximize the sensor coverage in less time and less movements (lower energy consumption)?*

Although the above problem is related to solving the global coordination issue, our algorithm will be completely distributed and each sensor makes the movement decision by itself on the basis of locally available information. The self-deployment algorithm will not require a prior map of a given ROI and will adapt to changes in the environment as well as the network itself.

In this paper, we evaluate the performance of our algorithm and the existing self-deployment algorithm with the following three metrics below:

2.1 Coverage Rate

The *coverage rate* is the most important metric necessary to evaluate the performance of self-deployment schemes. This refers to the level at which a random point in the whole ROI is observed by at least one sensor. This can be expressed by the following formula:

$$Coverage(\%) = \frac{\bigcup_{i=1}^{N} C_i}{A}. \tag{1}$$

where C_i is the region covered by a sensor s_i, N is the number of sensors, and A is the area of the given ROI. Here, the coverage area of each sensor is defined as the circular area within its sensing radius R_s. If a sensor s_i is located well inside the ROI and its complete coverage area lies within the ROI, the full area of that circle, i.e., πR_s^2, is included to calculate C_i. If a node is located near obstacles or the boundary of the ROI, then only the part of the ROI covered by that node is included in the calculation.

2.2 Moving Distance

For a sensor, the *moving distance* refers to the cumulative distance traveled until it moves to the final location after the random deployment at the beginning.

When D_i is the cumulative distance of a sensor s_i, the metric over all sensors can be exactly written as follows:

$$Moving\ Distance(m) = \frac{\sum_{i=1}^{N} D_i}{N}. \tag{2}$$

Since the motor is usually used for sensor locomotion, sensor movement results in significant energy consumption.

2.3 Deployment Time

The deployment of sensors in a short time is also important in time-critical sensor applications, such as emergency rescue and disaster recovery. The *deployment time* is defined in this paper as the elapsed time until all the sensors reach their final locations after the random deployment at the beginning. The deployment time as well as the moving distance is closely dependent on the completion criteria of the self-deployment algorithms.

3 The Proposed Self-deployment Scheme

3.1 Search for the Maximal Effective Position

A Voronoi polygon is composed of the set of edges that are equidistant from the neighboring nodes of the sensor which draws the Voronoid polygon, so that it represents the proximity information about a set of geometric points and nodes. It can be noted that a point p lies in the Voronoi polygon c_i corresponding to a sensor s_i if and only if $d(p, l_i) < d(p, l_j)$ for the location l_i of the sensor s_i and the location l_j of each sensor s_j, $j \neq i$ where $d(x, y)$ denotes the Euclidean distance between x and y. This is an important property in relation to the sensor coverage issue because if s_i cannot detect an expected phenomenon occurring in the polygon c_i, no other sensors can detect it. Therefore, each sensor should be responsible for the sensing task in its local Voronoi polygon.

By using the Voronoi polygon, the global coverage problem of all sensors can be translated into the local coverage problem for each sensor. The local coverage problem is described as follows. Let us assume that a sensor forms its Voronoi polygon by using the location information of neighboring nodes. The *effective region* is defined as the region obtained by the intersection of the Voronoi polygon and the coverage area (sensing circle) of the sensor. The *maximal effective position* within the polygon is defined as the location which can derive the largest effective region. The local coverage problem is to search for the maximal effective position inside the Voronoi polygon. In other words, each sensor uses only local information to calculate its Voronoi polygon and determine the maximal effective position in the Voronoi polygon to fix the coverage hole and expand the local coverage rate at the utmost. This procedure runs iteratively until a termination condition is satisfied. The conversion from the global problem to the local problem can reduce the complexity of the problem and allow a scheme to be executed concurrently by all sensor nodes.

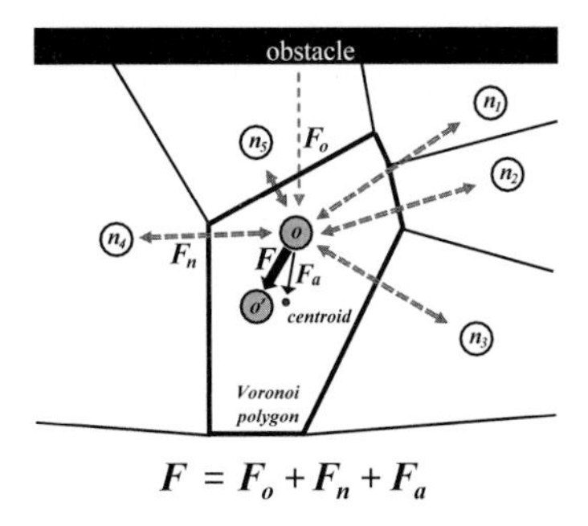

Fig. 1. Virtual force F in the proposed scheme

In our previous study [6], we demonstrated by simulation that the centroid of the local Voronoi polygon is close to the maximal effective position and that the proposed local coverage algorithm can efficiently increase the overall network coverage of a given ROI. The centroid of a convex polygon is the intersection point of all lines that divide the polygon into two parts of equal areas. Hence, the position is an average of all points of the polygon. More details can be found in [6]. In this paper, we adopt the strategy of fixing the coverage hole using the centroid, and then integrate it into the scheme to expand the sensor coverage using the traditional potential field.

3.2 Virtual Forces

We note that, in an earlier study [5], mobile sensor self-deployment using the potential field assumes that sensors gather in one place at the beginning, and that they spread throughout the entire region while simply pushing on one another. Here, each sensor may need an additional movement factor to fix coverage holes intellectually, while simultaneously pushing each other and finally being distributed in the entire ROI. In this paper, we think that the additional movement factor can be the attractive force generated from the centroid of each sensor's local Voronoi polygon. Thus, the proposed scheme enables sensors to spread out better and faster to expand the sensor coverage by utilizing the attractive force as well as the traditional repulsive force.

The proposed scheme uses the following three virtual forces independently exerted on each sensor: 1) F_o: the first repulsive force from surrounding external obstacles; 2) F_n: the second repulsive force from neighboring sensors; 3) F_a: the attractive force that draws to the centroid of the local Voronoi polygon.

Each of them has its individual magnitude and direction depending on the relative positions of obstacles, neighboring sensors, or centroid. A sensor is moved by F which is the sum of the three forces (see Fig. 1).

$$F = F_o + F_n + F_a. \tag{3}$$

This force F is called the *Centroid-directed Virtual Force.*

Let x denote the position vector of the current node. Let o_i, n_j, and c denote the position vectors which represent the positions of the obstacle i, the neighboring node j, and the centroid, respectively. Each of the virtual forces is presented as follows:

$$F_o = -K_o \sum_i \frac{1}{r_i^2} \cdot \frac{\mathbf{r_i}}{r_i} \tag{4}$$

$$F_n = -K_n \sum_j \frac{1}{r_j^2} \cdot \frac{\mathbf{r_j}}{r_j} \tag{5}$$

$$F_a = \begin{cases} 0 & \text{if the Voronoi polygon is fully covered.} \\ K_a \cdot r^2 \cdot \frac{\mathbf{r}}{r} & \text{otherwise.} \end{cases} \tag{6}$$

where $r_i(= |o_i - x|)$ is the Euclidian distance between the obstacle i and the current node, and $\mathbf{r_i} = o_i - x$. On the other hand, $r_j(= |n_j - x|)$ is the Euclidian distance between the neighboring node j and the current node, and $\mathbf{r_j} = n_j - x$. Finally, $r(= |c - x|)$ is the Euclidian distance between the centroid and the current node, and $\mathbf{r} = c - x$. K_o, K_n, and K_a are the force constants that describe the strength of each corresponding field.

After a sensor calculates the force F with Equations (3)~(6), its trajectory subject to the force is computed using the following equation of motion:

$$\ddot{x} = \frac{F - \nu \dot{x}}{m}. \tag{7}$$

where ν is the damping factor and m is the virtual mass of the sensor node.

3.3 Movement Adjustment

In the proposed scheme, a sensor moves after it decides on one of the two locations, i.e. either the present location or the calculated target location, on the basis of whichever is more beneficial towards expanding the effective region (and not just moving promptly to the target location). This means that a sensor maintains its present location if the present location is better for the local coverage than the calculated target location. Since the location of sensors is continuously changing, a sensor which maintains the present location can move again to the new location in the next round. This can increase the energy-efficiency by reducing the moving distance, and also suppresses unnecessary oscillation when a sensor repeatedly moves to the same locations.

3.4 Completion of Self-deployment

It is important to set the time when a sensor enters into the *equilibrium state* and ceases its own movement. The equilibrium state in this paper is classified into two cases as follows: 1) *Oscillatory Equilibrium State*: When a sensor has moved to the location that it has been at right before, it is decided that one 'oscillation' has occurred. If the 'oscillation' continuously occurs as frequently

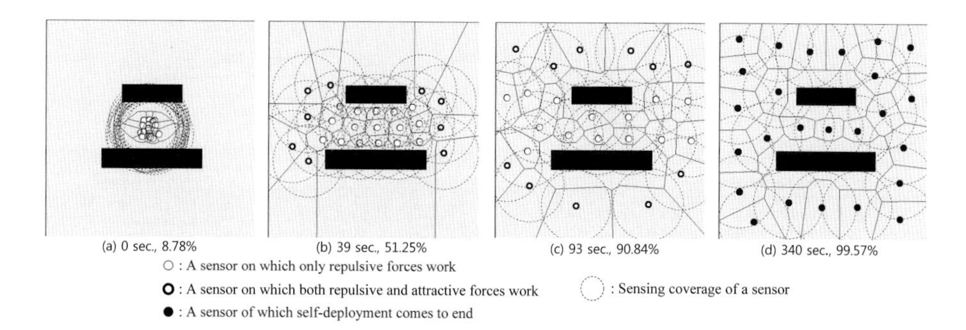

Fig. 2. Execution snapshots of the proposed scheme in an ROI with obstacles ($R_s = 6m$, $R_c = 20m$, and there are 25 sensors in the ROI measuring $60m$ by $60m$)

as the threshold parameter $threshold_o$, it is confirmed that the sensor has entered the *oscillatory equilibrium state*, and its self-deployment is then completed. 2) *Stationary Equilibrium State*: When a sensor's accumulated moving distance during a given time duration is smaller than the pre-defined value, it is decided that one 'stationary movement' has occured. If this stationary movement is continued as frequently as the threshold parameter $threshold_s$, it is confirmed that the sensor has entered the *stationary equilibrium state*, and its self-deployment is then completed.

When all sensors have entered into the oscillatory or stationary equilibrium state, the self-deployment over the whole ROI can be termed as complete. The time duration elapsed until all the sensors enter into the equilibrium state is defined as the deployment time.

The entire procedure of the proposed self-deployment scheme is as follows. While executing the following deployment procedures, the current location should be advertised periodically to the neighboring nodes.

- *Discovery phase*: At the beginning of each round, a sensor sets a timer to be the discovery interval and gathers the location information of the neighboring nodes and obstacles.
- *Determination phase*: Upon timeout, the sensor calculates its local Voronoi polygon and sensing coverage region. If the polygon is fully covered by the sensor's sensing coverage range, the sensor will not calculate the centroid and will only use the repulsive forces of the traditional potential field. If the polygon is not fully covered, the sensor then calculates the centroid of its local polygon. The centroid is used as one of the factors to generate the centroid-directed virtual force F which the sensor is subject to.
- *Moving phase*: Prior to the final decision, the sensor executes the movement adjustment to check whether the effective region increases by its moving to the new target location. If it increases at the new location, the sensor will move; otherwise, it will stay. After the movement, the sensor checks if it enters into the equilibrium state. If it enters the state, the deployment procedure comes to an end, or else the sensor repeats the procedure.

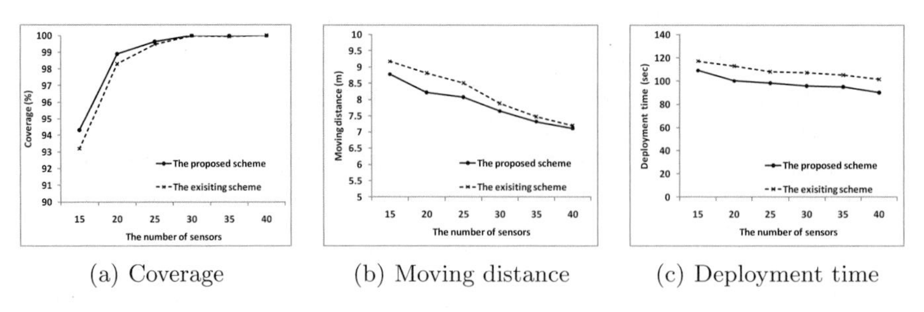

(a) Coverage (b) Moving distance (c) Deployment time

Fig. 3. Performance comparison versus the number of sensors ($R_s = 10m$, $R_c = 20m$, and the ROI measures $60m \times 60m$)

4 Performance Evaluation

We now compare the performance of the self-deployment scheme proposed in this paper to the existing potential-field-based scheme described in [5]. To begin, we randomly deployed $15 - 40$ sensors in a flat ROI measuring $60m \times 60m$. The default sensing range R_s and communication range R_c of a sensor were $10m$ and $20m$, respectively. In order to analyze the performance change according to the communication range, R_c was varied in the range of $14m$ to $26m$. The constant values K_o, K_n, and K_a are approximately calculated as 60.0, 15.0, and 0.003. We used 1.0 and 0.25 as the values of m and ν, respectively. Finally, we set the values of $threshold_o$ and $threshold_s$ as 3. Fig. 2 shows snapshots of our simulation in an ROI with obstacles. As shown in Fig. 2 (a), the 25 sensors were initially gathered in the center of the ROI. In the course of self-deployment, sensors encountered the obstacles and moved around them. As the simulation time passed from Fig. 2 (a) to Fig. 2 (d), sensors became more and more uniformly arranged, and thus we could see that the coverage rate in the ROI increased accordingly.

4.1 Coverage, Deployment Time, and Moving Distance in an Obstacle-Free ROI

Fig. 3 shows comparisons of the proposed scheme with the existing one in terms of (a) coverage, (b) moving distance, and (c) deployment time in an obstacle-free environment. For each number ($N = 15, 20, 25, 30, 35$ and 40) of sensors, we repeated the simulation 10 times with different initial random deployments. The mean values were then calculated and are shown in the figures. All the values were measured after the completion of every sensors' self-deployment. As shown in Fig. 3 (a), the proposed scheme yields better coverage performance than the existing one irrespective of the number of sensors. It can be noted that the performance gain diminishes with an increase in the number of deployed sensors. This is because the deployment of many sensors already results in good coverage of the given ROI even though a sophisticated self-deployment scheme is not supported. However, when a small number of sensors (15 or 20 sensors) are deployed, the proposed scheme well demonstrates its capability in the coverage expansion. Additionally, as shown in Fig. 3 (b) and (c), the proposed scheme

Table 1. Performance comparison in each ROI with different obstacle locations and shapes ($R_s = 6m$, $R_c = 20m$, and there are 25 sensors in each ROI measuring $60m$ by $60m$): C - Coverage Rate, D - Average Moving Distance, T - Deployment Time

	Location & shape of obstacles	The proposed scheme			The existing scheme		
		C (%)	D (m)	T (s)	C (%)	D (m)	T (s)
Case I		98.63	33.03	332.8	97.92	34.30	390.2
Case II		96.94	37.28	418.8	95.20	37.46	520.8
Case III		99.84	21.19	216.6	99.24	22.84	222.4

has a shorter moving distance and deployment time in all cases compared to the existing scheme. The reason for this is that the proposed self-deployment scheme using the additional attractive force moves sensors through a more optimal path with respect to coverage expansion than the existing scheme which uses only the repulsive force. Hence, the proposed self-deployment scheme is more energy-efficient than the existing scheme.

4.2 Performance Comparison in ROIs with Obstacles

Many real-world regions, such as a metropolitan area with buildings and structures, naturally have obstacles or holes. Therefore, a self-deployment scheme of mobile sensors should have an obstacle avoidance capability built-in. We simulated the proposed scheme in ROIs with obstacles to verify the obstacle-avoidance capability of the proposed scheme. Since our scheme is based on the traditional potential field, it inherits the obstacle-avoidance capability of the existing scheme using the potential field. Moreover, our scheme allows sensor nodes to move through a more optimal path. Thus, there is a difference in the performance between the proposed scheme and the existing scheme in ROIs with obstacles as well as an obstacle-free ROI.

Table 1 shows the results of the performance comparison for each ROI with different locations and obstacle shapes. We repeated the simulation 10 times for each ROI and the mean values of the coverage rate, moving distance, and deployment time were calculated. All the values were measured after the completion of the self-deployment of all sensors. It can be observed in the table that the performance of our proposed scheme is better than the existing scheme in all cases of ROIs. In the proposed scheme, a greater coverage rate of $0.6\% - 1.8\%$ and a reduced moving distance of $0.4\% - 7.6\%$ were obtained compared to the existing scheme. These results mean that the proposed scheme can maximize sensor coverage with less movement and in a shorter time even in an ROI with obstacles. Finally, it is worth noting that 2.6% – 19.5% of the deployment time was saved using the proposed scheme, implying that our scheme is more suitable to a sensor application that is to be used in an urgent situation.

5 Conclusion

We proposed a new energy-efficient mobile sensor self-deployment scheme which maximizes sensing coverage in a given sensing field. The major contribution of this paper is the proposal of a new virtual force to guide mobile sensors onto a more optimal path in terms of coverage expansion. This is achieved by incorporating the attractive force generated from the centroid of a sensor's local Voronoi polygon with the repulsive forces generated by obstacles and neighboring nodes. Our simulation results show that the proposed self-deployment scheme can successfully ensure higher coverage and less movement in a shorter time from initial uneven distributions in an energy-efficient manner.

References

1. Aitsaadi, N., Achir, N., Boussetta, K., Pujolle, G.: Potential Field Approach to Ensure Connectivity and Differentiated Detection in WSN Deployment. In: IEEE International Conference on Communications (ICC), pp. 1–6 (2009)
2. Boukerche, A., Fei, X.: A Voronoi Approach for Coverage Protocols in Wireless Sensor Networks. In: IEEE Globecom, pp. 5190–5194 (November 2007)
3. Du, Q., Faber, V., Gunzburger, M.: Centroidal Voronoi Tessellations: Applications and Algorithms. Society for Industrial and Applied Mathematics 41(1), 637–676 (1999)
4. Heo, N., Varshney, P.K.: Energy-Efficient Deployment of Intelligent Mobile Sensor Networks. IEEE Transactions on Systems, Man, and Cybernetics - Part A: Systems and Humans 35(1), 78–92 (2005)
5. Howard, A., Mataric, M.J., Sukhatme, G.S.: Mobile Sensor Network Delployment Using Potential Fields: A Distibuted, Scalable Solution to the Area Coverage Problem. In: The 6th International Symposium on DARS 2002, pp. 299–308 (June 2002)
6. Lee, H.J., Kim, Y., Han, Y.H., Park, C.Y.: Centroid-based Movement Assisted Sensor Deployment Schemes in Wireless Sensor Networks. In: The 70th Vehicular Technology Conference Fall (VTC 2009-Fall), pp. 1–5 (September 2009)
7. Lee, J., Dharne, A., Jayasuriya, S.: Potential Field Based Hierarchical Structure for Mobile Sensor Network Deployment. In: American Control Conference (ACC 2007), pp. 5946–5951 (2007)
8. Poduri, S., Sukhatme, G.S.: Constrained Coverage for Mobile Sensor networks. In: IEEE International Conference on Robotics and Automation, pp. 165–172 (May 2004)
9. Tan, G., Jarvis, S.A., Kermarrec, A.M.: Connectivity-Guaranteed and Obstacle-Adaptive Deployment Schemes for Mobile Sensor Networks. In: IEEE ICDCS 2008, pp. 429–437 (June 2008)
10. Wang, G., Cao, G., Porta, T.L.: Movement-assisted Sensor Deployment. In: IEEE Infocom, pp. 2469–2479 (March 2004)

ICSW^{2}AN : An Inter-vehicle Communication System Using Mobile Access Point over Wireless Wide Area Networks

Tae-Young Byun

School of Computer and Information Communications Engineering,
Catholic University of Daegu, Gyeonsan-si, Gyeongbuk, Rep. of Korea
tybyun@cu.ac.kr

Abstract. This paper presents a prototype of inter-vehicle communication system using mobile access point that internetworks wired or wireless LAN and wireless WAN anywhere. Implemented mobile access point can be equipped with various wireless WAN interfaces such as WCDMA and HSDPA. Mobile access point in the IP mechanism has to process connection setup procedure to one wireless WAN. To show the applicability of the mobile access point to inter-vehicle communication, a simplified V2I2V-based car communication system called ICSW^{2}AN is implemented to evaluate major performance metrics by road test. In addition, results of road test for traffic information service are investigated in view of RTT, latency and server processing time. The experimental result indicates that V2I2V-based car communication system sufficiently can provide time-tolerant traffic information to moving vehicles while more than two mobile devices in restricted spaces such as car, train and ship access wireless Internet simultaneously.

Keywords: Mobile Access Point, Wireless WAN, Car Communications, Inter-vehicle Communications.

1 Introduction

Advanced wireless WAN such as WCDMA, HSDPA and WiBro based on 3G+ of cellular networks increasingly are adopted to provide high speed data rate to subscriber in the world. Also, as demand of mobile device with 3G-capability is increasing, and revenue of carriers is rising compared to that of 2G cellular networks. In view of a variety of service, these advanced wireless WAN can provide more sophisticated services including U-healthcare, telematics and infotainment services by accessing advanced wireless Internet due to faster data rates than that of 2G+ cellular networks such as GPRS, CDMA etc.

This paper presents design and implementation of inter-vehicle communications system using mobile access point that provides Internet access over wireless wide area networks anywhere by interworking with Wi-Fi to offer wireless connectivity to two or more subscribers in restricted area such as car and train. Mobile access point is responsible for communicating with external mobile networks including CMDA,

T.-h. Kim, A. Stoica, and R.-S. Chang (Eds.): SUComS 2010, CCIS 78, pp. 355–366, 2010.

WCDMA, HSDPA and WiBro. When a vehicle moves, mobile devices within the vehicle can maintain connections to correspondent nodes because the whole car is a sub-network and mobile access point in a vehicle continuously provides wireless Internet connection to end-devices in vehicle.

Contributions. The goal of this paper is twofold : (a) to investigate the applicability of mobile access point in inter-vehicle communication over wireless wide area networks; (b) to provide a prototype of inter-vehicle communication system, and investigate the availability of inter-vehicle application over a variety of wireless wide area networks by experimental performance evaluation.

Roadmap. The remainder of this paper is organized as follows: next section surveys related work, and mentions the motivation to develop inter-vehicle communication system using mobile access point. Section 3 provides design details of inter-vehicle communication system. The effectiveness and performance are discussed in Section 5. Finally, Section 6 draws some conclusions.

2 Related Works and Motivation

3G telecommunication networks, specifically UMTS, are a relatively new player in this field [1, 2, 3] and offer a couple of benefits for TIS applications. Owing to their system design, in conjunction with much larger data rates compared to 2G networks, TIS operation based on 3G networks becomes economically feasible. In addition, unlike Vehicle-to-Infrastructure (V2I) solutions using WLAN or WiMAX, UMTS-based solutions can rely on readily-available infrastructure.

Compared to WLAN-based TIS solutions, however, the perceived strengths and weaknesses of UMTS networks are quite different. While, for example, the security of Vehicle-to-Vehicle (V2V) networks cannot easily be guaranteed [4], there already are strong security measures in place to guarantee 3G networks' integrity, which can be reused for V2I communication. As a second example, the distance between a message's sender and its intended receivers is almost a non-issue in 3G networks: its impact on the end-to-end delay is negligible. On the other hand, even for short distance messages the end-to-end delay is already quite high compared to that of direct radio links.

A key question to be asked about an infrastructure-based V2I communication system is therefore whether end-to-end delays will still be acceptable not only for common TIS applications, but also for the transmission of safety warnings. Another important question is whether such a system will scale better [5] than more traditional WLAN-based V2V solutions to accommodate high penetration rates, given that in this V2I solution all network traffic has to be routed through the available infrastructure. Together with obvious business reasons, both questions are at the core of the problems which hindered adoption of some of the early 2G-based approaches [6] to V2V and V2I communications via a cellular network proposed in the 1990s. Development of V2I solutions is now picking up again, with new approaches based on 3G networks or advanced wireless WAN.

Experimental approaches have accomplished post-hoc analysis of implemented testbeds using state of the art technology. In these setups, either detailed studies have been conducted [2] or complex extensible testbed architectures have been developed [7]. However, only the currently deployed UMTS versions could be tested and the size of the experiments was limited. Moreover, an evaluation of the environmental impact of TISs based on real world experiments is infeasible, and even simulative studies on this topic are rare [8].

3 Design and Implementation of ICSW^{2}AN

3.1 Major Components of ICSW^{2}AN

In this paper, the first experimentation results for a typical highway and urban scenarios are shown, based on real-world wireless wide area networks such as WCDMA, HSDPA and WiBro. The results clearly outline the capabilities of testbed and experimental evaluation results are consistent with all expectations. Figure 1 depicts an overview of testbed of ICSW^{2}AN, along with the various software modules that have been integrated to form the testbed that we will use for performance evaluation. Based on the testbed of ICSW^{2}AN, this paper describes modules of the testbed which is composed of and details how the modules interact with each other.

ICSW^{2}AN consists of three parts that are *MAP(Mobile Access Point)*, *client* and *LMS(Location Management System)*. Each component performs its own functionalities as followings.

MAP forwards packets which originated from several mobile devices such as Smartphone, PDA and Netbook in the moving vehicle to correspondent nodes simultaneously, or vice versa. *Client* periodically receives reliable location and time information from Global Positioning System (GPS) through GPS receiver that is attached to UMPC or mobile device. So, client has a library for parsing the GPS data format of NMEA standard to extract necessary information including current location, speed of vehicle and current time etc. In our implementation, client software is embedded in UMPC, provides the functionalities such as delivery of traffic information, displaying of safety-related information on satellite map provided by Google or Daum. Also, client provides running information of current moving vehicle to drivers based on GPS data. *LMS* is responsible to register safety-related traffic information such as car accident, traffic congestion and a trouble of car on the road. LMS selects some interested vehicles to avoid accident by delivery of safety-alarm to the vehicles as well as storing properties of each vehicle to database periodically.

3.2 Design and Implementation of MAP

This section explains the design and implementation details of MAP and specifications of it. MCU block contains MCU, NAND Flash and SDRAM in the layout of circuit board of MAP as shown in Fig. 1. MAP can support interworking between wireless WAN and wired or wireless LAN by attaching a variety of wireless WAN access modem such as CDMA, WCDMA, HSDPA etc. Two wired LAN interfaces for connecting mobile devices or another networking devices are provided, and additionally access modems can be attached by plugging into USB interfaces besides of internal access modem. In addition, LCD module can be equipped with LCD connector. Users can identify the operation status of MAP by the LEDs that indicates power on/off, normal operation, error occurrence, link status of LAN port etc. Also, user or developer can monitor details of operation through D-Sub interface which can sends current operation status of MAP to serial port of computer.

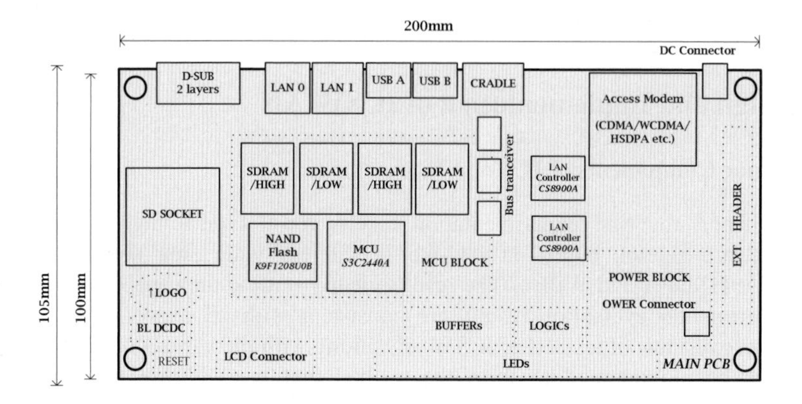

Fig. 1. Circuit board layout of mobile access point

Table 1. Specifications of MAP Prototype

Item		Description
MCU		SAMSUNG ARM920T S3C2440A(400MHz)
Memory	NAND Flash	64MByte, SAMSUNG K9F1208U0B
	SDRAM	64MByte, SAMSUNG K4S511632D
IP Address Allocation	NAT(masquerade) mode and Half-bridge mode	
3G+ Modem Type	Access Modem (Carrier, WWAN)	BSM860S(Korea Telecom, CDMA)
		CWE624K (Korea Telecom, WCDMA/HSDPA)
		WM210 (SK Telecom, HSDPA)
Wired LAN Spec.	Two LAN Ports : 100Base-TX, LAN Controller : CS8900A	

Two IP Assignment Methods of MAP. Implemented MAP provides two IP assignment methods, that is, *NAT* and *half-bridge mode* which optionally can be selected at programming stage by programmer. Fig. 2 illustrates the concept of two IP address assignment methods. *Network address translation* (NAT) is the process of modifying network address information in IP packet headers while in transit across a traffic routing device for the purpose of remapping one IP address space into another. Most often today, NAT is used in conjunction with *network masquerading* (or *IP masquerading*) which is a technique that hides an entire IP address space, usually consisting of private network IP addresses (RFC 1918), behind a single IP address in another, often public address space. This mechanism is implemented in MAP that uses stateful translation tables to map the "hidden" addresses into a single IP address and then readdresses the outgoing IP packets on exit so that they appear to originate from MAP.

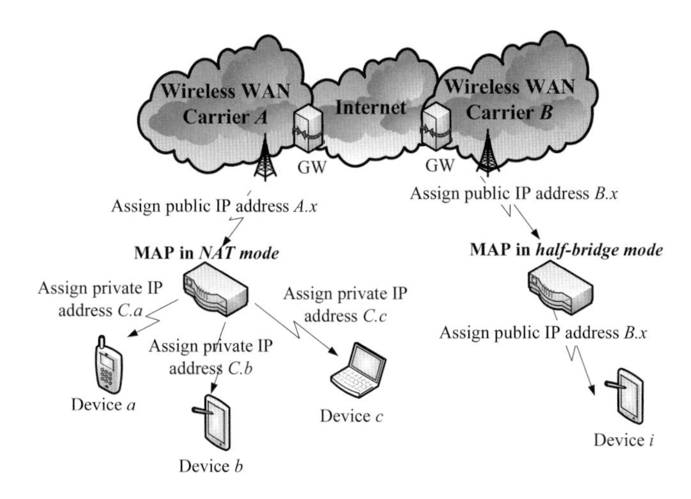

Fig. 2. Two IP Address Allocation Methods

In *half-bridge mode*, a public IP address assigned by carrier does not be modified within MAP, thus, original IP address is directly assigned to subscriber's device through MAP. In this mode, there is no need for modifying network address information in IP packet headers while in transit across MAP. However, in this mode, MAP can assign only one public IP address per WWAN interface within MAP to a mobile device. So, this mode optionally can be used in specific purposes or services.

MAP Iimplementation. MAP is implemented according to design details which already were shown in Fig. 1. MAP has several interfaces to connect peripherals such as external WWAN access modem and wired or wireless LAN devices.

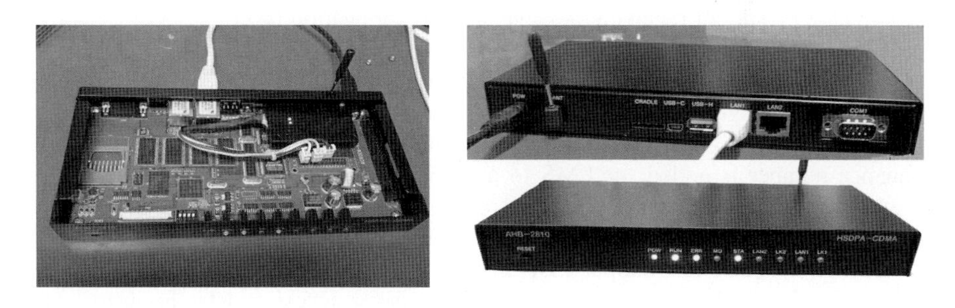

Fig. 3. Implemented MAP. The left figure shows internal circuit board, the right two figures shows the backside and the front of MAP respectively.

3.3 Design and Implementation of Client and LMS

Application Protocol. As explained in Section 3.1, major components of ICSW^{2}AP which consist of client, LMS and MAP communicates with each other as shown in

Fig. 4 which depicts overall protocol stacks of ICSW^{2}AP. Client and LMS exchange predefined application messages which indicate a variety of traffic information including registration of traffic congestion, car accident etc., and delivery of safety-alarm to some interested vehicles at application layer. Client software embedded in a variety of mobile devices has GPS device driver provided by manufacturer as well as legacy Internet protocol suites. GPS data acquired by GPS receiver are transmitted on USB interface or Blutooth wireless link to mobile device. Serial communication emulator pushes GPS data up client's application layer at interval of about one second. A link between client and MAP can be established through wired LAN(IEEE 802.3) or wireless LAN(IEEE 802.11g) link in the vehicle. MAP has two different datalink and physical layers for wireless WAN interface and wired & wireless interface to provide interworking between two heterogeneous networks. LMS & database server communicate with its peer correspondent layer of client by using application protocol.

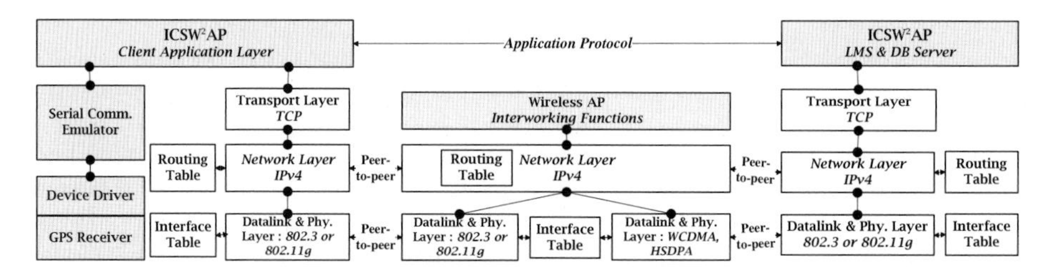

Fig. 4. Protocol Stacks of ICSW^{2}AP

Fig. 5 shows message format at application layer of ICSW^{2}AN between client and LMS. A message is a variable-length message consisting of two parts : header and data. The size of header depends on the size of each field in header and contains information essential to authentication and operation. All fields are filled with ASCII code to ease parsing of received message at LMS application layer.

Command field indicates the type of operation delivered from client. Operations include LOGIN, LOGOUT, GPS_Log, Alert etc. *Sequenc*e field means the sequence number of messages that are exchanged between client and LMS in a session. Although TCP is used to assure a reliable transmission at transport layer, additional sequence number is added for LMS to acknowledge successfully received message from client. Because of unreliable transmission of UDP, this sequence number can effectively be used to analyze frequencies of error occurrence, out-of-order delivery of message if ICSW^{2}AN adopts UDP instead of TCP as a transport protocol. LMS authenticates each driver by *Userid* field when driver access and try login to LMS. So, each driver has his or her own unique identifier. *Sessid* means session identifier between client and LMS, which is generated by combination connection establishment time and user identifier. The exact time that client sends routine message to LMS is stored at *Ctime* field in message. This field is needed to calculate round-trip-time (RTT) and latency between client and LMS. To assure the accuracy of calculation of time, all clients and LMS are synchronized to a specific time server periodically. *Data* filed contains supplementary data of *Command* field. If *Command* field contains a code of *GPS_Log*, this field conveys NMEA protocol data received from GPS satellite through GPS receiver. If *Command* field has a code of ALERT, *Data* field indicates one among predefined types of events such as car accident, construction, traffic congestion and out of order of vehicle. In all cases, *Data* field also provides vehicle's location information such as longitude and latitude.

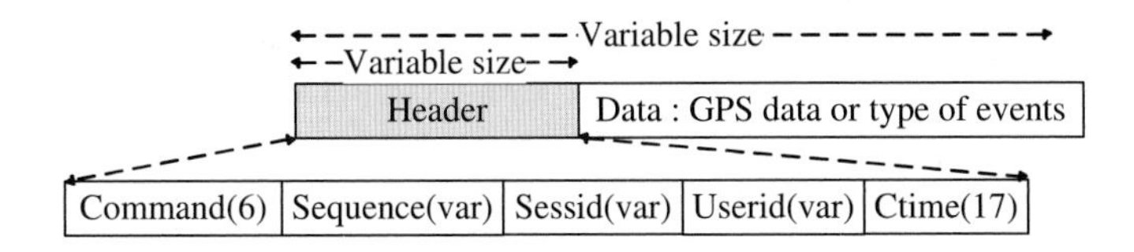

Fig. 5. Message format of application protocol

Operational Procedure. Operational procedures of application protocol between client and LMS are shown in Fig. 6. Also, state transition diagrams of client and LMS are shown in Fig. 7 respectively. To ease understanding of each state transition diagram, you can refer operational procedures among components of ICSW^{2}AN in Figure z and Fig. 6.

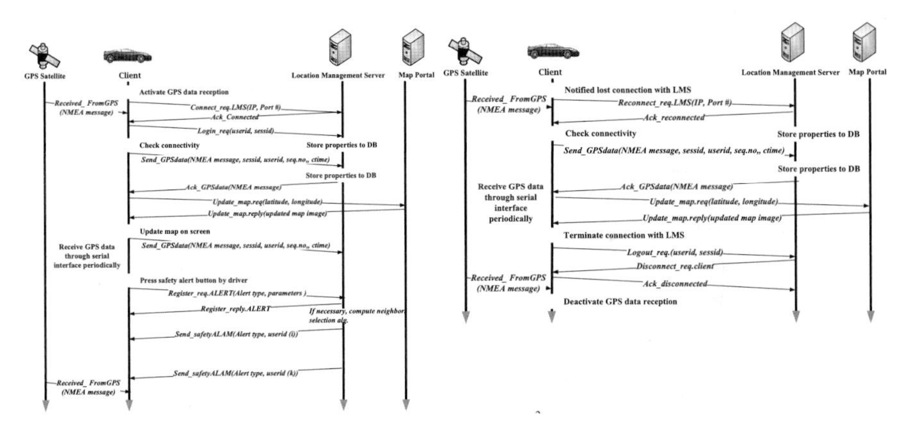

Fig. 6. Operational procedure among components of ICSW^{2}AN. The left figure illustrates procedures for connection establishment and data transfer, and the right figure depicts procedures for delivery of safety information and connection termination among components.

Software Architecture. Client software implemented in UMPC is composed of several modules according to its functionality. A class of software module and relations among them are depicted in Fig. 8. *GPS_Display* shows values of major fields of received GPS data on UMPC screen for providing driving information to driver periodically. *WebBrowser_Display* shows current location of moving vehicle on Google map or Daum map with satellite image at interval of one second as well as a driving course. *Log_Diplay* and *Log_Manager* store some data field into client's local database, then, process periodical GPS messages to measure performance of ICSW^{2}AN, collect statistics of performance metrics by combination with LMS's statistics at post-processing stage. *Connection_Manager* maintains current session with LMS by reconnecting to it automatically in case connection lost due to poor quality of signal over wireless WAN. If vehicle enters into tunnel etc. which prevents to receive a good quality of signal, the connection at datalink and physical layer can be lost due to poor signal quality. *Connection_Manager* try to reconnect to LMS and maintains the session continually. *NMEA Parser* analyzes incoming NMEA sentences and granting GPS data in data structures, also generates NMEA sentences. This module supports various types of sentences such as GPGGA, GPGSA, GPGSV, GPRMC and GPVTG.

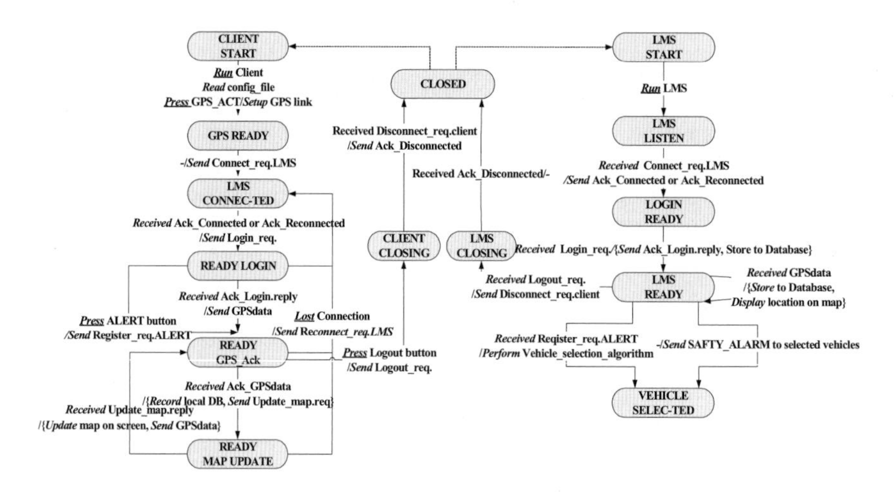

Fig. 7. State transition diagrams of both client and LMS

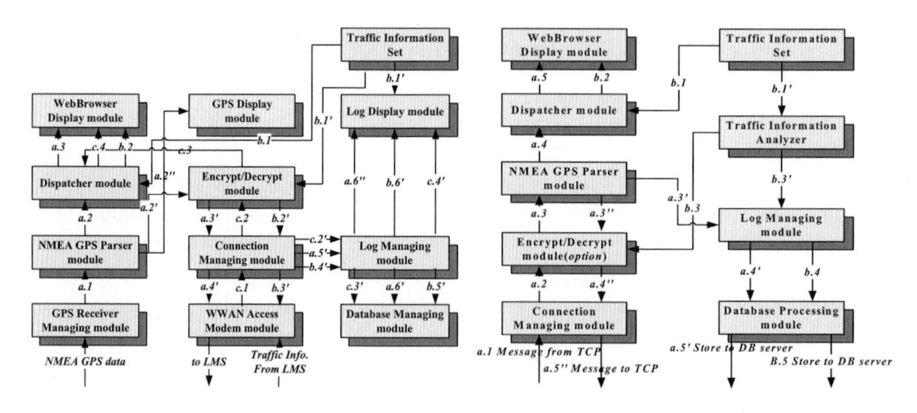

Fig. 8. Software architecture. Left and right figures respectively show the major modules both client and LMS, as well as data flows among major modules. Here, a step-by-step data flowing are designated by a.1, a.2 and a.3 etc., and different data sources are identified by different alphabet such as a.1, b.1 and c.1.

4 Test and Performance Evaluation

For performance measurement of ICSW^{2}AN, road test in real environment is performed for investigating major performance metrics such as round-trip-time, latency, standard deviation of latencies etc. as followings :

Performance Measurement of MAP. Performance of implemented MAP is experimentally measured by traffic quality measurement system which National Information Society Agency of Korea operates. This system provides a report of performance measurements which includes throughput, round-trip-time, latency, 95% latency, standard deviation and error ratio etc. Summarized results are shown in table 1. Comparing the result, performance of implemented MAP is nearly equivalent to that of commercial wireless WAN access modems.

Table 2. Summary of performance measurement of MAP (unit:Kbps[1)], ms[2)], %[3)])

Type of WWAN (Carrier)	Access Modem Model (Manufacturer)	Performance Metrics	NAT mode	Half-Bridge mode
WCDMA (Korea Telecom)	CWE624K (C-Mothech Ltd.)	Avr. DL Throughput[1)]	117.13	117.00
		Avr. UL Throughput[1)]	41.67	39.21
		Avr. Round Trip Time[1)]	215.81	242.32
		95% Delay[2)]	309.75	408.22
		Standard Deviation[2)]	42.69	79.18
		Loss Ratio[3)]	0.05	0.09
HSDPA (SK Telecom)	WM210 (M2MNET Ltd.)	Avr. DL Throughput[1)]	2300.78	2292.93
		Avr. UL Throughput[1)]	77.47	72.18
		Avr. Round Trip Time[1)]	118.23	132.22
		95% Delay[2)]	139.38	134.15
		Standard Deviation[2)]	15.15	23.76
		Loss Ratio[3)]	0.12	0.16

Performance Measurement of ICSW^{2}AN. Fig. 9 illustrates the network architecture and road test environment, and the leftmost picture shows equipments including client, MAP and GPS receiver in vehicle. Road tests are performed under two different environments of urban and rural areas over WCDMA, HSDPA and WiBro networks. In urban area, vehicles move at a speed of range from 0 Km/h to 60 Km/h around the Daegu, Korea. The Daegu city is densely populated city in the Korea, so, there are many pedestrian crossings, crossroads, buildings and another obstacles. Meanwhile, vehicles move at a speed of range from 0 Km/h to 130 Km/h on the expressway outside the Daegu boundary as a rural area. The measured performances are summarized in from Fig. 10 to Fig. 13. As vehicles move faster, as the RTT and latency are gradually increase in both rural and urban areas. On the other hands, LMS processing time is nearly constant to about 40 ms which is practically devoted to processing Database operation. Generally, measured values in the urban are superior to that on expressway in rural area in view of RTT and latency etc.

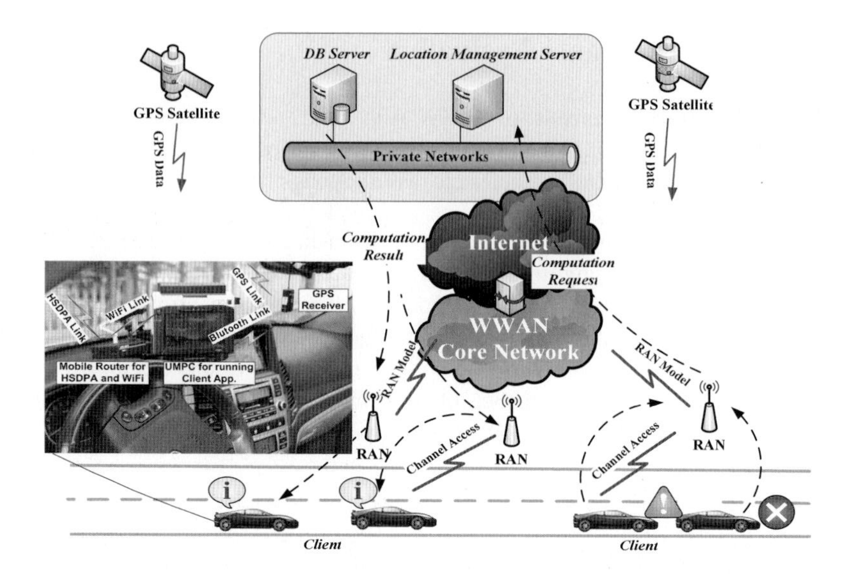

Fig. 9. Network architecture and road test environment

Table 3. Road test environment of ICSW^{2}AN

Item	Description
LMS and DB Server	Model : Desktop PC, OS : MS Windows XP professional, CPU : Intel Core2duo 2.66GHz, Memory : RAM 1GByte, Database :
Client	Model : Fujitsu U1010(UMPC), OS : MS Windows XP Home Edition, CPU : Intel A110 800MHz, Memory : RAM 1GByte, Bluetooth 2.0+EDR, 802.11g
WWAN Access Modem	Model : CBU-450D, WCDMA/HSDPA/WiBro supports
GPS Receiver	Model : GPS731 manufactured by Assem Ltd., MT3329 chipset-66channel, NMEA0183 v3.0support, Bluetooth v2.0 SPP, USB
Types of Wireless WAN	Tested with MAP over HSDPA networks. Tested without MAP over WCDMA, WiBro networks respectively
Road test time	Tested in urban area and on the expressway in rural area about 1 hour respectively

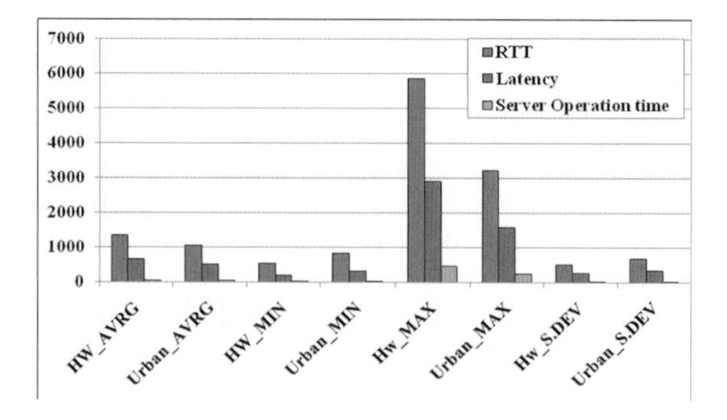

Fig. 10. Performance measurements over HSDPA network : the graph shows round-trip-time, latency between client and LMS, and LMS processing time in both highway(HW) and urban areas respectively

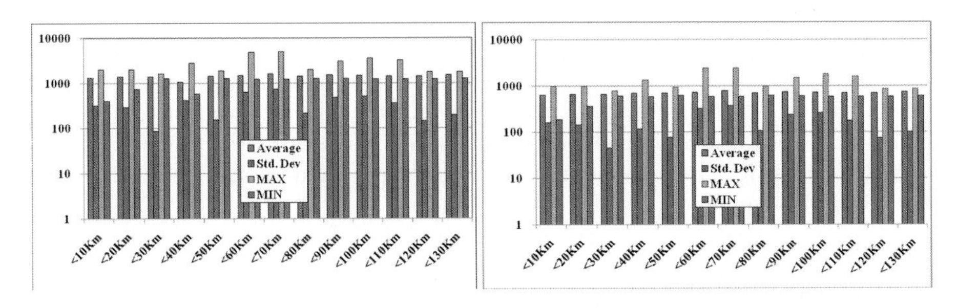

Fig. 11. Performance measurements on the expressway in rural area over WCDMA network (Y-axis : log scale)

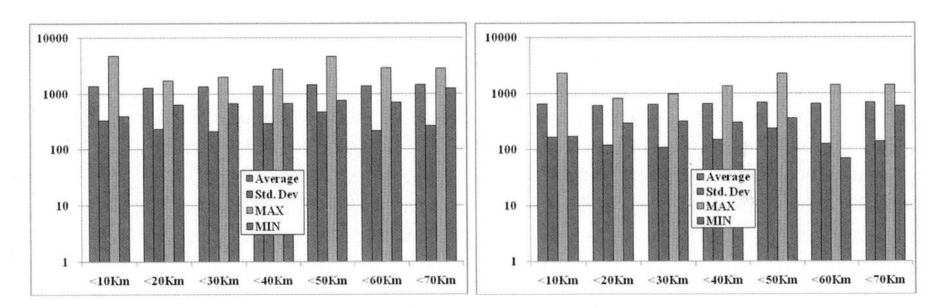

Fig. 12. Performance measurements in the urban area over WCDMA network (Y-axis : log scale)

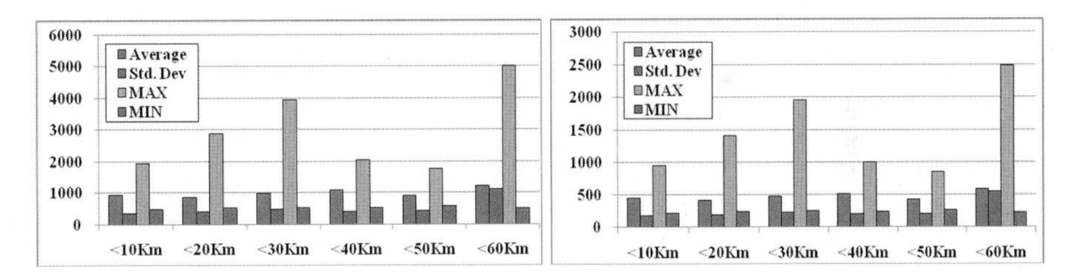

Fig. 13. Performance measurements in urban area over WiBro network : left graph shows round-trip-time between client and LMS, and right graph presents latency between client and LMS

5 Conclusion

In this paper, we presented experiments to help in the design and evaluation of WWAN-based vehicle communication systems. Such wireless WAN approaches might complement recent efforts to establish VANET-based traffic information system - basically because they are already widely deployed and provide capabilities such as inherent security measures and low latency communication independent of the current traffic density. The experimental performance evaluation is not limited by currently implemented WWAN infrastructure and thus able to use forthcoming technologies such as LTE and Advanced IMT-2000. We described in detail the each component implemented system were composed of and how these components interacted to form the system. The study demonstrated the availability of vehicle communication system over various WWANs. As a conclusion, vehicle communication system equipped with mobile access point sufficiently supports time-tolerant services such as time-tolerant traffic information and infotainment services in considering the experiment results of $ICSW^2AN$.

References

1. Sommer, C., Schmidt, A., Chen, Y., German, R., Koch, W., Dressler, F.: On the Feasibility of UMTS-based Traffic Information Systems. Ad Hoc Networks 8(5), 506–517 (2010)
2. Santa, J., Moragon, A., Gomez Skarmeta, A.F.: Experimental Evaluation of a Novel Vehicular Communication Paradigm based on Cellular Networks. In: IEEE Intelligent Vehicles Symposium, pp. 198–203 (2008)
3. Valerio, D., Ricciato, F., Belanovic, P., Zemen, T.: UMTS on the Road:Broadcasting Intelligent Road Safety Information via MBMS. In: 67th IEEE Vehicular Technology Conference, pp. 3026–3030 (2008)
4. Raya, M., Hubaux, J.P.: The Security of Vehicular Ad Hoc Networks. In: 3rd ACM Workshop on Security of ad hoc and Sensor Networks, Alexandria, VA, USA, pp. 11–21 (2005)
5. Gupta, P., Kumar, P.: The Capacity of Wireless Networks. IEEE Transactions on Information Theory 46(2), 388–404 (2000)
6. Zijderhand, F., Biesterbos, J.: Functions and applications of SOCRATES: a Dynamic In-car Navigation System with Cellular Radio based Bi-directional Communication Facility. In: Vehicle Navigation and Information Systems Conference, pp. 543–546 (1994)
7. Pinart, C., Sanz, P., Lequerica, I., Garcia, D., Barona, I., Sánchez-Aparisi, D.: DRIVE: a Reconfigurable Testbed for Advanced Vehicular Services and Communications. In: 4th International Conference on Testbeds and Research Infrastructures for the Development of Networks & Communities, Innsbruck, Austria (2008)
8. Collins, K., Muntean, G.-M.: A Vehicle Route Management Solution Enabled by Wireless Vehicular Networks. In: 27th IEEE Conference on Computer Communications, Poster Session, Phoenix, AZ (2008)

MAMODE: A Routing Scheme Using Mixed Address Mode for Wireless Sensor Networks

Jeongho Son[1] and Tae-Young Byun[2]

[1] Electronics and Telecommunications Research Institiute
2139 Daemyeong, Namgu, Daegu, Rep. Of Korea
[2] School of Computer and Information Communications Engineering,
Catholic University of Daegu, Gyeogsan-si, Gyeongbuk, Rep. Of Korea -
phdson@etri.re.kr, tybyun@cu.ac.kr

Abstract. As deployment of sensor networks are recently spread, a variety of services based on sensor network have been introduced. Nevertheless, prolonging of battery life becomes primary challenges to overcome for deploying sensor networks in various fields. Problems such as network bandwidth reduction, collision occurrence and performance deterioration due to broadcasting of message in large scale networks have become primary challenges. In this paper, we presented a new routing algorithm based on data-centric routing and address-based routing schemes, which is that query messages are delivered to specific area by using address-based routing scheme in unicast scheme, and then, broadcast scheme is used in specific area by inserting additional information of neighbor nodes to broadcast into the message payload. This method prevents severe broadcast storm caused by broadcast message, and also provides data reliability at reasonable level by utilizing address information in message. By computer simulation, our proposed scheme significantly reduces energy consumption caused by broadcasting of messages as well as it improves appropriate data reliability in wireless sensor networks.

Keywords: sensor networks, routing scheme, mixed address mode, broadcast.

1 Introduction

Sensor network consists of many sensor nodes which have capabilities of sensing, processing and communication. A specific sensor node is called *sink node* which collects interested data from sensor nodes.

As hardware technologies including MCU with low-power consumption, communication-related chip as well as MAC and routing protocols with low power consumption have been introduced, deployment of sensor networks in real world are increasingly promoted. Although services and applications deploying sensor networks greatly increase, there are still many challenges to develop it. A question of ‘How sensor networks can guarantee a reliable data delivery and stable operation & management for a long time’ has been the primary challenge in these s studies. Generally, sensor node has a limited power and sensor network has a difficulty of

T.-h. Kim, A. Stoica, and R.-S. Chang (Eds.): SUComS 2010, CCIS 78, pp. 367–375, 2010.

substitution of exhausted batteries across hazard zone in large area. It is very important to extend network's lifetime as well as sensor node's lifetime. Hardware technology to extend sensor node's lifetime is restricted because it is unlikely that performance of hardware will suddenly be improved. Thus, researchers have given attention to methods for extending sensor network's life time through software solutions recently.

In this paper, we present a simplified routing method that lessens traffic load caused by broadcast storm over sensor network, as a result, to reduce average power consumption of each sensor node. Our method prevents all nodes from participating in forwarding a query-message in flooding method. That is, a query message originated from a source is delivered to a designated node in a target area by unicast, then, designated node is responsible to broadcast the message to all nodes in target or interested area. Here, as the degree of branch out of node varies, the number of nodes of participating in flooding a message also varied. So, we investigate how many degree of branch out of a node in target area is sufficient to flood the message to all nodes in target area efficiently.

The rest of the paper is organized as follows. Section 2 reviews related works about sensor networks overall, and Section 3 explain our proposed scheme in detail. Simulations and results is presented in Section 4. Finally we conclude MAMODE scheme in Section 5.

2 Related Works

2.1 Wireless Sensor Network Standards

Several surveys [1][2][3][4][5], and [6] discussed various aspects on wireless sensor networks. In particular, this survey also deals with the increasing importance of the ZigBee/IEEE 802.15.4 standards, giving a review of these standards and comparing their solutions with the ideas emerged in the recent literature.

The ZigBee Alliance is an association of companies working together to develop standards (and products) for reliable, cost-effective, low-power wireless networking. ZigBee technology will probably be embedded in a wide range of products and applications across consumer, commercial, industrial and government markets worldwide. ZigBee builds upon the IEEE 802.15.4 standard which defines the physical and MAC layers for low cost, low rate personal area networks. ZigBee defines the network layer specifications for star, tree and peer-to-peer network topologies and provides a framework for application programming in the application layer.

2.2 Related Routing Scheme

In a WSN environment, where nodes can be deployed at random and in large quantities and the network topology may vary due to sensor failures or energy efficiency decisions, assigning and maintaining hierarchical structures is impractical. The message overhead to maintain the routing tables and the memory space required to store them is not affordable for the energy and resource constrained WSNs.

Reactive protocols such as AODV [7] and DSR [8] alleviate some of these problems (ZigBee actually uses an AODV-based protocol) but questionably scale to very large networks since they depend on flooding for route discovery. Furthermore, DSR requires the management of large route caches and large packet headers to store the path.

Routing protocols for WSNs should be lightweight in both processing power and memory footprint and should require minimal message overhead. Ideally they should be able to route packets based on information exchanged with its neighborhood and should be resilient to node failures and frequent topology changes. For these reasons most of the research on routing in sensor networks has focused on localized protocols which are tree-based or geography-based.

Routing Tree. Simple data gathering applications where readings collected by sensors are sent to the sink, possibly with some aggregation along the path, need trivial routing. As the query propagates through network, each node just remembers its parent toward the sink and later forwards it any messages it receives/originates. Directed Diffusion [9][10] is a variant that routes packets along the edges of a DAG rooted at the sink and allows for multipath data delivery. Routing trees are very easy to construct and maintain but this approach is not suitable for more complex applications that require end-to-end communication.

Geograph "greedy" routing. Geographic (or greedy) routing [11]naturally supports end-to-end communication. All nodes are assigned a location according to some flat (i.e., network-wide) coordinate system and a distance is defined for any two locations. Each node periodically broadcasts its location to neighbours. On the basis of the destination location (carried in each data packet) a node forwards the packet to the neighbor that minimizes remaining distance.

The first is a localization problem and consists in assigning a tuple of coordinates to each node. An obvious possibility is to use a physical (geographical) coordinate system with nodes equipped with GPS (or manually configured) or let nodes approximate their physical position from connectivity information with only a few GPS-equipped anchor nodes. An alternative to real coordinates is to run a protocol that assigns virtual coordinates to all nodes. Virtual coordinates are not bound to the physical position but only depend on relative position (i.e., node connectivity).

Hierarchical routing. Greedy routing is efficient in areas densely and regularly populated with nodes. It fails in the presence of voids or obstacles that introduce discontinuities in the topological connectivity structure. Recently developed alternatives to greedy routing consider taking a compact representation of the global sensor network topology structure and storing such representation at all nodes. The representation identifies and divides the network into a set of topologically regular regions. A local coordinate system is defined within each region and a greedy-like routing algorithm suffices to perform intra region packet forwarding. The role of the representation is to glue the regions together and drive long range routing across the network. Routing decisions within a given node consist of identifying an inter region path from the current node to the destination, and using local (greedy-like) routing to reach the next region in the path or the final destination (if it is in the current region).

One of the disadvantages of these approaches might lie in the complexity of deriving the high level topological structure of the whole network. Also the size of this representation must be small enough to be stored at each node, which precludes very articulated networks (e.g., sparse networks). Finally, local coordinate systems within regions tend to be a little more complex than integer tuples (as in flat greedy routing) and so are the corresponding greedy-like routing functions. Many studies have been performed to find proper solutions to various problems as mentioned above.

Especially, broadcast-based routing schemes [12][13] such as AODV, DSR and directed diffusion have a weakness of highly power assumption due to massive broadcast message which cause to deliver duplicated messages. Also, these duplicated messages reduce the efficient bandwidth over network, and frequent collisions of messages due to reduced bandwidth occur. As a consequence, a series of these events reduce the network lifetime overall.

Considering above problems, our contribution is find a proper routing method that as well as maintains reasonable efficient bandwidth.

3 Proposed Mixed Address Mode Scheme

We defined characteristics, problems and consumption energy for transceiver in large scale wireless sensor networks in section 2. Especially, we know that the broadcasting storm is a severe problem for networks life time.

In this section, we present a new mixed address mode routing scheme named MAMODE, which mixed an address based routing mode and data centric routing mode.

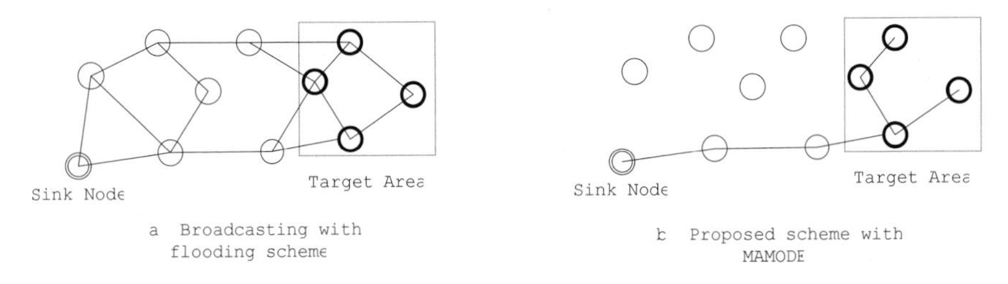

Fig. 1. Concept of mixed address mode

Fig. 1(a) is an example of full broadcasting. When each node received new message, the node sends the message using flooding scheme. At that time, the energy to receive is about:

$$E(total) = E(recv) * N * Avg(neighbors) \tag{1}$$

As equation (1), the total energy consumed for one broadcast is big so that it can rapidly reduce network's life time. Fig. 1(b) is our proposed scheme to limit the range of broadcast with a target region information that is included in the message. In MAMODE, it uses an address based routing scheme to deliver to specific region (especially target region). In the target region, it modifies the message into a broadcast message and inserts specific number of neighbors as in-message addresses. And then, the node broadcasts the message into the target area. The frame structure is shown in Fig. 2. The frame consists of MAC Frame Header (MHR), Network Frame Header (NHR), Payload, MAC Footer (MFR). The payload field contains a query message like Fig. 2.

MHR	NHR	Payload	MFR

Fig. 2. Frame structure of Zigbee

"Send me temperature, humidity information in region A in every 1 minutes"

The query message disseminates in Region A and sensor nodes in Region A send information to the sink node that generates original query message. Proposed MAMODE uses IEEE 802.15.4 and Zigbee protocol frames and includes a query message such as query message in directed diffusion. Query messages could be varied within frame length.

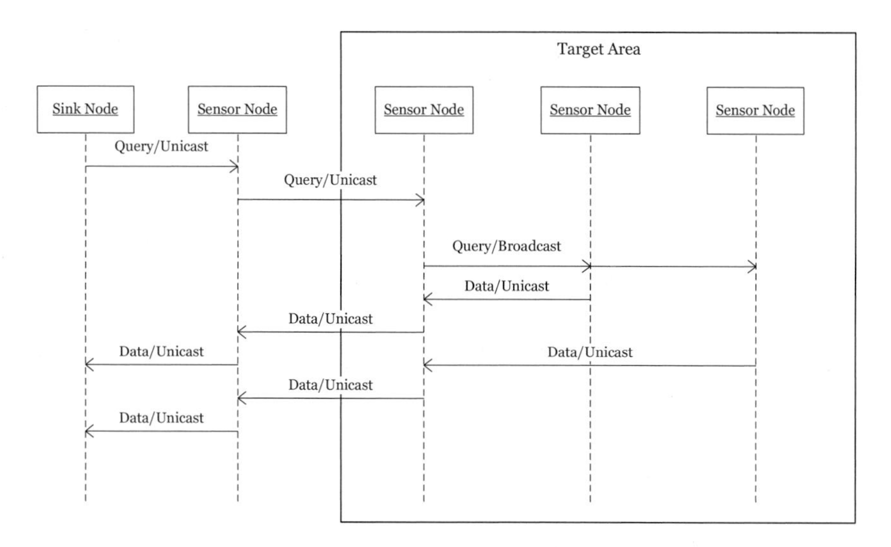

Fig. 3. Sequence diagram of processing in MAMODE

Fig.3 has shown a sequence diagram to process a MAMODE message in each node. Sink node initializes a query message and assembles a frame in address based mode. The query includes target area, interesting, and interval and inserted and it is encapsulated in Zigbee frame and IEEE 802.15.4 MAC frame. When it is forwarded out of target area, the address of MAC is filled out with a unicast address and routed using a routing table to deliver the message to a next hop sensor node. When it comes into target area, the received sensor node can disassemble the message and the node knows that it is in target area. Immediately, it makes the message into a broadcast message and selects relay nodes as much as the degree of broadcasting. The degree is selected as network density or desired coverage for broadcasting or to control the probability of collision occurrences. And then the selected node addresses are inserted into the payload field. When selected nodes receipt the message, these nodes repeat the previous task. Our basic selection scheme excludes the node itself, the node sent the message and neighbor nodes out of target area.

Although all nodes in target area cannot receive the message because the number of nodes is limited and collision can be occurred, these constraints can be in other protocol schemes using pure IEEE 802.15.4 MAC protocol. We have a positive point of view that it will reduce the collision occurrences and gather information appropriately from selected nodes in the target area. For example, if we want to know temperature in region A where it has 100 nodes, we can determine the temperature from only 20 nodes. The algorithm in this network is explaining the process when a node(N) has receipt a message(M) and N is not the destination node.

```
If (M is delivered in unicast mode)
   If (N is in the target area)
      Broadcast M with 2 neighbor nodes
   Else
      Deliver M to next node
   End if
Else (M is delivered in Broadcast mode)
   If (N is in the target area && N is in M's payload)
      Broadcast M with 2 selected neighbor nodes
   Else
      Drop M
   End if
End if
```

As we mentioned that broadcast storm affects to network life time. We need to reduce the broadcast messages so that we proposed a scheme to reduce the number of broadcast message transmissions with a mixed mode address scheme and selecting next broadcast nodes in sender based selection scheme. The degree of broadcasting that is the number of broadcasting candidates. The degree can be increased or decreased by network density. Hence, it limits the number of broadcasting messages with target area information using mixed address mode. In addition, our proposed scheme can make reverse route of a messages and can use in-network processing (fusion, compression etc.) to reduce a message transmission.

4 Simulation and Results

To evaluate our proposed scheme explained section 3 we use CSIM19 that is a discrete environment simulation tool. The network parameters are listed in Table 1. Basically, we use IEEE 802.15.4 MAC protocol parameters with recommended value in the standard specification.

Table 1. Network parameters for simulation scenarios

Metric	Value	Note
minBE	**3**	Minimum number of binary backoffs
maxBe	**5**	Maximum number of binary backoffs
Bandwidth	**250kb/s**	62.5ksymbols/s
CSMA backoffs	**4**	The number of retries

We randomly deployed fixed 150 nodes including 1 sink node in network size is 1000m * 1000m field. Transmission range of each node is 100m. And, we assume that each node knows its location information and neighbors. The neighbor information is collected by 'hello' message. The total message length to deliver is 100 bytes that include MAC header, MAC footer, Network header, payload, and node IDs of broadcast candidates. Each node operates in non-beacon mode with binary backoff algorithm so that it starts backoff algorithm and sends if channel is idle.

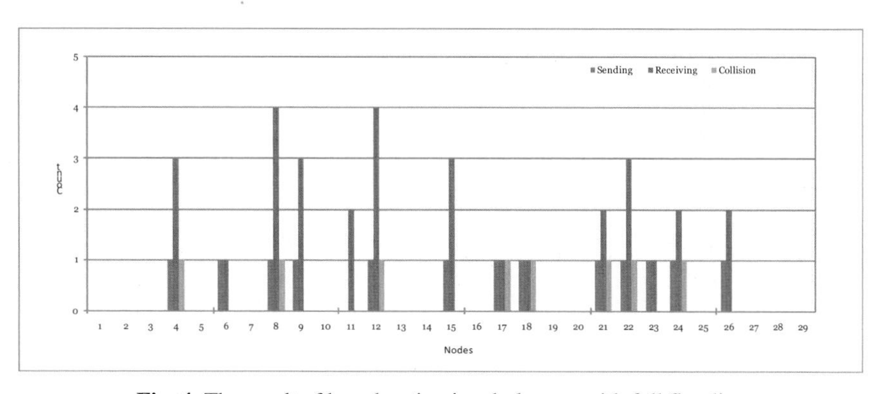

Fig. 4. The result of broadcasting in whole area with full flooding

We evaluate the number of sending, receiving, and colliding of broadcasting messages in whole area. We however present 29 nodes in the target area in Fig. 4. Only 13 nodes are received the query message. In the result of Fig. 4, we can find that collisions occurred and are interfere delivering a messages. Consequently, all nodes cannot receive the message.

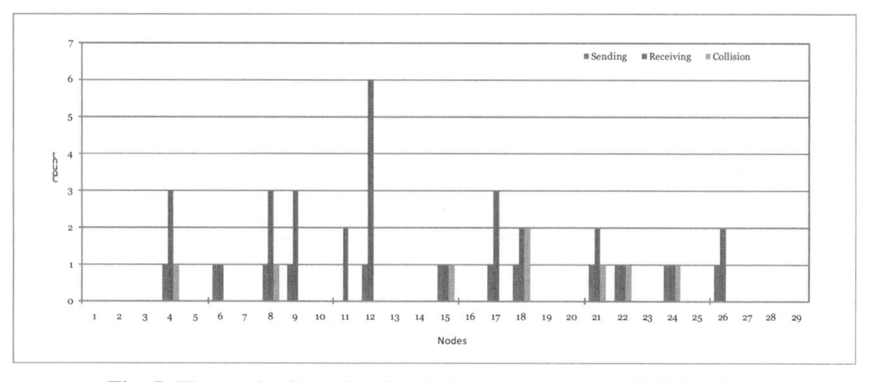

Fig. 5. The result of broadcasting in the target area with (full flooding)

In next scenario, we use MAMODE. We however use full flooding scheme in target area. As shown in Fig. 5, the number of received nodes is 12 that is 1 less than of previous scenario. It is a very valuable result that it doesn't need to broadcast into a whole network.

Table 2. Comparison of the number of transmissions, receptions, and collisions

Heading level	Whole area flooding	Target area flooding
The number of transmissions	**104**	**19**
The number of receptions	**232**	**41**
The number of collisions	**96**	**8**

In Table 2, our proposed MAMODE scheme with full flooding in target area can reduce significantly the number of transmissions, receptions, and collisions.

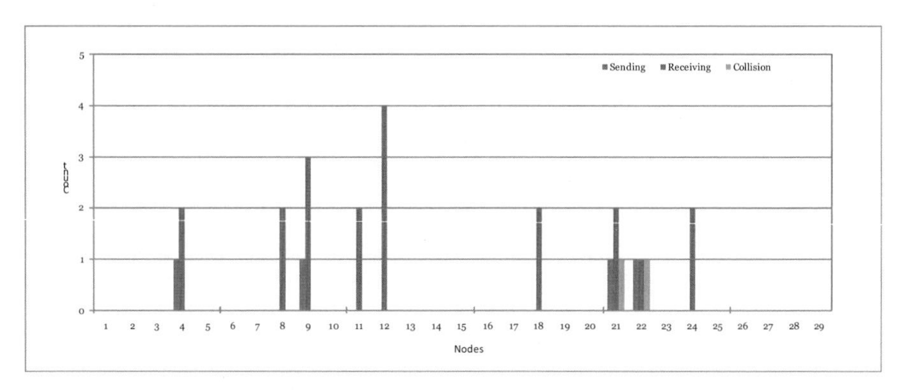

Fig. 6. The result of 3 nodes broadcasting in target area

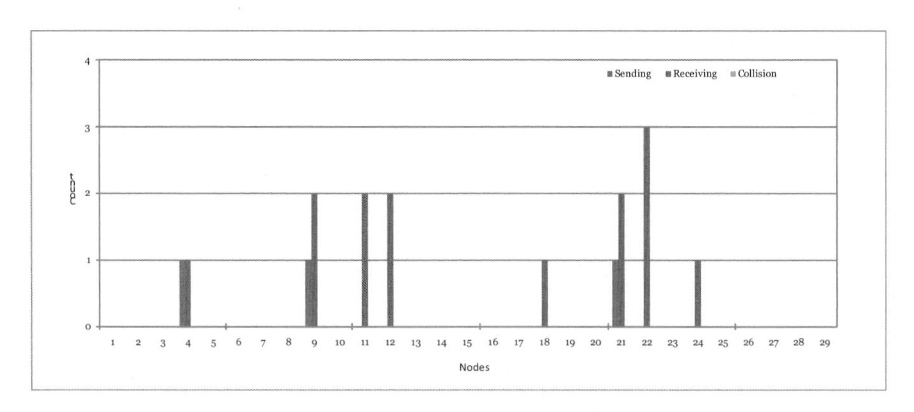

Fig. 7. The result of 2 nodes broadcasting in target area

Finally, our proposed MAMODE and limited broadcasting scheme is evaluated. In the results in Fig. 6 and Fig. 7, the number of receipt nodes is not much different and Fig. 7 shows only 8 nodes received the query message and collision doesn't occur.

According to results, we can reduce the number of transmissions and collision by reduced broadcasting with degree so that our proposed scheme can prolong the network life time. In addition, our proposed scheme can applied large scale sensor networks with varied broadcasting degrees as network density.

5 Conclusions

We proposed a routing protocol using mixed address mode (address based and data centric routing scheme) to reduce the number of broadcasts. It can prolong network life time and applicable precision value from target area. As changing the broadcast degree, our proposed scheme can be used in large scale networks. In the future works, we will evaluate the effectiveness of broadcast node degree and algorithms which broadcast candidate node is more applicable.

References

1. Gajbhiye, P., Mahajan, A.: A Survey of Architecture and Node deployment in Wireless Sensor Network. In: ICADIWT 2008, pp. 426–430. IEEE, Los Alamitos (2008)
2. Arampatzis, T., Lygeros, J., Maesis, S.: A Survey of Applications of Wireless Sensors and Wireless Sensor Networks. In: Intelligent Control 2005, pp. 719–724. IEEE, Los Alamitos (2005)
3. Fasol, E., Rossi, M., Widmer, J., Zorzi, M.: In-Network Aggregation Techniques For Wireless Sensor Networks: A Survey. Wireless Communications 14, 70–87 (2007)
4. Akyildiz, I.F., Su, W., Sankarasubramaniam, Y., Cayirci, E.: Wireless Sensor Networks: A Survey. Communications Magazine 40, 102–114 (2002)
5. Vieira, M.A.M., Coelho Jr., C.N., da silva Jr., D.C., da Mata, J.M.: Survey on Wireless Sensor Network Devices. In: ETFA 2003, vol. 1, pp. 537–544. IEEE, Los Alamitos (2003)
6. Ahmed, A.A., Shi, H., Shang, Y.: A Survey on Network Protocols For Wireless Sensor Networks. In: ITRE 2003, pp. 301–305. IEEE, Los Alamitos (2003)
7. Perkins, C.E., Royer, E.M.: Ad-hoc on-demand distance vector routing. In: Proceedings of the 2nd IEEE Workshop on Mobile Computing Systems and Applications (WMCSA 1999), New Orleans, LA, USA, pp. 90–100 (1999)
8. Johnson, D.B., Maltz, D.A.: Dynamic source routing in ad hoc wireless networks. Mobile Computing 353, 153–181 (1996)
9. Intanagonwiwat, C., Govindan, R., Estrin, D., Heidemann, J., Silva, F.: Directed Diffusion for Wireless Sensor Networking. IEEE/ACM Transactions Networking 11, 2–16 (2003)
10. Intanagonwiwat, C., Govindan, R., Estrin, D.: Directed diffusion: a scalable and robust communication paradigm for sensor networks. In: Proceedings of the 6th International Conference on Mobile Computing and Networking (MobiCom 2000), Boston, MA, USA, pp. 56–67 (2000)
11. Sanchez, J.A., Ruiz, P.M., Liu, J., Stojmenovic, I.: Bandwidth-Efficient Geographic Multicast Routing Protocol for Wireless Sensor Networks. Sensors Journal 7, 627–636 (2007)
12. Sabbineni, H., Chakrabarty, K.: Location-Aided Flooding: An Energy-Efficient Data Dissemination Protocol for Wireless Sensor Networks. IEEE Transactions Computers 54, 36–46 (2005)
13. Subramanian, S., Shakkottai, S., Araphstathis, A.: Broadcasting in Sensor Networks: The Role of Local Information. IEEE/ACM Transaction Networking 16, 1133–1146 (2008)

Processing of Large-Scale Nano-ink Data by Supercomputer

Sungsuk Kim[1] and Joon-Min Gil[2,⋆]

[1] Department of Computer Science and Engineering, Seokyeong University
16-1 ga, Jungnueng-dong, Sungbuk-gu, Seoul 136-701, Korea
sskim03@skuniv.ac.kr

[2] School of Computer and Information Communications Engineering,
Catholic University of Daegu
330 Geumnak, Hayang-eup, Gyeongsan-si, Gyeongbuk 712-702, Korea
jmgil@cu.ac.kr

Abstract. Recently, the simulation techniques using computers have been actively utilized to examine the effects of social or physical interactions. Those works often generate very large amounts of data, and to deal with them, it may be necessary to use high-performance supercomputers as well as to develop efficient algorithms. This paper presents the results of the simulation works to deal with the data generated in developing a printable board using nano-size, conductible inks. Our works are conducted with a supercomputer, based on parallel programming using message passing interface (MPI). The simulation results obtained from our works can be utilized to verify the findings obtained from the real physical experiments.

Keywords: Large-scale Data, Parallel Processing, Nano-ink, Supercomputer, Message Passing Interface.

1 Introduction

Recently, the amount of the data arising from various application areas has become so enormous that it is hardly possible to obtain the desired results within a given time period by the existing methods. In such applications, a high-performance computing such as supercomputer will be at least as efficient as the development of efficient algorithms.

In this paper, we implement a simulator for developing a nano-ink that is composed of conductive nano-sized particles. These nano-sized particles are defined as the components that have the dimensions of tens to hundreds of nanometers ($1nm = 10^{-9}m$). The results obtained from our simulation can be utilized to verify the findings obtained from the development of the trial product where a nano-ink is actually used [1].

Our simulator is designed for a 2D plane of 500×500 (which will be easily expanded to a 3D object of $500 \times 500 \times 100$ later on), and the size of the nano-sized

⋆ Corresponding author.

T.-h. Kim, A. Stoica, and R.-S. Chang (Eds.): SUComS 2010, CCIS 78, pp. 376–383, 2010.

particles is set to the values ranging from 0.01 (min) to 10 (max). In our simulator, diverse nano-sized particles are produced with various simulation factors and dropped on the 2D plane. After then, the density of the particles dropped on the whole board is calculated. Since the theoretical numbers of nano-sized particles are 2.5×10^9, our simulation will take a long time even if only simple simulation factors are considered. In our initial simulation through personal computers (PCs) as a computing tool, more than a week ware elapsed until the desired results are obtained. Thus, to obtain the results within a given time period, it is beneficial that the simulation is performed on a supercomputer with multiple processors. For this, our simulator is modified to suit the parallel processing of the supercomputer. In the actual simulation run on a supercomputer, we obtain the desired results within an hour under the same configuration as before.

The rest of this paper is organized as follows. In Section 2, we briefly describe background for the nano-ink and supercomputer that are the target and tool of our simulation, respectively. Section 3 presents the parallel algorithm to process large-scale nano-ink data by means of supercomputers. Simulation results are also given in this section. Section 4 concludes the paper.

2 Background

2.1 Nano-ink

Nano-ink is an ink-type electronic material made by spreading conductive nano-sized particles. In designing semiconductors' circuit wire, shaving methods have been commonly used. However, the nano-ink is capable of designing a more sophisticated circuit by actually "printing" the wires. Because of this, nano-ink technologies have been applied to a variety of fields, such as barcode, bio chip, anti-forgery money, etc.

Nano-ink technology can be largely divided into ink-jet and direct printing according to the way it is printed on the board. The ink-jet method has some flaws. Thus, recent research efforts have been focused on the direct printing one. The MDDW of nScrypt [2] and the M3D of Optomec [3] are representative technologies for the direct printing method. These technologies not only enhance the nano-ink's generality and compatibility, but also enable the design of a more sophisticated circuit.

Conductive ink is a key component in designing a nano-ink by the direct printing method and can be approached by various methods. Most researches have focused on manufacturing a metal particle by wet process and used this as ink [4]. The details of this process are beyond the scope of this paper and thus omitted.

As mentioned above, most researches are aimed at developing the method to enhance the conductivity of nano-inks. Fig. 1 shows the operation of direct printing method, in which nano-inks are printed on the simulation board from the head. In general, as packing density per unit area increases, the conductivity of nano-inks becomes higher. Thus, in this paper, we perform the simulation to mix nano-inks with various sizes and print them on the simulation board.

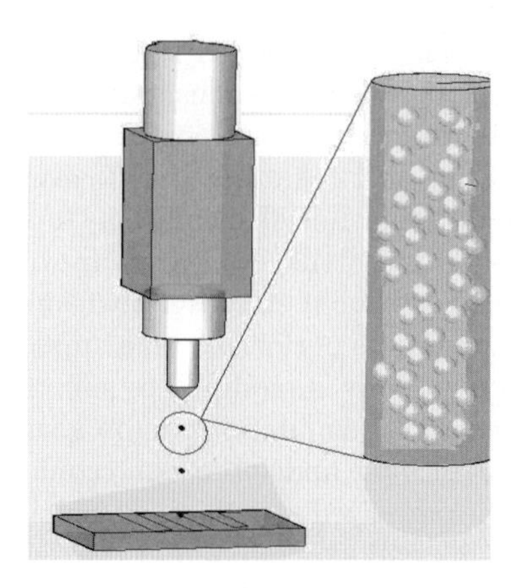

Fig. 1. Operation of direct printing method

2.2 Supercomputer

TACHYON (Sun Blade 6048) system [5], one of the supercomputers possessed by KISTI (Korea Institute of Science and Technology Information), is used to run our simulation. Table 1 summarizes its specification. TACHYON system is composed of many nodes that have separate memory spaces and are interconnected by high-bandwidth networks. In our simulation, we use MPI-2 (Message Passing Interface-2) library based on C language [6] for parallel processing in TACHYON system.

Table 1. TACHYON system and its specification

Processor	AMD Opteron 2.0GHz 16 nodes 3,008 (total)
Memory	32GB per node
External Storage Space	207TB (Disk), 422TB (Tape)
Inter-node Network	infiniband 4×DDR

3 Large-Scale Nano-ink Data Processing

3.1 Simulation Outline and Parallel Algorithm

The goal of the simulation is to continuously drop as many nano-sized particles as possible on the 2D plane and calculate theirs density. In the simulation, we assume that if a certain position in the 2D plane is already occupied by another nano-sized particle, a new nano-sized particle should be dropped at an empty position; *i.e.*, nano-sized particles should not be overlapped. When the new nano-sized particle is dropped at the position overlapping with other particles, it is again dropped, until it is dropped at an empty position or as many as a

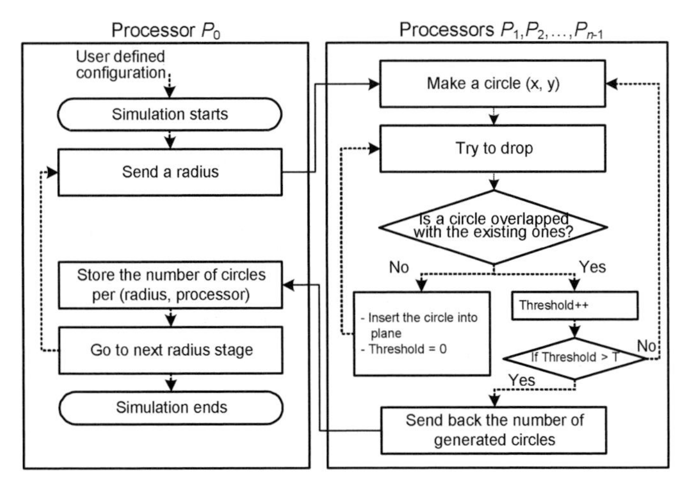

(a) Main flow for our simulation

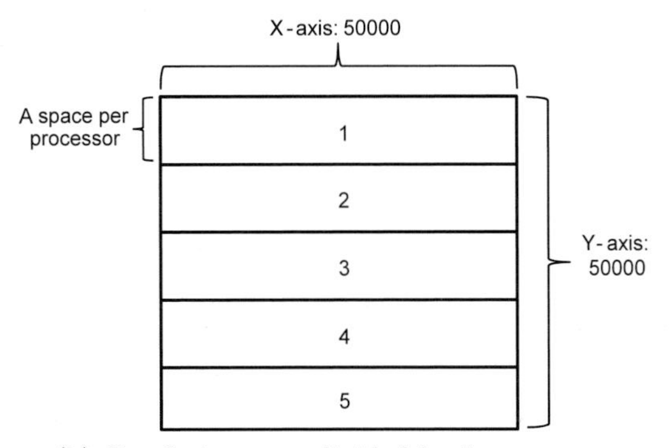

(b) Simulation area divided by 5 processors

Fig. 2. Simulation outline

predefined threshold. If the threshold becomes large, simulation time will extend in the proportion of it. The reason why this simulation is heavily affected by the amount of data is that the number of already generated nano-sized particles greatly increase as the simulation proceeds and so the numbers of comparison between existing and new nano-sized particles become more.

Fig. 2(a) shows the parallel algorithm to perform the simulation using n processors. In the figure, P_0 processor plays a role of a coordinator (the left node in this figure). The P_0 initially receives a user configuration with the simulation factors and transmits the information of particles to peer processors ($P_1 \sim P_{n-1}$). It also collects and corrects the results received from peer processors. The whole board is divided into $n-1$ parts, each of which is processed by a peer processor. Fig. 2(b) shows the board divided into 5 parts when $n = 6$. In our simulation, the whole board is set to 500×500, but the board in this figure is $50,000 \times 50,000$ because the minimum size of nano-sized particles is assumed to be 0.01.

In the actual simulation, nano-sized particles with 3 different sizes are dropped in decreasing order of the size. Nano-sized particles with the largest size are first dropped on the board, which is repeated until it continuously fails as many times as the threshold. The next smaller-sized particles are then generated and dropped.

P_0 sends the radius of the largest particle (user-defined) to the other processors, $P_1 \sim P_{n-1}$. Each of these processors then generates a point within its area and conducts simulation. $n - 1$ processors perform this processing in parallel. At this time, the generated point can be overlapped with the area controlled by another processor. Fig. 3 shows the case that a point is laid over two areas. In this figure, the particle generated by processor P_3 is overlapped with the area controlled by P_2 processor. In this case, P_3 should compare the particle not only with those printed in its own area but also with those in the area of P_2. The particles corresponding to this case are temporarily saved and then *corrected* by the following correction process.

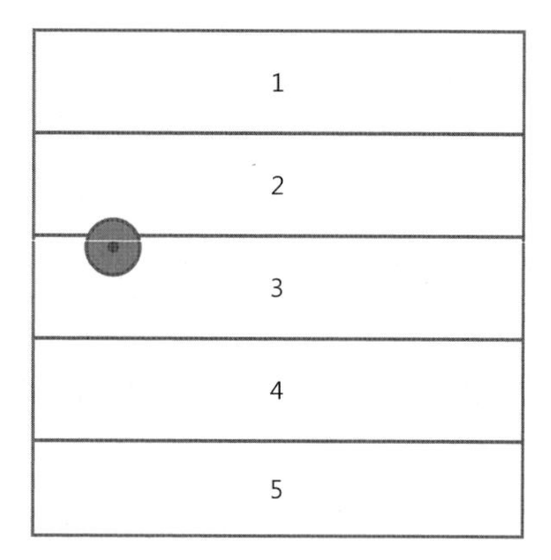

Fig. 3. The case that a particle is dropped at the boundary of two areas

In correction process, $P_1 \sim P_{n-1}$ processors send *primary task completion message* to P_0 upon completing comparison task. This message contains the particle information that each processor has to send to its neighboring processors. P_0 collects this information and sends it to each processor. The processors received this information from P_0 determine whether the points generated by neighboring processors are overlapped with the ones in its own area. Afterwards, the processors only send the information of "normal" particles to P_0. Finally, P_0 sends only such information to the other processors when sending the radius of the next particle.

3.2 Simulation Results

The amount of data generated in the simulation is heavily affected by the user-defined particle radius and threshold. The small radius causes the probability

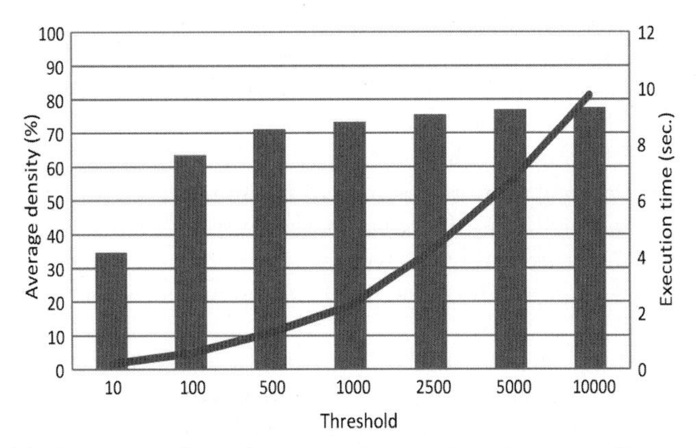

(a) Average packing density and execution time according to varying the numbers of processors

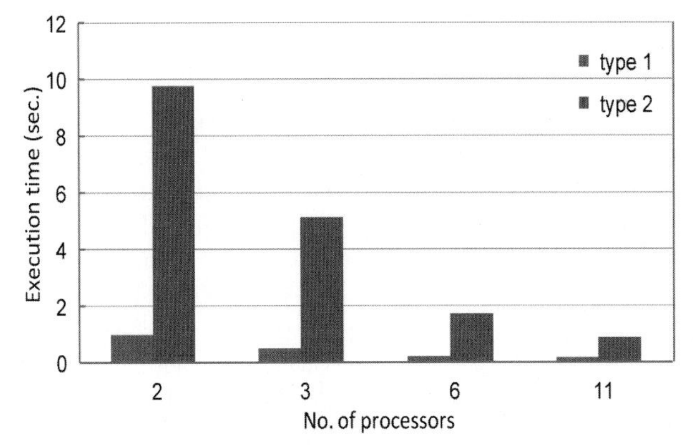

(b) Execution time according to varying the numbers of processors

Fig. 4. Simulation results

of overlapping with other particles to be lower. Meanwhile, the large threshold means that, even if the generated particle is overlapped with the existing ones, retries for dropping a particle at empty areas attempt much more so that as many particles as possible are printed on the whole board. The simulation can give us a lot of meaningful results under various configurations, but for simulation simplicity, only three kinds of particle radii $(0.025, 0.25, 1.0)$ are used in our simulation.

As aforementioned, our simulator is implemented with MPI-2 library for parallel processing. Simulation time will be greatly affected by the number of processors used. Thus, we have conducted the simulation to demonstrate the effect of the number of threshold and processor on execution time.

Fig. 4 shows simulation results obtained according to varying the number of thresholds and processors. Fig. 4(a) shows average packing density and execution time for the diverse numbers of thresholds when 6 processors are used. In our simulation, the particles are assumed to be circles, so even though they are densely packed on the whole board, there still exist empty spaces due to gaps between two neighboring circles. The average packing density is defined as the area of particles dropped on the board divided by total areas. We can see that the large threshold results in the increase of the number of the particles dropped on the board, leading to the increase of the average packing density as well. However, after 2,500 threshold, the average packing density converges to 76-77% which is the ideal value required for designing a conductive board with a metal nano-ink.

However, the large threshold means that the simulation is continuously proceeded even though the particle is overlapped with the existing ones. This causes the execution time to be longer. When the threshold is 10, the resulting execution time is about 0.07 sec. However, when the threshold exceeds 1,000, we can observe that the execution time abruptly increases. The execution time is 9.75 sec. in average when the threshold is 10,000.

In the next simulation, we have divided the whole board into equal parts and examined the effect of the number of processors on execution time. In this simulation, we use a value of 1000 as threshold. Since P_0 is used as a coordinator, only 10 processors of total 11 ones are actually used for the simulation. Fig. 4(b) shows the execution time according to varying the numbers of processors. In this figure, type 1 indicates the execution time when only one kind of radius for particles (*i.e.*, 1.0) is used; type 2 indicates the execution time when three kinds of radii for particles are used. Of course, if the various kinds of radii are used, the probability that the generated circles are not overlapped with the others already dropped on the board will be lower. This causes the execution time to be longer. Although it is expected that the use of many processors can be shorten the execution time, message exchanges between P_0 and each processor are relatively more. We can observe from this figure that when 2 processors are used, the execution time for type 2 is 9.6 sec. (max); 0.08 sec. when 11 processors are used. We can also observe that the execution time of type 1 is similar to that of type 2. It is apparently that many processors can lead to the reduction of the execution time.

4 Conclusions

In this paper, we implemented a simulator to process the large-scale nano-ink data when designing conductive printable board and presented the simulation results obtained by performing the simulator on a supercomputer. We expect that these simulation results will be used as the optimized standard in manufacturing the nano-ink later on.

The implemented simulator was limited to 2D board, but we plan to expand this to 3D simulations as well. Also, more detailed simulations such as conductivity test are needed as future works.

Acknowledgment

This work was supported by grant No. S1068259 from the Small and Medium Business Administration, Korea.

References

1. Maggs, B.M., Matheson, L.R., Tarjan, R.E.: Models of parallel computation: a survey and synthesis. In: Proc. of the 28th Hawaii Int. Conf. on System Science, pp. 61–70 (1995)
2. nScrypt Inc., http://www.nscryptinc.com
3. Optomec Inc., http://www.optomec.com
4. Jinag, H., Zhu, L., Moon, K.S., Wong, C.P.: Low temperature carbon nanotube film transfer via conductive polymer composites. Nanotechnology 18(12), Art. no. 125203 (2007)
5. TACHYON system, http://www.ksc.re.kr/eng/resources/resources1.htm
6. Gross, W., Huss-Lederman, S., Lumsdaine, A., Lusk, E., Nitzberg, B., Saphir, W., Snir, M.: MPI–The Complete Reference. The MPI-2 Extensions, vol. 2. MIT Press, Cambridge (1998)

Efficient Resource Management and Task Migration in Mobile Grid Environments*

DaeWon Lee[1], SungHo Chin[2], and JoonMin Gil[3,**]

[1] Division of General Education, SeoKyeong University, Korea
[2] Dept. of Computer Science Education, Korea University, Korea
[3] School of Computer and Information Communications Engineering, Catholic University of Daegu, Korea
daelee@skuniv.ac.kr, wingtop@comedu.korea.ac.kr, jmgil@cu.ac.kr

Abstract. Grid computing using wireless networks is receiving increasing attention and is expected to become a critical part of future grid computing. However, the inherent challenges of mobile environments such as mobility management, disconnected operation, device heterogeneity, service discovery, and resource sharing are significant issues in mobile grid computing. To achieve the best performance in a mobile grid computing environment, the mobile devices with the lowest probability of mobility should be selected for use first. We therefore focus on the idle state of mobile devices and use IP-paging scheme to identify the idle mobile devices. For this, we propose a user-defined checkpoint technique for task migration, which is based on the information of when a mobile device will leave the network or stop due to low battery. Our checkpoint technique performs checkpoints by two conditions: when a mobile device leaves its current cell and when it turns off due to low battery capacity.

Keywords: Mobile Grid, IP-Paging, Task Migration, Checkpoints.

1 Introduction

Grid computing is distinguished from conventional distributed computing by its focus on large-scale resource sharing, innovative applications, and high-performance orientation [1]. In the first stages of grid computing, most research has focused on fixed networks [1-5]. Because of improved Internet techniques, grid computing using wireless networks is now the subject of growing attention. It is expected to become a critical part of future grid computing involving mobile hosts to facilitate user access to grid networks to offer extended computing resources. Previous methods of resource selection in grid computing were suitable for resources based on the user requirement. However, because of the mobility in wireless networks, this resource selection

* This work was supported by the Korea Research Foundation Grant funded by the Korean Government (KRF-2008-331-D00447).
** Corresponding author.

T.-h. Kim, A. Stoica, and R.-S. Chang (Eds.): SUComS 2010, CCIS 78, pp. 384–393, 2010.

method will not result in the best performance for mobile grid computing. In a mobile grid computing environment, the mobile device that has the lowest probability of mobility should be selected as a resource first to maximize grid performance. We therefore focus on idle mobile devices. To achieve this purpose, this paper uses an IP-paging scheme [6-8] to find idle mobile devices and to gather information about them using an extended paging message.

The inherent challenges of mobile environments such as mobility management, disconnected operation, device heterogeneity, service discovery, and resource sharing are significant issues for grid computing. To overcome these challenges. we propose a task migration scheme using checkpoints for mobile grid computing [9-12]. This task migration scheme performs better than a re-execution scheme, although it does incur additional cost to perform a checkpoint procedure. For continuous grid services, it is useful to predict when a device will leave the network or stop due to low battery. In this paper, we use a user-defined checkpoint technique for task migration. This technique applies in two situations: when a mobile device leaves its current cell (determined by an analysis of its signaling strength) and when a mobile device turns off due to lower battery capacity.

The rest of this paper is organized as follows. Section 2 presents related works on mobile grid computing and the IP-paging scheme. Section 3 describes the proposed system architecture. Section 4 gives details of resource management for mobile devices. In Section 5, the task migration using the user-defined checkpoints is presented. Section 6 presents the performance evaluation of the proposed paging mobile grid system. Finally, we conclude the paper with future works in Section 7.

2 Related Work

2.1 Mobile Grid Computing

Two architectures exist for mobile grid computing: proxy-based and agent-based one. Proxy-based mobile grid architectures, designed to use mobile devices as resources for computational grid, have been proposed in [9-12]. Figure 1 shows a proxy-based mobile grid architecture. This architecture includes a cluster of mobile devices. The cluster is connected to a base station that acts as a router/node or a grid proxy server on the grid. The base station has two functions. It works as a wireless access point (AP) and as a proxy server that is responsible for data transfer, QoS, and resource access policy. In the proxy-based mobile grid architecture, mobile devices use a base station to access the grid, and base stations as proxy-servers allocate the tasks to mobile devices [9-14]. On the contrary, the agent-based mobile grid architecture [11, 12] uses mobile agents to provide, share, and access resources in grid networks.

Both architectures mainly focus on utilizing wireless networks for grid computing without considering how to manage mobile devices. A successful mobile grid architecture requires dealing with mobility management, disconnected operation, device heterogeneity, service discovery, and resource sharing.

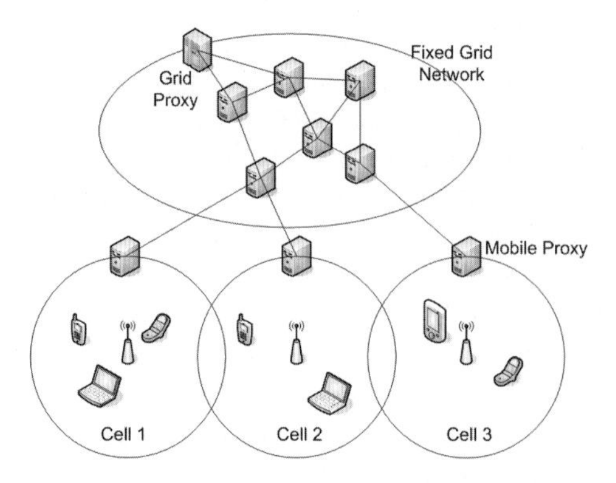

Fig. 1. Proxy-based mobile grid architecture

2.2 IP-Paging

Registration and paging techniques are important functions in cellular networks to minimize the signaling overhead and enhance the efficiency of mobility management. However, early mobile IP supports only registration function. Thus, mobile IP users do not actively communicate most of the time; *i.e.*, they are often in an idle state. IP-paging is one way to provide a more scalable and efficient location tracking scheme [6, 7]. It is a procedure that allows a wireless system to search an idle mobile device when a message is destined for it so that the mobile user does not need to register the precise location with the system whenever it moving. The IP-paging has two major benefits: reduction of signaling overhead and reduction of power consumption [6-8]. The reduction of power consumption is an important issue because grid tasks are executed on the mobile devices with limited battery capacity.

2.3 Checkpoint Technique in Grid Computing

Establishing an efficient task migration is very important in mobile grid environments. If one mobile device fails while executing a task, the task migration transfers the task to another mobile device and the task can be re-executed from the position of checkpoint. However, the task migration results in the increase of communication costs to save the state of the executing task on a stable storage. So, the task migration scheme does not guarantee better performance than the re-execution scheme [15-17]. The checkpoint technique in grid computing is divided into two types: system level and user-defined. The system level checkpoint is not suitable for heterogeneous resources. Meanwhile, two kinds of checkpoints for the user-defined checkpoint can be applied to the heterogeneous resources: coordinated checkpoint and independent checkpoint. However, coordinated checkpoint is not suitable for the mobile grid environment because the coordination process entails additional communication costs.

3 System Architecture

Mobile grid computing is based on a wireless network structure, in which each mobile device communicates with the destination through a base station. Wireless grid network is connected to a fixed grid network through a gateway router. A proxy server located on the gateway router manages send/receive tasks. In this paper, we propose a wireless grid computing architecture based on IP-paging to enable the use of idle mobile devices as grid resources and address the problem of limited battery capacity. Figure 2 shows the wireless grid computing architecture based on IP-paging.

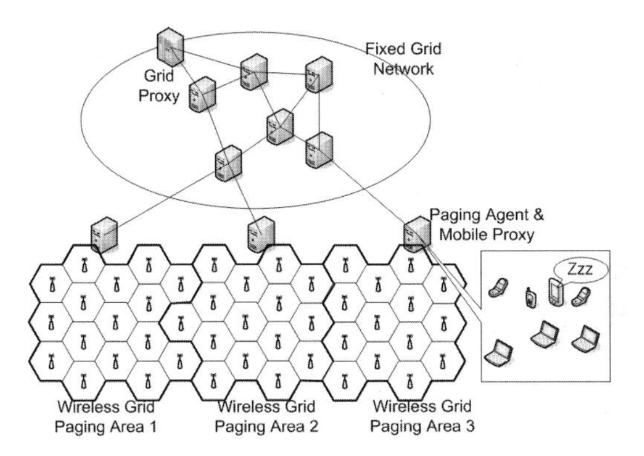

Fig. 2. Wireless grid computing architecture

The proposed wireless grid computing architecture consists of fixed grid networks and wireless grid networks. Grid users join a grid network through a grid proxy that operates as a pure meta-scheduler. It performs three operations: dividing a submitted job into small-size tasks, assigning the tasks to mobile proxies, and collecting finished tasks. A mobile proxy allocates tasks to mobile devices when they are assigned by the grid proxy. It has three important functions: a paging agent for resource management, a pure proxy for task distribution, and a stable storage for checkpoints.

4 Resource Management

In our architecture, resources are managed by the paging information of mobile devices. Each mobile device falls into one of two states: active and idle. Figure 3 shows the state transition diagram of mobile devices.

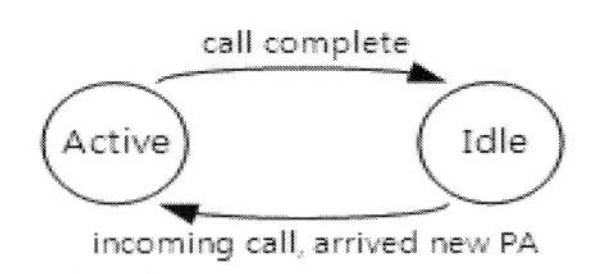

Fig. 3. State transition diagram of mobile device

```
 0                   1                   2                   3
 0 1 2 3 4 5 6 7 8 9 0 1 2 3 4 5 6 7 8 9 0 1 2 3 4 5 6 7 8 9 0 1
+-+-+-+-+-+-+-+-+-+-+-+-+-+-+-+-+-+-+-+-+-+-+-+-+-+-+-+-+-+-+-+-+
|     Type      |    Length     |        Sequence Number        |
+-+-+-+-+-+-+-+-+-+-+-+-+-+-+-+-+-+-+-+-+-+-+-+-+-+-+-+-+-+-+-+-+
|              the home address of paged mobile node            |
+-+-+-+-+-+-+-+-+-+-+-+-+-+-+-+-+-+-+-+-+-+-+-+-+-+-+-+-+-+-+-+-+
|CpuSpeed |Num|Ram  |P|Capa    |FreeSpace|Reserved              |
+-+-+-+-+-+-+-+-+-+-+-+-+-+-+-+-+-+-+-+-+-+-+-+-+-+-+-+-+-+-+-+-+
```

Fig. 4. Grid paging registration message format

The mobile device registers with the paging agent when it joins the paging area of a new wireless grid or when it change from the active to the idle state. The mobile device is removed from the paging agent when it leaves current paging area in a wireless grid or when it changes from the idle to the active state. In IP-paging scheme, a paging agent manages the address information of idle mobile devices. In this paper, we extend the IP-paging message format to manage resource status information as well as address information. Resource status information includes information on computing elements and storage elements. The information of computing elements includes class, CPU speed, number of CPUs, RAM size, and power. The information of storage elements includes capacity and free space. Figure 4 shows the grid paging registration message format. The resource information collected from mobile devices is stored in the paging cache. Figure 5 shows an example of the grid paging cache.

Num	Home address	Type	CpuSpeed	NumCpu	RamSize	Power	Capacity	FreeSpace	used
1	163.152.91.77	L	20	1	1024	C	10240	2048	A
2	163.152.91.78	L	27	1	2048	82	10240	2048	A
3	163.152.91.79	P	13	1	512	70	10240	2048	A
4	163.152.91.80	P	10	1	256	90	10240	2048	A
5	163.152.91.81	L	30	2	4096	C	10240	2048	U
-	------------	--	------	----	-----	---	------	------	--

Fig. 5. Grid paging cache example

```
 0                   1                   2                   3
 0 1 2 3 4 5 6 7 8 9 0 1 2 3 4 5 6 7 8 9 0 1 2 3 4 5 6 7 8 9 0 1
+-+-+-+-+-+-+-+-+-+-+-+-+-+-+-+-+-+-+-+-+-+-+-+-+-+-+-+-+-+-+-+-+
|     Type      |    Length     |        Sequence Number        |
+-+-+-+-+-+-+-+-+-+-+-+-+-+-+-+-+-+-+-+-+-+-+-+-+-+-+-+-+-+-+-+-+
|            the home address of paged mobile node              |
+-+-+-+-+-+-+-+-+-+-+-+-+-+-+-+-+-+-+-+-+-+-+-+-+-+-+-+-+-+-+-+-+
|           the care ofaddress of paged mobile node             |
+-+-+-+-+-+-+-+-+-+-+-+-+-+-+-+-+-+-+-+-+-+-+-+-+-+-+-+-+-+-+-+-+
|CpuSpeed |Num|Ram  |P|Capa    |FreeSpace|Reserved              |
+-+-+-+-+-+-+-+-+-+-+-+-+-+-+-+-+-+-+-+-+-+-+-+-+-+-+-+-+-+-+-+-+
```

Fig. 6. Grid paging reply message format

Network partitioning occurs in wireless environments because of the random movement of mobile devices. When tasks are assigned by a grid proxy, a mobile proxy creates a pre-candidate set of mobile devices based on the resource information in the paging cache. The mobile proxy sends a paging request message to the mobile devices in the pre-candidate set to confirm their state. The mobile proxy then receives paging reply messages from the mobile devices in pre-candidate set and decides upon a final resource set. The paging reply message contains additional information including the current address of the mobile device so that tasks can be sent to it. To prevent resource duplication, running mobile devices are so marked in the paging cache. Figure 6 shows the grid paging reply message format.

Although we use IP-paging scheme in this paper, this does nothing to address the problems with network partitioning and battery discharge in wireless grid networks. Therefore, we propose a task migration scheme using user-defined checkpoints in the next section.

5 Task Migration Using User-Defined Checkpoints

In this paper, we propose a user-defined checkpoint technique that performs checkpoints under two conditions: when a mobile device leaves its current cell (determined by analysis of the signaling strength) and when a mobile device turns off due to lower battery capacity.

5.1 Analysis of Wireless Signaling Strength

The wireless signaling strength is the strength of signal from the AP received at a mobile device. In wireless networks, handoff occurs when a mobile device moves to a new cell that is within signaling range of a new AP. When handoff occurs, mobile device's wireless connection with the current cell is terminated, and a new connection is established in the new cell. Depending on the condition of the wireless connection during handoff (*e.g.*, when moving into a fringe area), the mobile device can disappear. In mobile grid computing, the disappearance of a mobile device during operation is an important issue because any task assigned to it must be started over again from the initial state. Therefore, checkpoints are necessary to maintain efficient operation. We suggest performing a checkpoint before handoff occurs. Figure 7 shows the wireless signaling strength during handoff, and Table 1 shows checkpoint operation according to signaling strength.

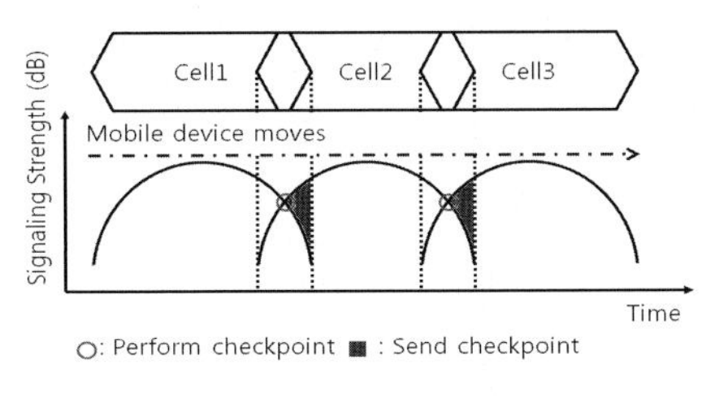

Fig. 7. Wireless signaling strength during handoff

Table 1. Checkpoint operation according to signaling strength

1: If current signaling strength is equal to new signaling strength
2: Perform checkpoint (current state)
3: Else if current signaling strength is less than new signaling strength
4: Send checkpoint to mobile proxy

5.2 Analysis of Battery Capacity

Mobile devices have limited battery capacity and can be classified as: charging and running. In a mobile grid environment, the charging state is the best case because a mobile device that is charging is not mobile, just like a wired device. It can be changed into the running state at any time. If a task is assigned to an idle mobile device in the running state, the battery consumption is nearly five times that of an idle mobile device without an assigned task. In this paper, we perform a checkpoint when the battery capacity is within the range defined by the provider. Figure 8 shows the battery capacity as a function of time, and Table 2 shows checkpoint operation according to battery capacity.

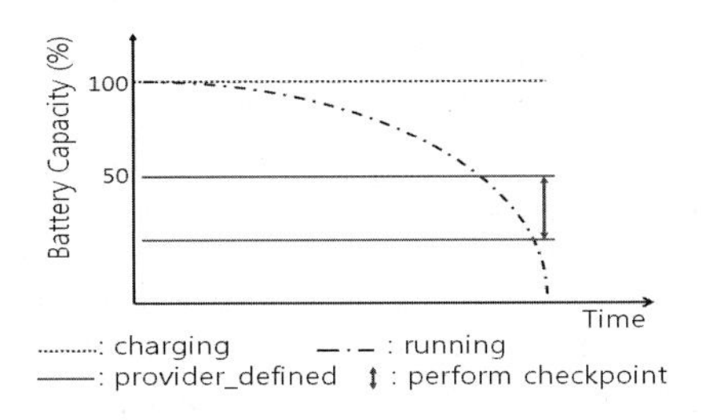

Fig. 8. Battery capacity as a function of time

Table 2. Checkpoint operation according to battery capacity

1: If current battery capacity is equal to battery capacity of provider defined
2: Perform checkpoint (current state)
3: Else if current battery capacity is less than battery capacity of provider defined
4: Send checkpoint to mobile proxy

5.3 Task Migration

Task migration occurs when a mobile device cannot execute the assigned task any longer. This can be due to network partition, discharged battery, or other reasons. The task migrates from one mobile device to another to minimize the completion time of the task. Table 3 describes broker service (task assignment), task migration, and re-execution phases along with an explanation of task completion time.

Table 3. Broker service, task migration, re-execution phases, and task completion time

Broker service phase	Task migration phase
1: Wait for tasks from grid proxy 2: Request grid paging cache 3: Decide pre-candidate set from paging cache 4: Send grid paging request current status to pre-candidate set 5: Wait for grid paging reply from pre-candidate set 6: Decide candidate set from pre-candidate set 7: Decide reserved mobile devices from candidate set 8: Send task to each mobile device 9: Request to change *Used fields* "U" and "N" in paging cache 10: If processing mobile device has disappeared, compare task completion times	11: If completion time$_{checkpoint}$ is less than completion time$_{re\text{-}execution}$ 12: Decide pre-candidate set from paging cache 13: Send grid paging request current status to candidate set 14: Wait for grid paging reply from candidate set 15: Decide reserved mobile device from candidate set 16: Migrate task to reserved mobile device 17: Request to change *Used fields* "U" and "N" in paging cache
Re-execution phase	Completion time
18: If completion time$_{checkpoint}$ is greater than completion time$_{re\text{-}execution}$ 19: Decide pre-candidate set from paging cache 20: Send grid paging request current status to candidate set 21: Wait for grid paging reply from candidate set 22: Decide reserved mobile device from candidate set 23: Send task to reserved mobile device 24: Request to change *Used fields* "U" and "N" in paging cache	The completion time = scheduling time + task assigned time + processing time + remaining time + collection time The completion time$_{checkpoint}$ = scheduling time + task assigned time + processing time + checkpoint time + migration time + remaining time + collection time The completion time$_{re\text{-}execution}$ = (scheduling time + task assigned time + processing time) $*$ re-execution + collection time

6 Performance Analysis

Figure 9 shows the network model used for our simulations. The network model includes six routers and three gateway routers on an Ethernet-based fixed grid network. Each wireless grid paging area has 14-18 cells. Mobile devices have a wireless link (11Mbps, 802.11b) with an AP in each cell. The AP and gateway router in the fixed grid network are connected via an Ethernet-based Internet.

The purpose of our simulations is to quantitatively evaluate the improvements in total task execution time and total job completion time in a system using the proposed enhancements, when compared with the architecture with no paging. Two parameters are studied: task execution time and job completion time. The task execution time is defined as the time that elapses between starting and finishing the task execution. We study the task execution time as a function of the number of available mobile devices and the probability of handoff. The job completion time is defined as time between scheduling a job and collecting the finished job. We study the job completion time as a function of the number of available mobile devices and the probability of handoff.

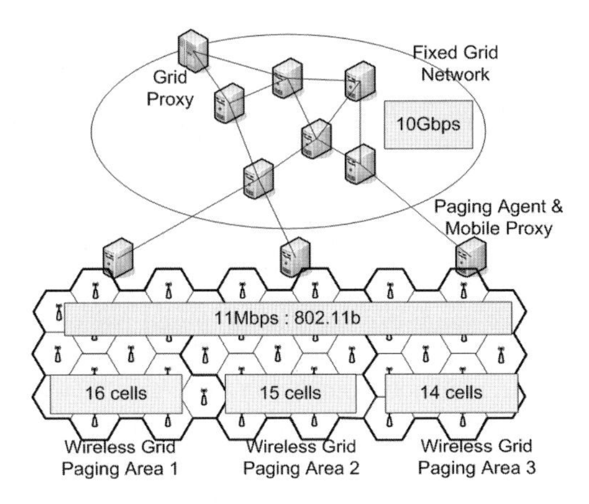

Fig. 9. Network model for performance analysis

For performance evaluation, a job that is divided into 20 tasks is executed on 20 mobile devices. We consider the following five cases:

- Case 1: 50 available mobile devices with 0% probability of handoff
- Case 2: 50 available mobile devices with 30% probability of handoff
- Case 3: 50 available mobile devices with 50% probability of handoff
- Case 4: 100 available mobile devices with 30% probability of handoff
- Case 5: 100 available mobile devices with 50% probability of handoff

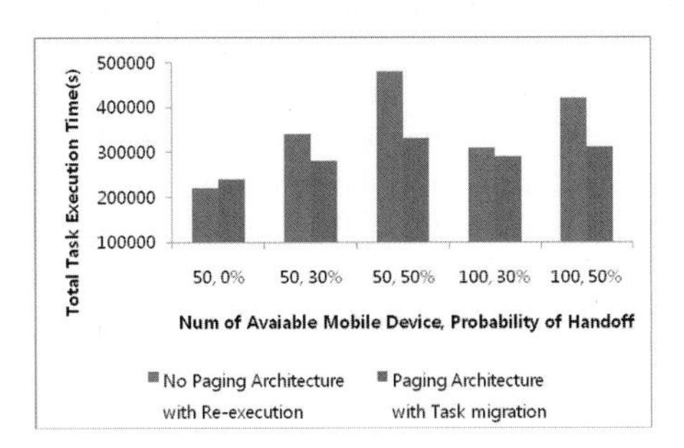

Fig. 10. Total task execution time as a function of the number of available mobile devices and the probability of handoff

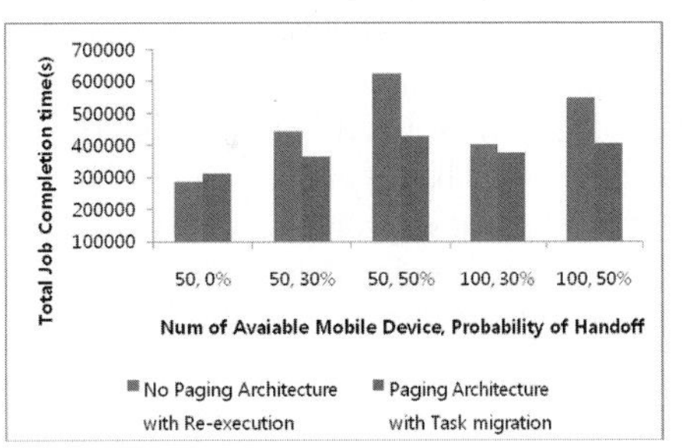

Fig. 11. Total job completion time as a function of the number of available mobile devices and the probability of handoff

Figure 10 shows the total task execution time as a function of the number of available mobile devices and the probability of handoff. The no-paging architecture was faster in Case 1 than the proposed paging architecture. This is because the proposed paging architecture spends additional time for paging-cache management, network delay due to paging request, and the paging reply message. As for Case 2, the proposed paging architecture was faster than the no-paging architecture. Because of the 30% probability of handoff, the tasks assigned to some mobile devices were re-executed. Additionally, our task migration scheme performed better than simple re-execution. The proposed paging architecture was faster in Case 3 than the no-paging architecture for the same reason as Case 2, and the difference in times was much greater than in Case 2. Because of the 50% probability of handoff, tasks assigned to many mobile devices were re-executed. The paging architecture was faster than the no-paging architecture in Case 4 as well, although the difference in times was much less than in Case 2. Due to the limited number of available mobile devices, neither paging architecture was able to find suitable new mobile devices easily. The proposed paging architecture had similar performance in Cases 2 and 4, due to the use of the pre-candidate set. The proposed paging architecture in Case 5 was faster than the no-paging architecture for the same reason as in Case 3.

Figure 11 shows the total job completion time as a function of the number of available mobile devices and the probability of handoff. The job scheduling time and network delay for gathering the finished job were added to the total task execution time. Because we used a static value for the job scheduling time and the network delay for gathering the finished job, the results of Fig. 11 have a similar tendency of these of Fig. 10.

7 Conclusion and Future Work

We have addressed the management of mobile devices in mobile grid environments by focusing on IP-paging, which is capable of managing idle mobile devices and grid resource status information. We proposed the IP-paging architecture for mobile grid environments to support mobile grid services. The proposed architecture can be used to deal with network partition problem, which is an important issue in creating a practical mobile grid environment with a task migration scheme using checkpoints. However, this imposed an additional cost due to task migration required to perform the checkpoint. In addition, being able to predict when a mobile device will leave the network or when its battery will fall below a certain level of charge is very useful in addressing these problems to provide continuous grid services. We focused on a user-defined checkpoint that is performed under two conditions: when a mobile device leaves its current cell and when a mobile device turns off due to lower battery capacity.

In this paper, we attempted to solve mobile resource management using IP-paging and task migration based on user-defined checkpoints. Many challenges still remain including disconnected operation, job scheduling, device heterogeneity, and security. We will tackle these issues as future works. We will also extend our investigation into what wireless networks can contribute to grid computing in an effort to provide effective support to mobile grid computing.

References

1. Foster, S.I., Kesselman, C., Tuecke, S.: The Anatomy of the Grid: Enabling Scalable Virtual Organizations. Int. Journal of High Performance Computing Applications 15(3), 200–222 (2001)
2. Foster, I., Kesselman, C.: The Grid 2: Blueprint for a New Computing Infrastructure. Morgan Kaufmann Publishers, San Francisco (2004)

3. Foster, I., Roy, A., Sander, V.: A Quality of Service Architecture that Combines Resource Reservation and Application Adaptation. In: 8th International Workshop on Quality of Service, pp. 181–188 (2000)
4. Foster, I.: The Physiology of the Grid: An Open Grid Services Architecture for Distributed Systems Integration. In: Global Grid Forum (2002)
5. Foster, I.: The Grid: A New Infrastructure of 21st Century Science. Physics Today 55, 42–52 (2002)
6. Zhang, X., Castellanos, J.G., Campbell, A.T.: P-MIP: paging extensions for mobile IP. Mobile Networks and Applications 7(2), 127–141 (2002)
7. Campbell, A.T., Gomez, J.: IP Micro-Mobility Protocols. ACM SIGMOBILE Mobile Computer and Communication Review (MC2R) 4(4), 45–54 (2001)
8. Castelluccia, C.: Extending Mobile IP with adaptive individual paging: A performance analysis, INRIA (1999), `http://www.inrialpes.fr/planete/people/ccastel/`
9. Ghosh, P., Roy, N., Das, S.K., Basu, K.: A Game Theory based Pricing Strategy for Job Allocation in Mobile Grids. In: Proceedings of 18th International Parallel and Distributed Processing Symposium, pp. 26–30 (April 2004)
10. Hingne, V., Joshi, A., Finin, T., Kargupta, H., Houstis, E.: Towards a Pervasive Grid. In: Proceedings of 17th International Parallel and Distributed Processing Symposium, pp. 22–26 (April 2003)
11. Fukuda, M., Tanaka, Y., Suzuki, N., Bic, L.F., Kobayashi, S.: A mobile-agent-based PC grid. In: Proceedings of Autonomic Computing Workshop Fifth Annual International Workshop on Active Middleware Services (AMS 2003), p. 142 (June 2003)
12. Barbosa, R.M., Goldman, A.: MobiGrid Framework for Mobile Agents on Computer Grid Environments. In: Karmouch, A., Korba, L., Madeira, E.R.M. (eds.) MATA 2004. LNCS, vol. 3284, pp. 147–157. Springer, Heidelberg (2004)
13. Phan, T., Huang, L., Dulan, C.: Challenge: Integrating Mobile Wireless Devices into the Computational Grid. In: Proceedings of 8th International Conference on Mobile Computing and Networking, pp. 271–278 (September 2002)
14. Wesner, S., Jähnert, J.M., Aránzazu, M., Escudero, T.: Mobile Collaborative Business Grids: A short overview of the Akogrimo Project, `http://www.akogrimo.org/download/White_Papers_and_Publications/Akogrimo_WhitePaper`
15. Allen, G., Angulo, D., Foster, I., Lanfermann, G., Liu, C., Radke, T., Seidel, E.: The Cactus Worm: Experiments with dynamic resource discovery and allocation in a grid environment. International Journal of High Performance Computing Applications 15(4), 345–358 (2001)
16. Roe, P., Szyperski, C.: Transplanting in Gardens: Efficient Heterogeneous Task Migration for Fully Inverted Software Architectures. In: Proceedings of the Fourth Australasian Computer Architecture Conference (January 1999)
17. Krishnan, S., Gannon, D.: Checkpoint and Restart for Distributed Components in XCAT3. In: Proceedings of the Fifth IEEE/ACM International Workshop on Grid Computing, Pittsburgh, Pennsylvania, pp. 281–288 (November 2004)

Group-based Scheduling Algorithm for Fault Tolerance in Mobile Grid*

JongHyuk Lee[1], SungJin Choi[2], Taeweon Suh[1], HeonChang Yu[1], and JoonMin Gil[3,**]

[1] Dept. of Computer Science Education, Korea University
Anam-dong, Sungbuk-gu, Seoul 136-701, Korea
{spurt,suhtw,yuhc}@korea.ac.kr

[2] Cloud Service Business Unit, KT
17 Umyeon-dong, Seocho-gu, Seoul 137-792, Korea
lotieye@gmail.com

[3] School of Computer and Information Communications Engineering, Catholic University of Daegu
330 Geumnak, Hayang-eup, Gyeongsan-si, Gyeongbuk 712-701, Korea
jmgil@cu.ac.kr

Abstract. Mobile Grid is a branch of Grid computing where the infrastructure includes mobile devices. Because mobile devices are resource-constrained, mobile Grid should provide new scheduling strategies considering its environment. This paper presents a group-based fault tolerance scheduling algorithm. The algorithm classifies mobile devices into several groups considering characteristic parameters of mobile Grid. Then, it uses an adaptive replication algorithm for enduring faults in an active manner. The experimental results show that our scheduling algorithm provides a superior performance in terms of execution times to the one without considering grouping and fault tolerance. Throughout the experiments, we found that the active fault tolerance (i.e., replication) is essential to improving performance in mobile Grid.

Keywords: mobile Grid, scheduling algorithm, replication.

1 Introduction

Grid [1] is a large-scale virtual computing environment where geographically distributed resources collaboratively provide a computing infrastructure. It is used for solving computing-intensive and data-intensive problems that are not practically feasible to run in traditional distributed computing environments. The early Grid was implemented mostly with physically fixed resources with high-performance, and the resources are connected through reliable networks

* This research was supported by Basic Science Research Program through the National Research Foundation of Korea (NRF) funded by the Ministry of Education, Science and Technology (No. 2009-0070556).

** Corresponding author.

T.-h. Kim, A. Stoica, and R.-S. Chang (Eds.): SUComS 2010, CCIS 78, pp. 394–403, 2010.

with high speed. Emerging Grids [2] are extending the scope of resources to mobile devices and sensors that are loosely connected through wireless networks, which is referred to as mobile Grid. Mobile Grid computing is a branch of Grid computing in which the foundational infrastructure includes mobile devices.

Mobile devices are not reliable because they have limited battery life and are casually disconnected from wireless network. In our prior work [3], we presented a balanced scheduling algorithm taking into account the mobility and availability in scheduling. The algorithm requires the active measurement of failures of mobile devices in execution to decrease execution time of tasks. Therefore, this paper proposes a group-based scheduling algorithm to provide the fault tolerance. The algorithm classifies mobile devices into several groups considering characteristic parameters of mobile Grid. Then, it uses an adaptive replication algorithm for enduring faults in an active manner.

The rest of the paper is organized as follows. Section 2 presents related work on scheduling algorithms in mobile Grid. Section 3 illustrates our group-based scheduling algorithm in mobile Grid. Experimental results are presented in Section 4. Finally, we conclude our paper with future works in Section 5.

2 Related Work

Studies on scheduling in mobile Grid mostly focus on power efficiency, communication availability, or job replication. For the power efficiency, Huang et al. [4] proposed a proxy-based hierarchical scheduling model taking mobility and power management into account in wireless environment. In this model, the scheduler is comprised of two levels (top level and proxy level) to efficiently utilize the energy of wireless node and guarantee QoS at the same time. There are several studies on the communication availability. Park et al. [5] proposed a scheduling algorithm with the processor and the communication availabilities. This algorithm confines the communication scope and is usable when the network link is broken due to mobility. However, this algorithm has a shortcoming in that it is applicable to specific job types with no communication among tasks during the job execution. For this, Lee et al. [3] proposed a balanced scheduling in mobile Grid, which takes the load balancing and users' mobility patterns into account. However, the algorithm does not provide fault tolerance. In the job replication front, Litke et al. [6] proposed a method that estimates a number of job replication using the Weibull reliability function and maximizes the resource utilization for workloads caused by replication with the knapsack formulation.

3 Group-Based Scheduling Algorithm for Fault Tolerance

Group-based Fault Tolerance Scheduling (GFTS) is a scheduling algorithm based on the mobile device group (MDG). The algorithm adaptively applies a different scheduling scheme and a fault tolerance algorithm according to the property of MDGs. In this section, we firstly introduce characteristic parameters of mobile Grid. Then, we describe how to construct MDG according to the property of a mobile device. Finally, we discuss the scheduling and replication algorithm.

3.1 Characteristic Parameters of Mobile Grid

A mobile device fails to execute a task when unexpected events happen in the middle of the task execution. We consider a computation failure and a communication failure as follows in this paper.

Definition 1. *Computation failure: The computation failure (Φ) is defined as the severe interruption caused by hardware and software problems of a mobile device.*

Definition 2. *Communication failure: The communication failure (Ψ) is defined as the severe interruption caused by a problem with the network and the user's movement.*

In our prior research, we introduced metrics to represent the characteristics of mobile Grid [5]. They are duplicated here for readability and expressed in terms of the computation and communication failures.

Definition 3. *Available time: Available time (Υ) is a time period during which a mobile device is able to provide its resources.*

$$\Upsilon = \Upsilon_F + \Upsilon_P \tag{1}$$

We classify the available time into full available time (Υ_F) and partial available time (Υ_P) according to the computation failure and the communication failure. Υ_F is defined as a time period during which a mobile device is able to execute tasks and return outcome via network link. In other words, Υ_F is a period of time during which Φ and Ψ do not occur. However, Υ_P is defined as a time period during which a mobile device is able to execute tasks, but is not able to return outcome because of Ψ.

Definition 4. *Availability: Availability (A) is the probability that a mobile device will be correctly operational and be able to return outcome.*

$$A = A_F + A_P = \frac{\Upsilon_F}{\Upsilon + \Upsilon^{-1}} + \frac{\Upsilon_P}{\Upsilon + \Upsilon^{-1}} \tag{2}$$

where the Υ^{-1} is unavailable time and the full availability (A_F) is a probability that a mobile device is in the full available time and the partial availability (A_P) is a probability that a mobile device is in the partial available time.

3.2 Construction of Mobile Device Groups

A MDG is a set of mobile devices with similar properties. We apply different policies on scheduling and fault tolerance techniques for each MDG rather than each individual mobile device, to reduce the management overhead. For the MDG classification, we use dynamic properties such as availability rather than static properties such as CPU, memory, storage, and network bandwidth. It is because mobile devices are typically equipped with less powerful resources. The

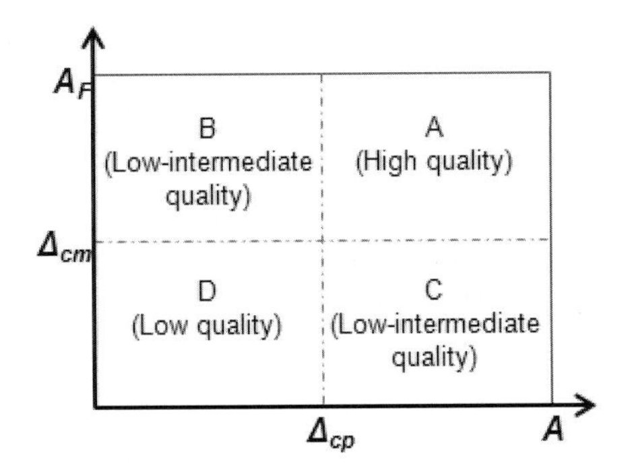

Fig. 1. Classification of mobile devices according to availability and full availability

```
// To construct mobile devices groups
if (Ni.A ≥ Δcp) then // Ni: one of the classified mobile devices
  if (Ni.AF ≥ Δcm) then
    MDGA ← Ni ; // ←: assign, MDGA: A mobile device group
  else
    MDGC ← Ni ; // MDGC: C mobile device group
  fi;
else // Ni.A < Δcp
  if (Ni.AF ≥ Δcm) then
    MDGB ← Ni ; // MDGB: B mobile device group
  else
    MDGD ← Ni ; // MDGD: D mobile device group
  fi;
fi;
```

Fig. 2. Group construction algorithm of mobile devices

dynamic properties provide more pertinent information on how long and how many the mobile devices can be used for the job execution.

Mobile devices are classified into four classes (A, B, C, and D), as shown in Fig. 1. Δ_{cp} is the expected availability of a mobile device for computation and Δ_{cm} is the expected availability of a mobile device for communication. MDGs are constructed using an algorithm shown in Fig. 2, where mobile devices are classified into MDG_A, MDG_B, MDG_C, and MDG_D according to the availability and the full availability.

In MDG_A, the chance that a task is executed and its outcome is returned within a predicted time is very high due to its high availability and full availability. In MDG_B, a task has a high chance of being stopped in execution due to its low availability. However, once the task is executed to its completion, the probability of returning its outcome is high since the full availability is relatively high. In MDG_C, a task is likely to be executed to its completion due to the high availability. However, the result of the task is not likely to be returned within a predicted time due to its low full availability. Therefore, MDG_B and MDG_C requires a fault tolerance mechanism for the reliable task execution. In MDG_D, the chance of failure is very high due to its low availability and full availability.

Table 1. Scheduling groups of mobile devices

Scheduling group	No. of task allocation	A_F compensation	A compensation	Description
$AD\&BC$	$AD \cong BC$ or $AD \geq BC$	O	O	The tasks are distributed to each scheduling group. A compensates for D, and C compensates for B.
$AB\&CD$	$AB \cong CD$ or $AB \geq CD$	X	O	The tasks are distributed to each scheduling group. As both C and D have low A_F, they do not compensate A_F.
$AC\&BD$	$AC \gg BD$	O	X	The tasks are mainly distributed to AC. Both B and D do not compensate A.

Since the availabilities of MDG_A and MDG_C are larger than its Δ_{cp}, most tasks are likely to be executed to their completion. For the fresh execution of a task, MDG_B is not suitable because of its low availability. However, when a failure occurs in MDG_A and MDG_C, MDG_B could be the right candidate for migration to execute the remaining execution of a task. Nevertheless, if the checkpoint mechanism is adopted for fault tolerance, MDG_B and MDG_D can be used for the fresh execution of non-time-critical tasks.

3.3 Task Scheduling via Scheduling Mobile Agent

After creating MDGs, a proxy server in a proxy-based mobile Grid environment allocates scheduling mobile agents (S-MAs) to MDGs. The lower the availability of an MDG, the MDG is less preferred for the S-MA allocation. Nevertheless, it is a waste of resources if completely excluded for allocation. Therefore, we use new scheduling groups (SGs) to effectively utilize all the system resources.

Table 1 shows the scheduling group candidates of MDGs. The 1st and 2nd groups are more appropriate than the 3rd one because SG_{BD} would complete fewer tasks than SG_{AC} due to the difference in availability. The 1st group would be superior to the 2nd one because SG_{AD} and SG_{BC} in the 1st group complement each other in terms of both availability and full availability. Whereas, SG_{AB} and SG_{CD} in the 2nd group complement each other in terms of only availability. Therefore, we make use of the 1st group for scheduling.

After S-MAs are allocated to SGs, each S-MA distributes task mobile agents (T-MAs) to scheduling group members. Each S-MA executes different scheduling and replication algorithm according to the SG type. The S-MA of SG_{AD} schedules tasks as follows.

1. Sort MDG_A in the descending order of A.
2. Distribute T-MAs to the sorted MDG_A.

3. When a T-MA fails, replicate it to other members of MDG_A or MDG_D by replication algorithm, or migrate it to other member of MDG_A or MDG_D if the task migration is permitted.

S-MA of SG_{BC} schedules tasks as follows.

1. Sort MDG_B and MDG_C in the descending order of A.
2. Distribute T-MAs to the sorted MDG_C.
3. When a T-MA fails, replicate it to other members of MDG_C by replication algorithm, or migrate it to other member of MDG_B or MDG_C if the task migration is permitted.

T-MAs are firstly distributed to SG_{AD} and then SG_{BC}. In addition, T-MAs are preferentially allocated to mobile devices with high A_F and high A in SG. If checkpointing is not used, T-MAs are not allocated to MDG_B and MDG_D because the communication failure in mobile devices with low A_F occurs frequently and the execution outcome is not returned within a proper time, resulting in the longer execution time. However if migration is permitted, MDG_B is used to compensate the main MDG (i.e., MDG_A and MDG_C) for low A_F. In addition, if the replication is used, S-MAs calculate the number of redundancy and choose replicas. Then, S-MAs distribute T-MAs to the selected replicas. When a T-MA fails, the T-MA is migrated or replicated to a new mobile device.

3.4 Adaptive Replication Algorithm

Replication is a well-known technique to improve reliability and performance in distributed systems. This paper uses replication to improve reliability by tolerating failures of a mobile device. Adaptive replication algorithm adjusts the number of redundancy automatically and selects a pertinent mobile device according to the property of each MDG.

We use the *Weibull distribution* to represent the reliability function of a mobile device and calculate the number of redundancy. The Weibull distribution can represent the case where failure rate is constant as well as the case where failure rate (or hazard rate) increases or decreases as time goes by, as opposed to the *Exponential distribution*. The general form of the probability density function f of a Weibull random variable x is as follows.

$$f(x) = \frac{k}{\lambda}(\frac{x}{\lambda})^{k-1}e^{-(\frac{x}{\lambda})^k} \quad (3)$$

where $\lambda > 0$ is the scale parameter and $k > 0$ is the shape parameter.

The cumulative distribution function F for t is as follows.

$$F(t) = \int_0^t f(x)dx = 1 - e^{-(\frac{t}{\lambda})^k} \quad (4)$$

We use the least squares parameter estimation (regression analysis) to estimate the parameters (i.e., λ and k) of the Weibull distribution. Eq. 5 is simple linear regression model.

$$y_i = \beta_0 + \beta_1 x_i + \epsilon_i \quad (5)$$

where β_0 denotes the intercept and β_1 is the slope of the regression line and ϵ_i , called the residual, is the vertical distance from a point (x_i, y_i) to the estimated regression line.

The cumulative distribution function $F(x)$ can be transformed as derived in Eq. 6 so that it is in the similar form to Eq. 5.

$$F(x) = 1 - e^{-(\frac{x}{\lambda})^k}$$
$$\Leftrightarrow \ln(\ln(\frac{1}{1 - F(x)}) = k \ln x - k \ln \lambda \tag{6}$$

Correlating Eq. 6 with Eq. 5, $\ln(\ln(\frac{1}{1-F(x)})$ is corresponding to y_i, $\ln x$ to x_i, k to β_1, and $-k \ln \lambda$ to β_0. By the least-squares estimators for β_0 and β_1, k and λ are calculated as follows.

$$k = \frac{n \sum_{i=1}^{n} x_i y_i - (\sum_{i=1}^{n} x_i)(\sum_{i=1}^{n} y_i)}{n \sum_{i=1}^{n} x_i^2 - (\sum_{i=1}^{n} x_i)^2} \tag{7}$$
$$\lambda = e^{-(\bar{y} - \beta_1 \bar{x})(n \sum_{i=1}^{n} x_i^2 - (\sum_{i=1}^{n} x_i)^2)/(n \sum_{i=1}^{n} x_i y_i - (\sum_{i=1}^{n} x_i)(\sum_{i=1}^{n} y_i))} \tag{8}$$

where n denotes the number of points.

This paper focuses a computation failure and a communication failure. In mobile Grid, the communication failure occurs more frequently than the computation failure due to the network instability and the user's movement (i.e., $\Psi \geq \Phi$). This paper calculates the number of redundancy with the communication failure. The number of redundancy r is calculated as follows.

$$R_s(t) = 1 - P[all\ mobile\ device\ fail] = 1 - \prod_{i=1}^{r} [1 - R_i(t)] \geq \gamma$$
$$\Leftrightarrow r \leq \frac{1 - \gamma}{e^{-(\frac{t}{\lambda})^k}} \tag{9}$$

where γ is the reliability threshold.

In this paper, each MDG has a different γ for the reliable execution. Therefore, each task has a distinct r. For example, a task allocated in MDG_A has a smaller r than one allocated in MDG_C.

4 Experiments and Analysis

4.1 Experimental Environment

We evaluated our scheduling algorithm using the SimGrid toolkit [7] with a real-life trace: WLAN trace [8] of the Dartmouth campus like our prior work [3]. The environmental change compared to prior work is a date range for the trace. We chose a trace from June 5, 2006 to June 7, 2006, which includes 1,031 APs and 4,840 mobile devices.

4.2 Experimental Result

We investigated the effect of four factors on execution time: prediction of availability (PA), mobile device grouping without failure detection (G), mobile device grouping with failure detection (GFD), and task replication (TR).

The first experiment evaluates the impact of the time limit used for prediction of availability. For example, if the execution time of a task is 1,000 seconds and the number of a multiple is 2, the time limit used for prediction of availability is 2,000 seconds (i.e., 1,000 * 2). In this experiment, failures are not detected. That is, S-MA does not take appropriate measures when a failure occurs in a mobile device. After the failure recovery, the task is executed again by T-MA. Then, based on the first experiment, we conducted the next experiment, investigating the impact of group using the full and partial availabilities. Then, we evaluated the impact of failure detection. In this experiment, when a failure is detected, the task is executed again from the beginning. Based on the experiment, we conducted the final experiment, evaluating the impact of task replication. The reliability thresholds γ of MDG_A, MDG_B, MDG_C, and MDG_D are 0.7, 0.8, 0.8, and 0.9 respectively.

Experiments were executed for five cases: The simulation start times are 0, 12, 24, 36, 48, 60 hours. Tasks were classified into two classes: small-size and large-size. Small-sized and large-sized tasks take 1,000 and 10,000 seconds to complete, respectively. The number of tasks is 500.

Fig. 3 and 4 show the average execution times of the four experiments according to the multiples used for prediction of availability in terms of large-size and small-size tasks, respectively. As shown in Fig. 3, when tasks are large with the prediction, the bigger multiples report the shorter execution time. In other words, the longer time limit used for prediction is, the higher the accuracy rate is. On the other hand, as shown in Fig. 4, when tasks are small with the prediction, the bigger multiples report the longer execution times. In other words, the accuracy rate is not increased even though the time limit is longer. Although the accuracy rate improves according to multiples, the execution time of tasks is much longer than its size. It is due to the network connection pattern of a specific mobile device. For example, a mobile device is connected to a network for 14 hours when a task is allocated (because availability is 1.0 at this time, this allocation is perfectly timed). It is disconnected after about 640 seconds (the task cannot execute anymore). After about 38,000 seconds, it is connected again (the task is executed again).

When the grouping is used without the failure detection, the execution time would not be distinguished from one with the prediction of availability. However, if the partial availability becomes higher, the difference in the execution times between them would be high. When the grouping is used with the failure detection, the execution time is shortened compared with the grouping without the failure detection. Meanwhile, when the task replication is used, the execution time become short compared with others and almost equal to task size. Therefore, a scheduling strategy using only prediction and only grouping would not improve the execution time. A scheduling strategy using the failure detection or task replication would enhance the execution time.

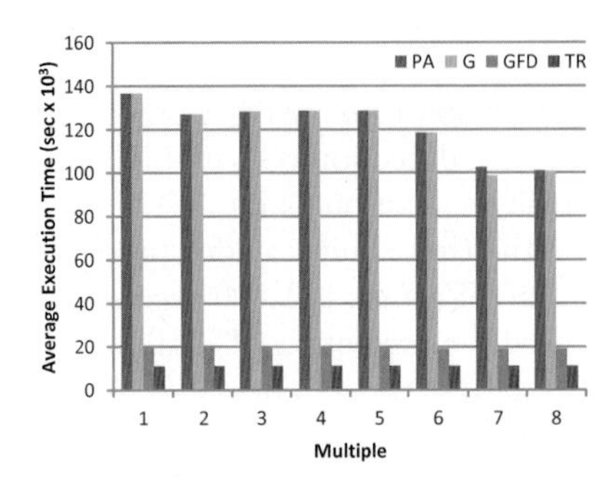

Fig. 3. Average execution times according to prediction, grouping, and replication in large-sized tasks

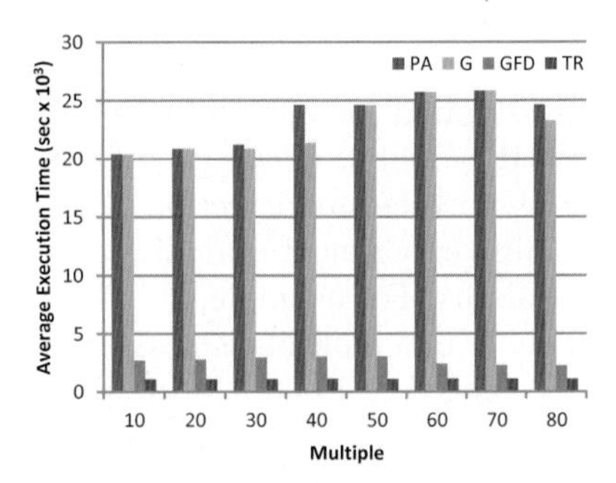

Fig. 4. Average execution times according to prediction, grouping, and replication in small-sized tasks

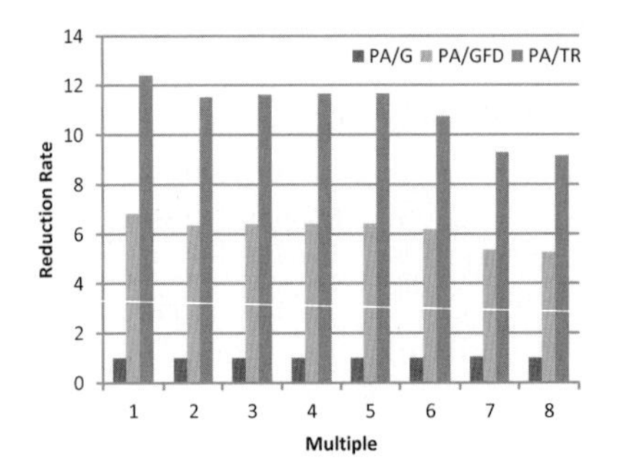

Fig. 5. Execution time reduction rate in large-sized tasks

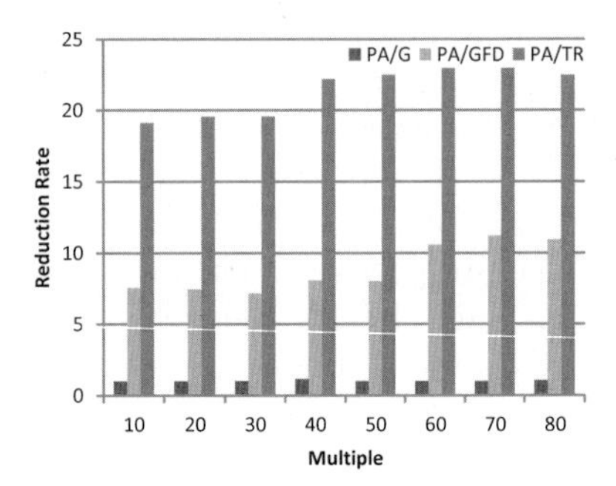

Fig. 6. Execution time reduction rate in small-sized tasks

Fig. 5 and 6 show the execution time reduction rate for large-sized and small-sized tasks, respectively. Reduction rate is defined as the ratio of execution time used with the general method to one used with the improved method. As shown in Fig. 5 and 6, we could expect that a scheduling strategy used by task replication is essential in mobile Grid.

5 Conclusions and Future Work

This paper presents a group-based fault tolerance scheduling algorithm in mobile Grid. Our algorithm classifies mobile devices into mobile device groups according to the full and partial availabilities and uses an adaptive replication algorithm for fault tolerance. The experimental results show that our scheduling algorithm provides a superior performance to the one that use only a prediction of availability. Throughout the experiments, we found that the active fault tolerance (i.e., replication) is essential to improving performance in mobile Grid. In the future, we have a plan to conduct a wider variety of experiments to study additional factors that contribute to performance in mobile Grid.

References

1. Foster, I., Kesselman, C.: The Grid 2: Blueprint for a New Computing Infrastructure. Morgan Kaufmann, San Francisco (2004)
2. Kurdi, H., Li, M., Al-Raweshidy, H.: A Classification of Emerging and Traditional Grid Systems. IEEE Distributed Systems Online 9 (2008)
3. Lee, J., Song, S., Gil, J., Chung, K., Suh, T., Yu, H.: Balanced Scheduling Algorithm Considering Availability in Mobile Grid. In: Proceedings of the 4th International Conference on Advances in Grid and Pervasive Computing, pp. 211–222. Springer, Heidelberg (2009)
4. Huang, C.-Q., Zhu, Z.-T., Wu, Y.-H., Xia, Z.-H.: Power-Aware Hierarchical Scheduling with Respect to Resource Intermittence in Wireless Grids. In: Proceedings of the Fifth International Conference on Machine Learning and Cybernetics, pp. 693–698 (2006)
5. Park, S.-M., Ko, Y.-B., Kim, J.-H.: Disconnected Operation Service in Mobile Grid Computing. In: Proceedings of the International Conference on Service Oriented Computing, pp. 499–513. Springer, Heidelberg (2003)
6. Litke, A., Skoutas, D., Tserpes, K., Varvarigou, T.: Efficient task replication and management for adaptive fault tolerance in Mobile Grid environments. Future Generation Computer Systems 23, 163–178 (2007)
7. Casanova, H., Legrand, A., Quinson, M.: SimGrid: a Generic Framework for Large-Scale Distributed Experiments. In: 10th IEEE International Conference on Computer Modeling and Simulation (2008)
8. A Community Resource for Archiving Wireless Data At Dartmouth, http://crawdad.cs.dartmouth.edu/dartmouth/campus/syslog/05_06

Context-Aware Hierarchy k-Depth Estimation and Energy-Efficient Clustering in Ad-hoc Network

Chang-min Mun[1], Young-hwan Kim[2], and Kang-whan Lee[1,*]

[1] Ubiquitous on Chip Laboratory, Advanced Technology Research Center, Korea University of Technology and Education Cheonan-city, Korea
{heroant,kwlee}@kut.ac.kr

[2] Laboratory of Intelligent Networks, Advanced Technology Research Center, Korea University of Technology and Education Cheonan-city, Korea
cherish@kut.ac.kr

Abstract. Ad-hoc Network needs an efficient node management because the wireless network has energy constraints. Previously proposed a hierarchical routing protocol reduce the energy consumption and prolong the network lifetime. Further, there is a deficiency conventional works about the energy efficient depth of cluster associated with the overhead. In this paper, we propose a novel top-down method clustered hierarchy, CACHE(Context-aware Clustering Hierarchy and Energy-Efficient). The proposed analysis could estimate the optimum k-depth of hierarchy architecture in clustering protocols.

Keywords: Ad hoc network, distributed clustering, hierarchy clustering, optimal depth.

1 Introduction

In the past decade, there are numerous papers about the overhead of routing clustered hierarchy. These hierarchical clustering could be classified into top-down method and bottom-up method according to the clustering procedure. Especially top-down method needs cluster head election procedure, then the network could know when to begin the procedure of clustering [1],[3]. Further, these both methods have different energy efficiency depending upon the depth of cluster. Because of energy constraint of WSN and MANET, finding optimal depth of cluster is critical problem. That is, we could decrease energy dissipation and prolong the network lifetime by estimating the optimal depth of cluster [4].

In [2], the authors have considered a level-2 hierarchical telecommunication network in which the nodes at each level are distributed according to two independent homogeneous Poisson point processes and the nodes of one level are connected to the closest node of the next higher level. In [5],[6], they provides a theoretical upper bound on the communication overhead incurred by a particular clustering algorithm for hierarchical routing in MANET. This paper addresses including k-depth coverage

* Corresponding author.

T.-h. Kim, A. Stoica, and R.-S. Chang (Eds.): SUComS 2010, CCIS 78, pp. 404–410, 2010.

the energy efficient for k-hop depth with respect to increase number of nodes and context-aware clustering hierarchy in MANET.

2 CACHE Structure

In this paper, we assumed that the network is hierarchical cluster-based. To find the optimal depth of hierarchy, we use CACHE algorithm represented below we need a energy model to estimate the energy consumption like to compare proposed clustering with [6].

2.1 Energy Model of Sensor with k-coverage

A typical sensor node consists mainly of a sensing circuit for signal conditioning and conversion, digital signal processor, and radio links [2]. The key energy parameters for communication in this model are the energy/bit consumed by the transmitter (α-trans), energy dissipated in the transmit op-amp (α-amp), and energy/bit consumed by the receiver electronics (α-recv). The energy consumption is:

$$E_{tx} = (\alpha_{trans} + \alpha_{amp} \times (d)^{\beta}) \times r \ . \quad (1)$$

Where r is the size of packet

$$E_{rx} = \alpha_{recv} \times r \ . \quad (2)$$

2.2 CACHE Algorithm

CACHE forms clusters by using a distributed algorithm, where nodes make autonomous decisions without any centralized control. A Level-0 Cluster Head(CH) selected arbitrarily. Let the Rt is a transmission range covered by Kc-times of Rc the coverage range, then Rc/Kc is the number of hop at Level-k clustering. To find optimal depth, we estimate the energy dissipation for u-level (1≤u≤n, n is number of node). The Level-0 CH could select a member node for u within the area of Rc/Kc and the member node be marked. Further, each member node be a Level-1 CH and this procedure can be recursively applied to build a u-level cluster hierarchy and determine the energy consumption for u-level cluster. This procedure could be repeated until u meets the n and find optimal depth.

3 Communication Overhead

After determining optimum depth of hierarchy architecture using CACHE algorithm, k-depth clustering is performed with different overhead. Each level cluster's size is not even, and 2 times of coverage range must be performed with its level(d) in k-depth neighbors at each level node. The average control packet(Pk) size at level(d) is

$$\begin{aligned} P_k &= N \times \left[\pi(R_C / K_C \times k)^2 - \pi(R_C / K_C \times (k-1))^2\right] \\ &= N \times \left[\pi\left(\tfrac{R_C}{K_C}\right)^2 (2k-1)\right], k \geq 1 \end{aligned} \ . \quad (3)$$

Where d is the depth of cluster, R_C is the communication range of level-0 cluster head and N is the number of nodes in the cluster.

Distance from level-0 cluster head to all nodes is $R_C / d \times k$, and according to Eq. 1, energy consumption of k-depth clustering formation is

$$E_k = (\alpha_{trans} + \alpha_{amp} \times (R_C / d \times k)^4) \times P_k \times r \ . \tag{4}$$

CACHE Algorithm

```
for d = 1, 2, ..., n
{  Ds = Round( Rc/d + 0.5 )
   {marked nodes} = NULL
   for u = 1, ..., d
   {  S(u) = {a set of nodes within Ds*u}
      S(u) = S(u) - {marked nodes}
      {marked nodes} = {marked nodes} + S(u)
      if v•1 then
      {  For v = 1, ··· , number of S(u)
         {  S(uv) means member of set S(u)
            S(uv) selects nearest one of the S(u-1) as a
            father node.
         }
      }
   }
   Sum(d) = (energy consumption of cluster)*We
            +(remained batter of each tx node)*Wb
   if minimal_sum > Sum(d) then
   minimal_sum = Sum(d)
   optimal_depth = d;
}
return optimal_depth;
end .
```

4 Analysis of CACHE

In this section, we analyze the CACHE theoretically with respect to energy dissipation and control packet overhead. And we compared these with the [6].

4.1 Experimental Setup

For our experiments, we used 20, 50 and 100 node network where nodes were randomly distributed in circle. The network area is assumed to be a circle and various radii were used for each experiment where radii are 15, 30 and 50. Each data message was 2048 bytes long. For the experiments described in this paper, the communication energy parameters are set as:

$$\alpha_{trans} = 50nJ / bit \, , \ \alpha_{amp} = 0.0013 pJ / bit / m^4 \, .$$

4.2 Energy Dissipation in Cluster with CACHE

With CACHE, we can analytically determine the optimal value of k in CACHE using the computation and communication energy models. According Eq.(1), the dissipated energy in the cluster is

$$E_{tx} = [\sum_{i=0}^{N-1}\sum_{j=1}^{d}(\alpha_{trans} + \alpha_{amp} \times (R_{ijCH})^4) \times r] \cdot \tag{5}$$

Where R_{ijCH} is distance from member node i to Cluster Head j.

$$E_{rx} = \sum_{i=0}^{N-1}\sum_{j=1}^{d}(\alpha_{recv} \times r) \tag{6}$$

$$E_{total} = E_{tx} + E_{rx} \tag{7}$$

$$Cost_{total} = E_{total} \times \omega_E + \sum_{i=0}^{N-1}\sum_{j=1}^{d}(Battery_{node\_i}) \times \omega_B \tag{8}$$

Where ω_B is weight value of remained energy of each node, ω_E is weight value of consumed energy and the $Battery_{node\_i}$ means the remained energy of each node.

In this approach, assuming N is the number of nodes in a cluster, we can get minimum value of total link cost with Eq. 8 and the level of cluster as well. Fig. 1 is the Energy dissipation upon level of cluster where the spatial density is 1.5, 5 and 10.

In order to verify the performance of protocol, we compare CACHE with [6]. The weight value of remained energy of each node is zero. In Fig. 1, show how the energy consumption decreases as the number of levels in the hierarchy increases. The energy savings by algorithm proposed in [6] is observed at level-1 cluster, and there is lesser energy savings after level-1 cluster. The energy savings with increase in the number of levels in the hierarchy are also observed to be more significant for CACHE than [6]. This can be explained that the proposed algorithm in [6] consume 1unit energy per 1unit datagram forwarding, but CACHE uses multiple mode for energy dissipation depending on distance between two nodes. The number of control packet depend on the level of cluster is given in Fig. 2.

4.3 Optimum Depth of Cluster

The throughput of network could be estimated by not only energy consumption but also network capacity about medium material. In this paper, we use the normalized value of energy and network latency as a indicator of throughput of network clustered hierarchy.

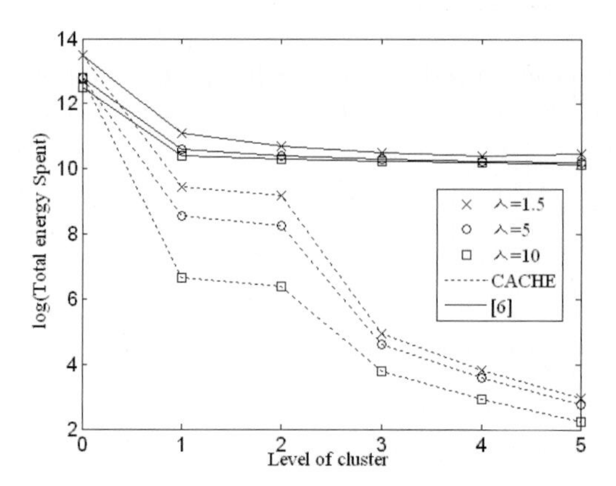

Fig. 1. Total energy consumption vs. depth of cluster in the clustering with three different density, 1.5, 5, 10 (marker is *'x', circle, square*). This shows that the total energy dissipation of proposed algorithm (*dotted lines*) is lower than [6] (*solid lines*).

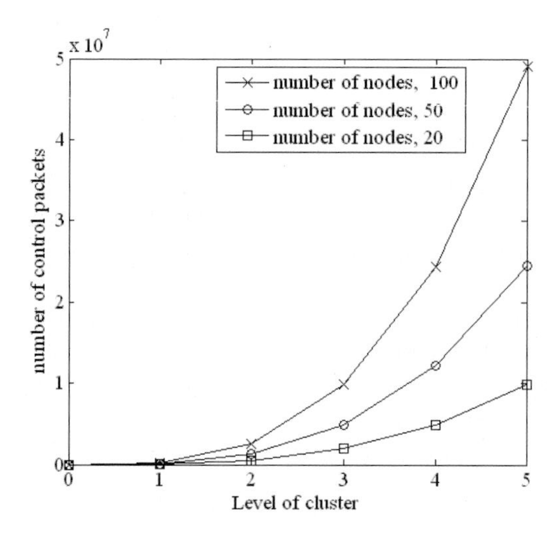

Fig. 2. The number of control packet depending on a depth of cluster. The number of control packet increase exponentially as the depth of cluster varies 1 to 5.

The Fig. 3 gives the assessment which normalized with Eq. 3 and Eq. 4. In the figure, the dissipated energy represents the total consumed energy in a cluster and the packet latency means the overhead by packet retransmit. In the fig. 3 we could find lowest value of the normalized assessment, and the level of this value will influence the network throughput better.

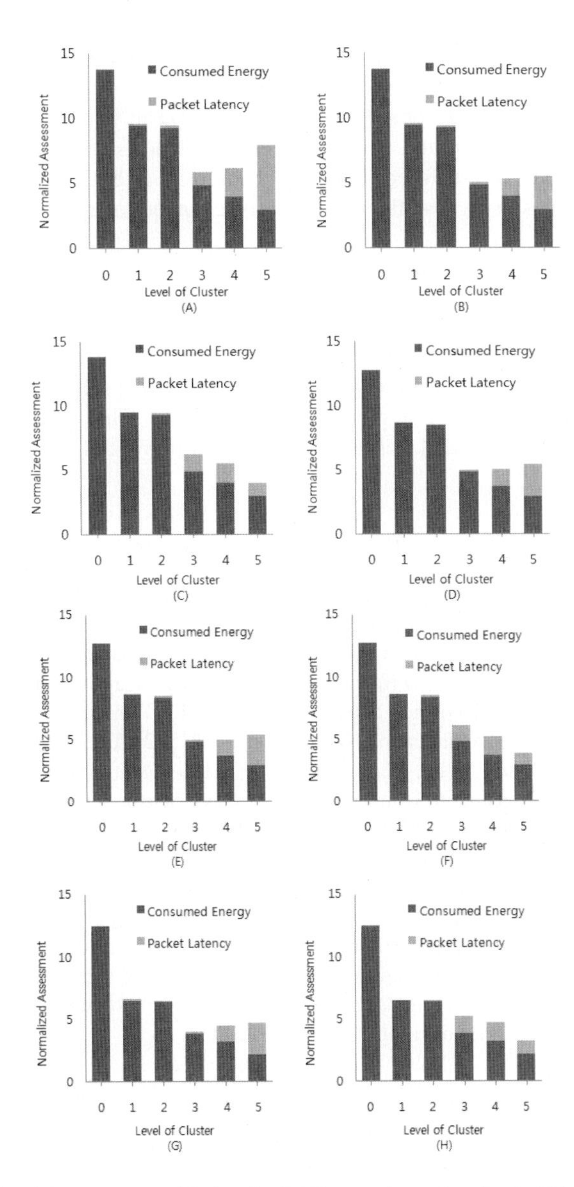

Fig. 3. Normalized Assessment with density λ. The number of nodes is 100 (*A, B, C*), 50(*D, E, F*), and 20(*G, H*).

5 Conclusions

A conclusion section is not required. Although a conclusion may review the main points of the paper, do not replicate the abstract as the conclusion. A conclusion might elaborate on the importance of the work or suggest applications and extensions.

In this paper, assuming the link cost in the hierarchical cluster, we try to find energy efficient level of clustering k-depth. The results from our model results show that increasing depth of hierarchy provide significant energy conserving, but the control packet overhead is increased as shown Fig. 2. Hence, we could assume the tradeoff between consumed energy and packet overhead. The minimum value of product of dissipated energy and control packet overhead is reasonable for the depth of cluster. The y-axis' are different both Fig. 1 and Fig. 2, but the number of control packets make re-transmission so the unit could be changed to energy.

Acknowledgments

This research is financially supported by the Ministry of Knowledge Economy(MKE) and Korea Institute for Advancement in Technology (KIAT) through the Workforce Development Program in Strategic Technology. And This paper was partially supported by the Education and Research Promotion Program of KUT.

References

1. Tashtarian, F., Honary, M.T., Haghighat, A.T., Chitizadeh, J.: A New Energy-Efficient Level-based Clustering Algorithm for Wireless Sensor Networks. In: 6th IEEE International Conference on Information, Communications and Signal Processing, pp. 1–5 (2007)
2. Handy, M.J., Haase, M., Timmermann, D.: Low Energy Adaptive Clustering Hierarchy with Deterministic Cluster-Head Selection. In: 4th IEEE International Conference on Mobile and Wireless Communication Networks, pp. 368–372 (2002)
3. Bansa, N., Sharma, T.P., Misra, M., Joshi, R.C.: FTEP: A Fault Tolerant Election Protocol for Multi-level Clustering in Homogeneous Wireless Sensor Networks. In: 16th IEEE International Conference on Networks, pp. 1–6 (2008)
4. Heinzelman, W.B., Chandrakasan, A.P., Balakrishnan, H.: An Application-specific protocol architecture for wireless microsensor networks. IEEE Transaction on Wireless Communications, 660–670 (2002)
5. Bandyopadhyay, S., Coyle, E.: An energy efficient hierarchical clustering algorithm for wireless sensor networks. In: INFOCOM, vol. 3, pp. 1713–1723 (2003)
6. Sucec, J., Marsic, I.: Clustering Overhead for Hierarchical Routing in Mobile Ad Hoc Networks. In: INFOCOM, vol. 3, pp. 1698–1706 (2002)

Two Scheduling Schemes for Extending the Lifetime of Directional Sensor Networks*

Joon-Min Gil[1], Chan-Myung Kim[2], and Youn-Hee Han[2,**]

[1] School of Computer and Information Communications Engineering,
Catholic University of Daegu
jmgil@cu.ac.kr

[2] School of Computer Science and Engineering,
Korea University of Technology and Education
{cmdr,yhhan}@kut.ac.kr

Abstract. Since directional sensor networks (DSNs) are composed of a large number of sensors equipped with a limited battery and limited angles of sensing range, maximizing network lifetime while covering all the targets in a given area is still a challenge problem. In this paper, we first address the MSCD (Maximum Set Cover for DSNs) problem that is known as NP-complete. We then present a new target coverage scheduling scheme to solve this problem with a greedy algorithm. We also present another target coverage scheduling scheme based on a genetic algorithm that can find an optimal solution for target coverage by evolutionary global search technique. To verify and evaluate these schemes, we conduct simulations and show that they can contribute to extending the network lifetime. Simulation results indicate that a genetic algorithm-based scheduling scheme has a better performance than a greedy algorithm-based one in terms of the maximization of the network lifetime.

Keywords: Target Coverage, Directional Sensors, Greedy Algorithms, Genetic Algorithms.

1 Introduction

Wireless sensor networks (WSNs) have been employed in various application fields, such as environmental monitoring, battlefield surveillance, smart spaces, and so on [1]. The networks are typically composed of a great number of sensors that have sensing, data processing, and communication functionalities. These sensing nodes should cover all the targets that have a fixed location in a given area. This is referred as the *target coverage* problem in WSNs [2].

* This research is financially supported by the Ministry of Knowledge Economy (MKE) and Korea Institute for Advancement in Technology (KIAT) through the Workforce Development Program in Strategic Technology, and supported by Basic Science Research Program through the National Research Foundation of Korea (NRF) funded by the Ministry of Education, Science and Technology (No. 2010-0015637).

** Corresponding author.

T.-h. Kim, A. Stoica, and R.-S. Chang (Eds.): SUComS 2010, CCIS 78, pp. 411–420, 2010.

For the target coverage problem, it is essential that sensors should monitor all the targets continuously for a long time as possible. Initially, each sensor has a limited battery. Once sensors are randomly scattered, it is hardly possible to replace their battery by a new one or be recharged [3]. Accordingly, under such circumstances, the problem of maximizing the network lifetime while covering all the targets is an important issue. To achieve this purpose, each sensor should minimize its battery consumption in an energy-efficient manner. Typically, the radio state of sensors falls into four kinds of states: transmit, receive, idle, and sleep [2]. We can denote transmit, receive, and idle states as an active state, because these three states consume more energy than a sleep one. Therefore, a scheduling scheme to properly alternate between active and sleep states will be a promising method to extend the network lifetime.

There have been a lot of works to maximize the network lifetime by alternating sensors between active and sleep states. In particular, these works have assumed that WSNs have omni-directional sensors, each of which can sense an omni-directional range at every instance [2,4]. Recently, directional sensors such as camera sensors, ultrasonic sensors, infrared sensors, etc. have emerged in sensor markets and the networks consisting of such sensors, *i.e.*, *directional sensor networks (DSNs)*, are widely used. Each sensor in DSNs has a sensing range as a sector. Unlike WSNs, target coverage in DSNs is determined by both location and direction of sensors. This feature of DSNs makes a target coverage scheduling more complex. As a result, maximizing the network lifetime of DSNs is still a challenge problem. Nevertheless, there are few works dealing with the target coverage problem in DSNs.

In this paper, we study a problem of target coverage scheduling in DSNs, in which directional sensors have limited battery capacity and are randomly and densely deployed to cover all targets. We describe the MSCD (Maximum Set Cover for DSNs) problem that finds the cover sets monitoring all the targets in an energy-efficient way and maximizes the network lifetime by assigning different scheduling time to each cover set. As referred in [5], this problem is known as NP-complete. To solve the problem, we first devise a greedy heuristic algorithm which has the advantage of finding a solution faster than other heuristic ones. Due to its local search, however, the greedy algorithm may fail to find an optimal solution for target coverage with the objective of maximizing the network lifetime of DSNs. As another solution, we secondly introduce a genetic algorithm, based on evolutionary global search techniques, into the target coverage problem in order to find optimal cover sets in DSNs. Simulation results verify that these two schemes can solve the MSCD problem. They also show that the genetic algorithm-based target scheduling scheme can find cover sets with more extended network lifetime as compared to the greedy algorithm-based one.

The rest of this paper is organized as follows. In Section 2, we formally define the MSCD problem. A target scheduling scheme based on a greedy algorithm to solve the problem is also given in this section. In Section 3, we propose another target scheduling scheme based on a genetic algorithm. This section also provides the detailed descriptions of our genetic algorithm. In Section 4, we present the performance evaluation of these schemes with simulations. Section 5 concludes the paper.

2 Maximum Set Cover for DSNs

In this section, we define the MSCD (Maximum Set Cover for DSNs) problem and present a greedy algorithm to solve the problem.

2.1 MSCD Problem

Let us consider a DSN composed of N sensors, each of which has W directions and operates only one direction with a uniform sensing range at any instance. We also consider that all sensors are randomly scattered to cover M targets in a two-dimensional plane. We define $S = \{s_1, s_2, \ldots, s_N\}$ as the set of N sensors and $R = \{r_1, r_2, \ldots, r_M\}$ as the set of M targets. Unlike a sensor network composed of omni-directional sensors, a DSN should additionally consider the definitions concerned with sensor directions.

- $D_{i,j}$: the jth direction of a sensor s_i ($i = 1, 2, \ldots N$ and $j = 1, 2, \ldots, W$). We assume that a sensor s_i has not any overlap between two neighbor directions.
- D: the collection of $D_{i,j}$ for $i = 1, 2, \ldots, N$ and $j = 1, 2, \ldots, W$.
- C_k ($\subseteq D$): the kth set of the directions that cover all targets in R such that every element in C_k covers at least one element in R and every two elements in C_k cannot belong to the same sensor in S. We call this set C_k a *cover set*.
- R_m ($\subseteq D$): the set of the directions that cover a target r_m ($m = 1, 2, \ldots, M$).
- L_i: the lifetime of a sensor s_i. We assume that a sensor s_i spends a uniform energy regardless of its direction when it is active.
- t_k: the allocated active time for the kth cover set ($0 \leq t_k \leq 1$).

We organize the directions in D into K cover sets, where K is the maximum number of cover sets for a given coverage relationship between S and R. Since $D_{i,j}$ can belong to multiple cover sets until the lifetime of a sensor s_i, L_i, completely runs down, we can define a boolean variable $x_{i,j,k}$ as in [5]:

$$x_{i,j,k} = \begin{cases} 1 & \text{if } D_{i,j} \in C_k \\ 0 & \text{otherwise.} \end{cases} \quad (1)$$

By the way presented in [5], we define the MSCD problem as follows.

$$\text{Maximize} \sum_{k=1}^{K} t_k \quad (2)$$

$$\text{subject to} \sum_{k=1}^{K} \sum_{j=1}^{W} x_{i,j,k} \cdot t_k \leq L_i, \forall s_i \in S \quad (3)$$

$$\sum_{j=1}^{W} x_{i,j,k} \leq 1, \forall s_i \in S, k = 1, 2, \ldots, K \quad (4)$$

$$\sum_{D_{i,j} \in R_m} x_{i,j,k} \geq 1, \forall r_m \in R, k = 1, 2, \ldots, K \tag{5}$$

$$\text{where} \quad x_{i,j,k} = \{0, 1\} \text{ and } t_k \geq 0$$

Greedy algorithm for the MSCD problem $(S,\ D,\ R,\ t)$

1: set L_i of each sensor to 1
2: $SENSORS = S$
3: $DIRECS = D$
4: $k = 0$
5: **while** each target is covered by at least one direction in $DIRECS$ **do**
6: $k = k + 1$
7: $C_k = \emptyset$
8: $TARGETS = R$
9: **while** $TARGETS \neq \emptyset$ **do**
10: $D_c = \emptyset$
11: find a critical target $r_c \in TARGETS$
12: find all directions $\in DIRECS$ that cover r_c and insert them into D_c
13: select a direction $D_{s,t} \in D_c$ with the greatest contribution
14: $C_k = C_k \cup \{D_{s,t}\}$
15: **for each** direction $D_{i,j} \in DIRECS$ **do**
16: **if** $i = s$ **then**
17: $DIRECS = DIRECS - \{D_{i,j}\}$
18: **end if**
19: **end for**
20: **for each** target $r_i \in TARGETS$ **do**
21: **if** r_i is covered by the direction $D_{s,t}$ **then**
22: $TARGETS = TARGETS - \{r_i\}$
23: **end if**
24: **end for**
25: **end while**
26: **for each** direction $D_{x,y} \in C_k$ **do**
27: $L_x = L_x - t$
28: **if** $L_x \leq 0$ **then**
29: $SENSORS = SENSORS - \{s_x\}$
30: **end if**
31: **end for**
32: $DIRECS = \cup_{j=1}^{W} \{D_{i,j}\}$ **for each** sensor $s_i \in SENSORS$
33: **end while**
34: **return** k-number of cover sets and the cover sets $C_1, C_2, \ldots, C_k$

Fig. 1. The greedy algorithm to solve the MSCD problem

2.2 Greedy Algorithm

Fig. 1 describes the details of the greedy algorithm devised to solve the MSCD problem by using a same active time t for all cover sets. It is similar to the one proposed in [2], but modified to capture the characteristics of DSNs. The algorithm consists of the following steps:

Step 1 Initialize the energy of each sensor and the variables $SENSORS$, $DIRECS$, and k. (lines $1 \sim 4$).

Step 2 Increase k by one and initialize the kth cover set and the variable $TARGETS$ (lines $6 \sim 8$).

Step 3 Initialize the variable D_c and a *critical target* r_c is selected (lines $10 \sim 11$). As the critical target, we select the target most sparsely covered in terms of the number of sensors.

Step 4 Once the critical target r_c has been selected, our algorithm selects the direction $D_{s,t}$ with the greatest contribution that covers the critical target (lines $12 \sim 13$). Various contribution functions can be defined. In this paper, we use the following function F:

$$F(D_{i,j}, r_c) = \alpha \cdot N_{i,j,c} + (1 - \alpha) \cdot L_i, \quad 0 \leq \alpha \leq 1. \tag{6}$$

$$D_{s,t} = \arg\max_{D_{i,j}} F(D_{i,j}, r_c). \tag{7}$$

where $N_{i,j,c}$ denotes the number of targets which the direction $D_{i,j}$ covers while $D_{i,j}$ already covers the target r_c. By choosing a proper value of α, the direction $D_{s,t}$ will be selected such that it covers a larger number of uncovered targets and the sensor s_s with the selected direction has more residual energy available.

Step 5 Once a direction $D_{s,t}$ has been selected, it is added to the current cover set C_k (line 14), and other directions of the same sensor s_i are removed from the $DIRECS$ set (lines $15 \sim 19$).

Step 6 All targets additionally covered by $D_{s,t}$ are removed from the $TARGETS$ set (lines $20 \sim 24$). When all targets are covered, the new cover set was formed. The condition in line 9 guarantees that a new cover set will cover all targets.

Step 7 After a cover set C_k has been formed, the lifetime of each sensor in C_k is updated (lines $26 \sim 31$). Once a sensor finishes its lifetime, it is removed from the set of available sensors, $SENSORS$.

Step 8 Before going to line 5 to find a new cover set, the set of available directions $DIRECS$ is updated based on the set $SENSORS$ (line 32).

3 Extending Network Lifetime by Genetic Algorithms

This section presents a genetic algorithm-based target coverage scheduling scheme which can solve the MSCD problem and get much more the network lifetime than the greedy-algorithm-based one.

3.1 Representation

Each chromosome in a population represents a candidate solution encoded as the direction of sensors for the MSCD problem. Fig. 2 illustrates the two-dimensional

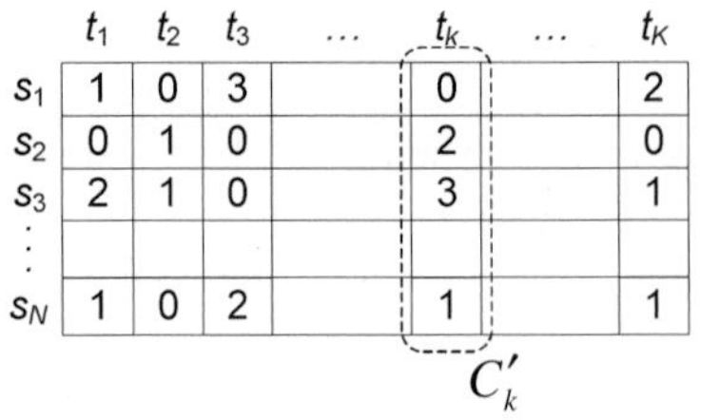

	t_1	t_2	t_3	...	t_k	...	t_K
s_1	1	0	3		0		2
s_2	0	1	0		2		0
s_3	2	1	0		3		1
⋮							
s_N	1	0	2		1		1

C'_k

Fig. 2. Chromosome representation

chromosome which represents a grid with N rows and K columns. In Fig. 2, a gene $g_{i,k}$, that means the ith row and the kth column ($i = 1, 2, \ldots, N$, $k = 1, 2, \ldots, K$) in the chromosome, is interpreted as

$$g_{i,k} = \begin{cases} 0 & \text{if sensor } s_i \text{ is sleep} \\ j & \text{if } D_{i,j} \text{ of sensor } s_i \text{ is active for } j = 1, 2, \ldots, W. \end{cases} \quad (8)$$

Every gene in the chromosomes of an initial population is randomly set to a integer value by using Eq. (8). One column in the chromosome of Fig. 2 corresponds to a candidate cover set C'_k. Such a candidate becomes a cover set C_k when the following three conditions are all satisfied: 1) all the targets should be covered by the directions in C'_k, 2) C'_k should have the directions, each of which can cover at least one target, and 3) when target sets T_1 and T_2 are respectively covered by any two directions D_1 and D_2 in C'_k, $T_1 \not\subseteq T_2$ and $T_2 \not\supseteq T_1$.

The total network lifetime is calculated as $K' \cdot t$ $(0 \leq K' \leq K)$, by counting C'_k satisfying the above conditions for $k = 1, 2, \ldots, K$.

3.2 Fitness Function

The two-dimensional chromosome presented in Fig. 2 is evaluated to find efficient cover sets with a fitness function. The evolutionary process of our genetic algorithm should achieve two kinds of objectives: 1) the network lifetime should be extended with as many cover sets as possible and 2) each sensor should have the residual energy as much as possible in the sense that it minimizes its energy consumption. To achieve these two objectives properly, we define a weighted fitness function f as follows.

$$o_1 = \kappa_1 \cdot K', \quad o_2 = \kappa_2 \cdot \sum_{i=1}^{N} L_i \quad (9)$$

$$f = \omega \cdot o_1 + (1 - \omega) \cdot o_2 \quad (10)$$

where o_1 and o_2 represent object functions for the network lifetime and the residual energy of all sensors, respectively. κ_1 and κ_2 represent scaling parameters for o_1 and o_2, respectively. ω represents a weighted parameter determining a significance of each object function.

In Eq. (10), the value of ω can be determined by the design goal of DSNs. If ω is set to 1, the final form of a DSN will be made focusing on only the maximization of network lifetime. With a value of 0.5 for ω, we will obtain a DSN whose lifetime can be extended while remaining residual energy much more.

3.3 Reproduction

Reproduction process is the core of a genetic algorithm. In the reproduction process, a selection mechanism is used to organize a new population from a current population [6]. Among various selection mechanisms, we use the following three ones.

- **Elitist:** The best runner and runner-up chromosomes are chosen in a current population and copied to a new population without any change; *i.e.*, these two chromosomes survive in the next generation.
- **Replace:** Two chromosomes with the lowest fitness in a current population are replaced by newly created chromosomes.
- **Roulette wheel:** The selection of chromosomes from a current population depends on the proportion of each chromosome's fitness to total fitness. As the fitness gets higher, the probability to be chosen as parents in a current population becomes more; *i.e.*, chromosomes with high fitness are more copied to the population of the next generation.

Given the population size P, $P-4$ chromosomes in a new population are made by the roulette wheel selection mechanism. The elitist and replace selection mechanisms are applied to produce the remaining four chromosomes.

3.4 Genetic Operators

Here, we describe crossover and mutation operators to achieve the extended network lifetime in a DSN. In general, a crossover is a process to take two parents and produce offstrings from them with an aim to get better chromosomes in the next generation. After a crossover point is randomly chosen, the part from the beginning of chromosome to the crossover point is copied from one parent. The rest is copied from the second parent [6]. We design two types of crossovers suitable for the structure of two-dimensional chromosome shown in Fig. 2.

- **Crossover for inter-cover sets:** This crossover is used to exchange cover sets in two chromosomes between each other. Fig. 3(a) illustrates an example of this crossover operation. By using this crossover, we expect the network lifetime to be extended.
- **Crossover for intra-cover sets:** This crossover is applied for two cover sets in a chromosome. Fig. 3(b) illustrates an example of this crossover operation. Directions in two cover sets are exchanged with each other to find a new cover set with the energy efficiency of directional sensors.

Mutation is used to maintain the genetic diversity in a population [6]. When a mutation operation is performed according to a mutation probability p_m,

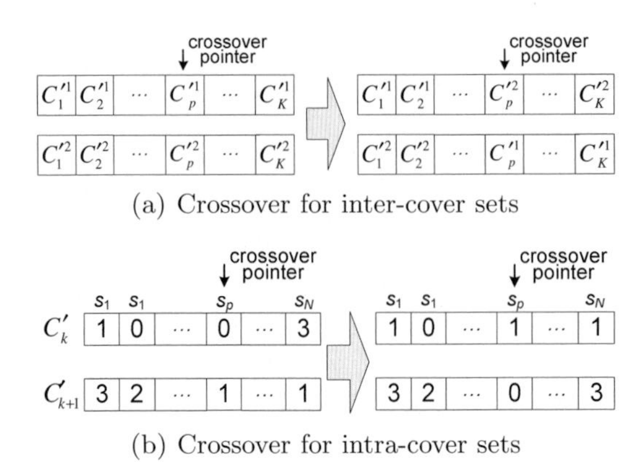

Fig. 3. Example of two crossover operations

it modifies a previous gene $g_{i,k}$ in chromosomes into a new gene $g'_{i,k}$ according to a random value β, as follows.

$$g'_{i,k} = \begin{cases} j & \text{if } 0 \leq \beta < \frac{1}{3} \\ (g_{i,k} - 1)\%(W+1) & \text{if } \frac{1}{3} \leq \beta < \frac{2}{3} \\ (g_{i,k} + 1)\%(W+1) & \text{if } \frac{2}{3} \leq \beta < 1 \end{cases} \tag{11}$$

where j is a random value ranging from 0 to W. In Eq. (11), according to a random value β, a sensor s_i can switch its current direction to the randomly chosen one. It can also switch to the left or right direction from the current one.

4 Performance Evaluation

In this section, we evaluate and analyze the performance of the proposed two schemes through simulations. The performance comparison for the schemes is also presented.

4.1 Simulation Environment

Our simulation environment assumes that the different numbers of targets ($M = 5$ and 10) are uniformly deployed in a region of $500m \times 500m$ and the different numbers of directional sensors ($N = 10, 15, 20, 25$, and 30) are randomly scattered in the region. It also assumes that all sensors have the same sensing range of $250m$ and face one of three directions, each of which has a direction angle of $\frac{2\pi}{3}$ ($W = 3$). Fig. 4 shows an example of a target and sensor deployment in our simulation environment when $M = 5$, $N = 10$, and $W = 3$.

In the greedy algorithm-based scheduling scheme, the parameter of a contribution function (α in Eq. (6)) is set to 0.5. In the genetic algorithm-based scheduling scheme, the chromosomes with 500 cover sets ($K = 500$) are encoded to represent candidate solutions. Each gene in these chromosomes is initially set to a random value with the range of $[0, 3]$. Table 1 summarizes the parameters and values used in our simulations.

Table 1. Parameters and values used in our simulations

Parameter	Value
Number of targets (M)	5, 10
Number of directional sensors (N)	10, 15, 20, 25, 30
Number of directions (W)	3
Sensing range	$250m$
Population size (P)	100
Number of generations	300
Crossover probability (p_c)	0.5
Mutation probability (p_m)	0.01
Weighted parameter for fitness function (ω)	0.9

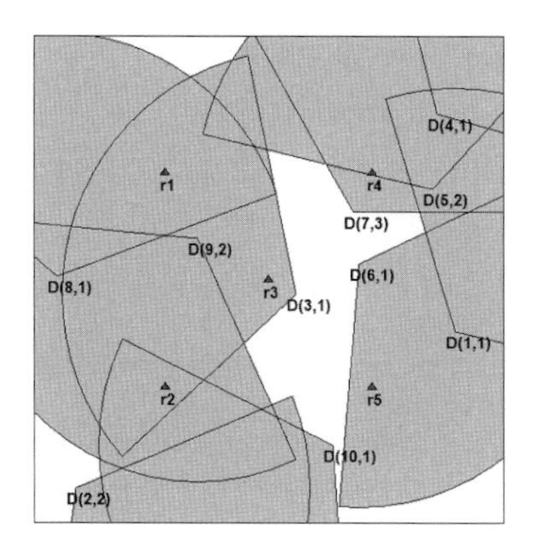

Fig. 4. An example of target and sensor deployment

Fig. 5. Comparison of network lifetimes

4.2 Simulation Results

Our simulations were conducted according to the diverse numbers of directional sensors when 5 and 10 targets with a fixed location were deployed, respectively. Each of the 10 simulations was run for the proposed two schemes. Fig. 5 shows the average network lifetimes obtained from these simulations.

For both schemes, the network lifetime for 10 targets is shorter than that for 5 targets. When more targets are deployed, they should be covered by a greater number of directions in cover sets, consuming the more energy of directional sensors. It is apparently shown that as the number of directional sensors increases, the network lifetime becomes higher. This is because the increase of directional sensors can lead to finding more cover sets. From this figure, we can also see that our genetic algorithm-based scheme has more extended network lifetime than our greedy algorithm-based one, regardless of the number of directional sensors. This result indicates that the genetic algorithm-based scheme can find a near-optimal solution to the MSCD problem by global search, compared with the greedy algorithm-based one that is dependent on a heuristic way.

5 Conclusions

In this paper, we have studied the target coverage scheduling for DSNs. Contrary to conventional sensor networks, DSNs are composed of a number of directional sensors with limited sensing ranges and directions, and thus the target scheduling to maximize the network lifetime needs a highly sophisticated optimization technique. We have presented two target scheduling schemes, the greedy algorithm-based and the genetic algorithm-based ones, to solve the MSCD problem that is known as NP-complete. Simulation results showed that these schemes can find the cover sets monitoring all the targets by switching directions in an energy-efficient way. They also showed that, by an evolutional global search technique, the genetic algorithm-based scheme gets the longer network lifetime than the greedy algorithm-based one. As a future work, we plan to extend the schemes to maximize the network lifetime of DSNs considering the communication ranges of sensors.

References

1. Akyildiz, I.F., Su, W., Sankarasubramaniam, Y., Cayirci, E.: Wireless sensor networks: a survey. Computer Networks 38, 393–422 (2002)
2. Cardei, M., Thai, M.T., Li, Y., Wu, W.: Energy-efficient target coverage in wireless sensor networks. In: IEEE INFOCOM 2005, pp. 1976–1984 (2005)
3. Zhou, G., He, T., Krishnamurthy, S., Stankovic, J.A.: Models and Solutions for radio irregularity in wireless sensor networks. ACM Transations on Sensor Networks 2(2), 221–262 (2006)
4. Pyun, S.-Y., Cho, D.-H.: Power-saving scheduling for multiple-target coverage in wireless sensor networks. IEEE Communications Letters 13(2), 130–132 (2009)
5. Ai, J., Abouzeid, A.A.: Coverage by directional sensors in randomly deployed wireless sensor networks. J. of Combinatorial Optimization 11(1), 21–41 (2006)
6. Sivanandam, S.N., Deepa, S.N.: Introduction to Genetic Algorithms. Springer, New York (2008)

Data Aggregation Using Mobile Agent Mechanism on Distributed Sensor Networks

Youn-Gyou Kook, Joon Lee, Ki-Seock Choi, Jae-Soo Kim, and R. Young-Chul Kim

Dept. of NTIS (National Science & Technology Information Service)
KISTI (Korea Institute of Science and Technology Information)
Daejeon, 305-806, Korea
{ykkook,rjlee98,choi,jaesoo}@kisti.re.kr
Dept. of Computer Information & Comm. Hongik University
Jochiwon, 339-701, Korea
bob@hongik.ac.kr

Abstract. In this paper, we propose mobile agent mechanism to efficiently aggregate data on distributed sensor networks (DSN), and describe the itinerary of mobile agent to establish by the sink node. The proposal of this mechanism is to reduce consumption of sensor network energy, network bandwidth and the limited resources. This mobile agent mechanism is based on mobile-agent-based DSN (MADSN) which the sink node migrate mobile agent to the sensor node on multicasting based on the binomial tree. It reduces time-cost to aggregate sensor data on MADSN.

1 Introduction

Since there are the limited resources, powers and network bandwidth on distributed sensor networks, it is important to aggregate sensor data in efficient on that environments. It needs to communicate with sensors for aggregating their data and transporting that to an application or some nodes. To efficiently aggregate sensor data in distributed sensor networks reduces consumption of sensor's power and network resources, and it can prolong life of sensor networks.

So many researchers proposed the methodology of prolonging sensor network's life, H. Qi suggested Mobile-Agent-based Distributed Sensor Networks (MADSN) as a solution of that[3, 5]. Mobile agent is small size programmed code which is able to migrate to the other nodes. That node has to be able to communicate with the other node and send/receive a mobile agent. MADSN is improved the traditional distributed sensor networks (DSN), that can migrate mobile agent in DSN. MADSN has the several benefits that are consumption of low power, extensibility, scalability and stability than DSN.

The sink node migrate mobile agent to the sensor nodes to aggregate sensor data in MADSN. There are the various methods of migration that the sink node migrate mobile agent to the all sensor nodes. The broadcasting is the sink node migrate mobile agent to the all sensor nodes one by one. The unicasting is same the ring topology that the sink node migrate mobile agent to the first sensor node with the itinerary to migrate the next node. And the multicasting is mixed migration method with

T.-h. Kim, A. Stoica, and R.-S. Chang (Eds.): SUComS 2010, CCIS 78, pp. 421–426, 2010.

broadcasting and unicasting[5,8,9]. The itinerary of multicasting is based on the binomial tree topology, the sink node migrate mobile agent some sensor nodes and a sensor node migrate that some sensor nodes. The binomial tree is the recursive tree structure, that is the optimization route path of mobile agent[1,8]. It reduces the time of migrating mobile agent and aggregating sensor data in efficient.

This paper proposes mobile agent mechanism based on the binomial tree. The itinerary of mobile agent is not static route path, and is to dynamically establish based on the location of sensor nodes and the priority of ones. To establish the dynamic itinerary of mobile agent, the sink node have to acquire the information of sensor node that are the level of tree, input node, input degree, sub tree, output degree and output nodes. This paper presents the method of finding the value of that to dynamically establish the itinerary of mobile agent and aggregating the sensor data in MADSN.

The organization of this paper is as follows. Section 2 descript MADSN and present the redesigned that. In section 3, we propose mobile agent mechanism based on the binomial tree. Section 4 presents the conclusion.

2 Mobile-Agent-based Distributed Sensor Network (MADSN)

In traditional DSNs, all of sensors collect sensor data and transmit data to the sink node or a higher-level processing element as like client-server paradigm. It decreases the life of sensor network to spend the power of sensors and to allocate the network bandwidth because large amounts of data are moved around the sensor network. And it is difficult to expand the scope of sensor networks that have to communicate sensors and sink node.

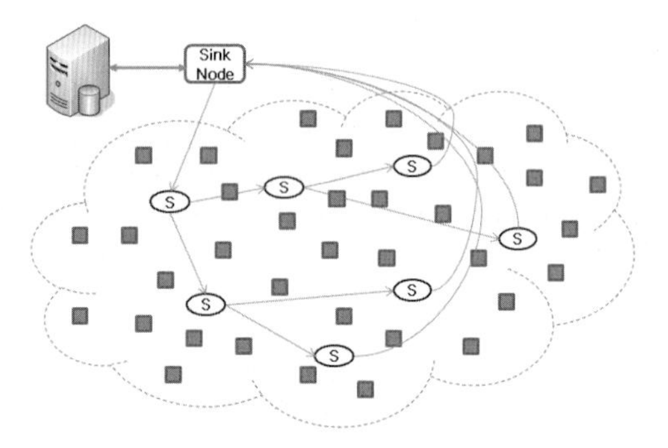

Fig. 1. The overview of mobile-agent-based distributed sensor network (MADSN)

So many researchers proposed the methodology of prolonging sensor networks, H. Qi designed MADSN and improved DSN architecture[5]. He presented the several important benefits as following that:

- Network bandwidth requirement is reduced. Mobile agent with small size moved around networks instead of round trip with large amount of raw data.
- Better network scalability. The architecture of MADSN is able to increase the number of sensor and expand the scope of network.

- Extensibility. Because mobile agent is small size code, we can improve the capability of sensor networks.
- Stability. If mobile agent migrates to the other node, then only the network connection is alive.

We redesigned the architecture of MADSN based on [5, 7], suggest mobile agent mechanism in those environments. The sink node communicates with the application server and launches mobile agent to the sensor node which is able to receive and send that. The sink node transmits the sensor data which is aggregated by mobile agent to the application server, and the application server sends a policy or the small size code (mobile agent) to aggregate sensor data.

3 Mobile Agent Mechanism

The itinerary of mobile agent in MADSN is based on the location of sensor and the priority of sensor nodes. It is important to order the route path of mobile agent which is comprised with sensor nodes topology based on the binomial tree to aggregate sensor data. Let n is the total number of sensor nodes in which have to visit for gathering sensor data. S is set of sensor nodes, as follows that: $S = \{S_{i-1} \mid i = 1,2,\cdots,n\},\ n < \infty$. S is the set of between S_0 to S_{n-1} that is composed of the sensor node's location. It is established dynamically the route path of sensor on which the sink node transports the mobile agent to aggregate sensor data. And P is set of the sensor node priorities, as follows that: $P = \{P_{i-1} \mid i = 1,2,\cdots,n\},\ n < \infty$. P is set of between P_0 to P_{n-1} that is composed of the sensor node priority which is established at the sink node to aggregate sensor data. Figure 2 shows the itinerary of mobile agent which is based on the binomial tree.

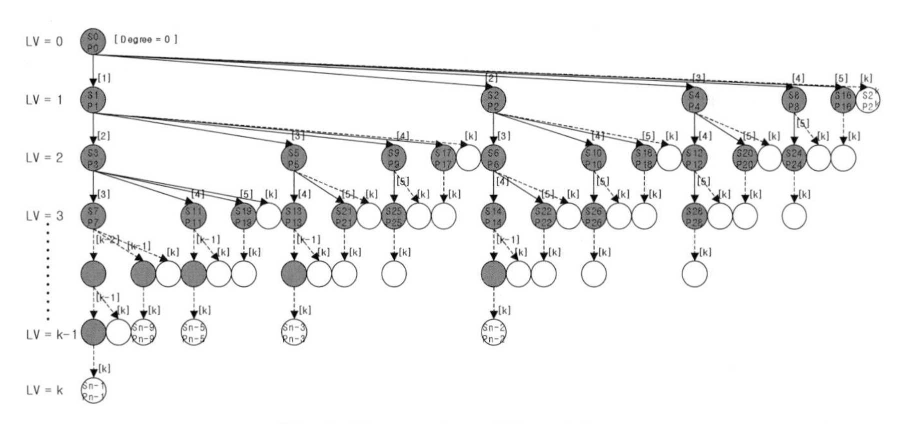

Fig. 2. The overview of binomial tree

So, S_i is composed of 6 elements which are LV_i, IN_i, ID_i, ST_i, OD_i, ON_i. The means of that is as follows: LV is a level of binomial tree, IN is an input node as the parent node, and ID is an input degree from the parent node. ST is a sub tree of node,

OD are an output degree to the child node, and *ON* is an output node to the child nodes. Where is a node S_i of the binomial tree *S*, $S_i \in S$, $S_i = \{LV_i, IN_i, ID_i, ST_i, OD_i, ON_i\}$. We defined the several variables to find the values of S_i's elements as follows that:

- n : the total number of sensor nodes to migrate the binomial tree
- l : the depth of a node at the binomial tree
- k : the maximum bit code size of a node
- m : the number of a node

If the total number of sensor nodes to migrate the binomial tree is *n* and the depth of that tree is *l*, then the formula of the binomial tree nodes is $2^l \leq (n-1) < 2^{l+1}$. We need to find the log value of that to define the maximum bit code size of sensor nodes.

$$2^l \leq (n-1) < 2^{l+1} \xrightarrow{\log} l \leq log_2(n-1) < l+1$$
$$\therefore k = l \tag{1}$$

And the order of mobile agent itinerary at the binomial tree is replaced by the number of a node *m*. The sink node establishes dynamically a set of *m* value, *m* value of S_i is not fixed. All of *m* value is progression 0 to *n-1* as below.

$$m = \prod_{i=0}^{k}(a_i * 2^i), \ \ a_i = [0,1] \tag{2}$$

Through the values of k, m from (1), (2), the sink node acquires the elements of S_i. When m's value is 0, m is the first sensor node to visit by mobile agent. LV is same the depth of a node. R of IN means root node which is the first sensor node.

$$LV = l$$
$$IN = \begin{cases} R\ (m=0) \\ m - 2^{k-1}\ (m \neq 0) \end{cases}$$
$$ID = \begin{cases} 0\ (m=0) \\ l\ (m \neq 0) \end{cases}$$
$$ON = \begin{cases} 0\ (m=0) \\ m + 2_{i=l}^{k}\ (m \neq 0 \end{cases}$$
$$ST = \begin{cases} k-l+1\ (m=0, ON \leq n) \\ k-l\ (m \neq 0, ON \leq n) \\ k-l-1\ (m \neq 0, ON > n) \end{cases}$$
$$OD = \begin{cases} 0\ (k-l=0) \\ log_2\left(m + 2_{i=0}^{k}\right)\ (k-l \neq 0, m=0) \\ log_2(m + 2_{i=l}^{k})\ (k-l \neq 0, m \neq 0) \end{cases}$$

After the sink node acquires the elements of S_i, the sink node establishes the itinerary of mobile agent. The itinerary is composed of 'Sensor Node', 'Priority', 'Level', 'Input Node', 'Input Degree', 'Sub Tree', 'Output Degree', 'Output Node'. When the total number of sensor nodes is 54 and the sensor node number is same the priority, the sink node establishes the itinerary of mobile agent shown as figure 3.

54 Sensor Nodes : P0 is the first of all priority =>

SensorN	PriorN	Level	InputN	InputD	SubTree	OutputD	OutputN
S0 [	P0,	0,	R,	0,	6,	6,	+1+2+4+8+16+32]
S1 [	P1,	1,	0,	1,	5,	6,	+3+5+9+17+33]
S2 [	P2,	1,	0,	2,	4,	6,	+6+10+18+34]
S3 [	P3,	2,	1,	2,	4,	6,	+7+11+19+35]
S4 [	P4,	1,	0,	3,	3,	6,	+12+20+36]
S5 [	P5,	2,	1,	3,	3,	6,	+13+21+37]
S6 [	P6,	2,	2,	3,	3,	6,	+14+22+38]
S7 [	P7,	3,	3,	3,	3,	6,	+15+23+39]
S8 [	P8,	1,	0,	4,	2,	6,	+24+40]
S9 [	P9,	2,	1,	4,	2,	6,	+25+41]
S10 [	P10,	2,	2,	4,	2,	6,	+26+42]
S11 [	P11,	3,	3,	4,	2,	6,	+27+43]
S12 [	P12,	2,	4,	4,	2,	6,	+28+44]
S13 [	P13,	3,	5,	4,	2,	6,	+29+45]
S14 [	P14,	3,	6,	4,	2,	6,	+30+46]
S15 [	P15,	4,	7,	4,	2,	6,	+31+47]
S16 [	P16,	1,	0,	5,	1,	6,	+48]
S17 [	P17,	2,	1,	5,	1,	6,	+49]
S18 [	P18,	2,	2,	5,	1,	6,	+50]
S19 [	P19,	3,	3,	5,	1,	6,	+51]
S20 [	P20,	2,	4,	5,	1,	6,	+52]
S21 [	P21,	3,	5,	5,	1,	6,	+53]
S22 [	P22,	3,	6,	5,	0,	0,	+T]
S23 [	P23,	4,	7,	5,	0,	0,	+T]
S24 [	P24,	2,	8,	5,	0,	0,	+T]
S25 [	P25,	3,	9,	5,	0,	0,	+T]

Fig. 3. An example of establishing the itinerary of mobile agent

We presented a mobile agent mechanism based on multicasting approach based on the binomial tree topology. The sink node send mobile agent to the sensor node of the itinerary. When the sink node migrate mobile agent, the itinerary of mobile agent is established dynamically by the sink node through the 6 elements of sensor node.

Let the total number of sensor node is 8 shown as figure 4. After the sink node establish the itinerary of mobile agent, the sink node transmit mobile agent to the first sensor node S_0. The sensor node S_0 migrate mobile agent to the other sensor nodes S_1, S_2, S_4 one by one. And sensor node S_1 migrate mobile agent to the child sensor node S_3, S_5 one by one. The leaf sensor nodes S_7, S_5, S_6, S_4 transmit mobile agent to the sink node shown as figure 4.

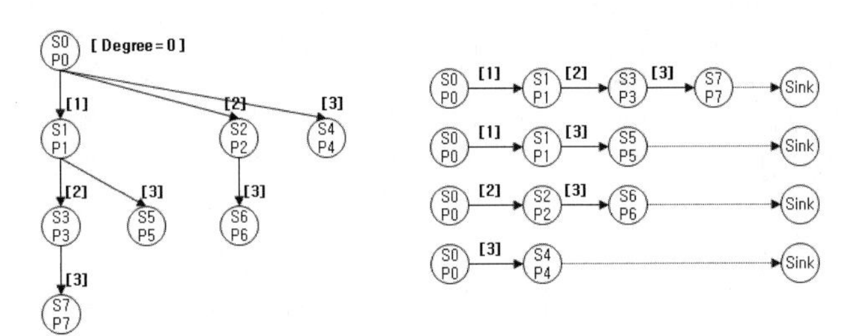

Fig. 4. Mobile agent mechanism based on the binomial tree

When a sensor node is fault, the itinerary of mobile agent is adjusted the priority of sensor nodes to guarantee the policy of fault-tolerance. A sensor node migrate mobile agent to the other node, after detecting the network channel. If a node S_j is fault, a node S_i adjusts the itinerary and migrate the next node S_k shown as figure 5.

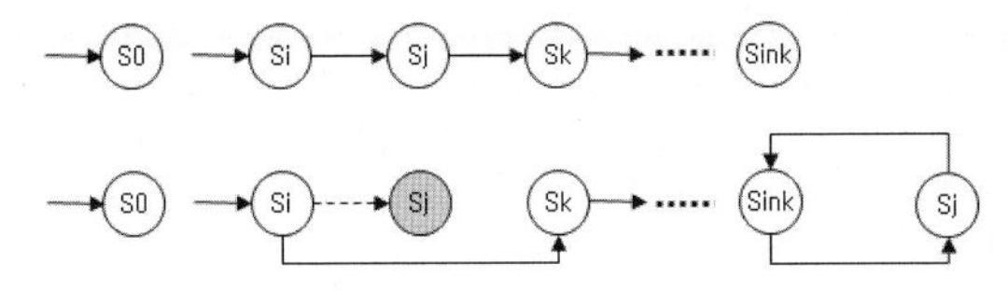

Fig. 5. Adjusting of sensor node priority

4 Conclusion

Since there are the limited resources, powers and network bandwidth on distributed sensor networks, it is important to aggregate data in efficient on that environments. It needs to communicate with sensors for aggregating their data and transport that to an application or some nodes. To effectively aggregate sensor data in distributed sensor networks reduces the power consumption of sensors and network resources, and it is able to bring sensor networks into all life-prolonging. Therefore, many researchers proposed the methodology of prolonging sensor networks, H. Qi suggested Mobile-Agent-based Distributed Sensor Networks (MADSN) as a solution of that.

So we suggested mobile agent mechanism to be reduced consumption of sensor network energy, network bandwidth and the limited resources. The proposed mobile agent mechanism is an approach of mobile agent migration which is multicasting based on the binomial tree. When the mobile agent launch to the MADSN, the sink node used the itinerary is composed of sensor nodes. All of the sensor nodes have the 6 elements to migrate mobile agent to the next sensor node. It is important to guarantee the stability of sensor networks because it provides the policy of fault-tolerance to migrate mobile agent.

References

[1] Kouvatsos, D., Mkwawa, I.-H., Awan, I.: One-to-All Broadcasting Scheme for Hypercubes with Background Traffic. In: Proceedings of the 36th Annual Simulation Symposium. IEEE CS, Los Alamitos (2003)

[2] Cappos, J., Hartman, J.H.: San fermın: aggregating large data sets using a binomial swap forest. In: USENIX Symposium on Networked System Design and Implementation (NSDI 2008), pp. 147–160 (2008)

[3] Qi, H., Wang, X., Sitharama Iyengar, S., Chakrabarty, K.: Multi-sensor Data Fusion in Distributed Sensor Networks Using Mobile Agents. Information Fusion (2001)

[4] Qi, H., Xu, Y., Wang, X.: Mobile-agent-based collaborative signal and information processing in sensor networks. Proceedings of IEEE, Special Issue on Sensor Networks and Applications 91(8), 1172–1183 (2003)

[5] Qi, H., Iyengar, S.S., Chakrabarty, K.: Multi-resolution Data Integration Using Mobile Agents in Distributed Sensor Networks. IEEE Transactions on Systems, Man, and Cybernetics-Part C: Applications and Rev. 31(3), 383–391 (2001)

[6] Roth, P.C., Arnold, D.C., Miller, B.P.: MRNet: A Software-Based Multicast/Reduction Network for Scalable Tools. In: Proceedings of the ACM/IEEE SC 2003 Conference (2003)

[7] Choi, S., Moon, S.J., Eom, Y.H., Kook, Y.-G., Jung, G.D., Choi, Y.G.: An Efficient Data Aggregation using Mobile Agent in Distributed Sensor Network. Proceedings of Korea Information Science Society 33(2)(B), 138–142 (2006)

[8] Cho, S.-H., Kim, Y.-H.: A Fast Transmission of Mobile Agents Using Binomial Trees. Korea Information Processing Society A 9-A(3) (September 2002)

[9] Kook, Y.-G., Young-Chul Kim, R., Choi, Y.-K.: Multi-Agent System Using G-XMDR for Data Synchronization in Pervasive Computing Environments. In: Lukose, D., Shi, Z. (eds.) PRIMA 2005. LNCS(LNAI), vol. 4078, pp. 297–309. Springer, Heidelberg (2009)

Issues on Selecting National R&D Project

Joon Lee*, Youn-Gyou Kook, Jae-Soo Kim, and Ki-Seock Choi

Dept. of NTIS (National Science & Technology Information Service)
KISTI (Korea Institute of Science and Technology Information)
Daejeon, 305-806, Korea
rjlee98@kisti.re.kr

Abstract. Because of the difficulties on selecting National R&D projects, it is expected that the errors of decision making can be reduced if the information on project (i.e. project duration, people, budget etc.), and its evaluation result is properly provided to stakeholders as a reference. In reality, however, the result of project evaluation is rarely utilized in its own purpose. One reason is that the information on evaluation result and budget is not shared amongst stakeholders at the right time. The other is that the interconnection between systems to support the R&D evaluation and budget information is not realized yet for further utilization. Therefore, this paper is focused on the improvement of decision process that reviews and selects the R&D project to provide relevant information through the data mapping approaches from different system domain and to suggest the enhanced process for the seamless interconnection between National R&D performance evaluation and budget information.

Keywords: National R&D, Performance Evaluation, Performance Budgeting, NTIS, dBrain.

1 Introduction

In Korea, the scale of budget for national R&D project has been rapidly increasing from 4 trillion won(4billion US dollars) in 2003 to 13.7 trillion won(13.7billion US dollars) in 2010 and this figure shows that the R&D budget continuously increases about 10% in annual average. As the investment is expanding, Korean Government introduced the evaluation system to enhance the efficiency of R&D expenditure. According to the law of Performance Evaluation and Performance Management for national R&D, government should apply the evaluation results of R&D project to the formulation of the following year's budget. Spending for R&D is an investment for uncertain future, budget formulators and policy makers have difficulty in knowing the effect of R&D budget they authorize. They need many detailed information and professional opinions about the projects and technical trends in regard to national R&D, in making their decisions in the process of budget formulation. Nonetheless, exact and sufficient information are hardly provided to them at the right time. Information of evaluation results for national R&D is not sufficiently utilized due to

* Corresponding author.

T.-h. Kim, A. Stoica, and R.-S. Chang (Eds.): SUComS 2010, CCIS 78, pp. 427–433, 2010.

the following reasons. One is that NTIS (National Technology Information service for R&D) that contains national R&D information (i.e. project duration, people involved in a project, research papers, patents, project evaluation results etc.) is not connected to dBrain(National digital accounting system) that contains all national financial information including R&D financial data. As a result, the R&D performance information loaded on NTIS is not able to be connected to dBrain. Second, the R&D evaluation results in the public sector is hardly utilized through the process of budgeting not just for Korea, but for almost any other countries. Therefore, the purpose of this study is to review current R&D performance evaluation system and the budget management system, and to diagnose the processes for the target systems (NTIS and dBrain) in terms of the possibility of connected utilization, and to propose the improvement solutions to enhance the information flow between performance evaluation and budget making. Consequently, the key issues from selecting national R&D projects which have to be concerned to do the decision making are identified.

2 The Status of Korean Government Performance Management

The current budget management system and the evaluation system for R&D were built respectively for its own sake, the compatibility between systems was not considered at all from the initial development. Furthermore, there have been many changes in accordance with the political changes. For examples, Korean government organization structures were changed in terms of starting new government in 2008 and national R&D functions are adjusted the changed structures accordingly. Ministry of Strategy and Finance (MSF) takes a responsibility of all performance evaluations of government based programs including R&D projects. This change brings about advantages and weaknesses compared with the former government process. The main advantage is to obtain a consistent decision process and a clear responsibility for decision making. Otherwise the weakness is a lack of consulting service of expert group that operates to compensate government officers for special knowledge of science and engineering domain. To fill the gap of this situation, MSF temporary utilize the expert committee which consists of experts in academic and private sector. The major concept of performance based budgeting was not changed, however, it was succeeded from the former government. The idea on the utilization through the connection between performance and budgeting information originated from performance budgeting of the United States. The Government Performance and Results Act (GPRA) established for promoting a focus on improving program performance and to provide greater accountability for results within federal government agencies in the U.S. [1]. The legislation explicitly requires agencies to develop measurable goals for outcomes and outputs and to report actual results. According to John Mercer [1], a major benefit of GPRA has been to bring greater transparency to the operations of the federal government, particularly to the relationship between investments and results. In order to assist program managers achieve their goals, government agencies should implement comprehensive performance management systems. An effective performance management system ensures that an agency's administrative and support functions such as budget, financial management, human resources, information technology, procurement etc. A system also encourages managers to operate their programs in ways that maximize performance while minimizing costs. Concerning the R&D projects, however, most researchers do not want to waste their time such an administrative and management work to maintain the system. The key issue lies on this area how to make researchers to recognize the benefits to be expected from using the dedicated system and to be willing to accept and advocate this transformation. R&D activities are affected by the complicated factors and their interactions. It makes difficult to measure output and outcomes from the result of R&D activities. For example, the interception of subjective factor during the measurement is inevitable even if the system was developed accurately [2]. Therefore, we assumed that the errors and risk of decision process can not remove at all, but can be reduced by the appropriate IT technology support. Based on this hypothesis, a framework to share information amongst stakeholders is proposed in the following section.

3 A Framework for Interoperability

Nowadays, interoperability is the most important issue facing necessary that need to access information from multiple information systems. The growing demand on integrating services from the different legacy systems in a network centric environment has introduced another type of complexity compared to develop large monolithic systems in a traditional way. While most research challenges focus on interoperability within the same domain, provision of cross-domain interoperability among collaborating domains is a new challenge that needs more attention from the research community. The interoperability requires data and service extraction to obtain common sets of information and services in collaborating domains.

To tackle the complexity of network centric interoperability, the trend is towards ease of use, vendor, program language, platform independently, and in general raising the level of communication abstraction from low-level techniques to provider-independent techniques, and finally to high-level abstractions such as service oriented architecture (SOA). These technologies to a large extent have diminished the problem of interoperability of heterogeneous data among distributed systems. However, a major challenge in interoperability among systems is interpretation of concepts from outside of one's domain of expertise. This emerging need must be addressed by cross-domain facilitators with enough knowledge from each participating domain to establish the required communications. Therefore, the first task is to extract both data and services from participating application domains to allow systems to perform mutual business. The next task is to provide the means for communication of information (syntactic interoperability) and communication of meaning (semantic interoperability) which are achieved through comprehensive and standard information and concept representations and communication through standard messages [3]. Achieving interoperability among heterogeneous systems is increasingly important in different domains. Concerning semantic interoperability, Jinsoo park and Sudha ram [5] categorized it into three broad areas: mapping-based, intermediary-based, and query oriented approaches. The following illustration of each approach is considered from their work for developing a research framework. The mapping-based approach attempts to construct mappings between semantically related information sources. It is usually accomplished by constructing a federated schema and by establishing mappings between federated schema and the participating local schemas [6][7][8]. Mappings are not limited to schema components such as entity classes, relationships, and attributes, but may be established between domains and schema components. The drawback of the federated schema approach is that it is not designed to be independent of particular schemas and applications. Explicit representations of semantics of information sources can help resolve the problems associated with interoperability when constructing mappings between them. Another approach is the intermediary-based approach. This approach depends on the use of intermediary mechanisms, for example, mediators, agents, ontologies, etc. to achieve interoperability [9][10]. Such intermediaries may have domain-specific knowledge, mapping knowledge, or rules specifically developed for coordinating various autonomous information sources. In most cases, such intermediaries use ontologies to share standardized vocabulary or protocols to communicate with each other. The advantage of using ontologies is its ability to capture the tacit knowledge within a certain domain in great detail in order to provide a rich conceptualization of data objects and their relationships. Its knowledge is domain-specific, but independent of particular schemas and applications. Even though such an approach may be theoretically valid, it is practically infeasible to develop and maintain an ontology in autonomous, dynamic, and heterogeneous databases due to the inherent complexities of the knowledge domain. Hence, this approach is typically applied only to a restricted application domain, which limits its general applicability in practice. The third approach, query-oriented approach, is based on interoperable languages, most of which are either declarative logic-based languages or extended SQL [7][11][12]. They are capable of formulating queries spanning several databases. In order to resolve semantic conflicts over data structure and data semantics, it is desirable to have high-order expressions that can range over both data and metadata. One of the main drawbacks of this approach is that it places too heavy a burden on users by requiring them to understand each of the underlying local databases. This approach typically requires users to engage in the detection and resolution of semantic conflicts, since it provides little or no support for identifying semantic conflicts [13]. However, these three approaches are rather mutually compensable than mutually exclusive. Some approaches based on intermediaries also rely on mapping knowledge established between a common ontology and local schemas. It is also often the case that mapping and intermediaries are involved in query-oriented approaches.

From the review of research so far, our framework for interoperability is proposed in Figure1. The framework provides the empirical basis for defining the various components required to interconnect different system domains, namely NTIS and dBrain. Main findings and suggestions from developing a framework are as follows. First, research project managers input information details about researches they are carrying out, the information is stored on NTIS on the form of Investigation and Analysis(I&A). Nonetheless, I&A is hardly utilized by budget makers. To solve this problem, information like I&A should be delivered to budget makers via NTIS-dBrain connection system. Second, Program managers of R&D ministries and agencies carry out self-evaluation system for the activities and projects. Meanwhile, budget managers of R&D ministires and agencies make budget request. Self-evaluation procedures utilize two systems like e-IPSES and NTIS. That is, program managers input same information twice to each of the information systems. This procedure could be enhanced by allowing them to use only one system such as NTIS, which has already initiated a work for connecting with dBrain. Third, MSF reexaminers do the self-evaluation, which is called upper level evaluation(ULE), being carried out off-line. Almost all the performance evaluation are done by networked system, but the last stage of it, ULE, is not processed electronically. Therefore, dBrain, should be upgraded by the enhanceing its sub-system called Business Management System. Through this process, bugdet makers are able to be supported by the information on the various aspects of national R&D projects. Fourth, budget officers of MSF require high quality information about R&D projects. If they could access to the information stored in NTIS via NTIS-dBrain Linking system, there will be great helpful for officers to make decisions associated with the national R&D budget.

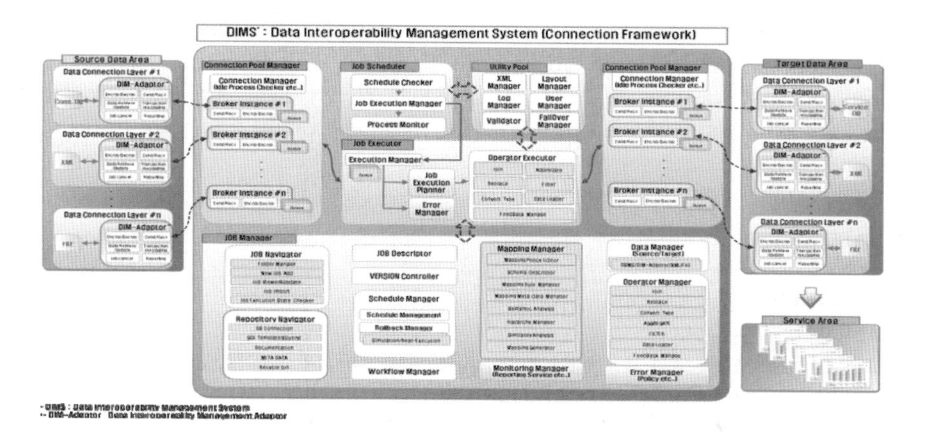

Fig. 1. A Research Framework

4 Key Issues in Project Selection

Key Issues in National R&D project selection are roughly classified into two categories, environmental and technical issues. Examples of environmental issues include efficiency/bureaucracy, transparency/control, centralization/distributed provision dilemmas in political situations [4]. On the other hand, technical issues are mainly associated with semantic conflicts during the developments of integrated service from different systems [5]. Some issues are related to data level and/or schema level conflicts. Data level conflicts are differences in data domains caused by the multiple representations and interpretations of similar data. On the other hand, schema level conflicts are characterized by differences in logical structures and/or inconsistencies in metadata of the same application domain. Examples of these conflicts are naming conflicts, entity-identifier conflicts, schema-isomorphism conflicts, generalization conflicts, aggregation conflicts and semantic discrepancies. Most of cases, developers tend to concentrate on technical issues rather than environmental issues. However, environmental issues are sometimes much more important than technical issues. For an example, it may be more important than other cases that the project is associated with the public sector which has to consider various requirements from stakeholders who behave in terms of their own interest that sometimes contains mutual exclusive.

5 Conclusion

A goal of research in this paper is to suggest the enhanced decision making process on selecting national R&D project through the seamless integration of R&D performance and budgeting information operated from different system. If current processes are improved by the integrated service, budget makers are able to be more likely to utilize various information resources associated with national R&D project, such as performance based feasibility reviews. Furthermore, such a service can be widely used by different stakeholders, i.e. the National Assembly, academic researchers, government officers of each government branches with their own interest and viewpoint. Consequently, this kind of interconnected system is able to provide a useful guideline for choosing national R&D project and can be a standard model case for service integration between relevant systems.

References

[1] Mercer, J.: Government Performance Management, `http://www.john-mercer.com/index.htm`

[2] Saaty, T., Vargas, L.G.: Models, methods, concepts & applications of the analytic hierarchy process. Kluwer Academic Publishers, Dordrecht (2001)

[3] Sartipi, K., Dehmoobad, A.: Cross-Domain Information and Service Interoperability. In: Proceedings of iiWAS 2008, November 24-26 (2008)

[4] Rossel, P., Finger, M.: Conceptualizing E-Government. In: ICEGOV 2007, December 10-13 (2007)

[5] Park, J., Ram, S.: Information Interoperability: What lies beneath? ACM Transactions on Information Systems 22(4), 595–632 (2004)

[6] Hayne, S., Ram, S.: Multi-user view integration system(MUVIS): an expert system for view integration. In: Proceedings of the 6th International Conference on Data Engineering, February 5-9 (1990)

[7] Krishnamurthy, R., Litwin, W., Kent, W.: Language features for interoperability of databases with schematic discrepancies. In: Proceedings of the 1991 ACM SIGMOD International Conference on Management Data, May 29-31 (1991)

[8] Navathe, S.B., Elmasri, R., Larson, J.A.: Integrating user views in database design. IEEE Computer 19(1), 50–62 (1986)

[9] Sciore, E., Siegem, M., Rosenthal, A.: Using semantic values to facilitate interoperability among heterogeneous information systems. ACM Trans. Database Syst. 19(2), 254–290 (1994)

[10] Parakonstantinou, Y., Garcia-Molina, H., Ullman, J.: Medmaker: A mediation system based on declarative specifications. In: Proceedings of the 12th IEEE International Conference on Data Engineering, February 26 - March 1, pp. 132–141 (1996)

[11] Arens, Y., Knoblock, C.A., Shen, W.M.: Query reformulation for dynamic information integration. J. Intell. Inf. Syst. 6(2), 99–130 (1996)
[12] Lakshmanan, L.V.S., Sadri, F., Subramanian, I.N.: Logic and algebraic languages for interoperability in multidatabase systems. J. Logic Programm. 33(2), 101–149 (1997)
[13] Goh, C.H., Madnick, S.E., Siegel, M.D.: Context interchange overcoming the challenges of large-scale interoperable database systems in a dynamic environment. In: Proceedings of the 3rd International Conference on Information and Knowledge Management, November 29 - December 2, pp. 337–346 (1994)

A Study on S-band Short-range Surveillance Radar Optimum Deployment Considering Frequency Interference

Bong-Ki Jang*, Young-soon Lee**, Byung-sam Kim***, and Ui-jung Kim***

Kumoh National Institute of Technology, Sanho-ro 77, Gumi, gyongbuk, Korea
{jang815,yslee,bskim,elijahkim}@kumoh.ac.kr

Abstract. Inter-radar interference can cause the important impact to the radar detection performance because radar operates with high transmitter power, sensitive receiver and wideband. Because short-range surveillance radar is deployed rather close to each other, inter-radar interference is more critical. In this paper, the international criteria for radar interference protection is reviewed based on the ITU-R and NTIA documents, and the radar analysis is presented by taking into account the S-band short-range surveillance radar operating environments. Finally, S-band short-range surveillance radar optimum deployment is presented with the interference impact analysis.

Keywords: Noise, Interference, Signal to Noise, Interference to Noise, FDR.

1 Introduction

Radar is an object detection system that uses electromagnetic waves to identify the range, altitude, direction or speed of both moving and fixed objects such as aircraft, ships, motor vehicles, weather formations and terrain. The term RADAR was coined in 1940 by U.S Navy as an acronym for RAdio Detection And Ranging[1].

Inter-radar interference can cause the important impact to the radar detection performance because radar operates with high transmitter power, sensitive receiver and wideband. Especially, In case of short-range surveillance radar, frequency interference is more important due to short available time.

ITU-R(International Telecommunication Union-Radiocommunication Sector) recommends radar operation band/bandwidth and provides interference protection criteria of some frequency band[6]. NTIA(National Telecommunications and Information Administration) studies various experiments and tests and comments radar range loss due to interference[2].

In this paper, international criteria for radar interference protection is reviewed based on the ITU-R and NTIA documents, and the radar analysis model is presented by taking into account the short-range surveillance radar operating environments. Finally, S-band short-range surveillance radar optimum deployments are presented with the interference impact analysis.

---

* Kumoh National University, Radio Communication Engineering(doctor's course)
** Kumoh National University, Radio Communication Engineering(professor)
*** Kumoh National University, Radio Communication Engineering(doctor)

T.-h. Kim, A. Stoica, and R.-S. Chang (Eds.): SUComS 2010, CCIS 78, pp. 434–439, 2010.

2 Radar Noise and Interference

2.1 Noise and Interference

Unwanted energy coupled into radar receivers from natural sources and from manmade devices that are not intentional radio transmitters is called NOISE. Noise degrades radar receiver performances and is generated by many sources. A natural source is the energy produced by thermal electrons within radar receiver circuitry, and it is one of the most fundamental limiting factors in radar receiver performance. Manmade noise across the spectrum from HF to lower microwave frequencies is coupled into radar receivers from unintentional emission sources such as electric motors, above-ground power lines and automotive ignition systems. Noise sources are unavoidable and generally uncontrollable[2].

In contrast to noise, INTERFERENCE is unwanted radio energy coupled into radar receivers from manmade, intentional radio transmitter sources such as communication devices. Like noise, interference degrades radar receiver performance. Interference, however, generally has statistical and spectrum characteristics that are different from noise. Another important contrast between noise and interference is that, while noise may be unavoidable and largely uncontrollable, interference is both avoidable and controllable through sound spectrum engineering and management procedures[2]. The types of interference are CW(Continuous Wave), Noise-like, Pulse, Impulse and so on.

2.2 Radar Interference

In case of no interference, signal-to-noise(S_N/N) of surveillance radar is given by[3]

$$\frac{S_N}{N}=\frac{P_t G_t}{4\pi R_N^2}\cdot\frac{\sigma}{4\pi R_N^2}\cdot\frac{G_r\lambda^2}{4\pi}\cdot\frac{1}{L_{bf}}\cdot\frac{1}{N} \tag{1}$$

Where S_N is radar receiver signal power, N is noise power of receiver, P_t is maximum transmitter power, R_N is target detection range, σ is target RCS(Radar Cross Section), L_{bf} is beam free space loss, G_t is transmitter antenna gain, G_r is receiver antenna gain, λ is operating wave length. R_N is given by

$$R_N^4=\frac{P_t G_t G_r\lambda^2}{(4\pi)^3}\cdot\sigma\cdot\frac{1}{\left(\frac{S_N}{N}\right)}\cdot\frac{1}{N\cdot L_{bf}} \tag{2}$$

If there is interference, target detection range(R_{N+I}) is given by

$$R_{N+I}^4=\frac{P_t G_t G_r\lambda^2}{(4\pi)^3}\cdot\sigma\cdot\frac{1}{\left(\frac{S_{N+I}}{N+I}\right)}\cdot\frac{1}{(N+I)\cdot L_{bf}} \tag{3}$$

Therefore, target detection range of with/without interference is given by[4]

$$\frac{R_{N+I}^4}{R_N^4}=\frac{\left(\frac{S_N}{N}\right)}{\left(\frac{S_{N+I}}{N+I}\right)}\cdot\frac{N}{N+I} \tag{4}$$

Both interference condition and no interference condition, detection probability (P_d) and false alarm probability(P_{fa}) is same and provided threshold level is set properly, equation (4) is given by

$$\frac{R_{N+I}}{R_N} = \left(\frac{N}{N+I}\right)^{(1/4)} \tag{5}$$

Detection range loss can be estimated by using equation (5), and the results show figure 1 and table 1.

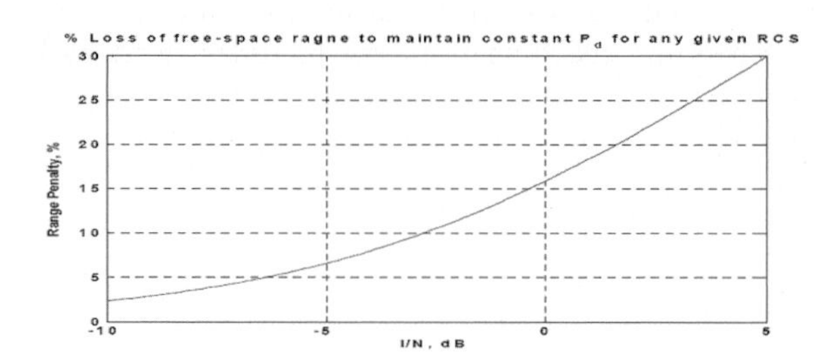

Fig. 1. I/N versus Surveillance range loss

Table 1. I/N versus Surveillance Range Loss

I/N dB(Numeric)	-10(0.1)	-6(0.26)	0(1.0)	3(2)
(I+N)/N dB(Numeric)	0.4(1.1)	1(1.26)	3(2.0)	4.8(3)
R_{N+I}/R_N $=[N/(N+I)]^{1/4}$	0.976	0.946	0.841	0.76
Range Loss	2.4(%)	5.4(%)	15.9(%)	24(%)

From the result, we can see this : as interference-to-noise(I/N) ratio increase, range loss also increase more and more. If I/N is -10dB, range loss is 2.4% and if I/N is -6 dB, range loss is 5.4%. According to ITU-R recommendations[6], CW radar is generally required I/N≤-6dB and in case of special purpose is I/N=-10dB. And NTIA test results of long-range air search radar, it happens target detection loss in case of I/N=-3 dB. Especially in case of I/N=-9dB, detection probability shows sudden degradation[3]. Additionally from test results of short-range air search pulse radar, detection probability suddenly degraded at the point of I/N=-10dB.

Considering interference impact, ITU-R recommendations, NTIA test results and specially short-range surveillance radar operating characteristics, I/N=-10dB needs in case of S-band short-range surveillance radar.

3 Radar Interference Power

3.1 Interference Power

For the purpose of enhancing operating and deployment, radio system needs frequency and distance separation trade-off. As comments previous chapter, interference degrades radar operating performance. Therefore optimum deployment is necessary by analyzing interference.

Interfered radar(radar #1) receive interference power(I) is given by[5]

$$I = P_t + G_t + G_r - L_{bf} - FDR(\Delta f) \tag{6}$$

Where P_t is interfering radar(radar#2) transmitter power, G_t is interfering radar antenna gain, G_r is interfered radar(radar#1) antenna gain. L_{bf} is beam free space loss between interfered radar and interfering radar. L_{bf}=20Log(4πd/λ) and d is distance and λ is wavelength. FDR is frequency-dependent rejection produced by the receiver IF selectivity curve on an unwanted transmitter emission spectra(dB) and is given by

$$FDR(\Delta f) = 10\,Log\frac{\int_0^\infty P(f)df}{\int_0^\infty P(f)|H(f+\Delta f)|^2 df} \tag{7}$$

FDR is summation OTR(On-Tune Rejection) and OFR(Off-Frequency Rejection): FDR(Δf)= OTR+OFR(Δf), Where OTR and OFR is given by

$$OTR = 10\,Log\frac{\int_0^\infty P(f)df}{\int_0^\infty P(f)|H(f)|^2 df} \tag{8}$$

$$OFR(\Delta f) = 10\,Log\frac{\int_0^\infty P(f)|H(F)|^2 df}{\int_0^\infty P(f)|H(f+\Delta f)|^2 df} \tag{9}$$

OTR is correction factor and approximately is $20Log(B_t/B_r)$. Where B_t is interfering transmitter 3dB bandwidth and B_r is interfered receiver 3dB bandwidth.

If interfering and interfered radar are same radar and uses same frequency, Δf of equation (9) is zero. Consequently OFR is zero(10Log1=0).

3.2 S-band Surveillance Radar Interference Power

In this paper, interfering radar(radar #2) and interfered radar(radar #1) are same S-band short-range surveillance radar and detail parameters are Table 2.

Table 2. S-band Surveillance radar parameter

parameters	Interfering Radar (Radar #1)	Interfered Radar (Radar #2)
Transmitter Power	1,200W	1,200W
Antenna Gain(Main Beam)	25dBi	25dBi
Frequency	F1(S-band)	F1(S-band)
Receiver Bandwidth	-	1MHz
Transmitter Bandwidth	1MHz	-
Noise Figure	2.7dB	2.7dB

The interference power(I) of interfered radar is,

$$I = P_t + G_t + G_r - L_{bf} - FDR(\Delta f) = 80.79 - L_{bf} - FDR(\Delta f)$$
$$I = -10 + N = -10 + KTBF = -151.25(dBw)$$
$$L_{bf} = 20Log(4\pi d/\lambda) = 32.4 + 20Logf + 20Logd \ (d = 10, 20, 50, 100Km)$$

When considering Table 2 parameters, in case of main-main beam, interference powers(I) of 10Km, 20Km, 50Km, 100Km are respectively -41.43(dBw), -47.45 (dBw), -55.41(dBw), -61.43(dBw). In case of mail-side beam are respectively -69.43 (dBw), -75.45(dBw), -83.41(dBw), -89.43(dBw). In case of side-side beam are respectively -97.43 (dBw), -103.45(dBw), -111.41(dBw), -117.43(dBw). Because all of cases are above than -151.25 (dBw), it can be estimated interference. Therefore we should choose frequency separation in stead of distance separation..

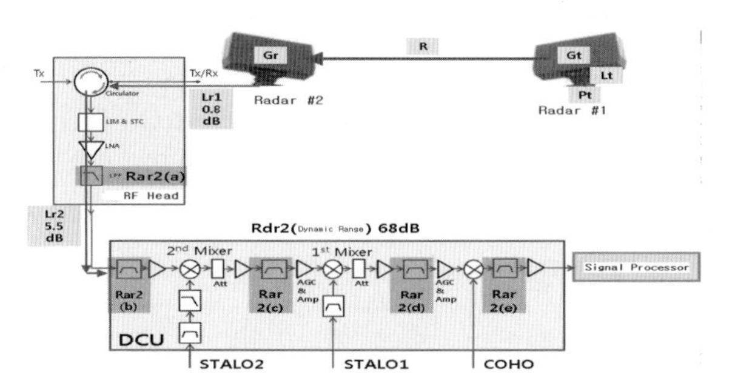

Fig. 2. Radar #1 and #2 Receiver Block Diagram

Table 3. Frequency-Distance Separation I/N Ratio : Main Beam-Main Beam

I/N	Frequency Separation(MHz)											
(dB)	8.0	16.0	24.0	32.0	40.0	48.0	56.0	64.0	72.0	80.0	88.0	96.0
X+0.5	63.6	40.5	33.4	26.4	19.7	12.0	4.9	-2.2	-9.3	-16.6	-23.7	-30.8
X+1.0	57.6	34.5	27.4	20.3	13.7	6.0	-1.1	-8.2	-15.4	-22.6	-29.7	-36.8
X+1.5	54.1	31.0	23.8	16.8	10.2	2.5	-4.6	-11.8	-18.9	-26.1	-33.2	-40.4
X+2.0	51.6	28.5	21.3	14.3	7.7	0.0	-7.1	-14.3	-21.4	-28.6	-35.7	-42.9
X+2.5	49.7	26.5	19.4	12.4	5.8	-2.0	-9.1	-16.2	-23.3	-30.5	-37.6	-44.8
X+3.0	48.1	24.9	17.8	10.8	4.2	-3.5	-10.7	-17.8	-24.9	-32.1	-39.3	-46.4
X+3.5	46.7	23.6	16.5	9.5	2.8	-4.9	-12.0	-19.1	-26.2	-33.5	-40.6	-47.7
X+4.0	45.6	22.4	15.3	8.3	1.7	-6.0	-13.2	-20.3	-27.4	-34.6	-41.8	-48.9
X+4.5	44.5	21.4	14.3	7.3	0.7	-7.1	-14.2	-21.3	-28.4	-35.7	-42.8	-50.8
X+5.0	43.6	20.5	13.4	6.4	-0.3	-8.0	-15.1	-22.2	-29.3	-36.6	-43.7	-51.6
X+5.5	42.8	19.7	12.6	5.5	-1.1	-8.8	-15.9	-23.1	-30.2	-37.4	-44.5	-52.4
X+6.0	42.0	18.9	11.8	4.8	-1.8	-9.6	-16.7	-23.8	-30.9	-38.2	-45.3	-53.1
X+6.5	41.4	18.2	11.1	4.1	-2.5	-10.3	-17.4	-24.5	-31.6	-38.8	-46.0	-53.7
X+7.0	40.7	17.6	10.5	3.4	-3.2	-10.9	-18.0	-25.1	-32.3	-39.5	-46.6	-54.3
X+7.5	40.1	17.0	9.9	2.8	-3.8	-11.5	-18.6	-25.7	-32.9	-40.1	-47.2	-54.9
X+8.0	39.5	16.4	9.3	2.3	-4.3	-12.1	-19.2	-26.3	-33.4	-40.7	-47.8	-55.4
X+8.5	39.0	15.9	8.8	1.8	-4.9	-12.6	-19.7	-26.8	-34.0	-41.2	-48.3	-55.4
X+9.0	38.5	15.4	8.4	1.3	-5.4	-13.1	-20.2	-27.3	-34.5	-41.7	-48.8	-55.9
X+9.5	38.1	14.9	7.8	0.8	-5.8	-13.6	-20.7	-27.8	-34.9	-42.1	-49.3	-56.4
X+10.0	37.6	14.5	7.4	0.3	-6.3	-14.0	-21.1	-28.2	-35.4	-42.6	-49.7	-56.8

4 Short-range Surveillance Radar Optimum Deployment

For enhancing operating and deployment, S-band short-range surveillance should consider frequency separation. After calculating FDR of Figures 2, I/N ratio shows according to frequency(8MHz steps)and distance(0.5Km steps).

5 Conclusion

So far radar performance is mainly focused radar single performance(detection range, detection altitude, detection velocity and so on). But as short-range surveillance radar increase, now inter-radar interference also should be considered. Referencing the international recommendations and test result, and considering S-band short-range surveillance characteristics, -10dB I/N ratio and optimum deployment plan is presented. Hereafter in designing radar systems, receiver FDR(frequency-dependent rejection) design is also considered properly to enhance radar operation efficiency.

References

1. Kwak, Y.-k., Yang, J.-y., Jung, J.-s.: Inter-radar interference anlysis based on radar interference criteria. KIEES (The Korean Institute of Electromagnetic Engineering Science) 19(6HO), 657–662 (2008)
2. NTIA Report TR-06-444 : Effects of RF Interference on Radar Receivers. U.S. Department Of Commerce (2006)
3. NTIA Report TR-05-432 : Interference Protection Criteria phase I – complication from existing sources. U.S. Department Of Commerce (2005)
4. ITU-R Report : Radar Protection Criteria in perspective (2005)
5. Recommendation ITU-R SM.337-5 : Frequency and distance separations (2007)
6. Recommendation ITU-R M.1465-1 : Characteristics of protection criteria for radars operating in the radiodetermination service in the frequency band 3,100~3,700MHz (2007)

Proposal of Secure VoIP System Using Attribute Certificate

Jin-Mook Kim[1], Young-Ae Jeong[1], and Seong-sik Hong[2]

[1] Sunmoon University, Department of IT Education
[2] Hyejeon college, Department of Internet security
{calf0425,yajung}@sunmoon.ac.kr,
sshong@hyejeon.ac.kr

Abstract. VoIP is a service that changes the analogue audio signal into a digital signal and then transfers the audio information to the users after configuring it as a packet; and it has an advantage of lower price than the existing voice call service and better extensibility. However, VoIP service has a system structure that, compared to the existing PSTN (Public Switched Telephone Network), has poor call quality and is vulnerable in the security aspect. To make up these problems, TLS service was introduced to enhance the security. In practical system, however, since QoS problem occurs, it is necessary to develop the VoIP security system that can satisfy QoS at the same time in the security aspect. In this paper, a user authentication VoIP system that can provide a service according to the security and the user through providing a differential service according to the approach of the users by adding AA server at the step of configuring the existing VoIP session is suggested. It was found that the proposed system of this study provides a quicker QoS than the TLS-added system at a similar level of security. Also, it is able to provide a variety of additional services by the different users.

Keywords: VoIP, Secure protocol, performance, QoS.

1 Introduction

All over the world is linked by development of internet and link with multimedia technology is accelerated accordingly. Demand about services such as VoIP(Voice over Internet Protocol) that transmit multimedia data including audio through IP (Internet Protocol) net, video etc, representatively is increasing fast[1].

But, while PSTN that is existent telephone envy can attack though approach physically, but QoS (Quality of service) problem and security vulnerability of multimedia data service can happen in internet environment.

Specially, VoIP has limitation that long-distance attacker can listen in variation of signal ring message and voice packet easily taking advantage of network technology. Insides such as SIP proposed to solve this. However, the alternative such as SIP has side effect that user's the convenience is damped [2].

So, we provide gradation service by user's access adding AA server in VoIP System establishment step that do not add in TLS base system in existing in treatise

T.-h. Kim, A. Stoica, and R.-S. Chang (Eds.): SUComS 2010, CCIS 78, pp. 440–447, 2010.

that see therefore, and wish to improve security and propose safe VoIP system that can guarantee QoS.

Composition of this treatise is as following. First, in related research - base technology of VoIP that accomplish base, SIP for Session initialization, PKI that is open height base certificate technology, attribute certificate and so on of internet telephone system - in chapter 2 because ocean describe. Chapter 3 explained about structure of proposal system and action process that wish to propose in treatise that see. And chapter 4 described contents about performance analysis. We described contents that compare about performance and function about method that propose VoIP system and treatise that see that add VoIP system, TLS technology that existent security technology is not added. Finally, described conclusion for our treatise.

2 Related Researches

2.1 VoIP

It is communication service that VoIP converts voice data to data packet and do so that currency may be possible in general telephone networks. Fare is inexpensive than existent telephone network service doing so and several person can use at the same time and extensity is excelled. It is SIP, H.323, MGCP, MEGACO etc, by representative protocol that is used for this. The VoIP's voice transmission characteristic digitalizes priority analog voice elements, and apply Data Compression Algorithm for efficiency of important duty use and create voice packet. And created packet attach real time protocol (RTP: Real-time Transport Protocol), UDP and IP header and transmit depending on transmission standard of relevant network [8].

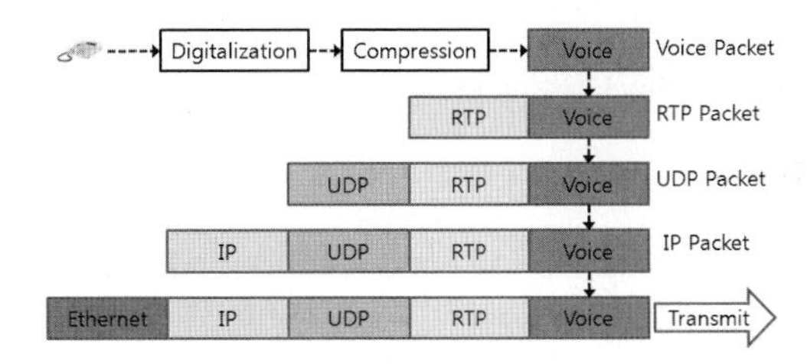

Fig. 1. VoIP Structure

2.2 SIP

As IP telephone communication signal protocol that SIP develops in IETF, applied security technology can be classed by Hop-by-Hop public safety division End-to-End security technology greatly. Technology of digest certification, TLS, IPSec etc. is included to Hop-by-Hop security, and End-to-End security applies S/MIME [6,8,9].

And SIP is Text based communication protocol. SIP message are consisted of request that send by client to server and response that send by sever to client. Invite messages that use common start-line and message header fields more than one, and SIP compose [7]. Figure 2 is shown about consist of messages.

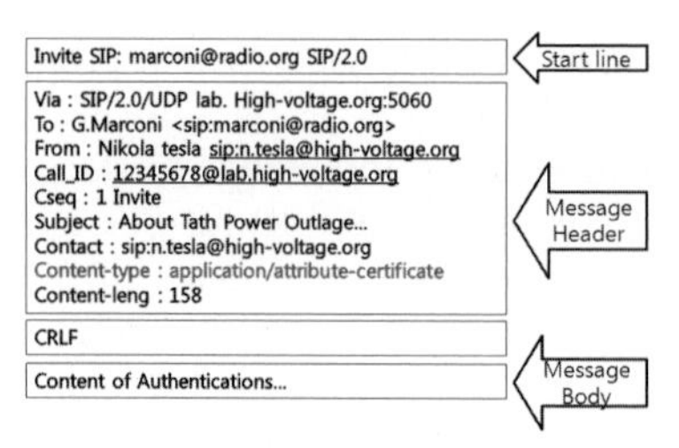

Fig. 2. SIP message

2.3 Attribute Certificate

Some lacking problems in that PKI system of X.509 base provides certification service about user were found. Research for attribute certificate proposed to solve these problems. Instead of attribute certificate has structure similar to open height certificate, but have open height information, it is structure including attribute about position of object, role, security level, competence information etc. Figure 3 is displaying example of attribute certificate.

Field name	Explain
Version	Version of certificate
Holder	Holder name or ID
Issuer	Issuer name or ID
Signature	Signature algorithm
Serial#	Serial number
Validate	Validate time
Attributes	Information of attributes
Extensions	Extension fileds

Fig. 3. Example of attribute certificate

2.4 TLS Methodology

Apply TLS to security mechanism to apply between SIP terminals. It can supply confidentiality and integrity for message that transmit if form TLS security channel

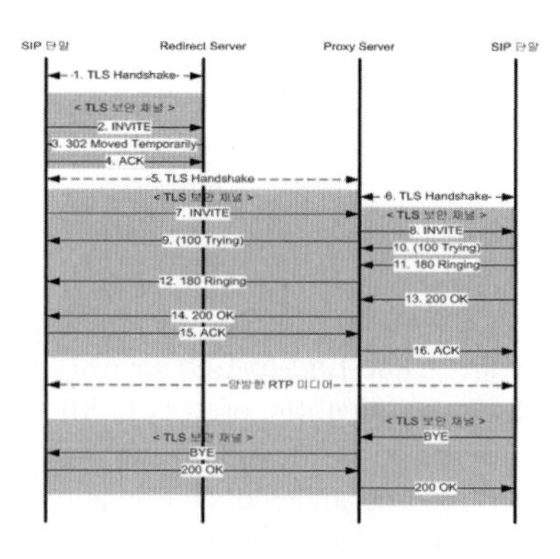

Fig. 4. TLS handshaking

beforehand to transmit SIP message. Figure 4 is expressing process for safe message transmission through TLS Handshaking.

However, have shortcoming that many calculations need because must pass through session connection process to form new security channel whenever new SIP terminal is added in TLS security techniques.

3 Proposal Method

We examined about various certification protocols that can apply to VoIP system so far. However, each research is fragmentary and there is difficult to apply in actuality. So, we wish to propose more effective and safe VoIP certification protocol in treatise that sees hereupon. And we established following virtue item for this:

① AA server and KMS server flow certification work beforehand and know public key of each other.

② User requests certificate issuance as user creates public key and individual key and register public key to KMS server based on PKI certification techniques.

③ KMS server transmits including AA server's public key information when behave certificate.

3.1 User Registration

First, precede user registration process to satisfy about 3 assumptions which present in front in proposal techniques. User registration process is as following.

1) User requests public key entry to registration server after creates Key pair (individual key, public key).

2) Registration server registers public key that user requires and store relevant contents to location server. In this time, registration and Location server composed by uniformity system.

3) Registration server issues certificate to user after seek location server's confirmation about user register truth.

It is appearing in figure 5 about this.

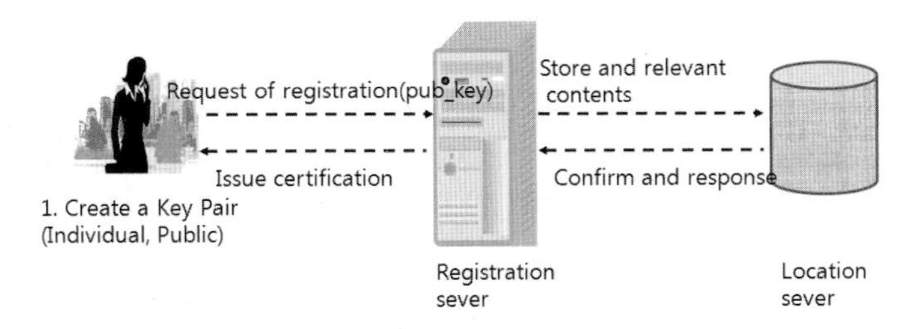

Fig. 5. Registration and Issues certification

3.2 Session Process

Between terminals in proposal techniques data to do send-receive as following Session connection process flow.

1) User own certificate transmission of a message proxy server of encrypt that exhibit and transmit to transmission of a message proxy server with Invite message.

2) Transmission of a message proxy server confirms user information and transmit user certification information and own certificate to AA server with invite message.

3) Issue attribute certificate and register to registration server after AA server confirms user information.

4) Registration server confirms attribute certificate that is issued to user and deliver reception side proxy server certificate and address information.

5) Encrypt invite message and transmit to reception side proxy server because transmission of a message proxy server spends open height value that get in certificate.

6) Reception side server accepts user's position information which send invite message in registration server last month.

7) Ringing

8) Confirm that was linked normally between transmission of a message proxy server and reception side proxy server.

It is displaying in Figure 6 about this.

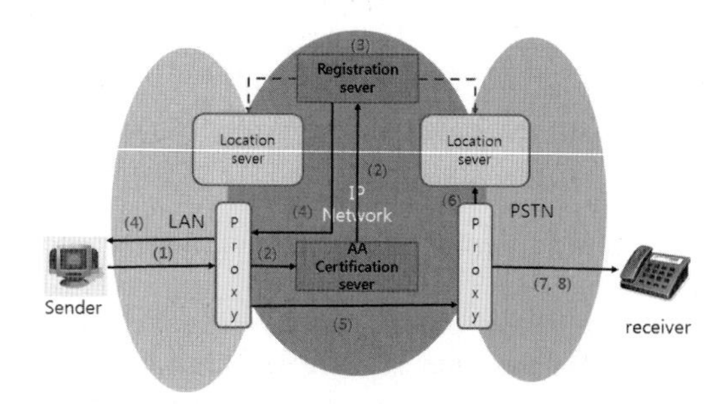

Fig. 6. Session process

3.3 Data Process

Data processing process between SIP terminals that Session connection consists is same with figure 7.

First, proxy server delivers with user information to AA certification server if transmit invite message to proxy in terminal. In midway IP network whether can be passed to reception proxy server through redirect servers while do case availability confirm. In case is verified attribute certificate from redirect server, invite message is passed by ocean reception proxy and when is not so, transmission is canceled. Reception side proxy informs to reception side terminal and transmits response message in transmission of a message in reply about invite message that is delivered from transmission of a message proxy. After confirm between each other that communication connection was confirmed and exchange ACK message that finish data transmission after transmit data lastly each other, finish data delivery.

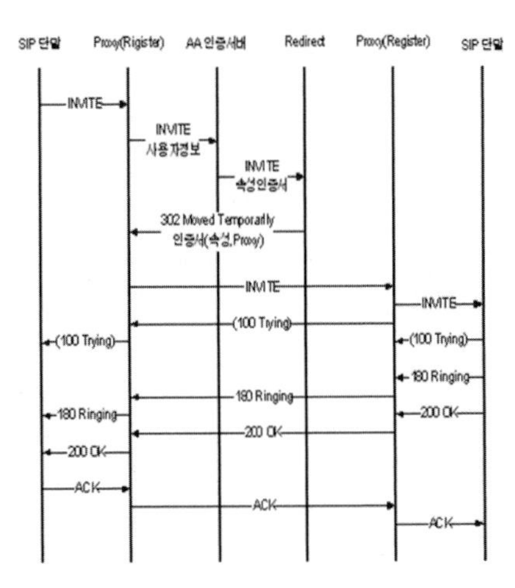

Fig. 7. Data procedure

4 Analysis of Our Proposal

We explained about action process of techniques and processing formality safe and effective VoIP message last month that use attribute certificate that wish to propose in treatise that see so far. For front, techniques to propose in these chapter wishes to describe performance results of measurements about existent VoIP system, and TLS system about response time, availability.

4.1 Response Time

We sent invite messages 20 times for performance analysis about response time for system and existent VoIP that propose and measure each response time. Additionally, we measured response time for TLS base system in equal condition. It is appearing with next time Table 1 about this.

Table 1. Response time

	1 round	5 round	10 round	20 round
SIP	10	30	50	95
Proposal	15	60	90	140
TLS	80	126	148	285

As is appearing in figure 8, when compare with existent SIP, response time of proposal techniques is little more late on the whole. However, it can know that show logical result seeing that is displaying smooth slant on the whole. However, in TLS's case, early response time is exposing by late thing about 1.6 times than SIP or proposal techniques. This displays in early setting of TLS that many overheads need. Also, we can know that TLS's response time appears late as is different in response time since 10th.

4.2 Availability Result

In this phrase, we transmit 500 invite messages and measure CPU usage for evaluate availability to SIP, proposal techniques, and TLS base system. This time, reached and measure last time to when get into situation that it is no solubility for system actually and compare solubility more than CPU amount used 90%. Through this, we can compare availability for 3 systems about DoS attack indirectly.

Proposal system expressed high solubility about 12% relatively than TLS. It shows Figure 8 about this result.

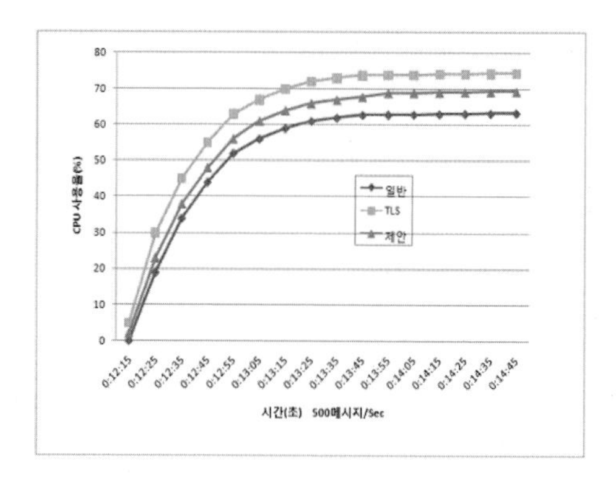

Fig. 8. Availability result

5 Conclusion

VoIP service is true effective thing in QoS side than existent PSTN by voice transmission technology based on internet. However, there was no preventive measure about security, and have much controversial points actually although security problems that use the SIP or TLS recently is offered in part VoIP service. For example, in SIP's case, was successful in early idea presentation, but is incongruent to realistic embodiment example. In the case of TLS, there is restriction point that early establishment expense is expensive and there are many difficulties in use.

So, in treatise that see hereupon attribute certificate to base safe VoIP service proposal techniques propose, and showed that excel actually as result that experiment about existent SIP, response time with VoIP system of TLS base, availability etc. about this. And response time that examine so far, will achieve addition research about secret nature, integrity, certification performance etc. in addition to urgent problem such as availability.

References

1. RFC 2617, HTTP Authentication: Basic and Digest Access Authentication, IETF (1999)
2. RFC 2402, IP Authentication Header, IETF IPSec WG (1998)
3. RFC 2246, The TLS Prototol Version 1.0, IETF TLS WG (1999)
4. RFC 3261, SIP: Session Initiation Protocol (June 2002)

5. Session Initiation Protocol(sip) Working Group, http://www.ietf.org/html.charters/sip-charter.html
6. Baugher, M., et al.: The Secure Real-time Transport Protocol, IETF draft-ietf-avt-srtp-09.txt (July 2003) (Work in Progress)
7. Housley, R., et al.: Internt X.509 Public Key Infrastructure: Certificate and CRL Profile, IETF RFC 3280 (April 2002)
8. Schulzrinne, H., Casner, S., Frederick, R., Jacobson, V.: RTP: A Transport Protocol for Real-time Application, RFC 1889, Audio/Video Transport Working Group (January 1996)
9. Handley, M., Schulzrinne, H., Schooler, E., Rosenberg, J.: SIP: session initiation protocol, Request for Comments (Proposed Standard) 2543, Internet Engineering Task Force (March 1999)

Cooperation System Design for the XMDR-Based Business Process

SeokJae Moon, GyeDong Jung, ChiGon Hwang, and YoungKeun Choi

Department of Computer Science and Engineering, Kwangwoon University 447-1
Wolgye-dong, Nowon-gu, Seoul, 139-701, South Korea
{msj8086,gdchung,duck1052,yhchoi}@kw.ac.kr

Abstract. This paper proposes a cooperation system for the XMDR-based business process. The proposed system solves the problem of heterogeneousness that may take place regarding interoperability of queries in a XMDR-based business process. Heterogeneousness in an operation of a business process may involve metadata collision, schema collision, or data collision. This can be handled by operating a business process by making use of XMDR-based Global Query and Local Query.

Keywords: Process, XMDR, Heterogeneous, Cooperative, Interoperability.

1 Introduction

As a way of operating various processed, a technology called business process is proposed[1,2]. For a process needed for cooperation in a business environment to be operated effectively, interoperability between legacies is vital. Legacy systems distributed individually, however, have different goals, and thus it might be difficult to operate them cooperatively. Besides, a service-based enterprise data integration is essential to operate processed between legacy systems in a business environment, but the processes might not be designed properly for data integration or might cause problems of heterogeneousness due to data interoperability. Heterogeneousness can be divided to schema collision and data collision in the semantic classification in terms of metadata information. Schema collision involves the semantic process of collision to semantics, structures, and expressions between database schemas, while data collision to units, formations, and validations between instance values. This paper proposes the XMDR-DAI[4] based system as a method for a business process to be effectively operated cooperatively. XMDR-DAI(Data Access & Integration) enables interoperability between legacy systems needed for data integration. In addition, collision between different units of metadata information among queries in the business process, that is, schema collision, is handled by making use of XMDR[5]. XMDR, in this paper, consists of global schemas that designate as the standard schemas used in legacy systems, and uses them to solve collision problems between units of metadata information upon business process execution in connection with local schemas. This paper states the XMDR-based business process in Chapter 2, and the cooperative system design and execution by means of XMDR-DAI in Chapter 3. The conclusion is in Chapter 4.

T.-h. Kim, A. Stoica, and R.-S. Chang (Eds.): SUComS 2010, CCIS 78, pp. 448–453, 2010.

2 Definition of XMDR and GQBP, LQBP

XMDR[5] is a technique to save relational database metadata in the object-oriented DB to solve the collision problems in the data integration. In other words, this is a system to combine MDR and ontology and integrate the data to solve the problem of collision of distributed data[7]. Thus, XMDR in this paper consists of MSO(MetaData-Semantic Ontology), InSO(Instance-Semantic Ontology), MLoc(MetaData-Location), and MDR(MetaData Registry). MDR consists of GS(Global Schema) and LS(Local Schema)[4]. This paper includes LQBP(Local Query Business Process) and GQBP(Global Query Business Process) with the inner queries of a business process as the basis for XMDR[4]. GQBP involves GQquery(GQ:Global Query) based on the global schema in a business process logic while GQ is classified to SELECT, INSERT, and UPDATE. It is defined by the generation rules according to ANSI. LQBP includes LQquery(LQ:Local Query) in which GQBP is reformed based on the local schema according to the conversion rule in connection with GQ in a business process logic. LQ as well is classified to SELECT, INSERT, and UPDATE. It is defined by the generation rule according to ANSI. GQBP adopts GQ generation rules 1,2, and 3 as defined in.

3 Proposed Systems

The system proposed in this paper enables the business process necessary for cooperation to be effectively executed. To this end, data integration and interoperability services are provided. Fig.1 shows the XMDR-based system for cooperation, which consists of Client Zone, Data Service, XMDR-DAI, Data Integration Service, Legacy DBS, and Analytic DBS. The major roles are as follows:

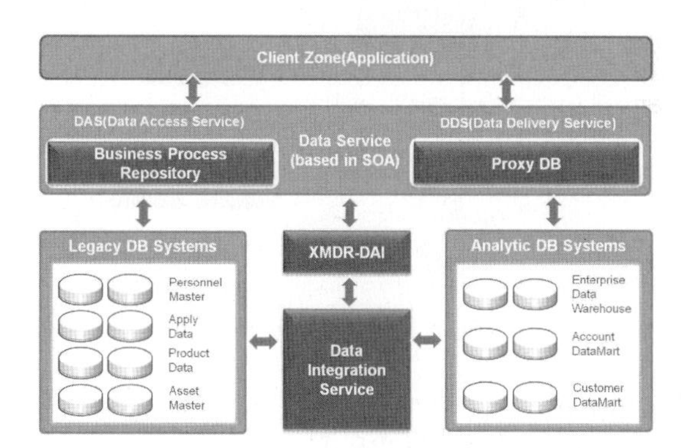

Fig. 1. A Cooperative System by Means of XMDR-DAI

- Clinet Zone: On This layer is an application area where a business process can be formed by means of a user interface.
- Data Service: This layer defines and executes a data service based business process, and stores the result data after the execution.

- XMDR-DAI: This element is the key part of this system that can solve collision problems of metadata information produced while the business process is being performed, and enhances data interoperability in the process between legacy systems by converting GQBP to LQBP.
- DIS(Data Integration Service): This element acquires the system access, performs transaction execution, and collects data so that the business process delivered in XMDR-DAI can be executed in the legacy DBS and Analytic DBS as the mediator between XMDR-DAI and DBS.
- Legacy DBS, Analytic DBS: This is an area where there are existing legacy DB and Analytic DB.

3.1 System Execution Process

In Fig.2, the Client requests BP(Business Process) of the BP Agent necessary for cooperation. BP Agent searches for information registered in BPR(Business Process Repository), and returns the selected BP information to the Client. The Client that received BP information prepares GQBP and delivers it to the DQP Agent. DQP Agent confirms whether there is FROM in the query after parsing GQBP, and executes the following step upon finding any FROM phrase. It inquires the Proxy Agent of generating a Temp Table if there is no FROM phrase. Then the DQP Agent requests the XMDR Agent of Global Meta information based on the parsed GQBP information. XMDR Agent returns the requested Global Meta information and related Local Meta information to the DQP Agent.

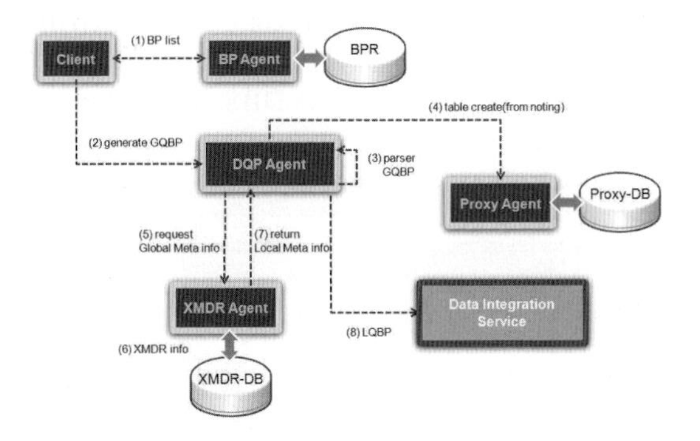

Fig. 2. Execution Process 1

Afterwards, the XMDR Agent refers to the returned Local Meta information and delivers it to DIS after remaking it LQBP. Fig.3 shows that the DQP Agent delivers LQBP to the BP Analyzer Agent. BP Analyzer Agent delivers to Connector Agent the connection information for LQBP parsing and DBS access.

The Connector Agent acquires the DBS access with regard to LQBP by means of the connection information, and delivers it to the Transaction Agent. BP Analyzer Agent delivers the BP queries in the parsed information to the transaction agent upon receiving DBS access from the Connector Agent. Transaction Agent executes the

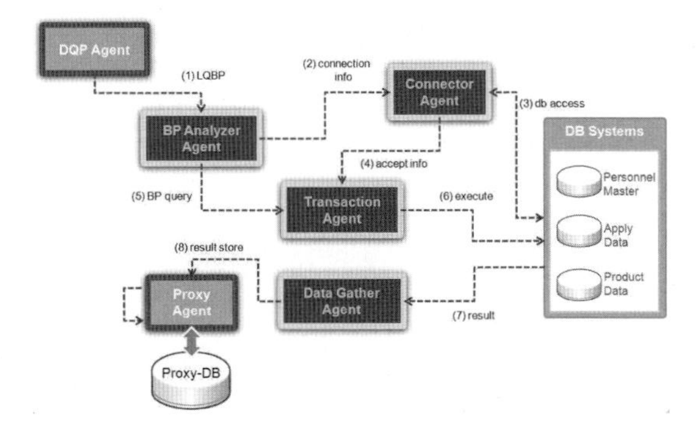

Fig. 3. Execution Process 2

execution queries by means of BP queries and access. Upon the completion of executing the BP queries, the execution results (data) are collected by the Data Gather Agent, delivered to the Proxy Agent, and stored in the Proxy DB.

3.2 System Execution Process

- BP Agent: When the business process requested by the user exists in BPR, the process is searched and the BP_list is provided accordingly. However, when the business process is to be prepared by the user, the global schema information provided through the XMDR Agent is utilized so that a newly defined business process can be adopted in a template form. Fig.4 shows the BP Agent execution algorithm. When the Input is (BP_list, *id) the parameter, the existing process is used by means of BPR_serach() when it is found to be in BPR. When it is not in BPR, however, BP_list parsing and extraction of parseTable, parseField, and parseValue follow

```
Algorithm BP_Agent
Input: request of BP_list, Output: LQBP
Begin
    If (BP_list exist in BPR) THEN  //BPR: Business Process Repository)
      BP := BPR_search(BP_list);
      return BP;
    Else parser(BP_list);
        gTable := XMDR_Agent(parseTable, 0);
        gField := XMDR_Agent(parseField, 1);
        gValue := XMDR_Agent(parseValue, 2);
        GQBP := GQBP_generate(gTable, gField, gValue, Bpformat);
        BRP_reg(*GQBP, *id); //register BPR
    End
End
```

Fig. 4. BP Agent Algorthm

The extracted values call XMDR_Agent, and they are provided with the global schema information. Then GQBP_generate() generates GQBP by means of global schema information and BPtemplate, and registers it by calling BPR_reg(). The generated GQBP is returned to the user.

- XMDR Agent: Fig.5 is an XMDR Agent execution algorithm. For Input, (itemList, handle) parameter values are inserted, and the hadle is divided into 0,1,2, and 3 to extract the table, field, value, and loc schema information from XMDR from XMDRExtract().

```
Algorithm XMDR_Agent
Input: request of itemLists, handle, Output: XMDR information
Begin
    Case handle of //Metadata info
        0 := xmdrTable = XMDRExtract(XMDR, 0); return;
        1 := xmdrTable = XMDRExtract(XMDR, 0); return;
        2:= xmdrTable = XMDRExtract(XMDR, 0); return;
        3 := xmdrTable = XMDRExtract(XMDR, 0); return;
End
```

Fig. 5. XMDR Agent Algorthm

- DQP Agent: Fig.6. shows the DQP Agent execution algorithm, which needs GQBP as the input, and GQBP is parsed by parser(). The results are divided to GQtable, GQfield, GQvalue, and GQloc respectively. The separate schema information calls createTable() to generate a table in the Proxy DB to save the GQBP execution results. Each GQtable, GQfield, GQvalue, and GQloc extracts LQtable, LQfield, LQvalue, and LQloc, which are part of the schema information needed to generate LQBP after XMDR_AGENT() performs the mapping with the local schema information. The extracted local schema information is conversed to LQBP by GQ2LQ(), then generated in an XML document form by SpliteXML(), and delivered to each legacy system.

```
Algorithm DQP_Agent
Input: GQBP, Output: LQBP;
Begin
    token[n] := parser(GQBP);
    For i:=1 to i<=n do
        Case token[i] of
            table[i] : GQtable[i] := token[i]; return;
            field[i]: GQfield[i] := token[i]; return;
            value[i]: GQvalue[i] := token[i]; reuturn;
            loc[i]: GQloc[i] := token[i]; return;
        End
    End
    gsTemp := (GQtable[i], GQfield[i], GQvalue[i], GQloc[i], id);
    createTable(gsTEMP, 1); //proxy-db table create
    LQtable := XMDR_Agent(GQtable, handle);
    LQfield := XMDR_Agent(GQfield, handle);
    LQvalue := XMDR_Agent(GQvalue, handle);
    LQloc := XMDR_Agent(GQloc, handle);
    LQBP := GQ2LQ(LQtable, LQfield, LQvalue, LQloc);
    SpliteXML(LQBP2XML(LQBP)); //send to legacy systems
End
```

Fig. 6. DQP Agent Algorthm

- Proxy Agent: Proxy DB is a temporary saving area for the executed data. Fig.7 shows the Proxy Agent execution algorithm, and with (resultXML, gsTEMP) parameter as the input value. ResultXML is the result of executing LQBP in the legacy while gsTEMP is the table information to be generated in the Proxy DB that includes the global schema delivered from the DQP Agent. The two elements - resultXML and gsTEMP - are divided by the y value. When the y value is 1, create_tabel() is called to generate a table in the Proxy DB while when it is 0, the result of processing the resultXML document in DOMProcoess() is saved in the Proxy DB. Then the user is informed of the result.

```
Algorithm Proxy_Agent
Input: LQBP execute resutl data(resultXML, gsTEMP, e, y);
Output: notice to UI;
Begin
    If ( y!=null ) Then
      create_table(gsTEMP); return;
    Else
        If ( e!=null ) Then
            resultXML := DataGather();
            DOMProcess(resultXML);
            return true; //notify UI
        Else
            return false;
        End
    End
End
```

Fig. 7. Proxy Agent Algorthm

4 Conclusion

This paper proposes a cooperative system for an XMDR-based business process in a business environment. This system is advantageous in that the user does not need to consider the metadata collusion when a business process is executed for cooperation between legacy systems. Besides, additional legacy systems do not need to modify the local schema for data access to other legacy systems in consideration of data interoperability and cooperation.

References

1. Smith, H., Fingar, P.: Business Process Management. In: The Third Wave. MK Press (2003)
2. Indulska, M., Chong, S., Bandara, W., Sadiq, S., Rosemann, M.: Major Issues in Business Process Management. In: ACIS 2006 (2006)
3. Dustdar, S., Treiber, M.: Integration of heterogeneous web service registries – the case of visr. WWW Journal (2006)
4. Moon, S., Jung, G., Choi, Y.: XMDR-DAI Based on GQBP and LQBP for Business Process. In: AST 2010 Proceedings, pp. 72–85 (2010)
5. Keck, K.D., McCarthy, J.L.: XMDR: Proposed Prototype Architecture Version 1.01 (February 2005), http://www.xmdr.irg
6. Arjuna, et al: Web service business activity framework (2005), http://specs.xmlsoap.org/ws/2004/10/wsba/
7. ISO/IEC IS 11179, Information technology -Specification and standardization of data elements (2003)

Architecture of the SDP in the Cloud Environment

Jae-Hyoung Cho and Jae-Oh Lee

Information Telecommunication Lab,
Dept. of Electrical and Electronics Engineering,
Korea University of Technology and Education, Korea
{tlsdl2,jolee}@kut.ac.kr

Abstract. Recently, cloud computing, which is Internet-based computing, is attracting type of business. Besides, network is integrated wire and wireless (including mobile) by the IP Multimedia Subsystem (IMS). Therefore, cloud computing needs to support the IMS and common platform in order to support between the IMS and cloud computing. So, in this paper we consider about deployment of the IMS and suggest Service Delivery Platform (SDP) that is common platform for cloud computing.

Keywords: SDP, Cloud Computing, OMA.

1 Introduction

Cloud computing is the computing equivalent of the electricity revolution of a century ago. Before the advent of electrical utilities, every farm and business produced its own electricity from freestanding generators. After the electrical grid was created, farms and businesses shut down their generators and bought electricity from the utilities, at a much lower price (and with much greater reliability) than they could produce on their own. Look for the same type of revolution to occur as cloud computing takes hold. Additionally, networks such as wired/wireless and mobile, is integrated by the IMS. The IMS is an architectural framework for delivering IP multimedia services. From a logical architecture perspective, services need not have their own control functions, as the control layer is a common horizontal layer. By integrating between networks, service development method which is moved by Service Provider is made up of the form of different structure and each separated vertical service structure. And these cause complication, intensiveness of resource and increase of cost management. These situations call for the creation of common service delivery platform that can develop new service based existing service and IMS into valued added services. Service providers not only move the appropriate cost and the new innovative service but also integrate and create the service in existence and demand the environment that moves these with the IMS and cloud computing. Therefore, in this paper, we suggest the common SDP for cloud computing. In chapter 2, we'll consider about cloud computing and the OMA as related works. And then, we suggest of the architecture of the SDP and do an experiment on the SDP using SOA BPEL and common service enablers.

T.-h. Kim, A. Stoica, and R.-S. Chang (Eds.): SUComS 2010, CCIS 78, pp. 454–459, 2010.

2 Related Works

2.1 OMA

The OMA is to facilitate global user adoption of mobile data services by specifying market driven mobile service enablers that ensure service interoperability across devices, geographies, service providers, operators, and networks, while allowing businesses to compete through innovation and differentiation. For creating service development and deploying environment, OMA specifies enablers that provide standardized components. These components and their interaction compose the OMA Service Environment (OSE) [6]. The structure of the OSE includes intrinsic and non-intrinsic functions that are possible to reuse enabler. Besides, these functions protect enabler and resource. For scalability, the OSE provides enabler with interfaces such as Web Service, Java, and .NET. It also applies to policy in order to control access enabler and also provides Life Cycle management on purpose to upgrade enablers.

Fig.1 represents the relationship between the OSE structures. Service enablers are technology that is used for service development, operation and deploying within service provider domain. The implementation of service enablers are accomplished to service and terminal domain, through life cycle management such as start, stop, and trace. The OSE context is a conceptual and structured environment that includes OMA enablers, interfaces to application that make use of OMA enablers, interfaces to service providers' Execution Environment (e.g., software life cycle management) and the interfaces to invoke and use underlying capabilities and resources for enabler implementations.

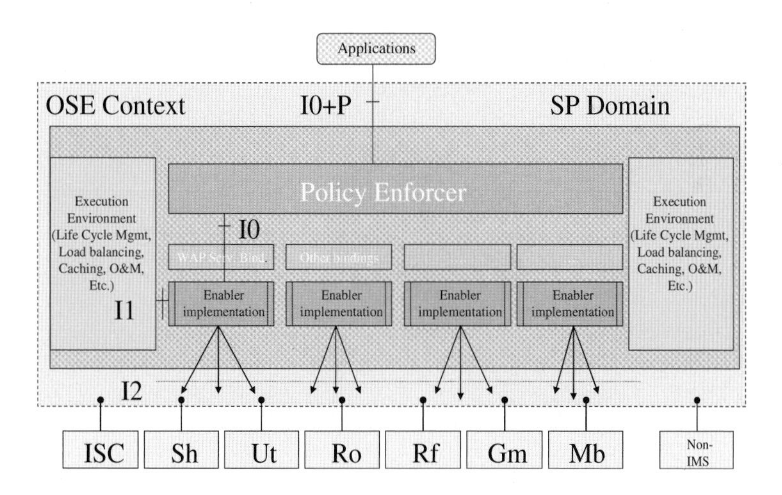

Fig. 1. The Structure of OSE

Execution Environment is accomplished by life cycle management eTOM-based software. Models include Develop, Sell, Provide, Bill, Service, Report, and Modify/Exit. Execution Environment includes life-cycle-management, system supporting (e.g., thread manage, load control, cashing), and running management as the enabler. Among which life-cycle-management executes software creation, deployment and management. Policy Enforcer protects resources from non-authority request and manages appropriate charging, logging, the secret of users and preference.

The capacities of the SDP are differed from the view point of vendors. But many capacities of vendor framework are able to contain OMA. Vendors consider the control of IMS and service layer component. As time goes by, the SDP structure that is different from vendors is more similar at present.

2.2 Cloud Computing

Cloud computing is the ability to provide IT resources over the Internet. These resources are typically provided on a subscription basis that can be expanded or contracted as needed. Services include storage services, database services, information services, testing services, security services, and platform services—pretty much anything user can find in the data center today can be found on the Internet and delivered as a service. In many respects, cloud computing is about abstracting the cloud computing resource from the underlying hardware and software, which are remotely hosted. Thus, user deals with the service and almost never with the needs and requirements of the platform, including maintenance, monitoring, and the cost of the hardware and data center space [1]. The concept is to leverage computing resources that you do not own or maintain, and thereby lower the cost of computing through economies of scale. The more that organizations share cloud computing resources, the less they should cost. Moreover, cloud computing can be leveraged computing resources that provide more pre-built component parts and thereby avoid having to build everything from scratch. Users can find many bits and pieces of an application within the clouds and get much farther down the road than if you built your system from scratch [2].

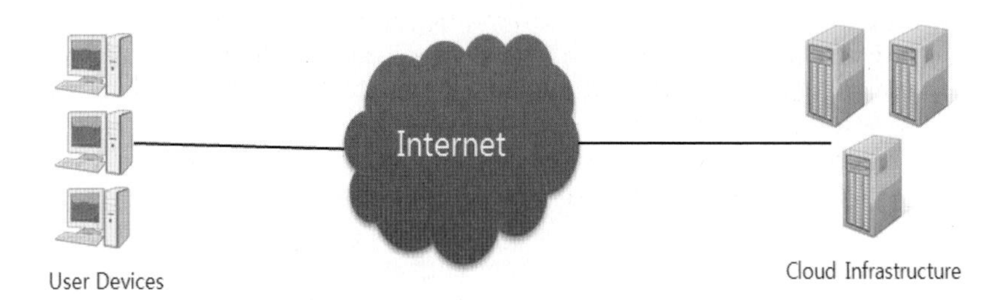

Fig. 2. Concept of Cloud Computing

Fig. 2 describes the basic concept of cloud computing. Users are able to access cloud infrastructure (database center, service and information and so on) through Internet. There are several different ways the infrastructure can be deployed. The infrastructure will depend on the application and how the provider has chosen to build the cloud solution. This is one of the key advantages for using the cloud. These needs might be so massive that the number of servers required far exceeds your desire or budget to run those in-house. Alternatively, you may only need a sip of processing power, so you don't want to buy and run a dedicated server for the job. The cloud fits both needs. Additionally, cloud computing is able to support full virtualization and para-virtualization. The former is a technique in which a complete installation of one machine is run on another, the latter allows multiple operating systems to run on a single hardware device at the same time by more efficiently using system resources, like processors and memory.

3 Architecture of the SDP and Orchestration Scenario

Fig. 3 represents that devices access to cloud infrastructure through the IMS. All devices (IP-based) are controlled by the IMS network through IMS session control scenario.

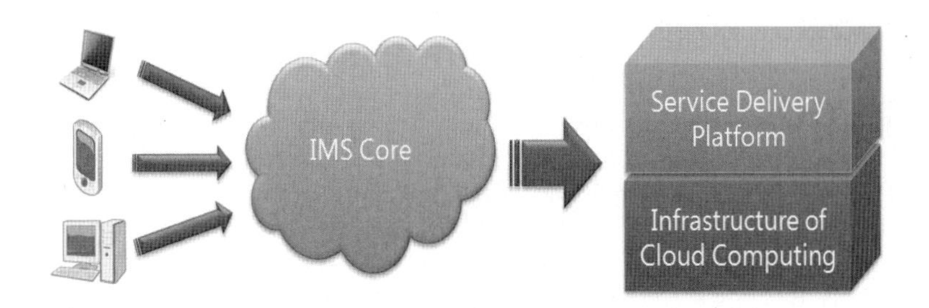

Fig. 3. Devices Access to Cloud Infrastructure through the IMS

The SDP supplies all the resources required to build applications and services completely from the IMS, without having to download or install software. The SDP offers some support to help the creation of user interfaces. The SDP also supports web development interfaces such as Simple Object Access Protocol (SOAP) and Representational State Transfer (REST), which allows the construction of multiple web services, sometimes called mashups. The interfaces are also able to access databases and reuse services that are within a private network [3]. Fig.4 represents composition elements and interfaces in the SDP. IMS services are able to access the SDP through SIP AS or OSA-SCS (through Parlay API). Also, the intra-bus, which composes the SDP, is defined by SOA/Web Service (WS). And SIP interface is defined for the inter-bus of each SDP [7]. To access the IMS network, a module is existed as various protocol gates in a middleware structure. The developer uses the middleware to provide high-level API about network protocol. The middleware carries out the necessary protocol conversion between high-level API command and protocol. Supported APIs can be defined as private one or standard ones (e.g., Parlay API, SS7-family SIP, MM7, LIP, SMPP, PAP, SMTP, POP3). Service Logic Execution Environment (SLEE), dealing with scalability, distribution, configuration of service, is a real-time environment for service providers. The candidate technologies are based on J2EE/.NET, SIP Servlet, JAIN SLEE and Web Service. Common service enabler supports a basis structure within the service domain. A service developer is able to control and prepare application service that uses the service capabilities provided by common service enablers. Accordingly, service providers don't have to re-create these service functions as stove-pipes. This service enabler includes contents management, devices management, policies control, logging, subscribers/profiles management and so on. Service composition and orchestration is used to create service in a fast way.

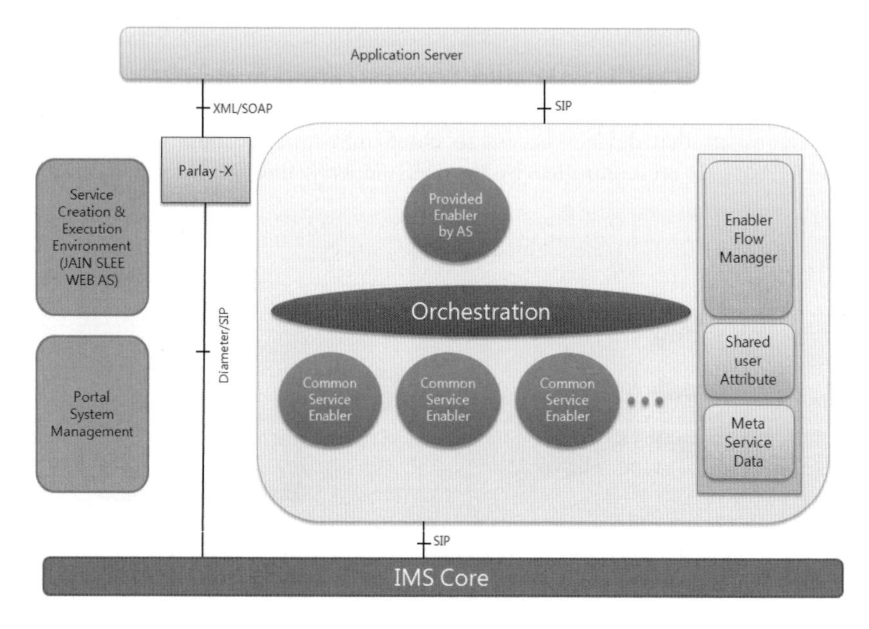

Fig. 4. Composition Element and Interfaces in the SDP

There is a modeling tool that might compose a desirable service. The communication among services is accomplished as Service Integration Bus (SIB) that is based on Business Process Engineering Language (BPEL). The SIB uses SOA/WS technology as an interface between one service enabler and other service enablers. The Service Creation & Execution Environment (SCEE) uses a GUI-based tool for making a real-time service bundling. Portal System is used as a third-party management tool. And resource adapter as a SDP service component takes a role of access gateway to the IMS network. Fig 5 represents orchestration using SOA/BPEL among service enablers. In this figure, we are used by two common service enablers (PRS, Pr) [5] and one service enabler (IPTV) which is provided by content provider. When users request or create service, enablers interact with each other through BPEL. For example, we suppose that user wants to see IPTV that is customized for them. Firstly, Presence enabler (Pr) invokes presentity of the user and the service provider including availability information. Additionally, Personalization Resource Service enabler (PRS) invokes personalized information of user including preference, offering and so on, and make new document (including presence and personalized information) through interaction. After this interaction, including presence and personalized information document interacts with IPTV enabler. IPTV enabler invokes the list of contents in the Electronic Program Guide (EPG) server. Finally, user is able to gain the list of customized IPTV.

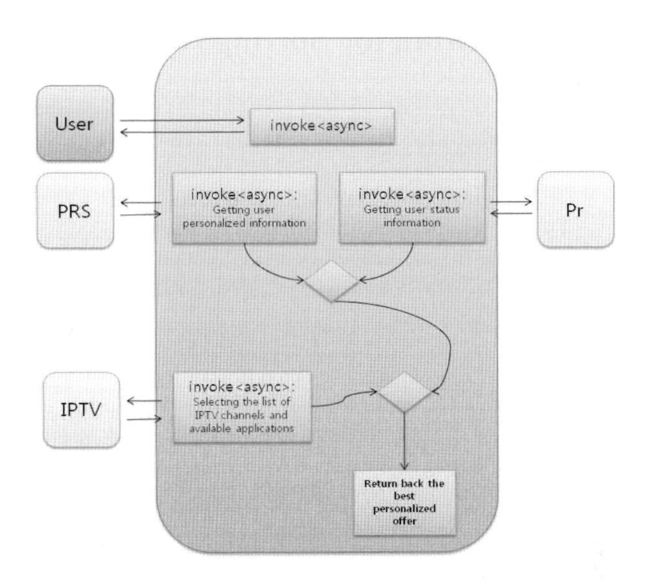

Fig. 5. The Orchestration using SOA/BPEL among service enablers

4 Conclusions and Future Works

We have suggested the structure of the SDP for integrated environment and cloud. Furthermore, we have made workflow using common service enablers through SOA/BPEL. In the Future, we would involve evaluating the node performance such as the IMS nodes (CSCFs), cloud infrastructure (servers) etc. Also we would implement more common service enablers through analyzing the OMA enablers. Later on, the IMS/SDP structure in the cloud computing will continue on in the future, further strengthening its leadership role, helping customers in their on demand business enablement and the growth, availability, and systems management.

References

1. Velte, A.T., Elsenpeter, R., Velte, T.J.: Cloud Computing A practical approach. Mx Graw Hill, New York
2. Miller, M.: Cloud Computing Web-Based Applications That Change the Way You Work and Collaborate Online. QUE (2008)
3. Linthicum, D.S.: Cloud Computing and SOA Convergence in Your Enterprise
4. Cho, J.-H., Lee, J.-O.: The Structure of the SDP Using Enablers in the IMS. In: Hong, C.S., Tonouchi, T., Ma, Y., Chao, C.-S. (eds.) APNOMS 2009. LNCS, vol. 5787, pp. 448–452. Springer, Heidelberg (2009)
5. 3GPP TS 23.141 v7.2.0, Presence Service; Architecture and functional description (Release 7) (September 2006)
6. OMA, Presence_SIMPLE-V1_1 (January 2008)
7. Cho, J.-H., Lee, J.-O.: The IMS/SDP Structures and Implementation of Presence Service. In: Ma, Y., Choi, D., Ata, S. (eds.) APNOMS 2008. LNCS, vol. 5297, pp. 560–564. Springer, Heidelberg (2008)

P-CSCF's Algorithm for Solving NAT Traversal

Jung-Ho Kim, Jae-Hyoung Cho, and Jae-Oh lee

Information Telecommunication Lab,
Dept. of Electrical and Electronics Engineering,
Korea University of Technology and Education, Korea
{jungho32,tlsdl2,jolee}@kut.ac.kr

Abstract. Many ways for efficient use of limited IP address of IPv4 are existed. The one of these ways is to construct the private network using Network Address Translator (NAT). NAT filtering rule makes network management easier. However, NAT Filtering rule makes NAT Traversal. Many solutions like Simple Traversal of UDP through NAT (STUN), Traversal Using Relay NAT (TURN) and Media Relay method exist. But these solutions require additional servers or devices. So, we suggest that P-CSCFs in the IP Multimedia Subsystem (IMS) change the packet's header and solve the NAT Traversal without any additional equipment.

1 Introduction

The increase of IP-based equipment and network users has caused the shortage of IP address. The way to overcome the shortage of IP address is building a private network using NAT devices. By connecting to public network using NAT device, multiple users with a single public IP address can connect to the internet. In the IMS, all network equipment is IP-based. The lack of shortage of IP address became more and more seriously. Therefore, using NAT devices is more important.

In private network using NAT, media transport using SIP occurs NAT Traversal. Because of difference between IP information of Session Description Protocol (SDP) and IP header of packet, NAT Traversal is occurred. [1]

There are many solutions for NAT Traversal. Typical solutions are using STUN server or TURN server and setting SBC equipment. In the IMS, Interconnection Border Control Function (IBCF) and Interconnection Border Gateway Function (IBGF) that is media relay server are used to solve NAT Traversal. IBGF hides the IMS network and acts as a firewall between the IMS networks. Most of the solutions like STUN, TURN and SBC are needed additional server or device. Therefore in this paper, we will suggest the solution for NAT Traversal using the existing P-CSCF in the IMS without adding devices.

2 Related Works

2.1 IMS

The IMS is developed to realize real-time or non-real-time multimedia service in the mobile environment. The service object of the IMS provides synthetic multimedia

T.-h. Kim, A. Stoica, and R.-S. Chang (Eds.): SUComS 2010, CCIS 78, pp. 460–465, 2010.

services using IP protocol such as voice, video and data services. Additionally, the IMS has advantages that are easy in service developments and modifications. The IMS can also improve a completive price as well as is easy to interlock with various 3rd party applications that are based on efficient session management. It enables mobile operator to support the structure based on Session Initiation Protocol (SIP) such as session management, security, mobility, Quality of Service (QoS) and charging.

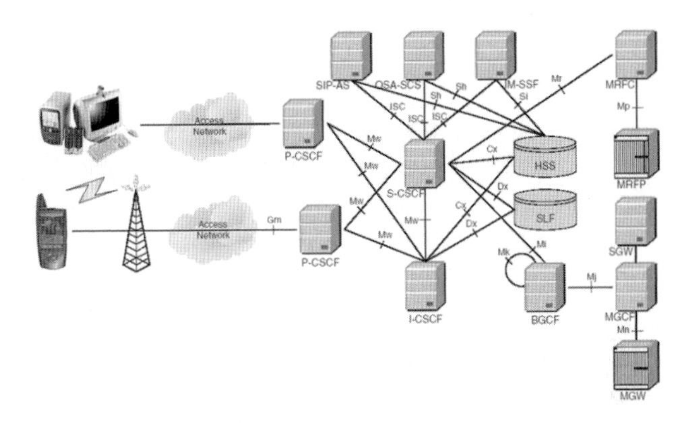

Fig. 1. The IMS Structure

Figure 1 shows the IMS structure. In the IMS network, Call Session Control Functions (CSCFs) process signal. The CSCFs based on SIP implement basic functions to control multimedia session with infra-system, and are divided into P(Proxy)-CSCF, I(Interrogating)-CSCF, S(Serving)-CSCF by roles. The CSCFs do the function of subscriber registration, authentication, charging, service triggering, routing of relevant application server, collating the receiving of user location, and compressing and decompressing of the SIP message. And profile information, authentication and related location information data of a subscriber are stored in Home Subscriber Server (HSS). Application Server (AS) has real service logic to provide services.

For multilateral service, the Multimedia Resource Function (MRF) is used to control the signal and process media mixing. Additionally, the IMS can interact with PSTN through one or more PSTN gateways. Media Gateway Control Function (MGCF) is responsible for inter-working with PSTN. The Breakout Gateway Control Function (BGCF) is the IMS element that selects the network in which PSTN breakout is to occur. If the breakout is to occur in the same network as the BGCF then the BGCF selects a MGCF. The MGCF then receives the SIP signaling from the BGCF. [2][3]

2.2 NAT and Solution for NAT Traversal

NAT devices exchange the packet header to its public IP address. So, a number of users in private network can connect to public internet using one public IP address. But, when the packet routes private network to public network through NAT devices, NAT devices make pinhole made according to a mapping table combined source

IP:Port and destination IP:Port. Because of these pinholes, unrelated packet, that is, does not correspond to the mapping table cannot pass the NAT device. NAT is divided 4 types by mapping rule: Full cone, restricted, port restrict and symmetric. [4]

There are many solutions for NAT Traversal. Solutions are distinguished by NAT Types.

STUN

STUN is a technology that realizes a NAT traversal by using UDP hole punching technology. A communication path is provided through the access to the STUN server on the Internet from the terminal behind the NAT. STUN has an advantage that the existing NAT router can be used. However, STUN server is required on the Internet, and its application is limited. Furthermore, it is not applicable in the case that Symmetric NAT is used.

TURN

TURN allows an end point behind a firewall to receive SIP traffic on either TCP or UDP ports. This solves the problems of clients behind symmetric NATs which cannot rely on STUN to solve the NAT traversal issue. TURN connects clients behind NAT to a single peer. Its purpose is to provide the same protection as that created by symmetric NATs and firewalls. The TURN server acts as a relay server; any received data is forwarded to client. The client on the inside can be received. However, if the number of user increases, TURN server has heavy traffics. So delay is occurred.

Media Relay

This method is similar to TURN about relaying the media packet. Unlike TURN, however, media relay method can change the Session Description Protocol (SDP). Typically, media relay method has server called NAT proxy changing SDP for passing the NAT. Moreover, instead connecting and flowing the media stream directly between two users of using SIP, by changing SDP, media session is established between NAT proxies. Then, NAT proxy is linked to users following the NAT's mapping table. [5]

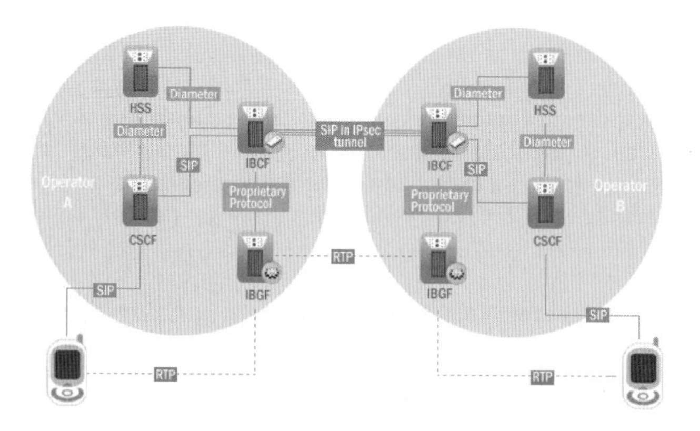

Fig. 2. Solution of NAT Traversal using IBCF

2.3 Solution in the IMS for NAT

Figure 2 shows present solution using IBCF and IBGF in the IMS. IBCF acts Border Controller between the IMS networks. And IBCF processes issues, like security, network hiding and firewall, occurred when the IMS networks are linked. If there are SDP offer and SDP answer when the session is established, IBCF inform the SDP's IP information to IBGF. IBGF received the SDP's IP information acts media relay server. The IMS also solves NAT Traversal by using SDP's IP information and additional media server. [6]

3 Solution for NAT Traversal Using P-CSCF

RTP transmission is based on SDP's IP information that is private IP address, however, public network can't find a user who is behind NAT device. In this paper, we will solve this problem using P-CSCF in the IMS. P-CSCF is first core meets users. Therefore, all packets must go through its P-CSCF. Because P-CSCF can perform over the 4 layers, it can read and change SIP and SDP. If we can add the functions that change packet's IP header and SDP's IP information at P-CSCF, we will be able to solve NAT Traversal without additional devices. [7]

3.1 P-CSCF's Algorithm for Solving NAT Traversal

Figure 3 is the algorithm, when P-CSCF receives the packet from its user and public network. In the SIP, configuration for media plane is defined SDP. Therefore, when a message arrives from the inside a user, firstly, P-CSCF checks the existence of SDP(left in Figure 3.). If there is no SDP, P-CSCF forwards the packet to public network. On the other hand if there is SDP, P-CSCF compares SDP's IP information

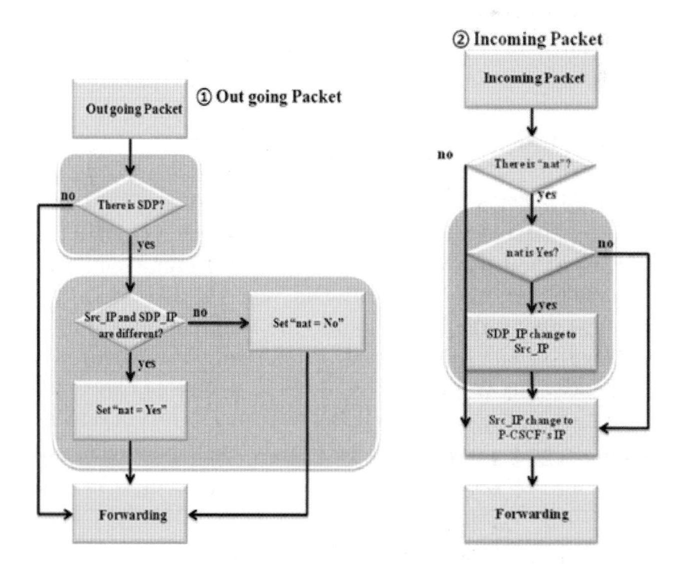

Fig. 3. P-CSCF's algorithm for outgoing and incoming packet

with packet header's source IP. The result of comparison identifies existence of NAT device. If a result of comparison is same, this means NAT device. So, P-CSCF sets SIP's nat parameter to "nat = No". If a result of comparison is not same, P-CSCF sets nat parameter to "nat = Yes". And then, P-CSCF forward to public network.

When Packet is arrived, P-CSCF checks existence of "nat" parameter (right in Figure 3). If there is no "nat" parameter, P-CSCF confirms that this message doesn't have the SDP and send packet that includes source IP replaced by P-CSCF's IP address. By doing this, the NAT passes the packets through the pinhole created in the process of registration. While if there is "nat" and "nat" parameter is "Yes", P-CSCF sets SDP's IP information to packet's source IP address and replaces source IP to P-CSCF's IP address. And, if "nat" parameter is "No", P-CSCF replaces the packet's source IP to its IP address. By processing this algorithm, media plane can pass the NAT through the pinhole made SDP offer, and other packets can pass the NAT through the pinhole made registration.

3.2 Scenario Solved NAT Traversal Using P-CSCF's Algorithm

Figure 4 describes the scenario of NAT Traversal. UA1 is behind the NAT, knows its P-CSCF address and sends registration message to P-CSCF. And, UA1 will make session for media transfer using INVITE message of SIP. At this point, packets are classified two kinds of part. One is outgoing packet, the other is incoming packet. In this scenario, P-CSCF algorithm for outgoing packet is represented ①, incoming packet is represented② .UA1 wants to transmit media. Before communication, UA1 should do registration. When (1) REGISTER message sent by UA1 arrived NAT, pinhole1 that is linking between P-CSCF and UA1 is created. Pinhole1 will be used to pass the packets except media to UA1. After that, NAT device forwards (2) REGISTER message to P-CSCF. And then, P-CSCF processes the packet according to ① algorithm shown figure 3. Because REGISTER messages don't have SDP, P-CSCF forwards (3) REGISTER message to the IMS network without any handling. After registering UA1 to the IMS, (4)200 OK message arrives at the UA1's P-CSCF. Then, P-CSCF modifies the (4)200 OK message according to ② algorithm shown figure 3. Because 200 OK messages don't have SDP body, there is no "nat" parameter. Therefore, P-CSCF forwards the packet after changes the packet's source IP to its IP address. (5)200 OK message can pass the NAT through Pinhole1 that links UA1 to its P-CSCF. After the SIP registration, UA1 sends (7) INVITE message that includes SDP for media transfer to UA2. UA2 is another user agent, who is not shown in figure 3. (7)INVITE message's source IP and SDP's IP address is same as A. When (7) INVITE message passes the NAT, source IP change to NAT's IP address N. So, (8) INVITE message's source IP is N and IP information of SDP for media plane is A. And then, Pinhole 2 that is linking UA1 to UA2 is created. This pinhole 2 will be used for flow the media plane. P-CSCF receiving the (8) INVITE message performs the ① algorithm. In this process, P-CSCF sets "nat" parameter to "Yes". This processing makes easy for P-CSCF to check the existence of NAT. Passing through the ① algorithm, (9) INVITE message has "nat = Yes". When (9) INVITE message is arriving at UA2's P-CSCF, UA2's P-CSCF does ② algorithm and forwards UA2's NAT. So, media session is established between both side NAT and flows through pinhole2. Other packets like (10) Ringing and (13)200 OK messages are processed ① algorithm and can pass the NAT using Pinhole1, too. By doing this, (19) Media plane flow between two NATs and NATs pass media packet to their user using pinhole2. We solve NAT Traversal without any media server or device.

4 Future Works

In this paper, we solve NAT Traversal without any devices. By adding P-CSCF's function handling the packets used session establishment. But, P-CSCF's overhead and SIP encryption problems will be emerged. We will solve these problems at simulation stage.

References

[1] Perea, R.M.: Internet Multimedia Communication Using SIP, pp. 467–493 (2008)

[2] Camarillo, G., Garcia-Martin, M.A.: The 3G IP Multimedia Subsystem (IMS), pp. 29–33 (2006)

[3] Cho, J.-H., Lee, J.-O.: The IMS/SDP Structure and Implementation of Presence Service. In: Ma, Y., Choi, D., Ata, S. (eds.) APNOMS 2008. LNCS, vol. 5297, pp. 560–564. Springer, Heidelberg (2008)

[4] M"uller, A., Klenk, A., Carle, G.: Behavior and Classification of NAT devices and implications for NAT Traversal. IEEE draft (2008)

[5] Park, J.-H., Park, C.-G., Seok, S.-H., Kim, T.-Y., Chung, B.-D.: Implementation of the OSGi-based Home SIP Proxy and UDP Relay for NAT Traversal. In: JCCI 2007 (2007)

[6] Vargic, R., Krhla, M., Schumann, S., Kotuliak, I.: IMS interworking using IBCF. In: Third 2008 International Conference on Convergence and Hybrid Information Technology (2008)

[7] Huang, T.-C.: Smart Tunnel Union for NAT Traversal. In: NCA 2005 (2005)

Model Based User's Access Requirement Analysis of E-Governance Systems

Shilpi Saha[2], Seung-Hwan Jeon[1], Rosslin John Robles[1], Tai-hoon Kim[1], and Samir Kumar Bandyopadhyay[3]

[1] Hannam University, Daejeon-306791, Korea
jeoninoldenburg@hanmail.net, roslin_john@yahoo.com, taihoonn@empal.com
[2] Computer Science and Engineering Department, Heriatge Institute of Technology, Kolkata-700107, India
shilpisaha7@gmail.com
[3] Department of Computer Science and Engineering, University of Calcutta, Kolkata-700009, India
skb1@vsnl.com

Abstract. The strategic and contemporary importance of e-governance has been recognized across the world. In India too, various ministries of Govt. of India and State Governments have taken e-governance initiatives to provide e-services to citizens and the business they serve. To achieve the mission objectives, and make such e-governance initiatives successful it would be necessary to improve the trust and confidence of the stakeholders. It is assumed that the delivery of government services will share the same public network information that is being used in the community at large. In particular, the Internet will be the principal means by which public access to government and government services will be achieved. To provide the security measures main aim is to identify user's access requirement for the stakeholders and then according to the models of Nath's approach. Based on this analysis, the Govt. can also make standards of security based on the e-governance models. Thus there will be less human errors and bias. This analysis leads to the security architecture of the specific G2C application.

Keywords: G2C, Broadcasting model, Critical flow model, Comparative analysis model, E-advocacy model, Interactive model.

1 Introduction

Using Information Communication Technology, e-governance is approaching the citizens to provide them web-based services at lower cost and higher efficiency. Among the various components of e-governance, using G2C the citizen can directly interact with Govt. Citizens will be reluctant to use the web based services offered by the government, due to their poor skill, lack of confidence, security and privacy concerns [Backus 2001]. Many of the citizens of the developing countries like India are illiterate. Also, presently there is a lack of inexpensive and easy-to-use security

T.-h. Kim, A. Stoica, and R.-S. Chang (Eds.): SUComS 2010, CCIS 78, pp. 466–471, 2010.

infrastructure in G2C applications. Moreover, to support all levels of the society, there must be a proper access control technology according to the models.

In our previous paper [5], G2C applications are classified based on the Egov models. The information flow follows the work flow in these models. The information provided to the citizens is to be reliable, but the information channels are almost always public and insecure. Moreover, there is a necessity for maintaining a secure environment for hosting the govt's data.

This paper looks at the access permissions of the users of G2C applications from a model driven approach. The users and their roles have been identified. Based on this analysis, access requirements of the roles have been identified. This analysis leads to the security architecture of the specific G2C application.

2 Previous Works

Based on different classes of information, their sources and frequency of updation and exchange, various models of E-governance projects can be evolved. The National E-governance Action Plan of the Government of India [NeGP] can act as a model for such projects. Other sources of information include and DigitalGovernance.org web page of Mr. Vikas Nath [Nath, 2005]. In the latter, Nath has classified the models into the following categories:

- Broadcasting model
- Critical flow model
- Comparative analysis model
- E-advocacy model
- Interactive model.

2.1 Broadcasting Model

The model is based on broadcasting or dissemination of useful governance information which already exists in the public domain into the wider public domain through the use of ICT and convergent media. The utility of this model is that a more informed citizenry is better able to benefit from governance related services that are available for them.

2.2 Critical Flow Model

The model is based on broadcasting or dissemination information of 'critical' value (which by its very nature will not be disclosed by those involved with bad governance practices) to targeted audience using ICT and convergent media. Targeted audience may include media, opposition parties, judicial bench, independent investigators or the wider public domain itself.

2.3 Comparative Analysis Model

Comparative Knowledge Model is one of the least-used but a highly significant model for developing country which is now gradually gaining acceptance. The model can be used for empowering people by matching cases of bad governance with those of good governance, and then analyzing the different aspects of bad governance and its impact on the people.

2.4 E-Advocacy Model

E-Advocacy / Mobilization and Lobbying Model is one of the most frequently used Digital Governance model and has often come to the aid of the global civil society to impact on global decision-making processes. The strength of this model is in its diversity of the virtual community, and the ideas, expertise and resources accumulated through this virtual form of networking.

2.5 Interactive Model

Interactive-Service model is a consolidation of the earlier presented digital governance models and opens up avenues for direct participation of individuals in the governance processes. Fundamentally, ICT have the potential to bring in every individual in a digital network and enable interactive (two-way) flow of information among them.

3 Our Work

Access to protected information must be restricted to people who are authorized to access the information. The computer programs, and in many cases the computers that process the information, must also be authorized. This requires that mechanisms be in place to control the access to protected information. The sophistication of the access control mechanisms should be in parity with the value of the information being protected- the more sensitive or valuable the information the stronger the control mechanisms need to be.

In this section, the roles of the stakeholders as the user will be identified according to the models. The stakeholders of egovernance systems are Government officials, LUB/LSG, private sector, NGO/civil society, organizations and citizens. The users can be classified as Database managers, Developers, Implementation officers, Information officers, Chief information security officers, officers and employees, Agents, Public.

We are considering here that at the client side only the information are being accessed with the help of some security checking. All access permissions according to the model are described below:

3.1 Broadcasting Model

In broadcasting model, generally data are for public use. So read permission must be granted to all the stakeholders. Other permissions for the stakeholders are described in Table 1:

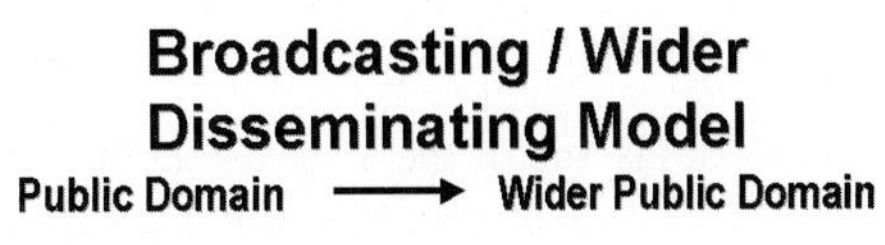

Broadcasting / Wider Disseminating Model

Public Domain ⟶ Wider Public Domain

Table 1.

Users	Read	Write	Delete	Execute	Modify	Append
DBM	Y	Y	Y	Y	Y	Y
Dev	Y	Y	N	Y	N	Y
IO	Y	N	N	Y	N	Y
Info	Y	Y	Y	Y	Y	Y
Ciso	Y	N	N	Y	N	Y
Emp	Y	N	N	Y	N	N
A	Y	N	N	Y	N	N
P	Y	Y	N	Y	N	Y

3.2 Critical Flow Model

In critical flow model, data must reach to the targeted domain not to all. All access requirements for the stakeholders are described in Table 2:

Critical Flow Model

Critical Domain ⟶ Targeted / Wider Domain

Table 2.

Users	Read	Write	Delete	Execute	Modify	Append
DBM	Y	Y	Y	Y	Y	Y
Dev	Y	Y	N	Y	N	Y
IO	Y	N	N	Y	N	Y
Info	Y	Y	N	Y	N	Y
Ciso	Y	N	N	Y	N	Y
Emp	Y	N	N	Y	N	N
A	Y	N	N	Y	N	N
P	Y	N	N	Y	N	Y

Table 3.

Users	Read	Write	Delete	Execute	Modify	Append
DBM	Y	Y	Y	Y	Y	Y
Dev	Y	Y	N	Y	N	Y
IO	Y	N	N	Y	N	Y
Info	Y	Y	N	Y	N	Y
Ciso	Y	N	N	Y	N	Y
Emp	Y	N	N	Y	N	N
A	Y	N	N	Y	N	N
P	Y	N	N	Y	N	Y

3.3 Comparative Analysis Model

In comparative analysis model, the analysis is done based on old records. All access requirements for the stakeholders are described in Table 3:

Comparative Analysis Model

Private / Public Domain + Public / Private Domain

⟶ Wider Public Domain

3.4 E-Advocacy Model

E-advocacy model has come to the aid of the global civil society to impact on global decision making process. All access requirements for the stakeholders are described in Table 4:

Mobilisation and Lobbying Model

Networking Networks for Concerted Action

Table 4.

Users	Read	Write	Delete	Execute	Modify	Append
DBM	Y	Y	Y	Y	Y	Y
Dev	Y	Y	Y	Y	Y	Y
IO	Y	N	N	Y	N	Y
Info	Y	Y	N	Y	N	Y
Ciso	Y	N	N	Y	N	Y
Emp	Y	N	N	Y	N	N
A	Y	N	N	Y	N	N
P	Y	Y	Y	Y	Y	Y

2.5 Interactive Model

In interactive model, information flows in two ways. All access requirements for the stakeholders are described in Table 5:

Service Delivery Model

Citizen ⇄ Government

The above tables provide the broad access control mechanisms for different stakeholders for different models. The government can also make standards of security based on the egovernance models. Thus there will be less human errors and bias.

Table 5.

Users	Read	Write	Delete	Execute	Modify	Append
DBM	Y	Y	Y	Y	Y	Y
Dev	Y	Y	Y	Y	Y	Y
IO	Y	N	N	Y	N	Y
Info	Y	Y	Y	Y	Y	Y
Ciso	Y	N	N	Y	N	Y
Emp	Y	N	N	Y	N	N
A	Y	N	N	Y	N	N
P	Y	Y	N	Y	N	Y

DBM → Database Manager
Dev → Developers
IO → Implementation Officers
Info → Information Officers
Ciso → Chief Information security officer
Emp → Officers and Employees
A → Agents
P → Public

4 Conclusion

In summary, this paper presents a methodology to formulate the access control mechanisms for different stakeholders of different G2C applications in a model driven manner. The methodology can be used for protecting all G2C applications from unauthorized user s.

This paper contains a general study but not from the point of individual application. Here model based access controls have been designed. Further work in this direction is the development of control directories as per the accepted ISO 27001 standards.

Acknowledgement. This work was supported by the Security Engineering Research Center, granted by the Korea Ministry of Knowledge Economy.

References

1. Ammal, Anantalakshmi, R.: E-Governance Application Life Cycle Management - Issues and Solutions. egovasia (2007)
2. Backus, M.: E-governance in Developing Countries. International Institute of Communication & Development (IICD), Research Brief No. 1 (March 2001), `http://www.ftpiicd.org/files/research/reports/report3.pdf`
3. Mazumdar, C., Kaushik, A.K., Banerjee, P.: On Information Security Issues in E-Governance: Developing Country Views. CSDMS Journal (July 6, 2009)
4. Nath, V.: Digital Governance Initiative, `http://www.DigitalGovernance.org`
5. Kaushik, A.k., Mazumdar, C., Bhattacharjee, J., Saha, S.: Model Driven Security Analysis of Egovernance Systems. In: eIndia 2008, November edn. (2008), `http://www.egovonline.net/Resource/eindia08-full-paper-for-abstract-173.pdf`

Energy Efficient, Chain Based Clustering Routing Protocol for Wireless Sensor Networks

Subhajit Pal[1], Debnath Bhattacharyya[2], and Tai-hoon Kim[2]

[1] Heritage Institute of Technology
Kolkata-700107, India
pal.subhajit77@gmail.com
[2] Hannam University
Daejeon, Korea
debnath@sersc.org, taihoonn@empal.com

Abstract. In this paper we try to improve the LEACH [1] protocol by implementing a chain among the cluster head (CH) nodes. Initially, the clusters are formed and the sensor nodes are kept under any one of the cluster heads according to the LEACH protocol. In LEACH, after sensing the environment, all the sensor nodes transmit the sensed data to its own CH at their time schedule. The CHs fuse all data and send them to base station (BS). The BS get information from all the cluster heads. In our approach, all the CHs form a chain among themselves and only one CH (called leader) nearest to BS sends data to this base station by implementing the concept of PEGASIS [2] protocol. This chain is started from the farthest CH from BS and ending at CH nearest to BS. Each CH gets data from its previous CH in the chain, fuses its own data and sends it to its next CH in the chain. The CH, nearest to BS within this chain only sends data to BS. LEACH, a clustering based protocol uses randomization among the CHs to distribute the energy load among the sensors and thus it reduces energy dissipation. In our protocol, in addition to this energy reduction, the CHs decrease the data transmission and energy among them by incorporating a chain based clustering approach.

General Terms: Wireless Sensor Networks, energy efficient routing protocols.

Keywords: Wireless Sensor Networks, routing protocols, LEACH protocol, PEGASIS protocol, clustering.

1 Introduction

A wireless sensor network (WSN) monitors physical or environmental conditions such as temperature, pressure, sound, vibration etc. It is consisting of hundreds to thousands of sensor nodes for collecting information from a predefined geographical region. Theses sensor nodes, having capabilities of sensing, computing and communicating are equipped with a radio transceiver or other wireless communication devices, a small micro controller and an energy source, usually a battery. A large number of such sensor nodes are distributed over an area of interest for collecting information. These

T.-h. Kim, A. Stoica, and R.-S. Chang (Eds.): SUComS 2010, CCIS 78, pp. 472–481, 2010.

nodes can communicate with each other either directly or through other nodes and thus forming a network. These nodes also can communicate with the base station (BS), which is located far away from this area. The BS then communicates with the users either directly or through the existing wired network. Since the sensor nodes are mobile, the topology of the sensor network changes very frequently. There are different routing techniques for sending data between sensor nodes and the base stations for data communication. A very essential constraint of sensor nodes is their limited energy resource. So the routing techniques should be aware of this limited resource and the routing technique should manage the energy so that data transmission among the nodes should be reduced and thus reducing the energy dissipation as well as increasing the lifetime of the whole network [3]. Different routing protocols [4] are proposed for wireless sensor network and these protocols can be classified depending on the several parameters. In some protocols, each and every sensor nodes directly send their data to base station. In that case, energy of the nodes situated far away from the BS, is drained out and dies quickly. In this type of protocols, a large number of data transmissions take place, creating data traffic and reducing the system lifetime.

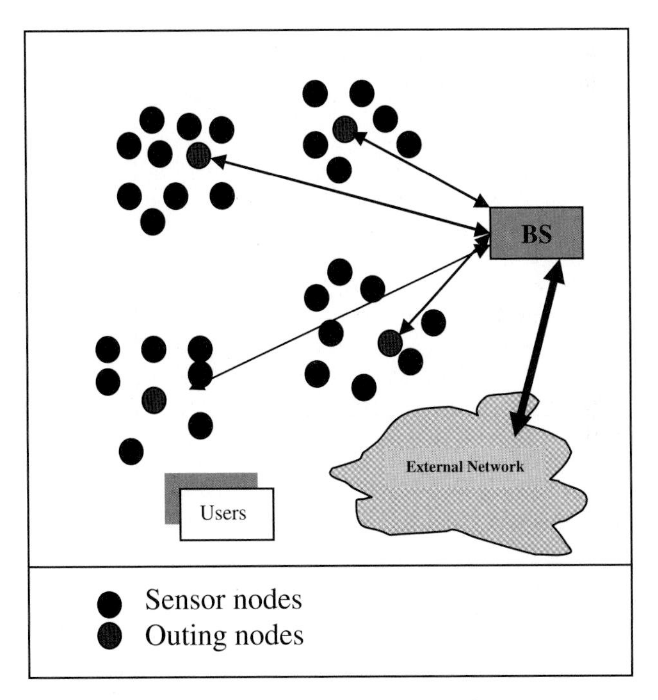

Fig. 1. Wireless sensor network

After getting data from nodes, BS fuses or aggregates data and sends to the user application. Therefore, nodes are not responsible for fusing data and thus increasing their energy. In another category, data communication is done through multihope transmission where each node gets data from its nearest node, fuses its own data and then transmits the data to its next nearest node. Ultimately nodes with maximum energy send fused data to BS directly. In that case, each node required a very small energy for transmitting data and the system lifetime will be greater than the previous case. Moreover since the data transmission is done between the nodes only, due to less number of data transmissions, there is no heavy data traffic in the system [5]. The only overhead in that case is data fusion or aggregation by each node.

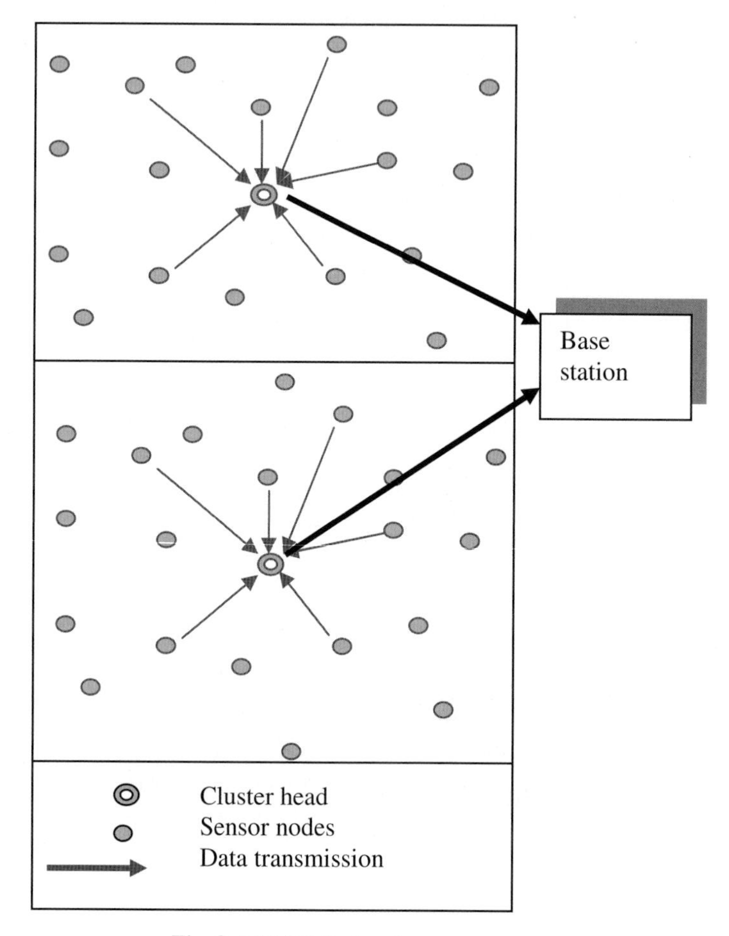

Fig. 2. LEACH Protocol

Another very good approach is clustering the whole networks proposed by LEACH [1]. According to this protocols, some clusters are formed throughout the region and each cluster contains a cluster head (CH) and some sensor nodes. Cluster heads are selected from the sensor nodes with highest energy and all sensor nodes within a cluster head send their data to their CH only. The CHs then fuse the collected data and send to BS. Since nodes only send data to its corresponding CH, they reduce energy by itself. Again the fusion or aggregation is done by CHs only, energy dissipation for each node within a cluster is reduced and thus LEACH reduces the energy of the whole network.

In the LEACH protocol, only the CHs fuse data and send data to BS and in that way LEACH reduces energy [6]. But our modified protocol can reduce more energy than LEACH. According to our protocol, the cluster heads form a chain among themselves and one of the cluster heads with high energy within the chain, send data to BS. As a result, the farthest CH from the BS sends data to its nearest CH within the chain rather than to BS and therefore this CH saves self-energy. In that way, all CHs fuses their data one by one in a chain and at last the CH, nearest from the BS, only send data to BS and thus reducing the energy of all other CHs except the last one. So the modified protocol can save more energy of the system by creating a chain among the cluster heads.

2 Leach Protocol

LEACH [1] is a self-organizing, adaptive clustering routing protocol that uses randomization for distributing the energy load among the sensor in the network. In the process of energy distribution high energy nodes are used to process and send the information and the low energy nodes are used to perform the sensing the area of interest. LEACH employs hierarchical routing as well as clustering approach. It has the following assumptions:

All nodes can take part for transmission of data to base station.

Each node has computational power to support different MAC protocols.

The adjacent nodes may have correlated data.

The total operation of LEACH is consisting of a number of rounds. Each round again has two phases: setup phase and steady-state phase.

2.1 Setup Phase

In each round, the setup phase starts by forming the clusters. The entire region is splitted into some small regions called clusters. For each and every cluster then cluster head is selected among the sensor nodes. After forming the clusters, cluster head is selected for each cluster depending on some factors: percentage of cluster heads for the network (previously determined) and the number of times the node already has been a cluster head.

For each node in a cluster, first a random number is generated ranging from 0 to 1 and then a threshold value T(n) is determined by the following expression:

$$T(n) = \frac{P}{1 - p(r MOD \frac{1}{P})} \quad \text{if } n \text{ e } G$$

$$= 0 \quad \text{otherwise,}$$

where P = the desired percentage of cluster heads, r = the current round and G is the set of nodes that have not been cluster heads in the last $\frac{1}{P}$ rounds. If this threshold value is higher than the generated random number, the corresponding node becomes a cluster head. Each cluster head being elected broadcast an advertisement message to the whole network using CSMA MAC protocol and using the same transmit energy.

All the other nodes must keep their receivers on during this phase. They receive the broadcast message and take decision in which cluster they belong to by using the signal strength of the broadcast message. Each non-cluster head nodes then transmit a message of existence inside the cluster to their corresponding cluster head by using CSMA MAC protocol. Thus cluster heads have the information of all the nodes of their own clusters. Cluster heads creates a TDMA schedule and broadcasts this schedule among the nodes of this cluster. Nodes in a particular cluster can transmit data to their cluster head only at their schedule time. Therefore Cluster head must keep its receiver always on and get data from sensor nodes one by one at a time.

2.2 Steady-State Phase

In the steady-state phase, data transmission among the nodes begins. All the non-cluster head nodes sense the environment and transmit the sensed data to their cluster heads at the time of their allocated schedule. All the nodes turns off their receivers except their schedule time and they transmit data only their corresponding cluster heads. Thus LEACH minimizes energy dissipation. After getting data from sensor nodes CH combines them to eliminate greatly redundant amount of raw data to the BS and executes a signal processing function for transforming it into a single signal. The single signal then is transmitted to the base station by the cluster head. Since it is a high-energy transmission, at the end of this round, the CH node dies. After a certain time (predetermined) interval, the next round begins as previously and continues.

3 Pegasis (Power Efficient Gathering Sensor Information System)

PEGASIS [2] is a near optimal chain based power efficient protocol based on LEACH. This protocol has the following assumptions:

i) The base station is fixed and it is located far away from the area of interest.
ii) The sensor nodes are homogeneous and energy constrained with uniform energy.
iii) The energy level for transmitting data depends on the distance of transmission.

According to this protocol, all the nodes have information about all other nodes and each has the capability of transmitting data to the base station directly. PEGASIS assumes that all the sensor nodes have the same level of energy and they are likely to die at the same time. Since all nodes are immobile and have global knowledge of the network, a chain can be constructed easily by using greedy algorithm using the sensor nodes. On the contrary, the base station can create the chain and broadcast this message to all the sensor nodes. Chain creation is started at a node far away from BS. Each node transmits and receives data from only one closest node of its neighbors. To locate the closest neighbor node, each node uses the signal strength to measure the distance from the neighbors and then adjusts the signal strength so that only one node can be heard. Node passes token through the chain to leader from both sides. Each node fuses the received data with their own data at the time of constructing the chain. In each round, a randomly chosen node (leader) from the chain will transmit the aggregated data to the BS. Node (i mod N) is the leader in round i. The chain consists of those nodes that are closest to each other and form a path to the base

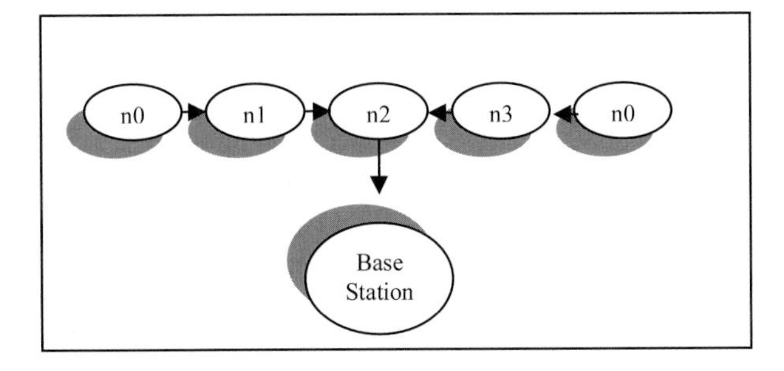

Fig. 3. Chaining in PEGASIS

station. The leader sends the aggregated data to the base station. If a node in the chain dies, the chain is reconstructed again in the same manner to avoid the dead node. Sometimes the distance between two nodes may be large compared with the other distance. In that case, the energy dissipation required for data transmission is high in each round. PEGASIS does not allow this type of nodes to become leaders and thus improving the performance. PEGASIS implements this using a threshold on neighbor distant.

PEGASIS outperforms LEACH by eliminating the overhead of dynamic cluster formation, minimizes the sum of distances and limits the number of transmission. Each node requires global information of the network. This is a drawback of this protocol because at any time it can be collected from the network. Simulation result shows that PEGASIS performs better than LEACH by about 100 to 300% when 1%, 20%, 50% and 100 % of nodes die for different network sizes and topologies [2].

4 Modified Protocol

In our modified protocol, we apply the above two protocols (LEACH and PEGASIS) for a wireless sensor networks. This is also a self-organizing, adoptive, clustering as well as hierarchical protocol as LEACH. We assume the following assumptions:

1) All nodes can transmit data to base station.
2) All nodes support MAC protocols for computation and transmission.
3) All the nodes have the global knowledge of the network.
4) The base station is fixed and located far away from the sensor nodes.
5) The sensor nodes are homogeneous and energy constrained with uniform energy.
6) The energy required for data transmission depends on the distance.

Since sensor nodes can use their limited supply of energy for computations and data transmissions, energy-conserving forms of communication and computation are essential for Wireless Sensor Networks. Moreover, WSNs Consists of hundreds to thousands of sensor nodes, there is a possibility of huge number of transmissions within the network. Therefore, the routing protocol should be aware of these points. The proposed protocol can solve these problems by applying a chain among the cluster heads and thus conserving energy and reducing the number of data transmissions. It is assumed that the local base station is a high energy node so that it has constant power supply.

The total process for this protocol consists of a number of rounds. At the end of each round, data is sent to base station. And then the base station transmits the required data to user through external networks. The operations of the proposed protocol are as follows:

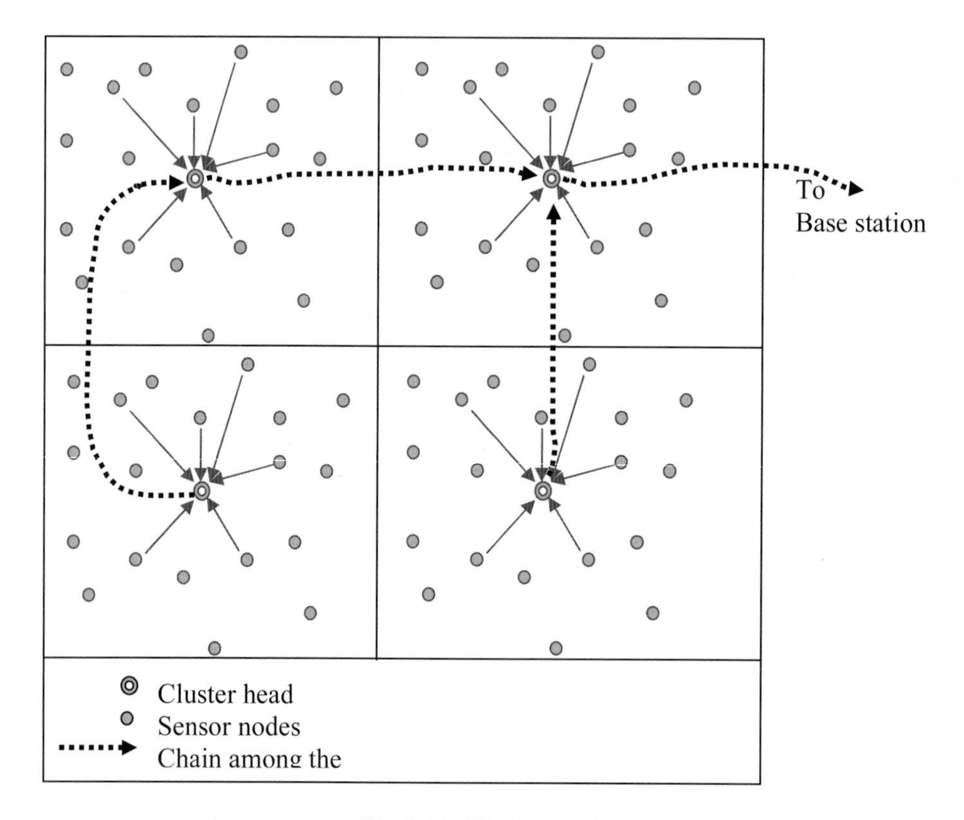

Fig. 4. Modified Protocol

4.1 Cluster Formation

In each round, first clusters are formed according to LEACH protocol and then the cluster heads are selected from the sensor nodes for each cluster. The cluster heads are nodes with highest energy for each cluster, i.e. the high-energy node is selected as cluster head in each cluster. After selecting the cluster heads, they broadcast an advertisement message throughout the network using CSMA MAC protocol and using the same transmits energy. All other nodes must keep their receivers on during this process and they receive this message and take decision in which cluster they belong to by using the signal strength of broadcast message. They also transmit a message to their corresponding cluster heads using CSMA MAC protocol, so that the cluster heads come to know which nodes exits under its own cluster. All cluster heads creates a TDMA schedule and broadcast this schedule among the nodes of their own cluster.

4.2 Data Transmission

All the non-cluster head nodes sense the environment continuously. They transmit the sensed data to their corresponding CHs at the time of their allocated time schedule only. This transmission requires a minimum amount of energy. They must turn off their receivers off except their schedule time. On the opposite side, i.e. the cluster head should keep its receiver always on for receiving data from all nodes within the cluster.

4.3 Chain Creation

A chain is created among the CHs starting from the farthest CH from the base station using the concept of PEGASIS protocol. This chain creation is established either by the sensor nodes themselves using a greedy algorithm starting from some node or the BS can compute this chain and broadcast it to all the sensor nodes. This cluster head first aggregates data coming from sensor nodes and send it to its next CH in the chain. The next CH also has data for transmitting. So after getting the previous fused data, it aggregates its own data with the previous one and sends it to the next CH in the chain. In this way data are transmitted from one CH to another. The nearest CH from the base station is selected as leader in the chain and it send aggregated data to base station directly. The leader can be selected either by the base station or can be chosen from the highest energy cluster head nodes. In each round, only one node (leader) transmits data to base station. Base station then send it to user.

The proposed protocol minimizes more energy than LEACH. In LEACH, in one round all the CHs send data to base station. Since this transmission is more powerful, all CHs losses their energy and drain out quickly. Therefore in each round a large number of nodes die. But in the modified protocol, only one node transmits data to base station in one round. So in each and every round, only one node may die. Again since this node is very close to base station with respect to other CHs, it may not die, it may loss some energy only.

In LEACH, there is an overhead [3] of selecting the cluster head where a threshold value and a random number are calculated, but this is resolved in our protocol by selecting the high-energy node as CH.

This protocol minimizes energy than PEGASIS protocol also. In case of PEGASIS, all nodes in the network involve in one chaining process and each node takes part in data fusion[7]. So it takes more time than the time required by the modified protocol, since in modified protocol, nodes in a cluster send their data to its CH and CH aggregate data.

In PEGASIS, one node sends data to its neighboring and the neighboring node then fuses its own data and then sends to its neighboring node. Thus the gathered data will be increased gradually and the nodes at the end of the chain would process these gathered data. As a result, the nodes at the end of the chain loss their energy quickly. But only the CHs aggregates data in the case of modified protocol and thus it reduces energy dissipation.

In our modified case, only the overhead is the creation of chain in each round. Since chain creation is done within the cluster heads only, it must not take more time and generate more complexity within the network.

5 Conclusion

This paper first analyzes the energy efficiency of existing wireless routing protocols (LEACH and PEGASIS) and then presents a energy preserving, chain based clustering protocol for Wireless Sensor Networks by modifying LEACH protocol with the concept of PEGASIS protocol. The proposed protocol preserves energy by implementing two modifications over LEACH protocol – one modification is the cluster head selection, and the other is forming a chain among the CHs. In LEACH protocol, the cluster head is selected by determining a threshold value and a random number for each sensor nodes [8]. If this threshold value is greater than the generated random number for a particular node then this node is selected as cluster head for a cluster. But in the modified approach the high-energy node within a cluster is selected as CH for that particular cluster. In the second modification, a chain is created among the CHs at the end of each round by using the concept of PEGASIS protocol. This chain creation is started from the farthest node from the base station and ended at the nearest node to the base station. A high energy node among the nodes in the chain (called leader) transmits fused data to base station. The closest CH from the base station must be chosen as leader of the chain so that the transmission energy should be less. The network can reduce energy dissipation and number of data transmissions using this modified protocol. Again there are no overheads for computation of threshold values at the time of cluster head selection and thus increasing the system life time.

We can further modify the proposed protocol by implementing a chain among the sensor nodes for each cluster i.e. chain is created cluster wise. Chain is created inside each cluster and leader of the chain transmits gathered data directly to the base station. In this case we must not construct chain among the cluster heads or leaders of the chains. So the base station gets data from all clusters and aggregates them. Finally it sends that data to user. In this case all the sensor nodes inside a cluster send data to its nearest node in the chain. Therefore each node saves energy and thus increasing the total energy of the network.

References

1. Heinzelman, W., Chandrakasan, A., Balakrishnan, H.: Energy –Efficient Communication Protocol for Wireless Microsensor Networks. In: Proceeding of the 33rd Annual Hawaii International Conference on System Sciences (HICSS), Big Island, USA, pp. 3005–3014 (January 2000)
2. Lindsey, S., Raghavendra, C.S.: PEGASIS: Power Efficient gathering in sensor information systems. In: Proceedings of IEEE Aerospace Conference (March 2002)
3. Routing Techniques in Wireless Sensor Networks: A Survey, Jamal N. Al-Karaki Ahmed E. Kamal, Dept. of Electrical and Computer Engineering, Iowa State University, Ames, Iowa 50011

4. Barati, H., Movaghar, A., Barati, A., Mazresh, A.a.: A Review of Coverage and Routing for Wireless Sensor Networks. World Academy of Science, Engineering and Technology 37 (2008)
5. Heinzelman, W., Kulik, J., Balakrishnan, H.: Adaptive Protocols for Information Dissemination in Wireless Sensor Networks. In: Proc. 5th ACM/IEEE Mobicom Conference (MobiCom 1999), Seattle, WA, pp. 174–185 (August 1999)
6. A survey on routing protocols for wireless sensor networks Kemal Akkaya *, Mohamed Younis Department of Computer Science and Electrical Engineering, University of Maryland, Baltimore County, Baltimore, MD 21250, USA Received 4 February 2003; received in revised form 20 July 2003; accepted 1 September 2003 Available online November 26 (2003)
7. Jiang, Q., Manivannan, D.: Routing Protocols for Sensor networks. IEEE, Los Alamitos (2004)
8. Wireless Sensor Networks, F. L. LEWIS, Associate Director for Research, Head, Advanced Controls, Sensors, and MEMS Group, Automation and Robotics Research Institute, The University of Texas at Arlington,7300 Jack Newell Blvd. S, Ft. Worth, Texas 76118-7115

Chain Based Hierarchical Routing Protocol for Wireless Sensor Networks

Subhajit Pal[1], Debnath Bhattacharyya[2], and Tai-hoon Kim[2]

[1] Heritage Institute of Technology
Kolkata-700107, India
pal.subhajit77@gmail.com
[2] Hannam University
Daejeon, Korea
debnath@sersc.org, taihoonn@empal.com

Abstract. Wireless Sensor Networks (WSNs) are very crucial network system in present day for getting information from an unattendant environment. The small tiny sensor nodes can sense environment's condition (temperature, pressure, humidity etc.) and send data to a particular location through wireless sensor networks. For collecting data from such an environment through these sensor nodes, there are different existing approaches in WSNs. LEACH[1] is one of them energy efficient cluster based routing protocol. In this paper we try to improve LEACH protocol by introducing a chain among the sensor nodes in each and every cluster as discussed in PEGASIS[2] protocol. The cluster heads get accumulated data or information from this chain and send it to its nearest base station (BS). Due to limited resources (energy, storage capacity, computing power etc) the lifetime of the sensor nodes are very restricted. LEACH protocol saves energy of the total network by using randomization for distributing the energy load among the sensor nodes and thus it enhances the lifespan of the network. But in the modified protocol, each sensor nodes except the cluster head sends sensed data to its nearest node, not to the cluster head like LEACH. As a result each node can preserve some energy and thus enhancing life span of the system. Moreover, since each node sends data to its nearest node, the modified approach can complete every round more quickly.

General Terms. Wireless Sensor Networks, chain based routing protocols, cluster based routing protocols,

Keywords: WSNs, routing protocols, modified LEACH protocol, PEGASIS protocol, energy efficient routing protocol, chaining in routing protocol, clustering in routing protocol.

1 Introduction

The recent development of micro sensor technology and wireless communication accelerate the evolution of Wireless Sensor Networks (WSNs) particularly for the hazardous environment for monitoring and collecting information from remote

T.-h. Kim, A. Stoica, and R.-S. Chang (Eds.): SUComS 2010, CCIS 78, pp. 482–492, 2010.

locations. A Wireless Sensor Network (WSN) consists of hundreds to thousands of sensor nodes having capabilities of sensing, computation, communications. The basic components[3] of a node are sensor unit, ADC (Analog to Digital Converter), CPU (Central Processing Unit) and a communication unit. The sensor unit is responsible for collecting required data from the area of interest. ATD converts the collected data by the sensor to digital form. CPU processes (data aggregation) data as requirement. The last unit i.e. communication unit transmits data to another node. Basically, a sensor node is micro-electro-mechanical device[4] and it can sense the environment periodically, fuse data if required and broadcast data to some other node. Wireless Sensor Networks are used for monitoring and collecting information from an unattendent environment and reporting events to user. Since the sensor nodes are constraints with their energy, a large number of sensor nodes are deployed over the area of interest. These nodes can communicate with each other by establishing a network among them. Since the sensor nodes may be mobile, the sensor network must be dynamic depending upon different wireless network protocol. According to some approach, nodes transmit their sensed data directly to base station which is far away from the sensor node. In another approach, data are sent to the nearest node and the nearest node fuses its own data, again sends it to its nearest nodes and in this way one particular node transmits data to base station- called *multi hope routing protocol*[5]. When data are transmitted by any node over the network, only its nearest node gets the signal with full energy, all other nodes receive this signal but discard them. In multi hope routing protocols each node acts as routers-they can receive and transmit signal[5].

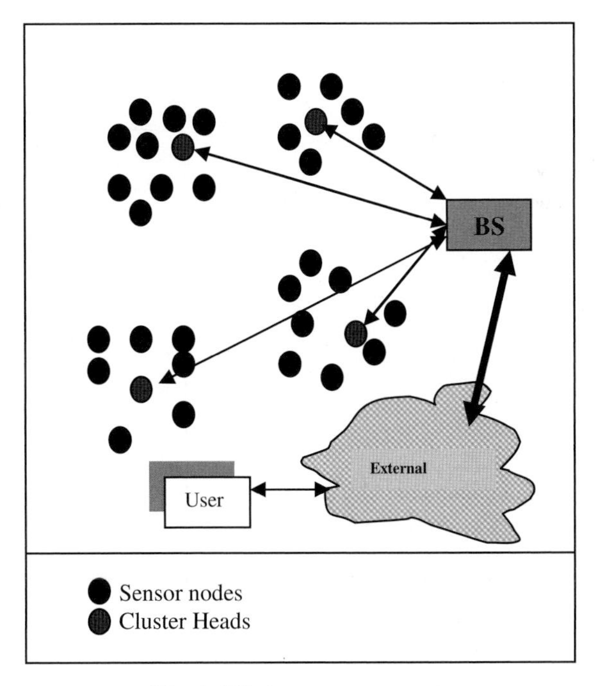

Fig. 1. Wireless sensor network

2 Leach Protocol

LEACH [1] is a self-organizing, adaptive clustering routing protocol that uses randomization for distributing the energy load among the sensor in the network. In the process of energy distribution high energy nodes are used to process and send the information and the low energy nodes are used to perform the sensing the area of interest. LEACH employs hierarchical routing as well as clustering approach. It has the following assumptions:

i) All nodes can take part for transmission of data to base station.
ii) Each node has computational power to support different MAC protocols.
iii) The adjacent nodes may have correlated data.

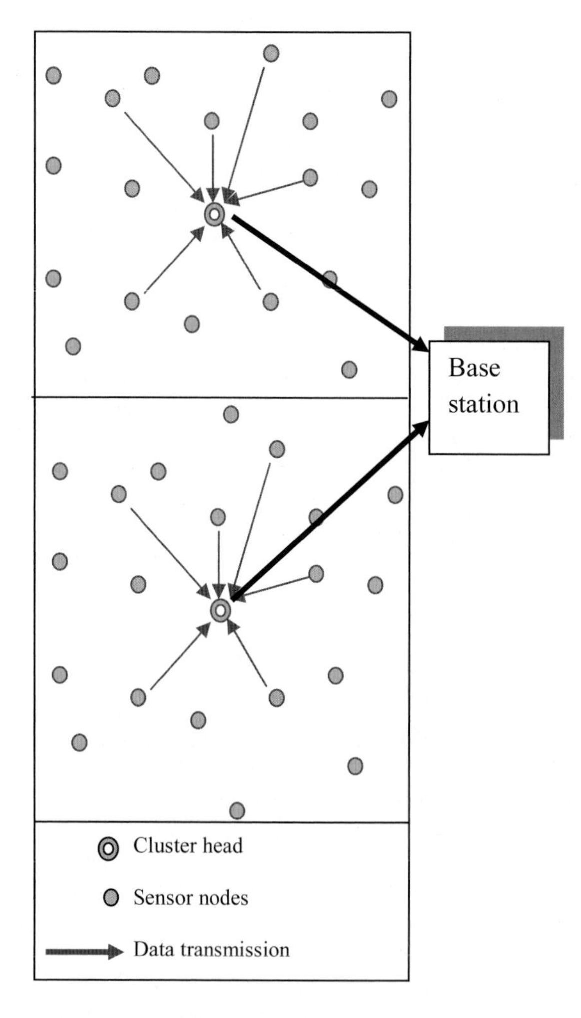

Fig. 2. LEACH Protocol

The total operation of LEACH is consisting of a number of rounds. Each round again has two phases: setup phase and steady-state phase.

2.1 Setup Phase[1][3]

In each round, the setup phase starts by forming the clusters. The entire region is splitted into some small regions called clusters. For each and every cluster then cluster head is selected among the sensor nodes. After forming the clusters, cluster head is selected for each cluster depending on some factors: percentage of cluster heads for the network (previously determined) and the number of times the node already has been a cluster head. For each node in a cluster, first a random number is generated ranging from 0 to 1 and then a threshold value T(n) is determined by the following expression:

$$T(n) = \frac{P}{1 - p(rMOD\frac{1}{P})} \qquad \text{if n e E G}$$

$$= 0 \qquad \text{otherwise,}$$

where P = the desired percentage of cluster heads, r = the current round and G is the set of nodes that have not been cluster heads in the last $\frac{1}{P}$ rounds. If this threshold value is higher than the generated random number, the corresponding node becomes a cluster head. Each cluster head being elected broadcast an advertisement message to the whole network using CSMA MAC protocol and using the same transmit energy. All the other nodes must keep their receivers on during this phase. They receive the broadcast message and take decision in which cluster they belong to by using the signal strength of the broadcast message. Each non-cluster head nodes then transmit a message of existence inside the cluster to their corresponding cluster head by using CSMA MAC protocol. Thus cluster heads have the information of all the nodes of their own clusters. Cluster heads creates a TDMA schedule and broadcasts this schedule among the nodes of this cluster. Nodes in a particular cluster can transmit data to their cluster head only at their schedule time. Therefore Cluster head must keep its receiver always on and get data from sensor nodes one by one at a time.

2.2 Steady-State Phase[1][4]

In the steady-state phase, data transmission among the nodes begins. All the non-cluster head nodes sense the environment and transmit the sensed data to their cluster heads at the time of their allocated schedule. All the nodes turns off their receivers except their schedule time and they transmit data only their corresponding cluster heads. Thus LEACH minimizes energy dissipation. After getting data from sensor nodes CH combines them to eliminate greatly redundant amount of raw data to the BS and executes a signal processing function for transforming it into a single signal. The single signal then is transmitted to the base station by the cluster head. Since it is a high-energy transmission, at the end of this round, the CH node dies. After a certain time (predetermined) interval, the next round begins as previously and continues.

3 Pegasis (Power Efficient Gathering Sensor Information System)[2][6]

PEGASIS is a near optimal chain based power efficient protocol based on LEACH. This protocol has the following assumptions:

i) The base station is fixed and it is located far away from the area of interest.
ii) The sensor nodes are homogeneous and energy constrained with uniform energy.
iii) The energy level for transmitting data depends on the distance of transmission.

According to this protocol, all the nodes have information about all other nodes and each has the capability of transmitting data to the base station directly. PEGASIS assumes that all the sensor nodes have the same level of energy and they are likely to

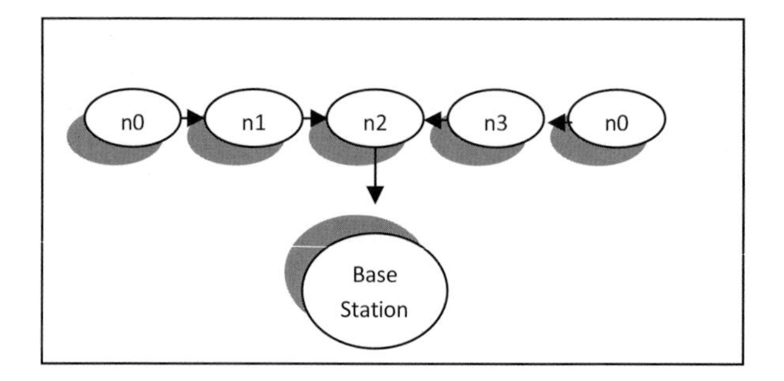

Fig. 3. Chaining in PEGASIS

die at the same time. Since all nodes are immobile and have global knowledge of the network, a chain can be constructed easily by using greedy algorithm using the sensor nodes. On the contrary, the base station can create the chain and broadcast this message to all the sensor nodes. Chain creation is started at a node far away from BS. Each node transmits and receives data from only one closest node of its neighbors. To locate the closest neighbor node, each node uses the signal strength to measure the distance from the neighbors and then adjusts the signal strength so that only one node can be heard. Node passes token through the chain to leader from both sides. Each node fuses the received data with their own data at the time of constructing the chain. In each round, a randomly chosen node (leader) from the chain will transmit the aggregated data to the BS. Node (i mod N) is the leader in round i. The chain consists of those nodes that are closest to each other and form a path to the base station. The leader sends the aggregated data to the base station. If a node in the chain dies, the chain is reconstructed again in the same manner to avoid the dead node. Sometimes the distance between two nodes may be large compared with the other distance. In that case, the energy dissipation required for data transmission is high in each round. PEGASIS does not allow this type of nodes to become leaders and thus improving the performance. PEGASIS implements this using a threshold value on neighbor distant. PEGASIS increases its performance further by implementing a threshold adaptive to the remaining energy levels in nodes. PEGASIS outperforms LEACH by eliminating the overhead of dynamic cluster formation, minimizes the sum of distances and limits the number of transmission. Each node requires global information of the network. This is a drawback of this protocol because at any time it can be collected from the network. Simulation result shows that PEGASIS performs better than LEACH by about 100 to 300% when 1%, 20%, 50% and 100 % of nodes die for different network sizes and topologies[2].

4 Chain Based Hierarchical Routing Protocol for WSNs

In this section, we lay out our modified protocol which is a self-organizing, adoptive, clustering and chain based routing protocol for wireless sensor networks. We assume the following assumptions for the modified protocol:

1) All nodes can transmit data to base station.
2) All nodes support MAC protocols for computation and transmission.
3) The base station is fixed and located far away from the sensor nodes.
4) The sensor nodes are homogeneous and energy constrained with uniform energy.
5) The energy required for data transmission depends on the distance.

4.1 Radio Model[1][2]

In modified protocol, we implement the first order radio model of LEACH. According to this model, for transmitting and receiving of energy, a radio dissipates E_{elec} =50 nJ/bit and for transmitter amplifier it dissipates E_{amp}= 100 pJ/bit/m^2 .We consider r^2 energy loss due to channel transmission. We also assume that the radio channel is symmetric in the sense that for a given signal to noise ratio, the required energy for transmitting a message from node i to node j is the same as the required energy for transmitting a message from node j to i. The following equations are used for calculating transmission costs and receiving costs for a k-bit message and a distance d:

For transmitting:-

$$E_{Tx} (k,d) = E_{Tx\text{-}elec} (k) + E_{Tx\text{-}amp} (k, d)$$

$$E_{Tx} (k, d) = E_{elec} *k + E_{amp} *k*d2$$

For receiving:-

$$E_{Rx} (k) = E_{Rx\text{-}elec} (k)$$

$$E_{Rx} (k) = E_{elec} *k$$

4.2 Algorithm

The algorithm consists of a number of rounds; each round again has the following steps:-

4.2.1 Cluster Formation

At the start of each round, the total area of interest is divided into small regions. Depending on some predefined parameters (total number of small regions, radius of each region) the base station splits the region into small segments. Each such segment is called cluster. The received signal strength by the base station also takes part in cluster formation.

4.2.2 Cluster Head Selection

After cluster formation, the high energy nodes for each cluster are selected as *cluster heads (CHs)* for the corresponding cluster i.e. for each cluster, nodes with more residual energy are selected as Cluster Head(CH). So each and every cluster has a cluster head and the local cluster head acts as router to base station. After being selected as cluster heads, they broadcast an advertisement message throughout the network using CSMA MAC protocol[5][7] and using the same transmits energy. All other nodes must keep their receivers on during this process and they receive this message and take decision in which cluster they belong to by using the signal strength of broadcast message. The cluster nodes also transmit a message to their corresponding cluster heads using CSMA MAC protocol, so that the cluster heads come to know which nodes exits under its own cluster. All cluster heads creates a TDMA schedule and broadcast this schedule among the nodes of their own cluster.

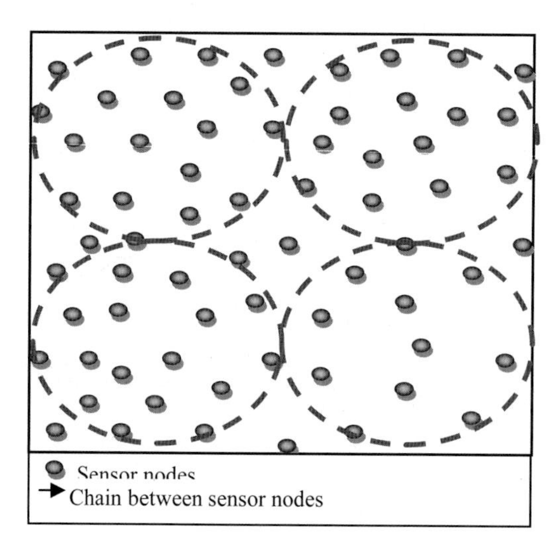

Fig. 4. Clusrter formation

4.2.3 Chain Creation

In each cluster, chains are created among the sensor nodes within the cluster starting from the farthest node from the cluster head using the concept of PEGASIS[2] protocol. This chain creation is established by the sensor nodes themselves using a greedy algorithm starting from some node or the CHs can compute this chain and broadcast it to all of its sensor nodes. The sensor nodes always sense the environment i.e. it always has data for transmission. First the farthest node from the CH starts the chain creation and sends its data to its nearest node within the cluster. The nearest

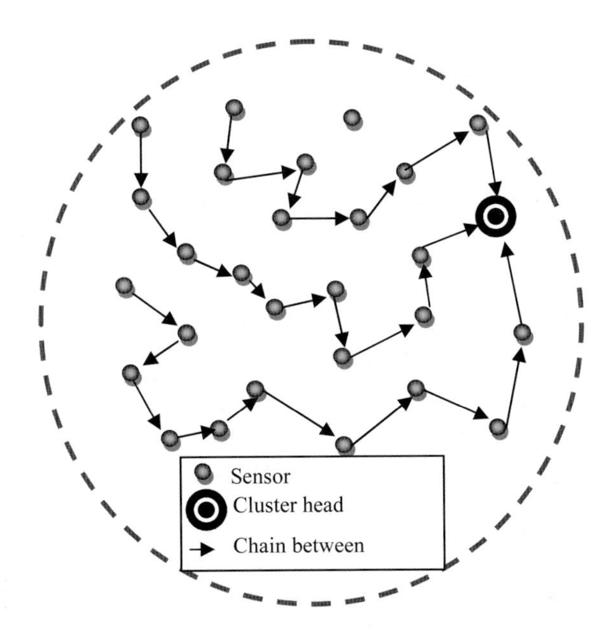

Fig. 5. Chain Creation in a Cluster

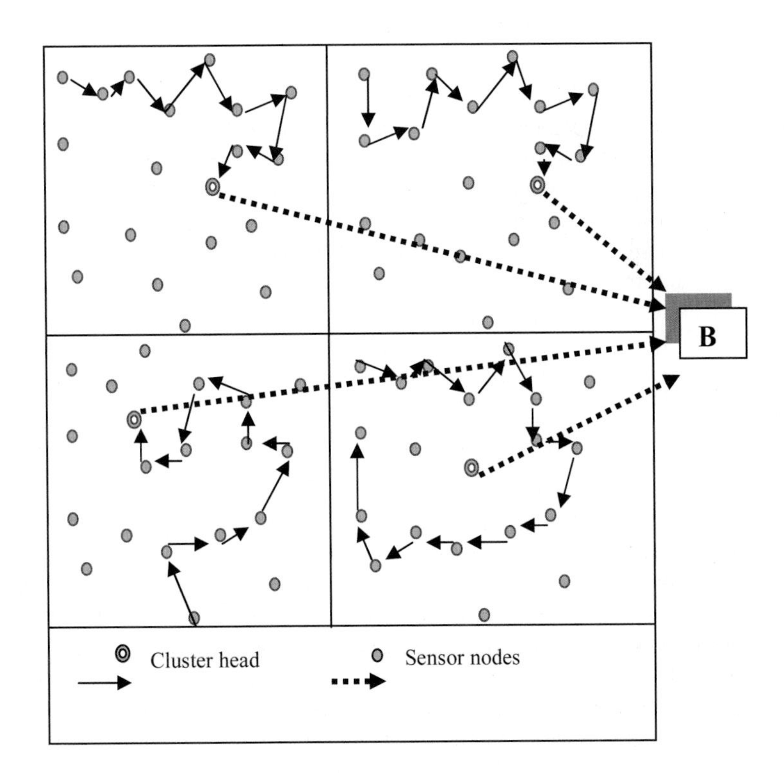

Fig. 6. Modified Protocol

node is computed by the signal strength for that node. After getting this data, the nearest node fuses its own data with it and the aggregated data is further transmitted to its nearest node and so on. In that manner chain is created and the data is transmitted from one node to another node within the chain.

4.2.4 Data Transmission between Chains and CHs

For every cluster, the node(called leader) having the highest remaining energy and nearest to CH within the chain transmits fused data to the cluster head. In one cluster more than one chain can be created and all the leaders from these chains transmit data to its CH at the time of their allocated time schedule only. This transmission requires a minimum amount of energy. The leaders always must turn their receivers on and the cluster head also should keep its receiver always on for receiving data from all nodes within the cluster.

4.2.5 Data Transmission between CH and BS

After getting data from leaders within a cluster, CHs aggreagate data and transmit the compressed data to base station. Since BS is far away from the CHs, the transmission between CH and BS requires high energy. The total process for this protocol consists of a number of rounds. At the end of each round, data is sent to base station. And then the base station transmits the required data to user through external networks.

5 Comparison

The proposed protocol minimizes more energy than LEACH. In LEACH, in each round all sensor nodes send their data to CH and thus they all losses their energy. But in the proposed model, only leader of a chain transmits energy to CH. Thus in the latter case, all nodes except the leader save energy and enhance the system lifetime. Again for the latter case sensor nodes broadcast their data to its nearest node in the chain; this requires low transmission cost.

In LEACH, there is an overhead[8] of selecting the cluster head where a threshold value and a random number are calculated, but this is resolved in our protocol by selecting the high-energy node as CH.

This modified protocol minimizes energy than PEGASIS protocol also. In case of PEGASIS, all nodes in the network involve in one chaining process[2][9]. So it takes more time than the time required by the modified protocol, since in modified protocol, nodes in a cluster form chain(s). In one round many chains are created and leaders of these chains send data to CH. So each round will be quicker than the previous case.

In PEGASIS, one node sends data to its neighboring and the neighboring node then fuses its own data and then sends to its neighboring node. Thus the gathered data will be increased gradually and the nodes at the end of the chain would process these gathered data. As a result, the nodes at the end of the chain loss their energy quickly. That possibility will not be created in modified approach, since there are a number of chains for each cluster.

In our modified case, the first overhead is the creation of chains inside the cluster in each round. Chain creation is done by the CHs within each cluster. Since this chain creation process starts simultaneously in each cluster, each round will be finished more quickly. The second overhead is the selection of leader for a chain. The high energy node and nearest to CH is selected as leader for a chain.

6 Conclusion

This paper first analyzes the energy efficiency of existing wireless routing protocols (LEACH and PEGASIS) and then presents a energy preserving, chain based clustering protocol for Wireless Sensor Networks by modifying LEACH protocol with the concept of PEGASIS protocol. The proposed protocol preserves energy by implementing two modifications over LEACH protocol – one modification is the cluster head selection, and the other is forming chain among the sensor nodes for each cluster. In LEACH protocol, the cluster head is selected by determining a threshold value and a random number for each sensor nodes. If this threshold value is greater than the generated random number for a particular node then this node is selected as cluster head for a cluster. But in the modified approach the high-energy node within a cluster is selected as CH for that particular cluster. In the second modification, chain is created among the sensor nodes inside the cluster in each round by using the concept of PEGASIS protocol. This chain creation is started from the farthest node from the CH and ended at the nearest node to the CH. A high energy node among the nodes in the chain (called leader) transmits fused data to CH. All the CHs transmit the aggregated data to base station. The network can reduce energy dissipation and number of data transmissions using this modified protocol. Again there are no overheads for computation of threshold values at the time of cluster head selection and thus increasing the system life time.

We can further modify the proposed protocol by implementing a chain among the CHs. The CHs form a chain among themselves. The chain creation is starts from the farthest CH from the BS and ends at the nearest CH to the BS. After getting accumulated data from leader(s) of a cluster a CH fuses its own data to it and transmit the gathered data to the next CH within the chain. The nearest CH to the BS ultimately sends the fused data to BS. In this case all the CHs can save energy, since all CHs except the last one transmit data to its closest CH and thus increasing the total energy of the network.

References

1. Heinzelman, W., Chandrakasan, A., Balakrishnan, H.: Energy–Efficient Communication Protocol for Wireless Microsensor Networks. In: Proceeding of the 33rd Annual Hawaii International Conference on System Sciences (HICSS), Big Island, USA, pp. 3005–3014 (January 2000)
2. Lindsey, S., Raghavendra, C.S.: PEGASIS: Power Efficient gathering in sensor information systems. In: Proceedings of IEEE Aerospace Conference (March 2002)
3. Wireless Sensor Networks, F. L. LEWIS, Associate Director for Research, Head, Advanced Controls, Sensors, and MEMS Group, Automation and Robotics Research Institute, The University of Texas at Arlington,7300 Jack Newell Blvd. S, Ft. Worth, Texas 76118-7115
4. Barati, H., Movaghar, A., Barati, A., Mazresh, A.a.: A Review of Coverage and Routing for Wireless Sensor Networks. World Academy of Science, Engineering and Technology 37 (2008)
5. Rappaport, T.S.: Wireless Communications. Prentice-Hall, Englewood Cliffs (1996)

6. Routing Techniques in Wireless Sensor Networks: A Survey, Jamal N. Al-Karaki Ahmed E. Kamal, Dept. of Electrical and Computer Engineering, Iowa State University, Ames, Iowa 50011
7. Heinzelman, W., Kulik, J., Balakrishnan, H.: Adaptive Protocols for Information Dissemination in Wireless Sensor Networks. In: Proc. 5th ACM/IEEE Mobicom Conference (MobiCom 1999), Seattle, WA, pp. 174–185 (August 1999)
8. A survey on routing protocols for wireless sensor networks Kemal Akkaya *, Mohamed Younis Department of Computer Science and Electrical Engineering, University of Maryland, Baltimore County, Baltimore, MD 21250, USA Received 4 February 2003; received in revised form 20 July 2003; accepted 1 September 2003 Available online November 26 (2003)
9. Jiang, Q., Manivannan, D.: Routing Protocols for Sensor networks. IEEE, Los Alamitos (2004)

Extraction of Features from Signature Image and Signature Verification Using Clustering Techniques

Samit Biswas[1], Debnath Bhattacharyya[2],
Tai-hoon Kim[2], and Samir Kumar Bandyopadhyay[3]

[1] Department of Computer Science and Engineering,
Bengal Institute of Technology,
Kolkata- 700150, India
samitbiet@yahoo.com
[2] School of Multimedia,
Hannam University,
Daejeon, Korea
debnathb@gmail.com, taihoonn@empas.com
[3] Department of Computer Science and Engineering,
University of Calcutta,
Kolkata, India
skb1@vsnl.com

Abstract. Humans are comfortable with pen and papers for authentication and authorization in legal transactions. In this case it is very much essential that a person's Hand written signature to be identified uniquely. The development of efficient technique is to extract features from Handwritten Signature Image and verify the signature with higher accuracy. This paper presents a method for off line hand written signature verification with higher accuracy. In this paper we have introduced a procedure to extract features from Handwritten Signature Images. That computed feature is used for verification. Here we used a clustering technique for verification.

Keywords: Segmentation, Morphological operation, Thinning, Region of Interest (ROI), Clustering Techniques.

I Introduction

Handwriting is a skill that is highly personal to individuals and consists of graphical marks on the surface in relation to a particular language.

Many researches have been done on this topic. Signatures of the same person can vary with time and state of mind [1]. A method proposed in [1] a signature verification system which extracts certain dynamic features derived from velocity and acceleration of the pen together with other global parameters like total time taken, number of pen-ups. The features are modeled by fitting probability density functions i.e., by estimating the mean and variance, which could probably take care of the variations of the features of the signatures of the same person with respect to time and state of mind.

T.-h. Kim, A. Stoica, and R.-S. Chang (Eds.): SUComS 2010, CCIS 78, pp. 493–503, 2010.

Handwritten signature is a form of identification for a person A method is introduced by Md. Itrat Bin Shams [2] where a signature image is first segmented (vertical and horizontal) and then data is extracted from individual blocks. Here these data is then compared with the test signature.

Signatures are composed of special characters and flourishes and therefore most of the time they can be unreadable. Also intrapersonal variations and the differences make it necessary to analyze them as complete images and not as letters and words put together [3].

The research work proposed by Debnath Bhattacharyya, Samir Kumar Bandyopadhyay, Poulami Das, Debashis Ganguly and Swarnendu Mukherjee [3] was based on the collection of set of signatures from which an average signature was obtained based on the stated algorithm[3] and then taking decision of acceptance after analyzing the correlation in between the sample signature and the average signature.

Baseline is the imaginary or invisible line, which a signature is assumed to rest on. A baseline is the line on which the letter sits. In our daily life, a baseline must be imagined when signing or writing in an unlined sheet of paper. The straightness and direction of the signature can be changeable features in a signature [5]. A method was introduced by Azlinah Mohamed, Rohayu Yusof, Shuzlina Abdul Rahman, Sofianita Mutalib [5] to extract the baseline feature.

The system developed by Alan McCabe, Jarrod Trevathan [6] analyses both the static features of a signature (e.g., shape, slant, size), and its dynamic features (e.g., velocity, pen-tip pressure, timing) to form a judgment about the signer's identity.

Support Vector Machine (SVM) can also be used to verify and classify the signatures [7].

For verification of signature image after extraction of features from handwritten signature images many authors use different mathematical formulas. Such as correlation is used for verification between sample and test signature [3, 11].

In this paper for verification of signatures after extraction of features we used clustering techniques.

Clustering involves dividing a set of data points into non-overlapping groups, or clusters, of points, where points in a cluster are "more similar" to one another than to points in other clusters. The term "more similar," when applied to clustered points, usually means closer by some measure of proximity. Any particular division of all points in a dataset into clusters is called a partitioning [11].

Image segmentation is the decomposition of a gray level or color image into homogeneous regions [13]. In image segmentation, cluster analysis is used to detect borders of objects in an image.

A method is already there [14] for image clustering based on a shared nearest neighbours approach that could be processed on both content-based features and textual descriptions (tags).

This paper is organized as follows. Previous work is surveyed in section I. Proposed Algorithm is introduced in section II. Simulation result and comparative study is shown in section III. Finally, we conclude in Section IV.

2 Proposed Algorithm

First we have to collect sample signatures that we have to keep in database. Here For every person we have collected n signature samples for database. It is better if we can collect more signature samples for database. Then for verification collect test signatures against the sample signatures. These test signatures we have to verify if it is genuine or forgery. Each of the signatures (Samples and corresponding test) has to take within a same sized rectangular area on paper by pen and collect the image of that rectangular area. The signed paper is then scanned at 500 dpi resolution by a gray scale scanner. Therefore, the signature image will be processed in four stages in order to determine if it is genuine or forgery. The following flow chart in Fig. 1 shows these stages which will be presented in the following subsections. Left side of fig. 1 shows the steps for extracting features from sample signature Image and right side of fig.1 shows the steps for extracting features from test signature image. The last step (for both cases) is for clustering the extracted data.

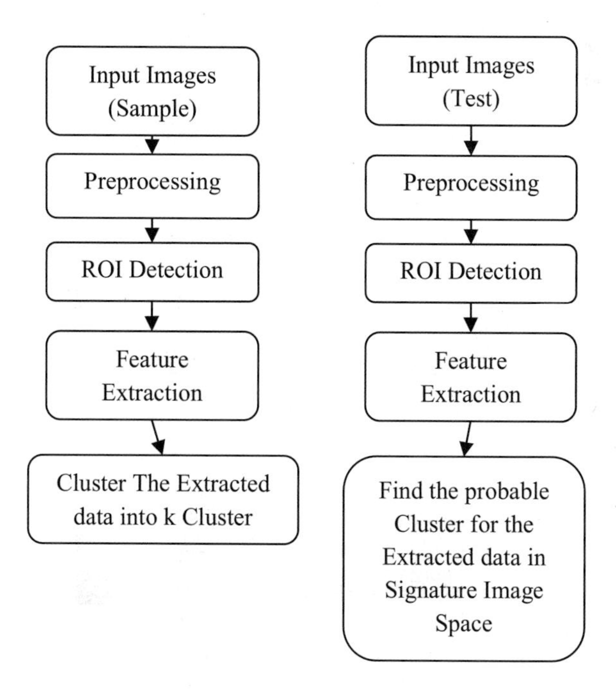

Fig. 1. Flow Chart for Data Extraction from Signature image (Sample and Test) And Clustering Data

2.1 Preprocessing Stage

The system deals with the static scanned image of the signatures. Unwanted images i.e. noises may include during the scanning process of the signature. The preprocessing of the signature images is related to the removal of noises, and thinning. The goal of thinning is to eliminate the thickness differences of pen by making the image one pixel thick. To remove noises and enhance, the images are preprocessed by filtering techniques [8]. For thinning Morphological operations [8] can be applied. Preprocessing have to be done for both sample and test signature images.

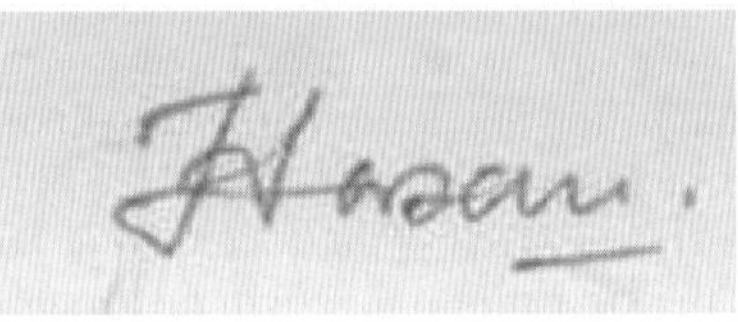

Fig. 2. Before Thinning the Signature Image

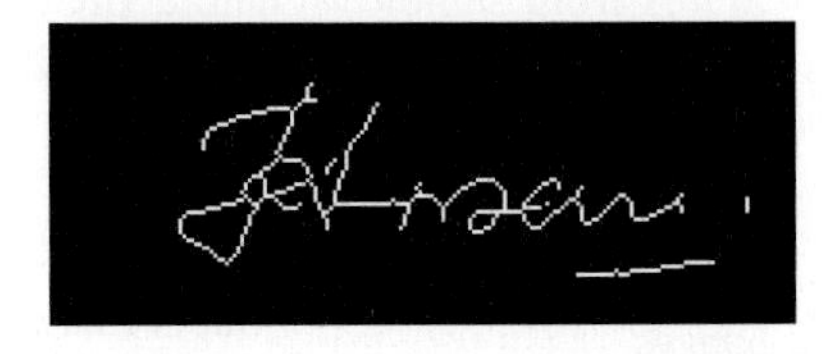

Fig. 3. After thinning the Signature Image

2.2 Region of Interest Detection and Scaling

In this step the signature area within the image is identified i.e. the region of interest (ROI) is identified. The Region of Interest detection of Fig. 3 is shown in Fig, 4. Region of Interest have to be identified from both Sample Signature and corresponding Test Signature. Then scaling is performed on both the sample and test signature. So stretching is performed to the input signature in case it is smaller than standard size or squeezing is done for being bigger. Normally all the signatures in the database are made to fit inside a rectangle of same height and width. Whole step is shown in figure.

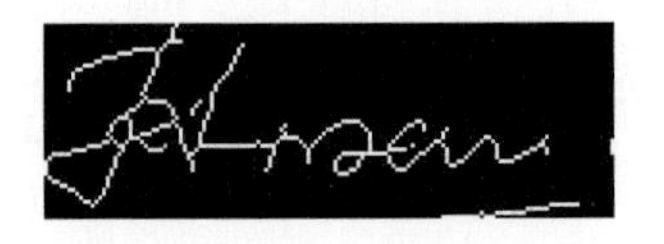

Fig. 4. Region Of Interest Detection

2.3 Feature Extraction Stage

Extracted features in this stage are used for clustering the signature images for verification stage. Features will have to be extract from both sample images and Test image. Three new Feature Extraction procedures for signature image verification is introduced here.

(a) *Signature Height Width Ratio:*

The ratio is obtained by dividing signature height to signature width. The height is the maximum length of the columns obtained from the cropped image.

Similarly the width is also calculated considering the row of maximum length. Signature height and width can change. But height-to-width ratios of an individual's signatures are approximately constant.

(b) Signature Occupancy Ratio:

It is the ratio of number of pixels which belong to the signature to the total pixels in the signature image. This feature provides information about the signature density. The Signature Density, D_i for i^{th} sample signature image can be calculated as follows:

$$D_i = \frac{I_i}{X_i}$$

If n is the total number of sample image for a person then i=1,2,3, ...n. I_i is the number of pixels which belongs to the signature of ith signature image sample. X_i is the total number of pixel in i^{th} sample signature. This feature provides information about the signature density.

(c) Distance Ratio calculation at boundary.

After cropping, the pixels in closest proximity to the boundaries (left, right, upper & bottom) are determined and their distance from the left & bottom boundaries are evaluated, i.e. for the upper leftmost pixel its distance from bottom boundary(L_1) & for the bottom left most pixel the distance from right boundary is calculated(L_2). These values are used later in verification process. Total procedure for calculation is shown in following figure.

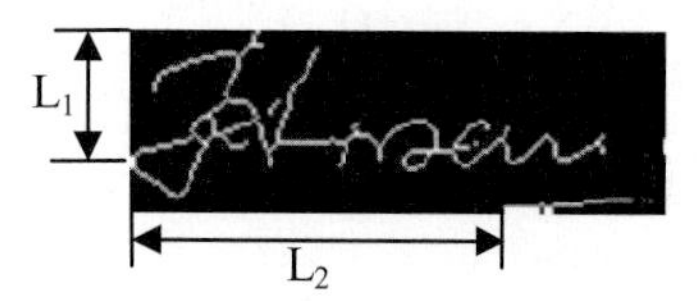

Fig. 5. Distance Calculation

The distance ratio, R is calculated as follows

$$R = \frac{L_1}{L_2}$$

The ratio should be approx same for all time for a person's signature.

(d) Compute the length and ratio of Adjacency Columns.

A new Feature Extraction procedure is introduced here. Features will have to be extract from both sample image and Test image. In this stage first we are computing the length of the adjacency columns from top and from bottom of the Sample signature image and store it in a 1D array L_T and L_B. Then compute the sum of all elements of L_T and sum of all elements L_B. Overall procedure for this feature extraction is shown in the following fig 6 . Upper half of fig.6 shows a portion of a signature Image. Lower half of fig. 6 shows the length of the corresponding adjacency columns from top and from bottom.

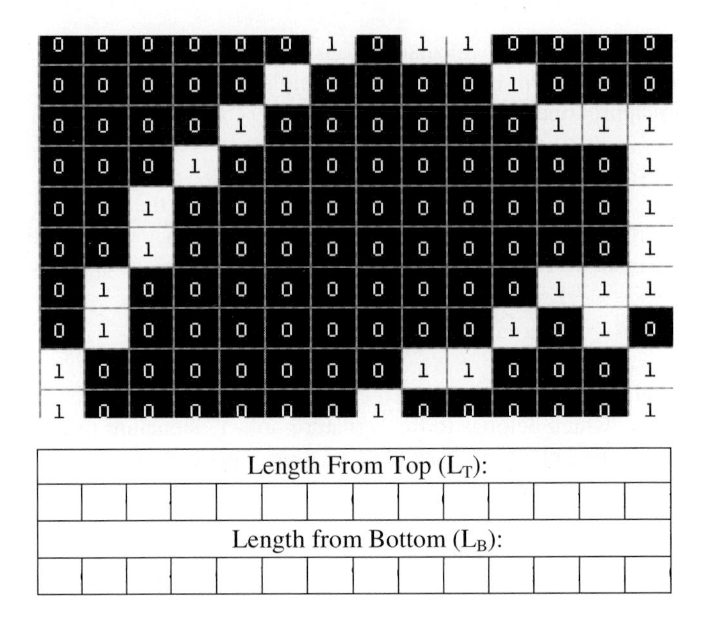

Fig. 6. Calculation of length of adjacency columns from top and bottom

Now Adjacency Ratio can be calculated as follows:

$$Adjacency\ Ratio = \frac{\mathrm{Sum}(L_T)}{Sum(L_B)} \text{ x Signature Occupancy}$$

For every signature of a person Adjacency ratio will be more or less same. If small differences are there that will be negligible.

(e) Compute the number of spatial symbols within the signature Image.

Every person in their signature uses some spatial symbols, such as they uses some 'x' marks (cross marks), star marks or other symbols. The total number of spatial symbols of a person's signature is unique. For calculating the total number of spatial symbols in a signature image we have to preprocess the image upto thinning. Then If we find that one pixel having more than two neighbors each of which get the values 1 then those pixels will form a Spatial symbol. Such types of pixels are shown in fig. 7

2.4 Signature Images Clustering

A number of methods have been proposed by many authors for clustering data. Hierarchical clustering, self-organizing maps, K-means, and fuzzy c-means have all been successful in particular applications. A person may have many sample signature images. We create separate clusters for set of sample signatures for each person. Here we use K-Nearest Neighbors' (KNN) clustering Technique for verifying a test signature belongs which cluster. The detailed algorithm for verification is as follows.

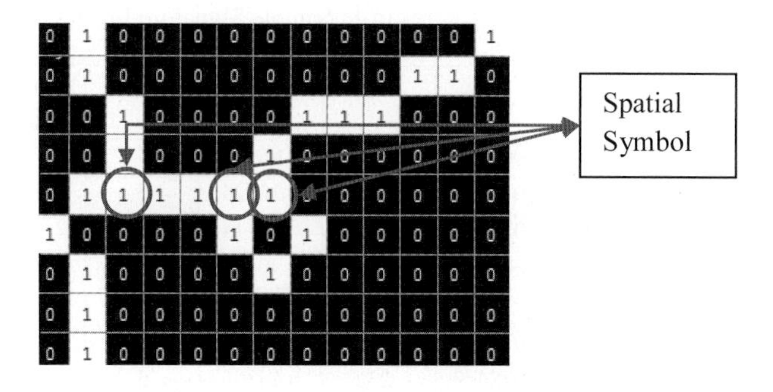

Fig. 7. Spatial Symbols in a Signature Image

Steps on how to compute k-Nearest Neighbors k-NN algorithm:

Input:

A clustered data set X (extracted feature from sample Signature image) of n points in d dimensional space in to K Clusters. Where K is the total number of persons, n is the total number of sample signature for each person and d is the total number of extracted feature.

$X : \{x_1, x_2, \dots \dots \dots \dots, x_n\}$

where $x_i = [x_{i1}, x_{i2}, \dots, x_{id}]$ for i=1,2,…,n.

A test data set T(extracted feature from Test Signature image)

T: $\{t_1, t_2, \dots \dots \dots \dots, t_n\}$

where $t_i = [t_{i1}, t_{i2}, \dots, t_{id}]$ for i=1,2,…,n.

Output:

Find the cluster for the dataset T.

Algorithm:

Step 1. Determine parameter K = number of nearest neighbors.

Step 2. Calculate the distance between the query-instance and all the training samples.

Step 3. Sort the distance and determine nearest neighbors based on the K-th minimum distance

Step 4. Gather the category of the nearest neighbors.

Step 5. Use simple majority of the category of nearest neighbors as the prediction value of the query instance

If $k = 1$, then the object is simply assigned to a new class i.e create a new cluster for this data point .

Here, another constant value is maintained, named as decision Value (distance between the query-instance and all the training sample), which is also calculated and set by professional statistician and security administrator deciding the security concern and policies of the organization after conducting surveys and testing over some sample experimental data set with already known results.

Table 1. Extracted features from Sample Signature Images

Person SL.	F_1	F_2	F_3	F_4	F_5	Cluster Index
1	0.23585	0.067937	26	0.06727	4	1
	0.18487	0.077536	23	0.10724	7	
	0.21138	0.072879	27	0.16114	9	
	0.22	0.087817	23	0.20664	8	
	0.17424	0.074561	24	0.16602	8	
	0.25472	0.073765	28	0.14931	8	
	0.18033	0.078473	23	0.14865	3	
	0.176	0.074189	23	0.11347	9	
	0.1811	0.070638	24	0.12843	7	
	0.17424	0.067043	24	0.13248	4	
2	0.33803	0.067778	1.0476	0.093693	2	2
	0.3	0.071111	1.2222	0.088208	8	
	0.34483	0.055352	1.1739	0.065907	1	
	0.26966	0.064	0.55556	0.085758	2	
	0.27174	0.052936	1.05	0.075477	1	
	0.30769	0.068354	0.5	0.086575	1	
	0.27059	0.064922	0.55556	0.054611	2	
	0.25263	0.06	0.95238	0.059183	1	
	0.31646	0.067788	1.2222	0.077703	6	
	0.27619	0.051887	0.52	0.065378	1	
3	0.41975	0.070732	0.52632	0.21338	6	3
	0.5	0.068052	0.37838	0.11818	6	
	0.3956	0.06698	0.51282	0.16905	2	
	0.34409	0.069632	0.34884	0.095108	4	
	0.34021	0.081032	0.3913	0.12189	7	
	0.41379	0.070946	0.32432	0.15692	7	
	0.35484	0.066959	0.18421	0.11855	4	
	0.29032	0.074088	0.39535	0.14497	4	
	0.29787	0.076588	0.33333	0.12362	3	
	0.4125	0.076616	0.7	0.14488	9	
4	0.42188	0.072527	2.2857	0.1207	4	4
	0.34375	0.07291	2.3333	0.099313	2	
	0.3375	0.053351	1.6667	0.085706	2	
	0.31034	0.064935	2.5714	0.078716	4	
	0.2907	0.067639	1.2727	0.091198	1	
	0.30263	0.079545	1.25	0.10304	5	
	0.2809	0.065385	2.125	0.093432	4	
	0.2375	0.083333	2	0.12449	3	
	0.28571	0.069176	1.375	0.13324	1	
	0.3913	0.076531	2.125	0.11829	3	

Table 1. (*continued*)

5	0.55385	0.067158	4.6	0.10148	3	5
	0.48052	0.070513	2.625	0.078737	7	
	0.5	0.074737	3.6	0.10549	3	
	0.47887	0.068651	1	0.073083	1	
	0.47826	0.069328	2	0.053906	5	
	0.58333	0.065626	1.75	0.064644	7	
	0.43678	0.066725	3.6667	0.12478	4	
	0.55224	0.071594	2	0.065406	3	
	0.44578	0.069549	2	0.069995	3	
	0.47945	0.061186	2.375	0.062193	5	

Signature Acceptance Percentage (%) can be calculated as follows:

$$Acceptance\ Percentage = \frac{M_s}{T_c} \times 100\ \%$$

Where M_s is the number of nearest neighbors in the majority cluster as the prediction value of the query instance and T_c is the total number of signature or data point in the majority clutser.

3 Simulation Result

In this section we show a few experimental results to illustrate the performance of the proposed method. The algorithm proposed in this paper consists of two distinct divisions. First features from the sample signature image and Test signature image have to be computed. Then a decision will be made based on the clustering results between the computed feature of Sample signature image and Test Signature Image.

Table 2. Extracted features from Genuine Test Signature Images and Acceptance Result.(Query Distance Considered:2)

Person SL.	F_1	F_2	F_3	F_4	F_5	Cluster & Acceptance Percentage (%)
1	0.19008	0.076161	24	0.14058	15	Cluster:1 % : 40
2	0.36905	0.055515	0.7368	0.055457	2	Cluster:2 %: 80
3	0.41667	0.071569	0.3437	0.12637	5	Cluster: 3 % : 50
4	0.47541	0.076882	2.4286	0.090737	3	Cluster: 4 %: 70
5	0.56	0.074737	3.6	0.10549	4	Cluster:5 %: 70

The detailed extracted data is shown in Table 1. Here we have considered five peoples signature. We have taken ten sample signatures from each of five peoples. Extracted data for genuine test signature and verification result is shown in Table 2 and for forgery signatures is shown in Table 3.

Table 3. Extracted features from Forgery Test Signature Images and Acceptance Result.(Query Distance Considered:2)

Person SL	F_1	F_2	F_3	F_4	F_5	Cluster & Acceptance Percentage (%)
1	0.235	0.067937	26	0.067	5	Cluster: 1 %: 10
2	0.34524	0.063922	1.091	0.080165	6	Cluster: 3 %: 40
3	0.42857	0.068362	5.25	0.059312	1	Cluster: 5 % : 0
4	0.37209	0.055033	2.572	0.11197	5	Cluster: 4 %: 40
5	0.51807	0.058983	3.25	0.060097	5	Cluster: 5 %: 30

Here meaning of F_1,F_2,F_3,F_4,F_5 will be as follows
F_1 = Signature Height Width Ratio
F_2 = Signature Occupancy Ratio
F_3 = Distance Ratio at boundary
F_4 = ratio of Adjacency Columns
F_5= number of spatial symbols in the signature image.

4 Discussion and Conclusion

A new method to extract features from handwritten signature and recognition of handwritten signature is presented here. Achieved results are encouraging and suggest the adequacy of the selected features. This proposed algorithm will help community in the field of signature verification, signature analysis and signature recognition. This work studies an image clustering process based on a k nearest neighbours approach enabling to handle clusters of different sizes and shapes. This types of Image clustering techniques can also be used in the field of Face recognition and Thumb impression recognition.

References

1. Kiran, G.K., Srinivasa Rao Kunte, R., Samuel, S.: On-line Signature Verification System Using Probabilistic Feature Modelling. In: International Symposium on Signal Processing and its Applications (ISSPA), Kuala Lumpur, Malaysia, August 13-I6 (2001)
2. Itrat Bin Shams, M.: Signature Recognition by Segmentation and Regular Line Detection. In: TENCON 2007 - 2007 IEEE Region 10 Conference, October 30 - November 2, pp. 1–4 (2007)

3. Bhattacharyya, D., Bandyopadhyay, S.K., Das, P., Ganguly, D., Mukherjee, S.: Statistical Approach for Offline Handwritten Signature Verification. Journal of Computer Science 4(3), 181–185 (2008) ISSN 1549-3636
4. Bandyopadhyay, S.K., Das, P., Bhattacharyya, D.: Statistical Analysis towards Image Recognition. International Journal of Multimedia and Ubiquitous Engineering 3(3) (July 2008)
5. Mohamed, A., Yusof, R., Rahman, S.A., Mutalib, S.: Baseline Extraction Algorithm for Online Signature Recognition. Wseas Transactions On Systems 8(4), 1109–2777 (2009) ISSN: 1109-2777
6. McCabe, A., Trevathan, J.: Handwritten Signature Verification Using Complementary Statistical Models. Journal Of Computers 4(7), 670–680 (2009)
7. Özgündüz, E., Şentürk, T., Elif Karslıgil, M.: Off-line signature verification and recognition by support vector machine. In: Eusipco 2005, Antalya, Turkey, September 4-8, pp. 113–116 (2005), http://www.eurasip.org/Proceedings/Eusipco/Eusipco2005/deferent/papers/cr2010.pdf
8. Gonzalez, R.C., Woods, R.E.: Digital image Processing, ISBN 81-7808-629-8
9. Chanda, B., Dutta Majumder, D.: Digital Image Processing and Analysis, ISBN: 81-203-1618-5
10. Biswas, S., Bhattacharya, S., Sahu, S.: A method to extract features from handwritten signature image for signature verification. In: Proceedings of National Conference on Emerging Trends and Application in Computer Science (NCETACS 2010), St. Anthony's College, Shillong, April 9-10, pp. 275–277 (2010) ISBN:978-81-910147-0-9
11. Faber, V.: Clustering And the Continuous K-means Algorithm. Los Alamos Science (22) (1994)
12. Bradley, P.S., Fayyad, U.M.: Refining Initial Points for K-Means Clustering. In: Shavlik, J. (ed.) International Conference on Machine Learning (ICML 1998), pp. 91–99 (1998)
13. Comaniciu, D., Meer, P.: Mean Shift: A Robust Approach Toward Feature Space Analysis. IEEE Transactions on Pattern Analysis and Machine Intelligence 24(5) (May 2002)
14. Moëllic, P.-A., Haugeard, J.-E., Pittel, G.: Image Clustering Based on a Shared Nearest Neighbors Approach for Tagged Collections. In: Proceedings of the 2008 International Conference on Content-Based Image and Video Retrieval, pp, pp. 269–278 (2008) ISBN:978-1-60558-070-8
15. Maulik, U., Bandyopadhyay, S.: Performance Evaluation of Some Clustering Algorithms and Validity Indices. IEEE Transactions On Pattern Analysis And Machine Intelligence 24(12) (December 2002)
16. Liu1, H., Li, J., Chapman, M.A.: Automated Road Extraction from Satellite Imagery Using Hybrid Genetic Algorithms and Cluster Analysis. Journal of Environmental Informatics 1(2), 40–47 (2003)

Medical Imaging: A Review

Debashis Ganguly[1], Srabonti Chakraborty[1], Maricel Balitanas[2], and Tai-hoon Kim[2]

[1] Banking and Capital Markets,
Infosys Technologies Limited
DebashisGanguly@gmail.com, srabonti.chakraborty@gmail.com
[2] Hannam University, Daejeon – 306791, Korea
maricel@sersc.org, taihoon@empal.com

Abstract. The rapid progress of medical science and the invention of various medicines have benefited mankind and the whole civilization. Modern science also has been doing wonders in the surgical field. But, the proper and correct diagnosis of diseases is the primary necessity before the treatment. The more sophisticate the bio-instruments are, better diagnosis will be possible. The medical images plays an important role in clinical diagnosis and therapy of doctor and teaching and researching etc. Medical imaging is often thought of as a way to represent anatomical structures of the body with the help of X-ray computed tomography and magnetic resonance imaging. But often it is more useful for physiologic function rather than anatomy. With the growth of computer and image technology medical imaging has greatly influenced medical field. As the quality of medical imaging affects diagnosis the medical image processing has become a hotspot and the clinical applications wanting to store and retrieve images for future purpose needs some convenient process to store those images in details. This paper is a tutorial review of the medical image processing and repository techniques appeared in the literature.

Keywords: bio-instruments, tomography, magnetic resonance, physiologic, anatomy, clinical.

1 Introduction

Medical imaging refers to the techniques and processes used to create images of the human body (or parts thereof) for various clinical purposes such as medical procedures and diagnosis or medical science including the study of normal anatomy and function. In the wider sense, it is a part of biological imaging and incorporates radiology, endoscope, thermograph, medical photography, and microscopy. Measurement and recording techniques such as electroencephalography (EEG) and magneto encephalography (MEG) are not primarily designed to produce images but which produce data susceptible to be represented as maps, can be seen as forms of medical imaging.

In the clinical context, medical imaging is generally equated to radiology or "clinical imaging". Research into the application and interpretation of medical images is usually the preserve of radiology and the medical sub-discipline relevant to medical

T.-h. Kim, A. Stoica, and R.-S. Chang (Eds.): SUComS 2010, CCIS 78, pp. 504–516, 2010.

condition or area of medical science (neuroscience, cardiology, psychiatry, psychology) under investigation. Many of the techniques developed for medical imaging also have scientific and industrial applications.

Although the mathematical sciences were used in a general way for image processing, they were of little importance in bio-medical work until the development of computed tomography (CT) for the imaging of X-rays (leading to computer-assisted tomography or CAT) and isotope emission tomography (leading to Positron Emission Tomography or PET scans and single Positron Emission Computed Tomography or SPECT scans), then MRI (magnetic resonance imaging) ruled over the other modalities in many ways as the most informative medical imaging methodology [1].

Besides all these well established techniques computer based methods are being explored in application of ultrasound and electroencephalography as well as new techniques of optical imaging, impedance tomography and magnetic source imaging. Though the final images obtained from many techniques have similarities but the technologies used and the parameters represented in the images are very different in characteristics as well as in medical usefulness, even different mathematical and statistical models have been used.

Several techniques have been developed to enable CT, MRI and ultrasound scanning software to produce 3D images for the physician. Traditionally CT and MRI scans produced 2D static output on film then to produce 3D images many scans are made and then produced a 3D model which can be manipulated by physician [2].

In this paper, we have tried to present a detail survey on Medical Imaging and we hope that this work will definitely provide a concrete overview on the past, present and future aspects in this field.

2 Overview

Medical imaging is considered as a part of biological imaging, which has been developed from 19th century onwards. A brief overview of medical imaging is as follows [3]:

In 1895 Roentgen accidentally discovered X-rays. Conventional radiography has been the most widespread medical imaging technique ever science. From 1896 radio-nuclides were for therapy and for metabolic tracer studies rather than imaging. Then γ- ray imaging rectilinear scanner was invented.

During World War 2 Sonar Technology and in 1970's ultrasound became widely available in medicine.

In 20th century the mathematical principles behind tomographic reconstruction have been understood and positron emission tomography (PET) and X-ray computed tomography (CT) have been developed. Nuclear magnetic resonance has been using for imaging in magnetic resonance imaging (MRI).

In 21st century X-rays, MRI, ultrasound kept dominating but more interesting techniques especially imaging is getting included with microscopic as well as macroscopic biological structures (thermal imaging, electrical impedance tomography, scanned probe techniques etc).

In future the emphasis will be increased on obtaining functional and metabolic information along with structural (image) information. This can be done to some extent with radioactive tracers (e.g. PET) and magnetic resonance spectroscopy [4].

3 Techniques and Applications

Advances in image technology; visualization technology and graphics workstation has initiated many different processes and ways of medical imaging. Among which the application of wavelet transform in medical images, segmentation of medical images, virtual medical imaging subsystems are of paramount importance.

⇒ *Medical image creation and capture techniques:*

A. *Application of Wavelet Transform Technology:*

The wavelet technology is widely applied to the domain of medical imaging and wavelet transform and inverse transform algorithms are introduced. Wavelet technology has been used in ECG signal processing, EEG signal processing, medical image compression, medical image reinforcing and edge detection, and medical image register.

- *Wavelet Transform:*

Wavelet families of functions generated from base function $\Psi(t)$, called an analyzing wavelet or mother wavelet.

$$\Psi_{a,\tau}(t) = 1/\sqrt{a}\ \Psi((t-\tau) / a) \quad a>0.$$

Where $\Psi(t)$ satisfies $\int \Psi(t)\, dt = 0$

a=scale parameter

τ = Translation parameter.

The wavelet transform has features of multi-resolution or multi scale. The scaling operation is just “stretching” and “compressing” operation. We restrict ourselves in binary scaling and use discrete wavelet transform (DWT) [5].

- *2D wavelet and inversion transform:*

The image can be expanded in terms of the 2D wavelets. At each stage of the transform, the image is decomposed into 4-quarter size images. For an N-by-N image, the image is decomposed into four N/2-by-N/2 images for that stage of the transform.

1. *Application of wavelet transform in ECG signal processing:*

As it is known, QRS complex, P waves and T waves of ECG contains plenty of information of human heart. The method of QRS waves detector ago mainly contain differential threshold method, slope method, areas method etc., but these has more e. m. r rate in the condition of seriously interference. Bradie [6] propose wavelet packet-based compression of signal lead ECG. Zheng Kaimei [7] proposed semi-loss less compression algorithm of ECG signal based on WT.

2. *Application of wavelet transform in EEG signal processing:*

The EEG signal is the mainly bases for the analysis of disease and symptom in neural system especially epilepsy. Kalayic [8] detect EEG spikes by using wavelet transform. Zhou Weidong [9] study the EEG signal singularity detection and denoising methods based on the dyadic WT modulus maxima, and the de-noising method can remove noise effectively as well as keep original EEG singularity.

3. *Applications of wavelet transform in medical image processing:*

Image Compression:

To meet the demand for high-speed transmission of image in efficient image storage and remote treatment, the efficient image compression is essential. Recently some new and very promising method emerges in the field of image compression algorithm based on WT, such as wavelet packet transform [10-11].

A.S.Tolba [12] discovers the hest design parameters for a data compression scheme applied to medical images of different imaging modalities. The proposed technique aims at reducing the transmission cost while preserving the diagnostic integrity.

Reinforcing medical image & edge detection:

According to the properties of its multi-scale, direction and local characteristic, determining the local maxima of wavelet coefficients provided the image edge features. Medical image enhancement and edge detection are very important in breast image.

Yuan ye [13] proposed a method of edge detection based on WT and fuzzy algorithm.

Register medical image:

Medical image registration is a pre-processing step in object identification and object classification.

Raj Sharman [14] et al. presents a fast, accurate, and automatic method to register medical image using Wavelet Modulus Maxima. It uses wavelets to obtain control points.

Several main aspects need further development study on its foundation theories and methods; the choice method of the best wavelet base; multi wavelet theory and its application; application of combined wavelet transform with neural network, application of combination of fractal and wavelet etc.

B. Segmentation of medical images (using LEGION method):

Computer vision literature typically identifies three processing stages before object recognition: image enhancement, feature extraction, and grouping of similar features. The last step, image segmentation, where pixels are grouped into regions based on image features, is an important feature, which we would discuss. The goal is to partition an image into pixel regions that together represent objects in the scene.

A recently proposed oscillator network called the locally excitatory globally inhibitory oscillator network (LEGION) whose ability to achieve fast synchrony with local excitation and desynchrony with global inhibition makes it an effective computational framework for grouping similar features and segregating dissimilar ones in an image. Using algorithms of LEGION dynamics and as results of the algorithm to two-dimensional (2-D) and three dimensional (3-D) (volume) computerized topography (CT) and magnetic resonance imaging (MRI) medical-image datasets.

• *LEGION model:*

LEGION was proposed by Terman and Wang [15, 16, 17], [32] as a biologically plausible computational framework for image analysis and has been used successfully to *segment binary and gray-level images* [33]. It is a network of relaxation oscillators, each constructed from an excitatory unit and an inhibitory unit. 2-D network architecture with four-neighborhood coupling is also used here. The global inhibitor, usually represented with a black circle, is coupled with the entire network.

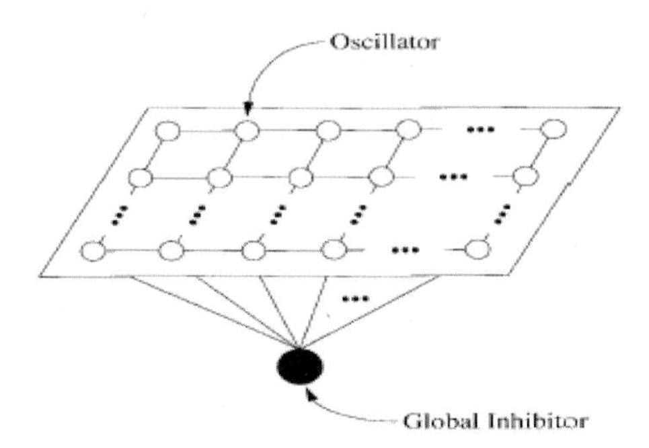

Fig. 1. A 2-D LEGION network with four-neighborhood connections

• *Segmentation algorithm:*

There are *three intuitive criteria* for defining segments (groups) on an image. The first is that leaders should be generated from both homogeneous and brighter parts of the image. Second, brighter pixels should be considered similar to wider ranges of pixels than darker ones. The third criterion stipulates that the boundaries of segments are given where pixel intensities have relatively large variations. For Three-dimensional segmentation is readily obtained by using 3-D neighborhood kernels.

Here the segmentation algorithm has been used on 2-D and volume CT and MRI medical datasets of the human head. The user provides six input parameters: the potential neighborhood; the recruiting neighborhood; the threshold; the power of the adaptive-tolerance-mapping function; and the tolerance range-variables.

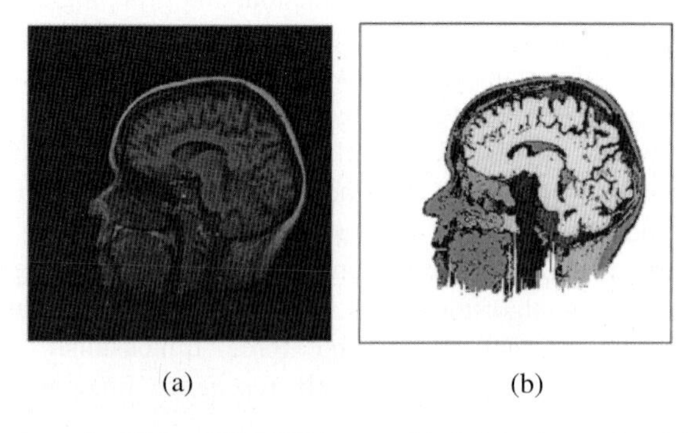

Fig. 2. Segmentation of a 256 _ 256 MRI image. (a) Original gray-level image of a human head. (b) A gray map showing the result of segmentation.

LEGION network, besides its biological plausibility, is especially feasible for parallel-hardware implementation, which would be important for real-time segmentation of volume datasets. *Manual segmentation* generally gives the best and most reliable results when identifying structures for a particular clinical task.

One objective of medical-image segmentation is to separate *white matter and gray matter*. This algorithm is intended to be more flexible for segmenting a variety of structures. However, it is interesting to note that once the brain is segmented, this algorithm in another pass can perform the separation of gray matter from white matter [18].

Layers of LEGION networks are effective computational framework that is capable of grouping and segregation based on partial results from preceding layers, and thus may further enhance segmentation performance. The network architecture is amenable to VLSI chip implementation, which would make LEGION a plausible architecture for real-time segmentation.

C. Superimposing a medical image within the subject:

For superimposition of medical image, within the subject itself, a virtual medical imaging system has been used in which 3-dimensional (3D), stereoscopic, motion (if necessary) medical image is superimposed within the subject via a see-through head mounted display [19]. For displaying more realistically, the 2D images are reconstructed in 3D images using computer graphics and for stereographic views a head mounted display providing separate images for each eye is useful. Virtual reality has been popular in representing data obtained using medical imaging devices much as echography, MRI, CT.

The medical imaging system consists of 3 subsystems: geometric information I/O, 3D object generations and image merging subsystem.

The geometric information I/O subsystem consists of polhemus receivers and polhemus transmitter. The see-through head-mounted display, shimadzu STV-E is also used.

The system is applied to echocardiogram. The original echo images in the 3D object generation subsystem was obtained by measuring 28 successive B-mode echo images of the heart, using a sector scanning Tran esophageal probe.

When the virtual imaging system was applied to echocardiogram, the virtual heart image changed its size, orientation and binocular parallax approximate for both the observer and the subject. The problem that the echo image was just pasted one onto the subject was somewhat improved in this system due to the stereographic image having reality along depth direction. This virtual medical imaging system will be beneficial to clinics for purposes such as surgical planning in the near future.

⇒ *Medical image repository and image categorization:*

In the last decade medical imaging has become an essential component of medical field. Moreover, the development of the Internet has made medical images available in large numbers in online repositories, atlases and other heath-related resources. These images represent a valuable source of knowledge and have significant importance for medical information retrieval. Medical image repository plays an important role in the hospital workflow. The operations in the hospital workflow such as image storage and retrieval, viewing, post processing can be saved as a generic repository component using which different modalities can develop their own clinical application.

A. COTS-Like Generic Medical Image Repository:

Commercial off-the-shelf software (COTS) is used for storage and retrieval of medical image data with the underlying storage being a commercial RDBMS. A generic medical image repository capable of serving multiple modalities can be treated similar to a COTS component.

The phrase "Medical Image Repository" is being used to describe a generic storage component consisting of a Data Access Layer (DAL) [20]. Clinical applications wanting to store and retrieve images use this storage component. Here in ref. [20] an innovative approach of how one single generic medical image repository subsystem (where the components satisfy characteristics of COTS product) can allow various modalities, viewing, work stations and PACS (Picture Archiving and Communications System) to perform 'store', 'view', 'query' operations have been explained. It also explains how this enables generic applications like basic viewing and printing of images to be reused across modalities and workstations.

The process for COTS software product evaluation [21] defines a COTS product as one that is:

a. Supported and evolved by the vendor, who retains the intellectual property rights.
b. Available in multiple, identical copies.
c. Used without modification of the internals.
d. Offered by a vendor trying to profit from it.
e. Sold, leased, or licensed to the public.

The medical image repository component described in ref. [20] satisfies all of the above except for (d) and (e) as only various product groups (modalities). Hence, it can be referred to as a COTS-like product.

In addition to the above-mentioned criteria [22] also talks of other criteria that are typically present in any COTS-like component. These criteria include:

- Interoperability: It defines Information Objects, which are abstractions of real information entities such as CT Image, MR Image etc.
- Diversity in requirements: A generic medical image repository must satisfy the following criteria:
  - Support for multiple modalities.
  - Different information models.
- Flexibility
- Usability
- Performance
- Reliability
- Portability

There was an assumption that generic repository means a trade-off between maintenance and performance. With the emergence of this COTS-like repository, this assumption has been proven to be a myth. Places, where the modalities as well as the PACS are from the same vendors, the generic repository component can be deployed in a client/server model, thereby giving a provision for an efficient proprietary protocol instead of DICOM.

B. Medical image categorization using a texture based symbolic description:

In the field of medical image indexation, automatic categorization provides the means for extracting, otherwise unavailable, information from images.

The content-based automatic medical image categorization methods, in the on-line context of the CISMeF health-catalogue are focused in ref. [23]. The compact symbolic image representation conveys enough of the initial texture information to obtain high recognition rates, despite the complex context of multimodal medical image categorization.

Medical image categorization architecture, based on a new type of image descriptor, aiming to accurately extract the modality (e.g. MRI, XRay1), and the anatomical region present in medical images.

When publishing (e.g. on the Internet) the images are suffering further transformations resizing, cropping, high-compression, superposed didactical drawings and annotations. Thus the image variability (already significant due to anatomical and pathological differences) is increased. The strong inter-class similarity between some classes (representing different modalities and anatomical regions) further increases difficulty in categorization.

The categorization approach consists of three stages:

1) The extraction of statistical and texture image-feature sets to describe the image visual content: Each image is represented by a vector of 16 blocks, and from each block features will be extracted to describe its content.
2) The description of these features using a symbolical representation: CLARA (Clustering Large Applications) [22], AGNES (Agglomerative Nesting) [24] are used respectively for clustering.
3) The classification of the description vectors into the defined classes: A k-Nearest Neighbor classifier is employed, using the first (INN), the first three (3NN) and the first five (5NN) neighbors (weighted by distance). For computing distances between nominal representations VDM (i.e. Value Difference Metric) is used, a metric introduced by ref. [25] to evaluate the similarity between symbolic (nominal) features.

The suggested feature representation/transformation method is close to Vector Quantification (VQ) where the blocks of pixels are labeled with the indexes of the prototype blocks [26]. The fixed block split, here used in ref. [23], is considered necessary to take into account the spatial distribution of the texture and statistical features. Unfortunately this approach is more sensitive to image rotations and translations. However, given that the modern digital acquisition equipments are following standard acquisition procedures, the images are rarely presenting significant variations to rotation and/or translation.

C. Web-based interactive applications of high-resolution 3D medical image data:

The demands for sharing medical image data on the Internet for computerized visualization and analysis are growing consistently such data sizes (ranging from several hundred megabytes to several dozen gigabytes [27,28]) severely stress storage systems and networks, and present many challenges in development of Web-based interactive applications.

Because of the limits imposed by the available bandwidth of the Internet, Web-based interactive applications are limited to low- or medium-resolution image data [29,30], which are often insufficient for reliable use in clinical diagnosis.

Gustafson et al. have developed a software package (stand alone system [32, 27, 31, 28]), named MACOSTAT, for assembly and browsing of 3D brain atlases [31].

A novel framework for Web-based interactive applications of high-resolution 3D medical image data has been proposed in ref. [33]. Specifically, first partition the 3D data into buckets, and then compress each bucket separately. Also an indexing structure for these buckets to efficiently support typical queries such as 3D slicer and region of interest (ROI), and only the relevant buckets are transmitted instead of the whole high resolution 3D medical image data.

The proposed framework in ref. [33] (see Figure 3) consists of three major components: *data storage structure, disk access optimization,* and *server query processing*.

◊ *Data storage structure:*

1) Partitioning the whole high-resolution 3D image data into buckets.
2) Each data bucket is a small subset of the whole 3D image data set and can be compressed/decompressed using any existing loss less or lossy compression technique [34, 27, 35], depending on the given application.
3) To facilitate disk access optimization, disk space is allocated to the data buckets in Hilbert curve order. Hilbert curve [36, 37] is a space filling technique, which maps a multi-dimensional data space into a one-dimensional data space; that is, it defines a linear order to visit every data bucket in the three-dimensional space exactly once.

◊ *Data access optimization:*

To reduce the demand on server in terms of disk access and communication costs, we use two techniques.

- Incremental transmission:

A client can send some bucket IDs along with a query to inform the server that those buckets are available locally and are not necessary to be retransmitted.

- Group Access:

In a multi-user environment, instead of retrieving data buckets for each of the queries independently, we allow them to share disk access by retrieving the data buckets inside the MBB which encloses all the ROI's, in one sequential access.

◊ *Server Query Processing:*

For ROI queries, first the MBB of the ROI is determined and then process the corresponding range query using the octree to identify the relevant data buckets that overlap with the MBB.

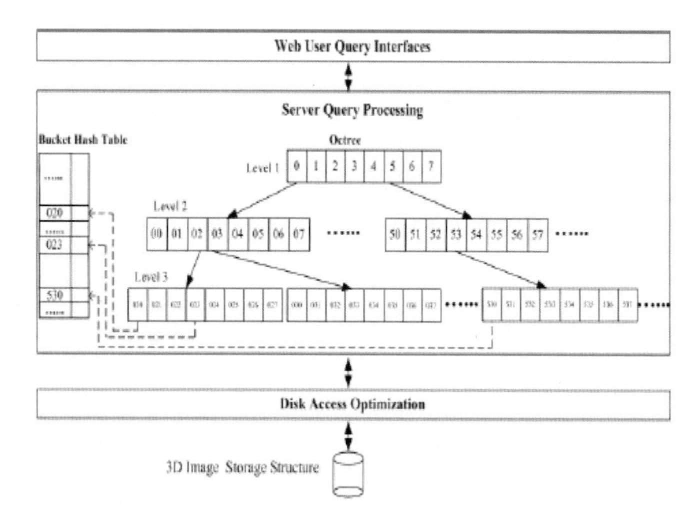

Fig. 3. An overview of the proposed framework

The problem of developing Web-based interactive applications of high-resolution 3D medical image data enables real-time interaction with remote high-resolution 3D medical images. This is achieved with the scalability to allow many concurrent users to receive the service simultaneously.

4 Discussion

So far we have seen various methods of medical image processing and the way of getting 3-dimensional images. In the wavelet technology for high-speed transmission of images efficient compression scheme is essential. Further development can be done in various aspects multi-wavelet theory and its application, application of combined wavelet transform with neural network, application of combination of fractal and wavelet etc.

Segmentation is a very difficult problem for general images, which may contain effects such as highlights, shadows, transparency, and object occlusion. On the other hand, sampled image datasets lack these effects with a few exceptions. Three broad classes that divide algorithms to segment sampled image data: are manual, semiautomatic, and automatic. The LEGION approach is able to segment volume datasets with appropriate parameter settings; produces results that are comparable to commonly used manual segmentation.

Other tolerance functions may also be used to better define pixel similarity.

For viewing the images more realistically, 3D representation is necessary, so a virtual imaging subsystem using a see-through head mounted system was used. It is beneficial for clinical purpose.

Generic medical image repository components catering all the important characteristics of the COTS like software products have been develop. Further, it has now spread its wing beyond the medical imaging repository and is now being used as a log repository and configuration repository.

High-resolution three-dimensional (3D) medical image data have become increasingly common with the advances and wide availability of medical image acquisition technologies.

Most existing Web-based 3D medical image interactive applications therefore deal with only low- or medium-resolution image data. But it is possible to download the whole 3D high-resolution image data from the server. An indexing structure for data buckets to efficiently support typical queries such as 3D slicer and region of interest (ROI), and only the relevant buckets are transmitted instead of the whole high resolution 3D medical image data. The study based on a human brain MRI data set indicates that the proposed framework can significantly reduce storage and communication requirements, and can enable real-time interaction with remote high resolution 3D medical image data for many concurrent users.

5 Conclusion

Biomedical imaging has seen truly exciting advances in recent years. Newly invented imaging methods can now reflect internal anatomy and dynamic body functions heretofore only derived from textbook pictures, and applications to a wide range of diagnostic and therapeutic procedures can be possible. Not only improvement in computer technology, but development will require continued research in physics and the mathematical sciences (e.g., artificial intelligence), fields that have contributed greatly to biomedical imaging and will keep continuing to do so.

The major topics of recent interest in the area of functional imaging involve the use of MRI and positron emission tomography (PET) to explore the activity of the brain when it is challenged with sensory stimulation or mental processing tasks. The emerging imaging methods have the potential to help major medical and societal problems, including the mental disorders of depression, schizophrenia, and Alzheimer's disease and metabolic disorders such as osteoporosis and atherosclerosis.

Although computing speed certainly has reached the point where iterative methods are clinically feasible for 2D problems, the focus is now on 3D PET where the size of A is 11-15 times larger than in 2D (after exploiting symmetries). Thus there is continuing need for new ideas in image reconstruction algorithm development.

Finally, it is worth mentioning, the explosion in the use and utility if the Internet including some resources of specific interest to the medical imaging community. The World Wide Web offers a great platform for education and teaching and has become the major method of sharing and communicating medical information. The creation of "digital departments" [38, 39] provides access to multimedia reporting (text, images, cines) from inexpensive client systems. Further, the WWW will allow use of Java applets to provide additional functionality such as analysis, to be implemented on Java compliant browsers.

Acknowledgement. This work was supported by the Security Engineering Research Center, granted by the Korea Ministry of Knowledge Economy.

References

1. Webb, S.: The Physics of Medical Imaging Medical Science Series, New York (1988)
2. http://en.wikipedia.org/wiki/Medical_imaging (visited on 01.04.2008)
3. http://www.lancs.ac.uk/ (visited on 01.04.2008)
4. Zaidi, H.: Medical Imaging: Current Status and Future Perspective, Division of Nuclear Medicine, Geneva University Hospital
5. Tian, D.-Z., Ha, M.-H.: Applications of Wavelet Transform in Medical Image Processing, Faculty of Mathematics and Computer Science, Hebei university, Baoding 071002
6. Bradie, B.: Wavelet packet-based compression of signal lead ECG J. IEEE Trans. BME 43(1), 49–60 (1994)
7. Kaimei, Z., Shengchen, y.: Semi-lossless compression algorithm of ECG signal based on wavelet transform. Shandong Joumal of Biomedical Engineering 22(2), 8–10 (2003)
8. Kalayci, T., Ozadmar, 0.: Wavelet processing for automated neural network detection of EEG spikes Jl. IEEE Eng. in Med. and Biol. 2, 160–166 (1995)
9. Weidong, Z., Yingyuan, L.: EEG spikes detection and denoising methods based on wavelet transform. Chinese Journal of Medical Physics 18(4), 208–210 (2001)
10. Meyer, E.G., Averbuch, A.Z., Stmmberg, J.O.: Fast adptive wavelet packet image compression. IEEE Trans. On Image processing 9(5), 792–800 (2000)
11. Xiong, Z., Ramchandran, K., Orchard, M.T.: Waveletpacket image coding using space-frequency quantization. IEEE Trans. On Image processing 7(6), 892–898 (1998)
12. Tolba, A.S.: Wavelet packet compression of Medical images. Digital Signal Processing 12, 441–470 (2002)
13. ye, Y., Zongying, O.: Method of edge detection. Letters 21, 447–462 (2000); Based on wavelet transform and fuzzy algorithm. Journal of Dalian University of Technology 42(4), 504–508 (2002)
14. Shman, R., Qler, J.M., Pianykh, O.S.: A fast and accurate method to register medical images using Wavelet Modulus Maxima. Pattem Recognition Letters 21, 447–462 (2000)
15. Terman, D., Wang, D.L.: Global competition and local cooperation in a network of neural oscillators. Physics D 81, 148–176 (1995)
16. Wang, D.L., Terman, D.: Locally excitatory globally inhibitory oscillator networks. IEEE Trans. Neural Networks 6, 283–286 (1995)
17. Wang, D.L., Terman, D.: Image segmentation based on oscillatory correlation. Neural Comput. 9, 805–836 (1997) (for errata see Neural Comput. 9, 1623–1626 (1997)
18. Haralick, R.M., Shapiro, L.G.: Image segmentation techniques. Comput. Graph. Image Processing 29, 100–132 (1985)
19. Matani, A., Ban, Y., Oshiro, O., Chihara, K.: A system for superimposing medical image within the subject. Nara Institute of Science and Technology
20. Chandrashekar, N., Gautam, S.M., Shivakumar, K.R., Srinivas, K.S., Vijayananda, J.: COTS-Like Generic Medical Image Repository
21. A Process for COTS Software Product Evaluation CMU/SEI-2003-TR 017 ESC-TR-2003-017 Santiago Comella-Dorda et al, http://www.sei.cmu.edu/pub/documents/03.reports/pdf/03tr017.pdf (visited on 01.04.2008)

22. Jaccheri, L., Sørensen, C.-F., Inge, A.: COTS Products Characterization Marco Torchiano, Wang Department of Computer and Information Science (IDI), Norwegian University of Science and Technology (NTNU) Sem Saelands vei 7-9, N-7491 Trondheim, Norway Phone: +47 735 94489, Fax: +47 735 94466 {Marco,letizia,carlfrs,alfw}@idi.ntnu.no, http://www.idi.ntnu.no/grupper/su/publ/marco.SEKE2002.pdf (visited on 01.04.2008)
23. Florea, E., Barbu, E., Rogozan, A., Bensrhair, A.: VBuzuloiu LITIS Laboratory, INSA de Rouen 1060 Av. de l'Universite. St. Etienne du Rouvray, France LAPI Laboratory Politehnica University Bucharest 1-3 Iuliu Maniu Blvd, Bucharest, Romania, Medical image catagorization using a texture based symbolic description
24. Kaufman, L.: Finding groups in data: an introduction to cluster analysis. In: Finding Groups in Data: An Introduction to Cluster Analysis. Wiley, New York (1990)
25. Stanfill, C., Waltz, D.: Toward memory based reasoning. Communications of the ACM 29(12), 1213–1228
26. Gersho, A., Gray, M.R.: Vector quantization and signal compression. Kluwer Academic Publishers, Boston (1992)
27. Bajaj, C.L., Ihm, I., Park, S.: 3D RGB image compression for interactive applications. ACM Transactions on Graphics 20(1), 10–38 (2001)
28. Udupa, J.K., Herman, G.T.: 3D Imaging in Medicine. CRC Press, Boca Raton (1999); Proceedings of the 19th IEEE Symposium on Computer-Based Medical Systems (CBMS 2006) 0-7695-2517-1/06 $20.00 © 2006 IEEE
29. http://www.hms.harvard.edu/research/brain/atlas.html (visited on 01.04.2008)
30. http://www.loni.ucla.edu/Research/Atlases/Data/monkey/MonkeyAtlasViewer.html (visited on 01.04.2008)
31. Gustafson, C., Tretiak, O., Bertrand, L., Nissanov, J.: Design and implementation of software for assembly and browsing of 3D brain atlases. Computer Methods and Programs in Biomedicine 74(1), 53–61 (2004)
32. 3D Slicer, http://www.slicer.org/ (visited on 01.04.2008)
33. Liu, D., Hua, K.A., Sugaya, K.: A Framework for Web-based Interactive Applications of High-Resolution 3D Medical Image Data
34. Aguirre, A., Cabrera, S.D., Lucero, A., Vidal Jr., E., Gerdau, K.: Compression of Three-Dimensional Medical Image Data based on JPEG 2000. In: Proceedings of the 17th IEEE Symposium on Computer-Based Medical Systems, pp. 116–121 (2004)
35. Tang, X., Pearlman, W.A.: Three-Dimensional Wavelet-Based Compression of Hyperspectral Images, Hyperspectral Data Compression. Kluwer Academic Publishers, Dordrecht (2005)
36. Moon, B., Jagadish, H.V., Faloutsos, C., Saltz, J.H.: Analysis of the clustering properties of the Hilbert space filling curve. IEEE Transactions on Knowledge and Data Engineering 13(1), 124–141 (2001)
37. Sagan, H.: Space-Filling Curves. Springer, Heidelberg (1994)
38. Wallis, J.R., Miller, M.M., Miller, T.R., Vreeland, T.H.: An Internet-based nuclear medicine teaching file. J. Nucl. Med. 36, 1520–1527 (1995)
39. Parker, J.A., Wallis, J.W., Halama, J.R., Brown, C.V., Ceadduck, T.D., Graham, M.M., Wu, E., Wagenaar, D.J., Mammone, G.L., Greenes, R.A., Holman, B.L.: Collaboration using Internet for the development of case-based teaching files: Report of the Computer and Instrumentation Council Internet Focus Group. J. Nucl. Med. 37, 178–184 (1996)

Hybridization of GA and ANN to Solve Graph Coloring

Timir Maitra[1], Anindya J. Pal[1], Minkyu Choi[2], and Taihoon Kim[2]

[1] Heritage Institute of Technology
Chowbaga Road, Kolkata 700107, India
timirmaitra@gmail.com, anindyajp@yahoo.com
[2] School of Multimedia
Hannam University
Daejeon, Korea
freeant7@naver.com, taihoon@empas.com

Abstract. A recent and very promising approach for combinatorial optimization is to embed local search into the framework of evolutionary algorithms. In this paper, we present one efficient hybrid algorithms for the graph coloring problem. Here we have considered the hybridization of Boltzmann Machine (BM) of Artificial Neural Network with Genetic Algorithms. Genetic algorithm we have used to generate different coloration of a graph quickly on which we have applied boltzmann machine approach. Unlike traditional approaches of GA and ANN the proposed hybrid algorithm is guranteed to have 100% convergence rate to valid solution with no parameter tuning. Experiments of such a hybrid algorithm are carried out on large DIMACS Challenge benchmark graphs. Results prove very competitive. Analysis of the behavior of the algorithm sheds light on ways to further improvement.

1 Introduction

To introduce our work we first discuss the classical problem Graph Coloring that we are going to solve. The problem is nothing but assigning colors to the vertices of graph with minimum number of colors. The constraint is adjacent vertices never be assigned with same color. There are two versions of the coloring problem, Optimization version where we have to find smallest number of color needed to color the graph, which is of NP-hard [1] type. Decision version where we have to decide whether the graph is colorable using at most k colors, where k is a positive integer. This problem is well known NP-complete problem [2]. We are considering optimization version of Graph Coloring, consequently heuristic methods must be used for large graphs. Heuristics in optimization is any method that finds an 'acceptable' feasible solution. Local search procedure based heuristics most of the time terminates with local optimum. Randomization and restarting approaches is in need to overcome the local optimum problem. Among the existing heuristic approaches for the graph coloring we propose a hybrid algorithm of Boltzmann Machine of Artificial Neural Network and Genetic Algorithm (GA) to implement. Both of this method is very suitable to implement randomization and restarting approaches. Genetic algorithm is an evolutionary approach which does not have the above said problem but it cannot

T.-h. Kim, A. Stoica, and R.-S. Chang (Eds.): SUComS 2010, CCIS 78, pp. 517–523, 2010.

guaranty the optimum solution. Hopfield neural network the most common and generally used artificial neural network itself has local optimum problem that is the reason to select Boltzmann Machine model of ANN.

The reason behind the selection of above mentioned problem is there are different areas of practical interest where coloration of graph with minimum color has direct influence on how efficiently a certain target problem can be solved. Such areas include Timetable Scheduling [3], Examination Scheduling[4], Register Allocation[5], Electronic Bandwidth Allocation[6], Channel Routing[7] etc.

The purpose of the selection of GA is the power of GA to give different partial or probable solutions. GA itself or with some modification can solve GCP. GA were introduced not to solve a particular problem but to investigate the effects of natural adaptation in stochastic search algorithms. It consists of a population of possible problem solutions that get refined over time through selection, crossover and mutation operator. Fitness function has a crucial role in GA that evaluates the quality of the solution calculating fitness of a solution. GA evaluates a population of N solutions; it also evaluates a much larger number of partial solutions. It does this job without spending more than N fitness function evaluations. The general scheme of a GA is discussed in different papers [8,9].

We initiate Neural Network Algorithm with the outcome of GA. The job of the ANN is to refine the solution to the desired level. The ANN is characterized by its pattern of connections between the neurons (called its architecture), its method of determining the weights on the connections (called its training or learning algorithm), and its activation function. There are several areas where ANN is already in use. Few to mention are Function approximation including fitness approximation, classification including pattern and sequence recognition and sequential decision making, data processing including filtering clustering, robotics etc. There are several models that include ANN, and we select Boltzmann Machine (BM). The Boltzmann distribution has some beautiful mathematical properties and it is intimately related to information theory. In particular, the difference in the log probabilities of two global states is just their energy difference. The equilibrium distribution is independent of the path followed in reaching equilibrium are what make Boltzmann machines interesting.

2 Previous Work

If we look into previous work on heuristic to solve GCP, we find metaheuristic approach multilevel cooperative search [10], traditional random walk with learning automata based learning capability encoding the GCP as a boolean satisfiability problem [11], ant colony optimization is also well-known metaheuristic [12], Q'tron Neural Network for GCP is one of the approaches that built as a known-energy system and can make Local-minima-free [13], Continuous Hopfield Network based on an energy or Lyapunov function that decreases as the system evolves until a local minimum value is attained also proposed [14], A model called the Integer Merge Model has been proposed which aims to reduce the time complexity of graph colouring algorithms. They also claim that the model provide information that will help to create heuristics [15]. Multiple-restart quasi-Hopfield network is also been proposed for this problem where only problem-specific knowledge is embedded in the energy function that the algorithm tries to minimize. It is an up-gradation of their previous work on this field [16], Iterated Local Search that is simple and powerful metaheuristic claimed as very good when the algorithm is implemented on hand benchmark graph [17], Genetic algorithm proved to be a good approach where a large pool of trivial solutions are created satisfying the constraints of the problem. Then they are improved (optimized) by genetic operations of selection, crossover and mutation [18], a new neuron structure to be implemented on Generalized Boltzmann Machine (GBM). In this neuron stochastic weights replaced the stochastic activation function, they termed this neuron as Multi State Bitstream Neuron (MSBSN). [19], a Graph coloring solution using Hopfield Neural Network, implemented using FPGAs, which is defined by VHDL [20], a combination of Hopfield type of neurons and Potts neurons to solve Graph Coloring problem proposed in [21], Channel Routing Problem in VLSI using Graph Coloring Problem solved using a Chaotic Neural Network algorithm [22], GCP solved after reducing the problem in Maximum Clique problem using Binary weights Hopfield net special case [23], simulated annealing [24], tabu search [25] etc.

Some hybrid algorithm like GA with different implementation of operator and Tabu search implemented for DIMACS [26,27,28], Previous work on BM are like approximate learning algorithm for BM [12], modeling categorical random variables using BM [13], Multi State Bitstream Neuron for BM [14], Monte Carlo version of the Hopfield network [15] etc.

3 Our Work

Our algorithm starts with input of a graph from text file. Job starts with creation of adjacency matrix from input. Following the Genetic Algorithm (GAGCP in our creation of initial population of chromosomes is the next job where we randomly create chromosome using n (no. of vertex) color. Algorithm contains a fitness function that checks fitness of the chromosome that counts number of color used to color the chromosome. Depending on the crossover probability we will implement crossover operator on two random pair of chromosomes to create new offspring. Depending on the mutation probability we will implement mutation operator on randomly selected chromosome. After some iteration of this process we will select some best chromosome based on there fitness values to create a good population to work further.

In our next phase of algorithm we implement artificial neural network, based on Boltzmann Machine model which we will call ANNGCP. Based on the fitness value we will select chromosomes one by one from the population. We need to create two matrixes for Boltzmann Machine are state matrix and another matrix for color to vertex mapping. We will calculate energy of network based on the state matrix that represents states of the vertices at a particular time. This will be considered as the parameter of terminating condition of the iteration. In different phases of the algorithm we introduced stochastic behavior like vertex selection for checking improper coloring, selection of the procedure to make proper coloring, selection of color. We introduced another procedure to minimize the color that is color compaction where minimum color value will be replaced with maximum color value and vice-versa for few repetitions.

We executed GA on this to have a set of possible solution, and then tried to get best coloration from each chromosome of the pool of chromosomes through ANN. Our algorithm is given below:

Algorithm GAGCP

INPUT: Total no. of vertices, total no. of edges and vertices that create edges from file and create adjacency matrix of the graph with number of vertices

Step 1) Create random initial pool of chromosomes (population), using n no. of colors

Step 2) Calculate the fitness of the pool

Step 3) Sort the chromosomes of the pool in ascending order of their fitness value

Step 4) P_{best} = best individual

Step 5) For generation = 1 to max_iteration

Step 5.1) If the crossover probability $p_c <= 0.4$

Step 5.1.1) Perform crossover between any two random pair of chromosomes

Step 5.2) If the mutation probability $p_m <= 0.6$

Step 5.2.1) Randomly choose one chromosome from the pool

Step 5.2.2) Apply n_mutation to the chromosome

Step 5.3) Calculate the fitness of the pool new offspring generated by crossover and mutation

Step 5.4) Take improved chromosomes for next generation

Step 6) Output the best coloration

Algorithm ANNGCP

Step 1) Select a possible solution from the set of coloration created by GAGCP

Step 2) Create a matrix for color to vertex mapping

Step 3) Create state matrix and energy calculation function

Step 4) Repeat until energy is minimized

Step 4.1) Repeat until valid coloration is achieved

Step 4.1.1) Select a vertex randomly

Step 4.1.2) Check adjacent vertex for same color

Step 4.1.2.1) Increment adjacent vertex color depending on probability, (if color value reach the maximum value, initialize with minimum value) or assign a random color to that vertex.

Step 4.1.2.2) Update weight vector, state matrix and matrix for vector color mapping

Step 4.1.2.2) Collect the values that we are replacing

Step 4.2) Count color and energy of new coloration

Step 4.3) Find highest replaced color and replace with selected vertex depending on the probability or replace with new randomly selected color

Step 4.4) Short weight vector

Step 4.5) Repeat color compaction to minimize no. of color

Step 4.5.1) Replace minimum color value of the vertices with highest color value and maximum color value with minimum color value alternatively.

4 Result

SL. No.	Instances	V	E	X(G)	Best Known	ANNGCP
1	1-FullIns 5.col.b, CAR	282	3247	?	6	6
2	2-FullIns 5.col.b, CAR	852	12201	?	7	11
3	3-FullIns 4.col.b, CAR	405	3524	?	7	7
4	5-FullIns 4.col.b, CAR	1085	11395	?	9	11
5	2-Insertions 3.col.b, CAR	37	72	4	4	4
6	2-Insertions 4.col.b, CAR	149	541	4	5	5
7	3-Insertions 5.col.b, CAR	1406	9695	?	6	6
8	anna.col.b, SGB	138	493	11	11	11
9	David.col.b, SGB	87	406	11	11	11
10	Games120.col.b, SGB	120	638	9	9	9
11	Homer.col.b,SGB	561	1629	13	13	13
12	Huck.col.b, SGB	74	301	11	11	11
13	Jean.col.b,SGB	80	254	10	10	10
14	Miles1000.col.b, SGB	128	3216	42	42	42
15	Queen8_8.col.b, SGB	64	1456	9	9	11
16	Queen7_7.col.b, SGB	49	952	7	7	7

5 Conclusion

Our algorithm gives the competitive result for different type of benchmark graphs. To strengthen conclusions made about the power of the algorithm, it is worth to test it on some other classes of large graphs. The main purpose of this paper is to study hybridization of GA and BM on graph coloring problem. In the future, we intend to refine ANNGCP and apply it to find minimal color of large and small but complex graphs in reasonable time.

Acknowledgement. This work was supported by the Security Engineering Research Center, granted by the Korea Ministry of Knowledge Economy.

References

1. Garey, M.R., Johnson, D.S.: Computers And Intractability: A Guide To The Theory Of Np-Completeness. W. H. Freeman and Co., New York (1979)
2. Baase, S., Gelder, A.V.: Computer Algorithms: Introduction To Design And Analysis. Addison-Wesley, Reading (1999)
3. Wood, D.C.: A technique for coloring a graph applicable to large scale time-tabling problems. Computer Journal 12, 317–319 (1969)
4. Leighton, F.T.: A graph coloring algorithm for large scheduling problems. Journal of Research of the National Bureau of Standards 84(6), 489–505 (1979)
5. Chow, F.C., Hennessy, J.L.: Register allocation by priority based coloring. In: Proceedings of the ACM SIGPLAN 1984 Symposium on Compiler Construction, New york, pp. 222–232 (1984)
6. Gamst, A.: Some lower bounds for class of frequency assignment problems. IEEE Transactions on Vehicular Technology 35(1), 8–14 (1986)
7. Sarma, S.S., Mondal, R., Seth, A.: Some sequential graph coloring algorithms for restricted channel routing. INT. J. Electronics 77(1), 81–93 (1985)
8. Reeves, C. (ed.): Modern heuristic techniques for combinatorial problems. Orient Longman (1993)
9. Goodman, S.E., Hedetnieni, S.T.: Introduction to Design and Analysis of Algorithm, MGH (1997)
10. Kennedy, J., Eberhart, R.C.: A discrete binary version of the particle swarm algorithm. In: Proceedings of the 1997 Conference on Systems, Man, and Cybernetics, pp. 4104–4109. IEEE Service Center (1997)
11. Dorigo, M., Maniezzo, V., Colorni, A.: Ant system: Optimization by a colony of cooperating agents. IEEE Transactions on Systems, Man, and Cybernetics-Part B 26, 29–41 (1996)
12. Dorigo, M., Gambardella, L.M.: Ant colony system: A cooperative learning approach to the traveling salesman problem. IEEE Transions on Evolutionary Computation 1, 53–66 (1997)
13. Blum, C., Aquilera, M.J.B., Roli, A., Sampels, M. (eds.): Hybrid Metaheuristics: An Emerging Approach to Optimization. Studies in Computational Intelligence. Springer, Heidelberg (2008)
14. Prakasam, P., Toulouse, M., crainic, T.G., Qu, R.: Design of a Multilevel Cooperative Heuristic for the Graph Coloring Problem. Interuniversity Research Centre (2009)

15. Bouhmala, N., Granmo, O.-C.: Solving Graph Coloring Problems using Learning Automata. Springer Link (2008)
16. Salari, E., Eshghi, K.: An ACO Alogrithm for the Graph coloring Problem. Int. J. Contemp. Math. Sciences (2008)
17. Yue, T.-w., Lee, Z.Z.Z.: A Q'tron Neural Network Approach to solve the Graph Coloring Problems. IEEE, Los Alamitos (2007)
18. Juhos, I., van Hemert, J.I.: Increasing the efficiency of graph colouring algorithms with a representation based on vector operations. Journal of Software (2006)
19. Blas, D., Jagota, A., Hughey, R.: Energy function –based approaches to graph coloring. IEEE, Los Alamitos (2002)
20. Liu, W., Zhang, F., Zu, J.: A DNA algorithm for the Graph Coloring Problem. ACS Publication (2002)
21. Chakraborty, G.: Genetic Algorithm for Graph Coloring Problem. Journal of Tree Dimensional Images (2000)
22. Jagota, A.: An adaptive, multiple restarts neural network algorithm for graph coloring. European Journal of Operational Research 93, 257–270 (1996)
23. Schwefel, H.P.: Numerical optimization of Computer models. John Wiley & Sons Ltd., Chichester (1981)
24. Kirkpatrick, S., Gelatt, C.D., Vecchi, M.P.: Optimization by simulated annealing. Science 220, 671–680 (1983)
25. Glover, F., Laguna, M.: Tabu Search. Kluwer Academic Publishers, Dordrecht (1997)
26. Dorne, R., Hao, J.: A new genetic local search algorithm for graph coloring. In: Eiben, A.E., Bäck, T., Schoenauer, M., Schwefel, H.-P. (eds.) PPSN 1998. LNCS, vol. 1498, pp. 745–754. Springer, Heidelberg (1998)
27. Tagawa, K., Kanesige, K., Inoue, K., Haneda, H.: Distance Based Hybrid Genetic Algorithm: An Application for The Graph Coloring Problem. In: Proceedings of 1999 Comgress of Evolutionary Computation, vol. 3, p. 2325–2332 (1999)
28. Galinier, P., Hao, J.K.: Hybrid Evolutionary Algorithms for Graph Coloring. Journal of Combinatorial Optimization 3, 379–397 (1999)

Alerting of Laboratory Critical Values

Sang Hoon Song[1], Kyoung Un Park[1], Junghan Song[1], Hyeon Young Paik[2], Chi Woo Lee[6], Su mi Bang[3], Joon Seok Hong[4], Hyun Joo Lee[5], In-Sook Cho[7], Jeong Ah Kim[8], Hyun-Young Kim[9], and Yoon Kim[9,10]

[1] Department of Laboratory Medicine
[2] Medical Information
[3] Internal Medicine
[4] Obstetrics & Gynecology
[5] Pediatrics, Seoul National University Bundang Hospital, Seongnam, Korea
[6] ezCaretech Co., Ltd., Seoul, Korea
[7] Department of Nursing, School of Medicine, Inha University, Incheon, Korea
[8] Department of Computer Education, Kwandong University, Chuncheon, Korea
[9] R&D Center for Interoperable EHR, Seoul, Korea
[10] Dept. of Health Policy & Management, College of Medicine, Seoul National University, Seoul, Korea

Abstract. Critical value is defined as a result suggesting that the patient is in danger unless appropriate action is taken immediately. We designed an automated reporting system of critical values and evaluated its performance. Fifteen critical values were defined and 2-4 doctors were assigned to receive short message service (SMS).Laboratory results in LIS and EMR were called back to the DIA server. The rule engine named U-brain in the CDSS server was run in real-time and decision if the laboratory data was critical was made. The CDSS system for alerting of laboratory critical values was fast and stable without additional burden to the entire EMR system. Continuous communication with clinicians and feedback of clinical performance are mandatory for the refinement and development of user-friendly CDSS contents. Appropriate clinical parameters are necessary for demonstration of the usefulness of the system.

Keywords: clinical decision support system; critical value.

1 Introduction

Laboratory data are estimated to affect about 70% of medical diagnoses [13]. It is known that clinicians interpret more than 35% of laboratory data incorrectly [9], which means that about 1/3 of the results may be thrown away without appropriate interpretation. Critical value is defined as a result suggesting that the patient is in danger unless appropriate action is taken immediately [6]. Reporting of critical values is part of the requirements for accreditation of Joint Commission on Accreditation of Healthcare Organization (JCAHO) [4] and College of American Pathologists (CAP) [2] in US. In many countries, including Korea, it is part of parameters of laboratory

T.-h. Kim, A. Stoica, and R.-S. Chang (Eds.): SUComS 2010, CCIS 78, pp. 524–531, 2010.

accreditation and service assessment of healthcare organizations. The most widely used method for alerting of critical values is fixed-line call [12]. The system is accurate in that it can read back the time, receiver, and the contents of the alert. However, there are possibilities of failure or delay of notification to doctors [9], who are responsible for the final clinical decision. In those cases, even if the alerts may be considered successful in clinical laboratory's perspectives, the data remains only in laboratory information system and is meaningful only when the clinicians voluntarily look for them. Recent development in medical informatics and mobile communication make it possible to automatically transmit alarm message depending on laboratory results. Therefore, more and more laboratories are using SMS service for alerting of laboratory critical values. In US, about 45% of the clinical laboratories are estimated to communicate with clinicians by electronic methods [12]. Recent studies have advocated the usefulness of the electronic alerting system in terms of successful notification [8] and clinical response time [7].

Type of laboratory tests and criteria of critical values differ depending on the policies and conditions of hospitals or laboratories [3, 12]. Each laboratory selects its own critical value lists by consulting clinicians or by referring to previous publications. Although many tests are used for critical value alerting in each laboratory, rules of alerting specific for doctors or specialties are rarely applied. For such individualized alerting to be possible, easily manageable and edible interface programs are desirable. We introduce a new critical value alerting system by introducing interoperable CDSS system and electronic critical value alerting system.

2 Design of Electronic Alerting System

Electronic alerting system consists of three major parts. The first is the contents of the system, the critical value lists and criteria. The second is interpretation of the laboratory results. The third is to show and express the results in appropriate method. The integrated design of the system cannot be successful by experts in single part as there are no experts who can understand and manage all of the components of the system. Participation of clinicians in various fields should be encouraged to define the critical value lists and criteria and evaluation of clinical performance. As experts in computer or network do not understand the medical significance and vice versa, continuous meeting and communication are essential.

2.1 Medical Information System

The institution which participated in this study was Seoul National University Bundang Hospital, a tertiary university hospital with about 930 beds and 4,000 outpatients a day. The institution is using electronic medical record (EMR) system and the laboratory information system (LIS) is integrated into the EMR system. The meeting of CDSS TFT is held per one or two weeks and experts in various fields of medicine, medical information system, medical record, and nurses participate in the meeting. This study was a main topic of the meeting and carried out by the members of the TFT. The technical aspects of CDSS rule engine was consulted to R&D Center for Interoperable EHR regularly.

2.2 Critical Value Lists and Criteria

Although various laboratory tests and criteria are used for critical values [3, 12], we defined 15 critical values after consulting clinicians in our hospital and referring to publications [5, 10]. Electrolytes, CBC parameters, and glucose were selected as most other laboratories. In addition, we selected transfusion related tests, antibody screening and Rh blood type, and tests related to neonatal health, newborn screening test (NST) and neonatal bilirubin. The critical value lists and criteria are shown in Table 1.

Table 1. Critical value lists and criteria used in this study

Test name	*Alert name*	*Criteria*
Na	Hyponatremia	< 115 mEq/mL
Na	Hypernatremia	> 150 mEq/mL
K	Hypokalemia	< 2.5 mEq/mL
K	Hyperkelamia	> 6.5 mEq/mL
Glucose	Hypoglycemia	< 40 mg/dL
Glucose	Hyperglycemia	> 400 mg/dL
Hematocrit	Falling Hct	Hct fallen >10% since last result and Hct < 26%
CBC blast	Leukemia screen 1	> 0%
WBC	Leukemia screen 2	> 50,000 /uL
ANC	Neutropenia	< 500 /uL
Plt	Thrombocytopenia	< 20,000 /uL
Rh	Rh	Negative
Antibody screening	Irregular antibody	Positive
Bilibrubin	Hyperbilirubinemia	> 18 mg/dL
NST	Newborn screening	Positive

As application of above lists to all the clinicians and patients is not practically possible nor necessary, we allocated each criteria to specific condition. The specific conditions were location of the patient, such as inpatient, outpatient, or neonatal intensive care unit (NICU), and the alert-receiving doctors. All the outpatients were included, but bilirubin and NST were exclued for inpatients, which could be thoroughly monitored without electronic alerting. For the critical results of outpatients, four clinicians participated in out study. For Na and K, all the four doctors with different specialties were assigned. For Hct, CBC blast, WBC, ANC, and Plt, a hematologist was assigned. For Rh and antibody screening, a surgeon in obstetrics, and for bilirubin and NST, a neonatologist was assigned. For inpatients, residents and fellows who were primarily responsible for the patients were assigned to receive alert messages as laboratory results are produced all day.

2.3 Electronic Critical Value Alerting System

The electronic critical value alerting system consisted of EMR/LIS and CDSS (Fig. 1). EMR/LIS consisted of EMR/LIS server which sends user information such as patient ID, name, location, laboratory test ordered, time of order, and result of the laboratory test to the CDSS and short message service (SMS) servers. The SMS server was used to send alert messages from CDSS server to the responsible users. The CDSS was made up of rule engine server, CDSS server, and CDSS DB. Rule engine (u-Brain) server was designed to make decisions about the contents and send appropriate messages. CDSS server was made to link to LIS data, to send them to rule engine server, to display critical value lists, to send the messages from rule engine server to SMS server, and to store data.

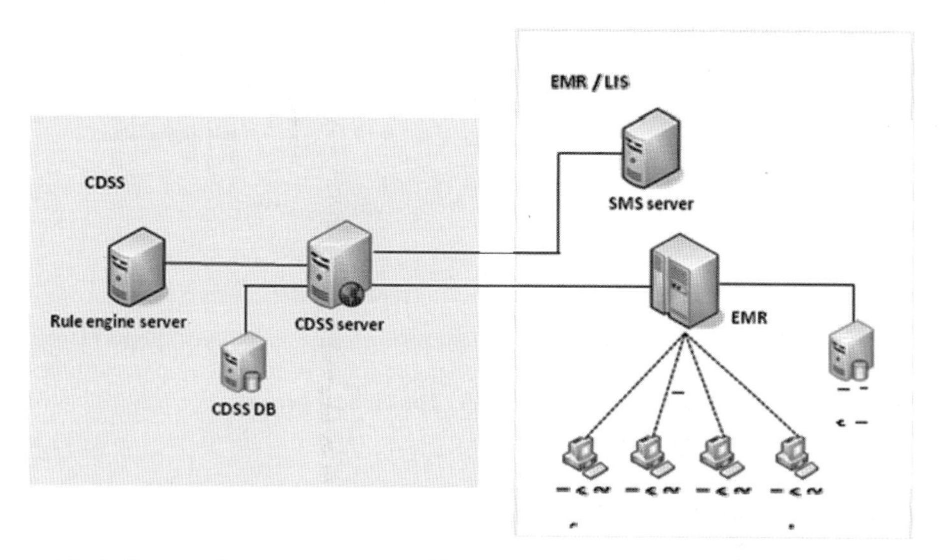

Fig. 1. The overall system components. The system is composed of EMR/LIS and CDSS.

When a patient is registered and his or her information exists in EMR, and if the order for laboratory tests is given, the laboratory test result is stored at LIS server. The data is linked to CDSS server via XML form, and stored at CDSS DB. The CDSS server is linked to DIA server and makes a document in XML form matching to the rule engine server receiving the alert message by linking to the u-Brain server. The message is stored at CDSS DB, sent to the corresponding users, and displayed as lists of critical values. The CDSS system process is presented in Fig. 2.

As it was necessary to evaluate the performance of the electronic critical value alerting system, we randomized the data to alert into two groups. One was SMS group for our new system and the other was non-SMS group, which was existing system that used fixed-line call to alert the critical values.

Finally, as there were possibilites of alert error in case of system failure or SMS failure, we decided to make lists of alert in the CDSS server. The lists were made to be retrieved in EMR for clinicians and in LIS for laboratorians. The clincians were encouraged to input the action they ordered into the lists after receiving alert message either by SMS or fixed-line call.

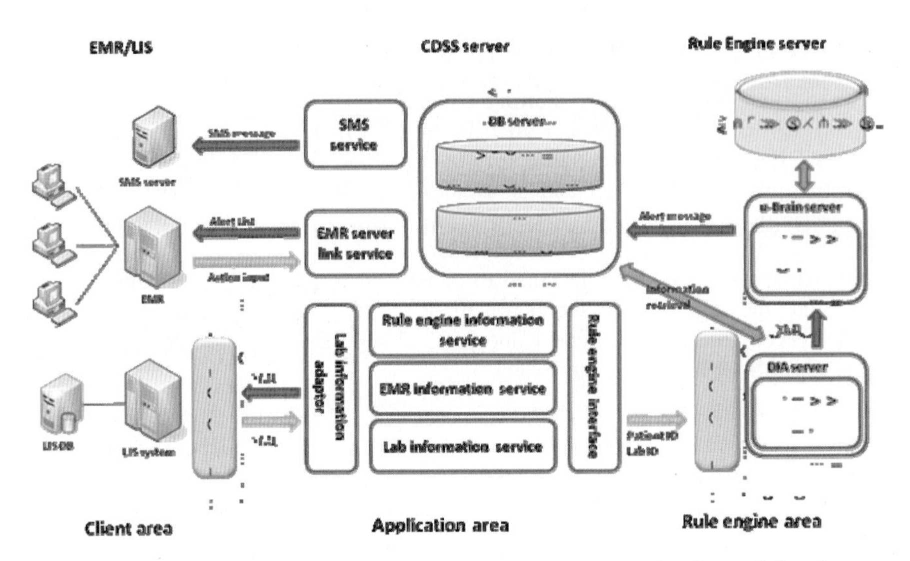

Fig. 2. Application architecture of the system. This figure shows the flow of data between EMR/LIS and CDSS.

2.4 Evaluation Parameters

The system was evaluated in terms of system performance and clinical performance. The system performance parameters were response times of interfaces, process times by stress test, CPU usage rate, and error rate. The clinical performance was evaluated by time to order and time to action.

3 Performance of the Electronic Alerting System

3.1 System Performance

The hardware environment of our system was 2.4 GHz Intel Pentium 4 CPU, 512 MB RAM, and MS Windows Server 2003 R2 Service Pack 2.

The average response time of LIS to CDSS (n=78,892) and CDSS to rule engine (n=14,981) were 0.059 and 6.292 seconds, respectively. The maximum response time for both interfaces were 68.329 and 91.438 seconds. The variable response times of LIS of CDSS were due to the variation of the laboratory results in a given time as CDSS server calls for LIS web service per one minute. Most laboratory tests are performed in daytime, especially in the morning. The large maximum respose time of CDSS to rule engine may be explained by following two reasons. One is that after the first data is received, it takes some time for DIA to link to DB and load the data to U-Brain for the first time. The other is when DIA fails to link to CDSS DB, the process time for the next data is delayed.

For evaluation of process times, stress tests with 10 and 500 data were performed. As the number of data increased from 10 to 500, the variation of the process times also increased. Thus, we decided to split the interface time between systems.

The overall average CPU usage rate was 33.5%, and for CDSS DEMON and rule engine, 2.7% and 3.1%, respectively. However, it also varied significantly according to the amount of the data, as the largest values reached over 94%.

For about 6 months of system operation, only 14 cases of DB connection errors, in which cases data processing was not performed.

3.2 Clinical Performance

From September 2009 to January 2010, there were 210,976 laboratory data whose test names coincied with the critical value lists. Of them, 24,708 (11.7%) data met the specific conditions of critical values. After all the data were sent to the U-Brain, 922 (3.7%) data were determined to be critical values to be sent to the responsible persons. Data not alerted due to the instability of CDSS system in early period and system error were not included. Totally, 838 (3.4%) data were analyzed for evaluation of clinical performance.

The age of the patients was 45±23 (mean±SD) years, and the time from order of test to receipt of the specimen was 49.2±85.6 hours. The seemingly delayed time of specimen receipt was due to the fact that much of the data used for analysis was from outpatients for whom laboratory tests are performed just before the next visit to hospital in general. Although the amount of data from outpatients is only 16.3%, the long follow-up periods significantly increase the overall time. The most frequent tests that produced critical values were ANC (48.8%), Plt (26.9%), blast (6.1%), and potassium (4.7%).

Table 2. Comparison of time taken at each step in terms of SMS groups

Parameters	SMS groups	No.	Mean	SD	Unit	P value
Specimen receipt	SMS	362	48.5	81.6	hr	0.837
	non-SMS	470	49.7	88.6	hr	
Performance of test	SMS	363	1.5	5.7	hr	<0.001
	non-SMS	473	6.1	22.4	hr	
Decision by U-brain	SMS	364	6.0	6.0	sec	0.090
	non-SMS	474	5.0	5.0	sec	
Order for action	SMS	122	3.4	8.3	hr	0.520
	non-SMS	151	2.9	4.0	Hr	
Taking action	SMS	218	5.3	7.7	Hr	0.781
	non-SMS	263	5.2	5.8	hr	

The time taken for each step was compared in terms of SMS groups. As shown in Table 2, the time to order and the time to actions were not significantly different between two groups. It was quite unexpected and it could be explained by the skewness of the frequency of the tests that produced critical values. The most frequently observed tests, ANC and Plt, tended to be ordered repeatedly in the same patient if the critical results had not been corrected. Furthermore, in many instances, leukemia patients are treated with predefined schedule and their laboratory test results can be predictable even if they are critical. In those cases, the order for appropriate actions such as administration of bone marrow stimulating drugs or transfusion is not mandatory but scheduled and predictable. The only parameter that differed significantly between the two groups was time taken for the performance of laboratory tests. The reason for the difference may be due to the delay of the verification of laboratory test results. When non-SMS alert was assigned, most laboratorians verified their critical results after calling to the nursing station. In those cases, although laboratory test results are already produced, it is not transmitted to LIS until the laboratorians finaly verify the results.

The time to order and the time to action were were shorter in outpatients than in inpatients. It may be due to the fact that many outpatients were treated in the same day they visited the hospital, whereas most inpatients were treated as they scheduled rather than on the basis of critical value alerts.

4 Discussion

In this study, we designed electronic critical vale alerting system by implementing CDSS rule engine. Although many laboratories are using the electronic alerting system, most of those systems are not interoperable and cannot be easily transplanted to other laboratories with different hardware and software systems. The advantage of interoperable CDSS system is that every laboratories with different environment can adopt the CDSS system with minor modification or development of interface programs only if they define critical value lists suitable for their environment. The system was evaluated only in one hospital, but there are possibilities that it can be implemented in other laboratories. For the least burden to the EMR system, we used independent CDSS server which called back the necessary information and made the most important decisions. The CDSS system provided, u-Brain and DIA, was fast and accurate enough to be used without additional burden to the entire EMR system. However, as the encoding of the CDSS system was done by the program provider, the flexibility of the system may be limited. The user-friendly interface and soft encoding environment are necessary for interoperable CDSS system.

The clinical performance which was measured by time to order and time to action was not improved. The main reason of the unexpected results is thought to be associated with the design of the contents. In our system, the vast majority of the critical values were CBC parameters, which tends to be ordered repeatedly in patients with hematological discorders. The large amount of data produced by the CBC parameters biased the overall results and the clinical performance of each critical value list was too small in number and might have been hidden. Thus, it would have been much better to exclude repeated laboratory tests in the same patient.

Another hurdle to the evaluation of clinical parameters were ambiguity of clinical data. It was not so easy to get exact time to order or time to action as the types of order or action were frequently obscure. When evaluating such clinical parameters, it is desirable to define each order and action and to select only clear cut ones. The more convincing and measurable clinical parameters such as prognosis, cost, and length of stay in hospital should be developed and tried for the CDSS system to be more popular and spread. Contents that can be easily implemented into CDSS system and that can help clinicians‘ decision is possible by considering the organization's environment and consulting the specialits [11]. Continuous communication and feedback by teamwork approach [1] may be the only solution to overcome above hurdles.

In conclusion, the electronic critical value alerting system was stable and accurate. Demonstration of clinical usefulness is mandatory for the system to be more widely used.

References

1. Bero, L.A., Grilli, R., Grimshaw, J.M., Harvey, E., Oxman, A.D., Thomson, M.A.: Closing the gap between research and practice: an overview of systematic reviews of interventions to promote the implementation of research findings. The Cochrane Effective Practice and Organization of Care Review Group BMJ 317, 465–468 (1998)
2. College of American Pathologists, `http://www.cap.org/apps/cap.portal`
3. Howanitz, P.J.: Errors in Laboratory Medicine: Practical Lessons to Improve Patient Safety. Arch. Pathol. Lab. Med. 129, 1252–1261 (2005)
4. Joint Commission on Accreditation of Healthcare Organizations, `http://www.jointcommission.org/PatientSafety/NationalPatientSafetyGoals`
5. Kuperman, G.J., Teich, J.M., Tanasijevic, M.J., Ma'Luf, N., Rittenberg, E., Jha, A., Fiskio, J., Winkelman, J., Bates, D.W.: Improving response to critical laboratory results with automation: results of a randomized controlled trial. J. Am. Med. Inform. Assoc. 6, 512–522 (1999)
6. Lundberg, G.D.: When to panic over abnornal values. MLO. Med. Lab. Obs. 4, 47–54 (1972)
7. Park, H., Min, W., Lee, W., Park, H., Park, C., Chi, H., Chun, S.: Evaluating the Short Message Service Alerting System for Critical Value Notification via PDA Telephones. Ann. Clin. Lab. Sci. 38, 149–156 (2008)
8. Piva, E., Sciacovelli, L., Zaninotto, M., Laposata, M., Plebani, M.: Evaluation of Effectiveness of a Computerized Notification System for Reporting Critical Values. Am. J. Clin. Pathol. 131, 432–441 (2009)
9. Piva, E., Plebani, M.: Interpretative reports and critical values. Clin. Chim. Acta. 404, 52–58 (2009)
10. Tate, K.E., Gardner, R.M.: Computers, Quality, and the Clinical Laboratory: A Look at Critical Value Reporting. In: Proc. Annu. Symp. Comput. Appl. Med. Care, pp. 193–197 (1993)
11. Trivedi, M.H., Daly, E.J., Kern, J.K., Grannemann, B.D., Sunderajan, P., Claassen, C.A.: Barriers to implementation of a computerized decision support system for depression: an observational report on lessons learned in "real world" clinical settings. BMC. Med. Inform. Decis. Mak. 9, 6 (2009)
12. Wagar, E.A., Richard, C.F., Ana, K.S.: Critical Values Comparison: A College of American Pathologists Q-Probes Survey of 163 Clinical Laboratories. Arch. Pathol. Lab. Med. 131, 1769–1775 (2007)
13. Wagar, E.A., Yuan, S.: The laboratory and patient safety. Clin. Lab. Med. 27, 909–930 (2007)

Accuracy and Performance Evaluation of a Laboratory Results Alerting

InSook Cho[1], JeongAh Kim[2], Ji-Hyeun Kim[1], Kyu Seob Ha[3], and Yoon Kim[4,5]

[1] Inha Univ., Seoul, Korea
Insook.cho@inha.ac.kr
[2] Kwandong Univ., Gangwon, Korea
clara@kd.ac.kr
[3] Seoul National Univ. Bundang Hospital, Gyeonggi, Korea
[4] Seoul National Univ. Seoul, Korea
[5] Institute of Health Policy and Management, Seoul National Univ., Seoul, Korea

Abstract. This study was designed to evaluate the accuracy and performance of clinical decision support (CDS) architecture which consisted of a knowledge engine and data interface adaptor (DIA) developed as knowledge sharing tools. A laboratory alerting system was selected for the test field and 10 rules represented by SAGE formalism were used. Laboratory results (N=323,455) captured from a tertiary and teaching hospital over three months were retrospectively evaluated. We were able to confirm accuracy, stability, and feasibility of the CDS application architecture and the knowledge engine. However, DIA throughput time suggested the need for further improvements.

Keywords: Clinical Decision Support System, quality evaluation.

1 Introduction

Electronic health records (EHRs), when used effectively, can improve the safety and quality of medical care. For maximum benefit, however, EHRs have to be paired with a clinical decision support system (CDSS) to effectively influence clinician behavior [1] and impact healthcare processes and outcomes. For widespread adoption of a CDSS, several issues must be tackled, including difficulties translating medical knowledge and guidelines into a form usable by EHRs, development of tools for doing so, and leveraging application of the system.

For several decades, the Arden Syntax[2], Asbru [3], GLIF3 [4], GEM [5], EON [6], SAGE [7], PRODIGY [8], and PRO*forma* [9] have been introduced as common shared models which facilitate direct interpretation or mapping to multiple implementation environments. SAGE (Standards-Based Active Guideline Environment) was derived from previous works such as PRO*forma*, GLIF3, EON, and PRODIGY [7]. It advanced the state of the art by focusing on requirements that previous models had not met simultaneously, including: A) incorporation of workflow awareness, B) employment of information and terminology standards, C) incorporation of simple flow-of-control standards, and D) attention to integration with a vendor CIS (clinical information system).

T.-h. Kim, A. Stoica, and R.-S. Chang (Eds.): SUComS 2010, CCIS 78, pp. 532–540, 2010.

However, SAGE execution is currently not available outside of the SAGE project. The authors (IC, JK) developed CDS application architecture, a knowledge engine, and an electronic medical record (EMR) data interface adaptor (DIA) which can execute SAGE formalism [10]. In this present study, we explored and evaluated the clinical aspect of performance by selecting laboratory alerts and converting the rules used in the laboratory alerting system into SAGE formalism. To date, most published cases have used a runtime database lookup strategy based on an existing database SQL Server platform and SQL code combined with a compiler [11-13]. This approach has practical advantages as far as operational aspects are concerned. From the knowledge sharing perspective, however, it causes dependency on a local database which is far from standard as well as not readable by both human and machine. Therefore, its interoperability among institutions is compromised.

It is important for physicians to respond to notifications of critical laboratory test results in a timely and appropriate manner. Due to the importance of critical values representing life-threatening situations, various types of automated wireless alert systems have been used to prevent medical errors and report critical laboratory values to physicians. Real setting response time and signal-to-noise ratio are truly important in these instances [14]; therefore, we have focused on these two measures of system performance.

2 Methods

The study was conducted within the development and testing environment of Bundang Hospital in Gyeonggi-province, which has the same environment as the real-time system. As a tertiary and teaching hospital with 900 beds, it has computerized physician order entry (CPOE), EMR, and LIS (laboratory information system) with callback. The daily volume of lab tests is approximately 55,000 including the inpatient, outpatient, and emergency departments.

The knowledge evaluated consisted of 10 alerting rules; nine based on the values of a single laboratory result, and one that detected changes in laboratory results over time. These rules were selected from the review of previous work [13], expected prevalence of use, and potential morbidity associated with failure to perform appropriate laboratory monitoring by the CDSS at Bundang Hospital. Transforming and encoding the rules into SAGE format was introduced in our previous work [14] in detail. Briefly, we used 18 concepts including concept qualifiers and seven activity graphs.

Figure 1 shows the system architecture of a test environment, based on the CiEHR CDSS architecture developed by the authors (IC, JK, YK) [15]. We used a testing tool to simulate a lab alerting application programmed to trigger the system to get data from a CDR (clinical data repository) every day. The CDR also received lab data from the LIS on a daily basis. A DIA requested CDR data using standard terminology and mapped the data model used in encoded rules with patient data from the CDR. Testing was conducted for nine days in September, 2008 using a total of 323,445 retrospective lab results collected over three months.

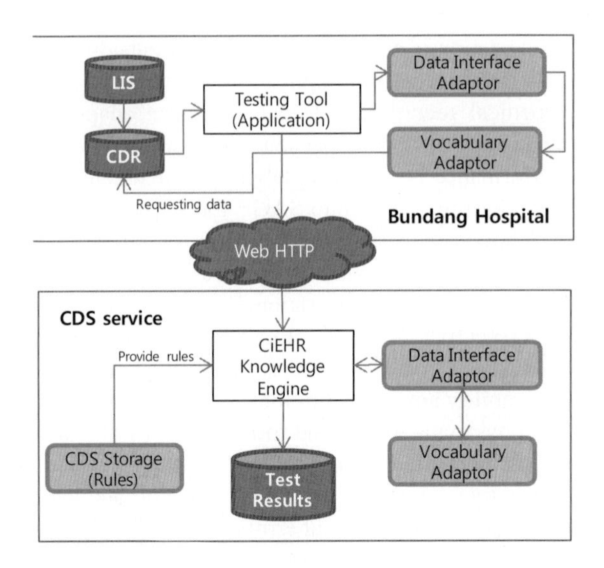

Fig. 1. System architecture for testing a lab alerting system
LIS: laboratory information system, CDR: clinical data repository

We analyzed the tracking data accumulated in the test results database to evaluate the accuracy of each alert. The test results database kept the data delivered from the CDR and the outputs created from the knowledge engine. Therefore, we could compare the input and output data of the engine as well as failure cases which did not create any results. Figure 2 shows the transaction intervals measured to detect overall response time: ① lab alerting system, ② engine throughput, and ③ DIA output. The test was performed on a personal computer with 1.86 GHz CPU, 1.5 GB memory, and Windows XP operating system.

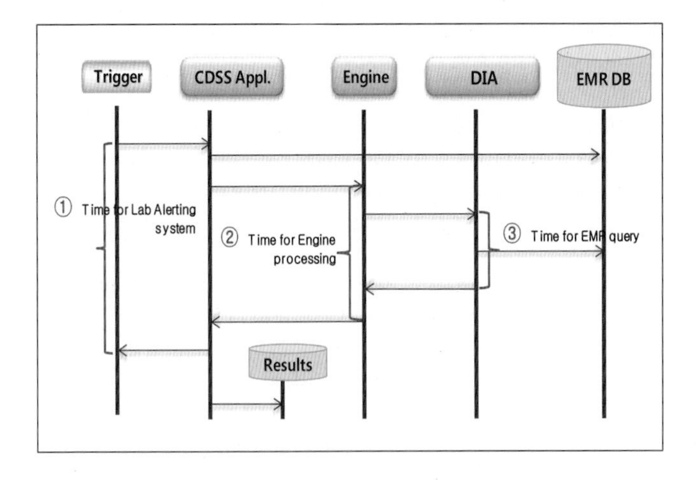

Fig. 2. Time measurement intervals for performance testing

3 Results

3.1 Alert Volume and Distribution

Based on the 323,445 lab results, 1,650 alerts were created: 1,164 (70%) for inpatients, 222 (13%) for outpatients, and 264 (16%) for emergency department patients. For inpatients, alerts for falling hematocrit and hyperkalemia had the highest frequencies of 764 (46.3%) and 203 (12.3%), respectively. Alert frequencies divided by the physician number of each department, to estimate the amount of alerts a physician received per day, ranged from 1.0 for pediatrics to 6.0 for internal medicine, while the emergency department showed 2.9 alerts per day per physician. In the outpatient setting, we were not able to adequately estimate the average alert frequency per physician due to daily variations in the number of physicians who saw patients. Daily test distribution had a typical bell shape with a peak point around 10 to 11 AM, except for Rh typing which had a peak point around 5 to 6 PM. Sodium, potassium, and WBC showed the highest test frequencies during the peak time period.

3.2 Alert Accuracy

Accuracy refers to the completeness of LIS data mapping and correctness of an alert triggered based on rule criteria. During testing, six points of the data mapping profile in the DIA were found and fixed. Several incorrect outputs caused by the converter transforming encoding knowledge into executable formats were also found and corrected. Through the white- and black-box validation process, the accuracy rate was raised to 100%.

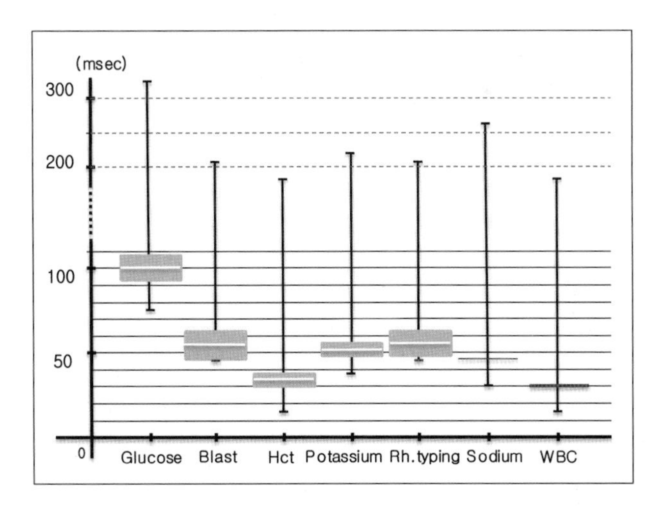

Fig. 3. Average response time of the knowledge engine by alerts

3.3 System Performance

The average time of overall system response (interval ① in Fig. 2) was 398.06 milliseconds (SD = 799.59; range = 191.50-842.04). Specifically, the rule for Rh typing consumed the highest amount of time and variation. The average time for the knowledge engine (interval ② subtracted from ③) was 51.90 milliseconds (SD = 24.59; range = 31.66-102.07). Glucose rules had the longest time (Fig. 3).

The DIA took an average 346.16 milliseconds (SD = 800.55; range = 787.31-136.86) and the Rh typing rule showed the highest response time.

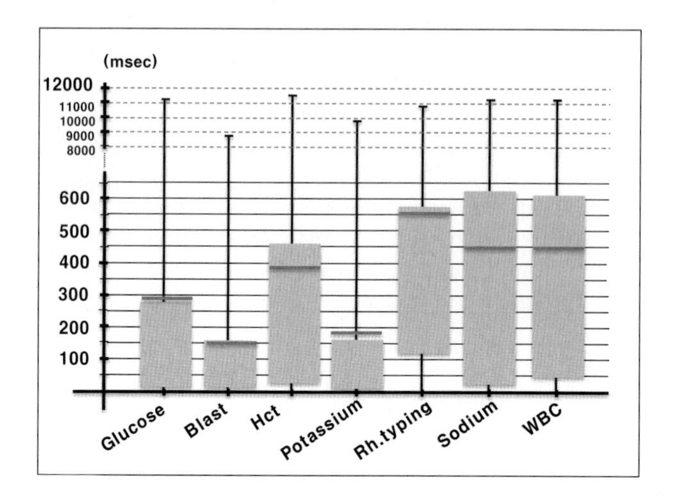

Fig. 4. Average response time of the data interface adaptor (DIA) by alerts

Analysis of peak time response revealed a mean of 475.23 milliseconds (SD = 660.21) except in the case of the Rh typing rule which was typically used from 5 to 6 PM with relatively low frequency and had the longest response time (Table 1).

Table 1. Average response time during the day at peak time by rule. The a, b, c, d superscripts indicate groups with significant differences at the .05 level of multiple comparisons.

Rule	Mean freq./ day	Mean freq. of peak time/ hr	Mean overall system response ms (SD)
Sodium	563.79	101.78	621.57 (835.66)[a]
Potassium	568.50	101.90	295.99 (532.43)[b]
Glucose	488.15	74.19	632.11 (727.21)[c]
Hematocrit	687.25	101.07	428.81 (807.52)[a]
Rh typing	114.69	25.05	762.46 (1,244.59)[d]
CBC blast	438.19	87.36	237.81 (490.03)[b]
WBC	685.58	101.26	586.57 (791.57)[c]

According to average response time analysis by clinical setting, inpatient and emergency cases showed similar overall and DIA times (Table 2). Engine times showed consistent and stable turnaround with an average of 50.0 milliseconds regardless of department.

Table 2. Average response time by clinical setting

Clinical Department	N	Mean response time (SD)		
		Overall	Engine	DIA
Outpatient	122,949	197.36 (251,46)	56.00 (29.02)	141.36 (249.17)
Inpatient	149,282	626.22 (853.22)	48.02 (17.14)	578.20 (853.80)
Emergency	51,214	193.98 (247.97)	53.10 (25.16)	140.89 (251.32)

4 Discussion

It is important for physicians to respond to notifications of critical laboratory test results in a timely and appropriate manner. Because critical values may represent life-threatening situations, therapeutic management by physicians of patients with critical values could serve as a valuable measure of laboratory outcomes [15]. Specifically, the point-of-care reminders delivered when busy clinicians are reviewing or renewing medications in the hospital are useful and the benefits of such have been reported by previous studies [13, 16]. The current consideration is how such systems or services could be implemented and delivered widely.

In the health care industry, the knowledge assets underlying CDSS are time-consuming and expensive to generate, as well as voluminous, and subject to change, so sharing and reuse once they are created, would be highly advantageous [16]. To facilitate sharing, the knowledge should be represented in standard form so that it can be disseminated and widely used and it needs to be updated on a regular basis as well. Also, an implementation environment is important to realize knowledge sharing in the form of CDSS. Therefore, in the present study, we explored and evaluated the tools – CDS architecture, knowledge engine, and data interface adaptor – developed by authors in previous works, with regards to system performance. Rules used in the laboratory alerting system were a simpler form than those of other CDS categories, so it was easier to implement and test using these.

With regard to alert volume and distribution, the system was heavily stressed at a specific time period during the day. Considering the work processes at hospitals, it is natural that tests are concentrated in the morning when ambulatory settings are open and work is starting for the day. As for the number of alerts a physician is expected to receive per day by department, internal medicine was the highest at 6.0 alerts.

Considering that most rules are relevant to internal medicine, the high number may be responsible for alert fatigue described in previous studies as a cause of system failure. For example, if a physician already knows that a patient's hematocrit is low or expected to be low, repeated alerts would not be meaningful, and may even be disruptive. Potential users recommended that the rules should be more sophisticated based upon each patient's specific clinical context, and not just depend on a value itself.

As for accuracy, at the beginning of our evaluation we found incorrect cases due to concept mismatches in condition expressions and LIS data. However, the rules were simple and explicit, so it was easy to refine the mappings and achieve 100% accuracy, assured by black-box validation as a post-hoc approach.

System performance showed feasible results around 400 milliseconds. During peak time, the mean frequency was about 25 to 100 per hour and the overall system response was 240 ~ 760 milliseconds. Sometimes response time was more than 1 minute. These outcomes were not unexpected for notification of laboratory results considering that users are not standing at a computer monitor waiting for the results. Comparing messaging time of an average of 15 seconds by pager, and 10 seconds by mobile phone according to the study of Chen et al. [12], 240 to 760 milliseconds represented quite an acceptable performance. However, there were wide variations in response times by test, to which the DIA's performance contributed. The role of the DIA was to map the concepts in knowledge representation with relevant data items in an EMR or LIS using a standard terminology code system. Thus, the DIA must pass several mapping steps in run time, which require time delays. The study of Goldberg et al. [17] which reported the performance of a commercial engine for a CDS service, also pointed out that delivery of data is the major bottleneck in rule service performance. In order to minimize the number of round trips between a rule service and an external repository, the rule service should be primed with a large swath of patient data. It will be necessary to further investigate methods for bulk data retrieval in order to assure a scalable infrastructure for an enterprise CDSS.

We analyzed the data in detail by medical department, clinical settings (inpatient, outpatient, and emergency department), months, days, and times to examine the effect of other variables, but no typical patterns or trends were uncovered. In contrast to the DIA, the knowledge engine showed quite stable and consistent performance at approximately 50 milliseconds.

5 Conclusion

System accuracy measured by signal-to-noise ratio easily achieved 100% due to the simple and explicit rules and system architecture. Response times measured by overall system time, engine throughput time, and DIA throughput time were reasonably sound and feasible in the domain of laboratory alerts; however, we detected an opportunity for improvement in DIA throughput time for other applications of CDSS.

Acknowledgments. This study was supported by a grant of the Korea Health-care Technology R&D Project, Ministry for Health, Welfare & Family Affairs, Republic of Korea (No. A050909).

References

[1] Reichley, R.M., Seaton, T.L., Resetar, E., Micek, S.T., Scott, K.L., Fraser, V.J., Dunagan, C., Bailey, T.C.: Implementing a Commercial Rule Base as a Medication Order Safety Net. Journal of the American Medical Informatics Association (JAMIA) (ScienceDirect) 12, 383 (2005)

[2] de Clercq, P.A., Blom, J.A., Korsten, H.H., Hasman, A.: Approaches for creating computer-interpretable guidelines that facilitate decision support. Artificial Intelligence in Medicine 31, 1–27 (2004)

[3] Shahar, Y., Miksch, S., Johnson, P.: The Asgaard project: a task-specific framework for the application and critiquing of time-oriented clinical guidelines. Artificial Intelligence In Medicine 14, 29–51 (1998)

[4] Boxwala, A., Peleg, M., Tu, S., Ogunyemi, O., Zeng, Q., Wang, D., Patel, V., Greenes, R., Shortliffe, E.: GLIF3: a representation format for sharable computer-interpretable clinical practice guidelines. J. Biomed. Inform. 37, 147–161 (2004)

[5] Shiffman, R., Karras, B., Agrawal, A., Chen, R., Marenco, L., Nath, S.: GEM: A proposal for a more comprehensive guideline document model using XML. J. Am. Med. Inform. Assoc. 7, 488–498 (2000)

[6] Musen, M., Tu, S., Das, A., Shahar, Y.: EON: a component-based approach to automatiuon of protocol-directed therapy. J. Am. Med. Inform. Assoc. 3, 367–388 (1996)

[7] Tu, S.W., Campbell, J.R., Glasgow, J., Nyman, M.A., McClure, R., McClay, J., Parker, C., Hrabak, K.M., Berg, D., Weida, T., Mansfield, J.G., Musen, M.A., Abarbanel, R.M.: The SAGE Guideline Model: achievements and overview. Journal of the American Medical Informatics Association 14, 589–598 (2007)

[8] Purves, I., Sugden, B., Booth, N., Sowerby, M.: The PRODIGY project - the iterative development of the release one model. Presented at AMIA Annual Symposium (1999)

[9] Fox, J., Johns, N., Rahmanzadeh, A.: Disseminating medical knowledge: the PROforma approach. Artificial Intelligence In Medicine 14, 157–181 (1998)

[10] Kim, J., Cho, I., Kim, Y.: Knowledge Translation of SAGE-based guidelines for Executing with Knowledge Engine. Presented at Proceedings of the Annual Symposium of the American Medical Informatics Association Washington, DC, USA (2008)

[11] Iordache, S.D., Orso, D., Zelingher, J.: A comprehensive computerized critical laboratory results alerting system for ambulatory and hospitalized patients. Studies In Health Technology And Informatics 84, 469–473 (2001)

[12] Chen, H.-T., Ma, W.-C., Liou, D.M.: Design and implementation of a real-time clinical alerting system for intensive care unit. In: Proceedings / AMIA... Annual Symposium. AMIA Symposium, pp. 131–135 (2002)

[13] Park, H.-i., Min, W.-K., Lee, W., Park, H., Park, C.-J., Chi, H.-S., Chun, S.: Evaluating the short message service alerting system for critical value notification via PDA telephones. Annals Of Clinical And Laboratory Science 38, 149–156 (2008)

[14] Schedlbauer, A., Prasad, V., Mulvaney, C., Phansalkar, S., Stanton, W., Bates, D.W., Avery, A.J.: What evidence supports the use of computerized alerts and prompts to improve clinicians' prescribing behavior? Journal of the American Medical Informatics Association 16, 531–538 (2009)

[15] Howanitz, P.J., Steindel, S.J., Heard, N.V.: Laboratory critical values policies and procedures: a College of American Pathologists Q-Probes study in 623 institutions. Archives of Pathology & Laboratory Medicine 126, 663–669 (2002)

[16] Kuperman, G.J., Teich, J.M., Tanasijevic, M.J., Ma'Luf, N., Rittenberg, E., Jha, A., Fiskio, J., Winkelman, J., Bates, D.W.: Improving response to critical laboratory results with automation: results of a randomized controlled trial. Journal of the American Medical Informatics Association 6, 512–522 (1999)

[17] Goldberg, H.S., Vashevko, M., Pastilnik, A., Smith, K., Plaks, N., Blumenfeld, B.M.: Evaluation of a commercial rule engine as a basis for a clinical decision support service. In: AMIA... Annual Symposium Proceedings / AMIA Symposium. AMIA Symposium, pp. 294–298 (2006)

National Medical Terminology Server in Korea

Sungin Lee, Seung-Jae Song, SoonJeong Koh, Soo Kyoung Lee, and Hong-gee Kim

Information Knowledge Engineering Laboratory, School of Dentistry,
Seoul National University, Seoul, Korea
{sunginlee,kernel}@snu.ac.kr, smilekoh@gmail.com,
soo1005s@gmail.com, hgkim@snu.ac.kr

Abstract. Interoperable EHR (Electronic Health Record) necessitates at least the use of standardized medical terminologies. This paper describes a medical terminology server, LexCare Suite, which houses terminology management applications, such as a terminology editor, and a terminology repository populated with international standard terminology systems such as Systematized Nomenclature of Medicine (SNOMED). The server is to satisfy the needs of quality terminology systems to local primary to tertiary hospitals. Our partner general hospitals have used the server to test its applicability. This paper describes the server and the results of the applicability test.

Keywords: Electronic Health Records, Terminology, Medical Records, Information System, Database, Medical Informatics Application.

1 Introduction

A terminology server [1] is a core constituent of EHR systems that offers flexible integration of clinical information services with authorized access mechanisms to repositories of standardized medical knowledge.

The purpose of the national medical terminology server (LexCare Suite) is to be used as a reference repository to serve local hospitals for their terminology needs. In order to accommodate multiple terminology systems into the server, we needed a data meta-model that would aid in converting and importing of the terminology systems we wanted to use. The model of choice was the LexGrid Model[2]. We adapted the LexEditor to add several novel modules: facet-browsing, new search options, keyword auto-completion, and mapper. The mapper was developed to ease the burden of inter-terminology mapping of concepts by terminologists. In all, the LexCare Suite consists of 1) a LexGrid-based integrated modules, called LexCare Editor, that provide managerial functionalities for medical terminologies, and 2) a terminology repository packaged together with the LexCare Editor.

The Suite has been tested for its usability at two general hospitals (Korea Cancer Center Hospital, and Gachon University Gil Hospital). Part of the test was to see if the Suite could help the hospital terminologists with their daily terminology chores, such as mapping local terms to those from international standard terminology systems such as SNOMED-CT. In order to do that, the hospitals were asked to import their local

T.-h. Kim, A. Stoica, and R.-S. Chang (Eds.): SUComS 2010, CCIS 78, pp. 541–544, 2010.

lists of medical terms into the Suite to check the coverage of the repository, and to see how easily they can perform mapping activities using the mapper.

The remainder of the paper is organized as follows: Section 2 gives the system architecture of the LexCare Suite. Section 3 presents the components of the Suite. Section 4 provides concluding comments.

2 System Architecture

The LexCare Suite is part of a larger initiative for developing infrastructure for interoperable HER as shown in Fig. 1, which shows a high-level architectural scheme for interoperable HER and the role of the LexCare Suite with other components.

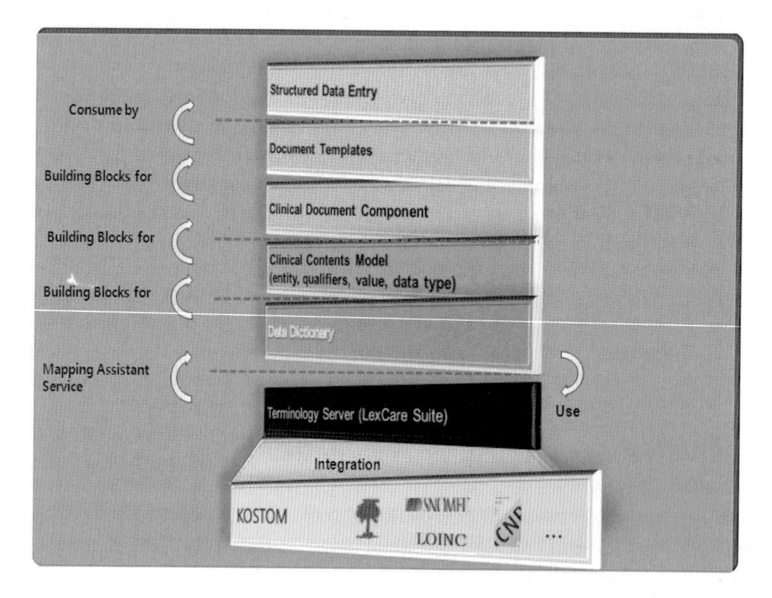

Fig. 1. Interoperable EHR infrastructure

The LexCare Suite houses eight terminology systems: ICNP, ICD-10, RxNORM, SNOMED-CT, LOINC, KOSTOM (Korean equivalent of UMLS), KCD-5 (Korean equivalent of ICD-10) and a locally developed data dictionary called CiDD, as show on the left-hand pane of Fig. 2.

3 LexCare Suite Components

There are two main components in the Suite: LexCare Editor and Terminology Repository. Most of the functionalities of the original LexGrid Editor remain in the LexCare Editor. The most important extensions are the concept mapper and new search and browse modalities.

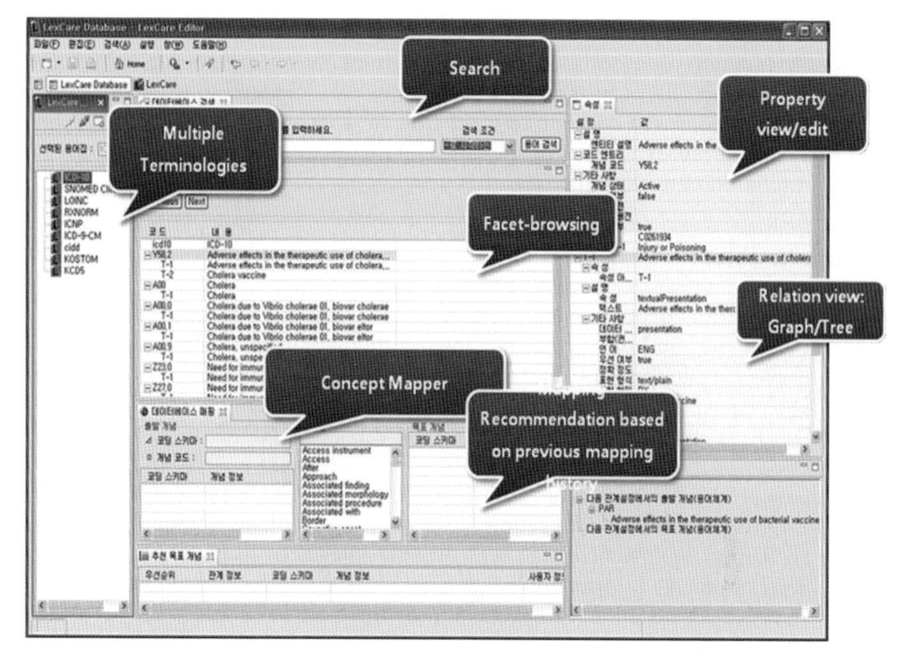

Fig. 2. LexCare Suite

3.1 Concept Mapper

The LexCare Editor is incorporated with a concept mapper that matches two (sets of) concepts or terms, which may originate from different terminology systems (that is, the mapper support 1:1, 1:N, and N:N concept mapping). Our concept mapper, shown in Fig. 2, is an in-house developed module.

Local terminology specialists in local hospitals have great demand for an easy-to-use, (semi-) automatic, semantics-enabled concept mapper, which will facilitate their day-to-day terminology related tasks. Our approach to this demand is to use collective intelligence, which in our case, supports collection and storage of previous mapping activities of terminology specialists and use them to guide users to the most 'correct' concepts for mapping. The historical data of mapping activities are stored at a remote server. And when a user wants to find a matching concept for a source concept, the previously matched target concepts are automatically loaded, with the activities with higher weights showing up first, for the user to choose. Each mapping activity is given a weight based on several parameters: activity recency, organization he/she works for, his/her job title as terminology specialist, etc. We believe that, with a reasonable amount of mapping activities recorded and with enough resources such as time and users, this will work like an automatic concept mapper.

3.2 Search and Browse Modes

Most of the search and browse modalities of the LexGrid Editor remain in the LexCare Editor. In order to accommodate local user requirements, however, we have added 1) Korean-based search; 2) more search options such as "start with", "end with", and "start and end with" the keyword entered; and 3) keyword auto-completion function. In terms of browsing, a new browsing mode, called facet browsing, is added. Facet-browsing is offered in two modes: concept or relation. When the user chooses a concept or a relation in a terminology in facet browsing mode, the system retrieves a list of the concepts or relations that contain the chosen concept or relation in the terminology.

4 Conclusions

The LexCare Suite is mature enough in terms of its contents and functionalities to be applied in local hospital environments. Two participating hospitals have tested the Suite in terms of its applicability and usability within their hospitals.

The XML-based source data has limit system performance, specifically when SNOMED-CT is loaded in the Suite. At present, RxNORM, ICD-10, ICNP, SNOMED-CT, LOINC, and CiDD have been converted into MySQL databases in accordance with the LexGrid database schema, except for KOSTOM. The Suite is at the final stage of source code revamping to accept the new data source format. Web services are being implemented that will provide interfaces to external applications such as LIS (Laboratory Information Systems) and local clinical data dictionaries

As a technical application of practical use, the LexCare Suite has become an important technical advancement and testing ground for a national terminology server that aims to services the medical terminology needs of local primary to tertiary hospitals, public and private clinics. The LexGrid model has been used in other contexts including NCI's Cancer Biomedical Informatics Grid(caBIG[3], https://cabig.nci.nih.gov/), the National Center for Biomedical Ontology(NCBO, http://www.bioontology.org), UK CancerGrid[4]; the LexCare Suite, however, appears to be the first attempt to use the LexGrid model and its accompanying editor, as the generic terminology server model for the purpose of servicing the terminology needs of all the hospitals in a nation. The LexCare Suite is by no means a complete endeavor, and further collaboration with terminology specialists and ensuing customizations to the Suite is expected – specifically, support for post-coordination of concepts.

Acknowledgments. The work is supported fully by the Ministry of Health and Welfare, South Korea, under Ontology-Based EHR Interoperability Development Project A05-0909-A8-405-06A2-15010A.

References

1. Rector, A., Solomon, W., Nowlan, W., Rush, T., Zanstra, P., Claassen, W.: A terminology server for medical language and medical information systems. Methods of Information in Medicine-Methodik der Information in der Medizin 34(1), 147–157 (1995)
2. Pathak, J., Solbrig, H., Buntrock, J., Johnson, T., Chute, C.: LexGrid: A framework for representing, storing, and querying biomedical terminologies from simple to sublime. Journal of the American Medical Informatics Association 16(3), 305–315 (2009)
3. Von Eschenbach, A., Buetow, K.: Cancer Informatics Vision: caBIG^{TM}. Cancer Informatics 2, 22 (2006)
4. Reddington, F., Wilkinson, J., Clark, R., Parkinson, H., Kerr, P., Begent, R.: Cancer informatics in the UK: the NCRI Informatics Initiative. Cancer Informatics 2, 389 (2006)

A Typology for Modeling Processes in Clinical Guidelines and Protocols

Samson W. Tu and Mark A. Musen

Stanford Center of Biomedical Informatics Informatics, Stanford University
Stanford, CA 94305-5479, USA

Abstract. We analyzed the graphical representations that are used by various guideline-modeling methods to express process information embodied in clinical guidelines and protocols. From this analysis, we distilled four modeling formalisms and the processes they typically model: (1) flowcharts for capturing problem-solving processes, (2) disease-state maps that link decision points in managing patient problems over time, (3) plans that specify sequences of activities that contribute toward a goal, (4) workflow specifications that model care processes in an organization. We characterized the four approaches and showed that each captures some aspect of what a guideline may specify. We believe that a general guideline-modeling system must provide explicit representation for each type of process.

1 Introduction

In recent years, professional organizations, government agencies, and health-care institutions have promoted clinical guidelines to reduce practice variations, contain costs, and improve quality of care. Most guidelines are published in narrative formats supplemented by diagrams depicting clinical algorithms. The diagrams organize the narrative in a form that allows clinicians to quickly grasp guideline recommendations. Similarly, in clinical trial protocols, schema diagrams sketch the broad outline of a study.

To provide of guideline-based computerized decision support, the medical knowledge embodied in guidelines must be formalized in computer models. The models must applicable to coded patient data to generate patient-specific recommendations. A number of guideline models that attempt to formalize clinical practice guidelines use diagrammatical methods for depicting control-of-flow relationships among recommendations [2-8]. The assumption is that a diagrammatic representation can be a bridge between informal narrative guidelines and a formal computable representation useful to clinicians not trained in computer science. Most guideline modelers, however, find that flowcharts cannot be translated easily into computable formalisms. Often the diagrams don't represent actual steps, but an idealized view. The diagrams may contain gratuitous sequencing of steps which would impede real implementation, but not the quick grasp of the gestalt of the guideline. Furthermore, often it is not clear what process is being depicted in diagrams.

T.-h. Kim, A. Stoica, and R.-S. Chang (Eds.): SUComS 2010, CCIS 78, pp. 545–553, 2010.

This paper clarifies the processes that we are trying to capture in our diagrammatic formalisms and characterize their properties. After studying several guideline-modeling methodologies and drawing upon our experiences in modeling guidelines and protocols, we developed a typology of four stereotypical processes modeled by these methodologies. This typology allowed us to analyze different intended uses of diagrams in narrative guidelines. It informs guideline developers of the distinctions that they should make explicit. Finally, it suggests that developers of guideline-modeling formalisms need to accommodate these processes in their models.

2 Methods

In this study, we defined a *process description* to mean a specification of a collection of activities occurring over notional or real time. Thus, an algorithm may define a step-by-step problem-solving process that takes place mentally, while a care plan may describe a set of activities to be performed by health-care providers throughout a patient's hospitalization. Activities may be performed sequentially or concurrently, and performing them may depend on certain conditions being true.

We analyzed the diagramming formalism proposed by the Society for Medical Decision Making [1] and formalisms of a number of different guideline modeling methodologies. They included EON [6], an architecture for creating guideline-based decision-support system developed at Stanford University; Asbru [9], a task-based guideline modeling and execution architecture jointly developed in Austria and Israel; GUIDE [2], a guideline modeling methodology developed at University of Pavia; PRODIGY3 [3], the third phase of the University of Newcastle's primary-care guideline modeling and execution system; PRO*forma* a guideline-modeling system developed at Imperial Cancer Research Fund; GLIF3 [5, 10], a product of the Intermed Collaboratory, which is a joint project of researchers at Columbia, Stanford, and Harvard Universities; and SAGE (**S**tandards-Based **A**ctive **G**uideline **E**nvironment) [7] [8], the work of a consortium consisting of medical informatics groups at GE Healthcare, the University of Nebraska Medical Center, Intermountain Health Care, Apelon, Inc., Stanford University, and the Mayo Clinic. We also drew upon our experiences in modeling guidelines and protocols.

3 Results

This paper proposes a typology of four modeling formalisms and the processes they typically model: (1) flowcharts for capturing problem-solving processes, (2) disease-state maps that link decision points in managing patient problems over time, (3) plans that specify sequences of activities that contribute toward a goal, and (4) workflow specifications that model care processes in an organization.

The four formalisms in our typology may overlap in the processes they model. A plan of actions, for example, can be mapped to a workflow process when its actions are assigned to healthcare providers playing different roles. A flowchart may depict decisions and actions that are performed over time. In that case, a flowchart takes on

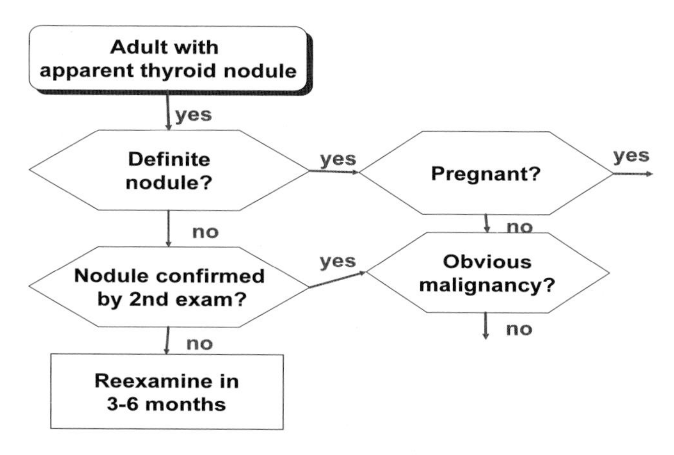

Fig. 1. Part of a flowchart for depicting an algorithm for managing patients with thyroid problems. (Adapted from first figure of [1])

the character of a disease-state map as defined here. Nevertheless, we believe that it is useful to distinguish these formalisms and the processes they model in conceptually distinct stereotypical forms.

3.1 Flowcharts

The Society for Medical Decision Making (SMDM) has published a proposal for standardizing the notation for encoding clinical algorithms [1]. In that proposal, a clinical algorithm is a step-by-step procedure for solving a problem that contains conditional logic. Elements of the procedure consist of clinical states, binary decisions that can be formulated as questions with yes/no answers, and sets of actions that should be performed (Figure 1). The proposal outlines a diagramming convention for clinical algorithms of this type. Shadowed boxes represent clinical states; hexagonal boxes represent decisions; and rectangular boxes represent actions. Arrows connecting boxes are labeled "yes" or "no." Link boxes, represented by ovals, connect regions of the algorithm that may be separated by page breaks or intervening boxes and arrows. Textual annotations can be associated with each node in the algorithm.

As these conventions indicate, this diagramming formalism is designed for paper-based dissemination of guidelines. Its conception of an algorithm consists of a rigid sequence of if-then-else statements that impose a specific ordering. It is limited to a simple decision model of yes/no questions. Nuances of decision making are represented as textual annotations. As a formalism for describing a sequence of decisions and actions that should be taken to solve a problem, the SMDM formalism has the advantage of simplicity and apparent clarity. Furthermore, the choice of decision variables and the sequence of decisions may reflect the sensitivity, specificity, and predictive power of the variables to classify patients into subpopulations for which a particular management strategy is appropriate [11] [Because flowchart is designed for human understanding, an optimal sequence of decision variables may allow a clinician to skip steps.] A trained clinician, with the

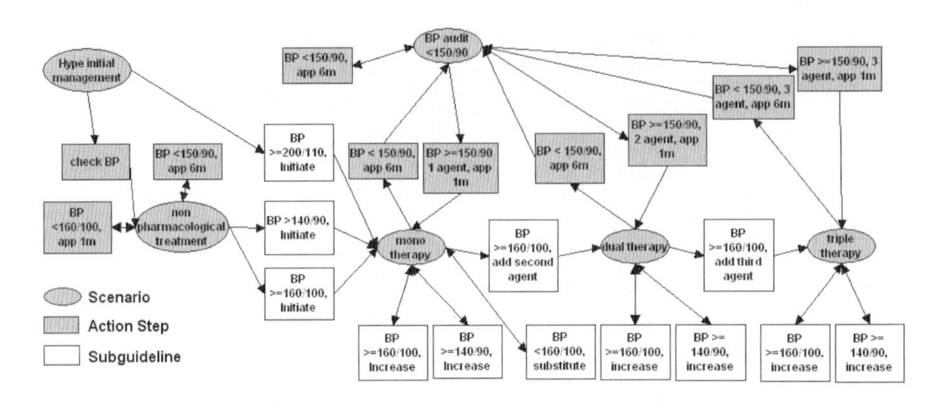

Fig. 2. Part of a PRODIGY3 hypertension management guideline. (Adapted from Figure 1 of [3]

aid of textual annotations associated with each node, can interpret the intent of the algorithm and adjust for the occasional arbitrary ordering of the steps imposed by its linear sequencing requirement. However, a naïve computer-based implementation of the algorithm following this rigid sequencing of steps in its interactions with clinicians is unlikely to be acceptable. Such sequencing may differ from a clinician's own way of thinking and may take up valuable time unnecessarily.

3.2 Disease-State Maps

The PRODIGY3 guideline model presents a very different approach for modeling the flow of decision making in clinical guidelines. In the third phase of a project supported by the UK National Health Service to provide clinical decision support in primary care, PRODIGY3 developers designed their formalism to model chronic disease management guidelines. In this approach, the flow of a guideline is defined by a *disease-state map* that enumerates a set of patient *scenarios* requiring decisions and management alternatives available at the decision points (Figure 2). Scenarios in a disease-state map represent combinations of disease states and treatment settings (e.g., hypertension treated with monotherapy). In each scenario, a clinician has a choice of management alternatives. Each alternative consists of a set of actions, such as adding an additional anti-hypertensive agent to a patient's drug regimen or increasing the dose of a prescribed medication. A decision model that ranks the alternatives according to an argumentation structure aids the choice among alternatives. This argumentation structure consists of absolute and relative rule-in criteria and absolute and relative rule-out criteria.

Each management alternative can be refined into more specific decision-making processes. For example, the top-level choice may suggest prescribing an additional anti-hypertensive agent. The choice is refined into detailed decisions regarding particular agents based on a patient's current medication and on the patient's co-morbidities.

In the disease-state map approach, choosing a management strategy and acting on it lead to a new patient scenario. Thus, the set of all scenarios and management strategies form a transition network in which a patient's location changes over time. Unlike a flowchart that emphasizes modeling simple decisions embedded in a problem-solving procedure, the disease-state map explicitly models the aggregate process of making decisions over multiple clinician/patient encounters. The model defines a notional process in which a patient may start from a diagnosis state and progress through a sequence of clinical states and management strategies represented by the disease-state map. For example, if the patient's blood pressure is not controlled, she may progress to the state of being prescribed one or more antihypertensive medications. However, there is no presumption that a patient necessarily start from a particular start node, nor is there necessarily a terminal node in the process description.

In PRODIGY3, the tasks of information management—acquiring and displaying information relevant for decision-making—are modeled separately in scenario-specific *consultation templates* that specify conditional actions. These actions are applicable before the provider makes a choice among treatment alternatives.

3.3 Plans

In contrast to PROIDGY3, where modeling decision-making in alternative scenarios is paramount, Asbru [9] adopts a model of processes for guidelines that emphasizes the flow of activities. Inspired by artificial intelligence work on planning, Asbru considers guidelines to be *skeletal plans* that are recursively refined into subplans. Plans are annotated by *intentions* (temporal patterns of actions and patient states to be achieved, maintained, or avoided). They also have set-up preconditions that should be achieved, filter preconditions that must be true for the plan to be executed, and post-conditions that hold after a plan is executed. Furthermore, they have activate, complete, suspend, and abort conditions that manage the state of plan execution. The plan is *skeletal* because the partial ordering among its subcomponents is pre-specified. Substituting subplans is allowed if they have effects similar to the original subplan. Each plan can be recursively refined.

The planning approach does not model patient scenarios or complex decision making based on an argumentation structure. Plan selection can be based on satisfying filter and setup conditions, on matching the effects of a plan to target states, or on other parameters, such as required resources, the duration, and plan complexity [12]. Instead of modeling decision-making, Asbru's planning approach models guideline objectives in detail. It also models temporal constraints on actions to be performed, and various organization of subplan (sequential, parallel, any-order, and cyclic) that are used to achieve plan objectives.

3.4 Workflow Specifications

A fourth approach to describing processes in guidelines and protocols is exemplified by the University of Pavia's guideline-based *careflow systems* [2]. In this approach, clinical guidelines provide the workflow process logic that, when combined with organizational knowledge, allows the use of workflow management systems to implement guideline-based patient workflow. Workflow management is concerned with automating procedures in which documents, information, or tasks are passed

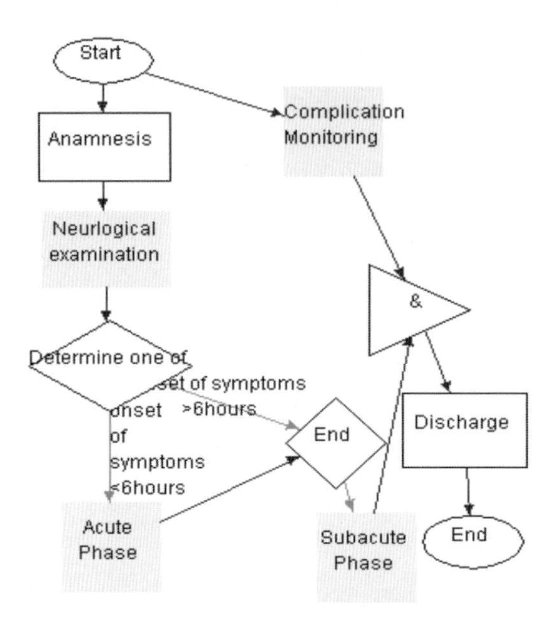

Fig. 3. Workflow-oriented modeling of an ischemic stroke guideline. Clear rectangles are atomic tasks; shaded rectangles represent subprocesses; diamonds represent decision nodes, and the triangle means waiting for concurrent processes to converge. (Adapted from Figure 3 of [2]).

between participants according to a set of rules to achieve or contribute to an overall business goal. Workflow management represents business logic in terms of a process model that emphasizes the sequencing of activities, their possible concurrent execution, and roles played by organizational actors.

The Workflow Management Coalition defines a process as a network of activities and their relationships, criteria to indicate the start and termination of the process, and information about individual activities, such as participants, associated IT applications and data, etc. [13]. In this activity-centric view of guidelines, decision-making is one type of activity in the overall context of guideline-based patient management. Thus, a guideline for ischemic stroke modeled by the Pavia group is characterized by a process that includes concurrent monitoring of complications and a sequence of neurological examinations and treatments for acute and subacute phases of the stroke. The two concurrent processes must end before a patient is discharged from the hospital. (Figure 3).

Like the planning approach exemplified by the Asbru language, the workflow model allows guideline authors to encode the timing and conditions for scheduling activities which may be concurrent. While the workflow approach emphasizes the coordination of participants in the healthcare process, it provides no facility for explicitly modeling the goals, intentions, and expected effects of activities.

3.5 Hybrid or Integrated Approaches

Most modeling methodologies take a hybrid approach to modeling processes in guidelines and protocols. At *indeterministic-decision* nodes, Pavia's guideline model

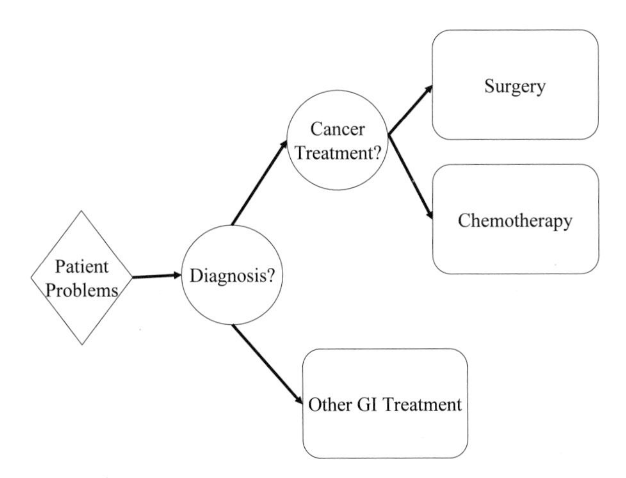

Fig. 4. A PRO*forma* task network that starts with an enquiry into patient problems, successive decisions to determine a diagnosis and, if a patient has cancer, the treatment plan. (Adapted from Figure 37 of [1).

specifies a URL through which a decision tree or an influence diagram can be invoked to assist a clinician. EON and GLIF3 incorporate the scenario and decision model of PRODIGY3. In addition, they provide a *Branch Step* that allows concurrent or unordered actions and decisions, and a *Synchronization Step* that provides the basis for controlling continuation of the thread of execution when multiple sequences of actions converge. EON goes further to associate goals with guidelines and subguidelines and to drive decision-making based on satisfaction of these goals. EON also introduces the notion of specializing modeling primitives for different classes of guidelines and protocols [6]. Thus, it can simulate PRODIGY3 disease-state maps by creating a specialized guideline model that uses only the scenario and decision-making primitives in its clinical algorithm. It can also simulate the workflow-oriented model by augmenting its guideline model with an organization model that assigns care activities to roles played by agents in the organization.

Like the Pavia group's careflow methodology, PRO*forma* [4] integrates explicit models for decision making and task scheduling that informs the diagramming conventions used by a guideline modeler. Instead of using a workflow process model as the foundation, the PRO*forma* approach extends the for-and-against argumentation structure of the Oxford System of Medicine [14] to include clinical tasks executed over time in a co-ordinated way. It uses an ontology of standard tasks consisting of decisions, actions, enquiries (data requests), and plans (recursive decomposition of tasks). Each task in PRO*forma* can have a goal, a precondition and a postcondition similar to Asbru plans. *Scheduling constraints* specify the execution order of the tasks (Figure 4). Thus, the graphical depiction of PRO*forma* tasks has the semantics of a constraint graph that is similar to a skeletal plan. At the same time, it has an integrated decision model based on an argumentation approach.

The SAGE project [7] [8] integrates decision making and workflow processes in another way. It proposes that the recommendations in a guideline can be organized as *recommendation sets* consisting of either *Activity Graphs* that represent guideline-directed care processes or *Decision Maps* that represent recommendations involving decisions at a time point. An Activity Graph is a specialization of the workflow process model for specifying clinical and computational activities designed to implement guideline recommendations. Activity Graphs represents related sets of actions and decisions as recommended in a particular clinical and organizational context. A Decision Map is a collection of recommendations that do not need to be organized and executed as a process. One use of the Decision Map is the encoding of a collection of asynchronous alerts and reminders that are not organized as a connected process of activities. Alternatively, a Decision Map may be used as the decomposition of a high-level action or as the workflow-independent decision making module that generates medically appropriate conclusions that are used to generate workflow-appropriate clinical information system interventions. For example, in encoding the comprehensive recommendations for immunization issued by the Center for Disease Control [15], Decision Maps generate the due dates of appropriate immunizations for a given patient. Two Activity Graphs, one encoding the workflow process for a primary care encounter and the other the periodic EHR monitoring of patient panels, invoke the Decision Maps and use their conclusions to alert primary care nursing staff of immunizations that are due and to generate summary message for primary care physicians respectively.

4 Discussion

In this paper, we report the result of analyzing guideline-modeling methodologies to create a typology of four kinds of processes that guidelines depict. We argue that a guideline may describe four semantically distinct entities, namely an algorithmic problem-solving process; a map of disease and therapeutic states for which a guideline suggests considerations in making management decisions; a care plan that achieve certain goals; and a workflow process that coordinates multi-providers patient care.

A guideline algorithm may give recommendations on problem-solving and decision-making for an end user. Another may give recommendations on the managing patients in a collection of clinical scenarios. A third guideline may specify treatment goals for different classes of patients, and a fourth one may describe a care plan that involves communication and coordination of care providers. Typically a guideline will contain elements from all four process models, with different emphasis on each depending on its target audience. We see the central roles that decision-making and activity coordination play in the guidelines. We see a variety of decision models: (1) a sequence of Boolean decision variables that end in recommended actions, (2) a for/against argumentation structure that weighs the benefits and evidences of a treatment strategy, and (3) decision-theoretic models that take patient preferences into account. We see the importance of modeling goals and care plan that may require coordination of multiple care providers. The challenge for developers who wish to create computer-interpretable guidelines is to create a formalism that allows these concepts and processes to be modeled coherently and intuitively.

Acknowledgments

The authors wish to thank Mor Peleg for her valuable comments and Valerie Natale for her editorial assistance. This work has been supported by the U.S. Department of Commerce, National Institute of Standards and Technology, Advanced Technology Program, Cooperative Agreement #70NANB1H3049, by grant LM05708 from the National Library of Medicine, by DARPA contract #N66001-94-D6052, a contract supported by NCI, and by grant LM06594 with support from the Department of the Army, Agency for Healthcare Research and Quality, and the National Library of Medicine.

References

1. Society for Medical Decision Making(SMDM) Committee on Standardization of Clinical Algorithms (CSCA), Proposal for Clinical Algorithm Standards. Medical Decision Making 12(2), 149–154 (1992)
2. Quaglini, S., Stefanelli, M., Cavallini, A., et al.: Guideline-Based Careflow Systems. Artif. Intell. Med. 5(22), 5–22 (2000)
3. Johnson, P.D., Tu, S., Booth, N., et al.: Using scenarios in chronic disease management guidelines for primary care. In: Proc. AMIA Symp., pp. 389–393 (2000); 2244127
4. Fox, J., Das, S.K.: Safe and Sound. MIT Press, Cambridge (2000)
5. Peleg, M., Boxwala, A.A., Tu, S.W., et al.: The InterMed Approach to Sharable Computer-Interpretable Guidelines: A Review. J. Am. Med. Inform. Assoc. 11(1), 1–10 (2004)
6. Tu, S.W., Musen, M.A.: A flexible approach to guideline modeling. In: Proc. AMIA Symp., pp. 420–424 (1999); 2232509
7. Tu, S.W., Campbell, J., Musen, M.A.: The structure of guideline recommendations: a synthesis. In: AMIA Annu. Symp. Proc., pp. 679–683 (2003); PMC1480008
8. Tu, S.W., Campbell, J.R., Glasgow, J., et al.: The SAGE Guideline Model: achievements and overview. J. Am. Med. Inform. Assoc. 14(5), 589–598 (2007)
9. Miksch, S., Shahar, Y., Johnson, P.: Asbru: A Task-Specific, Intention-Based, and Time-Oriented Language for Representing Skeletal Plans. In: 7th Workshop on Knowledge Engineering: Methods & Languages (KEML 1997). Milton Keynes, UK (1997)
10. Boxwala, A.A., Peleg, M., Tu, S., et al.: GLIF3: a representation format for sharable computer-interpretable clinical practice guidelines. J. Biomed. Inform. 37(3), 147–161 (2004)
11. Hadorn, D.C.: Use of algorithms in clinical guideline development, in Clinical Practice Guideline Development: Methodology Perspectives, pp. 93–104. Agency for Health Care Policy and Research, Rockville (1995)
12. Seyfang, A., Kosara, R., Miksch, S.: Asbru Reference Manual: Asbru Version 7.3 (January 2002), http://www.ifs.tuwien.ac.at/asgaard/asbru/asbru_7_3/asbru_7.3_reference.pdf
13. Workflow Management Coalition, Interface 1: Process Definition Interchange, Process Model. Workflow Management Coalition: Lighthouse Point (1999)
14. Huang, J., Fox, J., Gordon, C., et al.: Symbolic decision support in medical care. Artif. Intell. Med. 5(5), 415–430 (1993)
15. Kroger, A.T., Atkinson, W.L., Marcuse, E.K., et al.: General recommendations on immunization: recommendations of the Advisory Committee on Immunization Practices (ACIP). MMWR Recomm. Rep. 55(RR-15), 1–48 (2006)

How Process Helps You in Developing a High Quality Medical Information System

Yoshihiro Akiyama

Next Process Institute Ltd., 102-3-3024 Nogawa Miyamae Kawasaki Kanagawa 216-0001 Japan
y.akiyama@next-process.com

Abstract. A medical information system is one extreme in using tacit knowledge that patricians and medical experts such as medical doctors use a lot but the knowledge may include a lot of experience information and be not explicitly formulated or implied. This is simply different from other discipline areas such as embedded engineering systems. Developing a mechanical system critically depends on how effectively such various knowledge is organized and integrated in implementing a system. As such, the development process that customers, management, engineers, and teams are involved must be evaluated from this view point. Existence of tacit knowledge may not be sensed well enough at project beginning, however it is necessary for project success. This paper describes the problems and how the Personal Software Process (PSP[1]) and Team Software Process (TSP[2]) manage this problem and then typical performance results are discussed. It may be said that PSP individual and TSP team are CMMI[2] level 4 units respectively.

Keywords: Knowledge Management, Systems Engineering, Conceptual Integrity, Personal Software Process, Team Software Process, High quality software development process, self-directed team, CMMI.

1 Introduction

A medical information system is one extreme of the systems to be developed in that it needs extensively integration of tacit knowledge with experiences. Medical patients, doctors, nurses, technical and administrative staffs, and engineers and development teams need to deal with such tacit but (probably) unique to them. A big question is how these unique knowledge and experiences are integrated into a system to be developed.

The knowledge (and experiences) learned by a medical doctor may be formulated sufficiently but they may not be shared for such a system development easily. This difficulty become more serious when different medical doctors have different

---

[1] Personal Software Process, PSP, Team Software Process, and TSP are the services marks of Carnegie Mellon University.

[2] CMMI, Capability Maturity Model, CMM, and Carnegie Mellon are registered in the U.S. Patent & Trademark Office by Carnegie Mellon University.

T.-h. Kim, A. Stoica, and R.-S. Chang (Eds.): SUComS 2010, CCIS 78, pp. 554–559, 2010.

opinions. In addition, engineers and project teams work with such medical doctors, nurses, and other staffs and find a large variation in what they say and present.

In this paper, we describe first what problems are associated with a medical system development with tacit knowledge. Then we overview how such key problems could be overcome based on the PSP and TSP which are the SEI's process technology for the last 15 years to make them usable for real use.

2 The Problem

Knowledge (and experiences) may not be fully collected when starting development of a system. For example, a formulated (or stated) requirement should be either explicit or implied (not described but can be formulated) and should describe what it means or covers. However, there exists the third form of requirement, called "tacit" requirement which is neither stated nor implied. The tacit knowledge is triggered when a new requirement, new system design, new implementation, and new testing method etc. are need to be identified and required to achieve in system development, where specialized and talented professionals need to work on. Extensive interviewing with customers, for example, may not produce the tacit requirement easily while the explicit or implied requirements may be possible.

All tacit requirements must be revealed and identified to formulate the system to be developed. It is very clear that the development process that does not include the mechanism of revealing tacit knowledge may lead project failure.

Key tacit knowledge should be revealed at the planning phase so that a better project plan can be developed at the beginning. In addition, some tacit knowledge should be revealed during the project execution which may require updating the project plan, team member work assignment, or negotiating with customers as needed. This means that a project may have high risk if the team does not pay attention on and do not work on identifying tacit knowledge during the project lifecycle.

3 The Process

The most important goal for engineers is *the conceptual integrity* which must be satisfied in developing a system [4]. The integrity and completeness on design concepts, quality, and delivery are essential. A question is how you define and achieve the integrity and completeness and how you plan them where the tacit knowledge must be expected for such as a medical system development.

The Personal Software Process (PSP) and Team Software Process (TSP) include such as process mechanisms at the individual engineer level and the team level respectively.

Fig 1 shows the PSP planning framework. After describing the user requirements which are not complete, engineer *creates the conceptual design* which must represent sufficiently the completeness of the project work and objectives such as the components to be developed, reusing, and quality levels.

It is noted that a conceptual design is the result of transforming all *necessary* tacit knowledge into a concrete structure of project product which satisfies the conceptual integrity to the requirements and goals. Then the engineer moves to the estimation and development phases. The conceptual design phase is the time for engineers to identify much tacit knowledge to develop the complete systems concept.

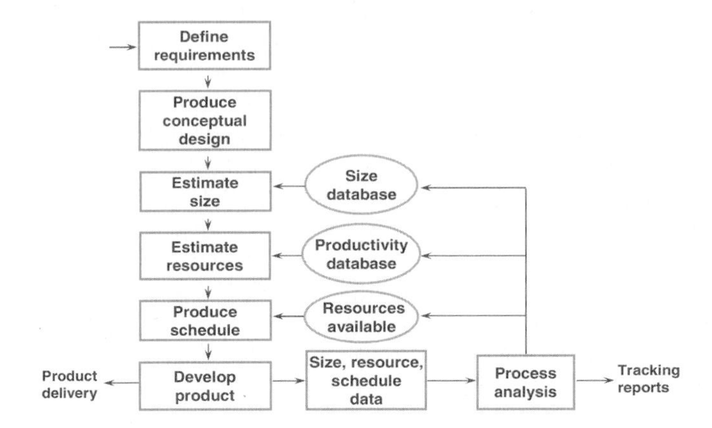

Fig. 1. PSP Planning Framework

Similarly in the TSP, before or at a team launch, the first step is that the customers and the organization's management state (explain) all requirements and goals that they want the team to meet, where the stated requirements and goals are not fully organized, formulated, and consistent. The team members then examine and identify what products must be produced, how they are built, and find how the stated requirements and goals will be satisfied with them. During this exercises, all team members work together, identify tacit knowledge, i.e., many ideas!, where the team creativity is required in identifying those missing requirements, design elements, and implementation throughout the project lifecycle.

It is noted that this revealing tacit knowledge and transforming them to the stated knowledge is driven by the engineer's creativity and professional willingness to stay the "conceptual integrity." This is essentially important for the project success. In order to make this happen, every team member must have been trained by the PSP course to establish such mind and skill before becoming a team member. A team member is required very good communication with customers, management, and team members. Also the team is required to be disciplined, i.e., faithfully follow the process with consistency, completeness, and improvement over and over on what they do and what they make.

4 The Results

The PSP has been provided mid 1990s and the TSP provided for industry or academic use (different the TSP versions) in early 2000s by the Software Engineering Institute USA.

More than 80,000 software engineers have been trained using the PSP and showed excellent performance results. Following is the list of the typical results:

1. Engineers liked transforming tacit knowledge into the conceptual design so that the entire picture of a project work is made explicit. This is a means to ensure the conceptual integrity of the engineering products and the project effort.
2. Setting the goal and making plan of project quality is a new experience for many students. This skill is used to manage the quality improvement. Fig 2 shows the defect density identified at the unit test along with the assignments from 1 through 10. The worst quality performers at the course beginning improved their performance toward the course end. It is impressive that the improved quality performance is better than the best performance at the course beginning. This is by the power of skills that can identify and transform tacit knowledge into the explicit knowledge.
3. Fig 3 shows that the productivity from defining requirements through the unit test (as the PSP provides) stays unchanged or improved even when serious design review and code review process are added to the assignments 7 through 10.
4. Engineers have confidence in using the PSP process to produce high quality work products as planned.

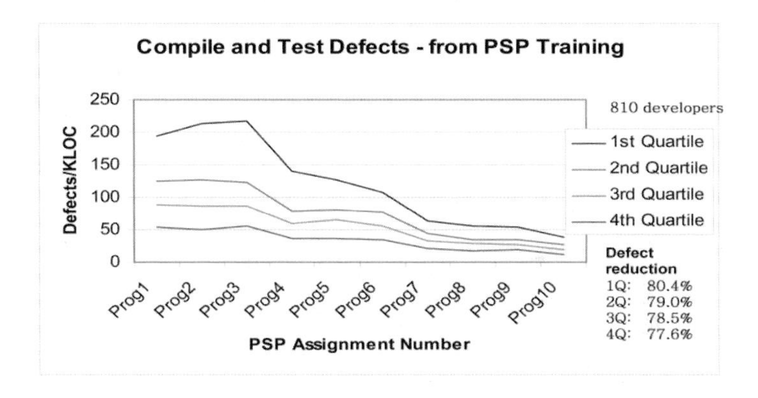

Fig. 2. The Quality Improvement during PSP courses for Engineers course

The TSP has been introduced to industry organizations such as Microsoft, Intuit, Softech, and many other companies. Typical messages of TSP results are as follows:

1. A TSP team accepts the planning session naturally that includes the conceptual design phase through which all team members know what and how to be done regarding the system elements, work to be done, skills needed, the delivery and quality goals, and risk handling.
2. Through these activities, the team members have tendency to contribute to the project at their best effort, which result in strong team building.
3. Every TSP team has weekly meeting and examine any change demanded or desired for short or long term of the project. One example is to manage a late work item by taking over it willingly by other member who completes work early. This works to keep the project on schedule.

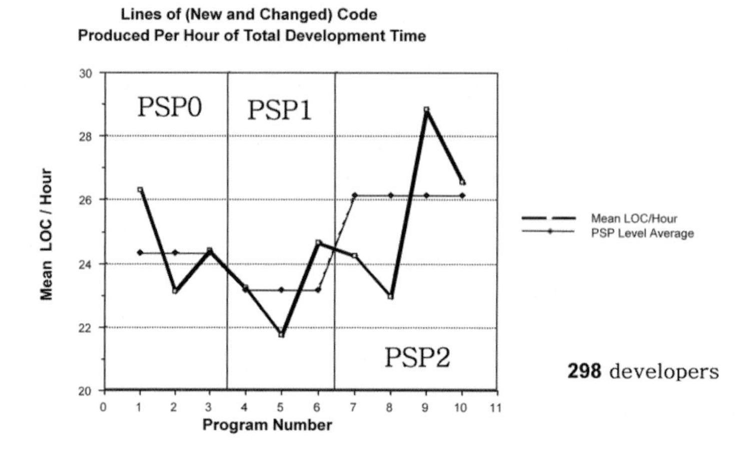

Fig. 3. The Productivity performance during PSP for Engineers courses

4. The schedule accuracy shows a few percent variation to the original estimate.
5. The workload for the integration and system test phases come down drastically.
6. The defect density at the integration and system test phase came down to 10 – 20% of non-TSP projects. The quality has been improved 5 to 10 times.
7. The schedule and quality of a team are tracked and managed with also the individual level progress and quality work, which help the team to identify any problem potentially impacting to the team goals and to solve them in early enough.
8. All team members are happy to work for better excellence together and to achieve the team commitment.

5 Summary

Identifying and transforming tacit knowledge on requirements, design, and implementation in early project phase is critical to the project success and must be addressed during the development lifecycle. This skill and mind of an engineer have been developed and established when it completed the PSP training. Such engineer is the smallest unit with CMMI Level 4 practices.

It is important to reveal tacit knowledge at the project initiation time and to communicate with other stakeholders to make the knowledge formulated and the stated knowledge augmented. The quality of a system is determined by the worst quality component of the system. Such a component is developed by a engineer who might not know or might not be informed the knowledge derived from tacit knowledge necessary to implement the component. This situation must be avoided.

Engineers must be trained by the PSP course because most of current organizational process does not address this yet. Once engineers learn the importance of the the process disciplines and the conceptual integrity on the requirements, design, implementation, and the project plan through the PSP training, they can easily achieve high quality goals. With the requirements and goals stated by customers and management, every engineers and team can do high quality work with the process which includes the conceptual design phase.

References

1. Humphrey, W.S.: PSP – A Self-Improvement Process for Software Engineers. Addison-Wesley, Reading (2005)
2. Humphrey, W.S.: TSP – Leading a Development Team. Addison-Wesley, Reading (2006)
3. Humphrey, W.S.: TSP – Coaching Development Teams. Addison-Wesley, Reading (2006)
4. Brooks Jr., F.P.: The Design of Design. Pearson Education Inc., London (2010)
5. Musson, B.: Deploying Disciplined Methods. In: 21st IEEE-CS CSEET 2008 (2008), `http://www.csc2.ncsu.edu/conferences/cseet/special.php`
6. Fagan, E., Davis, N.: TSP/PSP at Intuit SEI SEPG (2005), `http://www.sei.cmu.edu/library/assets/tsppspatintuit.pdf`
7. De La Maza, A.: TSP Implementation for Outsoruced Application Development Projects. In: SEPG-NA 2010 (2010), `http://www.sei.cmu.edu/sepg/na/2010/index.cfm`

An Implementation Strategy of Evidence-Based Application Lifecycle Management

Jeong Ah Kim[1] and SeungYong Choi[2]

[1] Computer Education Department, 522 NaeKok Dong, KangNung
clara@kd.ac.kr
[2] Computer Engineering Department, 522 NaeKok Dong, KangNung
boromi@gmail.com

Abstract. Application lifecycle management (ALM) facilitate and integrate facilitate and integrate requirements management, architecture, coding, testing, tracking, and release management. Process automation and seamless traceability among tools are getting important. In this paper, we suggest new approach to achieve the seamless traceability and process automation. For these features, knowledge formalization is critical. Evidence-based medicine is good reference architecture to ALM2. Software Engineering ontology is powerful solution to semantic gap between tools and guideline-based process execution environment can be practical solution to process automation.

Keywords: Application Life Cycle Management, Evidence-based software engineering, ontology, guideline.

1 Introduction

Essentially sets of tools encompassing requirement management, architecture, coding, testing, tracking, release management and more, early ALM offerings were promoted as the unified approach to the entire life cycle[1]. At any time, analysts, coders, testers and other participants in the application development process could see what others were doing, and where a project stood—or so the vision went. Which requirements have been coded, tested and released? Which are still works in progress? There are 3 main pillars combining an ALM solution: traceability, process automation, and reporting and analytics. ALM doesn't support specific lifecycle activities; rather, it keeps them all in sync [2].

Early ALM is just set of soiled tools so that it was not easy to share the data and process among the tools and stakeholders. In ALM 1.0 tool was integrated as role-based but roles are anything but uniform, varying by company, by business unit, by development team, and even by individual.

ALM 2.0 is next generation of ALM and it is a platform for the coordination and management of development activities, not a collection of life-cycle tools. Key process automation, integration and traceability between activities and all development departments are key features in ALM 2.0. ALM 2.0 is about transforming tools from being discipline-centric to being role-focused and process-centric. In this

T.-h. Kim, A. Stoica, and R.-S. Chang (Eds.): SUComS 2010, CCIS 78, pp. 560–566, 2010.

perspective, ontology and the guidelines are essential knowledge asset for successful adoption of ALM 2.0.

We can find many approaches of knowledge-based process supporting in evidence-based medicine. Evidence-based medicine (EBM) integrates the best practices and well-validated knowledge with medical information system to give best recommendation or alerting at right time. Knowledge formalism as ontology and guideline is critical elements in EBM. In this paper, we suggest software engineering ontology with UML and software engineering guideline with BPML for knowledge formalism. Also we suggest guideline-based ALM as knowledge execution and process environment

In section 2, we introduce the evidence-based software engineering and evidence-based medicine both. In section 3, we suggest the formalization method for software engineering knowledge. We describe the simple guideline-based ALM for verifying our approach in section 4.

2 Backgrounds

2.1 Evidence-Based Software Engineering

Software Engineers and managers have challenges to select what technologies or methods to adopt on their development projects. They should know the problems of their current practices and the solution to them. EBSE(Evidence-based Software Engineering) aims to improve decision making related to software development and maintenance by integrating current best evidence from research with practical experience and human values [3,4].

EBSE involves five steps: 1) Convert a relevant problem or information need into an answerable question, 2) Search the literature for the best available evidence to answer the question, 3) Critically appraise the evidence for its validity, impact, and applicability, 4) Integrate the appraised evidence with practical experience and the customer's values and circumstances to make decisions about practice, 5) Evaluate performance and seek ways to improve it. EBSE has focused on SPI (Software Process Improvement) and not published specific evidences and not define how to implement these concepts.

2.2 Evidence-Based Medicine: Guideline-Based Decision Supporting

Evidence-based medicine aims to apply the best available evidence gained from the scientific method to medical decision making [5]. Evidence-based guidelines (EBG) are the practice of evidence-based medicine at the organizational or institutional level. Clinical guidelines are "systematically developed statements to assist practitioner and patient decisions about appropriate health care for specific clinical circumstances.[6]" Clinical practice guidelines define recommended strategies for managing health care in specific clinical circumstances. The goal is to reduce practice variation and foster adoption of best practices. For this, in medical information engineering fields, implementing the computer-interpretable guidelines in health care applications is so important. Many research results have achieved in guideline modeling formalism such as GLIF, SAGE so on, in guideline execution architecture as EON, CLEE so on.

The success of computer-interpretable guideline in clinical decision supporting system gives a new direction to implement the evidence-based application lifecycle management

3 How to Formalize the Software Engineering Knowledge

In software engineering fields, there are so many evidences in various styles and there are several information sources we apply. Methodologies, processes, practices, pattern, tactics are the useful evidences to use. Also, SWEBOK[11] is also useful resources since it categorizes the domain of software engineering as 6 KA(Knowledge Areas). Even though, we have many useful resources, they have different format and written in text. Computer cannot interpret these knowledge assets so that it is not easy to share among the tools to be integrated into ALM 2.0.

To make knowledge assets to be computer-interpretable, we need formalize the knowledge. In this paper, we suggest the conceptual framework (Fig 1) to formalize the knowledge of software engineering.

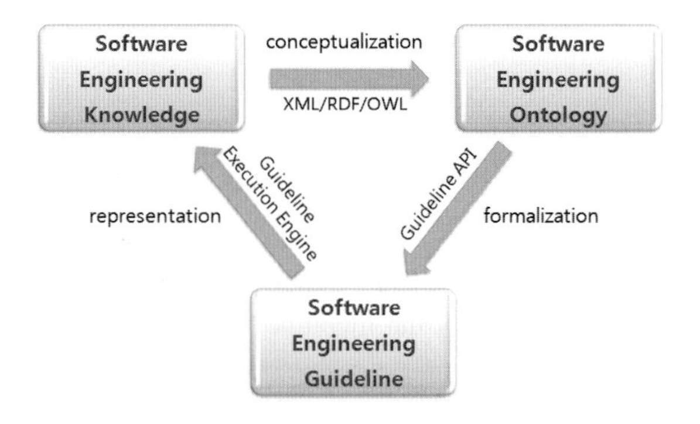

Fig. 1. Conceptual Framework to Formalize the Software Engineering Knowledge

First step to formalize the knowledge is constructing the ontology. Several efforts to make software engineering ontology have been tried [7,8]. Like the approaches in [8], we used object modeling methodology for modeling the SE ontology. Previous research addressed that UML has some limitation in modeling the several properties of ontology such as constraints or restrictions but OCL and several extension mechanism of UML can solve these problems. Especially structural diagram such as class diagram and instance diagram can be useful modeling tools to construct the SE ontology. One more reason to apply this approach, we have several meta model what already formalize the knowledge assets. SPEM (Software Process Engineering Metamodel), MOF(Meta Object facility), Common Warehouse Metamodel (CWMP) from OMG(Object Management Group) and Software Engineering -- Metamodel for Development Methodologies from ISO are examples of metamodel to reuse. With this meta model, we can start to construct the SE ontology in UML and translate into OWL.

Next step is to formalize the guideline in process perspectives. Fig 2 shows our basic conceptual model for software engineering guidelines. The goal of formalizing the software engineering knowledge is to achieve the executable knowledge for supporting the software engineers in the development process. To meet this goal, we apply the business process modeling approach to make a guideline. In medical guideline, workflow or activity-based formalisms are adopted also. In software development context, several roles are involved in one practice or one activity and process management is importance as well as leading into correct direction. Business Process Modeling Language (BPML) is a meta-language for the modeling of business processes, like XML, is a meta-language for the modeling of business process. Also, GEM (Guideline Element Model) [8] is good reference model to define the guideline model.

We define the Meta model for software engineering guideline in Fig 2.

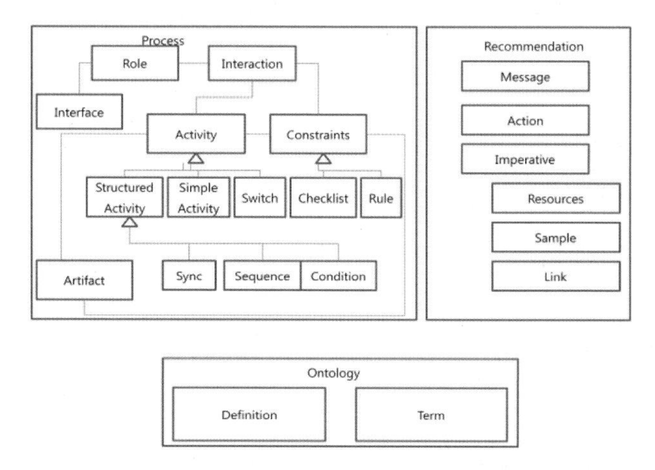

Fig. 2. Metal model for software Engineering guideline (partial)

4 How to Execute the Software Engineering Knowledge

As shown in Fig 3, general ALM architecture defines so many tools to integrate.

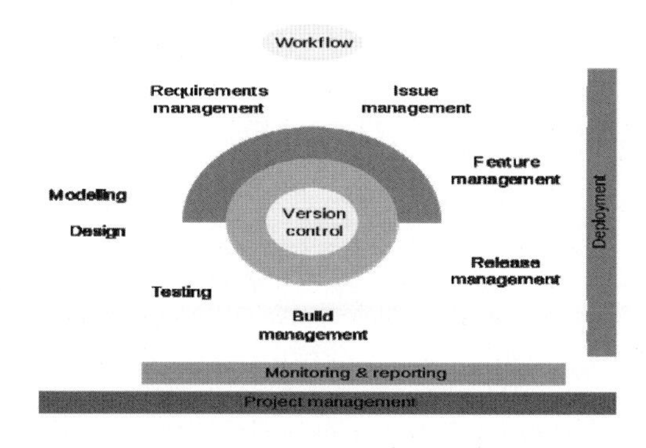

Fig. 3. General ALM Architecture

Each tool should be connected but each has different definitions. This can be solved with software engineering ontology and profiling mechanism. Software Engineering defines the meta model for each knowledge area of software engineering and define standard definition. With software engineering ontology, profile for each tool to integrated should be created. Profile is generic mechanism adapted in UML to provide the extension mechanism.

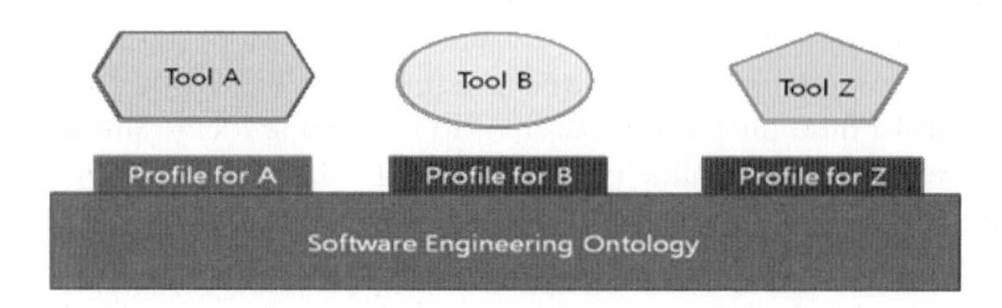

Fig. 4. Mapping from the tool to SE ontology

To execute the guideline, each software engineering guidelines should be modeled in process such as in Fig 5.

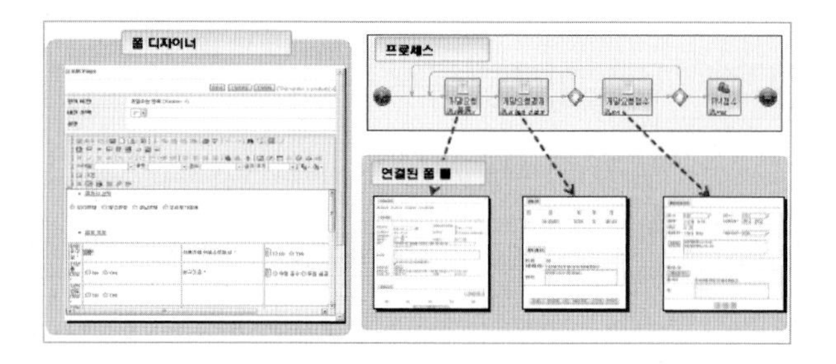

Fig. 5. Process modeling to formalize the guidelines

So far, to verify the approach, we constructed small ALM based on Eclipse. Eclipse is the most popular development environment. Also, Eclipse provides API to integrate with several tools and to customize the tools.

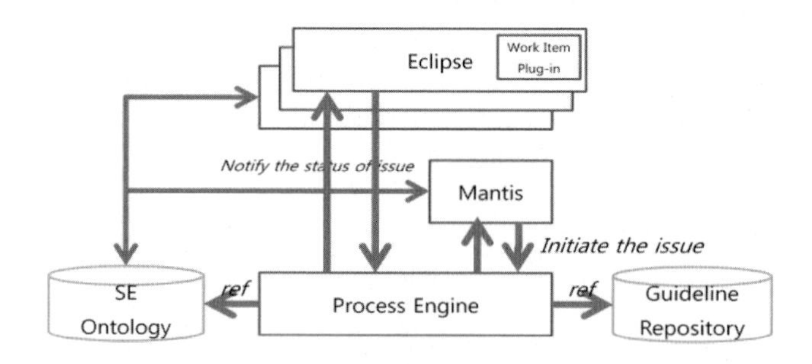

Fig. 6. Simple version of Guideline-based ALM

We tried to integrate the issue management process with Eclipse. As shown in Fig 3, issue management is defined as important element in ALM. Many organizations have their issue tracking system to register new issue, to allocate issue to resources, to monitor the progress of issue. Even though issue management is very critical process to improve the process quality and almost organization have tried to apply tools but the capability of issue management is still low and developers underestimate its importance. Issue management process should be embedded in development project in integrated development environment.

For this, first we define the issue management ontology in UML and define the basic issue management process in BPML as guideline. Next we translate the UML into XML and author the process with process modeling tool. We decided to integrate the issue management system Mantis what is bug tracking system and open-source software. To integrated, we developed work item component for plug-in Eclipse to understand the software engineering ontology. With software engineering guidelines, whenever new issue is registered at Mantis, process instance is created and new work item for assigned resource is displayed in Eclipse. When issue is closed by resource, process instance is removed and change the status of Mantis. When the issue is delayed, process engine perform the activity defined in the guidelines, for example, notification to allocated resources or related roles.

5 Conclusion and Future Direction

Application Lifecycle Management deals with approaches, methodologies and tools for integrated management of all aspects of software development. Its goal is to making software development and delivery more efficient and predictable by providing a highly integrated platform for managing the various activities of the development lifecycle from inception through deployment [10]. Attempts to create such an integrated platform for software development are not new. A number of systems and solutions were developed in the past to address this problem. However, all of them have failed to provide an industrial-strength solution adequately addressing the needs of ALM. But what we need now is systematic and automatic approach of ALM platforms.

In this research, we suggest the implementation strategy of knowledge formalizing and guideline-based process automation for ALM. Formalized knowledge as ontology can provide the semantic bridge among tools or practices. Formalized guideline as process make possible to process automated integration of tools. Also, we provide the sample small ALM integrating Eclipse and issue tracking system.

In the future, we will construct more detail process and method for formalizing the software engineering knowledge. Also concrete architecture for knowledge execution platform for ALM will be studied.

Acknowledgments. This research was supported by National IT Industry Promotion Agency (NIPA) under the program of Software Engineering Technologies Development and Experts Education.

References

1. deJong, J.: Mea culpa, ALM toolmakers say. SDTimes, http://www.sdtimes.com/SearchResult/31952 (retrieved 2008-11-22)
2. Application Lifecycle Management - ALM 2.0 Advantages, http://www.orcanos.com/why_alm_20.htm
3. Kitchenham, B.A., Dybå, T., Jørgensen, M.: Evidence-Based Software Engineering. In: Proc. 26th Int'l Conf. Software Eng. (ICSE 2004), pp. 273–281. IEEE CS Press, Los Alamitos (2004)
4. Kitchenham, B.A., Dybå, T., Jørgensen, M.: Evidence-Based Software engineering for practitioners. IEEE Software (2005)
5. Timmermans, S., Mauck, A.: The promises and pitfalls of evidence-based medicine. Health Aff. 24(1), 18–28 (2005)
6. Field, M.J., Lohr, K.N. (eds.): Clinical practice guidelines: directions for a new program. National Academy Press, Washington (1990)
7. Mendes, O., Abran, A.: Software Engineering Ontology: A Development Methodology in Metrics News, vol. 9 (2004)
8. Wongthongtham, P., Chang, E., Dillon, T., Sommerville, I.: Development of a Software Engineering Ontology for Multisite Software Development. IEEE Transactions On Knowledge And Data Engineering 21(8) (2009)
9. Shiffman, R.N., et al.: GEM: A Proposal for a More Comprehensive Guideline Document Model Using XML. J. Am. Med. Inform. Assoc. 7, 488–498 (2000)
10. Kravchik, M.: Application Lifecycle Management Environments: Past, Present and Future, Thesis of M.Sc. degree in Computer Science, The Open University of Israel (2009)
11. Bourque, P., Dupuis, R.: Guide to the Software Engineering Body of Knowledge: 2004 Version. IEEE, Los Alamitos (2005)

Architecture and Workflow of Medical Knowledge Repository

HyunSook Choi[1], Jeong Ah Kim[2], and InSook Cho[3]

[1] Graduate School of Information & Technology, Sogang University, Seoul, South Korea
[2] College of Computer Education, Kwandong Unifersity, South Korea
[3] Dept. of Nursing, Inha University, Incheon, South Korea
ezhschoi@gmail.com, clara@kd.ac.kr, insook.cho@inha.ac.kr

Abstract. Recently, clinical field builds various forms of computerized medical knowledge and tries to use it efficiently. In general, to build and reuse knowledge easily, it is needed to build a knowledge repository. Especially, the credibility of knowledge is important in clinical domain. This paper proposes methods for supporting it. To perform it systematically, we propose the method of the knowledge management processes. The methods for knowledge management can serve equal quality, usability and credibility of knowledge. Knowledge management methods consist of 2 methods. They are the knowledge management processes and the specification of the management targets. And this paper proposes the requirement of a knowledge repository and the architecture of the knowledge repository.

Keywords: architecture, workflow, process, knowledge repository.

1 Introduction

CDSS team in CiEHR (Center for Interoperable Electronic Healthcare Record in South Korea) has researched into CDS (Clinical Decision Support) System. This research has been being conducted for 5 years from 2006 to 2010. In this research, we made various forms of medical knowledge. But the knowledge has been made in the ad hoc ways, hence the difficulty in the management. When we make a similar knowledge as well as a new knowledge, we should make extra efforts since there is no process. The knowledge made by a person can only be used for that one's purpose. But the other people cannot use the knowledge and should remake the knowledge.

In order to solve the problem, we suggest a method which is building the knowledge repository. The way of digitalizing medical knowledge has already been done through the other researches. And the research using the knowledge has been conducted. Consequently, various forms of knowledge have been made. The researchers recognize the need of the knowledge repository to manage the accumulated knowledge. But there doesn't exist a knowledge repository suitable for medical knowledge while the other fields like business administration already have and use one.

T.-h. Kim, A. Stoica, and R.-S. Chang (Eds.): SUComS 2010, CCIS 78, pp. 567–573, 2010.

Therefore we suggest medical knowledge repository in the way of sharing and reusing medical knowledge. And the knowledge repository satisfies the requirements as follows: medical knowledge repository should have a knowledge management process and the process should suggest a workflow which can orderly fulfill the activities assigned to the role. And it is needed to specify the knowledge, the management target, so as to manage the knowledge systematically. Systematic knowledge management process ensures equal quality and credibility of the knowledge and specifying knowledge makes the knowledge reused easily.

This project has been progressing and will be finished in October 2010. Thus, the validation regarding this research is to be reported after the field research.

This paper is comprised of five sections. Section 1 introduces this paper. Section 2 reviews existing work related to knowledge asset infrastructure. Section 3 explains requirements of knowledge repository. Section 4 presents medical knowledge repository architecture. Section 5 concludes this paper.

2 Related Work

PHS (Partners Healthcare System) researching into Medical knowledge is supporting a lot of hospitals for clinical information system. In addition, knowledge management is being researched for Clinical Decision Support. PHS proposed knowledge management process for the purpose of constant learning and dissemination of knowledge.

The knowledge asset management research, a part of knowledge management process, is the research that studies the process and environment to control the created knowledge efficiently. In this research, PHS suggests the infrastructure for managing knowledge asset and a variety of requirements for collaboration environment for knowledge management.

Also, this research proposes some requirements of ideal knowledge management infrastructure to achieve cost-reduction, fast acquirement of knowledge, care quality and business performance. The requirements are as follows:

- Knowledge reuse via a centralized repository
- Support for content management and collaboration processes
- Support for Knowledge Authoring and Creation Environments

What the requirements mean is the need of specifying the knowledge stored and reused in the repository and the need of process for management, collaboration, authoring and auditing.

If you reuse the knowledge built in the knowledge repository when authoring knowledge, building knowledge becomes easier and faster. Furthermore, if you specify and reuse the knowledge according to the knowledge-authoring level, it becomes easier to reuse according to the phase. Therefore, it is necessary to define and specify the knowledge, i.e. the target of knowledge management, according to the phases.

Also, the process for management and collaboration of knowledge is needed. It is inevitable to collaborate with many workers in order to manage knowledge. It is essential to assign the role to the users participating in the work and assign the work to the role for the collaboration. Furthermore, an assigned work flow needs a certain definition. This leads naturally to the need of the management processes.

3 Requirements of Knowledge Repository

Based on section 2, this paper suggests two concrete methods. The first method is the specification of the knowledge management targets and the second is the definition of knowledge management process. The definition of process for knowledge management supports the usability of the establishment and the management for knowledge and the guarantee of the quality and the safety for knowledge.

3.1 Knowledge Management Process

Knowledge management process is divided into three parts: the definition of the role of the user, the definition of the activity to perform according to the role, the progress management according to the process, which mean knowledge management process should define a role of the user who conduct process, grant the role an activity to perform and manage to execute in accordance with the process.

Users can define four roles in total: Repository Manager managing the entire knowledge repository, Knowledge Engineer authoring and submitting knowledge, Knowledge Manager reviewing the submitted knowledge, Registered User who search for and use the registered knowledge. All the users who use knowledge repository can be assigned a role and perform an activity according to the role. Fig. 1 is the diagram that shows the activities in accordance with the user's role.

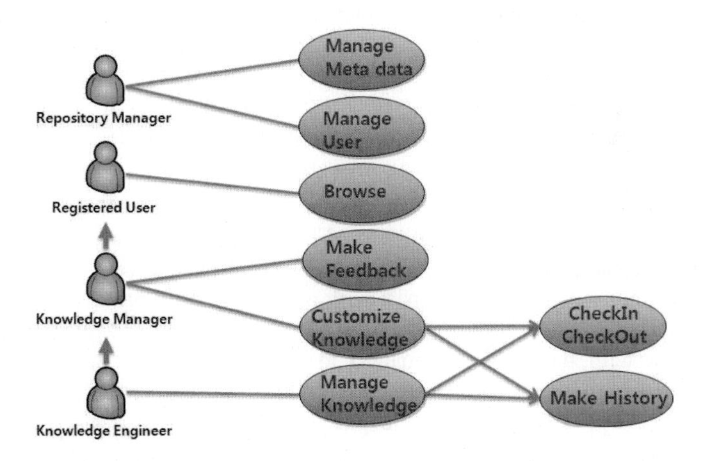

Fig. 1. Knowledge Repository Process Activity for user role

Knowledge repository aids an activity per role in proceeding in sequence and allows Knowledge management to be performed. Fig. 2 describes the workflow of knowledge repository. First of all, Repository Manager makes and registers Metadata. Knowledge Engineer then submits the knowledge made with the knowledge-authoring tools to knowledge repository. Submitted knowledge is reviewed by

Knowledge Manager. If the target knowledge is not adequate, Knowledge Manager gives his feedback to Knowledge Engineer and Knowledge Engineer restart submit process after refining the knowledge. On the contrary, if the target knowledge is proper, the knowledge is approved and registered in knowledge repository. Registered User can thereafter search for and use the knowledge in knowledge repository. All the process mentioned above is controlled by the workflow managers.

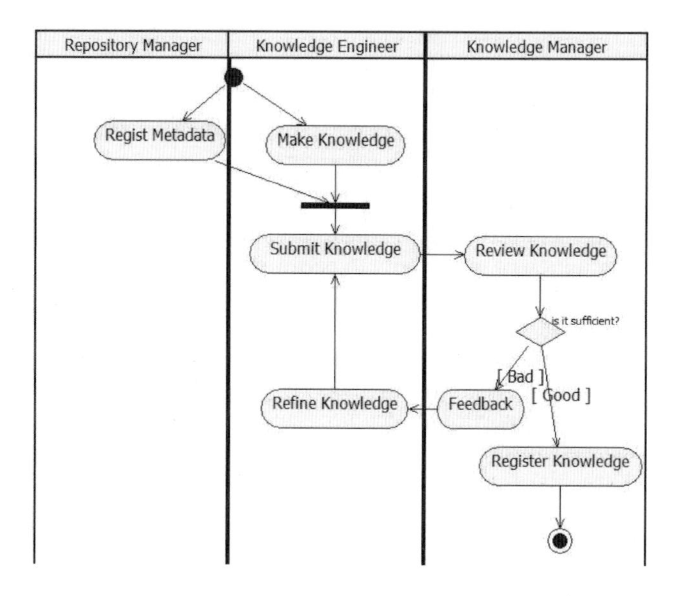

Fig. 2. Knowledge Repository Process Workflow

3.2 Specify Management Target

Management Target, a type or a form of the knowledge which is the target of knowledge management, refers to knowledge object that knowledge management process controls. This research proposes the following structure for the sake of specifying the management targets. Metadata structure is classified into the part with metadata information and the other part with the information based on the metadata. The part of metadata is comprised as follows: metadata schema to define the schema, metadata field with the field information to compose the schema, metadata field constraints to impose restrictions per field, field value to place limitations on the domain of the field input. Information part is made up of the knowledge part to identify the knowledge and the metadata value to have the metadata information of the knowledge. And the categorization of knowledge is comprised of metadata field constraints and field value. The type of metadata field constraints is the definition of how to use the field and the field is able to be chosen from character string, number or categorization (i.e. the form to choose from domains). If the type is categorization, the knowledge authored can be selected to belong to what categorization through inputting the value that can be chosen from the field values when authoring the knowledge.

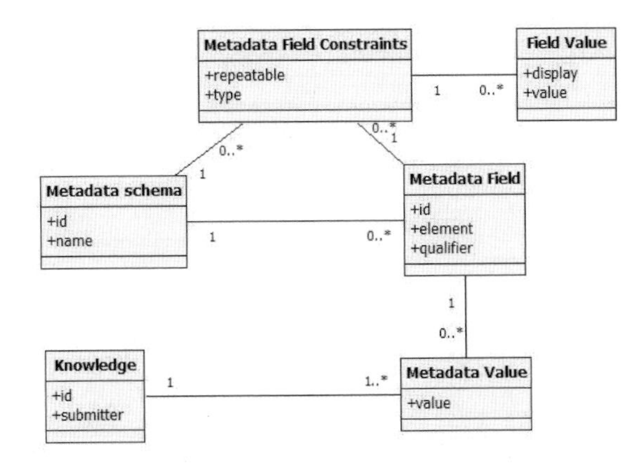

Fig. 3. Metadata structure

4 Knowledge Repository Architecture

Section 4 explains logical and physical architecture of knowledge repository for the purpose of meeting the requirements for knowledge repository.

4.1 Logical Architecture

Logical architecture mentioned here expresses the techniques to achieve logical entity and entity in the form of layered view. Knowledge Repository is comprised of 3 Layers in total: the top layer is Interface Layer in charge of web and interface of application, the middle layer is Business Layer responsible for business logic, the bottom layer is Storage Layer dealing with database and file system. Fig. 4 and Table 1 describe the logical architecture diagram of knowledge repository and the specification for the components of the architecture respectively.

4.2 Physical Architecture

Physical architecture describes the distribution of the main components and package software used. Physical architecture of knowledge repository is made up as follows. The place where the real data of knowledge repository is stored is RDB (Relational Database) and PostgreSQL is used in this research. Database in the architecture is independent from a certain vendor and can therefore be replaced by RDB such as Oracle besides PostgreSQL. Knowledge repository is formed based on Java (JDK1.5). Accordingly, if the circumstances support the JVM, the architecture has the portability that can be performed anywhere. Knowledge Repository provides two user interfaces: web, application (client). Web Based Knowledge Portal, that is to say web interface, provides a service through web browser. Web interface allows users to access to the service easily. CKMS (Clinical Knowledge Management System) Client, an application, furnishes the integrated knowledge management environment in which the knowledge can be authored, tested and executed. Users are able to select the interface in accordance with the purpose and manage the knowledge.

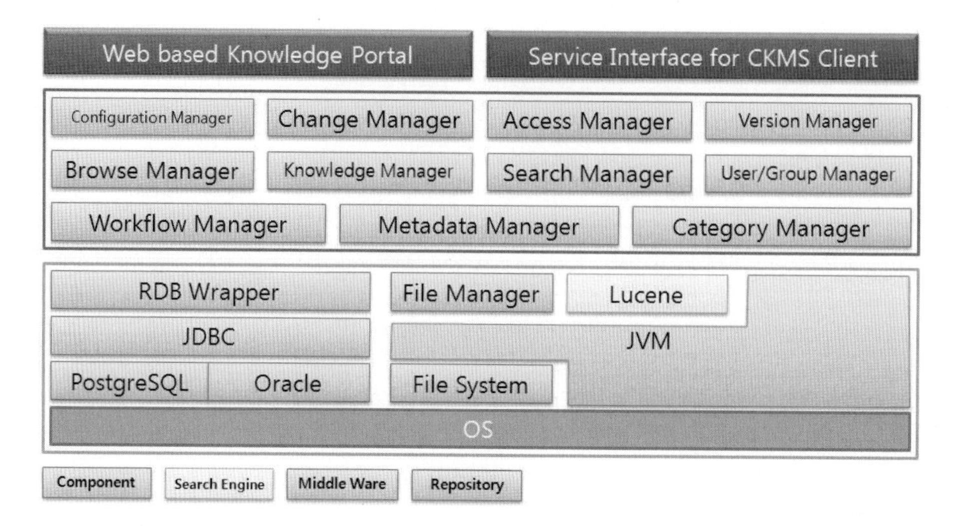

Fig. 4. Knowledge Repository Logical Architecture

Table 1. Architecture compoent description

Name	Description
Configuration Manager	Deploy package management component
Change Manager	Knowledge change history management component
Access Manager	User Authority management component for knowledge repository
Version Manager	Knowledge change version management component
User/Group Manager	User, Group, ones role and Authentication management component
Category Manager	Category management component
Workflow Manager	Knowledge management workflow(process control) component
Metadata Manager	Metadata management component
Knowledge Manager	Knowledge management component
Browse Manager	Searching knowledge repository component
RDB Wrapper	Data Management component for RDBMS
Browse Manager	File Management Supporting Component
Lucene	Java based search engine
PostgreSQL	A kind of RDBMS
Oracle	A kind of RDBMS
File System	File System
JVM	Java Virtual Machine
OS	Operating System

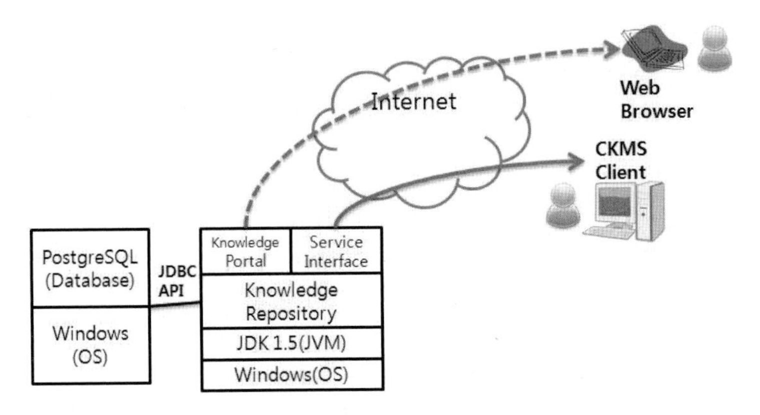

Fig. 5. Knowledge Repository Physical Architecture

5 Conclusion

This paper suggests medical knowledge repository architecture for CDS (Clinical Decision Support). We propose logical and physical architecture to build a medical knowledge repository which is reusable and capable of collaborating. And the final goal of this study is to show the architecture suggested through building a real system is practical. The validation of this project is to be tackled because the project is still ongoing. Therefore, it is necessary to confirm and discuss the effects on a specific field.

Acknowledgement

This study was supported by a grant of the Korea Health-care Technology R&D Project, Ministry for Health, Welfare & Family Affairs, Republic of Korea (No. A050909).

References

Greenes, R.A.: Clinical Decision Support: The Road Ahead, pp. 447–467 (2007)

Kashyap, V., Hongsermeier, T.: Towards a national health knowledge infrastructure (NHKI): The role of semantics-based knowledge management Tech Report No. CIRED-20005112301 (2005), http://www.partners.org/cird/pdfs/White_Paper.pdf

Fischer, G., Otswald, J.: Knowledge management: problems, promises, realities, and challenges. IEEE Intelligent Systems, 60–72 (2001)

O'Leary, D.E.: How knowledge reuse informs effective system design and implementation. IEEE Intelligent Systems, 44–49 (2001)

SW Architecture for Access to Medical Information for Knowledge Execution

Suntae Kim[1], Bingu Shim[2], Jeong Ah Kim[3], and InSook Cho[4]

[1] Dept. of Computer Science and Engineering, Sogang University, Seoul, South Korea
[2] Dreamer Corporation, Seoul, South Korea
[3] College of Computer Education, Kwandong Unifersity, South Korea
[4] Dept. of Nursing, Inha University, Incheon, South Korea
jipsin08@sogang.ac.kr, bin.shim@dreamercorp.com,
clara@kd.ac.kr, insook.cho@inha.ac.kr

Abstract. Recently, many approaches have been studied to author medical knowledge and verify doctor's diagnosis based on the specified knowledge. During the verification, intensive access to medical information is unavoidable. Also, the access approach should consider modifiability in order to cover diverse medical information from the variety of hospitals. This paper presents an approach to generating query language from medical knowledge, and shows software architecture for accessing medical information from hospitals by executing generated query languages. Implementation of this architecture has been deployed in a hospital of South Korea so that it shows the feasibility of the architecture.

Keywords: Software Architecture, Query Generation, Query Execution, SAGE, Electronic Medical Record.

1 Introduction

There have been many studies on describing clinical guidelines in reusable form and executing the guidelines in order to verify medical decisions [1][2], which are so-called computerized decision support system(CDSS) for medical information. These approaches are intended to help prevent patients and doctors from misdiagnosing and prescribing medicine that embodies potential problems. For reusing these guidelines throughtout various hospitals, CDSS should be built on a architecture that can handle the following heterogeneity over a number of clinical databases: datamodel heterogeneity, structural heterogeneity, naming heterogeneity and semantic heterogeneity[5]. Furthermore, the manual translation or creation of queries for a given EMR(Electronical Medical Record) should be minimized, since thses jobs are costly and error-prone.

Broad range of studies on handling the heterogeneous nature of existing clinical databases have been carried out. Most of these works introduce their own abstraction models and mapping models that bridge the gap between the concepts in the guidelines and the concepts in the clinical databases. Based on these mapping the

T.-h. Kim, A. Stoica, and R.-S. Chang (Eds.): SUComS 2010, CCIS 78, pp. 574–580, 2010.

previous works proposed their own approaches that generate queries automatically. However, the previous works do not handle the performance aspect of fetching clinical data using the generated queries.

This paper presents an approach to generating query syntax from CDSS and introduces SW architecture to access databases deployed in hospitals by executing the generated query syntax. This paper is based on our previous work [2] that proposes an approach to converting clinical guideline knowledge specified in SAGE(Standard-based Sharable Active Guideline Environment) [3] into the executable format, and SW architecture to execute the converted knowledge using rule engine system[4]. This paper is more focusing on access to diverse legacy hospital systems to complete execution of converted clinical guideline knowledge. Also the approach introduces tactics to handle performance issue in the software architecture.

The implementation of this architecture is applied to a legacy database system in a hospital of South Korea as a case study. It shows the architecture tackles modifiability to cover database schema changes of legacy systems. Also, the architecture achieves acceptable performance.

This paper is comprised of five sections. Section 2 introduces related work for accessing legacy data base systems. Section 3 presents an approach to generating queries and executing the queries to access information of diverse hospitals. Case studies are described in Section 4 and Section 5 concludes this paper.

2 Related Work

Handling the heterogeneity of clinical databases is one of the active areas of research in medical informatics. Most researches suggested their own framework that generates queries to a clinical database based on their own mapping model between guidelines and entities in the clinical database. Mor *et al.* [6] developed Knowledge-data ontology mapper(KDOM) framework that maps guidelines on different EMRs. This framework enables one to define abstract terms in ontology format that can bridge gap between abstract data in guidelines and those in an EMR. They evaluated the framework by mapping between a GLIF3-encoded guideline and two different EMR schemas. Boaz and Shahar [7] proposed a mediator based approach, called IDAN, that answers raw-data and abstract queries by integrating the appropriate clinical data with the relevant medical knowledge. Sujansky and Altman [5] suggested semantic data model based approach. They defines four components, a semantic data model, a high-level query language, a mapping language and a query-translation module to handle the heterogeneous nature of existing clinical databases.

Bilykh *et al.*[8] suggested an open interface model for medical data exchange. In this model, the HL7 Clinical Documents Architecture(CDA) are used for data exchange between EMR and CDSS components, and the EMR pushes required data in the specified document format to the CDSS compoments. They evaluated the model by applying it to three immunization guidelines.

3 SW Architecture for Generating and Executing Queries

This section presents our previous work in short, and then introduces how to generate query syntaxes from clinical guidelines authored in SAGE. In addition, this section is devoted to describing runtime architecture to execute generated query syntaxes.

3.1 Knowledge Engine Architecture and Data Access Components

We introduced SW architecture for translating clinical guideline knowledge specified in SAGE into executable format in our previous work [2]. The architecture addresses runtime behaviors to execute the translated guidelines and provides results of verification for conducted diagnosis. Fig. 1 depicts SW architecture and workflow among components.

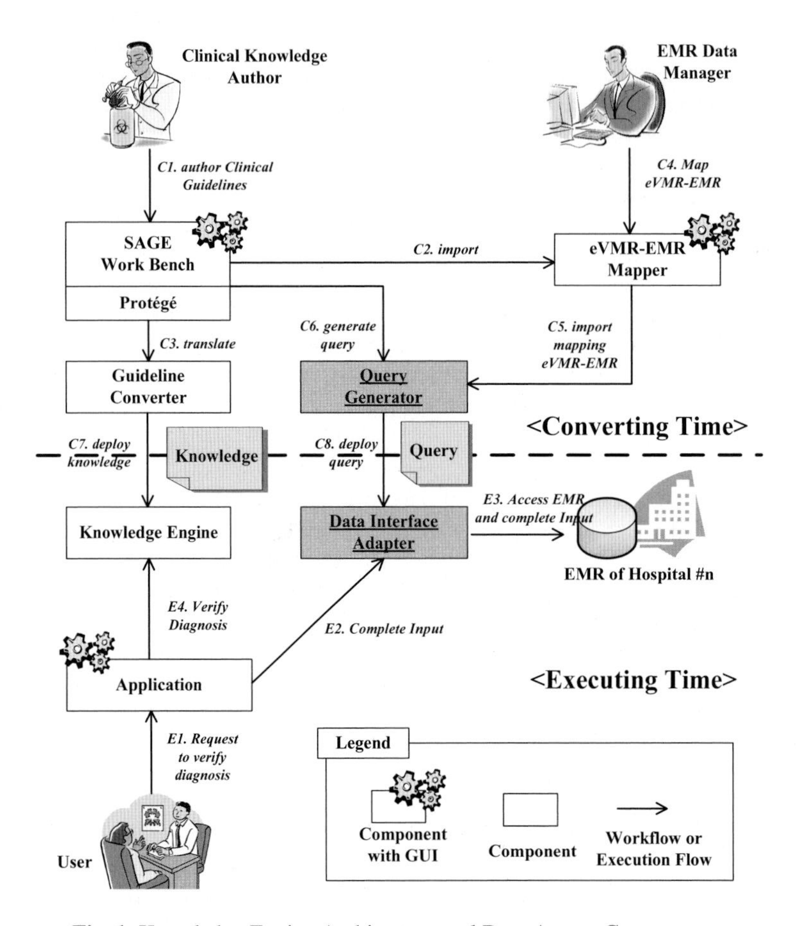

Fig. 1. Knowledge Engine Architecture and Data Access Components

The architecture is described depending on the time that components are running: converting and executing time. Converting time addresses that knowledge authors specify clinical guidelines through SAGE and EMR(Electronic Medical Record, so-called hospital data) data manager describes relationships between data part of SAGE and EMR. The data parts of SAGE are represented in eVMR that stands for extended virtual medical record in the figure. Finally the knowledge is translated into the format that knowledge engine [4] can execute and the queries are generated into the format that query executor component (depicted in *Data Interface Adapter* in the figure) to access EMR data. The results are deployed in the place which knowledge engine and data interface adapter can read. *Query Generator* handling query generation is discussed further in subsection 3.2.

Executing time denotes a time that users' diagnosis is verified automatically according to translated knowledge and queries by collaborating with *Application, Knowledge Engine* and *Data Interface Adapter.* After the application receives a request for verifying user's diagnosis, the data interface adapter accesses to users' (patients) data in EMR of a hospital by executing the generated queries, and builds input data to the rule engine at first. Then, the input data is delivered to knowledge engine to verify users' diagnosis in accordance with translated clinical guidelines. Structure and behaviors of data interface adapter are discussed in subsection 3.3 in more detail.

3.2 Query Generation

Query generation is composed of two steps: generation of eVMR query and EMR query. eVMR is the data model defined in SAGE, which is an extended virtual medical record that defines a set of hierarcal common clinical entites not specific to hospitals. eVMR quary is the the temporal output generated from our approach, which is a query which defines variables that EMR schema information is replaced. Table 1 presents a sample eVMR query. In eVMR query, the variable having '$' denotes a variable where EMR schema information should be bound.

EMR is a set of data defined in a hospital specific databse schema. EMR query is the secondly generated query from our approach based on eVMR and eVMR-EMR Schema Mapping. Schema mapping information is authored by EMR Data Manager, containing relationships between eVMR entity and EMR schema. The schema mapping in the table below shows that *PatientTable.CurAge* should be replaced with the variable *$Patient.Age* in eVMR query. In this manner, the EMR queries that access the patient data are generated. By redefining the schema mapping information, the query can be generated flexibily so that it can cover diverse hospital database systems and changes of database schema of a hospital.

Table 1. Major Outputs and Contents

Output Name	Sample Contents
eVMR Query	QID #1: Select $Patient.Age from $Patient where $PID = 'X'
EMR Query	QID #1: Select PatientTable.CurAge from PatientTable where PatientTable.ID = 'X'
eVMR-EMR Schema Mapping	PatientTable → $Patient PatientTable.CurAge → $Patient.Age

Fig. 2 (a) shows the inside structure of *Query Generator. eVMR Query Generator* gets eVMR data(see Fig. 2(b)) from SAGE and generates eVMR query. In order to generate EMR query, *EMR Query Generator* obtains schema mapping data from *eVMR-EMR Mapper.* In addition, *Query Generator* produces *Input Model Template* that constraints structure of input model to Knowledge Engine. Later, the input model template is filled out with results of query execution so that it is used as an input of knowledge execution. Fig. 2(c) shows the structure of the input model. In the figure, 'X' denotes a value that should be filled in during execution time. Because all accesses to the legacy database are conducted before executing knowledge engine, the knowledge engine does not need to access database at runtime, which is a tactic to increase execution performance.

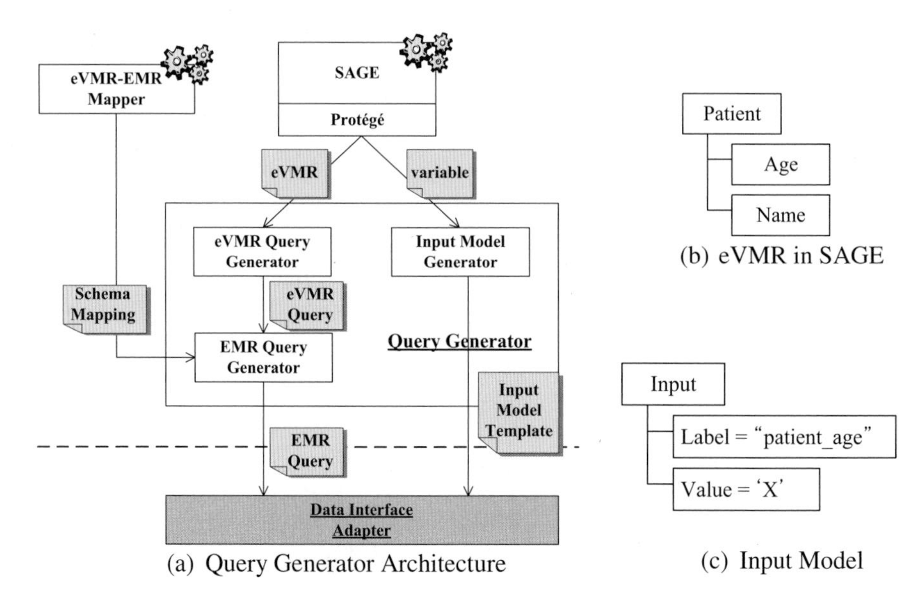

Fig. 2. Query Generator Architecture, eVMR and Input Model

3.3 Query Execution

Query execution is a process that executes generated EMR queries for EMR of hospital and sets the results of execution into the input model template. Finally, the output of this process is the input model containing patient's data. DIA(Data Interface Adapter) component is in charge of query execution (see. Fig. 3). Because DIA is a component that should interact with application and knowledge engine at runtime, performance must be important issue because accessing database intensively through network generally consumes much time. Thus, the component contains several components such as *Query Queue*, *DB Connection Pool* and *Pooled Thread* to increase performance of query execution.

In Fig. 3, Step 4-7 are executed in one thread sequentially. Thus, many EMR queries are executed in several threads simultaneously and the query results are set to the input model right after the execution. The input model then delivered to knowledge engine to execute translated guidelines.

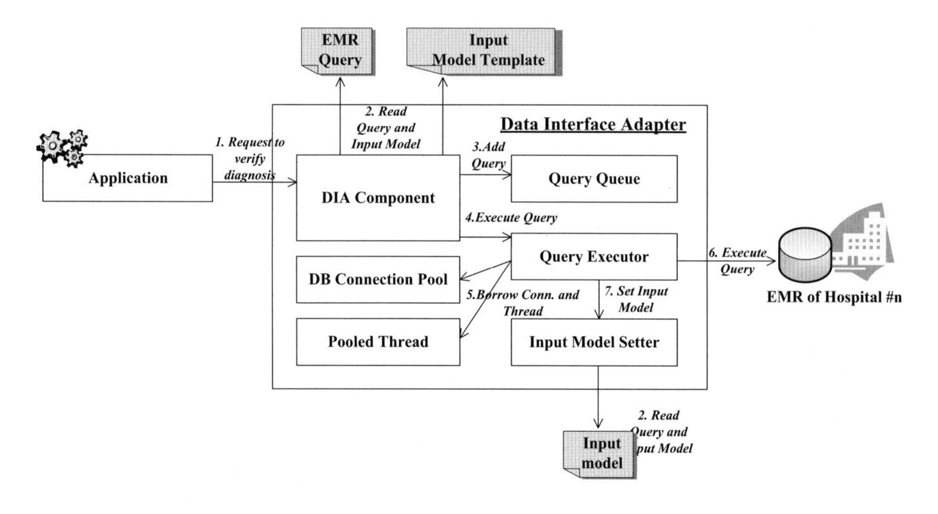

Fig. 3. Data Interface Adapter SW Architecture

4 Case Studies

The implementation of the proposed query generation approach and SW architecture are deployed in Seoul National University Hospital of South Korea as a proof of concept. Knowledge author described the clinical guideline knowledge on LAB Alerting in SAGE. Initially, the query generator produced 14 queries to get patient data. Later, the several queries are added and regenerated due to changes of the table schema. Finally, DIA executes 18 regenerated queries at runtime.

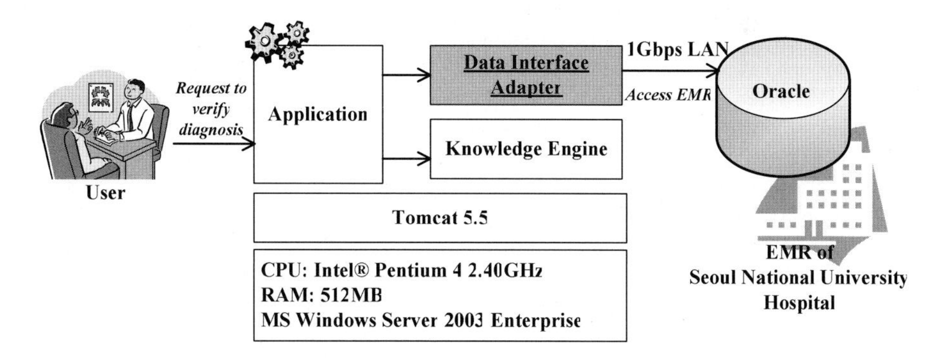

Fig. 4. Execution View of SW Architecture

Fig 4 shows execution view of software architecture. An application for Lab Alerting, DIA and knowledge engine are deployed in Apache Tomcat, and DIA accesses EMR of Seoul National University Hospital. Although the system is running on the relatively poor hardware, average execution time is recorded 500ms on average.

5 Conclusion

This paper presents an approach to generating queries for accessing patient's data in diverse legacy database of hospitals. Also, SW architecture to enhance performance of query execution is proposed. The architecture has been deployed in the real hospital to show feasibility. Based on this approach, CDSS system can access EMR data by just mapping database schema between eVMR and EMR. In order to extend this approach, we have ongoing research to generate XML query syntaxes and access XML based database to broaden the coverage for various types of legacy data base systems.

Acknowledgments. This study was supported by a grant of the Korea Health-care Technology R&D Project, Ministry for Health, Welfare & Family Affairs, Republic of Korea (No. A050909).

References

1. Kashyap, V., Morales, A., Hongsermeier, T.: On Implementing Clinical Decision Support: Achieving Scalability and Maintainability by Combining Business Rules and Ontologies. In: The American Medical Information Association (AMIA), Washington, DC, pp. 414–418 (2006)
2. Kim, J., Shim, B., Kim, S., Lee, J., Cho, I., Kim, Y.: Translation Protégé Knowledge for Executing Clinical Guidelines. In: 11th International Protégé Conference, Amsterdam, Netherlands (2009)
3. SAGE Guideline Model, `http://sage.wherever.org/model/model.html`
4. Kim, J., Taek, J., Hwang, S.: Rule-based Component Development. In: Third ACIS Int'l Conference on Software Engineering Research, Management and Applications (SERA 2005), Mt. Pleasant, MI, pp. 70–74 (2005)
5. Sujansky, W., Altman, R.: An evaluation of the TransFER model for sharing clinical decision-support applications. In: AMIA Annual Fall Symposium, pp. 468–472 (1996)
6. Peleg, M., Keren, S., Denekamp, Y.: Mapping computerized clinical guidelines to electronic medical records: Knowledge-data ontology mapper (KDOM). Journal of Biomedical Informatics 41, 180–201 (2007)
7. Boaz, D., Shahar, Y.: A framework for distributed mediation of temporal-abstraction queries to clinical databases. Artificial Intelligence in Medicine 34(1), 3–24 (2005)
8. Bilykh, I., Jahnke, J., McCallum, G., Price, M.: Using the Clinical Document Architecture as open data exchange format for interfacing EMRs with Clinical decision support systems. In: 19th IEEE Symposium on Computer-Based Medical Systems, Utah, pp. 855–860 (2006)

Implementation of Wireless Sensor Networks Based Pig Farm Integrated Management System in Ubiquitous Agricultural Environments

Jeonghwan Hwang, Jiwoong Lee, Hochul Lee, and Hyun Yoe*

School of Information and Communication Engineering,
Sunchon National University, Korea
{jhwang,leejiwoong,hclee,yhyun}@sunchon.ac.kr

Abstract. The wireless sensor networks (WSN) technology based on low power consumption is one of the important technologies in the realization of ubiquitous society. When the technology would be applied to the agricultural field, it can give big change in the existing agricultural environment such as livestock growth environment, cultivation and harvest of agricultural crops. This research paper proposes the 'Pig Farm Integrated Management System' based on WSN technology, which will establish the ubiquitous agricultural environment and improve the productivity of pig-raising farmers. The proposed system has WSN environmental sensors and CCTV at inside/outside of pig farm. These devices collect the growth-environment related information of pigs, such as luminosity, temperature, humidity and CO2 status. The system collects and monitors the environmental information and video information of pig farm. In addition to the remote-control and monitoring of the pig farm facilities, this system realizes the most optimum pig-raising environment based on the growth environmental data accumulated for a long time.

Keywords: WSN, Ubiquitous, u-IT, Agriculture, Pig farm.

1 Introduction

The wireless sensor networks technology based on low power consumption is one of the important technologies in the realization of the ubiquitous society. It is applied and utilized in various fields such as environment monitoring, disaster control, logistic management and home network [1][2]. As the ubiquitous technology such as wireless sensor network technology has brought big change to other industries and daily living, it can also be utilized in the agriculture and livestock industry in various ways [3].

The ubiquitous agriculture has its purpose on enhancing the productivity by combining IT technology with agriculture, examining the safety of agricultural crops by systematically managing the distribution/consumption of crops and making the process of distribution/consumption transparent [4].

* Corresponding author.

T.-h. Kim, A. Stoica, and R.-S. Chang (Eds.): SUComS 2010, CCIS 78, pp. 581–590, 2010.

For instance, an unmanned control device is installed in the facility house producing agricultural products. The device automatically measures the environmental change factors of the crop, such as temperature, humidity, ammonia gas, CO2 and the weather. The measured information is saved in database and utilized in the agriculture. Through this process, it is possible to reduce the material input for crops growth such as fertilizers and chemicals, decrease the production cost by making the most optimum growth environment and enhance the productivity [5][6].

The ubiquitous technology is also effectively used in the livestock-raising such as pig-raising or chicken-raising. For instance, RFID technology and WSN technology are used in the pig-raising to manage the feeding of pig-individual, pigsty environment and pigs growth tracking. Mobile device gives alarm when there is an abnormality in the pig-individual so that farmers can take immediate action [6].

Recently, domestic pig-raising industry faces a face-to-face duel with pig-raising advanced countries because of rise in the feed cost and the execution of FTA [7]. Also, the mortality rate caused by wasting diseases increases and production cost goes up together, giving double difficulties to pig-raising farmers. Accordingly, the productivity increase and high quality pork production became the essential tasks of pig-raising industry [8].

In order to cope with this issue, it is urgently required to secure the scientific and systematic pig-raising technology combining current u-IT technology with pig-raising industry, which is the primary industry.

This paper proposes ‘Pig Farm Integrated Management System’ based on wireless sensor networks, which is an integrated management system applying the WSN technology to the pig-raising environment management and control.

The proposed system enables the monitoring of pig farm environment and the control of pig farm facilities. Through these, the optimum pig-raising environment can be maintained, productivity can be increased and producer convenience through remote/automatic control can be also achieved.

This research paper is comprised of followings. Chapter 2 will explain the system structure and service process of pig farm integrated management system based on wireless sensor network. Chapter 3 will explain the system operation result. Chapter 4 will compare and analyze the experiment result. Finally, Chapter 5 will give conclusion to close the research paper.

2 Design of Pig Farm Integrated Management System

The Pig Farm Integrated Management System collects pig farm environment information and video information through environmental sensors measuring environmental elements and CCTV. It also supports the monitoring and control of pig farm status.

2.1 System Architecture

The proposed pig farm integrated management system is comprised of physical layer, middle layer and application layer. The physical layer is comprised of sensors, CCTV and pig farm facilities as in Figure 1. The middle layer supports the communication between physical layer and application layer. It makes the pig farm information into database and provides with monitoring and control service, maintaining the growth environment of pigs at optimum status. The application layer is comprised of interfaces which support the pig farm environment monitoring and pig farm facilities control service.

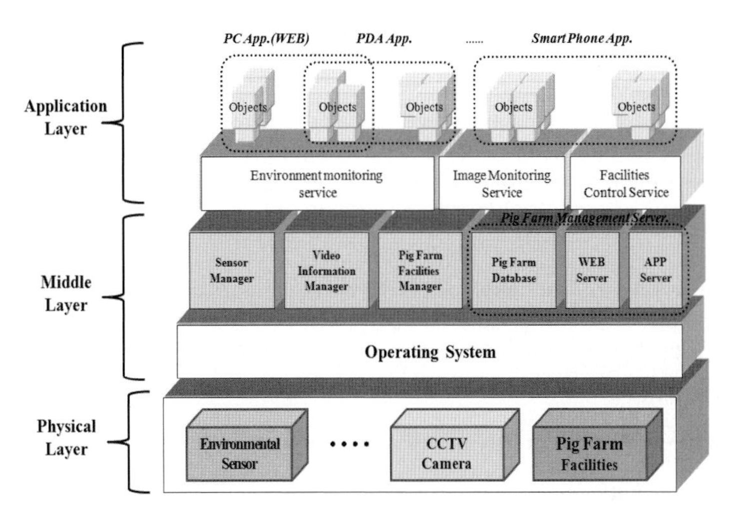

Fig. 1. Pig farm integrated management system architecture

Physical Layer. The physical layer is comprised of environmental sensors, collecting the pig farm environment information, CCTV, collecting the video information of pig farm and pigs, and pig farm facilities, making the optimum growth environment for pigs.

The environmental sensors can be classified into pig farm internal sensors and pig farm external sensors. Internal sensors measure the pig farm internal environmental information such as luminosity, temperature, humidity and CO2. External sensors measure the external environment change of pig farm.

CCTV in the pig farm collects the video information of pig farm and pigs. Pig farm facilities are comprised of lightings, humidifier, air conditioner and ventilator which control the pig farm environment that gives impact on the growth of pigs such as luminosity, temperature, humidity and CO2.

Middle Layer. The middle layer is comprised of sensor manager, video information manager, pig farm facilities manager and pig farm management server. The sensor manager manages he environmental information collected at the sensors in the physical layer. Video information collected at CCTV is managed by video information manager. Pig farm management server, with pig farm database comprised of pig farm information, monitors and controls pig farm facilities.

The sensor manager does 'format processing', which is changing the pig farm environmental information collected at the environmental sensors of physical layer into a format that can be saved in the pig farm database, ‘units change’, which is changing the units to meet with measurement elements, and ‘update query’, which is processing the data to save in the pig farm database.

The pig farm facilities manager receives the control signal and operates/manages pig farm facilities. It also saves the pig farm facilities status in the pig farm database. The video information manager provides web with stream data.

The pig farm database saves ‘pig farm facilities environment data’, collected at the sensors installed inside/outside of pig farm such as luminosity, temperature, humidity and CO2, ‘video data’, collected at CCTV, ‘pig farm facilities status/control data’ and ‘environmental standard values’ for the automatic control and status notification, in the tables allocated to each of them.

The pig farm management server is located between the producer and pig farm database. It examines the environmental data saved in the pig farm database in fixed cycle, reports them to the producer and controls the pig farm facilities by comparing them with the environmental standard values saved in the pig farm facilities control table.

Application Layer. The application layer is comprised of application services supporting various platforms such as laptop, web, PDA and smart phones. It provides producer with 'pig farm environment monitoring service', 'pig farm video monitoring service' and 'pig farm facilities control service'.

2.2 Services of Pig Farm Integrated Management System

This system provides with ‘pig farm environment monitoring service’, enabling the observation of internal/external environmental information of pig farm, ‘pig farm video monitoring service’, providing with pig farm video in real time, ‘pig farm facilities control service’, enabling the automatic control and manual control of pig farm facilities by producer based on the environmental standard values, and ‘danger alarm service’, giving notification of dangerous situation at the pig farm.

Pig Farm Environment Monitoring Service. The pig farm environment monitoring service shows the pig farm environmental data, collected at the environmental sensors measuring the environmental elements, such as luminosity, temperature, humidity and CO2, to producer through GUI so that producers can identify the environment changes of internal and external of the pig farm.

The detail of this service is that it collects pig farm internal/external environmental information giving impacts to pigs growth such as luminosity, temperature, humidity and CO2 from the environmental sensors installed at inside/outside of pig farm and transmits the information to sensor manager periodically.

The sensor manager will analyze the received data and extract each sensing value. Their formats will be changed and they will be saved in each table of pig farm database. The pig farm management server transmits pig farm internal/external environmental information saved in the pig farm database to producer and the producer can monitor the environmental information of pig farm through this information.

Pig Farm Video Monitoring Service. The pig farm video monitoring service provides producer/consumer with video of pig farm/pig-individuals through CCTV installed in the pig farm.

The CCTV sends the pig farm video to video information manager and the video information manager provides with this information by web through Internet. Users can confirm the pig farm video information through Internet.

Pig Farm Facilities Control Service. The pig farm facilities control service enables the pig farm management server automatically control the pig farm facilities, or, the producer manually control the pig farm facilities based on the collected information at the CCTV and environmental sensors installed at inside/outside of pig farm.

The automatic control service saves the information collected from pig farm at pig farm database. The pig farm management server calls up the information and compares it with the environmental standard values saved in the pig farm database. If it is more than or short of standard value, it will confirm whether the pig farm facilities are operating as saved in the pig farm database. Then it will send the control signal to pig farm facilities manager and control the pig farm facilities.

When pig farm facilities operate, the pig farm facilities status information is saved in the pig farm database and it will be notified to user.

The manual control service saves the information collected from pig farm in the pig farm database and the pig farm management server sends the information to the user in real time.

If the user wants to control the pig farm at this time, the user will send the pig farm facilities control signal to pig farm management server through GUI. The pig farm management server will check whether the pig farm facilities are operating through pig farm database and send the control signal to pig farm facilities manager to control the pig farm facilities.

Danger Alarm Service. The danger alarm service tells the weather change and pig farm status change to farmers in real time and takes emergency measure to prevent danger in advance. The data sensed at the environmental sensor is sent to the sensor manager. The sensor manager extracts the sensing values from received data and saves them in the pig farm database. The saved sensing values will be periodically monitored by pig farm management server. If it would be more than or less than the standard value, it will be notified to the element where the event had occurred.

3 Implementation of Pig Farm Integrated Management System

3.1 Components of Pig Farm Management System

Environmental Sensors. In order to collect the environmental information of pig farm, WSN environmental sensors were installed at inside/outside of pig farm. These sensors will form the wireless network together with WSN sensor gateway in the pig farm. The sensors are classified into 'integrated sensor node', measuring the temperature, humidity and luminosity, and 'CO2 nodes', measuring CO2.

The integrated sensor node receives the sensor data from temperature, humidity sensors. It processes the data at MSP430 MCU and transmits them to relay node and gateway, using CC2420 RF chip. In order to reduce the heat impact the sensor receives from the node, the node and the sensor will maintain certain distance from each other.

MSP430 is 16bit RISC with 48Kbyte 'program memory' and 10Kbyte RAM inside. It can process multiple sensor data at high speed. CC2420 is RF chip supporting Zigbee. It supports the frequency band of 2400~2483.5 MHz. It operates in DDDS method, supports O-QPSK modulation method and 250k bps baud rate. It enables real time wireless communication with small power consumption.

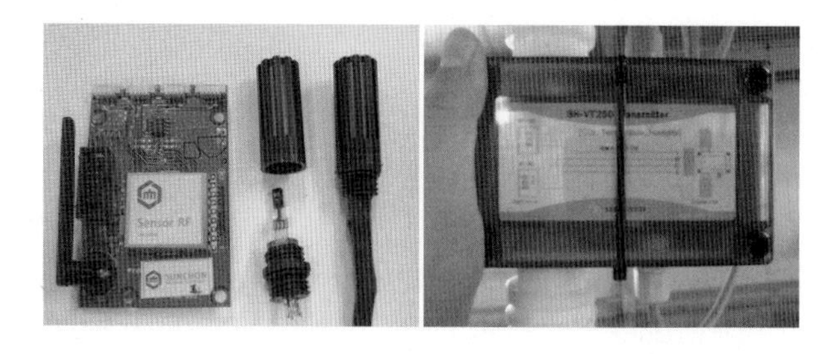

Fig. 2. Integrated sensor node and CO2 sensor

SHT71 was used for temperature/humidity sensors. SHT 71 temperature/humidity sensor has temperature sensor and humidity sensor in one body. It works on relatively small power source of 2.4V~5.5V and has small power consumption of average 28 uA. Having correction memory, it has 14bit A/D converter and digital 2-wire interface. It measures temperature from '-40°' to '120°' with 0.5° error accuracy. Humidity can be measured between 0% and 100% with 3.5% error accuracy.
3.3V operating voltage was connected to integrated sensors node and digital 2-wire was connected to MSP430 circuit to process the temperature and humidity information of pig farm.

CO2 sensor uses NDIR measurement method. It measures the range of 0 to 3,000 ppm with 3% error accuracy. RS485 method was used for communication method.

CCTV. In order to monitor the pig farm by 24 hours video, monitoring camera based on IP was installed as in Figure 3. This camera can monitor the pig farm status in real time and is also used to find out the cause of accident, in case there was an accident such as theft or accident in the pig farm, by monitoring and recording the pig farm inside 24 hours. The recorded video information is transmitted to the 'pig farm management server', where they are saved in the database after classification by pig farm ID and camera number.

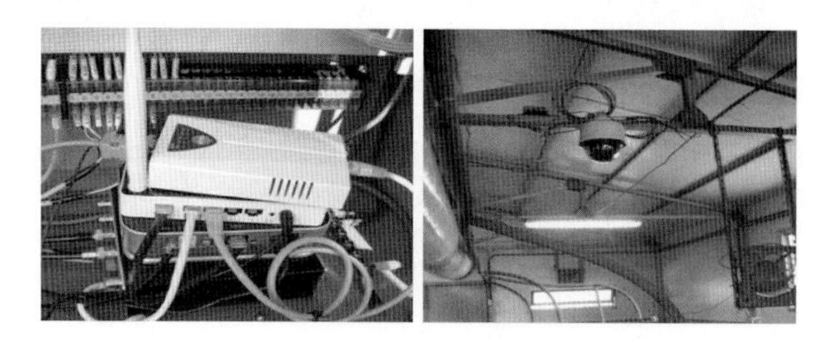

Fig. 3. DVR and CCTV camera

Pig Farm Facilities and Environmental Control Device. Luminosity, temperature, humidity and CO2 give impacts on the growth of pigs. Figure 4 shows the environmental control devices in the pig farm, which enables the control of pig farm facilities such as lighting, humidifier, fan heater, air conditioner and ventilator for those. Through these environmental control devices, it is possible to maintain pleasant pig-raising environment in the pig farm.

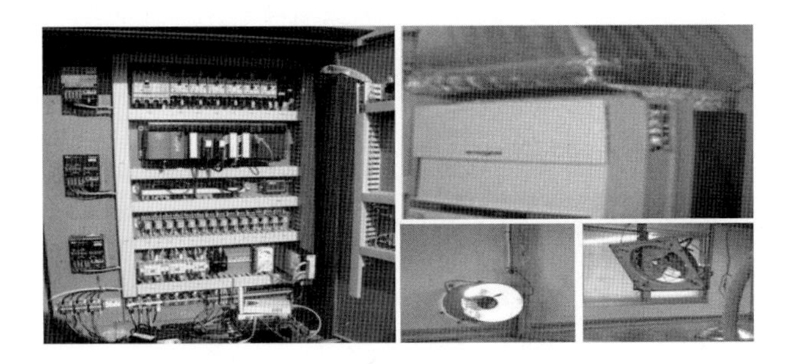

Fig. 4. Environmental control device and Pig farm facilities

Pig Farm Management Server. The environment measurement data in stream-form transmitted from pig farm is parsed and saved in the database. At the same time, it is sent to the manager in charge of relevant pig farm so that he/she can know the environment change in real time.

The process of pig farm environmental status of pig farm is classified into two processes. First one is automatically controlling the environmental status from system in reference to the designated environmental standard data. Second is directly controlling the system dependent on the necessity of the manager.

Applications. GUI for manager is developed for web environment. Tomcat-6.0.20 is used for WAS and 'mysql' is used for database. The latest released version 5.0 was used. It was possible to collect the pig farm environmental information and video information of pig farm through sensors and video monitoring camera and constantly monitor/control the pig farm status through user-intuitive GUI by way of above result. Figure 5 shows the Web GUI of pig farm integrated management system.

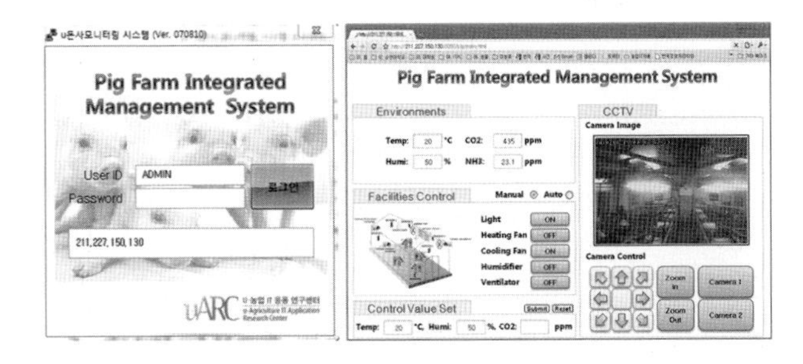

Fig. 5. Web Graphic User Interface of Pig Farm Integrated Management System

4 Results

4.1 Measurement Environment

In order to complete the design and evaluate the performance of this system, two test-beds were established in two pigsties. Pig farm A had sensors and WSN. Pig farm B

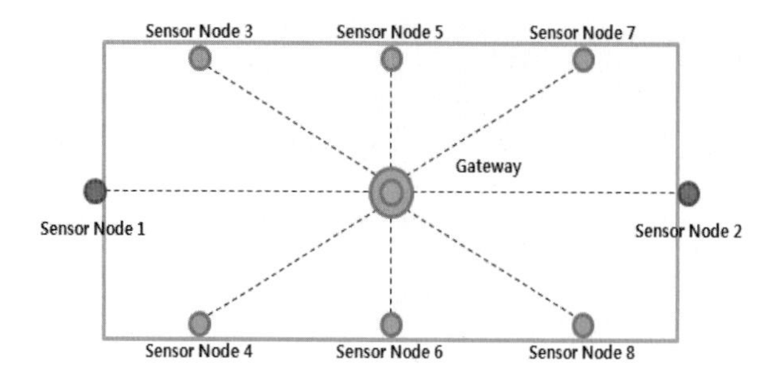

Fig. 6. Wireless sensor networks topology

had sensors, WSN and 'pig farm integrated management system'. Both pig farms were located in the same area with 5m distance from each other.

Figure 6 is the structure of sensors and gateways installed in the two pig farms. Sensor 1 and sensor 2 are located outside of pig farm to measure the external temperature and humidity. Sensors 4 to 6 were installed inside pig farm to measure the internal temperature and humidity.

4.2 Measurement Results and Analysis

The environment sensors installed at pig farm send measured data to server every 10 minutes. The server applies 'the pig farm environment decision making method' to the measured data and control the pig farm environment. Measurement period was from 00:00 hours of April 1st, 2010 until 23:00 hours of April 3rd. The measured environmental data are shown in graph with one hour interval.

Figure 7 is the variations of temperature and humidity in pig farm A using existing control method.

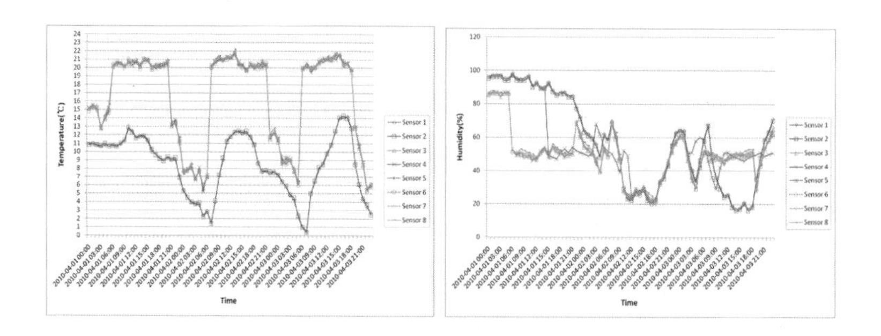

Fig. 7. Variations of temperature and humidity in pig farm A

The measurement result of pig farm A suggests that the temperature and humidity of pig farm inside keep constant level from 06:00 hour to 20:00 hour, because the producer directly controls the pig farm facilities. However, from 20:00 hour to 05:00 hour, there are rapid changes in the temperature and humidity caused by absence of proper pig farm facilities control. Such rapid environmental change in the pig farm gives severe stress to pigs, which can lead to the deaths of pigs.

Figure 8 is the variations of temperature and humidity in pig farm B which has operating 'pig farm integrated management system'.

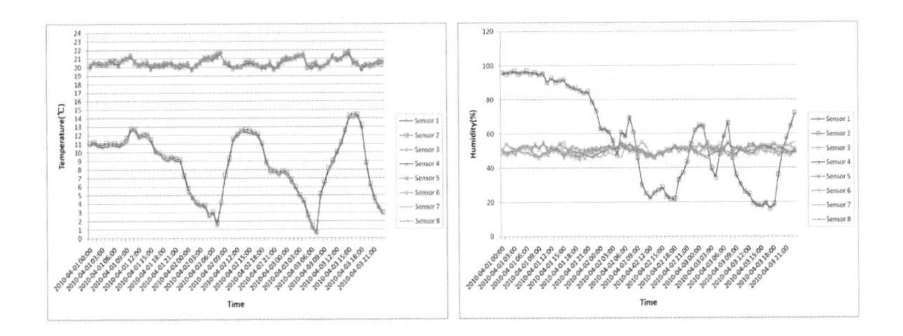

Fig. 8. Variations of temperature and humidity in pig farm B

Pig farm B generates the estimated data based on the measured data by way of 'pig farm integrated management system' proposed by this research. It activates the pig farm internal control devices and it was possible to maintain the pig farm internal temperature and humidity near to the environmental standard values. Pig farm B showed uniform temperature and humidity status compared to pig farm A using existing control method.

This measurement result suggests that the pig farm operation by proposed pig farm integrated management system is more effective than the pig farm operation by existing control method.

5 Conclusions

This research proposed 'Pig Farm Integrated Management System' based on wireless sensor network as the system to manage the pig farm environment in integrated way in the ubiquitous agricultural environment.

The pig farm integrated management system is comprised of three layers. The roles and services provided by these three layers have been explained. The comprising elements of each layer provide user with organic information by collecting and managing the environmental elements in the pig farm. The system provides with management devices proper to pig farm, saves the provided data, makes a 'control manual' and keeps the record, so that there would not be trial-and-error situation even if the person in-charge would be replaced. It is expected that the mortality rate of pigs could be reduced substantially.

The pig farm integrated management system based on wireless sensor network will contribute in the saving of labor force in the pig-raising farmers, production of high quality pork and further contribute in the securing competitiveness of pig-raising industry, by way of joining pig-raising industry with ubiquitous technology.

Acknowledgements. This research was supported by the MKE(The Ministry of Knowledge Economy), Korea, under the ITRC(Information Technology Research Center) support program supervised by the NIPA(National IT Industry Promotion Agency) (NIPA-2010-(C1090-1021-0009)).

References

1. Akyildiz, I.F., et al.: A survey on Sensor Networks. IEEE Communications Magazine 40(8) (2002)
2. Chong, C.-Y., Kumar, S.P., Hamilton, B.A.: Sensor networks: evolution, opportunities, and challenges. Proc. IEEE 91(8), 1247–1256 (2003)
3. Pyo, C.-S., Chea, J.-s.: Next-generation RFID / USN technology development prospects. Korea Information and Communication Society, Information and Communication, 7–13 (2007)
4. Shin, Y.-s.: A Study on Informatization Model for Agriculture in Ubiquitous Era. MKE Research Report (2006)
5. Lee, M.-h., Shin, C.-s., Jo, Y.-y., Yoe, H.: Implementation of Green House Integrated Management System in Ubiquitous Agricultural Environments. Journal of KIISE 27(6), 21–26 (2009)
6. Yoo, N., Song, G., Yoo, J., Yang, S., Son, C., Koh, J., Kim, W.: Design and Implementation of the Management System of Cultivation and Tracking for Agricultural Products using USN. Journal of KIISE 15(9), 617–674 (2009)
7. Lee, J.-h.: [Livestock industry research series 11] What is a threat to South Korea's livestock industry? Focus Attention GSnJ (55) (2008)
8. Yoo, Y.-h., Kim, D.-h.: The current state of automation in pig house establishment and prospection. Korea Society for Livestock Housing and Environment, 29–47(19) (2006)

A Study on Energy Efficient MAC Protocol of Wireless Sensor Network for Ubiquitous Agriculture

Ho-chul Lee, Ji-woong Lee, Jeong-hwan Hwang, and Hyun Yoe*

School of Information and Communication Engineering,
Sunchon National University, Korea
{hclee,leejiwoong,jhwang,yhyun}@sunchon.ac.kr

Abstract. Various technologies are used in the agricultural sites now. Especially, the recent application of sensor network related technology is quite notable. Considering the efficiency of MAC protocol of WSN is being researched in various aspects, it is believed that a research on how to apply the MAC protocol to agriculture would be also required. This research is based on the sensor node developed by Sunchon University ITRC. Once the sensor nodes are effectively located in the farm, they operate for a long time and they are rarely relocated once installed. The concentration of multiple sensor nodes in a narrow area is another characteristic the sensor node. The purpose of this research is to select a sensor network MAC protocol, which would be most proper to agricultural site with good energy efficiency and excellent transmission delay performance. The applicable protocols such as S-MAC and X-MAC were set up for the installation environment. They were compared and a methodology to select the most optimum protocol to agricultural site is suggested.

Keywords: WSN, Ubiquitous, u-IT, cultivation facility, Paprika.

1 Introduction

The recent innovation in IT technology is accelerating the fusion between industries. The fusion between IT and traditional industries continuously goes on. The application of ubiquitous technology to agriculture, which is a primary industry, is getting expectation that the convergence technology would enhance the added-value and productivity of agriculture [1]. In order to establish such u-agriculture environment successfully, the core ubiquitous technology development optimized to agriculture, such as sensor hardware, middleware platform, routing protocol and agriculture environment application service, would be essentially required [2].

For the development of such ubiquitous technology, various energy-efficient MAC protocols were studied in the wireless sensor network. S-MAC[3], applying "sleeping, stand-by", was suggested to improve the energy efficiency of MAC protocol. T-MAC[4] was suggested to reduce the unnecessary waking hours even a little bit more. Adaptive S-MAC [5] was developed to avoid the transmission delay phenomenon occurring when applying the duty cycle and hybrid type Z-MAC [6] was developed

* Corresponding author.

T.-h. Kim, A. Stoica, and R.-S. Chang (Eds.): SUComS 2010, CCIS 78, pp. 591–599, 2010.

combining CSMA and TDMA. There is also the X-MAC [7], which preoccupies the channel using preamble during the sleeping period in asynchronous method.

This research chooses the MAC protocol, which demonstrates the most efficient energy performance when WSN would be applied to paprika cultivation in a cultivation facility. Further, a methodology to choose MAC protocol proper to certain cultivation method or stock raising method will be suggested. Actual cultivation facility was taken as the model for the research and sensors were located proper to the cultivated crop. The network topology of the sensors was configured and sensors performance will be measured by a simulator.

Paprika cultivation facility was chosen because paprika takes an important role among Korean major exporting horticultural products. Paprika is being exported to Japan and United States, Russia and Taiwan are potential export markets [8]. Paprika is a tropical garden fruit. The harvest quantity of paprika shows big variation, dependent on sunlight, temperature and humidity environments in the cultivation facility, in addition to cultivation and management technology [9]. Especially, the harvest-cycle variation range of paprika is very big, dependent on the number of fruit-setting caused by interaction between luminosity and temperature [10]; therefore, very precise control of luminosity and temperature is required. When the productivities of paprika in the plastic film house and glass greenhouse were compared, the glass greenhouse showed twice productivity of plastic film house [11]. The productivity of paprika in Korea were 6.8kg/m² in year 2000 and 9.4kg/m² in year 2007. Even there was 38% productivity increase during seven years; it is still 30% of average productivity in Netherlands. [12].

This research paper comprised of followings. The MAC protocols to be compared and analyzed in this research will be introduced in Chapter 2. In Chapter 3, the methodology of selecting MAC protocol to be used at cultivation site is suggested and candidate MAC protocols are compared. Then there will be a conclusion section.

2 Relate Works

2.1 S-MAC

SMAC is a representative synchronous MAC protocol. It periodically repeats inactivated “sleep mode” and activated “listen mode” with fixed lengths [3]. In the “listen mode”, the data transmission between two nodes is possible. In the “sleep mode”, power waste at each sensor node is reduced by providing with minimum power to maintain the sensor node, while main power is shut-off. However, there will be “listen-sections” without communication caused by fixed lengths and power is wasted because of these unused “listen-sections”. Also, there is a disadvantage in the “sleep-section”, which is data transmission delay caused by inability to receive signal in the “sleep-section”.

2.2 X-MAC

The X-MAC protocol is suggested to resolve the problem of overhearing caused by the long preamble used in the B-MAC [13] protocol. It reduces the preamble overhearing of B-MAC protocol by repeated transmission of minimum preamble for

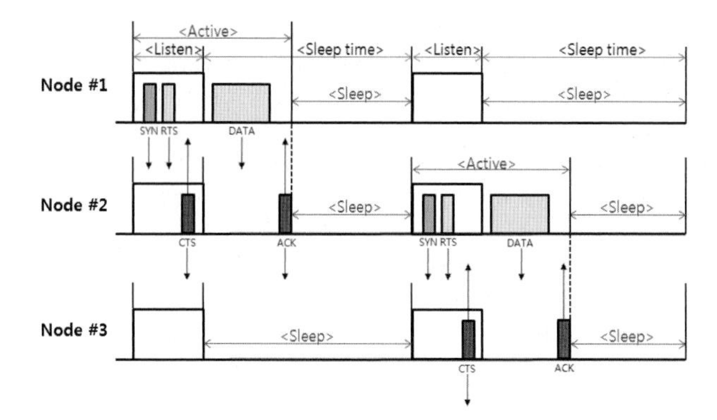

Fig. 1. S-MAC

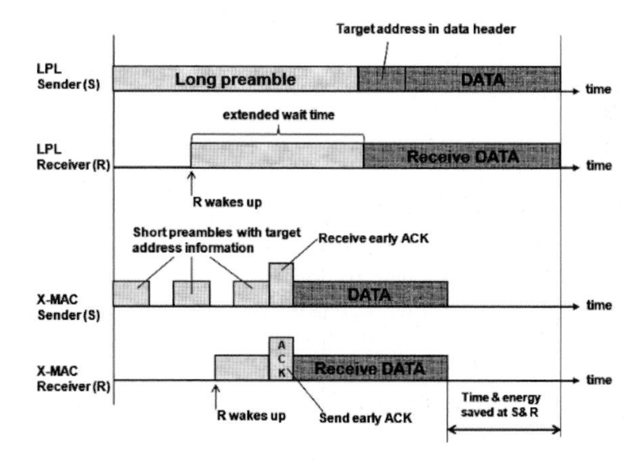

Fig. 2. Comparison of the timelines between LPL's extended preamble and X-MAC's short preamble approach[7]

synchronization and the "strobed preamble" containing the destination address. When there is data to transmit, the node operating with X-MAC transmits the minimum preamble and the "short preamble", containing the destination address, in order to tell nearby nodes that it has data to transmit. Then the node maintains "stand-by" mode of data reception for a sufficient period to receive early ACK [7].

3 Method of Protocol Adaptation

3.1 Cultivation Environment of Paprika

This research takes the actual paprika greenhouse in Gwangyang, Chollanam-do. Sensor nodes are located in the greenhouse and the network performance was examined in advance so that the sensor network can be installed for efficient operation by choosing the efficient MAC at site.

In the facility cultivation, paprika seeds are sown to the rock wool trays and they are planted temporarily on the rock wool cubes. When there would be the first branch

Fig. 3. LED Lamp, Sun Shield and Warm water supply

stem, they will be permanently planted to the culture medium for cultivation. Then culture solution made for the best paprika growth will be supplied. At the lower part of the culture medium, boiler pipe way will be installed and warm water will be supplied to maintain the temperature at the paprika rooting zone constant. LED lightings will be installed in the upper part to enhance the growth of paprika and ventilation equipment will be also installed to mix the upper air and lower air for constant temperature. Mobile screens will be installed inside of greenhouse roof to shut off the strong sunlight.

The paprika rock wool cubes will be located at 30cm distance with each other. There will be four rock wool cubes for each culture medium. Each culture medium will make 50m length in parallel with the rail installed at greenhouse floor. Each group of culture media will be located at 50cm distance having the rail between them. Other than moving and working space in the greenhouse, the whole greenhouse will be filled with culture media planted with paprika. For growth environment, daytime 22~25℃, nighttime 18~20℃ and humidity 70~75% will be maintained.

3.2 Hardware Description

Fig. 4. Sensor Node

The sensor node developed by Sunchon National University ITRC Research Center will be applied to this research. This sensor node can collect the information of leaf wetness, leaf temperature, greenhouse temperature/humidity and control the relay by one sensor. MSP430 MCU is applied to the CPU and CC2420 RF module of Chipcon Co. is used as the data transmission/reception device. The MSP430 microprocessor has 16 bit RISC structure and it works in very fast speed with its 48 Kbyte program memory and 10Kb RAM. 3.6V battery is used for power supply [14].

3.3 Network Topology

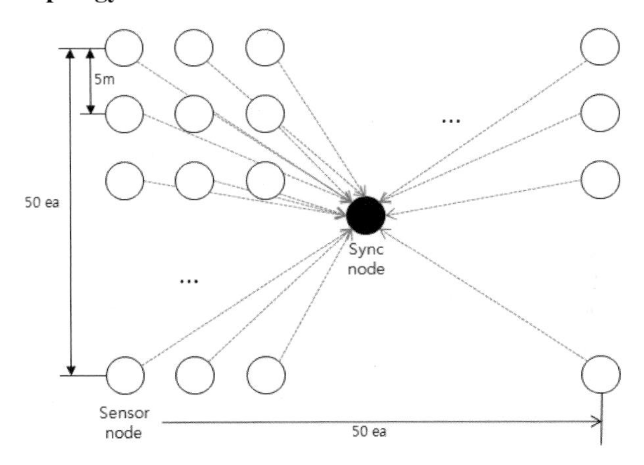

Fig. 5. Network Topology

The sensor nodes will be installed at every 5m along the culture media lined up in reference to the paprika at the most outer side. They will be installed alternately for the root zone parts and upper parts. Installation will continue to the culture media with 5m distance in reference to the culture media with sensors installed. The overall location shape of sensor nodes is grid-type with 5m distance. The sink node to transmit collected data to the server will be located in the center of 50 * 50 grids. The shape of sensor nodes location is grid shape; however, the network topology is a star topology in reference to the sink node in the center of the grid.

3.4 Duty Cycle

The sensor of sensor node measures leaf temperature, leaf wetness and greenhouse temperature/humidity. They are measured in 3 minutes cycle and transmitted to the server. The relay control port of the sensor node will not be used. The measurement cycle can be different dependent on the characteristic of the crops.

Now, the duty cycle to be applied to sensor node will be determined. The generated data and the number of nodes to transmit data to sink node, during the measurement cycle of crop environment data, will be estimated. Total number of data which can be generated during the measurement cycle will be estimated and the data quantity which can be processed for each duty cycle will be deduced. The duty cycle will be chosen

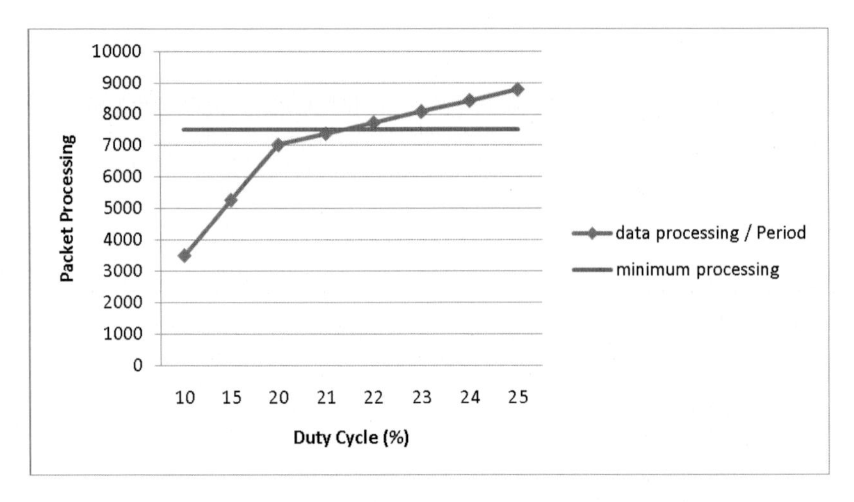

Fig. 6. Available Duty cycle of S-MAC

so that it can process more data than the data generated during the measurement cycle. 10% allowance will be given so that waste data would not be generated. If the measured data would be missed, the on-time responding to the change of crops growth environment change will be difficult and the quality, quantity of the harvested crop will get negative impact.

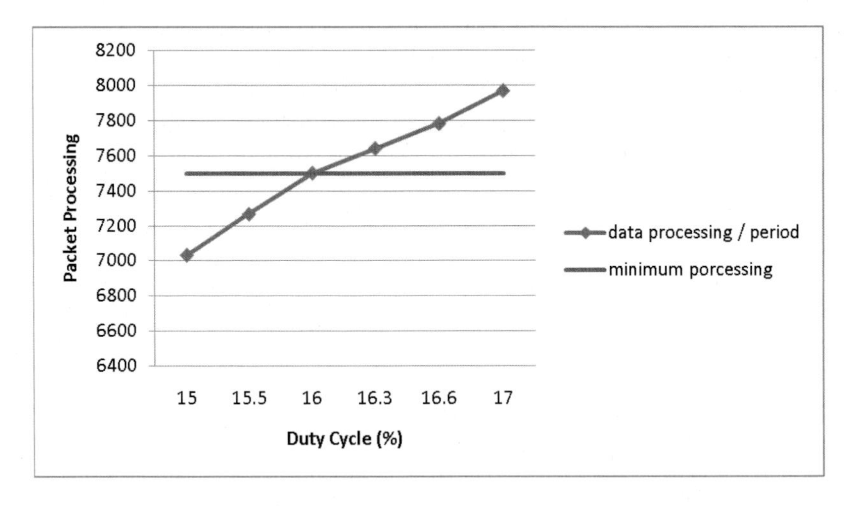

Fig. 7. Available Duty cycle of X-MAC

Figure 6 and Figure 7 shows the packets which can be processed by S-MAC and X-MAC, dependent on duty cycles. In the established situation, the effective duty cycle of S-MAC is 22%, including 10% allowance. For X-MAC, the proper duty cycle is shown as 16.3%.

3.5 Simulation

A simulation will be done to measure the energy performance of the MAC considered for the application. The performances of S-MAC and X-MAC will be examined. For this, a simulation environment was made using NS-2 [15]. [Table 1] is the system parameters for the simulation.

Table 1. Simulation Parameter

	S-MAC	X-MAC
NS2 Version	NS-2.34	
Simulation Time	5000 Second	
Packet Size	40 byte	
Packet Interval	1 minute / node	
Node Count	2500	
Routing Protocol	DSDV	
Duty Cycle	22%	16.3%
Bandwidth	250Kbps	
Initial Energy	30,000 J	

Total 2,500 nodes comprised the network topology as in [Figure 7]. The physical shape of nodes is grid; however, the shape of network topology is a star-shape with sink node in the center. Each node generates 40 bytes per minute of sensing data when the number of the nodes is fixed and the energy consumption at this time is measured. Each node measures leaf temperature, leaf wetness and greenhouse temperature/humidity. They are measured in 3 minutes cycle and transmitted to the server; however, it will be assumed in the simulation that the measurement items will be data with same size and the data is generated in one minute cycle.

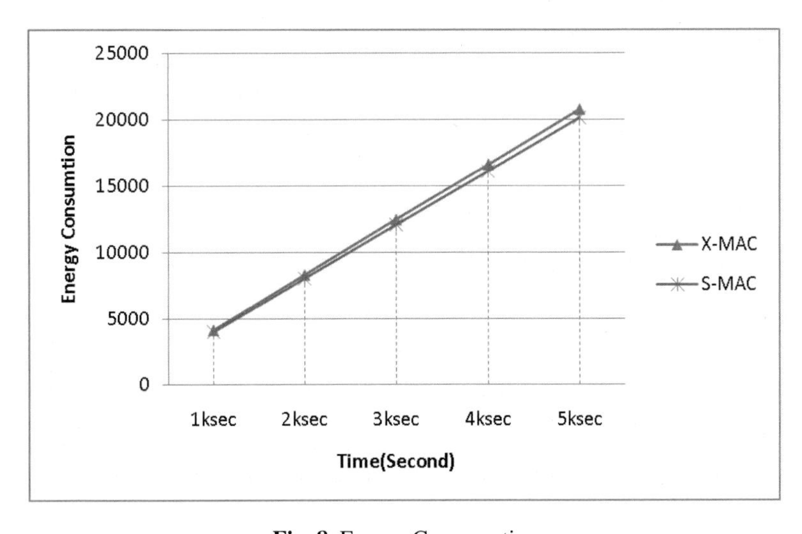

Fig. 8. Energy Consumption

The simulation measurement result in Figure 8 suggested that the energy performance of S-MAC is slightly better than X-MAC. It is believed that this phenomenon occurs when there are many nodes and the number of packets being transmitted is very small. As we can learn from the simulation result, when we apply sensor network to agricultural environment, the protocol proper to the operation environment can be chosen through the comparison and analysis of MAC protocol to be applied, together with the design of network topology in advance.

4 Conclusion

When applying a wireless sensor network to agriculture, the first thing to determine is whether agricultural site is dynamic or static. Next will be the sensors proper to the environment information of the crops to be measured, number of sensors for each sensor node and the data measurement cycle. If one sensor node will have multiple sensors, there will be more power consumption to operate sensor, in addition to the data transmission/reception. The capacity of the sensor to be used would be chosen considering all these.

Then the sensors will be located at the position of the crop to be measured. The location of sink node will be determined by the methodology of data collection. Then the data generation cycle of sensor and the shape of network topology will be determined. After that, the MAC protocol will be determined, applying the MAC protocol determining methodology suggested by this research. It was found that the S-MAC protocol is proper as the MAC to be applied to the facility cultivation of paprika.

This research result suggested the methodology to deduce the most efficient protocol which can be applied to the facility cultivation of paprika. However, it is believed that this result can be also applied to the outdoor cultivation, outdoor stock-raising and cultivation of other crops than paprika in the determination of proper MAC protocol for the situation and subsequent efficient operation.

Acknowledgements. This research was supported by the MKE(The Ministry of Knowledge Economy), Korea, under the ITRC(Information Technology Research Center) support program supervised by the NIPA(National IT Industry Promotion Agency) (NIPA-2010-(C1090-1021-0009)).

References

1. Lee, K.H., Ahn, C.M., Park, G.M.: Characteristics of the Convergence among Traditional Industries and IT Industry. Electronic Communications Trend Analysis 23(2), 13–22 (2008)
2. Lee, M.-h., Shin, C.-s., Jo, Y.-y., Yoe, H.: Implementation of Green House Integrated Management System in Ubiquitous Agricultural Environments. Journal of KIISE 27(6), 21–26 (2009)
3. Ye, W., Heidemann, J., Estrin, D.: An Energy-Efficient MAC Protocol for Wireless Sensor Networks. In: 21st Conf. of the IEEE Computer and Communicaions Societies (INFOCOM), vol. 3, pp. 1567–1576 (2002)

4. van Dam, T., Langendoen, K.: An adaptive energy efficient mac protocol for wireless sensor networks. In: 1st ACM Conference on Embedded Networked Sensor Systems (SenSys), pp. 171–180 (2003)
5. Ye, W., Heidemann, J., Estrin, D.: Medium access controlwith coordinated, adaptive sleeping for wireless sensor networks. ACM Transactions on Networking 12(3), 493–506 (2004)
6. Rhee, I., Warrier, A., Aia, M., Min, J.: ZMAC: a Hybrid MAC for Wireless Sensor Networks. In: Proc. of 3rd ACM Conference on Embedded Networked Sensor Systems, SenSys 2005 (2005)
7. Buettner, M., Yee, G.V., Anderson, E., Han, R.: X-MAC: A Short Preamble MAC Protocol For Duty-Cycled Wireless Sensor Networks. In: Conference On Embedded Networked Sensor Systems, pp. 308–320 (2006)
8. Korea Agricultural Trade Information (KATI), The state of sweet pepper industry in korea, Korea Agro-Fisheries Trade Corporation (2009)
9. Dorais, M.: The use of supplemental lighting for vegetable crop production: Light intensity, crop response, nutrition, crop management, cultural practices. In: Canadial Greenhouse Conference (2003)
10. Heuvelink, E., Marcelis, L.F.M., Korner, O.: How to reduce yield fluctuations in sweet pepper. Acta. Hort 633, 349–355 (2004)
11. Jeong, W.-J., Lee, J.H., Kim, H.C., Bae, J.H.: Dry Matter Production, Distribution and Yield of Sweet Pepper Grown under Glasshouse and Plastic Greenhouse in Korea. Journal of Bio-Environment Control 18(3), 258–265 (2009)
12. Jeong, W.J., Kang, I.K., Lee, J.Y., Park, S.H., Kim, H.S., Myoung, D.J., Kim, G.T., Lee, J.H.: Study of dry and bio-mass of sweet pepper fruit and yield between glasshouse and plastic greenhouse. The Kor. Soc. Bio-Environ. Control 17(2), 541–544 (2009)
13. Polastre, J., Hill, J., Culler, D.: Versatile low power media access for wireless sensor networks. In: The Second ACM Conference on Embedded Networked Sensor Systems (SenSys), pp. 95–107 (2004)
14. Park, D.-H., Kang, B.-J., Cho, K.-R., Sin, C.-S., Cho, S.-E., Park, J.-W., Yang, W.-M.: A Study on Greenhouse Automatic Control System Based on Wireless Sensor Network. Wireless Pers. Commun. (2009)
15. Network Simulator, `http://www.isi.edu/nsnam/ns`

An u-Service Model Based on a Smart Phone for Urban Computing Environments*

Yongyun Cho and Hyun Yoe

Information and Communication Engineering, Sunchon National University,
413 Jungangno, Suncheon, Jeonnam 540-742, Korea,
{yycho,yhyun}@sunchon.ac.kr

Abstract. In urban computing environments, all of services should be based on the interaction between humans and environments around them, which frequently and ordinarily in home and office. This paper propose an u-service model based on a smart phone for urban computing environments. The suggested service model includes a context-aware and personalized service scenario development environment that can instantly describe user's u-service demand or situation information with smart devices. To do this, the architecture of the suggested service model consists of a graphical service editing environment for smart devices, an u-service platform, and an infrastructure with sensors and WSN/USN. The graphic editor expresses contexts as execution conditions of a new service through a context model based on ontology. The service platform deals with the service scenario according to contexts. With the suggested service model, an user in urban computing environments can quickly and easily make u-service or new service using smart devices.

1 Introduction

Recently, along with the rapid growth of context-aware technologies using situation or location information in ubiquitous computing, there have been various researches for smart services in urban computing environments toward metro cities, which have filled with smart technologies such as sensors, RFID/USN, and hand-held devices [1–3]. One of the characteristics of the urban computing environments might be that the services should be based on the intercommunication between humans and environments or devices and devices. Now, the urban computing environment has produced various new smart service models in many of modern cities around the world. Generally, a smart service in urban computing environments, like in ubiquitous computing environments, may be also based on contexts to describe the situation information of an entity that consists of any domain such as u-home, u-office, or u-city, in which the interaction between humans and their environments frequently happens. The contexts

* This research was supported by the MKE(Ministry of Knowledge Economy), Korea, under the ITRC(Information Technology Research Center) support program supervised by the IITA(Institute of Information Technology Advancement)" (IITA-2009-(C1090-0902-0047)).

T.-h. Kim, A. Stoica, and R.-S. Chang (Eds.): SUComS 2010, CCIS 78, pp. 600–605, 2010.

can be defined as the knowledge about the user's and IT device's state, including surroundings, situation, and to a less extent, location" [4–6].

There have been many of researches about context-aware services in various application domains. In this paper, we propose a smart service model based on various smart devices for urban computing environments. The suggested model uses web-service technologies and workflow technologies to support an independent and automatic context-aware service. The workflow technology with context-awareness uses a context in a workflow service model as conditions for service execution. So, the suggested service model can offer an context-aware or smart service to humans with various contexts which can occur in urban computing environments. Section 2 in this paper describes the related works about context-aware service models or service systems in ubiquitous computing environments or urban computing environments. Section 3 describes the architecture of the suggested u-service model. Section 4 proposes the experimental service scenario using the suggested service model with a context-aware workflow and implementation. Section 5 mentions conclusion and future work.

2 Related Work

[7] is a recent interesting research about an urban computing management system using mobile phone, which is designed with wireless ad hoc networks. The system includes complete software which can make all kinds of autonomous devices communicate with each others in the urban computing environments.

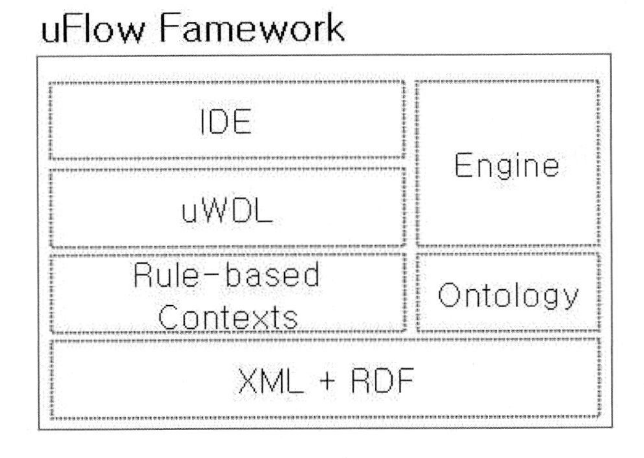

Fig. 1. A platform structure of uFlow

uFlow [8] is a web service-based framework to support a context-aware service using uWDL [2], which is a context-aware workflow language. uFlow can express independent services as a context-aware service flow and provide the functionalities to select an appropriate service based on high-level contexts, profiles, and events information, which are obtained from various sources and structured by ontology [9]. In Figure 1 shows the platform structure of uFlow. uWDL in Figure 1 can define low/high-level contexts using the rule-based contexts, and use the contexts as conditions for service execution. To do this, uWDL has a context model consisting of the triplet of <subject>, <verb>, and <object> based in RDF [10].

3 A Suggested u-Service Model Using Smart Devices

3.1 An Architecture of the Suggested Model

Figure 2 shows the architecture of the suggested u-service model using smart devices in urban computing environments.

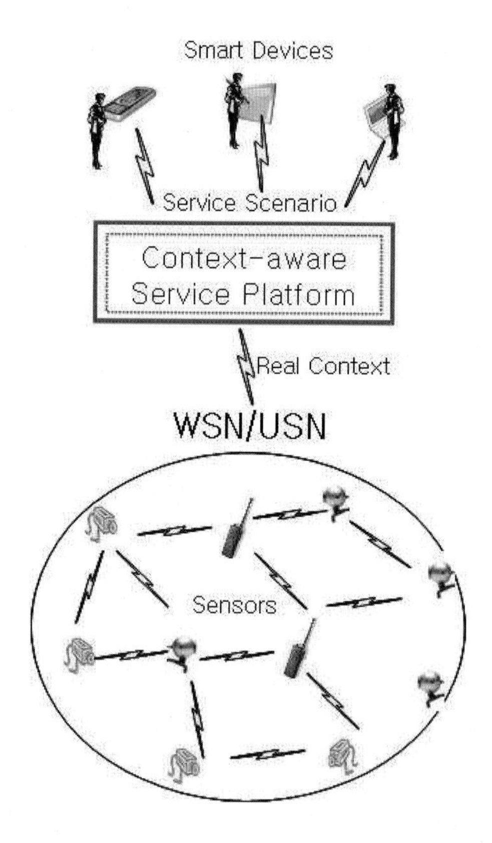

Fig. 2. A conceptual architecture of the suggested system

As shown in Figure 2, the architecture of the suggested service model supports a context-aware service using an user's situation information described through various smart devices. When a user makes a context-aware service scenario, the service scenario can be transmitted to the context-aware service platform through wireless communications. After then, the context-aware service platform performs a proper context-aware service through the communication with the sensors and devices in real-time. Figure 3 describes the structure of the context-aware service platform in the Figure 2.

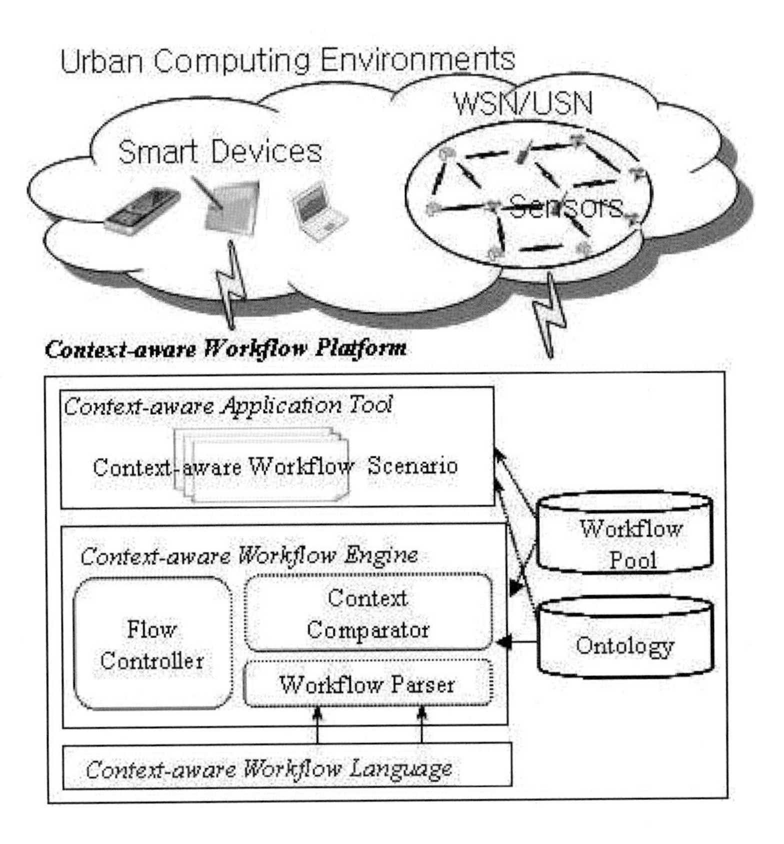

Fig. 3. A structure of the context-aware service platform

Figure 3 shows the more detail architecture of the suggested model for context-aware service in urban computing environments. As shown in Figure 3, the architecture for the suggested service model supports client application which an end-user's situation information is described as a service execution condition. When an end-user writes a service scenario in a context-aware workflow language, like uWDL, the service scenario is transmitted to a context-aware workflow scenario parser in Figure 3. The parser represents contexts described in a scenario as RDF-based tree data structure through parsing. And then, the flow controller and context comparator in Figure 3 performs context-aware service according to the results of the comparison with the user-made service scenario and the contexts produced from real urban computing environments.

3.2 A Context-Aware Service with a Smart Device

An user in urban computing environments wants to make a workflow service scenario through smart devices easily and quickly. That is, an user can easily reserve services that he/she want in anytime and anyplace. To edit workflow scenario in smart device's edit environment, that is better and more convenience than general computers. Therefore, context-aware service model has to offer a method with which a user can make a context-aware service scenario using a smart device. The environment may be based on graphical user interfaces to describe a context as conditions for a service transition through ontology dictionary.

Figure 4 shows the possible editing environment based on uFlow for a hand-held device such as a smart phone, smart note, and smart PDA.

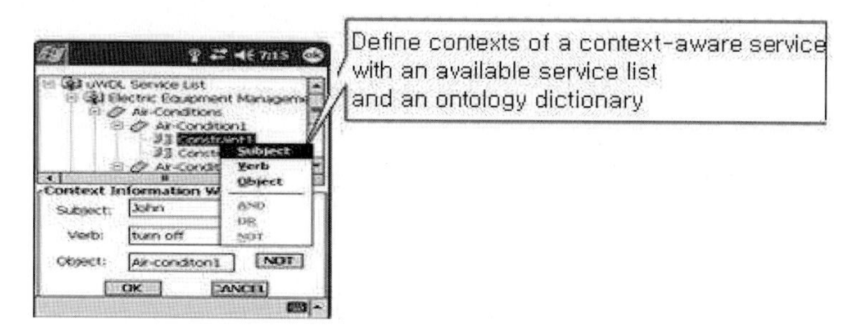

Fig. 4. A possible context-aware service editing environment for smart devices

In Figure 4, the possible environment to edit context-aware scenario can be include an available service list window, an ontology dictionary window, and an ontology-based context information window. So, an user can select various types of services which she/he wants from the workflow service list, and define contexts for the execution of the service.

4 Conclusion

This paper suggested an u-service model with smart devices for urban computing environment, which supports a context-aware or smart service based on contexts. For this, the suggested service model has a systematic architecture, which consists of the infrastructure based on the WSN/USN and sensors in urban computing environment, a context-aware service scenario editing environment using smart device, and a context-aware service framework to process the scenario. In the structure of the suggested service model, it includes a workflow technology, an ontology technology, and web service discovery technology. So, using the suggested service model, a developer can easily and quickly make a new u-service using various smart devices. Therefore, it can raise development efficiency of various u-service applications in urban computing environments. As a future work, we will research the method to implement the suggested conceptual architecture using the available context-aware service platform like uFlow.

References

1. Cho, Y., Yoe, H., Kim, H.: CAS4UA: A Context-Aware Service System Based on Workflow Model for Ubiquitous Agriculture. In: Kim, T.-h., Adeli, H. (eds.) AST/UCMA/ISA/ACN. LNCS, vol. 6059, pp. 572–585. Springer, Heidelberg (2010)
2. Han, J., Cho, Y., Choi, J.: Context-Aware Workflow Language based on Web Services for Ubiquitous Computing. In: Gervasi, O., Gavrilova, M.L., Kumar, V., Laganá, A., Lee, H.P., Mun, Y., Taniar, D., Tan, C.J.K. (eds.) ICCSA 2005. LNCS, vol. 3481, pp. 1008–1017. Springer, Heidelberg (2005)
3. Cho, Y., Moon, J., Yoe, H.: A Context-Aware Service Model Based on Workflows for u-Agriculture. In: Gervasi, O. (ed.) ICCSA 2010, Part III. LNCS, vol. 6018, pp. 258–268. Springer, Heidelberg (2010)
4. Tang, F., Guo, M., Dong, M., Li, M., Guan, H.: Towards Context-Aware Workflow Management for Ubiquitous Computing. In: Proceedings of ICESS 2008, pp. 221–228 (2008)
5. Dey, A.: Understanding and Using Context. Personal and Ubiquitous Computing 5(1) (2001)
6. Choi, J., Cho, Y., Choi, J.: The Design of a Context-Aware Workflow Language for Supporting Multiple Workflows. Journal of Korean Society for Internet Information 11(1), 145–158 (2009); 30-39 (2006)
7. Mitra, K., Bhattacharyya, D., Kim, T.: Urban Computing and Informaiton Management System Using Mobile Phone in Wireless Sensor Network. Internation Journal of Control and Automation 3(1), 18–26 (2010)
8. Han, J., Cho, Y., Kim, E., Choi, J.: A Ubiquitous Workflow Service Framework. In: Proceedings of the 2006 International Conference on Computational Science and its Application, pp. 30–39 (2006)
9. Lauser, B., Sini, M., Liang, A., Keizer, J., Katz, S.: From AGROVOC to the Agricultural Ontology Service / Concept Server - An OWL model for creating ontologies in the agricultural domain. In: Proceedings of the OWLED 2006 Workshop on OWL: Experiences and Directions, Athens, Georgia (USA), pp. 10–11 (2006)
10. Beckett, D.: W3C: RDF/XML Syntax Specification. W3C Recommendation, University of Bristol (2004)

An Incremental Join Algorithm in Sensor Network

Hyun Chang Lee[1,*], Young Jae Lee[2], and Dong Hwa Kim[3]

[1] 344-3 Sinyong-Dong, Iksan-si, Jeonbuk, Division of Information and e-Commerce, Wonkwang University, 570-749, Korea
hclglory@wku.ac.kr
[2] 45 Baekma-gil, Jeonju-si, Jeonbuk, School of Media and Information, Jeonju university
leeyj@jj.ac.kr
[3] 16-1, Deokmyung-Dong, Daejeon, Control Instrumentation Engineering Major, Hanbat National University, 305-719, Korea
kimdh@hanbat.ac.kr

Abstract. Given their autonomy, flexibility and large range of functionality, wireless sensor networks can be used as an effective and discrete means for monitoring data in many domains needed to gather and process real-time state information which is one of the sensor network features. But sensor nodes are generally very constrained, in particular regarding their energy and memory resources. While simple queries such as select and aggregate queries in wireless sensor networks have been addressed in the literature, the processing of join queries in sensor networks remain to be investigated. Previous approaches have either assumed that the join processing nodes have sufficient memory to buffer the subset of the join relations assigned to them, or that the amount of available memory at nodes is known in advance. Therefore, in this paper including these assumptions, we describes an Incremental Join Algorithm(IJA) in Sensor Networks to reduce the overhead of moving a join tuple to the final join node or minimize the communication cost that is the main consumer of the battery when processing the distributed queries in sensor networks environments. The simulation result shows that the proposed IJA algorithm significantly reduces the number of bytes to be moved to join nodes compared to the popular synopsis join algorithm.

Keywords: sensor network, incremental maintenance, wireless environment, query processing.

1 Introduction

Technological advances, decreasing production costs and increasing capabilities have made sensor networks suitable for many applications, including environmental monitoring, warehouse management and battlefield surveillance. As well, sensor networks have been adopted in various scientific and commercial applications [1,2,3].

* Corresponding author.

T.-h. Kim, A. Stoica, and R.-S. Chang (Eds.): SUComS 2010, CCIS 78, pp. 606–613, 2010.

Gathering data from sensors is achieved by modeling it as a distributed database where sensor readings are collected and processed using queries [4,5,6].

Especially, sensor nodes consisting of sensor networks are obtaining the state information from sensor device parts on those nodes and store those data. Accordingly, each sensor node in sensor networks is regarding as distributed database system generating data stream and has studied as a sensor database [7,8,9,10,11]. In query processing of sensor networks, Join operation cost much in sensor networks for correlating sensor readings like distributed database environments [12]. Therefore, many researches have studied on reducing the cost in sensor environment.

In this paper, we propose an Incremental Join Algorithm (IJA) as an in-network join strategy which is an efficient join processing in sensor network and minimizes communication cost. The IJA strategy is capable of reducing communication cost and utilizing data by gathering real-time state information from sensors which is one of sensor network features. In sensor network environments, it is hard to send all data stored at each node into server located in the center. Therefore it needs to be filtered whether data are sent or not. Therefore assume previous join results are stored in temporary repository to process efficiently. To evaluate the performance, we compare IJA strategy to synopsis strategy.

The remainder of the paper is structured as follows. In the next section, we describe typical join strategies in sensor networks including synopsis to compare. In section 3, we introduce and explain an incremental join algorithm. And we analyze the performance and compare IJA to typical join strategy including synopsis algorithm. In section 5, we conclude with future works.

2 Related Work

In sensor network, the value obtained from sensor nodes is lack of expressing all the information about event or entity. It needs join operation for that problem. To process a join query, we first have to decide which join queries are used. In this paper, we consider binary equi-join (BEJ). A BEJ query for sensor networks is defined as follows.

Definition. Given two sensor tables $R(A_1, A_2, \ldots, A_n)$ and $S(B_1, B_2, \ldots, B_m)$, a binary equi-join (BEJ) is

$$\mathbf{R} \infty_{\mathbf{Ai=Bj}} \mathbf{S} \quad (i \in \{1,2, \ldots., n\}, j \in \{1,2, \ldots.., m\},)$$

Where A_i and B_j are two attributes of R and S respectively, which have the same domain.

Consider a sensor network covering a road network from [12]. Each sensor node can detect the ID's of vehicles in close vicinity, record the timestamps at which the vehicles are detected. Suppose N_R and N_S represent two sets of sensor nodes located at two regions of a road segment, Region1 and Region2, respectively. To gather the necessary data for determining the speeds of vehicles traveling between the two regions, the following join query can be expressed.

SELECT R.autoID, R.time, S.time
FROM R, S
WHERE R.location IN Region1 AND S.location IN Region2 AND R.autoID = S.autoID

To evaluate the above query, sensor readings from Region1 and Region2 need to be collected and joined on the autoID attribute. Typical join strategies of sensor networks are classified into Naïve join, Sequential join and Centroid join according to the join location and shown in figure 1 [13].

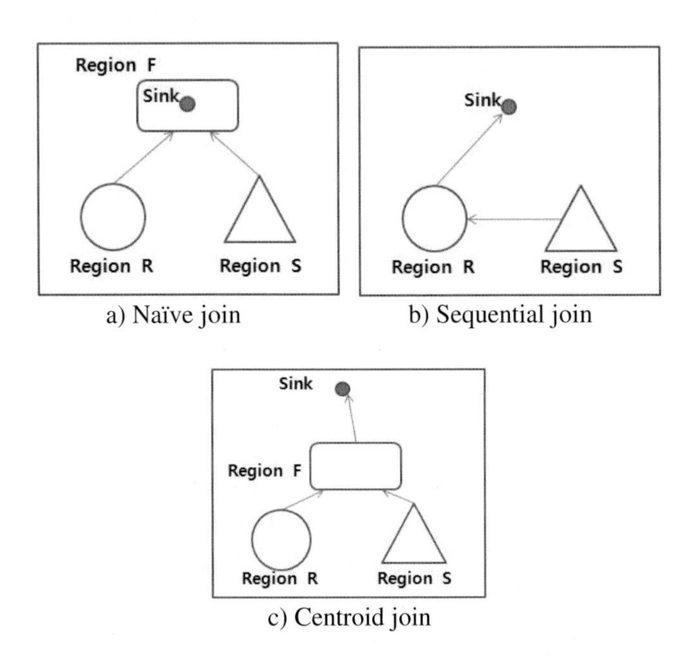

a) Naïve join b) Sequential join

c) Centroid join

Fig. 1. General join strategies

One of general join problems is a lower join selectivity of query regarding overhead in communication. For instance, given there are two tables, R and S, tuples which is not participated in joining operation between R and S tables are possible to be sent to other region F to join with other table.

To solve the problems above, synopsis strategy join (SNJ) was suggested. After reducing the number of tuples in R and S tables using synopsis to remove the rest tuples not to participate join operation, SNJ sends the tuples to join with others. The means of synopsis is an abstract of a table to process join operation. In addition, the size of the synopsis table is smaller than original table size. Therefore, each sensor creates its synopsis [13].

Synopsis strategy consists of 3 steps. First is synopsis join step. The second step is notification and third is final join operation. The first join operation of synopsis is following.

Each node, $n_i \in N_R$, stores local table R_i which is one of local tables consisting of table R. Also each node n_i creates local synopsis Si (R_i) by extracting join attributes A_j, and counting the frequency of the same value in the table. Synopsis join region N_L is selected to get a final join candidate tuples from joining table R and S synopsis. The synopsis join nodes after receiving synopsis from N_R and N_S synopsis process synopsis join operations.

The second step, a notification of synopsis join strategy, notifies final join candidate tuples to the N_R and N_S nodes. For this, synopsis join node n_l stores sensor ID of local synopsis originated. At the third step, each node of N_R or N_S notified from synopsis join nodes n_l sends join attribute v to final join node n_f. The final join node n_f joins with $R_v \infty S_v$, and then sends the results to query sink node.

3 IJA Algorithm

In this section, we propose an incremental join algorithm (IJA) to gather and process real-time state information which is one of the sensor network features. First, we describe general environment components including terms in next section [12] and the algorithm later.

3.1 General Environments

Suppose a sensor network consisting of *N* sensor nodes. We assume there are two virtual tables in the sensor network, *R* and *S*, containing sensor readings distributed in sensors. Each sensor reading is a tuple with two mandatory attributes, timestamp and sensorID, indicating the time and the sensor at which the tuple is generated. A sensor reading may contain other attributes that are measurements generated by a sensor or multiple sensors, e.g., temperature, autoID. We are interested in the evaluation of static one-shot binary equi-join queries in sensor networks. We assume that *R* and *S* are stored in two sets of sensor nodes N_R and N_S located in to distinct regions known as R and S respectively. A BEJ query can be issued from any sensor node called query sink, which is responsible for collecting the join result. A set of nodes is required to process the join collaboratively, referred to as join nodes.

When a join query is issued, a join node selection process is initiated to find a set of join nodes N_F to perform the join. *R* tuples are routed to a join region *F* where the join nodes N_F reside in. Each join node $n_f \in N_F$ stores a horizontal partition of the table *R*, denoted as R_f. *S* tuples are transmitted to and broadcast in *F*. Each join node n_f receives a copy of *S* and processes local join $R_f \infty S$. The query sink obtains the join results by collecting the partial join results at each n_f.

The selection of N_F is critical to the join performance. Join node selection involves selecting the number of nodes in N_F, denoted by $|N_F|$, and the location of the join region *F*. To avoid memory overflow, assuming *R* is evenly distributed in N_F, $|N_F|$ should be at least $|R|/m$, where $|R|$ denotes the number of tuples in *R* and *m* denotes the maximum number of *R* tuples a join node n_f can store.

3.2 IJA Strategy

The following figure 2 shows the flow of incremental join algorithm.

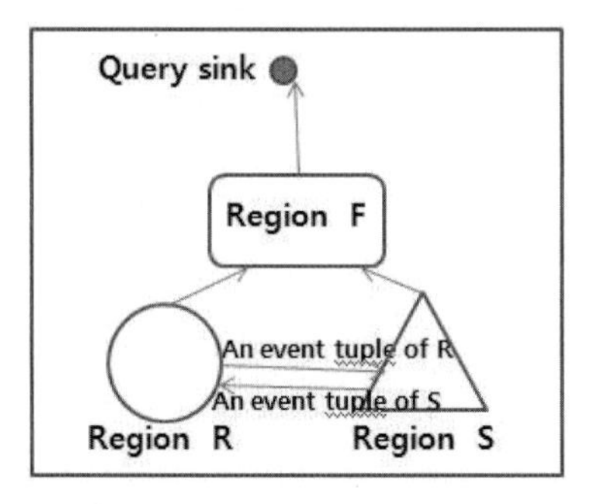

Fig. 2. Incremental join strategy

The steps for IJA are followings.

1. Send an event tuple of *R*(or *S*) to other part.
   Only send the tuple to Region S to make a semi table P_R (or P_S)at the counter part.
2. Operate join operation at each region to produce a semi table.
   Send the semi table to region F
3. Operate join operation with semi tables from R and S repectively.
   Send the join results to the query sink.
4. At the query sink node, the query can get the result within region F not to compute all of R and S computations.

The incremental join algorithm is objective to produce a smaller semi table by processing join operation at the counter region because it needs to be decided whether the event tuple is useful to join at the region or not. Another operations to process at the regions such as selecting a center location of the regions, routing protocol etc, is based on [12],[14],[15]. The number of the join nodes at semi table join region is decided by P_R and P_S arrived at N_H region. Given memory m for a node, the node number at semi table region is following.

$$|N_H| = (|P_R| + |P_S|) / m$$

The different point with other algorithms is just sending the tuple whenever a tuple is occurred with insert or update operation. To compute the communication cost, $|N_F|$ is following.

$$|N_F| = \Big(\sum_{n_i \in N_R} |C(R_i)| + \sum_{n_j \in N_S} |C(S_j)| \Big) / m$$

Where $|C(R_i)|$ is the number of join candidate tuples arrived from node $n_i \in N_R$. $|C(S_j)|$ is the number of join candidate tuples arrived from node $n_j \in N_S$.

4 Performance Evaluation of IJA

In this chapter, we evaluate the performance of incremental join algorithm (IJA) and compare it with typical join strategies including synopsis strategy, including naïve join, sequential join and centroid join. The experiment is mainly measured by the total number of messages incurred for each join strategy. Other comparisons for performance evaluation will be included in our near future work.

4.1 Experiment Environments

We created a simulation result the same as synopsis strategy [12] done for comparing naïve join, sequential join, centroid join and also including synopsis strategy. In case of the number of sensor node, this experiment has done with 10,000 sensor nodes uniformly placed in a 100*100 grid. Each grid contains one sensor node located at the center of the grid. The regions R and S are located at the bottom-right and bottom-left corners of the network region, respectively, each covering 870 sensor nodes. Table R consists of 2000 tuples, while S consists of 1000 tuples. We set a message size of 40 bytes, which is equal to the size of a data tuple. A tuple in the join result is 80bytes since it is ac concatenation of two data tuples. The messages for synchronization and coordination among the sensors are negligible compared to the data traffic for communication caused by large tables.

4.2 Performance Evaluation

We first varied the join selectivity and the synopsis selectivity for synopsis strategy. Join selectivity δ is defined as $|R \infty S| / (|R| * |S|)$. The join attribute values are uniformly distributed within the domain of the attribute. Figure 3 shows the total communication cost for different join selectivities while keeping the memory capacity and synopsis size fixed at 250 * 40 bytes and 10 bytes respectively.

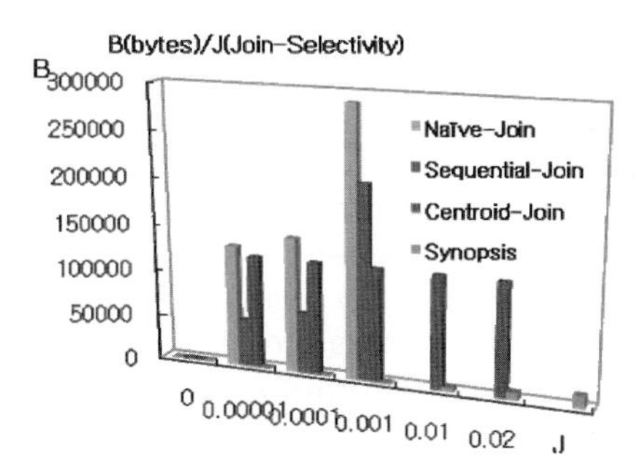

Fig. 3. Impact of selectivity

As shown in the figure, naïve join performs worse than all others due to the high cost of broadcasting S to all nodes in N_R. In addition, sequential join performs worse than centroid join and synopsis as well. Therefore we exclude them from when join selectivity is greater than 0.01. Synopsis strategy is lower than others and outperforms because non-candidate tuples can be determined in the synopsis in the synopsis join state, and only a small portion of data are transmitted during the final join. However, IJA performs than all algorithms though not to be shown in the figure. Therefore figure 4.a shows the comparison synopsis algorithm to incremental join algorithm.

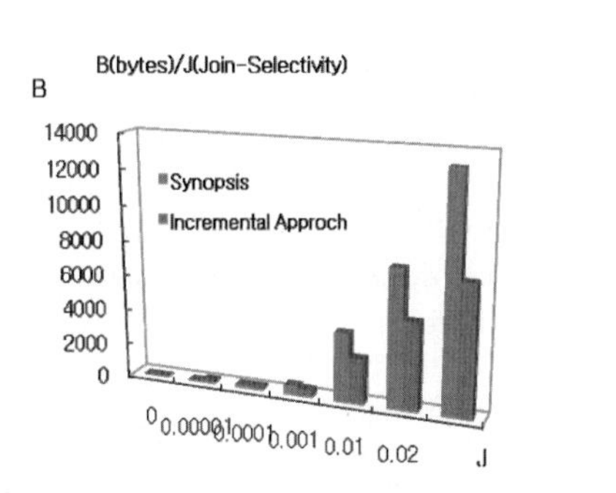

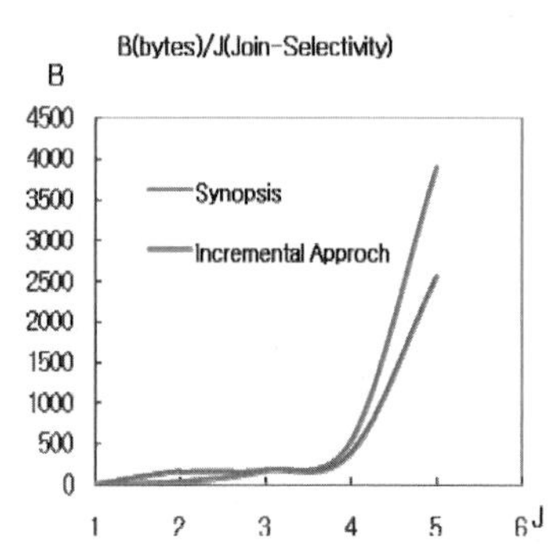

a)Comparison synopsis to IJA b)Magnification with join selectivity <=0.01

Fig. 4. Comparison IJA to synopsis algorithm

Figure 4.b shows the enlargement of lower selectivity than 0.01 in the axis of figure 4.a. For that case, synopsis has lower communication cost than IJA. This is because synopsis strategy has both join selectivity and synopsis selectivity parameters which have a strong effect on communication cost. However we have shown the result of the experiments for comparing the IJA to synopsis strategy.

5 Conclusion

Sensor networks have been adopted in various scientific and commercial applications. Gathering data from sensors is achieved by modeling it as a distributed database where sensor readings are collected and processed using queries. Sensor nodes are generally very constrained, in particular regarding their energy and memory resources. While simple queries such as select and aggregate queries in wireless sensor networks have been addressed in the literature, the processing of join queries in sensor networks remain to be investigated. Previous approaches have either assumed that the join processing nodes have sufficient memory to buffer the subset of the join relations assigned to them, or that the amount of available memory at nodes is known in advance.

Therefore, in this paper including these assumptions, we describes an Incremental Join Algorithm(IJA) in Sensor Networks to reduce the overhead of moving a join tuple to the final join node or minimize the communication cost that is the main consumer of the battery when processing the distributed queries in sensor networks environments.

To evaluate the experiments, we compare the IJA to the typical algorithms including synopsis algorithm which is representative strategy in sensor network to process query. We also show the result of comparisons. As a future works, we will vary and perform the parameters, such as network density, node memory capacity and synopsis size including communication cost.

Acknowledgments. This paper was supported by wonkwang university in 2010.

References

1. Mainaring, A., Culler, D., Plastre, J., Szewczyk, R., Anderson, J.: Wireless sensor networks for habitat monitoring. In: Proceedings of WSNA 2002 (2002)
2. Estrin, D., Govindan, R., Heidemann, J.S., Kumar, S.: Next century challenges: Scalable coordination in sensor networks. In: Proceedings of MobiCom (1999)
3. Estrin, D., Govindan, R., Heidemann, J.S. (eds.): Special Issue on Embedding the Internet. Communications of the ACM 43 (2000)
4. Bonnet, P., Gehrke, J., Seshadri, P.: Towards sensor database systems. In: Tan, K.-L., Franklin, M.J., Lui, J.C.-S. (eds.) MDM 2001. LNCS, vol. 1987, pp. 3–14. Springer, Heidelberg (2001)
5. Madden, S., Franklin, M.J., Hellerstein, J.M., Hong, W.: TAG: A Tiny AGregation service for ad-hoc sensor networks. In: Proceedings of OSDI 2002 (2002)
6. Yao, Y., Gehrke, J.E.: The cougar approach to in-network query processing in sensor networks. SIGMOD Record 31(3), 9–18 (2002)
7. Chowdhary, V., Gupta, H.: Communication-efficient implementation of join in sensor network. In: Zhou, L.-z., Ooi, B.-C., Meng, X. (eds.) DASFAA 2005. LNCS, vol. 3453, pp. 447–460. Springer, Heidelberg (2005)
8. Yao, Y., Gehrke, J.: Query processing for sensor networks. In: Proc. of Int'l Conference on Innovative Data System Research (2003)
9. Sun, J.Z.: An Energy-Efficient Query Processing Algorithm for Wireless Sensor Networks. LNCS, pp. 373–385.Springer, Heidelberg (2008)
10. Zhang, Z., Gao, X., Zhang, X., Wu, W., Xiong, H.: Three TApproximation Algorithms for Energy-Efficient Query Dissemination in Sensor Database System. LNCS, pp. 807–821, Springer, Heidelberg (2009)
11. Coman, A., Nascimento, M.A.: A distributed Algorithm for Joins in Sensor Networks. In: Proc. of Int'l Conference on SSDBM (2007)
12. Yu, H., Lim, E., Zhang, J.: In-network Join Processing for Sensor Networks. In: Zhou, X., Li, J., Shen, H.T., Kitsuregawa, M., Zhang, Y. (eds.) APWeb 2006. LNCS, vol. 3841, pp. 263–274. Springer, Heidelberg (2006)
13. Coman, A., Nascimento, M., Sander, J.: On join location in sensor networks. In: Proc. of MDM (2007)
14. Greenberg, I., Robertello, R.A.: The three factory problem. Mathematics Magazine 38(2), 67–72 (1965)
15. Karp, B., Kung, M.J.: GPSR: Greedy perimeter statelss routing for wireless networks. In: Proc. of MobiComm (2000)

Interface of Augmented Reality Game Using Face Tracking and Its Application to Advertising

Young Jae Lee[1] and Yong Jae Lee[2]

[1] Jeonju University, 1200 Hyoja Dong Wansan-Gu Jeonju Jeonbug 560-759, Korea
leeyj@jj.ac.kr
[2] Tongmyung University, 179 Sinsunro Nam Gu Busan, 680-711, Korea
leeyj@tu.ac.kr

Abstract. This paper proposes the face interface method which can be used in recognizing gamer's movements in the real world for application in the cyber space so that we could make three-dimensional space recognition motion-based game. The proposed algorithm is the new face recognition technology which incorporates the strengths of two existing algorithms, CBCH and CAMSHIFT and its validity has been proved through a series of experiments. Moreover, for the purpose of the interdisciplinary studies, concepts of advertising have been introduced into the three-dimensional motion-based game to look into the possible new beneficiary models for the game industry. This kind of attempt may be significant in that it tried to see if the advertising brand when placed in the game could play the role of the game item or quest. The proposed method can provide the basic references for developing motion-based game development.

Keywords: face interface, CBCH, CAMSHIFT, advertisement.

1 Introduction

With revolutionary growth of computer-related and sensor technologies, sensory and emotive technologies are being established as the new technological trend in the global game market. Especially, in the mainstream are the three-dimensional space-recognizing motion-based games and functional games which incorporate educational, training, and therapeutic purposes.[1-16] The main focus of the AR(Augmented Reality) game is on increasing fun by making the virtual interaction more like the real-world experiences. Attempts have steadily been made to apply 'Augmented Reality Technology' to games. The instances are 'Car Race Game on the Table' and 'Bowling Game'[1]. “Eye of Judgement", the card game for PlayStation 3, is also the AR game which has been released by Sony in 2008 and it shows three-dimensional virtual characters on the table by recognizing a specific card[2-3]. Another game provides interactive experiences by recognizing the hand of the user by installing many markers on the ceiling in limited space.[4-5] Augmented Reality can provide the user with more real experiences since the user interacts with the virtual object in the environment of the real world.

However, there are not many motion-based games using AR technology because of its complex structure(HMD and AR apparatus, tracking systems and etc.) and

T.-h. Kim, A. Stoica, and R.-S. Chang (Eds.): SUComS 2010, CCIS 78, pp. 614–620, 2010.

inconvenient user interface[6]. To address this kind of problem, this paper tried to track and analyze the human face by using the web camera. The experiments have been done after realizing the human friendly interface on the game screen which is controllable and its validity has been proven. And the possibility has been looked into whether and how this kind of program could be applied to advertisement as part of the interdisciplinary study consideration.

2 Augmented Reality

2.1 Augmented Reality

Augmented reality means the kind of technology which integrates the real and the virtual information in real time to augment the human senses. In other words, it is the technology which shows things which cannot be seen with eyes or cannot be felt with hands in the real world. One example is "Scouter" which appears in the popular Japanese animation, "Dragon ball Z". Augmented Reality derives from Virtual Reality Technology and we can add virtual experience to the real world experiences by using this technology. We can create anything in the virtual world but within it the user cannot feel exactly like in the real world. Augmented Reality is gaining more attention because it helps the user maintain the sense of the real world while enjoying the benefits of the virtual world[6].

2.2 Motion Based Game

In motion-based game[7] the user interface is very similar to the real human motions. Usually generic games use keyboards, mouse or other button-pressing mechanism, but the motion-based game is based on the real-life physical motions which happen in the real world not in the game world, such as controlling the steering wheel, aiming and firing of a gun, dancing and etc. Nintendo Wii is a motion-based game for household use which was released in December, 2006. The speed sensor is installed in the controller which can capture the motions of the gamer to be stored as data. Its applications are expected to be comprehensive since it was designed to be used in various kinds of motion-based games and can also be connected with the joystick.

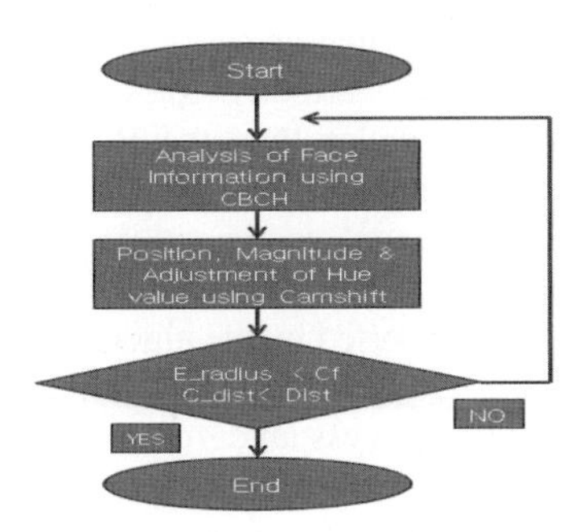

Fig. 1. Proposed Algorithm

The proposed algorithm uses the CBCH algorithm(Cascade of Boosted Classifier Working with Haar-like Feature Algorithm)[9] in the openCV[8] environment, to detect and analyze the front face and identify the center point and the circle size for face recognition. This information is used to automatically set up the ROI of CAMSHIFT algorithm.[10] Based on the circle size and the central coordinate given, CAMSHIFT algorithm is performed. After that, the Hue value is modified to get elliptical shape which is approximate to the pre-calculated circle of face size. If we apply the value of elliptical shape, we can realize the face recognition program which is automatically modified according to each face area size. Figure 1 shows the process on the flowchart.

3 Face Recognition

3.1 CBCH Algorithm

With CBCH, we generate classifier by using Harr-like Feature based on Boost technology, and search the results by using the Cascade structure. CBCH algorithm is the Gray-based algorithm and constitutes the collection of very simple features to be studied. To extract these features, Haar-like Feature which use s the light-darkness distribution and has been proposed by Papa-georiou et al[9-11], has been utilized as the feature collection for the purpose of face recognition. That is to say, integral image makes rapid calculation of face detection possible.[12]

3.2 CAMSHIFT Algorithm

The existing image-processing methods include differential image processing and the optical flow vector processing, which, however, mainly focus on tracking the movements of the object on the video streaming captured by the fixed camera, making it difficult to apply when the camera is not fixed because the object is moving. So the need arises to keep tracking the movements of the object as it moves by changing the position of the camera along. CAMSHIFT algorithm is the color-based, area-tracking algorithm which is based on the HSC color space. CAMSHIFT algorithm helps quickly calculate and track the position, rotating angle and even size by using the color information of the object, namely, the color probability distribution. So it allows for the benefits of being able to keep tracking the object in real time with the same minimal information about the search area, as long as another object does not overlap with the currently tracked object. Another benefit of CAMSHIFT is that it is much smaller than the RGB color space with complex calculating, since it uses the Hue channel in the HSV color space. It is also relatively resistant to the color change since it ignores the extreme brightness or darkness thanks to the intrinsic characteristics of the color probability distribution model. But there is the prerequisite that it has to be supported by the algorithm which guarantees a certain amount of light because of the characteristics specific to the color information. [12]

Considering the time the whole process takes to process the complex algorithm and to eliminate noise real time in the area in which more than 30 images of frames have to be processed, CAMSHIFT algorithm is more convenient than Template, Active Contour Model, MeanShift, and etc. But when we try to apply the exiting CamShift algorithm only, we have difficulties in automatically detecting and tracking the moving objects because we have to manually set up the ROI. CamShift algorithm tracks the object with the color value of the object, Hue and there can be many pixel values with the same Hue value. Therefore, if we do not control the area of the tracked object in an appropriate way, we can face the problem of the tracking area expanding unnecessarily.(Figure 2-(b))

(a) manual setup of ROI (b) expansion of the tracked area (c) the proposed algorithm

Fig. 2. Problems of CAMSHIFT algorithm

Let's address this problem. We analyze the face information derived from CBCH algorithm to identify the size of face circle and the center. We then get the circle inside the ellipsis which was gained after the CAMSHIFT algorithm calculation and modify the Hue value to get the new circle through meticulous modification based on the size and position of the previous circle. This new circle and the circle derived from the CBCH algorithm are continuously compared and if the error (E_radius, C_dist) increases, we re-perform the CAMSHIFT algorithm.

$$x^2 + y^2 = r^2$$

$$(x_2 - r_1)^2 + (y_2 - r_2)^2 = r_3^2$$

$$r_3^2 + r_4^2 = r_r^2$$

$$fr = r^2 / r_r^2 < E_{radius}$$

$$D = (r_3 - x)^2 + (r_4 - y)^2 < C_{dist}$$

The Circle in Formula (3) is the mathematical formula which is the approximation to circle of Formula (2). The confirm factor in Formula (4) establishes the new area and performs CANSHIFT if its radius comparison with the one in the circle made in Formula (3) shows less than E_radus. If the position is greater than CI_dist, CANSHIFT is re-performed to prevent the face area from being enlarged.

4 Experiment

Experiment was done on the face interface by applying the proposed algorithm to the Head Tracking library provided by nighsoft.com[14]. The background was created by using Quake 3 map and simple interface has been realized based on face movements between the background and Farie and Ninja characters. For this, CBCH (Method I) which is known for speedy recognition and the proposed modified CAMSHIFT algorithm(Method II) have been used to recognize face in various cases in order to test the face recognizing performance and applications. In Experiment 5, we dealt with the interdisciplinary studies between the 3D motion-based game and advertisement and derived an example of beneficiary models for synergy effect.

4.1 Experiment 1

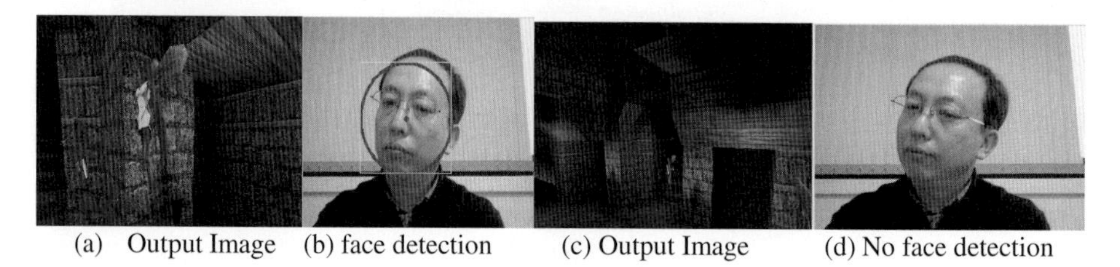

(a) Output Image (b) face detection (c) Output Image (d) No face detection

Fig. 3. Experiment 1

Experiment 1 uses Method I to detect the face(refer to Figure(b)) and shows the results of 3D camera application in the cyber game area(Figure(a)). But the problem arose when we tried to recognize the face when it was distorted(Figure(d)) and we got the remote image(Figure(c)).

4.2 Experiment 2

In Experiment 2, the face directly faces ahead (refer to Figure(b) and Figure(c)), and face was recognized with both Method I and Method II, which was then applied to realize the movement of the camera in the virtual world. In this case, the two algorithms worked just fine and face interface was possible.

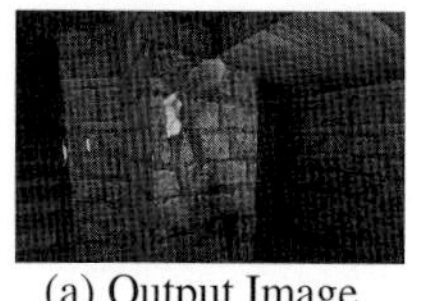

(a) Output Image

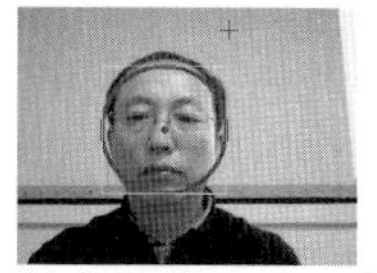

(b) face detection I

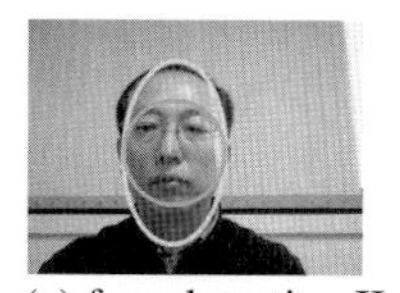

(c) face detection II

Fig. 4. Experiment 2

4.3 Experiment 3

(a) Output Image

(b) face detection I

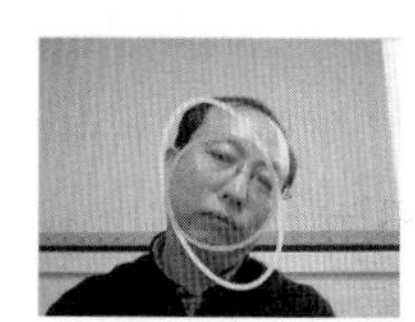

(c) face detection II

Fig. 5. Experiment 3

In Experiment 3, there was face distortion and when Method I was applied, the face was not to be recognized, making face interface impossible as shown in Figure(b). In this case, we have no other choice but to show the remote distance image like in Figure(a) of Experiment 1. If we use the proposed Method II, however, we can trace the face as in Figure(c). By moving the camera in proportion to the variations of the face, we could verify it could well express the 3D cyber space.

4.4 Experiment 4

Fig. 6. Brand exposure game

Experiment 4, looked into the possibility of interdisciplinary studies between game and advertising by placing particular brand in the three dimensional motion-based game. As a result, we could find out the possibility that the synergy effect could be maximized by integrating the two fields. It can also mean that advertising can be developed as a new beneficiary model for the game industry in the future. The beneficiary model we can think of is that when we make a certain brand appear in the game, we can get rewards whenever the user clicks on the brand. This kind of model can act as an effective marketing strategy since it can serve as an opportunity to raise the brand awareness for the sponsor and the game company can reap profits. It can also raise the enjoyment level and benefits for the user, making its realization highly probable. Prior researches also show that this kind of approach is highly implementable and beneficial for the game industry. [15-16].

5 Conclusion

This paper proposes recognition interface method for face which moves in the cyber space and verified its effectiveness through experiments. To do that, the face movements of the gamer in the cyber space has been tracked to get the movement factor, which has then been applied to the three dimensional camera. The conventional CBCH algorithm(Method I) has the benefits of rapidly recognizing face but presents a problem of not being able to deal with face distortion. To solve this problem, we may think of CAMSHIFT algorithm but it also presents a problem. This method tracks the object with the color value and hue and since there are comprehensive pixel values which have the same hue value, the tracking area gets enlarged and we have to manually set the ROI of algorithm. To solve this problem, this paper proposes Method II, considering the benefits of CBCH and CAMSHIF. Five experiments have proven its validity and possibility. This paper also looked at the design for the new beneficiary model for the game industry through interdisciplinary research between game and advertising. Further researches will be necessary for the development of various interface algorithms using fingers and hands and also for new beneficiary models for the game industry.

References

[1] Dong, Q., Sun, Z., Namee, B.M.: Physics-based table-top mixed reality games. In: Conference of the International Simulation & Gaming Association (2008)
[2] http://www.eyeofjudgement.com
[3] http://www.google.co.kr/search?hl=ko&newwindow=1&complete=1&q=Eye+of+Judgement&btnG=%EA%B2%80%EC%83%89&lr=&aq=f&aqi=g6g-m4&aql=&oq=&gs_rfai=
[4] Kim, K.Y., et al.: ARPushPush: Augmented Reality Game in Indoor Environment. In: KHCI, pp. 354–359 (2005)
[5] Park, J.S., Jeon, Y.J.: Design and Implementation of Motion-based Interaction in AR Game. Korea Game Society 9(5), 105–115 (2010)
[6] Handheld Augmented Reality Game System Using Dynamic Environment Kang, Won Hyung, Handheld Augmented Reality Game System Using Dynamic Environment, KAIST, Thesis for Master's Degree (2007)
[7] http://www.kmobile.co.kr/k_mnews/news/news_view.asp?tableid=IT&idx=278751
[8] http://sourceforge.net/projects/opencvlibrary/
[9] Viola, P., Jones, M.: Rapid object detection using a boosted cascade of simple features. In: IEEE Conf. on Computer Vision and Pattern Recognition, Kauai, Hawaii, USA (2001)
[10] Allen, J.G., Xu, R.Y.D., Jin, J.S.: Object Tracking Using CamShift Algorithm and Multiple Quantized Feature Spaces. In: Pan-Sydney Area Workshop on Visual Information Processing, pp. 3–7 (2003)
[11] Lienhart, R., Maydt, J.: An extended set of Haar-like feeature for rapid object detection. In: International Conference on Image Processing (2002)
[12] http://cafe.naver.com/opencv.cafe?iframe_url=/ArticleRead.nhn%3Farticleid=1328
[13] Lee, M., Jeong, S.: Computer vision programming using OpenCV. HongReung Press (2007)
[14] http://irrlicht.sourceforge.net/phpBB2/viewtopic.php?t=29408&sid=802b4e0b6d0c25ec6beead2ea867b4c2
[15] Cornwell, B.T., Schneider, P.L.: Cashing on crashes via brand placement in computer games. International Journal of Advertising 24(3), 321–343 (2005)
[16] Nelson, R.M., McLeod, E.L.: Adolescent brand consciousness and product placements. International Journal of Consumer Studies 29, 515–528 (2005)

Bankruptcy Problem Approach to Load-Shedding in Agent-Based Microgrid Operation

Hak-Man Kim[1], Tetsuo Kinoshita[2], Yujin Lim[3], and Tai-Hoon Kim[4]

[1] Dept. of Electrical Engineering, Univ. of Incheon, Korea
hmkim@incheon.ac.kr
[2] Graduated School of Information Science, Tohoku Univ., Japan
kino@riec.tohoku.ac.jp
[3] Dept. of Information Media, Univ. of Suwon, Korea
yujin@suwon.ac.kr
[4] Division of Multimedia, Hannam Univ., Korea
taihoonn@empas.com

Abstract. Research, development, and demonstration projects on microgrids have been progressed in many countries. Furthermore, microgrids are expected to introduce into power grids as eco-friendly small-scale power grids in the near future. Load-shedding is a problem not avoided to meet power balance between power supply and power demand to maintain specific frequency such as 50 Hz or 60 Hz. Load-shedding causes consumers inconvenience and therefore should be performed minimally. Recently, agent-based microgrid operation has been studied and new algorithms for their autonomous operation including load-shedding has been required. The bankruptcy problem deals with distribution insufficient sources to claimants. In this paper, we approach the load-shedding problem as a bankruptcy problem and adopt the Talmud rule as an algorithm. Load-shedding using the Talmud rule is tested in islanded microgrid operation based on a multiagent system.

Keywords: microgrid, microgrid operation, load-shedding, bankruptcy problem, Talmud rule.

1 Introduction

Microgrids, eco-friendly small-scale power grids, are expected to introduce into power grids in the near future. Microgrids are operated in the grid-connected mode and in the islanded mode. Balance between power supply and power demand is an important requirement for maintaining specific frequency such as 50 Hz or 60 Hz. Especially, microgrids isolated electrically from power grids should be operated to meet the balance. Although energy storage devices are used for imbalance of power generally in the case of supply shortage, it is difficult to resolve the imbalance completely. In this case, load-shedding, which is an action reducing load intentionally to resolve supply shortage is used generally. Especially, load-shedding should be performed minimally because load-shedding causes consumers inconvenience.

T.-h. Kim, A. Stoica, and R.-S. Chang (Eds.): SUComS 2010, CCIS 78, pp. 621–628, 2010.

Recently, the multiagent system has been studied for autonomous microgrid operation [1-4]. Especially, for load-shedding in autonomous operation based on a multiagent system, a fair and effective scheme is required because of consumers' competing and conflicting claims for their loads.

The bankruptcy problem deals with how to divide insufficient estate among all claimants. The origin of the problem is found from the Talmud literature. Bankruptcy problems have been used in many areas such as mathematics, finance, economics, and communications. The Talmud rule was extended by Aumann and Maschler as a popular method for bankruptcy problems [5-8].

Load-shedding results from an insufficiency of power supply. The load-shedding problem is a bankruptcy problem because it is related to distribute insufficient power to consumers requiring more power.

In this paper, we approach the load-shedding problem as a bankruptcy problem and adopt the Talmud rule as an algorism. Load-shedding using the Talmud rule is tested in microgrid operation based on a multiagent system.

2 Islanded Microgrid Operation

Microgrids are small-scale power grids composed of distributed generation (DG), distributed storage devices (DS), and loads. Fig. 1 shows the configuration of a microgrid. Microgrids are operated in the grid-interconnected mode and in the islanded mode.

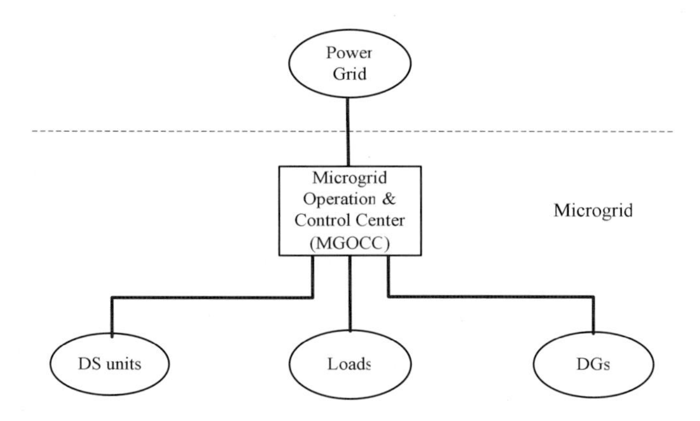

Fig. 1. Configuration of a microgrid [4]

Microgrids can be electrically isolated from a power grid on abnormal operation conditions such as fault occurrence at an interconnected power grid or by geographical environment like a distant island. Microgrid operation electrically isolated from power grids is called islanded operation. In the islanded mode, microgrids should operate without interconnection with any power grid and the following actions can be performed to meet power balance between power supply and power demand;

- The decrease in generation and the charge action of DS for solving power supply surplus
- Load-shedding and the discharge action of DS for solving power supply shortage.

Especially, load-shedding should be minimally performed not to reduce supply reliability. Fig. 2 shows the typical operation procedure according to power balance in the islanded microgrid.

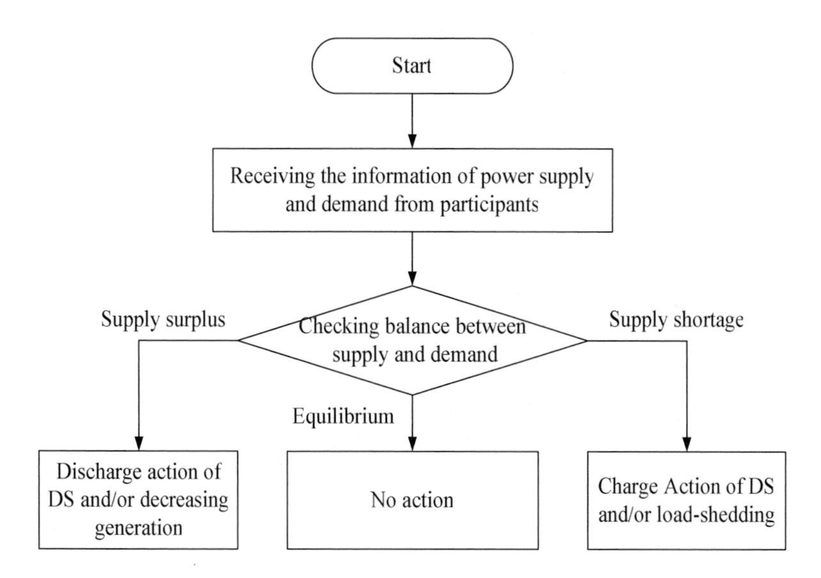

Fig. 2. Typical operation procedure in the islanded microgrid

3 Load-Shedding Using the Talmud Rule

3.1 Bankruptcy Problem

A bankruptcy problem is defined as a pair (c, E), where E is an amount to be divided and $c = (c_1, \cdots\cdots, c_N)$ is a set of claims of N agents and is described as

$$0 \le c_1 \le \cdots\cdots \le c_N \text{ and } 0 \le E \le c_1 + \cdots\cdots + c_N. \quad (1)$$

A solution of a bankruptcy problem is a vector of real numbers, $x = (x_1, \cdots\cdots, x_N)$ with $x_1 + \cdots\cdots + x_N = E$. Let $\mathcal{C}^N$ denote the class of all problems [6].

3.2 Talmud Rule

The Talmud rule is widely used to bankruptcy problems of many areas such as finance, economics, and communications. Table 1 shows the results of a well-known the contested garment problem in the Babylonian Talmud [7].

Table 1. Results of the contested garment problem in the Talmud

Estate\Debt	100	200	300
100	$33^1/_3$	$33^1/_3$	$33^1/_3$
200	50	75	75
300	50	100	150

3.3 Load-Shedding Using the Talmud Rule

Load-shedding is an action which intentionally reduces load to meet balance between power supply and power demand when supply shortage occurs. In this paper, we approach the load-shedding problem as a bankruptcy problem. For solving the load-shedding problem, the Talmud rule is adopted as follow;

$$Ti(l,P) = \begin{cases} min\{li/2, \lambda\} & if\ \Sigma\,(li\,/2) \geq P \\ li - min\{li/2, \lambda\} & other \end{cases} \tag{2}$$

where l is the vector of claims of loads, and $\lambda$is chosen so that $\Sigma\min\{l_i/2, \lambda\} = P$ and is chosen so that $\Sigma\,[l_i - min\,\{l_i/2, \lambda\}] = P$, respectively.

Finally, the load-shedding of each load (LS^*) is obtained from $LS^*=l - l^*$, where l^* is the vector of allocated loads by the Talmud rule. Additionally following additional rule is used for dealing with an infinite decimal of a result.

Computation Rule) Truncating number below the decimal point and the number is added to a consumer requiring the largest load.

4 Agent-Based Islanded Microgrid Operation

A multiagent system for islanded microgrid operation is defined as

$$Ag = \{Ag_{MGOCC}, AG_{DG}, AG_S, AG_L\}, \tag{3}$$

where Ag_{MGOCC} is the MGOCC agent and has sufficient knowledge and information as a manager of total operation procedures. AG_L is a set of load agents (Ag_L), and Ag_L operates and controls its load including load-shedding. AG_{DG} is a set of DG agents (Ag_{DG}) and Ag_{DG} governs a DG or a group of DGs located at same place. AG_S is a set of storage device agents (Ag_S) and Ag_{DS} takes charge of a storage devices or a group of storage devices located at same place [3]. Fig. 3 shows an agent-based islanded microgrid.

A multiagent system for islanded microgrid operation is designed as follows;

- Ag_{MGOCC} is a manager of islanded operation.
- Information Exchange Protocol and a modified CNP are used as protocols for interactions among agents.
- A modified version of the KQML is used as the ACL [3].
- Decision making of Ag_{MGOCC} is designed to follow the typical operation procedure as shown in Fig. 2.
- The final contractors among DGs are decided by the merit order algorithm [3].

- Ag_{MGOCC} decides load-shedding using the Talmud rule mentioned in Sec. III in the case of supply shortage.
- Distributed Agent System based on Hybrid Architecture (DASH) as a multiagent platform, Interactive Design Environment for Agent Designing Framework (IDEA) as a GUI-based interactive environment for the DASH, and JAVA [3,10-12].

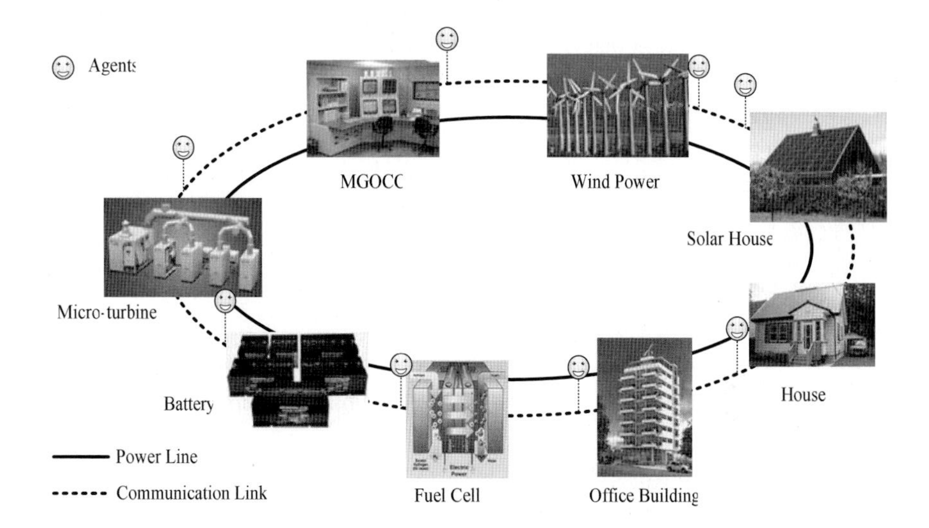

Fig. 3. Agent-based islanded microgrid

Fig. 4 shows the designed message flow among the agents for cooperative distributed problem solving.

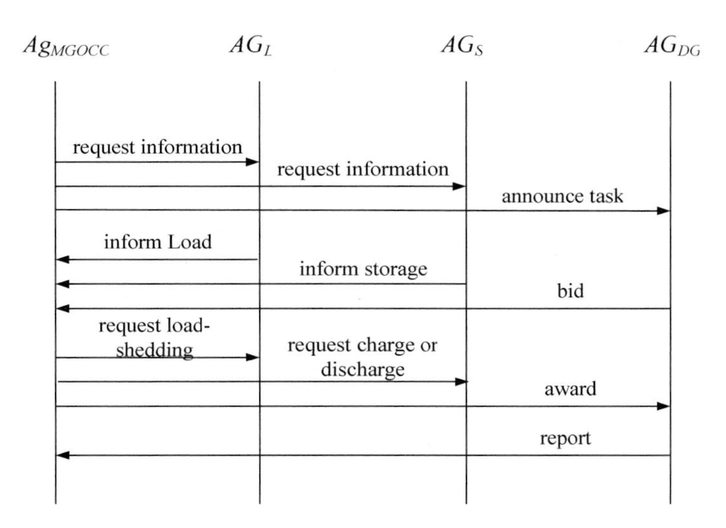

Fig. 4. Message flow among agents

5 Experiment

To test load-shedding using the Talmud rule for agent-based islanded microgrid operation in the case of supply shortage, an agent-based islanded microgrid is constructed as shown in Fig. 5. Here, IMG means the islanded microgrid. The multiagent system has seven agents: an MGOCC agent (Ag_{MGOCC}), three DG agents (Ag_{DG1}, Ag_{DG2}, Ag_{DG3}), a storage agent (Ag_{S1}), and three load agents (Ag_{L1}, Ag_{L2}, Ag_{L3}).

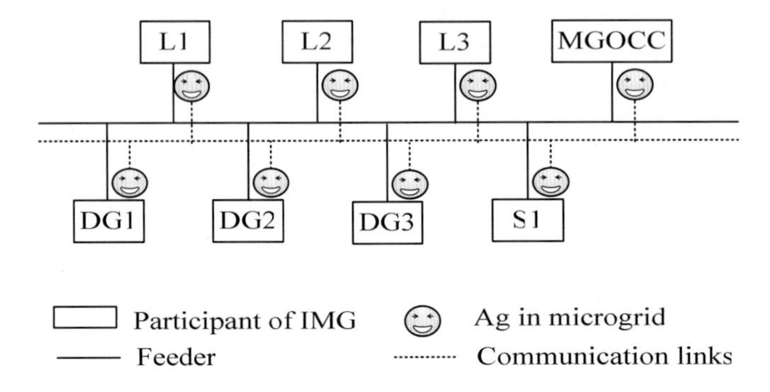

Fig. 5. Configuration of agent-based islanded microgrid

Two cases are considered according to power imbalance. The following shows information of a storage device and load for test.

- Capacity (S1): 10 [kWh]
- Initial charged amount (S1): 5 [kWh]
- L1: 100 [kWh]
- L2: 150 [kWh]
- L3: 200 [kWh]

Table 2 shows production costs and capacities of three DGs using three cases. It is assumed DGs bid their production costs as bid prices [3].

Table 2. DG information

Case	DG	Production Cost [¢/ kWh]	Capacity [kWh]	Power Imbalance [kWh]
Case 1	DG1	50	45	50
	DG2	70	150	
	DG3	80	200	
Case 2	DG1	50	45	100
	DG2	70	100	
	DG3	80	200	
Case 3	DG1	50	45	200
	DG2	70	100	
	DG3	80	100	

Table 3 and Table 4 illustrate allocated loads and load-shedding results, respectively as test results. From the result of Case 1 and Case 2, the effect of the computation rule mentioned in Sec. 3 appears.

Table 3. Allocated loads

DG	L1 [kWh]	L2 [kWh]	L3 [kWh]
Case 1	83	133	184
Case 2	66	116	168
Case 3	50	75	125

Table 4. Load-shedding results

DG	L1 [kWh]	L2 [kWh]	L3 [kWh]
Case 1	17	17	16
Case 2	34	32	34
Case 3	50	75	75

We can see that the load-shedding is successfully performed by the proposed scheme.

6 Conclusion

In this paper, we suggested load-shedding as a bankruptcy problem and load-shedding scheme using the Talmud rule in islanded microgrid operation based on a multiagent system. A multiagent system was constructed to test the proposed approach. The proposed approach was tested successfully in load-shedding in agent-based islanded microgrid operation.

We expect that the proposed approach can be widely used in autonomous agent-based microgrid operation. In future work, we plan to study additional algorithms for more effective microgrid operation based on the multiagent system.

References

1. Dimeas, A.L., Hatziargyriou, N.D.: Operation of a Multiagent System for Microgrid Control. IEEE Trans. on Power Systems 20(3), 1447–1455 (2005)
2. Dimeas, A.L., Hatziargyriou, N.D.: A Multiagent System for Microgrids. IEEE Power Society General Meeting, 55–58 (June 2004)
3. Kim, H.-M., Kinoshita, T.: A Multiagent System for Microgrid Operation in the Grid-interconnected Mode. Journal of Electrical Engineering & Technology 5(2), 246–254 (2010)
4. Kim, H.-M., Kinoshita, T.: Multiagent System for Microgrid Operation based on a Power Market Environment. In: INTELEC 2009, incheon, Korea (October 2009)

5. Yaiche, H., Mazumdar, R.R., Rosenberg, C.: A game theoretic framework for bandwidth allocation and pricing in broadband networks. IEEE/ACM Trans. on Networking 8(5), 667–678 (2000)
6. Thomson, W.: Axiomatic and game-theoretic analysis of bankruptcy and taxation problems: a survey. Mathematical Social Science 45, 249–297 (2003)
7. Aumann, R.J., Michael, M.: Game theoretic analysis of a bankruptcy problem from the Talmud. Journal of Economic Theory 36, 195–213 (1985)
8. Moreno-Ternero, J.D., Villar, A.: The Talmud rule and the securement of agents' awards. Mathematical Social Sciences, 245–257 (2004)
9. Kinoshita, T. (ed.): Building Agent-based Systems, The Institute of Electronics, Information and Communication Engineers (IEICE), Japan (2001) (in Japanese)
10. Kinoshita, T., Sugawara, K.: ADIPS Framework for Flexible Distributed Systems. In: Ishida, T. (ed.) PRIMA 1998. LNCS (LNAI), vol. 1599, pp. 18–32. Springer, Heidelberg (1999)
11. Uchiya, T., Maemura, T., Xiaolu, L., Kinoshita, T.: Design and Implementation of Interactive Design Environment of Agent System. In: Okuno, H.G., Ali, M. (eds.) IEA/AIE 2007. LNCS (LNAI), vol. 4570, pp. 1088–1097. Springer, Heidelberg (2007)
12. IDEA/DASH Tutorial, `http://www.ka.riec.tohoku.ac.jp/idea/index.html`
13. Smith, R.G.: The Contract Net Protocol: High-level Communication and Control in a Distributed Problem Solver. IEEE Trans. on Computer C-29(12) (December 1980)
14. Wooldridge, M., Jennings, N.R.: Intelligent Agents: Theory and Practice. The Knowledge Engineering Review 10(2), 115–152 (1995)
15. Wooldridge, M.: An Introduction to Multiagent Systems, 2nd edn. A John Wiley and Sons, Ltd., Chichester (2009)
16. Weiss, G. (ed.): Multiagent Systems: A Modern Approach to Distributed Artificial Intelligence. The MIT press, Cambridge (1999)

State of the Art of Network Security Perspectives in Cloud Computing

Tae Hwan Oh[1], Shinyoung Lim[2], Young B.Choi[3], Kwang-Roh Park[4], Heejo Lee[5], and Hyunsang Choi[5]

[1] Dept. of Networking, Security and Systems Adminstration
Rochester Institute of Technology
Rochester, NY, 14623 U.S.A
[2] Dept. of Rehab Sci & Tech
University of Pittsburgh
Pittsburgh, PA, 15260 U.S.A
[3] Dept. of Natural Science, Mathematics and Technology
Regent University
Virginia Beach, Virginia, 23464 U.S.A
[4] Electronics and Telecommunication Research Institute (ETRI)
138 Gajeongno, Yuseong-gu, Daejeon, 305-700
Rep. of Korea
[5] Korea University
Anam-dong Seongbuk-gu, Seoul, 136-701
Rep. of Korea

Abstract. Cloud computing is now regarded as one of social phenomenon that satisfy customers' needs. It is possible that the customers' needs and the primary principle of economy – gain maximum benefits from minimum investment – reflects realization of cloud computing. We are living in the connected society with flood of information and without connected computers to the Internet, our activities and work of daily living will be impossible. Cloud computing is able to provide customers with custom-tailored features of application software and user's environment based on the customer's needs by adopting on-demand outsourcing of computing resources through the Internet. It also provides cloud computing users with high-end computing power and expensive application software package, and accordingly the users will access their data and the application software where they are located at the remote system. As the cloud computing system is connected to the Internet, network security issues of cloud computing are considered as mandatory prior to real world service. In this paper, survey and issues on the network security in cloud computing are discussed from the perspective of real world service environments.

Keywords: Cloud computing, cloud security guidance, network security, risk analysis.

1 Introduction

When Google's Christophe Bisciglia proposed concept of cloud computing in 2006 [1], it has to wait until 2008 for most global enterprises' recognition of cloud

T.-h. Kim, A. Stoica, and R.-S. Chang (Eds.): SUComS 2010, CCIS 78, pp. 629–637, 2010.

computing as one of futuristic business services. Cloud computing provides customers with service in the form of virtualized computing resources [2][3]. Customers are able to acquire computing resources (i.e., software, storage, server, and network) based on their demands. Available cloud services are Software as a Service (SaaS) [4], Platform as a Service (PaaS) [5], and Infrastructure as a Service (IaaS) [6]. SaaS focuses on multiple leases of application software packages, PaaS focuses on providing software developer's environment, and IaaS focuses on providing service infrastructure such as storage or computing power over the Internet.

Although there are problems with the cloud security issues, there are a number of security benefits that comes with cloud computing. All security measures are cheaper if they are implemented on a large scale. Customers of cloud computing have security as their primary concern, and thus as the cloud provider provides more security, they will be more acceptable to customers as preferred suppliers. Large cloud providers are able to offer standardized interfaces for managing the clouds; this can help reduce the migration time, and also identifies certain problems. Having all the data on the same place is dangerous as it could be a single point of failure, but on the other hand, if all the data is on the same place this means it will be easier for monitoring. Cloud providers provide algorithms for hashes and checksums for saving files, thus any incident happens will produce a backup copy and will be provided instantly. You may have unaware of doing it yourself. Most companies (especially small ones) do not have a 24/7 or they can't provide them. Security advantages that come from virtualization also apply to cloud computing.

In this paper, we focus on network security perspectives in cloud computing. Section 2 discusses related work of technical aspects and commercially available cloud computing service, section 3, security threats of cloud computing, section 4, status of cloud computing security, and section 5, network security for cloud computing are discussed followed by conclusion in section 6.

2 Related Work

In this section, topics on deployment models of cloud computing and commercially available cloud computing are discussed to find out system architecture without or less security features.

2.1 Deployment Models of Cloud Computing

We can categorize clouds based on their visibility as follows: Public clouds, private clouds, hybrid clouds, and community cloud based on its customers and service policy.

Clouds have a five following unique characteristics.

Multi-tenancy (shared resources): In cloud computing environment, multiple users use the same resources in which resources are shared in network level, host level and application level, rather than dedicate to single host to server.

Massive scalability: Cloud computing provides the ability to scale thousands of systems, bandwidth and storage space.

Elasticity: The cloud computing provides the ability to increase or decrease their computing resource by their needs.

Pay as you go: This feature is to provide the user to pay for actual resource for their usage, which depends upon computing power, bandwidth and storage use.

Self–provisioning of resources: Having additional resources, like processing capability, software, storage network resources.

2.2 Commercially Available Cloud Computing Services

Currently there are a few companies that provide cloud computing services: Amazon, Google, Microsoft, Sales-force.com, IBM, HP, and VMware. In this paper, we summarize cloud computing services of Amazon and Google.

Amazon, known as one of world's largest online sellers, has the Internet online sale regarded as cloud services. Amazon provides cloud services in S3 and EC2 [7][8]. S3 stands for Simple Storage Services and users can access storage in stored objects in S3 from any place in the internet. EC2 is Elastic Computer Cloud. It's a virtual infrastructure that is able to run a lot of applications from web-hosting, emails, all the way to simulations. The control over the data is in the user's hands. Users can create their own virtual image to include customizable features such as OS, security and network access controls, and API. Some of their main security issues currently include availability. Another issue is the threat of attackers to leverage Amazon and their processors to a level that will be hard for detection (for example using multiple servers as a super computer to brute force encryption attack).

Google's App Engine (GAE) is the companies cloud services [9][10]. They give users possibility to build their own virtual application to run on web applications in either Java or Python. Resources used by applications are free up to 500MB in addition to the bandwidth. In Google's GAE, customers will not get any privileges as opposite to Amazon's services. Google's core business is in the cloud; all of its services like search, emails, online mapping, office productivity, and social networking are available in the clouds. Users can subscribe to those services for free or pay a little for more services and support. Google's Electronic Privacy Information Center (EPIC) has filed a complaint for the FTC about security standards in Google's Cloud computing services.

3 Security Threats in Cloud Computing

Among the requirements of disseminating the cloud computing services, acquiring reliability, availability, and compatibility are in active discussion in the community. As the cloud computing has different type of system architectural models and service models, the security risks are also different from other models. Cloud computing is about gracefully losing control while maintaining accountability even if the operational responsibility falls upon one or more third parties. But even though, as clouds do have benefits, they still have security concerns that need to be addressed. Some of security topics are being discussed by Gartner, European Network and Information Security Agency (ENISA) [14], and Cloud Security Alliance (CSA) [11].

Gartner announced seven cloud-computing security risks [19] and ENISA also announced 10 security risk assessments [20]. Gartner pointed out seven security risks in cloud computing as follows:

1. Privileged user access: Sensitive data processed outside the enterprise brings with it an inherent level of risk, because outsourced services bypass the 'physical, logical and personnel controls' IT shops exert over in-house programs. Get as much information as you can about the people who manage your data. As providers to supply specific information on the hiring and oversight of privileged administrators, and the controls over their access, customers should know as much as they know about how is their data being processed and handled so sensitive data should not be exposed to un-privileged users.

2. Regulatory compliance: Customers are ultimately responsible for the security and integrity of their own data, even when it is held by a service provider. Traditional service providers are subjected to external audits and security certifications. Cloud computing providers who refuse to undergo this scrutiny are signaling that customers can only use them for the most trivial functions. It is also necessary for the customers to pay precautious attention to the terms and conditions that service providers should give the services of customer compliance according to their policies.

3. Data location: Service provider will be responsible for storing sensitive data and whole process for customer but the customer or client not aware of process are running and where the data are stored. Cloud service providers might commit to storing and processing data in specific jurisdictions, and whether they will make a contractual commitment to obey local privacy requirements on behalf of their customers. Customer should inquiry the service provider about commitment for protects their sensitive data on behalf the customer and should obey their policies.

4. Data segregation: Data in the cloud is typically in a shared environment alongside data from other customers. Encryption is effective but isn't a cure-all. It should find out what is done to segregate data at rest. The cloud provider should provide evidence that encryption schemes were designed and tested by experienced specialists. Encryption accidents can make data totally unusable, and even normal encryption can complicate availability.

5. Recovery in the case of disaster: Even if customers do not know where their data is, a cloud provider should tell the customers what will happen to their data and service in case of a disaster. Any offering that does not replicate the data and application infrastructure across multiple sites is vulnerable to a total failure. Customers should ask their provider if it has the ability of a complete restoration, and how long it will take.

6. Investigative support: Investigating inappropriate or illegal activity should be impossible in cloud computing if these investigation or activities are against of user service and policy. Cloud services are especially difficult to investigate, because logging and data for multiple customers may be co-located and may also be spread across an ever-changing set of hosts and data centers. If customers are unable to get a contractual commitment to support specific forms of investigation, along with evidence that the vendor has already successfully supported such activities, then their safe assumption is that investigation and discovery requests will be impossible.

7. Long-term viability: Ideally, cloud computing provider will never go broke or get acquired and swallowed up by a larger organization. But customers must be sure their data will remain available even after such an event. It is required to make an inquiry to the potential providers how customers would get their data back and if it would be in a format that the customers could import into a replacement application.

The ENSIA security risk assessments are summarized as follows: When in using cloud infrastructures, the client necessarily cedes control to the Cloud Provider, thus leaving a gap in security defenses. There are no standards those are available so, it will be hard for customers to migrate data between providers as well moving them back to in-house IT departments. This risk category covers the failure of mechanisms separating storage, memory, routing between different tenants. However, the attacks on resource isolation mechanisms are much more difficult for an attacker compared on traditional operating systems. An investment in achieving certification or certain compliance regulations may be put at stake due to migration to the cloud, and sometimes the use of a public cloud infrastructure implies that certain kinds of compliance cannot be achieved. Customer management interfaces of a public cloud provider are doing so using the internet, and thus this is a more risk as it's publically available. It can be difficult for the customer to check the data handling correctly, and thus if it's a lawful way or not. As customers may not be sure of the way the cloud providers get rid of the data, this can be a security risk if they are not deleted in a lawful manner.

4 Improving of Cloud Computing Security

To improve security in cloud computing, companies and academia joined together and formed several groups and alliances to address the security issues for cloud computing.

The common goals for those groups and alliances are to enhance and improve security for cloud computing through education and by encouraging the use of best practices for providing security for clouding computing. The following list describes the different security organization for cloud computing.

Cloud Security Alliance (CSA): This is a non-profit organization and promotes best practices of security assurance for clouding computing. Also, this alliance allocates resources for awareness campaigns and education programs to encourage appropriate use clouding computing security solutions. Additionally, strong research activities are encouraged as well [11].

Open Cloud Consortium (OCC): This consortium supports standard development for cloud computing as well as framework development to address interoperable between different clouds. Additionally, the bench marks for cloud computing are supported as well as the reference implementations. Lastly, the consortium sponsors events that related to cloud computing [12].

Storage Networking Industry Association (SNIA): This non-profit organization assists members to store and manage large amount of information. This is important for cloud computing to address the storage issues including the security issues related to storage [13].

European Network and Information Security Agency (ENISA)**:** This organization is a centre of excellence for the European Member States and European institutions in network and information security, giving advice and recommendations and acting as a switchboard for information on good practices. The ENISA has presented few recommendations for providing security in the clouds and the cloud customers need assurance that providers are providing good security practices for cloud computing [14]. Therefore, following action items are check to ensure the security between the customers and providers.

- Assess the risk of adopting cloud services
- Compare different cloud provider offerings
- Obtain assurance from selected cloud providers
- Reduce the assurance burden on cloud providers.

National Institute for Standards and Technology (NIST):
NIST mission is promoting technical guidance and standards to provide effective and secure use of cloud computing technology. NIST wants to promote cloud security standards by proposing roadmaps for needed standard as well as catalysts to help industry to formulate their own standards. Also, NIST encourage government and industry to adopt the cloud standards [15]. The goals of NIST Cloud standards are fungible clouds that have following features:

- Mutual substitution of services
- Data and customer application portability
- Common interfaces, semantics, programming models
- Federated security services
- Vendors compete on effective implementations

Also, enable and foster values add on services for advanced technology and vendors compete on innovative capabilities.

5 Network Security for Cloud Computing

There are some recommend priority areas of research to improve the security of cloud computing. The following are the categories being considered [16]:

Building Trust in the Cloud

- Effects of different forms of breach reporting on security
- End-to-end data confidentiality in the cloud and beyond
- Higher assurance clouds, virtual private clouds etc

Data protection in large scale cross-organizational systems

- Forensics and evidence gathering mechanisms.
- Incident handling - monitoring and traceability
- International differences in relevant regulations including data protection and privacy

Large scale computer systems engineering:

- Resource isolation mechanisms, data, processing, memory, logs etc
- Interoperability between cloud providers
- Resilience of cloud computing

The Open Cloud Consortium (OCC) [12] is a group formed by universities and IT companies looking to investigate new ways of improving computing and storage costs across various cloud platforms and integrate communication standards among different providers. This is a relatively new group formed in the mid-2008, which confirms the novelty of the field. The OCC has undertaken the following goals:

- Development of standards for cloud computing and frameworks for interoperating between clouds
- Develop benchmarks for cloud computing
- Support reference implementations for cloud computing, preferably open source reference implementations.
- Manage a test bed for cloud computing - the Open Cloud Testbed
- Sponsor workshops and other events related to cloud computing

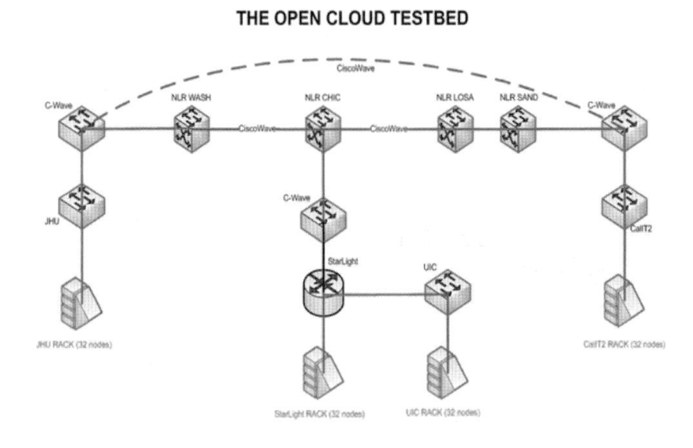

Fig. 1. OCC Network Testbed

The architecture in the Figure 1 above shows the OCC network and the connection among server racks at University of Illinois at Chicago, StarLight in Chicago, Calit2 in La Jolla and John Hopkins University in Baltimore to the switches, routers and wide area routers in between [12]. Using the aforementioned architecture, the OCC published has worked towards implementing high traffic flow design and protocols among several locations [12]."

According to the 'Security Guidance for Critical Areas of Focus in Cloud Computing V2.1' published by Cloud Security Alliance (CSA) in December 2009,

CSA focused on operating in the cloud and identified the following as the factors to consider in network security aspects in cloud computing [17]:

- Traditional Security, Business Continuity, and Disaster Recovery
- Data Center Operations
- Incident Response, Notification, and Remediation
- Application Security
- Encryption and Key Management
- Identity and Access Management
- Virtualization

Another view about the network security aspects in cloud computing can be found in the research performed by RSA on the role of security in trustworthy cloud computing. First, they identified the challenges of the cloud security and found that security is the big question mark because of the factors such as changing relationships, standards, portability between public clouds, confidentiality and privacy, viable access controls, compliance, and security service levels. Bay analyzing those factors RSA suggested the three principles for securing the cloud computing as 1) Identity security, 2) information security, and 3) infrastructure security [18].

6 Conclusion

To satisfy increasing needs of customers' about security in cloud computing, it is important to identify the outstanding issues, existing technologies, and future directions of network security in cloud computing.

In this paper, we surveyed the state of the art of network security perspectives in cloud computing. We researched related work in development models of cloud computing and commercially available cloud computing services. Considering security threats in cloud computing identified by Gartner, efforts to improve cloud computing security by various organizations including CSA, OCC, SNIA, ENISA, and NIST were explained respectively as major technical development efforts. Finally, major network security paradigms for cloud computing by OCC, CSA, and RSA were introduced to forecast future directions of network security perspectives for cloud computing.

References

1. Kimball, A., Michels-Slettvet, S., Biscigilia, C.: Cluster Computing for Web-Scale Data Processing. In: SIGCSE 2008, Portland, Oregon, pp. 116–120 (2008)
2. Wikipedia, http://en.wikipedia.org/wiki/Cloud_computing
3. Vision, Hype, and Reliability for Delivering IT Services as Computing Utilities, HPCC 2008 Keynote (2008)
4. Thomas, D.: Enabling Application Agility-Software as a Service, Cloud Computing and Dynamic Languages. Journal of Object Technology 17(4) (May-June 2008)

5. Lawton, G.: Developing Software Online with Platform-as-a-Service Technology. Computer (June 2008)
6. Armbrust, M., et al.: Above the Clouds: A Berkeley View of Cloud Computing (February 2009), http://radlab.cs.berkeley.edu
7. Amazon Elastic Compute Cloud (Amazon EC2), http://aws.amazon.com/ec2
8. Amazon Simple Storage Service (Amazon S3), http://aws.amazon.com/s3
9. Valdes, R.: Google App Engine Goes Up Against Amazon Web Services. Gartners (April 2008)
10. Mitchell, A.: Google Apps: Education Edition Overview Webinar, http://www.google.com
11. Cloud Security Alliance, http://cloudsecurityalliance.org/
12. Open Cloud Consortium, http://opencloudconsortium.org/
13. Storage Networking Industry Association, http://www.snia.org/home/
14. European Network & Information Security Agency (ENISA), http://www.enisa.europa.eu/
15. National Institute of Standards and Technology, Computer Security Resource Center, http://csrc.nist.gov/groups/SNS/cloud-computing/
16. Andrei, T., Jain, R.: Cloud Computing Challenges and Related Security Issues. Project report, Washington University in St. Louis (April 2009)
17. Security Guidance for Critical Areas of Focus in Cloud Computing V2.1, Cloud Computing Alliance (December 2009)
18. The Role of Security in Trustworthy Cloud Computing, White Paper, RSA
19. Brodkin, J.: Gartner: Seven cloud-computing security risks (July 2, 2008), http://www.infoworld.com
20. European Network and Information Security Agency (ENISA), Cloud Computing: benefits, risks and recommendations for information security (November 2009)

Design and Implementation of Wireless Sensor Networks Based Paprika Green House System

Jiwoong Lee, Hochul Lee, Jeonghwan Hwang,
Yongyun Cho, Changsun Shin, and Hyun Yoe*

School of Information and Communication Engineering,
Sunchon National University, Korea
{leejiwoong,hclee,jhwang,yycho,csshin,yhyun}@sunchon.ac.kr

Abstract. This research paper suggests the 'Paprika green house system' (PGHS), which collects paprika growth information and greenhouse information to control the paprika growth at optimum condition. The temperature variation range of domestic paprika cultivation facilities are relatively quite big and the facility internal is kept at relatively dry condition. In addition, the concentration of CO2 is not uniform, giving bad impact on the growth of paprika. In order to cope with these issues, the 'Paprika green house system' (PGHS) based on wireless technology was designed and implemented for the paprika cultivating farmers. The system provides with the 'growth environment monitoring service', which is monitoring the paprika growth environment data using sensors measuring temperature, humidity, illuminance, leaf wetness and fruit condition, the 'artificial light-source control service', which is installed to improve the energy efficiency inside greenhouse, and 'growth environment control service', controlling the greenhouse by analyzing and processing of collected data.

Keywords: USN, Paprika, Green house.

1 Introduction

Recently domestic horticultural industry achieved substantial growth both in quantity and quality with its technology and capital-intensive industry characteristic. Now it became a competitive industry with big potential in overseas export demand, in addition to existing domestic demand [1].

Paprika is one of horticultural products that create high added-value. The production quantity of paprika varies dependent on sunlight quantity, illuminance and sunlight hours [2]. The cultivation cost of paprika is comprised of heating cost, agricultural material cost and labor cost. Among them, the weights of heating cost and agricultural material cost are very high, giving difficulty to the cultivating farmers [3].

This research paper suggests the establishment of a 'Paprika green house system' (PGHS) in the paprika-cultivating green houses, which need precise growth management. 'Paprika green house system' (PGHS) utilizes IT technology in collecting

---

* Corresponding author.

T.-h. Kim, A. Stoica, and R.-S. Chang (Eds.): SUComS 2010, CCIS 78, pp. 638–646, 2010.

the crops-growth environmental-information in real time and controls the environmental system in the cultivation facility. ‘Paprika green house system’ (PGHS) reduces the deviations in growth, development, production-quantity and quality of crops. It also maintains optimum environment in the cultivation facility using biometric data and creates optimum condition at paprika root zone. The system optimizes the management of production elements and reduces the loss of energy, fertilizer and water, which will result in the decrease of production cost. The artificial light-source from artificial lighting makes pleasant growth-environment so that continuous supply of high quality, fresh vegetable would be possible. Farmers will be able to increase the productivity and income by having their cultivation facilities as continuous supply source of high quality fresh vegetable to clients. ‘Paprika green house system’ (PGHS) is designed and realized to enable all above based on wireless sensor network.

This research paper is comprised of followings. Chapter 2 introduces the technologies related to the monitoring system applied to the agricultural environment in Korea and overseas. Chapter 3 explains the configuration elements and services provision by ‘Paprika green house system’ (PGHS) suggested by this research. Chapter 4 explains the implementation content of ‘Paprika green house system’ (PGHS). Chapter 5 will be the conclusion.

2 Related Researches

2.1 Agricultural Monitoring System Using Integrated Sensor Module

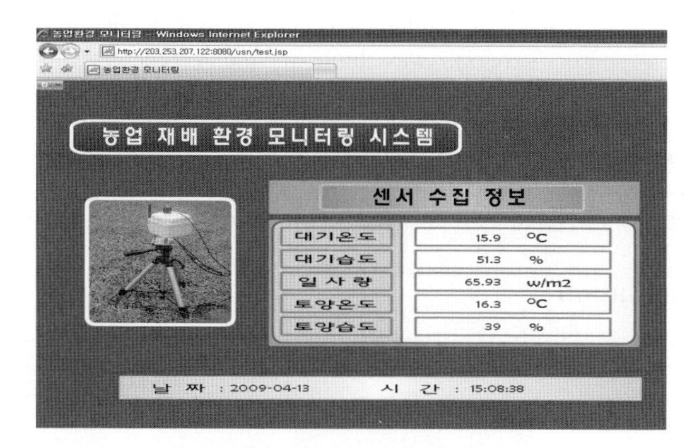

Fig. 1. Monitoring System

This system uses various environmental sensors to collect information required for the cultivation environment of crops. It is a real-time agricultural environment monitoring system based on sensor network. Most of existing wireless sensor nodes based on sensor network need separate conversion/control module for each sensor characteristic. To overcome this issue, an integrated sensor module was developed, which can integrate various sensors used in getting the information required for the crops cultivation, into a single node. New sensors and network monitoring system were also developed to let them fit with the new integrated sensor module and they were integrated to the test environment, in order to examine the operation of newly developed system [4]. Sensor node is also installed to measure the environmental information so that real-time monitoring would be possible [4].

2.2 'Greenhouse Environment Integrated Management System'

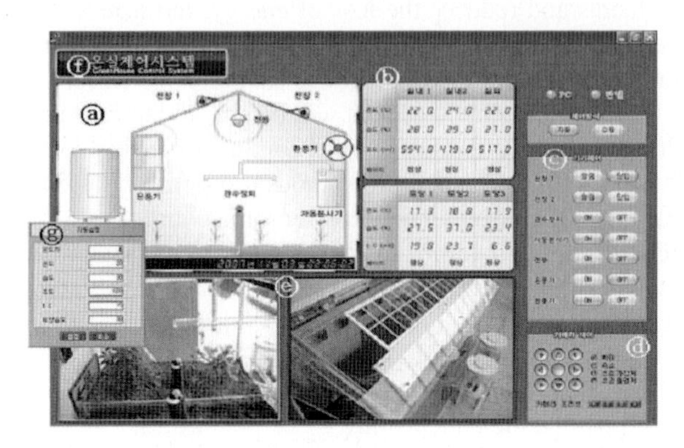

Fig. 2. Monitoring System GUI

The 'greenhouse environment integrated management system' enables the monitoring of greenhouse status in real time through Internet. With its remote control capability, it enables users to manage their farms without restriction of time and place, as long as Internet connection is available[5]. In order to make ubiquitous agricultural environment, a sensor network was built in the greenhouse to measure the environmental elements affecting cultivation environment, such as temperature, humidity, amount of insolation, CO2, ammonia, wind velocity and precipitation. Also, a 'greenhouse environment monitoring system' – comprised of ventilators, windows, heaters, humidifiers, lightings and video processors – is suggested to control the devices activated by the change of measured environmental elements [5].

3 Implementation

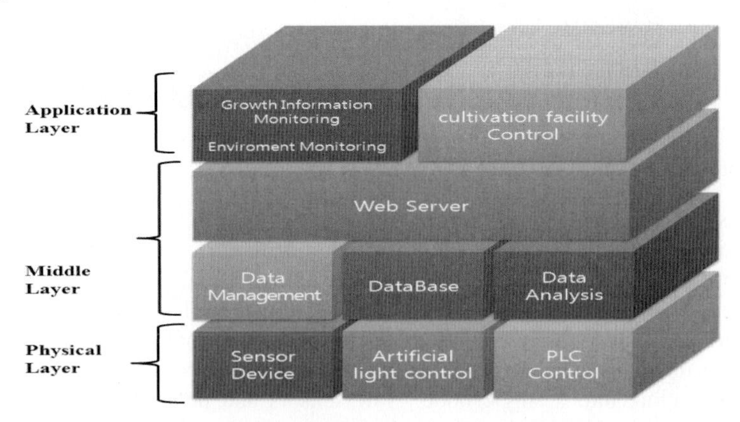

Fig. 3. Paprika Green House System Structure

3.1 System Configuration

'Paprika green house system' (PGHS) is comprised of following three layers. The physical layer has 'environmental sensor', collecting environmental information, 'artificial light-source growth control device' and PLC. The middle layer has data analysis and system control. The application layer has GUI and control.

3.1.1 Physical Layer

Physical layer has a sensor device which collects information from paprika culture media and sends the raw data to middle layer. It also has 'artificial light-source growth control device' which controls the wavelength and light quantity of LED light-source for the most optimum growth of paprika. PLC controller controls the temperature, humidity and growth-environment at the root zone based on the collected environmental information. It collects environmental information of cultivation location and makes a control platform to perform the monitoring and control of 'Paprika green house system' (PGHS) through each module.

3.1.2 Middle Layer

The middle layer is comprised of 'data filtering module', 'data analysis module', 'environment control module', 'artificial light-source control module', 'database' and 'web server'.

'Data filtering module' processes the raw data transmitted from sensors and saves the temperature, illuminance, humidity and root zone environment data in the database. 'Data analysis module' analyzes the cultivation location environment and crop status based on the information saved in the database. 'Environment control module' transmits the control signal to the PLC of physical layer. 'Artificial light-source control module' transmits the control signal to the artificial light-source controller. 'Database' saves the environment data and analysis data of paprika cultivation location. 'Web server' distributes the service to users through WEB-GUI.

3.1.3 Application Layer

The application layer is comprised of WEB-GUI, which provides user with service from 'Paprika green house system' (PGHS).

3.2 Service Provided

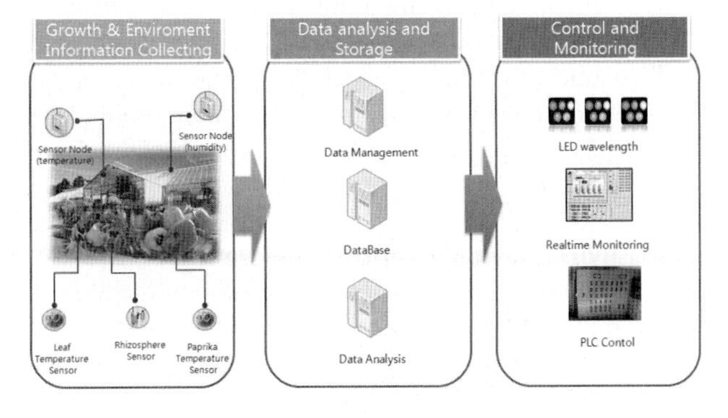

Fig. 4. Providing Service of Paprika Green House

There are 'paprika growth information monitoring service', 'growth environment monitoring service', 'root zone environment monitoring service' and 'artificial light-source control service' and 'cultivation environment control service'. Details of them are as following.

3.2.1 'Growth Environment Monitoring Service'

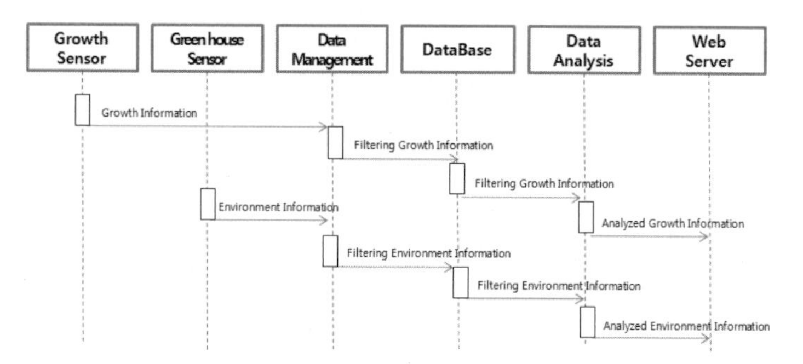

Fig. 5. Growth Environment Monitoring Service Sequence Diagram

'Growth environment monitoring service' provides with paprika growth information and Greenhouse information. During paprika cultivation, if the temperature difference between paprika and atmosphere would be more than 4oC, there will be dew condensation and paprika will suffer diseases like grey fungus. To cope with this problem, sensors to measure the temperature of paprika fruit and paprika leaves are located at the fruit surface and within 5cm of the leaves rear-side. The sensors will collect the temperature data in 2 minutes cycle. User can know the temperature difference between paprika and atmosphere through these sensors and can cope with the temperature difference problem caused by temperature difference between crop and atmosphere, in an active way.

Additional sensors for temperature, humidity and illuminance are installed in the greenhouse so that the user can know the situation through web and identify the growth environment of paprika culture media.

The activation process of 'growth environment monitoring service' is as following. Sensors installed in the greenhouse and on the crops collect the raw data. Data management will extract the leaf temperature, leaf wetness and greenhouse environment information (temperature, illuminance, humidity) and save them in the database. The saved data are analyzed and provided to user through web server in the form of web page.

3.2.2 'Root zone Environment Monitoring Service'

Paprika cultivation is mostly done by culture medium; therefore, the root zone management is very important because it gives big impact on the absorption efficiency of culture medium. The root zone of the crop means the soil environment, which changes as the growing roots absorb nutrient and save it. EC and PH are

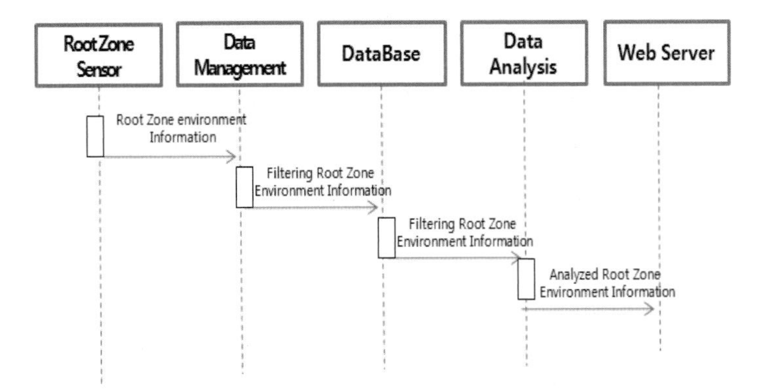

Fig. 6. Root Zone Environment Monitoring Service Sequence Diagram

especially important. If EC in soil is not enough, fruit does not grow compared to the leaves growth. If EC is too high, production decreases. If the fertilizer content in the soil becomes higher, PH decreases [6].

This service utilizes such characteristics of EC and PH and monitors the root zone environment. The activation process is same as 'growth environment monitoring service'.

3.2.3 'Artificial Light-Source Control Service'

LED can save 80% energy than existing incandescent light bulbs. LED prevents vermin and adjusts the growth velocity of crops so that shipping timing can be adjusted. As seen in following table, the wavelengths of LED give various impacts on the crops. This service applied those impacts [7] to the 'artificial light-source control service' and let it contribute in the control of paprika growth-speed and quality improvement.

Table 1. Artificial light-source Impact

Wavelengths	Impact
1400~1000 (IR-A)	No specific impact on crops. Gives heat impact
780(IR-A)	Promotes specific elongation effect on crops
660(red)	Maximize chlorophyll reaction (655)
610(red yellow)	Not good for photosynthesis. Prevents vermin (580~650)
430~440(blue)	Maximize photosynthesis (430), Maximize chlorophyll reaction (440), Entice vermin.
400 ~ 315(UV-A)	In general, makes leaves thick. Encourages the coloring of pigments. Entice vermin.
280(UV-B)	Important reaction in many synthetic processes (makes antibody), Harmful if too strong
100(UV-C)	Let crops wither rapidly.

3.2.4 'Cultivation Environment Control Service'

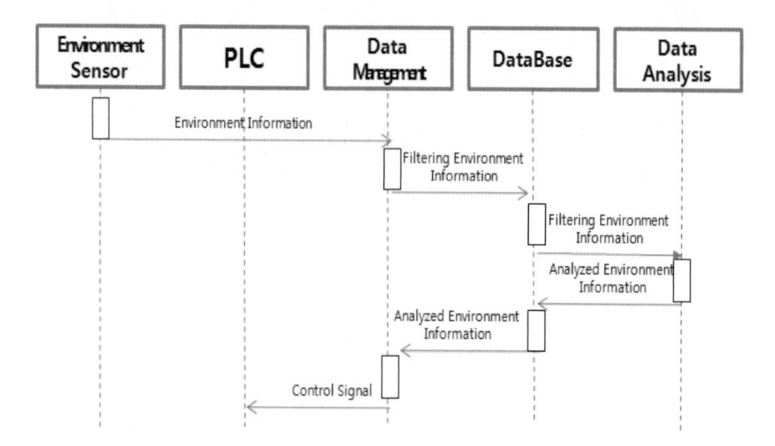

Fig. 7. Cultivation Environment Control Service Sequence Diagram

'Cultivation environment control service' controls the devices installed in the paprika greenhouse based on the data collected from sensors and saved in the database. The service maintains the optimum environment for the growth of the crop.

The activation environment is as following. The environmental information sent from cultivation location is transmitted to the data management in the middle layer. They are saved in the database after the correction of overlapping or wrong data. Then the saved data is sent to the 'data analysis module', where optimum control information for paprika growth would be analyzed. That information is saved again in the database and signals will be sent to PLC so that it would automatically control the devices such as ventilators and fan heaters.

4 Implementation

Various devices such as sensors, ventilators and fan heaters were installed in the paprika greenhouse to examine the performance capability of 'Paprika green house system' (PGHS).

Fig. 8. Fruit Sensor

Fig. 9. Leaf Sensor

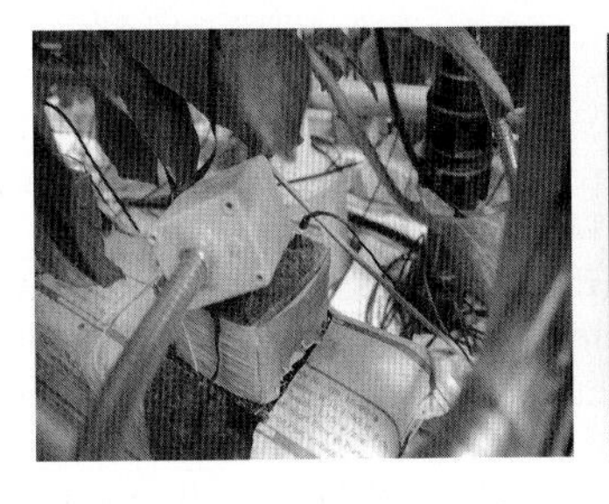

Fig. 10. Root Zone Sensor

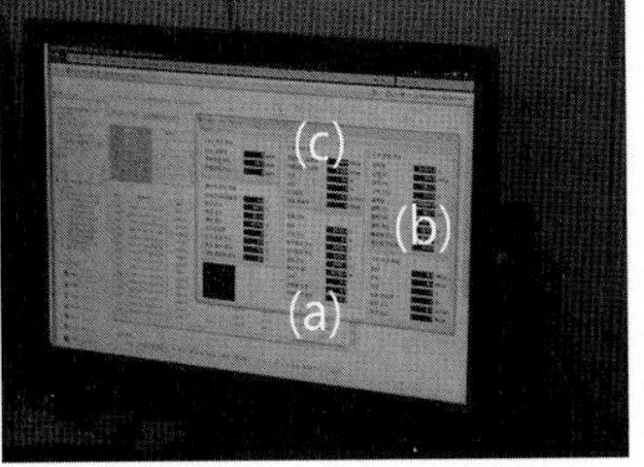

Fig. 11. Web-GUI

Figure 8 and Figure 9 are sensors measuring the fruit temperature, leaf temperature and leaf wetness.

Figure 10 is sensors collecting the root zone environment information.

Figure 11 is Web-GUI, providing users with 'Paprika green house system' (PGHS) service. Information collected from paprika fruits and leaves in Figure 8, Figure 9 can be confirmed in (b) of Figure 11. Root zone information collected in Figure 10 can be confirmed in (c) of Figure 11.

The greenhouse environment information values measured in Figure 12 can be confirmed in (a) of Figure 11. Data collected at the sensors go through server and the most optimum growth data will be sent to Figure 13. Then the server sends signals to artificial light-source controller of Figure 13 and PLC controller of Figure 14.

It has been demonstrated that the 'Paprika green house system' (PGHS) shows optimum control performance for the best growth of paprika.

Fig. 12. Environment Sensor

Fig. 13. Artificial Light Controller

Fig. 14. PLC controller

5 Conclusion

This research paper realized the paprika greenhouse environment monitoring system for the precise growth-management of high added-value crop, paprika. The suggested system 'Paprika green house system' (PGHS) makes a network comprised of sensors measuring temperature, humidity, illuminance and others. The system also controls ventilators, humidifiers, lightings and video-processing through web-based GUI by analyzing the measured data.

The suggested 'Paprika green house system' (PGHS) will contribute in the farmers' income increase, which is the top priority task in domestic agricultural industry. The system will create other research results by providing with the growth environment information from numerous paprika cultivation locations. Enhancement of price competitiveness of domestic horticultural industry can also be achieved by improving the distribution rate of new paprika species.

As the next research subject, the occurring conditions of paprika diseases and vermin will be made into database and a 'diseases and vermin forecast system' will be designed and implemented, so that farmers can actively cope with the diseases and vermin in advance.

Acknowledgements. This research was supported by the MKE(The Ministry of Knowledge Economy), Korea, under the ITRC(Information Technology Research Center) support program supervised by the NIPA(National IT Industry Promotion Agency) (NIPA-2010-(C1090-1021-0009)).

References

1. Yooun-il, N.: Present Status and Developmental Strategy if Protected Horticulture Industry in Korea. The KCID Journal 10(2), 191–199
2. Dorais, M.: The use of supplemental lighting for vegetable crop production: light intensity, crop response, nutrition, crop management, cultural practices. In: Canadial Greenhouse Conference (2003)
3. http://jindo.jares.go.kr/
4. Lee, E.-J., Lee, K.-l., Kim, H.-S., Kang, B.-S.: Development of Agriculture Environment Monitoring System Using Integrated Sensor Moudle. Korea Contents Association 10(2)
5. Lee, M.-h., Shin, C.-s., Jo, Y.-y., Yoe, H.: Implementation of Green House Integrated Management System in Ubiquitous Agricultural Environments. Journal of KIISE 27(6), 21–26 (2009)
6. http://www.kati.net/index.jsp
7. http://cafe.naver.com/chled
8. Jeong, W.-J., Myoung, D.-J., Lee, J.-H.: Comparison of Climatic Conditions of Sweet Pepper's Greenhouse between Korea and the Netherlands. Journal of Bio-Environment Control 18(3), 244–252 (2009)

An Implementation of the Salt-Farm Monitoring System Using Wireless Sensor Network

JongGil Ju, InGon Park, YongWoong Lee, JongSik Cho, HyunWook Cho, Hyun Yoe, and ChangSun Shin*

Dept. of Information and Communication Engineering, Sunchon National University, Korea
{jake,crescent1,cho1318,chohyunwook, pig9004,hyoe,csshin}@sunchon.ac.kr

Abstract. In producing solar salt, natural environmental factors such as temperature, humidity, solar radiation, wind direction, wind speed and rain are essential elements which influence on the productivity and quality of salt. If we can manage the above mentioned environmental elements efficiently, we could achieve improved results in production of salt with good quality. To monitor and manage the natural environments, this paper suggests the Salt-Farm Monitoring System (SFMS) which is operated with renewable energy power. The system collects environmental factors directly from the environmental measure sensors and the sensor nodes. To implement a stand-alone system, we applied solar cell and wind generator to operate this system. Finally, we showed that the SFMS could monitor the salt-farm environments by using wireless sensor nodes and operate correctly without external power supply.

Keywords: Solar Salt, USN, Salt-farm, Environment Monitoring System, Renewable Energy.

1 Introduction

The information technologies, wireless sensors, ubiquitous computing and communication devices techniques are applied to various industrial fields. But solar salt industry didn't receive above IT technologies yet. The solar salt is very sensitive to salt-farm environments. If we collect precision salt-farm environment data, the productivity and quality rate of products will be improved.

For producing high quality salt, we propose the Salt-Farm Monitoring System (SFMS) that uses hardware and software IT technologies. This system monitors and collects the information of salt-farm environments with a renewable power supply.

The rest of this paper is organized as follows: Section 2 is related works, Section 3 describes the system architecture of the SFMS and Section 4 presents implementation of the system. Finally, we discuss conclusions and future works in Section 5.

2 Related Works

There are five requisites for growing corps [1]. They are temperature, light, air, water and soil. A project of plant growth monitoring system was developed by Go-heung

* Corresponding author.

T.-h. Kim, A. Stoica, and R.-S. Chang (Eds.): SUComS 2010, CCIS 78, pp. 647–655, 2010.

agriculture technology & extension center in Korea. They applied environment monitoring sensors and software to Hanabong farm [2]. This system measures the greenhouse environment status by using sensors adhered to plant and sends information to grower's home via internet.

Fig. 1. The Hanabong monitoring system by Go-heung agriculture technology & extension center

Floating buoy is developed to monitor ocean environments via Orbcomm satellite and a method is proposed to increase measurement accuracy of sea water temperature with common low price temperature sensor [3].

The feasibility of the developed node was tested by deploying a simple sensor network into Martens Greenhouse Research Foundation's greenhouse in Närpiö town in Western Finland. They are the number of wireless sensor networks with tree structure form integrating sensors of the same category. In addition, three commercial sensors capable to measure four climate variables [4, 5].

By using above technologies, agricultural sensors and nodes are proposed, and various applications are developed.

3 Salt-Farm Monitoring System (SFMS) Architecture

The Salt-Farm Monitoring System (SFMS) is a collect real time environmental field data from various sensors. Figure 2 shows the architecture of the SFMS.

The SFMS divided into three layers. The physical device layer includes sensors and facilities. Sensors are temperature, humidity, solar radiation, wind direction, wind speed, rain and salt water level sensor. Facilities are salt water reservoir sluice, salt-farm sluice and salt water reservoir pump. SFMS is a self-charging stand alone system using renewable energy. We supply solar-power, wind generator into the system without any electric power.

The middle layer has the sensor manager, control manager, salt-farm database. The sensor manager manages the information from environments sensors [6]. The control manager controls the facilities device using the salt-farm database. The sensor data

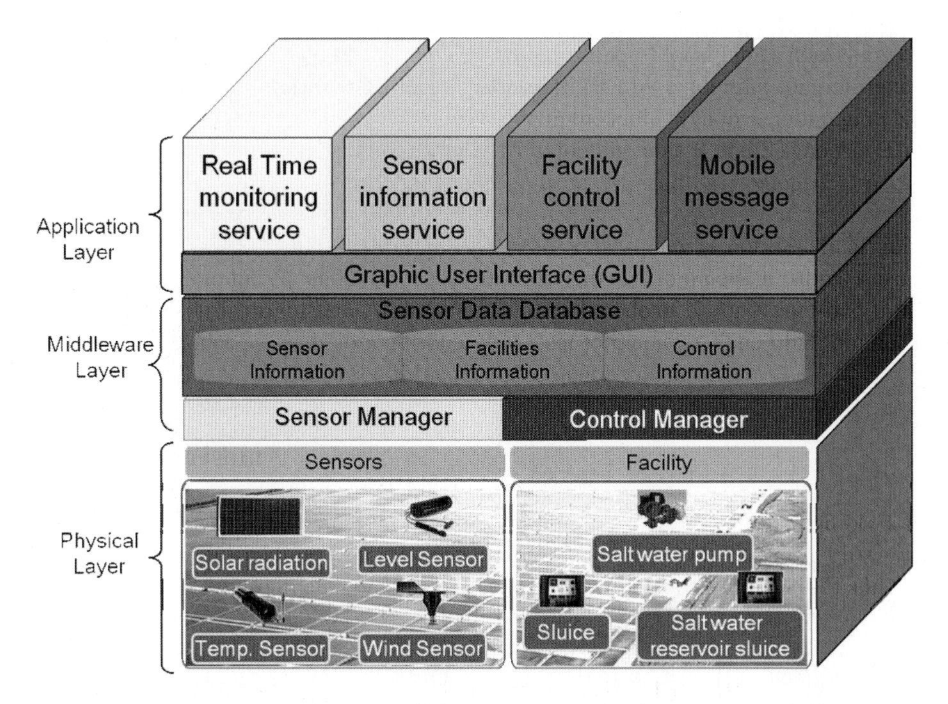

Fig. 2. Salt-Farm Monitoring System's architecture consisting of three layers

database provides environment information from physical layer devices to application layer via sensor, facilities control information.

The application layer provides with the real time monitoring service, sensor information service, facility control service and mobile message service [7]. These are provided with laptop GUI (Graphic User Interface), web GUI and mobile phone GUI. Three layers are integrated into the SFMS. By interacting with each layer, the system provides users with salt-farm environment information.

3.1 Environment Monitoring Service

The salt-farm should be real-time sensing, because it is very sensitive to environment element such as temperature, humidity, illumination, etc.

Figure 3 shows procedure of the environment monitoring service. First, this service sends the raw data of environment sensors to the sensor manager. The raw data are temperature, humidity, salinity and intensity of illumination information. Sensor manager verify which the raw data is error data or not. Wireless sensor network has virtually the low reliability in an open area. In order to reduce the risk of collecting environment information it needs to data filtering. A simple filtering such as compared with previous raw data through the classification and then stored in the database. A GUI obtained the sensor data from data storage then offers the environment information to users.

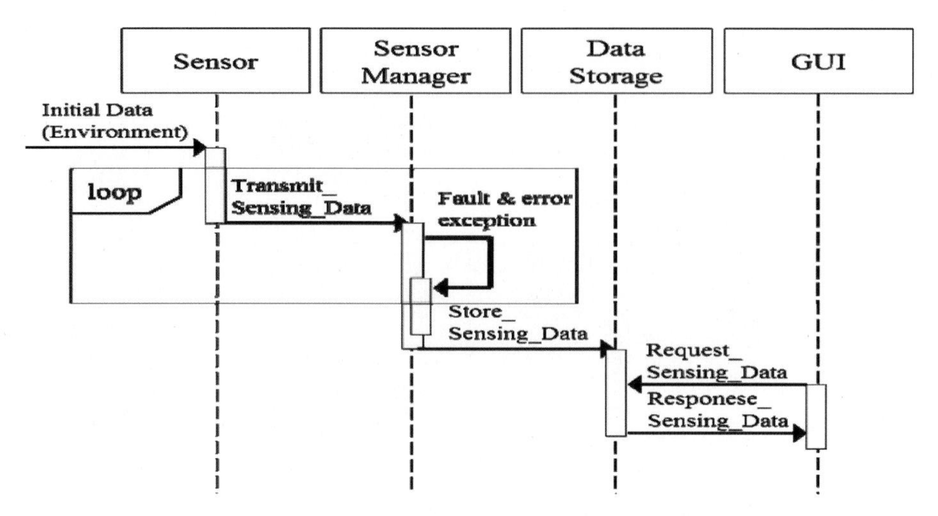

Fig. 3. Environmental Information Service sequence diagram

3.2 Control Services

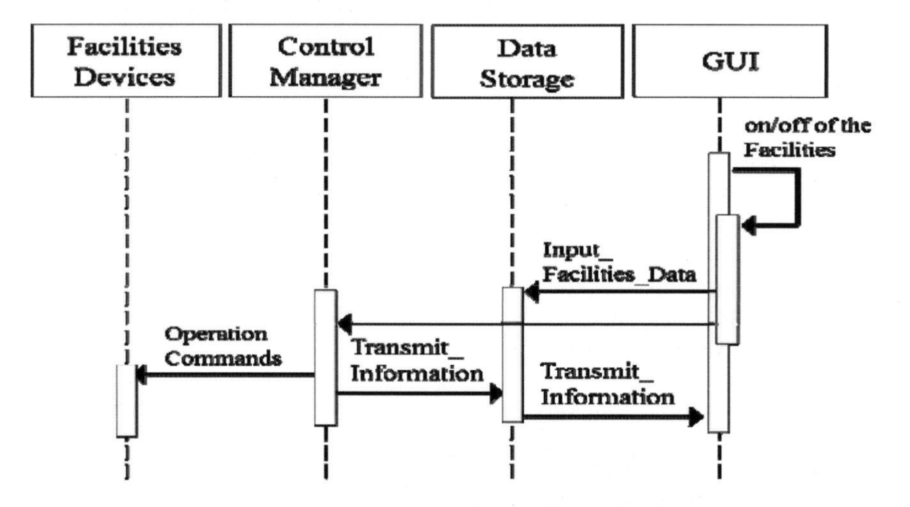

Fig. 4. Facilities Automatic Control Services Sequence Diagram

Figure 4 shows procedure of automatic control services facilities. Database provides the factors information such as facilities state and sensing data to GUI. GUI send the on/off command signal to control manager through database by Logical analysis of response factors data [8].

4 Implementation of the SFMS

In this chapter, we implement the SFMS by implementing the system's components. Figure 5 is the system model.

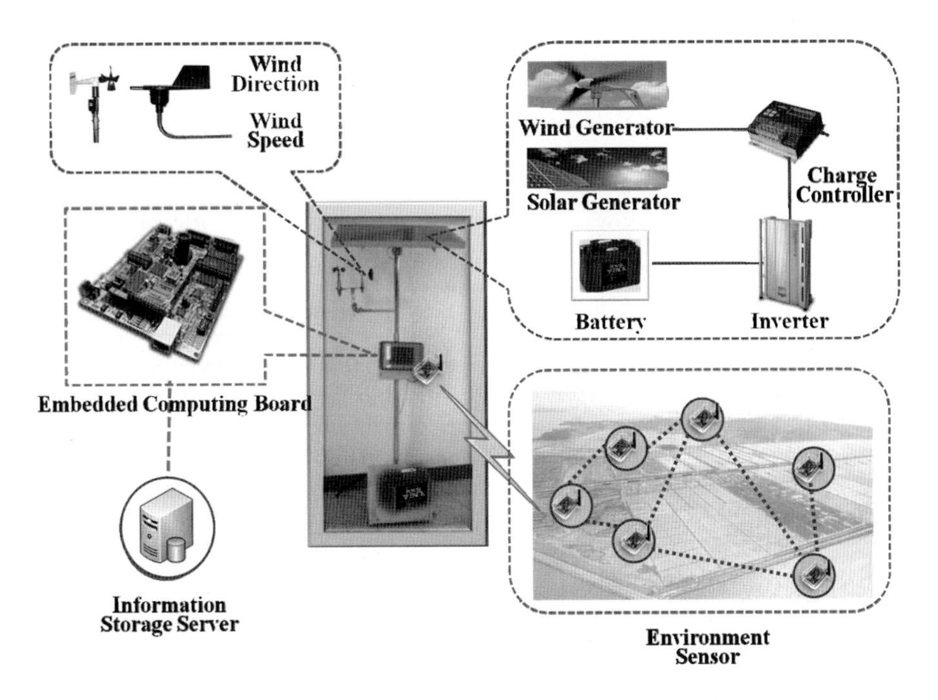

Fig. 5. SFMS model including embedded board, renewable charging devices and sensors

The SFMS was applied a renewable energy system. The system has solar cell, wind generator and storage battery. The system stores power in the daytime and using in the night time.

4.1 System Components

This system includes of physical devices and software modules. The physical devices have sensing and information gathering devices. You can see the devices in Figure 6.

Fig. 6. Wind turbines, Solar cell, Network sensor node, temperature sensor in the SFMS

Table 1. Power consumption of each module and Power supply of charging battery

<table>
<tr><th rowspan="2">Module</th><th colspan="3">Power consumption</th></tr>
<tr><th>Voltage</th><th>Current</th><th>Power</th></tr>
<tr><td>Embedded Board</td><td>DC 9V</td><td>500mA</td><td>5W</td></tr>
<tr><td>Environment Sensor</td><td>DC 3V</td><td>2.3A</td><td>6.9W</td></tr>
<tr><td>Salinity Sensor</td><td>DC 5V</td><td>10mA</td><td>0.05W</td></tr>
<tr><td>TOTAL</td><td>DC 17V</td><td>2.81A</td><td>11.95 W</td></tr>
<tr><th>Module</th><th colspan="3">Supply Power</th></tr>
<tr><td></td><td>Voltage</td><td>Current</td><td>Power</td></tr>
<tr><td>Solar Cell</td><td>DC 26.4V</td><td>7.6A</td><td>200W</td></tr>
<tr><td rowspan="2">Wind generator</td><td>DC 24V</td><td>7.7A</td><td>200W</td></tr>
<tr><td colspan="3">Avr Wind Speed 12.5m/s</td></tr>
<tr><td rowspan="2">Battery</td><td>Voltage</td><td>Capacity(20HR)</td><td></td></tr>
<tr><td>DC 12V</td><td>64A</td><td></td></tr>
</table>

Table 1 is showing the power consumption of equipped modules and power supply of solar and wind in the SFMS. The total power consumption of equipped modules like embedded board, salinity sensor and environment sensor is 11.95W. The solar cell and wind generator supplies with electric power of the maximum 200W for each in the 25℃ test environment. This is enough to operate the SFMS. You can see the main system installed sensors' data receiver and database in Figure 7 and a charging battery in Figure 8.

Fig. 7. Embedded board including environment sensor receiver and database

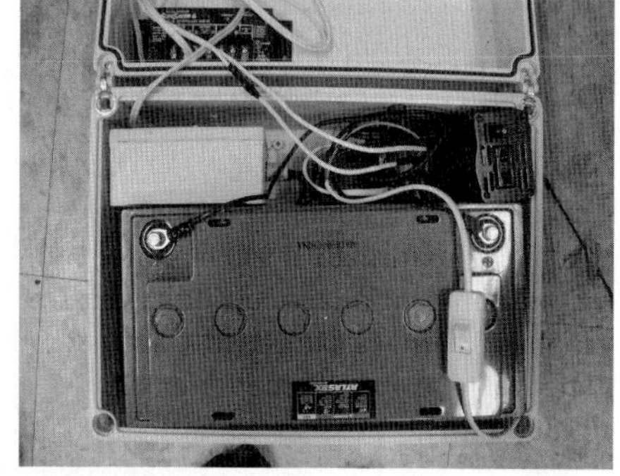

Fig. 8. salinity sensor receiver, integrated battery

Now, we integrate above components into the system. Figure 9 shows the SFMS's prototype including the software modules. The SFMS can apply various environments such as precision agriculture, aquaculture, fishing industry, livestock industry, greenhouse monitoring and salt farm monitoring.

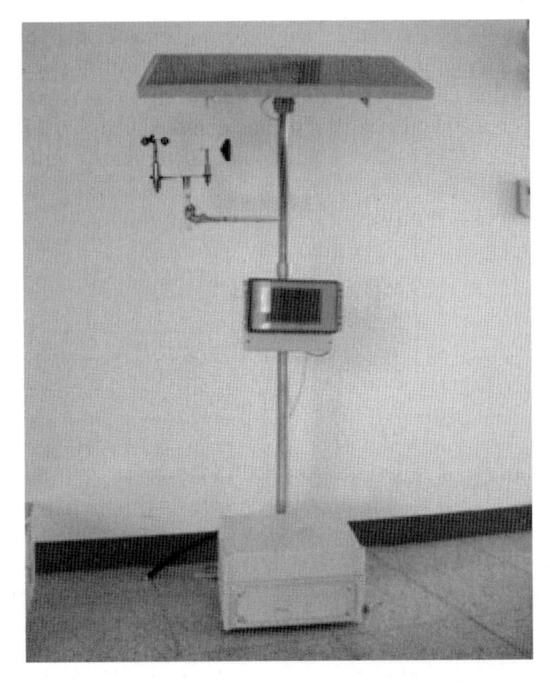

Fig. 9. Prototype of the SFMS

4.2 Implementation Results

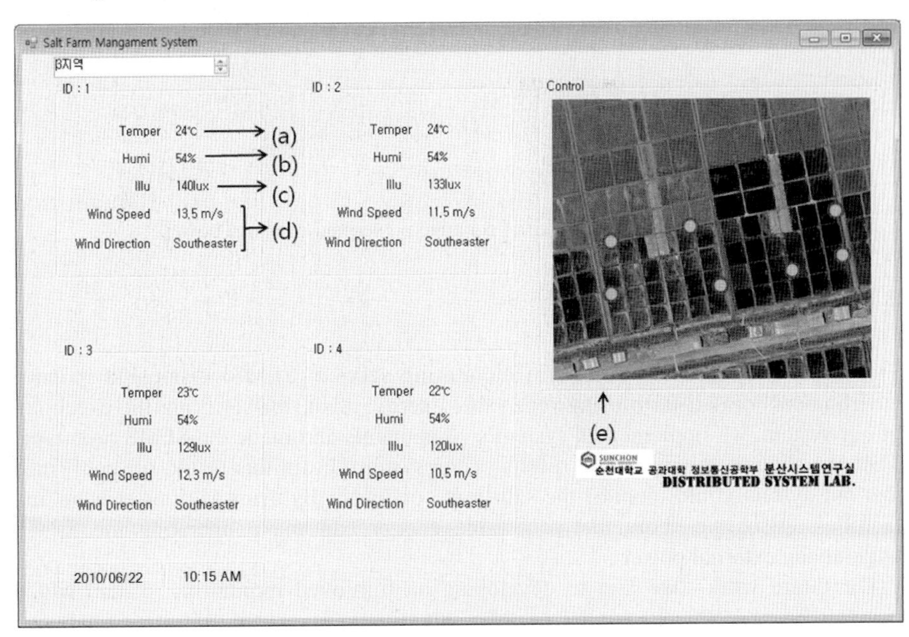

Fig. 10. FMSS's GUI

Figure 10 is the SFMS's GUI. The (a) shows the sensing value from the temperature sensor. The (b) is sensing value for humidity. The (c) showing the solar radiation sensing value, and the (d) is the sensing value from the wind speed, direction sensors. The (e) shows control of floodgate.

To confirm the successful operation of the SFMS using self-supply of electric power, we perform field test on a sunny day with a mean temperature of 25□ degree and wind speed 12.5m/s. As a result, during the daytime solar cell, wind generator generated power together and night time only wind generator generated the power. If there is windless and cloudy day, SFMS could be supplied power by recharged battery.

Hence, our SFMS can operate with the support of solar cell and wind power in the field without power supply from wired link or additional recharging process. Figure 11 shows a graph of field test result in power consumption.

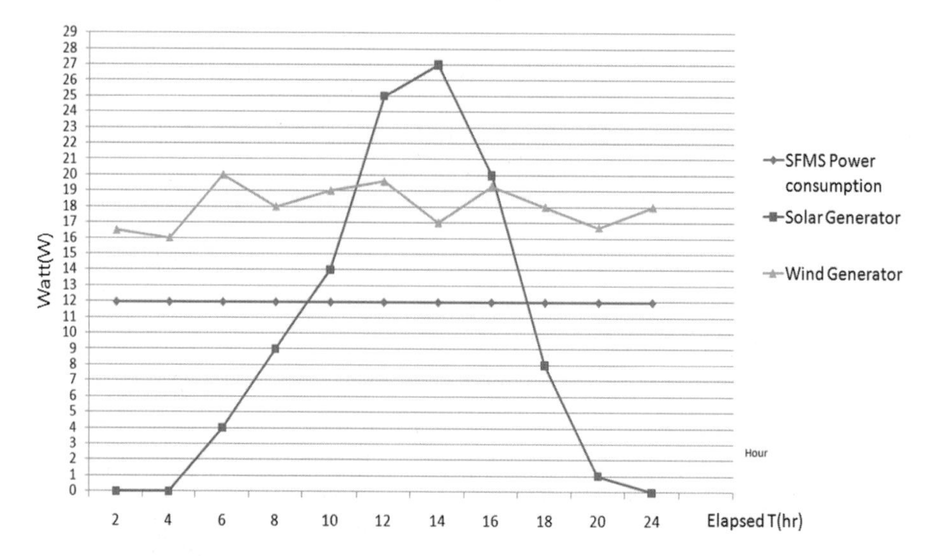

Fig. 11. Field test result of generate power and consume power

5 Conclusions

This paper proposed the Salt-Farm monitoring service (SFMS) that could monitor environments of salt-farm using renewable energy. Also, for verifying the execution of our system, we implemented system's components and made the SFMS prototype. Then we showed the executing results of the system. From this result, we confirmed that our system could monitor the salt-farm conditions by using various sensors and facilities. Also, we show that renewable energy can make operating the SFMS without any external power

For future works, we aim to developing an improved monitoring system which operates based USN and applies into the salt storage inventory. Also, it's challenge for us to keep a good condition from salt wind, salt water and extreme weather.

Acknowledgements. This research was supported by the MKE (Ministry of Knowledge Economy), Korea, under the ITRC(Information Technology Research Center) support program supervised by the IITA (Institute of Information Technology Advancement) (IITA-2009-(C1090-0902-0047)).

References

1. Shin, C.S., Joo, S.C., Lee, Y.W., Sim, C.B., Yoe, H.: An Implementation of Ubiquitous Field Server System Using Solar Energy Based on Wireless Sensor Networks. Studies in Computational Intelligence 209 (2009)
2. Lee, Y.W., Cho, C.J., Ju, J.K., Shin, C.S., Lee, J.H., Shin, H.H., Yum, Y.C., Yoe, H.: Implementation of System for a Ubiquitous Farming-diary. Journal of the Korean Society of Agricultural Engineers 52(2), 35–42 (2010)
3. Yu, Y., Gang, Y., Lee, W.: Development of a Floating Buoy for Monitoring Ocean Environments. Journal of the Korean Society of Marine Engineering 33 (July 2009)
4. Sensinode, OEM Product catalog (2007), http://www.sensinode.com/pdfs/sensinode-catalog-20071101.pdf
5. Ahonen, T., Virrankoski, R.: Greenhouse Monitoring with Wireless Sensor Network. In: IEEE/ASME International Conference on, pp. 403–408 (2008)
6. Delin, K.A., Jackson, S.P., Burleigh, S.C., Johnson, D.W., Woodrow, R.R., Britton, J.T.: The JPL Sensor Webs Project: Fielded Technology. In: Space Mission Challenges for IT Proceedings, Annual Conference Series, pp. 337–341 (2003)
7. Shin, C.S., Kang, M.S., Jeong, C.W., Joo, S.C.: TMO-based Object Group Framework for Supporting Distributed Object Management and Real-Time Services. In: Zhou, X., Xu, M., Jähnichen, S., Cao, J. (eds.) APPT 2003. LNCS, vol. 2834, pp. 525–535. Springer, Heidelberg (2003)
8. Kang, B.J., Park, D.H., Cho, K.R., Shin, C.S., Cho, S.E., Park, J.W.: A Study on the Greenhouse Auto Control System based on Wireless Sensor Network. In: International Conference on Security Technology, pp. 41–44 (December 2008)
9. Tilman, D., Cassman, K.G., Matson, P.A., Naylor, R., Polasky, S.: Agricultural sustainability and intensive production practices. Nature 418, 671–677 (2002)
10. Burrell, J., Brooke, T., Beckwith, R.: Vineyard computing: sensor networks in agricultural production. IEEE Pervasive Computing 3(1), 38–45 (2004)
11. Yoe, H., Eom, K.-b.: Design of Energy Efficient Routing Method for Ubiquitous Green Houses. In: Szczuka, M.S., Howard, D., Ślęzak, D., Kim, H.-k., Kim, T.-h., Ko, I.-s., Lee, G., Sloot, P.M.A. (eds.) ICHIT 2006. LNCS (LNAI), vol. 4413, Springer, Heidelberg (2006)
12. Lee, M.H., Eom, K.B., Kang, H.J., Shin, C.S., Yoe, H.: Design and Implementation of Wireless Sensor Network for Ubiquitous Glass Houses. In: 7th IEEE/ACIS International Conference on Computer and Information Science, pp. 397–400 (May 2008)

Author Index

Printing: Mercedes-Druck, Berlin
Binding: Stein+Lehmann, Berlin